SUSTAINABILITY
IN THE **MINERAL** AND **ENERGY SECTORS**

SUSTAINABILITY
IN THE **MINERAL** AND
ENERGY SECTORS

EDITED BY

SHEILA **DEVASAHAYAM**

KIM **DOWLING**

MANOJ K. **MAHAPATRA**

CRC Press
Taylor & Francis Group
Boca Raton London New York

CRC Press is an imprint of the
Taylor & Francis Group, an **informa** business

CRC Press
Taylor & Francis Group
6000 Broken Sound Parkway NW, Suite 300
Boca Raton, FL 33487-2742

First issued in paperback 2019

ISBN-13: 978-1-4987-3302-1 (hbk)
ISBN-13: 978-0-367-87380-6 (pbk)

Library of Congress Cataloging-in-Publication Data

Names: Devasahayam, Sheila, editor. | Dowling, Kim, editor. | Mahapatra, Manoj K., editor.
Title: Sustainability in the mineral and energy sectors / Sheila Devasahayam, Kim Dowling, and Manoj K. Mahapatra, editors.
Description: Boca Raton : Taylor & Francis, 2016. | "A CRC title." | Includes bibliographical references and index.
Identifiers: LCCN 2016006101 | ISBN 9781498733021 (alk. paper)
Subjects: LCSH: Mines and mineral resources. | Mineral industries. | Power resources. | Energy industries.
Classification: LCC TN23 .S87 2016 | DDC 338.2/7--dc23
LC record available at https://lccn.loc.gov/2016006101

Visit the Taylor & Francis Web site at
http://www.taylorandfrancis.com

and the CRC Press Web site at
http://www.crcpress.com

Contents

SECTION I Mining and Mineral Processing

SECTION II Metallurgy/Recycling

SECTION III   Environment

SECTION IV   Energy

SECTION V   Socio-Economic, Regulatory

SECTION VI   Sustainable Materials, Fleets

Preface

The market values of the global mining and the energy sectors in 2015 were estimated to be US$1 trillion and US$1.3 trillion, respectively. As the mining and energy sectors are pivotal to many world economies, the sustainability issues facing these sectors assume greater significance. China's economic rebalancing, fluctuating commodity prices that are taking longer to recover than anticipated, rising energy costs, and the environmental and socioeconomic impacts of mining and processing of the mined product have led the mineral sector to focus on and invest in innovative and bold sustainable practices. The high dependency of the mining sector on energy has paved the way for the development of technologies promoting renewable energy resources easing the reliance on nonrenewable resources as well as identifying energy efficiency opportunities and reducing the carbon footprint. Safety, security, and health all have huge impacts on mining productivity. A need for a more uniform and effective global regulatory regime for the mineral and energy sectors is still felt following the recent collapse of an iron ore tailings dam in the Brazilian state of Minas Gerais and Fukushima nuclear disaster in 2011.

This peer-reviewed book examines the worldwide sustainability practices in the mineral and energy sectors. The contributors are eminent representatives of academia, industry, and government bodies from Australia, Canada, India, Greece, Korea, Papua New Guinea, South Africa, Spain, Macedonia, Mexico, the United Kingdom, and the United States.

The book addresses six major themes:

Section I: Mining and Mineral Processing
Section II: Metallurgy/Recycling
Section III: Environment
Section IV: Energy
Section V: Socio-Economic, Regulatory
Section VI: Sustainable Materials, Fleets

The need for an integrated transdisciplinary approach with innovative interdisciplinary research to address the sustainability challenges facing the minerals and the energy sectors is emphasised throughout this book. This book deliberates on lifting the mining productivity and optimising energy efficiency across the mining and mill-to-melt operations and downstream in terms of innovation, step-change technologies. The need for integrated waste management practices including zero waste concepts and cost-effective environmentally friendly technologies to recover valuable resources from wastes are emphasised. Economic, environmental, and social parameter analyses are used to identify the areas for improvement in iron and steel production, as well as aluminium, lead, zinc, copper, and gold.

This book provides an overview of water and energy sustainability in mining, and analytical perspectives on nuclear energy, solar (photovoltaics), fuel cells, natural gases, and unconventional gases, carbon capture and storage, Limacon waste energy recovery, and energy from urban wastes. It discusses how they can be integrated to meet the current and future energy demands of the mineral sectors in order to mitigate the footprints of carbon, water, heat, and other anthropogenic products.

Environmental management, essential for sustainable mining, including water treatment and remediation, managing the mine-induced water, mine planning, rehabilitation and remediation of contaminated brownfields, and landfills are examined extensively in this book.

This book features critical analyses of the sustainable and innovative energy solutions for the mining and energy sectors. The use of mining wastes in novel energy materials is also considered. In essence, this book provides the insight and the current and future perspectives on sustainability in mineral and energy sectors, which will benefit the students, industrialists, academics, policy makers, and regulators of these sectors.

Editors

Sheila Devasahayam, PhD (Metallurgy) and PhD (Materials Science) is a faculty member of the Federation University Australia, Australia with a strong research background in both metallurgy and materials science. Sheila earned a PhD in materials science from the University of Queensland, Australia, with post-doctoral experiences from the University of Wollongong, University of Sydney, and University of New South Wales. Sheila has expertise in many aspects of industry-based materials science ranging from coatings, surface modifications, plasma and radiation chemistry, nanocomposites, super absorbents, super hydrophobic materials, and energy materials. Sheila previously earned a PhD in extractive metallurgy from the National Metallurgical Laboratory, CSIR-Madras University, India. She is involved in many industrial projects in the mineral sector including acid mine drainage, hydrometallurgy, pyrometallurgy, mineral processing, and coal processing with a focus on green chemistry and engineering. Sheila gained expertise in soil and water science and aquifer characterisation while working at the Centre for Water Resources, Anna University, India. She served in the Commonwealth Government, Australia as a scientific policy officer in the Department of Innovation. Sheila's inter-disciplinary, industry-focussed expertise has enabled her to provide innovative sustainable solutions to many immediate industrial problems and has motivated her to edit this important book of transdisciplinary nature transcending individual disciplines.

Kim Dowling, a senior academic in the Faculty of Science and Technology, works in the Faculty of Science and Technology at Federation University Australia, Australia. She has been an active researcher for over 30 years, and her current interest is in the identification, characterisation, and remediation of contaminated sites, with an emphasis on historical mining environments. She earned her PhD from James Cook University by developing a new and innovative method for assessment of likely gold deposits and also holds a Graduate Diploma in Environmental Management. She has extensive experience working in both the mining industry and the tertiary education sector, and actively contributes to both local and international discussions on environmental health and the emerging multidisciplinary field of medical geology.

Manoj K. Mahapatra, PhD, is a faculty member in the Department of Materials Science and Engineering at the University of Alabama at Birmingham (UAB), USA. His research focusses on ceramic and glass materials for energy applications. He has authored and coauthored over 35 peer-reviewed publications and holds 2 US patents. Dr. Mahapatra earned a PhD in Materials Science and Engineering from the Virginia Polytechnic Institute and State University. He worked at the Center for Clean Energy Engineering at the University of Connecticut prior to joining UAB.

Authors

Komal Babu Addagatla
School of Information Technology and
    Engineering
Federation University Australia
Ballarat, VIC, Australia

Alexander Agrios
Department of Civil and Environmental
    Engineering
Centre for Clean Energy Engineering
University of Connecticut
Storrs, Connecticut

Shahzada Ahmad
Abengoa Research
Sevilla, Spain

Ali Alhelal
Mechanical Engineering
Federation University Australia
Ballarat, VIC, Australia

Jim Avraamides
Korea Institute of Geoscience and Mineral
    Resources (KIGAM)
Daejon, Republic of Korea

Sri Bandyopadhyay
School of Materials Science and Engineering
University of New South Wales
Sydney, NSW, Australia

Dee Bradshaw
Department of Chemical Engineering
University of Cape Town
Cape Town, South Africa

Jennifer Broadhurst
Department of Chemical Engineering
University of Cape Town
Cape Town, South Africa

Geoffrey Brooks
Wealth from Waste Research Cluster
and
Swinburne University of Technology
Melbourne, VIC, Australia

Sandip Chatterjee
Department of Electronics and Information
    Technology
Government of India
New Delhi, India

Tapiwa Chenje
JKTech Pty Ltd
Indooroopilly, QLD, Australia

Sagar Cholake
School of Materials Science and
    Engineering
University of New South Wales
Sydney, NSW, Australia

Glen Corder
Wealth from Waste Research Cluster
and
University of Queensland
Brisbane, QLD, Australia

Michael J. Daniel
CMD Consulting Pty Ltd
Bellbowrie, QLD, Australia

Sheila Devasahayam
School of Applied and Biomedical Sciences
Faculty of Science and Technology
Federation University Australia
Ballarat, VIC, Australia

K. Karthika Devi
Department of Materials Engineering
Indian Institute of Science
Bengaluru, Karnataka, India

L. C. S. Dharmasiri
Chemical and Metallurgical Engineering and
    Chemistry
School of Engineering and Information
    Technology
Murdoch University
Perth, WA, Australia

Kevin Dodds
Australian National Low Emissions
    Coal R&D
Canberra, ACT, Australia

Augustine Doronila
Analytical and Environmental Chemistry
    Research Group
School of Chemistry
University of Melbourne
Melbourne, VIC, Australia

Kim Dowling
School of Biomedical and Applied Science
Federation University Australia
Ballarat, VIC, Australia

Divya Eratte
School of Biomedical and Applied Science
Federation University Australia
Ballarat, VIC, Australia

Tim Evans
Department of Environmental Sciences
Macquarie University
Sydney, NSW, Australia

Nimesha Fernando
School of Biomedical and Applied Science
Federation University Australia
Ballarat, VIC, Australia

Singarayer K. Florentine
School of Biomedical and Applied Science
Federation University Australia
Ballarat, VIC, Australia

Jean-Paul Franzidis
Department of Chemical Engineering
University of Cape Town
Cape Town, South Africa

Maryam Ghodrat
Wealth from Waste Research Cluster
and
Swinburne University of Technology
Melbourne, VIC, Australia

Juan Gimenez
School of Materials Science and Engineering
University of New South Wales
Sydney, NSW, Australia

Patrick Graz
School of Biomedical and Applied Science
Federation University Australia
Ballarat, VIC, Australia

Rajender Gupta
Chemical and Materials Engineering
University of Alberta
Edmonton, Alberta, Canada

Dharmappa Hagare
School of Computing, Engineering and
    Mathematics
Western Sydney University
Penrith, NSW, Australia

Yotamu R. S. Hara
The Institute for Materials Research
School of Chemical and Process
    Engineering
The University of Leeds
Leeds, United Kingdom

Susan T. L. Harrison
Department of Chemical Engineering
University of Cape Town
Cape Town, South Africa

Samanthala Hettihewa
Finance, Business School
Federation University Australia
Ballarat, VIC, Australia

Michael Hightower
Energy Systems Analysis Department
Sandia National Laboratories
Albuquerque, New Mexico

Boxun Hu
Materials Science and Engineering
Centre for Clean Energy Engineering
University of Connecticut
Storrs, Connecticut

Shihao Huang
School of Materials Science and
    Engineering
University of New South Wales
Sydney, NSW, Australia

Samudra Jayasekera
School of Engineering and IT
Federation University Australia
Ballarat, VIC, Australia

Animesh Jha
The Institute for Materials Research
School of Chemical and Process Engineering
The University of Leeds
Leeds, United Kingdom

Manis Kumar Jha
Metal Extraction and Forming Division
CSIR-National Metallurgical Laboratory
Jamshedpur, Jharkhand, India

Sarma S. Kanchibotla
Julius Kruttschnitt Mineral Research Centre
Sustainable Minerals Institute
University of Queensland
Brisbane, QLD, Australia

Meri Karanfilovska
Institute of Communication Studies
Jurij Gagarin, Skopje, Republic of
    Macedonia

S. Komar Kawatra
Department of Chemical Engineering
Michigan Technological University
Houghton, Michigan

Luke Keeney
CRC ORE
Queensland, Australia

Dong-Jin Kim
Korea Institute of Geoscience and Mineral
    Resources (KIGAM)
Daejon, Republic of Korea

Sarah Kim
Higher Education Faculty
Holmes Institute
Melbourne, VIC, Australia

Alex Kouznetsov
Peter Faber Business School
Australian Catholic University
Melbourne, VIC, Australia

Archana Kumari
Metal Extraction and Forming Division
CSIR-National Metallurgical Laboratory
Jamshedpur, Jharkhand, India

Gretchen T. Lapidus
Departamento de Ingeniería de Procesos
    e Hidráulica
Universidad Autónoma Metropolitana
Unidad Iztapalapa, Mexico City, Mexico

Wen Lee
School of Materials Science and
    Engineering
University of New South Wales
Sydney, NSW, Australia

Manoj K. Mahapatra
Materials Science and Engineering
University of Alabama at Birmingham
Birmingham, Alabama

Liza Maimone
PwC Australia
Canberra, ACT, Australia

Kenaope Maithakimako
School of Materials Science and
    Engineering
University of New South Wales
Sydney, NSW, Australia

Venkata Manthina
Fraunhofer Center for Energy Innovation
Storrs, Connecticut

Gail H. Marcus (Retired)
Office of Nuclear Energy, Science and
    Technology
U.S. Department of Energy
Washington, DC

D. Marinos
School of Mining and Metallurgical
    Engineering
National Technical University of Athens
Zografou, Athens, Greece

Natasa Markovska
Macedonian Academy of Sciences and Arts
Republic of Macedonia

Rachael Martin
School of Biomedical and Applied Science
Federation University Australia
Ballarat, VIC, Australia

Syed Masood
Wealth from Waste Research Cluster
and
Swinburne University of Technology
Melbourne, VIC, Australia

Karsten Michael
Energy—Oil, Gas & Fuels Program
Commonwealth Scientific and Industrial
    Research Organisation
Perth, WA, Australia

Brajendra Mishra
Metal Processing Institute
Worcester Polytechnic Institute
Worcester, Massachusetts

K. A. Natarajan
Department of Materials Engineering
Indian Institute of Science
Bengaluru, Karnataka, India

Kyung-Ho Park
Korea Institute of Geoscience and Mineral
    Resources (KIGAM)
Daejon, Republic of Korea

Dora C. Pearce
School of Biomedical and Applied Science
Federation University Australia
Ballarat, VIC, Australia

Joe Pease
CRC ORE
Queensland, Australia

Jochen Petersen
Department of Chemical Engineering
University of Cape Town
Cape Town, South Africa

Truong H. Phung
Mechanical Engineering
Federation University Australia
Ballarat, VIC, Australia

Deepak Pudasainee
Chemical and Materials Engineering
Clean Coal Technologies
University of Alberta
Edmonton, Alberta, Canada

Moshfiqur Rahman
Chemical and Materials Engineering
Clean Coal Technologies
University of Alberta
Edmonton, Alberta, Canada

Muhammad Muhitur Rahman
Institute for Infrastructure Engineering
Western Sydney University
Penrith, NSW, Australia

M. Akbar Rhamdhani
Wealth from Waste Research Cluster
and
Swinburne University of Technology
Melbourne, VIC, Australia

Kerryn Schrank
PwC Australia
Melbourne, VIC, Australia

Gamini Senanayake
Chemical and Metallurgical Engineering and
    Chemistry
School of Engineering and Information
    Technology
Murdoch University
Perth, WA, Australia

Alexander Senaputra
Chemical and Metallurgical Engineering and
    Chemistry
School of Engineering and Information
    Technology
Murdoch University
Perth, WA, Australia

Greg Shapland
CRC ORE
Queensland, Australia

Sandeep Sharma
Carbon Projects Pty Ltd
Perth, Australia

S. M. Shin
Korea Institute of Geoscience and Mineral
    Resources (KIGAM)
Daejon, Republic of Korea

Mohan Singh
Mineral Resources Authority
Papua New Guinea

Prabhakar Singh
Centre for Clean Energy Engineering
University of Connecticut
Storrs, Connecticut

Raman Singh
Department of Mechanical and Aerospace
    Engineering
and
Department of Chemical Engineering
Monash University
Melbourne, VIC, Australia

Raghu N. Singh
Department of Civil Engineering
Nottingham University
Nottingham, United Kingdom

Muttucumaru Sivakumar
Department of Civil, Mining and
    Environmental Engineering
University of Wollongong
Wollongong, NSW, Australia

Jeong-Soo Sohn
Korea Institute of Geoscience and Mineral
    Resources (KIGAM)
Daejon, Republic of Korea

Vladimir Strezov
Department of Environmental Sciences
Macquarie University
Sydney, NSW, Australia

Ibrahim A. Sultan
Mechanical Engineering
Federation University Australia
Ballarat, VIC, Australia

Jim Underschultz
Centre for Coal Seam Gas
The University of Queensland
Brisbane, QLD, Australia

Terry Wall
School of Engineering
The University of Newcastle
Callaghan, NSW, Australia

Stephen Walters
CRC ORE
Queensland, Australia

Steve Whittaker
Illinois State Geological Survey
Prairie Research Institute
University of Illinois at Urbana-Champaign
Champaign, Illinois

Dakshith Ruvin Wijesinghe
School of Engineering and IT
Federation University Australia
Ballarat, VIC, Australia

Christopher S. Wright
Higher Education Faculty
Holmes Institute
Melbourne, VIC, Australia

Greg You
School of Engineering and IT
Federation University Australia
Ballarat, VIC, Australia

Referees

1. Dr. Abhoy Kumar, Defence Metallurgical Research Laboratory, Hyderabad, India.

2. Dr. Aleks Nikoloski, senior lecturer, Metallurgical Engineering, Murdoch University, Perth, Australia.

3. Alex Kouznetsov, senior lecturer, Holmes Institute, Melbourne, Australia.

4. Dr. Ashish Aphale, Center for Clean Energy Engineering, University of Bridgeport, Bridgeport, Connecticut.

5. Associate Professor Anandhan Srinivasan, Department of Metallurgical and Materials Engineering, National Institute of Technology, Karnataka, India.

6. Dr. Andrew Scott, Mine Consulting Services Pty Ltd. Brisbane, Australia.

7. Associate Professor Brett Robinson, Soil and Physical Science, Lincoln University, Lincoln, New Zealand.

8. Dr. Chris Boreham, petroleum geochemist, Geoscience Australia, Canberra, Australia.

9. Corinne Unger, senior research officer, Sustainable Minerals Institute, The University of Queensland, Brisbane, Australia.

10. Dr. Daniel Riley, senior materials engineer, ANSTO, Sydney, Australia.

11. Professor Gautam Sarkhel, head, Department of Chemical Engineering and Technology, Birla Institute of Technology, Ranchi, India.

12. Associate Professor Gamini Senanayake, Metallurgical Engineering, Murdoch University, Perth, Australia.

13. Dr. Harpreet Kandra, lecturer, School of Engineering and IT, Federation University Australia, Ballarat, Australia.

14. Associate Professor Ibrahim Sultan, School of Engineering and IT, Federation University Australia, Ballarat, Australia.

15. Adjunct Professor John Rankin, Faculty of Science, Engineering and Technology, Swinburne University of Technology, Melbourne, Australia.

16. Adjunct Professor John Grandfield, Swinburne University of Technology, and director, Grandfield Technology Pty Ltd., Melbourne, Australia.

17. Dr. Manoj Khandelwal, lecturer, School of Engineering and IT, Federation University Australia, Ballarat, Australia.

Acknowledgments

We wish to acknowledge the authors of this book who provided their valuable insight, current, and future perspectives on sustainability in mineral and energy sectors. Our eminent referees, both listed and the anonymous, added value to the book's content through their suggestions and expert comments and we gratefully acknowledge them. We are extremely grateful to the entire CRC/ Taylor & Francis Publishing team, especially Allison Shatkin (senior acquisitions editor, Books, Materials, Chemical, and Petroleum Engineering) who has made this journey pleasant through her invaluable help and input, and Arlene Kopeloff, Jennifer Ahringer, Mikaela Kursell, Judith Simon, Mohamed Hameed, and Florence Kizza for their friendliness and cooperation throughout the entire process. Sheila Devasahayam, the lead editor, wishes to acknowledge her parents who instilled in her the courage, wisdom, and the positive core values to accomplish her goals.

Section I

Mining and Mineral Processing

1 New Paradigms for Sustainable Mineral Processing

S. Komar Kawatra

CONTENTS

ABSTRACT

Sustainable mineral processing is the ultimate goal driving mineral processing plants to generate zero waste. This encompasses: (1) clean and reusable water—recycle all water used in the plant and remove any metal particulates in the water; (2) solid waste utilization—use solid tailings for infrastructure instead of impounding, and efficiently extract all useful minerals from the tailings; and (3) clean air—zero release of dust, nanoparticles, or toxic gases, and capture of emissions such as CO_2 and H_2SO_4 to convert into useful products. By striving to achieve zero-waste production in mineral processing, less hazardous compounds will be released into the environment, less material will be wasted, and companies will increase profit by processing old "waste" streams into marketable products. In the progression toward zero waste, sustainable mineral processing will address the issues of reducing carbon emissions, minimizing environmental effects, and maintaining the societal "license to operate."

INTRODUCTION

Sustainability in the mineral processing industry is of paramount importance as resources continue to decline and regulations become more stringent. The ultimate goal of a sustainable mineral processing operation is to maintain a zero-waste strategy in which the plant emits clean air, recycles all water, and uses solid wastes to create a value-added product. SUStainable OPerations (SUSOP) is a theoretical approach to the mining industry that relates to Michigan Technological University's (MTU) innovative concept of a zero-waste strategy. Initiated by the Co-operative Research Centre for Sustainable Resource Processing, this method is being developed with input from research and industry participants, and focuses on sustainable development (SD) in all areas of mineral processing from corporate to plant operations (Corder and Green, 2011). The need for SD in mining is directly related to increasing regulations on common water and airborne pollutants. Water is currently regulated for mineral content such as Sb, As, Be, Cr, Cu, Pb, Ag, Sn, Cn, and asbestos, and air is regulated for emissions such as CO_2, O_3, Pb, dust, and airborne mercury.

Zero-waste production is essentially recovering each waste stream from a mining plant's unit process, and converting it into a feedstock for another process. The challenge is determining in which market each reprocessed end product will be most profitable and advantageous. Some materials, such as cement kiln dust (CKD) and low-grade fly-ash, are difficult to market because of their chemical properties (extreme alkalinity), but there are ways to benefit from this. For example, the Advanced Sustainable Iron and Steelmaking Center at MTU is researching innovative ways to sequester and reduce CO_2 emissions using CKD and low-grade fly-ash as CO_2 absorbents.

Iron ore mineral processing plants are one of the major driving forces behind discovering and implementing SD processes. What markets are available to use tailings for profit, and how can the long-term environmental footprint of tailings processing and storage be reduced? Certain mineral plants also consume vast quantities of water, a highly coveted resource in many parts of the world. How can plants increase water recovery from tailings streams to 95% or more? Questions like these need to be answered to transform mineral processing into a sustainable industry.

BACKGROUND: ZERO-WASTE CONCEPT

As the global demand increases for iron ore and other minerals, major environmental, health, and socioeconomic concerns arise. Sustainable production in mineral plants can help prevent and reduce threats to the environment, public health, and our sense of security. This ideally means elimination of all waste, whether solid, liquid, or gaseous, as shown in Figure 1.1. The zero-waste approach encompasses clean air output, clean water output and maximum water recycling, and full utilization of solid waste. This chapter focuses on ways to reduce mining waste output in each of these categories.

SD principles need to be applied to the design phase of mineral processing, in addition to operations. Sustainability objectives must be a priority in the design and implementation of new mineral

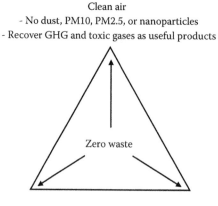

FIGURE 1.1   Three main components of the zero-waste approach. Ultimately, it is the repurposing of all solid wastes, zero air emissions, and recovery of water and effluents in water. Materials are used for internal recycling or conversion to profitable materials. (Adapted from Kawatra, S. K. 2011. Zero emissions in iron ore processing plants, Fray International Symposium, December 4–7, 2011, Cancun, Mexico.)

processing plants. McLellan et al. (2009) argue that the opportunities to reduce environmental impact and increase sustainability stem from bridging the gaps between production, recycling chains, and supply chains. They propose a two-stage review to drive mineral processing toward sustainability. The first stage engages the industry with the government and local community to develop pragmatic approaches for (1) tangibly improving the impacts on the "triple bottom line" (social, environmental, and financial aspects), (2) designing production cycles and phases which focus on sustainability, and (3) addressing the challenges from existing minerals processing infrastructure. The second stage focuses on building an economic environment where the mining industry supports SD, while continuing the approaches described in the first stage.

AIR

CARBON DIOXIDE

According to the Environmental Protection Agency (EPA), the greenhouse gas emissions from the minerals sector of industry were 107.5 million metric tons of carbon dioxide in 2012 from facilities in the United States. The largest contributor to these emissions is cement production, followed by lime manufacturing. From 2010 to 2012, emissions increased by 6% due to increased reported emissions from the cement manufacturing subsector. Figure 1.2 represents what areas of mineral processing produce the largest amount of CO_2 emissions.

Table 1.1 shows a summary of the total greenhouse gas emissions (carbon dioxide, methane, and nitrous oxide) released for years 2010 through 2012. As the number of mineral processing plants increased, the amount of carbon dioxide released to the environment has increased. Cement production is consistently the largest contributor to greenhouse gas emissions, and the emission rate is increasing every year. The need to drive the mining industry toward zero emissions of dust vapors and air pollutants is greater than ever.

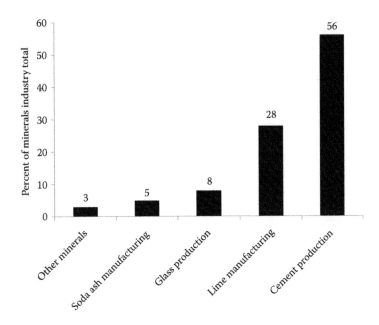

FIGURE 1.2 Visual representation of which mining industry subsectors contributed the most to global CO_2 emissions in 2013. Cement production was the highest, rendering 57% of the total CO_2 emissions in the mining industry. (Adapted from EPA, 2014. 2010-2011-2012 GHGRP Industrial Profiles, Minerals Sector Report, https://www.epa.gov/ghgreporting/ghgrp-minerals-sector-industrial-profile.)

TABLE 1.1

Greenhouse Gas Emissions by Mining Subsector

Sector	Reporting Year		
	2010	**2011**	**2012**
Number of facilities	359	367	369
Total emissions (MMT CO_2 equivalent)	101.1	103.2	107.5
Emissions by GHG			
Carbon dioxide (CO_2):			
• Cement production	55.2	55.3	59.8
• Glass production	8.1	8.4	8.2
• Lime manufacturing	28.9	30.5	30.2
• Soda ash production	5.1	5.1	5.1
• Other materials	3.4	3.6	3.8
Methane (CH_4):			
• Cement production	0.1	0.1	0.1
• Glass production	<0.05	<0.05	<0.05
• Lime manufacturing	<0.05	<0.05	<0.05
• Soda ash production	<0.05	<0.05	<0.05
• Other materials	<0.05	<0.05	<0.05
Nitrous oxide (N_2O):			
• Cement production	0.1	0.1	0.1
• Glass production	<0.05	<0.05	<0.05
• Lime manufacturing	0.2	0.1	0.1
• Soda ash production	<0.05	<0.05	<0.05
• Other materials	<0.05	<0.05	<0.05

Source: Adapted from EPA, 2014. 2010-2011-2012 GHGRP Industrial Profiles, Minerals Sector Report, https://www.epa.gov/ghgreporting/ghgrp-minerals-sector-industrial-profile.

The most commonly proposed method of carbon dioxide capture and storage (CCS) is to feed CO_2 contaminated air into an absorption column, where the CO_2 is absorbed by an amine solution. A CO_2 free gas stream leaves through the top of the column and a CO_2-rich amine solution leaves through the bottom. The CO_2 loaded solution is then sent to a stripping column where it is heated to desorb high-purity CO_2 and regenerate the amine absorption solution. The high-purity CO_2 may then be compressed for transport to an underground storage reservoir. This proposed method is not ideal due to the costs associated with it. The amine used for absorption is monoethanolamine, which is highly hazardous, volatile, corrosive, and expensive ($1.25/kg). The stripping stage of the process is energy costly due to the high energy consumption for heating large volumes of feed and compressing the high-purity CO_2 to a supercritical state for transportation to storage.

Researchers at MTU, along with industry leaders, are proposing a low-cost method to capture CO_2 emissions using an inexpensive absorbent chemical to convert it into mineral waste products (Spigarelli, 2013). The process not only increases the value of the waste products (as they can be used in construction applications rather than being simply immobilized as a part of natural mineral formations), but also reduces CO_2 emissions, converts CO_2 into a solid, stable form, and regenerates the absorbent needed to capture the CO_2.

The proposed chemical absorbent to replace the amine solution is a Na_2CO_3 solution. After the CO_2 is captured with Na_2CO_3 in the absorption column, the CO_2 loaded solution reacts with an alkaline industrial waste material to both regenerate the Na_2CO_3 absorbent and to store the CO_2 as

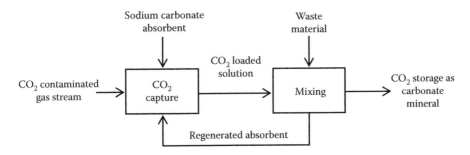

FIGURE 1.3 Block diagram of proposed CCS method. (Adapted from Spigarelli, B. P. 2013. A novel approach to carbon dioxide capture and storage. Dissertation, Michigan Technological University, Houghton, Michigan, http://digitalcommons.mtu.edu/etds/633.)

a carbonate mineral. Figure 1.3 represents a basic overview of the entire proposed CCS process. The main reaction associated with this method is the conversion of aqueous sodium carbonate to sodium bicarbonate.

$$Na_2CO_3 + H_2O + CO_2 \rightarrow 2NaHCO_3 \qquad (1.1)$$

Sodium carbonate is already in industrial use as an absorbent in some sulfur dioxide scrubbers according to this reaction:

$$Na_2CO_3 + H_2O + SO_2 \rightarrow NaHSO_3 + NaHCO_3 \qquad (1.2)$$

There are many benefits to using sodium carbonate as the absorbent in CCS as opposed to using an amine solution. The carbonate-based raw material cost is significantly lower at $0.11/kg and the process occurs at ambient conditions, reducing the energy costs. It is also considered nonhazardous, is nonfouling, has a low equipment corrosion rate, and has low volatility. Most substantially, the sodium carbonate solution can capture CO_2 emissions in addition to other airborne pollutants such as sulfur oxides (SO_x) and mercury.

The proposed method of CCS also makes use of mining solid waste materials. The CO_2 loaded solution requires alkaline industrial waste materials for the final sequestration reaction, such as materials rich in calcium or magnesium. Reacting the CO_2 saturated absorbent with the waste materials regenerates the absorbent, in addition to converting the CO_2 into a stable form using mine tailings, high CaO manufacturing wastes, and metallurgical slags. Another material well suited for CO_2 sequestration is CKD. Cement plants can use their own stockpiled CKD to absorb and reduce their CO_2 emissions.

Dust Pollution

Dust emission is another major concern of mineral processing plants and is considered air contamination. Also known as particle pollution (or particulate matter, PM), this encompasses fine mineral particulates/metals, soot, aerosol chemicals, and small particles. Inhalable course particles have diameters between 2.5 and 10 μm, and fine particles have diameters smaller than 2.5 μm.

The EPA regulates PM standards for the United States. The regulations have become more rigorous since the first standard was released as the understanding of inhalation and health hazards from PM has evolved. The first standard was released in April 1971, and dictates the total suspended particulates (TSPs) allowed in the air. This standard was a result of studies indicating that exposure to airborne dust can cause serious health problems such as chronic bronchitis, aggravated asthma,

TABLE 1.2

Total Suspended PM Standards Enacted in April 1971 by U.S. EPA

Standard Designation	24 h Average	Annual Geometric Mean Value
Primary TSP standard	260 µg/m³	75 µg/m³
Secondary TSP standard	150 µg/m³	N/A

TABLE 1.3

Total Suspended PM Standards Enacted in December 2012 by U.S. EPA

Standard Designation	Particle Size	24 h Average Limit	Three-Year Arithmetic Mean Value
Primary TSP standard	$PM_{2.5}$	35 µg/m³	12 µg/m³
Secondary TSP standard			15 µg/m³
Primary TSP standard	PM_{10}	150 µg/m³	N/A
Secondary TSP standard			

decreased lung function, nonfatal heart attacks, and premature death (EPA, 2015). The TSP standard regulated the total concentration of airborne dust (µg/m³) using a primary and secondary classification. The primary TSP standards were aimed at improving occupation and community health while the secondary TSP standards were purposed to protect the surrounding environment. The original TSP standards implemented in 1971 are summarized in Table 1.2.

The U.S. EPA PM standards as of December 2012 are summarized in Table 1.3. It is very clear that EPA standards are becoming much more stringent over time, and are focusing on smaller particulate sizes. It is logical to conclude these PM standards will continue to evolve as our understanding of the health risk of exposure increases.

Nanoparticulate regulations (particles smaller than 1 µm) will most likely be included in the next TSP standard. These fine particles are able to deeply infiltrate the lungs, and possibly enter the bloodstream and cause serious health issues for other tissues. The EPA is currently studying the effects of nanoparticulates, and will revise TSP standards once a clearer understanding of the environmental and health hazards is reached.

PM SUPPRESSION

As more stringent PM standards are enacted, more effective and lasting suppression methods will be required. The challenge is deciding which suppressant is most effective for a given application. The current most common practice is to wet the dust particles with a liquid, resulting in adhesion to coarse, heavier particles, thus preventing these particles from drifting into the air. Certain dust materials, like iron ore, are easily wetted and suppressed using water; others, like coal, are hydrophobic and require other wetting materials. Another key factor is the timing of adding the suppressant, which relates to how the dust is generated. If an operation produces dust quickly, then fast suppression is required. If dust is generated in the act of transportation and shipping of products, then a long-lasting suppressant is required. It is critical to select a suppressant based on the exact needs of the application.

Coal particulates are difficult to wet due to their hydrophobic nature; therefore, water sprays have little or no effect when suppressing coal dust. Surfactants lower the surface tension of water, however, allowing a water–surfactant mixture to easily engulf coal particles. This overcomes their hydrophobic nature to readily allow wetting and suppression.

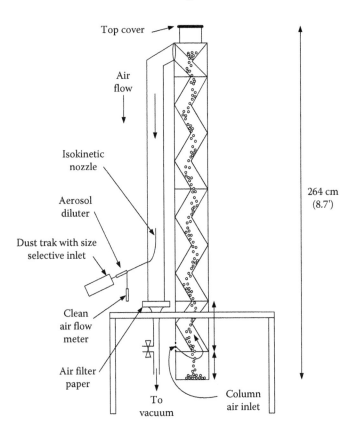

FIGURE 1.4 MTU dust tower apparatus. This apparatus is capable of simulating industrial activities that lead to PM release, and measuring the effect of suppressants. Tower air flow rate was 4.5 L/s, with a mean flow velocity of 45 cm/s. (Adapted from Copeland, C. R. and S. K. Kawatra. 2005. *Minerals and Metallurgical Processing*, 22(4): 177–191.)

Hydrophilic materials (like taconite) wet easily with water and have no need for surfactants. The suppressant is only effective until it evaporates. Hygroscopic materials can be used to slow the evaporation process.

To study the effects of hygroscopic reagents on dust suppression, a freefall dust test was conducted at MTU, using a custom-built dust tower and taconite pellet feed as shown in Figure 1.4. The test requirements included reproducibility, large representative samples, the use of dust suppressants, and mimicking pellet handling conditions in a plant as much as possible. Several different hygroscopic materials were used, with distilled water as the control suppressant. The results of the test are summarized in Figure 1.5. The PM_{10} concentration over time significantly decreased when $CaCl_2$, $MgCl_2$, and Na_2SiO_3 were used as suppressants, proving that the evaporation rate of wetting agents can be slowed with hygroscopic materials (Copeland and Kawatra, 2005).

SOLIDS

Tailings are a major waste in the mining industry, and economical and sustainable methods of repurposing them must be implemented in order to drive toward zero solid waste. The United States produces over 125 million metric tons of fine mine waste tailings every year. At one Michigan mine, over 418 million metric tons of mine tailings have accumulated in just 38 years. Because of the worldwide abundance of tailings, they are attractive as a raw material that can be recovered and utilized in infrastructure construction.

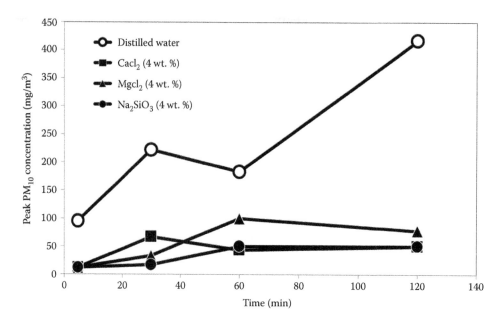

FIGURE 1.5 Effects of hygroscopic materials on the generation of PM_{10}. The different hygroscopic materials were more effective than water at reducing the PM_{10} concentration of taconite dust over time. The test used a mixture of 1 kg plant D taconite pellets, 10 g of plant D tumble test fines (P90 = 40 μm, P80 = 27 μm, and P10 = 3 μm), and 4 kg/tonne treatment. (Adapted from Copeland, C. R. and S. K. Kawatra. 2005. *Minerals and Metallurgical Processing*, 22(4): 177–191.)

Taconite Tailings

Taconite is a low grade of iron ore which must be processed to extract a high grade concentrate, which results in high volumes of tailings. At one Minnesota iron mine, over 130 million metric tons of taconite tailings were produced, and this quantity is approximately 1/12th of the United States' annual production of commercial crushed stone for construction applications. Taconite tailings, among others, have high potential for use in industrial applications, such as road construction and repairs (Zanko, 2011).

Lawrence Zanko, a senior research fellow at the University of Minnesota, Duluth, has conducted research on ferrous and nonferrous minerals for the Natural Resources Research Institute (NRRI) since 1988. One of his main projects focuses on using waste materials, such as fine mine tailings, for value-added aggregate products. These products include taconite-based repair compounds for use in road pothole patching, coarse taconite tailings for high friction airport runway surfacing aggregates, taconite aggregate materials in pavement design of upper wear courses, and taconite materials paired with microwave technology to repair road damage more sustainably and efficiently (NRRI, 2015).

Cold Bonding Fine Tailings for Aggregate

The biggest challenge with tailings is characterizing their physical and chemical properties in order to determine the best future application of the material. The properties of the tailings determine if additional milling and separation processes are required to obtain the desired material. Past methods to recover tailings required expensive sintering temperatures of 1300°C, which is not energy efficient or economical for processing waste materials.

How will research and industry accomplish driving mine solid waste toward sustainability? Cold bonding technologies are the most promising answer to converting variable mineral tailings to applicable and economical materials. The "cold bonding" technique hardens agglomerates in low-temperature and high-humidity, or elevated-pressure environments. This is an attractive alternative to the current high-temperature/high-energy processes used to harden materials in the iron and steel industry. MTU is currently researching the novel use of cold bonding fine mine tailings for repurposement, especially those of the taconite/magnetite nature. Not only do certain mine tailings have a higher bond work index than currently used construction materials, but also they do not require further processing before use (Walkiewicz et al., 1991).

The cold bonding technique polymerizes waste tailings at economically attractive low temperatures (200–500°C) via carbonation and hydrothermal processes, creating pellets that can withstand high compressive forces. Cold bonding utilizes calcium hydroxide as the binding agent. This material is then cured in a humid, CO_2-rich environment, at an elevated pressure (300 psig), thus reducing the time required to harden the agglomerates.

The cold bonding technique is a highly advantageous method of repurposing mining tailings. The technology has proved to significantly decrease the curing times of cementitious or hydraulic minerals waste. It also lowers energy consumption because it requires lower temperatures than the sintering method. The resultant product of cold bonding mineral tailings is a material that can withstand high compression, improves the friction resistance of road surfaces, and will be economically superior to natural minerals that are currently used in cement and concrete production.

WATER

On January 30, 2000 in Baia Mare, Romania, the dam encircling a tailings pond broke, resulting in a spill of 100,000 m^3 of liquid and suspended waste (Soldan et al., 2001). This spill contained 50–100 tonnes of cyanide, as well as copper and heavy metals. The spill caused tremendous damage to major parts of Romania, Hungary, and Yugoslavia. Rivers including Sasar, Lapus, Somes, Tisza, and Danube were affected, and the contamination eventually reached the Black Sea. Approximately 2000 km of the Danube water catchment area was affected, and Romania reported the spill interrupted the water supply to 24 municipalities. Hungary estimated the spill killed 1240 tonnes of fish within the country.

Aurul SA was the company responsible for the Baia Mare plant and the spill that occurred. The plant was designed to process 2.5 million tonnes per year of gold and silver mining tailings. These tailings were produced by previous mining activities in the area. The plant recovered 1.6 tonnes of gold and 9 tonnes of silver per year. The dam surrounding the tailings pond broke because it was not designed for the fast rise in water level from heavy rainfall and rapidly melting snows. Neither the company nor local government had adequate accident/unplanned event response protocols for what occurred (Regional Environmental Center for Central and Eastern Europe, 2000).

Water is a precious commodity, becoming increasingly more regulated around the world. In the case of Baia Mare, large quantities of important water resources were contaminated due to poor mineral processing operating discipline and design. Smart water use and clean water output are key in keeping the global ecosystem uncontaminated by industrial processes. Effective water management, recycling, and treatment are and will continue to be crucial in developing sustainable operations. The two fundamental issues the mining industry will face with regard to water are

1. Conserving the water currently used to minimize the need to draw more from available, but possibly limited supplies.
2. Controlling accumulation of heavy metals, sulfates, nitrates, acids, and process chemicals in process water, and proper treatment and removal of these substances to reduce or eliminate contamination to water supplies.

GENERAL WATER PROCESSING

Paste thickeners are used in mining tailings to thicken and concentrate the solids sent to surface stacking. Paste tailings have a higher viscosity and higher concentration of suspended (nonsettling) solids than conventionally thickened tailings. Surface stacking is a method of disposing of tailings on the ground surface. When thickened tailings are impounded via surface stacking, minimal process water is lost and dry time of the impoundment tailings decreases. Ponding, which uses significantly more water, is the alternative method to stacking. Stacking thickened tailings allows for a smaller impoundment area, less water usage, improved water and chemical recovery throughout the plant, and lower risk of contaminant breach and ground water contamination. Conventional ponds also require construction of a new dam once the tailings pond has reached capacity. By switching to use of paste thickeners to generate thickened tailings, the tailings stream can deposit into the existing pond and displace any free water for recovery. As less water is used due to the paste, the tailings/solids concentration in the pond will increase, thereby increasing the drying speed and decreasing the pond size (Westech, 2014).

Deep cone paste thickeners are engineered to process high volumes of fines using minimum energy. The cone thickeners are fed from the top, allowing for free-settling of solids. Pickets or rakes rotate within the tank to release water in the slurry recovery. The output stream of the cone thickener is a paste which negates the need for further separation of free water. Multiple companies exist today which specialize in building deep cone thickeners and filters for mining companies (McLanahan, 2015).

Thickener design is an important concept of water recovery in tailings. According to O'Brien et al., a paste thickener by itself will allow the plant to recover 80%–90% of the water in the tailings underflow slurry. To increase this yield, the underflow can also pass through a filter press to output a dry cake which is then sent to the tailings "pond." This significantly improves water recovery, and essentially eliminates the need for a tailings pond, as the waste transferred to the impoundment is dry (Walker, 2014).

In order to decrease fresh water needed for a tailings pond, and increase recovery of water used in the process, mineral processing plants should focus on implementing paste thickener chemistry, combined with deep cone thickener and filter press technology. The methodology of processing and drying tailings aligns with environmental regulations, decreases the land area needed for the impoundment, reduces water losses in the process, reduces the risk of water table contamination, and extends the lifetime of the impoundment.

CONCLUSIONS

The zero-waste concept in mineral processing is a key part of achieving sustainability and full recyclability in the mining industry. It is essential to examine all aspects of a product's lifecycle in a plant to determine where and how the input and output streams change from initial creation to ultimate disposal or recycle. Zero discharge cannot be achieved until recycle streams are optimized and "waste" streams are analyzed for repurposing. MTU is determined to expand the zero-waste concept to all aspects of mining, and with further research and development of new technology and markets, this concept will become a reality.

REFERENCES

Copeland, C. R. and S. K. Kawatra. 2005. Dust suppression in iron ore processing plants. *Minerals and Metallurgical Processing*, 22(4): 177–191.

Corder, G. D. and S. Green. 2011. Integrating sustainability principles into mineral processing plant design. *CEEC the Future*, http://www.ceecthefuture.org/publication/integrating-sustainability-princi ples-into-mineral-processing-plant-design/.

EPA, 2014. 2010-2011-2012 GHGRP Industrial Profiles, Minerals Sector Report, https://www.epa.gov/ ghgreporting/ghgrp-minerals-sector-industrial-profile.

EPA, 2015. Particulate matter, http://www.epa.gov/ncer/science/pm/.

Kawatra, S. K. 2011. Zero emissions in iron ore processing plants, Fray International Symposium, December 4–7, 2011, Cancun, Mexico.

McLanahan. 2015. Deep cone and paste thickeners, http://www.mclanahan.com/p/Deep_Cone_and_Paste_Thickeners/509/346/327.

McLellan, B. C., G. D. Corder, D. Giurco, and S. Green. 2009. Incorporating sustainable development in the design of mineral processing operations—Review and analysis of current approaches. *Journal of Cleaner Production*, 17(16): 1414–1425.

Regional Environmental Center for Central and Eastern Europe. 2000. The Cyanide Spill at Baia Mare, Romania. http://documents.rec.org/publications/Cyanide_spill_June2000_EN.pdf.

Soldan, P., M. Pavonic, J. Boucek, and J. Kokes. 2001. Baia Mare accident—Brief ecotoxicological report of Czech experts. *Ecotoxicology and Environmental Safety*, 49(3): 255–261.

Spigarelli, B. P. 2013. A novel approach to carbon dioxide capture and storage. Dissertation, Michigan Technological University, Houghton, Michigan, http://digitalcommons.mtu.edu/etds/633.

UMD Natural Resources Research Institute. 2015. Larry Zanko, senior research fellow, http://www.nrri.umn.edu/staff/lzanko.asp.

Walker, S. 2014. Dewatering: An increasingly important mineral process. *Engineering and Mining Journal*, http://www.e-mj.com/features/3823-dewatering-an-increasingly-important-mineral-process.html#.VPdtbPnF-Sq.

Walkiewicz, J. W., A. F. Clark, and S. K. McGill, 1991. Microwave assisted grinding, *IEEE Transactions in Industrial Applications*, 27: 239–243.

WesTech. 2014. "Paste and High Density Thickeners," brochure. Salt Lake City, Utah. http://www.westech-inc.com/public/uploads/global/2014/4/Brochure%20Paste%20and%20High%20Density.pdf.

Zanko, L. 2011. Research, development, and marketing of taconite aggregate. TERRA Innovation Series, *MnROAD Research Conference*, Minneapolis, Minnesota.

2 Assessment of Sustainability of Mineral Processing Industries

Vladimir Strezov and Tim Evans

CONTENTS

ABSTRACT

Mineral processing activities pose considerable challenges associated with the emission of pollutants to the environment. On the other hand, these industrial activities have a significant positive impact on the economy. The balance between the economic input versus the environmental footprint of industrial mineral processing determines its sustainability. This chapter provides a sustainability assessment of the production of iron and steel, aluminium, copper, lead, zinc, nickel and gold and summarises their contribution to the economy and environmental impacts. The environmental indicators assessed in this work were consumption parameters, priority volatile pollutants, particulate matter (PM), greenhouse gas emissions, emissions of priority metals, organic air toxics and the properties of the produced non-recyclable waste. These parameters were assessed based on their environmental risk scores and based on the economic contribution of the final product. The results indicate that lead production ranks as the least sustainable technology, due to significant emissions of priority volatiles, toxic metals and high depletion rates, relative to the economic input of this industry. The other low ranking industry was nickel production due to the highest $PM_{2.5}$ emissions, high amount of produced toxic waste and low economic input. This work reveals the priority areas for improvement for each industry as it gives a relative comparison of the mineral processing industries based on environmental risk and economic contribution.

INTRODUCTION

Mineral processing industries have considerable input to the global economies and provide the backbone for a nation's wealth. Production of metals is one of the most important branches of mineral processing, which is a significant contributor to economic development. However, these activities also pose considerable environmental risks through atmospheric emissions of pollutants and the management of toxic waste, often left as a legacy for future generations long after closure of the facilities.

The balance between the economic input of the industry and its environmental footprint is the fundamental expression of its contribution to sustainable development (Strezov et al., 2013). Industrial sustainability should aim to address the three pillars of sustainable development; economic, environmental and social. Singh et al. (2007) provided a comprehensive approach to developing a methodology for sustainability assessment of steel companies and highlighted the importance of applying sets of indicators that quantify the progress towards sustainable development. A range of indicators was proposed that defined the emission of pollutants and greenhouse gases, consumption of power, raw materials and water, hazardous waste generation and noise pollution. Fruehan (2009) further defines the environmental industrial sustainability as: (i) conservation of natural resources; (ii) reduction of greenhouse gas emissions; (iii) reduction of volatile emissions; (iv) reduction of landfill waste and (v) reduction of hazardous waste. Norgate et al. (2007) evaluated the environmental impact of metal production processes based on greenhouse gas emissions, energy requirement for production, acidification potential and solid waste burden. Arena and Azzone (2010) added the social dimension of industrial development and tested the preliminary set of indicators across different steel production processes. Worldsteel Association (2012) provided a list of eight sustainability indicators based on environmental sustainability (greenhouse gas emissions, energy intensity, material efficiency and environmental management systems), social sustainability (lost time injury frequency rate and employee training) and economic sustainability (investment in new processes and products and economic value distributed). These indicators were based on global average figures derived from the members of the World Steel Association with the aim to compare the historical trends of global indicator performances. From an industry perspective (Naylor, 2005), the sustainability of the steel industry is challenged by a declining knowledge transfer with lower student numbers enrolled in science and engineering university courses across the developed nations.

The relative contribution of metal production industries to sustainable development requires a system-wide analytical approach to quantify and compare different production processes. This will enable continuous economic and environmental improvements and contribute to social wellbeing. The aim of this work is to provide a detailed review of environmental indicators for industrial ecology and sustainability, and to comparatively analyse the performance of the most important metal production industries, iron and steelmaking, aluminium, lead, zinc, nickel, copper and gold production.

ECONOMIC INDICATORS FOR INDUSTRIAL SUSTAINABILITY

Metal production industries contribute to the global economy through financial and trade benefits, and through local direct and indirect employment. Production of the seven most important metals, iron/steel, aluminium, copper, lead, zinc, nickel and gold, contribute close to 5% to the global economy (Table 2.1). Production of iron and steel is the most important of all with annual income of almost $2.5 trillion US$. Aluminium is the second most important metal for the global economy, followed by gold and copper. Lead and zinc have the smallest contribution to the global economy.

ENVIRONMENTAL INDICATORS FOR INDUSTRIAL SUSTAINABILITY

The environmental performance of industrial activities can be assessed based on their: (i) consumption and contribution to depletion of natural resources, (ii) greenhouse gas emissions, (iii) pollutant emissions to air, (iv) pollutant emissions to water and (v) type and management of the produced waste. Table 2.2 summarises the set of parameters applicable for assessment of industrial environmental performance.

CONSUMPTION PARAMETERS

The consumption parameters include withdrawal of freshwater, land use for mining, plant operation and contamination and use of depleting resources, such as fossil fuels. Metals can be recycled and

TABLE 2.1

Economic Performance of Metal Production Industries for 2012

	Price (US$/t)	World Income from Metal Production (Billion US$/a)	Contribution to World Gross Domestic Product (%)
Iron and steel	$826	$2,435	3.35
Aluminium	$2,017	$520	0.72
Copper	$7,948	$140	0.2
Lead	$2,060	$11	0.02
Zinc	$1,947	$25	0.04
Nickel	$17,508	$43	0.06
Gold	$58,828,200	$163	0.22

Source: Adapted from Armitage, G. 2013. *Resources and Energy Statistics.* Canberra, Australia: BREE, 4 December.

potentially reused infinitely, which is one of the most important parameters for their sustainability, as society can ultimately reach a level when global demands for metal can be met through recycling of waste and scrap metal. However, human society is still far from reaching this goal as meeting the current demands for metals necessitates mining and production through processing of ores and minerals, which needs exploration and use of the natural ore reserves.

Table 2.3 summarises the global reserves of ores and annual production of the selected metals. The depletion rates were calculated based on the 2013 estimates of ore reserves and production rates. The largest relative reserves of ores are for bauxite ore used for aluminium production at over

TABLE 2.2

Environmental Criteria and Indicators for Industrial Sustainability

Criteria	Indicators
Consumption	Water use
	Land use
	Use of non-renewable and non-recyclable resources
Greenhouse gas emissions	CO_2, CH_4, N_2O, CFCs and HCFs, CF_4 and SF_6
Emissions to air	*Priority pollutants:*
	SO_2, NO_x, CO, Pb, PM (PM_{10} and $PM_{2.5}$) and O_3
	Air toxics:
	Metals and metal compounds
	As, Cd, Hg, Cr6+, Pb and *Ni*
	Polycyclic aromatic hydrocarbons
	Benzo(a)pyrene (BaP)
	Volatile organic compounds
	Benzene
	Persistent organic pollutants
	Aldrin, chlordane, DDT, dieldrin, dioxins, furans, endrin, hexachlorobenzene, heptachlor, mirex, polychlorinated biphenyls and *toxaphene*
Emissions to water	pH, dissolved oxygen, turbidity, fecal coliforms, biochemical oxygen, nitrates, total suspended solids and thermal pollution
Waste	Inert, solid, industrial and hazardous

TABLE 2.3

Global Reserves and Production of Metals for 2013

Ore	Reserves (t)	Annual Production (2013) (t)	Depletion Rate at 2013 Production Yields (Years)	5 Year Change in Production (2008–2012) (%)	Adjusted Depletion Rate Based on Annual Production Change (Years)
Iron ore	1.7×10^{11}	2,950,000,000	57.5	17	33
Bauxite	2.8×10^{10}	259,000,000	108	16	48
Copper	690,000,000	17,900,000	38.5	8.5	30
Lead	89,000,000	5,400,000	16.5	38	12
Zinc	250,000,000	13,500,000	18.5	7	17
Nickel	74,000,000	2,490,000	30	10	24
Gold	54,000	2,770	20	−13.5	28

Source: Adapted from U.S. Geological Survey. 2014. Mineral Commodity Summaries 2014, Washington DC; Armitage, G. 2013. *Resources and Energy Statistics*. Canberra, Australia: BREE, 4 December.

100 years estimated ore reserves, followed by iron ore and copper reserves. The largest depletion rates are estimated for lead, zinc and gold ores. All metal production rates increased in the 5 years period between 2008 and 2012, except for gold production, which decreased by more than 10%. The largest increase in production was evidenced for lead, the metal that has the least projected reserves. When taking into consideration the annual production increase based on an average 5 year production rates, the depletion rates of the ore reserves will further decline for all ores, except for gold, which has somewhat prolonged projected life expectancy. The depletion rates are based on the projected reserves of ores with the ore quality mined at the time of projection. The future of the sustainability of the industry will rely on new and more efficient extraction methods of lower quality and low concentration ores, which can substantially extend the life expectancy of the global ore production and processing, but are not included in the projections in Table 2.3.

Production of metals is an energy-intensive process that also needs withdrawal of freshwater to sustain production. Table 2.4 summarises the average estimates of water consumption for metal production, the energy requirement to produce a tonne of metal and subsequent greenhouse gas emissions. Where pyrometallurgical and hydrometallurgical processing of the ores is possible, such as in case of copper and nickel, the values for the pyrometallurgy were selected for assessment. Water consumption is an important parameter for countries with scarce water availabilities, such as Australia. Mining industries and metal production in Australia consume approximately 2% of Australia's total water consumption, however, this consumption is in geographical locations with very low natural water availability (Norgate and Lovel, 2004). Water consumption for various metal production industries ranges from 2.9 m³/t for steel production to 252,087 m³/t for production of gold. These figures are estimates accounting for various stages of water withdrawals, ranging from mining, ore processing, smelting and metal refining. Water and energy consumptions are directly dependent and related to the metal content in the ore that needs to be extracted. Norgate and Lovel (2004) provide an expression for correlation between water consumption (W in m³/t refined metal) and the grade of ore (G in percentage of metal in the ore)

$$W = 167.7 \times G^{-0.9039} \tag{2.1}$$

Mineral processing and production of metals are significant energy consuming processes that require energy for mining, processing, smelting and refining of the final metal product. The energy

TABLE 2.4

Water Consumption, Energy Consumption and Greenhouse Gas Emissions from Production of Metals Based on Metric Tonne of Produced Metal

Metal	Water Consumption (m³/t)	Energy Consumption (GJ/t)	Greenhouse Gas Emissions (tCO$_{2e}$/t)
Iron and steel	2.9	20.9	2.1
Aluminium	35.9	211.5	21.8
Copper	25.9	33	3.3
Lead	12.6	19.6	2.1
Zinc	21.2	48.4	4.6
Nickel	79	113.5	11.4
Gold	252,087	200,000	18,000

Source: Adapted from Norgate, T. E. and W. J. Rankin. 2000. *Proceedings, Minprex 2000, International Conference on Minerals Processing and Extractive Metallurgy*, September, pp. 133–138, Melbourne, Australia; Norgate, T. E. and R. R. Lovel. 2004. Water use in metal production: A life cycle perspective. CSIRO Report DMR-2505, Clayton, Australia; Price, L. K. et al. 2011. *ACEEE Industrial Summer Study*, American Council for an Energy-Efficient Economy, New York, July; Rankin, J. 2012. *4th Annual High Temperature Processing Symposium 2012*, Swinburne University, Melbourne, Australia, pp. 7–9, 6–7 February; Strezov, V., A. Evans, and T. J. Evans. 2013. *Journal of Cleaner Production*, 51: 66–70.

is consumed either directly for the process, for instance coke used for sintering and ore reduction, or indirectly through electricity use during processing. The energy intensity and greenhouse gas emissions for production of metals stated in Table 2.4 are for the full life cycle from ore mining to metal refining. The greenhouse gas emissions are estimates of the direct and indirect greenhouse gas emissions assuming production of electricity from black coal at 35% efficiency (Rankin, 2012). The main energy intensive and greenhouse gas emitting industry is the production of gold, followed by aluminium production. Gold production requires large energy input due to the lowest grade of metal in the ore. Aluminium production has high energy input due to the use of electricity in the electrolytic process. Iron and steel and lead production have the lowest energy densities and greenhouse gas emissions.

EMISSION PARAMETERS

The emission parameters of metal production for the priority pollutants to air and non-recyclable waste stored on the site or landfilled were determined for each industry. The estimates were performed for individual and representative metal production industries in Australia, using the emission reports for 2012/2013 by the Australian National Pollutant Inventory (NPI, 2015) for each industry and the individual industrial annual performance reports and publicly available information for the individual annual productions.

Table 2.5 shows the selected representable industries for each of the produced metals and their corresponding publicly available sites for sourcing the information for annual production used to calculate the emission of pollutants per tonne of metal, shown in Table 2.6. The data presented in Table 2.6 are based on direct emissions, as reported for the national emission pollutant inventory. When the industry produces more than one type of metal, for instance, the specific nickel industry also produces a smaller fraction of cobalt, the iron and steel industry produces both iron and steel, all manufacturing proportions were included in the assessment, but the values are assigned to the major metal produced by the specific industry.

TABLE 2.5

Individual Industries Selected as Representatives for Calculation of the Metal Production Emission Rates

Metal	Represented Industry	Public Information Sources
Iron and steel	Onesteel Whyalla	http://www.onesteel.com/businesses. asp?id = 8&pageSource = business https://www.sa.gov.au/__data/assets/pdf_file/0016/14812/ OneSteel_Whyalla_sector_agreement.pdf
Aluminium	Tomago Aluminium Company Pty Ltd	http://www.tomago.com.au/about-us
Copper	Copper Refineries	http://www.mountisamines.com.au/EN/Publications/ Sustainability%20Documents/North%20Queensland%20 2012%20Sustainability%20Report.pdf
Lead	Nyrstar Port Pyrie	http://www.nyrstar.com/operations/Documents/NYR%20 Hydra%20Brochure%20EN.pdf
Zinc	Sun Metals	http://www.northernstevedoring.com.au/projects/sun-metal- case-study.html http://www.sunmetals.com.au//aboutus/company.aspx
Nickel	Queensland Nickel	http://www.brisbanetimes.com.au/queensland/toxic- queensland-nickel-dam-full-environment-minister-20140414- 36mob.html http://www.qni.com.au/about-qn/Documents/Our%20 Refinery.pdf
Gold	Kalgoorlie Consolidated Gold Mines	http://ec.europa.eu/environment/integration/research/newsalert/ pdf/302na5_en.pdf

The estimated emissions from each metal producing industry are significantly variable. Production of nickel emits the largest amount of acidic gases NO_x and SO_2, followed by lead production, which also contributes to the largest emissions of CO per tonne of produced metal. Production of aluminium is a significant emitter of SO_2 and CO to the environment, relative to the weight of produced metal. Production of nickel is the single largest emitter of PM (PM_{10} and $PM_{2.5}$), far outweighing iron and steel manufacturing, which is the second largest emitter of particles to the environment. Gold production has no contribution to the emission of volatiles, but has the largest contribution of emissions of air toxics to the environment, specifically toxic metals. Emission of lead and nickel from gold production far outweighs the relative emissions from the individual industries that specialise in production of these metals. The atmospheric emissions of highly toxic mercury and arsenic from gold production, as well as their concentrations in the produced waste are considerable, prompting the importance of and opportunities for integrating metal recovery technologies to reduce these emissions. Iron and steelmaking emit the largest amounts of organic air toxics, such as dioxins and furans, polycyclic aromatic hydrocarbon (PAH) and volatile organic compounds (VOCs), compared to the other industries.

Each produced metal offers different levels of economic value to the society, as shown previously in Table 2.1. For this reason, the relative comparison of emissions between the different metal production industries can be better represented if expressed based on the value they contribute to the economy. Table 2.7 summarises each individual emission of pollutants expressed in grams of emitted pollutant per US$ value of produced metal. The organic air toxics are emitted to the environment with the highest concentration relative to the economic input of the produced product. For instance, iron and steel manufacturing emits close to 2 t of VOCs and close to 28 kg of benzene per US$ of contribution to the economy. The aluminium, copper, lead and zinc industries also have significant emissions of VOCs relative to the income from the produced metal.

TABLE 2.6

Emission Rates for the Priority Air Pollutants, Metals, Air Toxics and Waste Expressed in Kilograms

	Iron/Steel	Aluminium	Copper	Lead	Zinc	Nickel	Gold
			Emissions to Air (kg/t)				
NO_x	1.8	0.63	0.05	1.1	0.06	293	0
SO_2	0.75	17.5	4.5×10^{-4}	352	1.5	140	7×10^6
CO	50	75	0.037	145	0.038	5.7	0
PM_{10}	2.2	0.16	3.6×10^{-6}	0.84	0.064	40	0
$PM_{2.5}$	0.01	0.07	3.6×10^{-6}	0.07	5×10^{-3}	0.77	0
Pb	7×10^{-5}	2×10^{-4}	5×10^{-6}	0.26	4.8×10^{-3}	5×10^{-3}	2.4
Hg	1.2×10^{-6}	1×10^{-5}	1.1×10^{-7}	3.5×10^{-4}	1.2×10^{-6}	1.2×10^{-4}	129
As	2.8×10^{-5}	8.4×10^{-5}	8.2×10^{-6}	0.006	8×10^{-5}	4×10^{-4}	39
Cr^{6+}	3.2×10^{-6}	1.3×10^{-5}	0	7.3×10^{-5}	0	2.5×10^{-4}	0
Ni	6.2×10^{-5}	3.6×10^{-4}	1.1×10^{-5}	9.5×10^{-3}	4.4×10^{-6}	0.9	153
Cd	2.8×10^{-6}	6.2×10^{-5}	1×10^{-6}	5.2×10^{-3}	1.1×10^{-4}	2.1×10^{-4}	0
Dioxins and furans (TEQ)	1.1×10^{-6}	4.8×10^{-9}	0	5×10^{-10}	3×10^{-10}	0	0
PAH (B[a]Peq)	0.56	0.032	1.8×10^{-4}	8×10^{-4}	3.6×10^{-4}	0	0
VOCs	1937	7.4	1.1	2.5	1	0	0
Benzene	27.8	0	0	0	0	0	0
			Waste (kg/t)				
As	2.7×10^{-4}	0	0.75	0	1×10^{-6}	0	7055
Cd	5.9×10^{-4}	0	0	0	0	0	0
Pb	5.1×10^{-3}	0	0.05	0.78	8.8×10^{-3}	1.37	99.6
Hg	9.2×10^{-5}	0	0	0	0.056	4×10^{-4}	66.4
Ni	2.1×10^{-3}	0	0.3	0	2.2×10^{-5}	153	913

Each substance poses different risks to the environment. The Australian NPI Technical Advisory Panel (NPI, 1999) estimated the risk to the health and environment for approximately 400 substances. Each substance was evaluated for human health effects, environmental effects and exposure using Equation 2.2:

$$\text{Risk} = \text{Hazard}(\text{Human Health} + \text{Environment}) \times \text{Exposure} \qquad (2.2)$$

The equation also takes into account the acute and chronic toxicity on health and environment, including reproductive toxicity, carcinogenicity, persistence and bioaccumulation. The exposure in Equation 2.2 was evaluated based on point sources, diffuse sources and bioavailability of each substance. Table 2.8 summarises the normalised health (H), environmental (En), exposure (Ex) and total NPI risk scores estimated according to Equation 2.2 for the priority volatile and air toxic pollutants. VOCs as a group and $PM_{2.5}$ were not included in the NPI (1999) assessment and in further comparisons are only weighed as a risk factor of 1. NO_x (NO and NO_2) is the highest ranked pollutant substance with the highest risk score of 13.5. The second substance is hexavalent chromium followed by carbon monoxide and sulphur dioxide.

Normalised emission rates based on the amount, risk and the contribution to the economy are summarised in Table 2.9 for the priority volatiles (NO_x, SO_2 and CO), metals (Pb, Hg, As, Cr^{6+}, Ni and Cd), organic air toxics (dioxins/furans, PAH, VOCs and benzene) and waste. The normalised

TABLE 2.7

Emission Rates for the Priority Air Pollutants, Priority Metals, Air Toxics and Waste Based on the Value Created for the Economy

	Iron/Steel	Aluminium	Copper	Lead	Zinc	Nickel	Gold
			Emissions to Air (g/US$)				
NO_x	2.2	0.3	6×10^{-3}	0.5	0.03	16.7	0
SO_2	0.9	8.6	5×10^{-5}	170	0.8	8	120
CO	60	37	0.004	70	0.02	0.32	0
PM_{10}	2.6	0.08	4.5×10^{-4}	0.4	0.03	2.3	0
$PM_{2.5}$	0.013	0.035	4.5×10^{-4}	0.03	2.5×10^{-3}	0.04	0
Pb	8×10^{-5}	10×10^{-5}	6×10^{-7}	0.12	2×10^{-3}	3×10^{-4}	4×10^{-5}
Hg	1.5×10^{-6}	5×10^{-6}	1×10^{-8}	1.7×10^{-4}	6×10^{-7}	7×10^{-6}	2×10^{-3}
As	3.4×10^{-5}	4×10^{-5}	1×10^{-6}	3×10^{-3}	4×10^{-5}	2×10^{-5}	6.6×10^{-4}
Cr^{6+}	3.8×10^{-6}	6×10^{-6}	0	3.5×10^{-5}	0	1.5×10^{-5}	0
Ni	7.5×10^{-5}	1.8×10^{-4}	1.4×10^{-6}	4.6×10^{-3}	2.2×10^{-6}	0.051	2.6×10^{-3}
Cd	3.4×10^{-6}	3×10^{-5}	1.3×10^{-7}	2.5×10^{-3}	5.7×10^{-5}	1.2×10^{-5}	0
Dioxins and furans (TEQ)	1.1×10^{-3}	4.8×10^{-6}	0	5.2×10^{-7}	3.5×10^{-7}	0	0
PAH (B[a]Peq)	557	32	0.18	0.8	0.36	0	0
VOCs	1.9×10^6	7380	1090	2540	1000	0	0
Benzene	27.8×10^3	0	0	0	0	0	0
			Waste (Non-Recyclable) (mg/US$)				
As	0.32	0	94	0	5.3×10^{-4}	0	120
Cd	0.7	0	0	0	0	0	0
Pb	6.2	0	6.6	300	4.5	72	1.7
Hg	0.11	0	0	0	29		
Ni	2.5	0	38	0	0.011	8060	15.5

values are useful for comparison purposes to determine the highest risks from grouped pollutants and the largest risk category relative to the industrial activity. For instance, the production of lead poses the largest relative risk to the environment from emission of priority volatiles and priority metals, when compared to the other metal production industries, while iron and steel production pose the largest environmental risks from organic air toxics emissions, largely associated with production of coke. The nickel industry produces solid waste with the most significant risk factor.

SOCIAL CONSIDERATIONS FOR INDUSTRIAL SUSTAINABILITY

The contribution of industrial activities to local communities, employment and social wellbeing is of significant importance to their sustainability. Table 2.10 summarises the number of employed people per million of US$ of annual income estimated for the representative case studies from Table 2.5 and excludes production costs. Table 2.10 shows that lead, nickel and iron/steel manufacturing industries employ the largest number of people per income. Copper, followed by gold production, employ the lowest number of people relative to the income of these industries.

SUSTAINABILITY RANKINGS OF MINERAL PROCESSING INDUSTRIES

Sustainability of industrial activities can be assessed based on a set of indicators to compare relative weightings of individual indicators between different industrial activities. In this work, the

TABLE 2.8

Normalised Health (H), Environmental (En), Exposure (Ex), Total NPI Risk Scores for the Priority Volatile and Air Toxic Pollutants, and Their National Rank

	H	En	Ex	Risk Score	Rank
NO_x	1.5	3.0	3.0	13.5	1
SO_2	1.5	1.3	3.0	8.5	3
CO	2.0	0.8	3.0	8.5	3
PM_{10}	1.2	1.3	3.0	5.5	9
Pb	1.7	1.5	2.2	5.3	14
Hg	1.7	2.0	1.3	4.6	35
As	2.3	1.7	1.8	7.0	9
Cr^{6+}	2.5	3.0	1.8	7.3	2
Ni	1.2	1.0	1.7	3.6	54
Cd	2.3	2.0	1.8	7.6	6
Dioxins and furans	2.6	0.7	0.8	2.6	83
PAH	1.3	1.5	2.3	6.4	18
Benzene	2.3	1.0	2.0	6.7	14

Source: Adapted from NPI. 1999. Technical Advisory Panel, Final Report to National Environment Protection Council, National Pollutant Inventory, Canberra, Australia, January.

industrial activities are ranked based on economic, social, consumption (ore depletion, energy and water consumption) and a set of environmental parameters, which include greenhouse gas emissions, PM ($PM_{2.5}$) emissions, emissions of priority volatile pollutants, priority metals, organic air toxics and priority metals in the non-recyclable waste. Table 2.11 summarises the sustainability indices where each industrial activity is given a rank between 0, for the highest ranked industry per parameter, and 1 for the least ranked industrial activity. The final sustainability score was estimated as an average of all 10 selected sustainability parameters and each industry was ranked in order of the sustainability score. The assessment shows that iron and steelmaking industries have the highest ranking, closely followed by gold production. Lead production ranks the lowest due to the lowest contribution to the economy, highest depletion, highest emissions of priority volatiles, metals and

TABLE 2.9

Emission Rates of Priority Volatiles, Metals, Organic Air Toxins and Waste Based on the NPI Risk Factors and Economic Input (Risk kg/$)

	Priority Volatiles	Priority Metals	Organic Air Toxins	Waste
Iron and steel	0.54	1.2×10^{-6}	2130	0.06
Aluminium	0.4	1.9×10^{-6}	7.6	0
Copper	1.2×10^{-4}	1.8×10^{-8}	1.1	0.85
Lead	2	9.2×10^{-4}	2.5	2.1
Zinc	7×10^{-3}	1.8×10^{-5}	1	0.16
Nickel	0.3	1.9×10^{-4}	0	29.5
Gold	1	2.5×10^{-5}	0	0.9

TABLE 2.10

Number of Employees per Annual Income for Mineral Processing Industries

	Iron/Steel	Aluminium	Copper	Lead	Zinc	Nickel	Gold
Number of employees per million US$	1.8	0.85	0.12	2	0.7	1.9	0.38

TABLE 2.11

Comparison of Sustainability Parameters and Ranking of Mineral Processing Industries

Sustainability Index	Iron/Steel	Aluminium	Copper	Lead	Zinc	Nickel	Gold
Economic index	0	0.79	0.95	1	0.99	0.99	0.94
Depletion	0.42	0	0.5	1	0.86	0.67	0.56
Water consumption	0.02	1	0	0.2	0.52	0.09	0.07
GHG	0.21	1	0.01	0.07	0.2	0.03	0
$PM_{2.5}$	0.3	0.8	0.01	0.74	0.06	1	0
Priority volatiles	0.26	0.19	0	1	0	0.15	0.49
Priority metals	0.001	0.002	0	1	0.02	0.2	0.03
Organic air toxins	1	0.004	0.0005	0.0012	0.0005	0	0
Waste	0.002	0	0.03	0.07	0.006	1	0.03
Social index	0.007	0.086	1	0	0.11	0.004	0.27
Sustainability score	0.22	0.39	0.25	0.51	0.28	0.41	0.24
Rank	1	5	3	7	4	6	2

substantial emissions of PM. Nickel production ranks the second lowest, due to low global economic contribution, high PM emissions, high non-recyclable waste and considerable depletion rates. Aluminium production has the largest impact on water consumption and greenhouse gas emissions.

Assessment of the sustainability indicators of the mineral processing industries gives a better understanding of the areas for improvement of each industrial activity. For instance, the iron and steelmaking industry needs to address the substantial emissions of organic air toxics as their environmental impact score relative to the economic input weighs orders of magnitude higher than the other mineral processing industries. Similarly, the environmental risk from toxic metal emissions from lead production is significantly higher than the other metal production industries. Some parameters, such as greenhouse gas emissions from aluminium production, are well-known problems that the industries are trying to address. One possible solution is integration of renewable energy sources, such as wind, solar or hydropower, for production of aluminium, which would significantly improve the sustainability ratings. The social parameters, however, specifically the employment rates per income of the industry, have been overlooked and need to be also addressed. Copper and gold production industries can improve their social contribution through employment and social growth.

CONCLUSIONS

Mineral processing industries provide strong input into the world economy, but also have substantial environmental impacts. This work presents assessment of seven mineral processing industries, iron and steel production, aluminium, lead, zinc, copper and gold production and provides assessment of their contribution to sustainable development through analysis of economic, environmental and social parameters. The environmental parameters considered here were broad and involved consumption parameters, greenhouse gas emissions, emissions of PM, atmospheric emissions of priority volatiles, metals, organic air toxics and non-recyclable waste produced during metal production. Australian representative industries of each metal production technology were used as case studies for the evaluation of environmental performance. The results revealed lead manufacturing as the lowest ranked industry, due to the largest depletion rates of the ore and largest emissions of priority volatiles and metals to the environment, relative to the contribution to the economy and environmental risk. Nickel production also ranked low due to the largest quantity of $PM_{2.5}$ emissions and waste generated during the process. Aluminium production has the largest impact on greenhouse gas emissions and water consumption. Iron and steel manufacturing had the highest sustainability ranking, but showed the largest emission of organic air toxics, a category that appears to be least addressed, as the emitted amounts are the largest relative to their risk and economic contribution. This assessment reveals the areas of consideration for each industry that can be addressed to improve their contribution to sustainable development.

REFERENCES

Arena, M. and G. Azzone. 2010. Process based approach to select key sustainability indicators for steel companies. *Ironmaking and Steelmaking: Processes, Products and Applications*, 37: 437–444.

Armitage, G. 2013. *Resources and Energy Statistics 2013*. Canberra, Australia: BREE, 4 December.

Fruehan, R. J. 2009. Research on sustainable steelmaking. *Metallurgical and Materials Transactions A*, 40B: 123–133.

NPI. 1999. Technical Advisory Panel, Final Report to National Environment Protection Council, National Pollutant Inventory, Canberra, Australia, January.

NPI. 2015. Department of Environment, National Pollution Inventory Australian Government, Canberra, Australia. Accessed January 22, 2015.

Naylor, D. 2005. Industry perspective: Sustainability of ferrous metallurgy and the steel industry. *Ironmaking and Steelmaking: Processes, Products and Applications*, 32: 97–100.

Norgate, T. E., S. Jahanshahi, and W. J. Rankin. 2007. Assessing the environmental impact of metal production processes. *Journal of Cleaner Production*, 15: 838–848.

Norgate, T. E. and R. R. Lovel. 2004. Water use in metal production: A life cycle perspective. CSIRO Report DMR-2505, Clayton, Australia.

Norgate, T. E. and W. J. Rankin. 2000. Life cycle assessment of copper and nickel production, *Proceedings, Minprex 2000, International Conference on Minerals Processing and Extractive Metallurgy*, September, pp. 133–138, Melbourne, Australia.

Price, L. K., A. Hasanbeigi, N. T. Aden, Z. Chunxia, L. Xiuping, and S. Fangqin. 2011. A comparison of iron and steel production energy intensity in China and the US, *ACEEE Industrial Summer Study*, American Council for an Energy-Efficient Economy, New York, July.

Rankin, J. 2012. Energy use in metal production. *4th Annual High Temperature Processing Symposium 2012*, Swinburne University, Melbourne, Australia, pp. 7–9, 6–7 February.

Singh, R. K., H. R. Murty, S. K. Gupta, and A. K. Dikshit. 2007. Development of composite sustainability performance index for steel industry. *Ecological Indicators*, 7: 565–588.

Strezov, V., A. Evans, and T. J. Evans. 2013. Defining sustainability indicators of iron and steel production. *Journal of Cleaner Production*, 51: 66–70.

U.S. Geological Survey. 2014. Mineral Commodity Summaries 2014, Washington DC.

Worldsteel Association. 2012. *Sustainable Steel Policy and Indicators 2012*, Worldsteel Association, ISBN 978-2-930069-71-5, Brussels, Belgium.

3 A Step Change in Mining Productivity
Time to Deliver the Promise[*]

Joe Pease, Stephen Walters, Luke Keeney, and Greg Shapland

CONTENTS

ABSTRACT

The mining industry has long sought a step change in productivity by integrating operations from mine to market. While there have been some success stories, in general the promise has not been delivered due to some crucial gaps in technology and systems. Many of those gaps have now been closed, or at least recognized, meaning the tools are now available to deliver the benefits of integration.

THE MINING PRODUCTIVITY CHALLENGE

The minerals industry faces a productivity and investment crisis. The "Millennium Super Cycle" from 2003 to 2011 was an unprecedented period of growth and investment. Throughput was increased and lower grade resources were developed to meet demand. The urgency to bring production to market quickly stretched people, project, and management resources. But the "boom" ended and prices have declined, with the industry left with a legacy of high costs, declining ore quality, and less efficient operating practices. Step changes to practice and productivity are needed to sustainably produce the minerals society needs.

Groups including the Cooperative Research Centre for Optimising Resource Extraction (CRC ORE) are working with the global resources industries to reverse the trend of declining feed grade and quality through novel approaches and innovative solutions (Figures 3.1 and 3.2).

TRANSFORMING MINING PRODUCTIVITY: AN INTEGRATED APPROACH

CRC ORE was established in 2010 to address these productivity challenges. It is a large scale, industry-led initiative that brings together orebody knowledge, mass mining, mineral processing,

---

[*] This Chapter is based on an article first published in the Australasian Institute of Mining and Metallurgy Bulletin of March 2015.

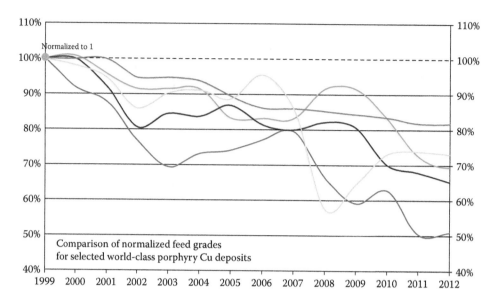

FIGURE 3.1   Decline in feed grades for copper operations.

and resource economics. CRC ORE provides a bridge between technology development and site implementation, working with a consortium of global mining companies, mining equipment, technology and services providers, and research organizations. The aim is to achieve a step change in productivity by adopting an integrated, manufacturing-style approach to the production of metals from drill core to product. In particular, there is a focus on improving feed quality early in the production value chain.

The minerals industry has long sought improvement by integrating operations from mine to metal. In the 1990s, this was the driving force behind "Mine-to-Mill." The disappointing thing was not that Mine-to-Mill failed to deliver value. The disappointing thing was that it did so successfully

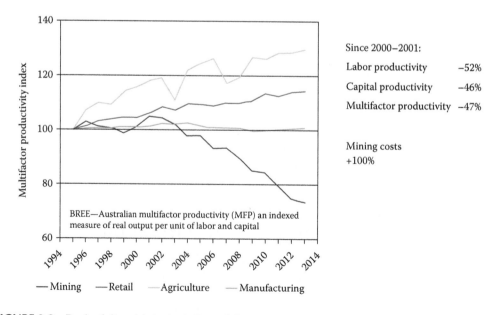

FIGURE 3.2   Productivity crisis in Australian mining.

and yet then withered at many sites. It was widely accepted—but not as widely adopted. Even worse, many of the successful "poster" sites gradually reverted to the traditional "silo" approach of managing operations.

CRC ORE has worked to understand why those good intentions failed to deliver or be sustained. It was not because industry did not try, but rather because crucial gaps in technology and systems meant solutions were not "robust" enough to survive the operating environment. CRC ORE seeks to close those gaps to enable robust, long term, and integrated systems for minerals production.

CLOSING THE GAPS

CRC ORE believes that the barriers to integration were *measurements, ore heterogeneity, integrated systems,* and *supporting management systems.* Therefore, the industry should identify and develop solutions in these areas. Tools have to be combined, assembled, and "ruggedized" on other sites, and management and organization systems must support and "lock in" the changes.

MEASUREMENTS

A manufacturing plant controls its feed within strict limits, setting specifications and measuring to ensure compliance. In contrast, in mining the feed quality is variable. In most cases, we cannot measure quality as it enters the production process. We measure and control so many variables on a mining truck that we can operate it autonomously from the other side of the world. But we have almost no measure of the most important thing, the quality of the payload, the feed to the metal manufacturing (Figure 3.3).

Providing routine on-line feed quality measurements of coarse run of mine (ROM) ores or mill feed is technically challenging. However, ongoing developments in sensor technologies in fields such as neutron activation, magnetic resonance, and laser-induced breakdown spectroscopy offer potential solutions. In some cases, these technologies are already being used at feed belt scale in rock-based industries where maintaining feed quality specification is crucial to generating saleable

FIGURE 3.3 Measurement problem.

FIGURE 3.4   Vision for early measurement and separation.

product, for example, cement manufacture. The next step is to prove their application and business value for bulk base and precious metal mining. Measurement of grade early in production opens new fields of possibility for coarse separation.

A medium-term vision is to sense the grade in every loader bucket or transfer point, to enable a separation decision at multiple points. This requires the development of more robust and compact measurement hardware. But the technology already exists to measure grade on conveyors. This could identify intervals of high-grade or low-grade ore that could be diverted to separate destinations. This will enable some important applications of grade engineering. An ultimate manifestation would be an in pit conveying and separating system as shown below, yet there are many less ambitious options that can be applied now (Figure 3.4).

HETEROGENEITY: FRIEND OR FOE?

Mineral deposits are heterogeneous. The common response is to attempt to smooth or blend the feed to downstream processing. This smoothing process often starts early in data collection, so even mine block models do not capture the full heterogeneity of the orebody. While smoothing makes sense once a flow sheet is settled, perversely it has also hindered the adoption of manufacturing principles in the design of new flow sheets.

CRC ORE believes that a manufacturing approach to integration must start with an acceptance of the things that fundamentally differ between mining and manufacturing—measurement difficulty and feed heterogeneity. Rather than try to *eliminate* heterogeneity by smoothing, the early stages of mining, and processing should embrace and *exploit* it. This is the principle behind "Grade Engineering™." It means identifying and removing low-grade unprofitable materials as early as possible, whenever possible, however possible.

The concept of removing uneconomic material rather than smoothing it through plant feed is not new—it was practiced by our forebears as hand-picking; is practiced in some plants as dense medium separation, and in others as ore sorting. In a well-reported example, Bougainville Copper upgraded sub-marginal ore by screening to remove coarse low-grade rocks, exploiting a natural preferential deportment of copper values to fines. The impact can be enormous—the dense medium plant at Mount Isa lead zinc removes about 35% of coarse and hardest feed before the fine grinding treatment process. That increases the throughput, reduces the capital intensity, and reduces energy requirement by over 40%.

Yet, it seems the principles of early waste removal are not always considered in the design phase of a mine. Many operations design the "standard" circuit of stockpile-conveyor-semi-autogenous grinding (SAG) mill-ball mills-flotation. Once selected, this circuit prefers a smoothed feed, and the materials handling and feed sizing prevents the application of most coarse separation options. This flow sheet may well be the best solution for an ore; but that can only be judged after options to exploit coarse ore heterogeneity have been examined. An interesting question is this: if your company was developing Bougainville or Mount Isa lead zinc today, would it consider a coarse screening plant or dense medium plant? Or would it just accept the higher capital, higher operating cost of the "standard" circuit?

COARSE UPGRADING OF ORES

Every ore is different. Different areas of the same orebody are different. Therefore, any solution must recognize that there is not a general solution, but there can be a general approach. One way to categorize and compare solutions is the concept of Grade Engineering® developed by CRC ORE. This is a toolbox of analytics and techniques to assess the potential to apply coarse upgrading to any ore. A range of possible separation techniques can be assessed, and the response of that ore ranked relative to other orebodies. CRC ORE has characterized and assessed a large database of global ores using this approach. On some sites, the heterogeneity does not support a business case; on other sites, a significant business case has emerged. The increasing database of industry studies means faster and more accurate "desktop" assessment for new operations.

Five potential coarse separation mechanisms are:

- *Induced sized deportment by preferential blasting*—for example, blast higher grade zones fine and low-grade zones coarse, then separate by coarse screening.
- *Natural size deportment*—that is, exploit natural tendency of valuable minerals to concentrate in fines; upgrade ore by screening out coarse low-grade rocks.
- *Coarse gravity separation*—exploit coarse gangue liberation by removing it before grinding, for example, dense medium, jigs.
- *Sensor-based mass sorting*—measure (or infer) grade, divert low-grade batches or conveyor intervals to waste.
- *Sensor-based particle sorting*—measure distinctive characteristic of valuable ore gangue, and eject individual particles.

CRC ORE has tested a wide range of ores and has developed protocols to place an ore on "response ranking curves" to assess coarse separation potential to each of these mechanisms relative to other ores (Figure 3.5).

Any orebody may respond to one or more (or none) of these levers, and they may combine to increase effect. For example, *induced* size deportment does not rely on *natural* size deportment but may be enhanced by it. Induced size by differential blasting will also increase mill throughput because of the finer mill feed. The most appropriate mechanism(s) will be determined by the characteristics of the mineralization and the heterogeneity of the deposit. Dense medium separation or natural size deportment will not suit disseminated mineralization. Yet, the orebody may exhibit significant variation in grade across the production bench. This could be exploited by grade sensing in belt or bucket and diverting low-grade intervals to waste. Alternatively, differential blasting can

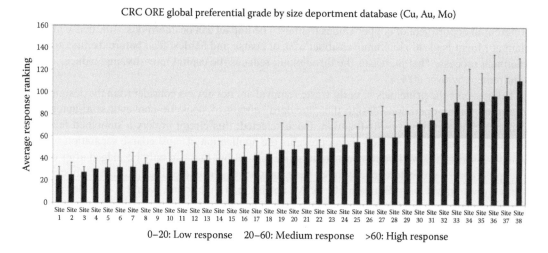

FIGURE 3.5   Response ranking curve for natural grade deportment by size.

induce a size difference between high- and low-grade zones in the pit, with the low-grade coarse fraction removed by screening. The results of a full scale site demonstration of this are depicted in Figures 3.6 through 3.8.

This technique increases mill feed grade by diverting below-cut-off-grade material. While this reduces metal feed to the mill, this can be recovered by similarly recovering small high-grade areas from waste benches. Thus, metal production rate can be maintained or increased with a higher mill

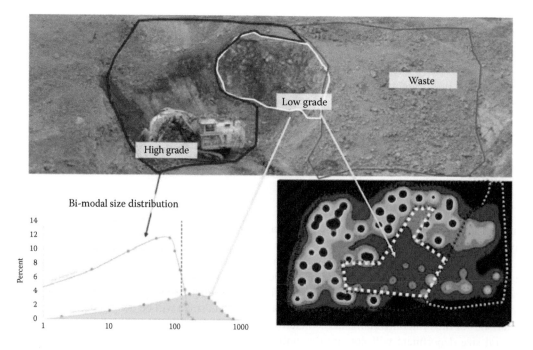

FIGURE 3.6   Differential blasting. The high-grade zone is blasted to a fine size by close blast-hole spacing and high powder factor. Minimal blasting in the lower grade zone produces a coarser sizing. The high- and low-grade zones can be mined together and separated by screening.

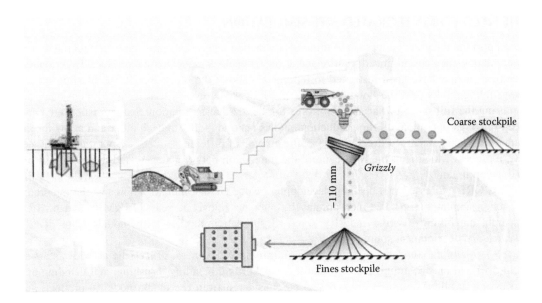

FIGURE 3.7 Flow sheet for site demonstration of selective blasting.

feed grade. Economic streams can be generated from previously sub-marginal ore. This is a step change in productivity using simple technology—blasting, belt sensing, diverters, and screening. The technology components are available, but first the business case for various options must be assessed. Then, the components must be assembled and engineered into a robust solution. These are the objectives of CRC ORE.

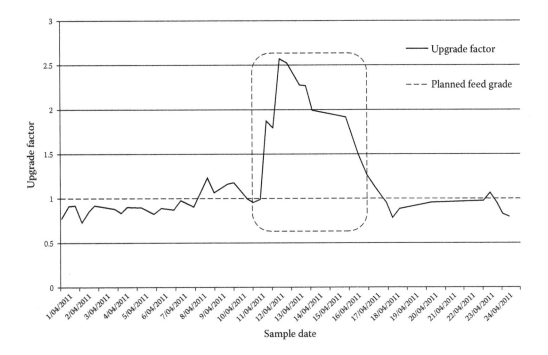

FIGURE 3.8 Results of site demonstration of selective blasting and screening.

THE NEED FOR INTEGRATED SITE SIMULATION

To support the integrated approach to minerals extraction, operations must take an integrated view to simulating, then optimizing, the entire operation, from block model to final product. Traditionally, technical models have been restricted to individual "silos": the block model, the blasting model, the grinding model, and the flotation model. Some of those models were single dimensional, not capturing the full range of variability of ores (grade, hardness, flotation characteristics, etc.). Yet to optimize an operation, all the production models need to "talk" to each other and allow for the multidimensional nature of ore. Otherwise, the outputs of different models are put together manually, a process which is unlikely to capture the interaction between stages, and therefore will not find the optimum.

CRC ORE addressed this need by developing a value chain simulator known as the integrated extraction simulator (IES). IES incorporates the outcomes of over 50 years of JKMRC and AMIRA research, combining existing industry standard simulation models with models from diverse research and development sources.

IES is a mining simulator that integrates all mining and mineral processing activities starting from drill and blast, through loading, hauling, stockpiling, blending, crushing, and grinding and processing. It allows multiple ore types to be considered from the block model to product, and allows for the interaction between those steps. For example, IES allows operators to assess future ore sources for the effect that changes in blasting will have on grinding and how this will impact flotation. Therefore, it can be used to assess changes in the design, layout, and operating steps to optimize metal production and environmental footprint. It provides a model development environment that allows the user to access or input models that suit their specific equipment.

SUPPORTING MANAGEMENT AND ORGANIZATION SYSTEMS

The technology and techniques for a step change in industry productivity are well within reach. The next phase of the program is to assemble and demonstrate them in high value site applications.

But to achieve the value and to lock in the gains, the technology must be supported with appropriate management and control systems. The integrated approach to production must be matched with a similar management approach. Though every organization supports this principle, often existing management and reward systems inadvertently hinder it. Careful design of key performance indicators (KPIs), targets, and incentives is required to ensure that individual efforts combine to optimize the overall site, and not isolate activities into "silos." For example, the drill and blast crew should be rewarded (not penalized) for increasing their unit cost if it increases site productivity. This principle is well understood, yet KPIs remain insidious barriers to genuine integration. Researchers, technical staff, senior management, and system providers need to work together to develop the tools and business support for integrated operations.

THE TIME IS RIGHT

The Mine-to-Mill initiatives of the 1990s showed that better coordination of activities will yield significant productivity gains. Yet, the changes were not robust enough to be maintained.

After an era of major expansion to meet the "supercycle," the minerals industry now desperately needs to increase productivity as both feed quality and prices decline.

The tools, both technical and analytical, are now much better developed. They can now be combined to quickly assess ores and options, and in many cases to demonstrate significant improvement. The technology is simple. Further engineering and site demonstration is needed to make it robust and reliable. Then, it needs to be supported with appropriate organization and management systems.

Finally, the time has arrived and the tools are within reach to truly integrate mining operations.

4 An Approach to Improve the Energy Efficiency of Mining Projects

Sarma S. Kanchibotla and Tapiwa Chenje

CONTENTS

ABSTRACT

As more nations strive to become part of the developed world, the demand for commodities continues to increase rapidly. However, the high-grade and easy-to-mine deposits are mostly on the decline and future mineral deposits are likely to be deeper with higher stripping ratios, lower grade and in more remote locations with little infrastructure and skilled labour pools. The reduction in feed grades will increase the comminution energy demand per unit of metal produced. In addition to decreasing grades, the amount of waste and haul distances will increase as the deposits get deeper thus significantly increasing the haulage energy requirement. Deeper pits also increase the risks of slope stability and wall failure and if not managed in a timely manner slope failures not only pose risk to regular production but also significantly increase the energy signature of the operations by increasing unplanned haulage. Producing commodities from such deep and low-grade deposits is likely to substantially increase the energy (fuel, electricity and embodied) requirements per unit of mineral/metal produced as well as the carbon footprint of the operations. Embodied energy is the quantity of energy necessary for the fabrication of material or consumable used in the process. In addition to higher energy requirements and operating expenses (OPEX), mining deeper and lower grade deposits also require higher capital expenditure (CAPEX), which has become a significant impediment to investment.

The value chain in most mining operations involves a number of complex inter-dependent processes in which each process is sensitive to a range of rock mass attributes within the mine block that are transformed by the previous processes. The value created per ton of rock within that mine block is the difference between the price it commands when sold as the final product and the cost to produce it. Even though it is obvious total mine-to-mill processes are interdependent, traditionally, mining and milling are managed separately right from feasibility studies to day-to-day operations. Such an approach will not only increase the risk of the project but also may ultimately lead to sub-optimal operations.

During the past 15 years, the Julius Kruttschnitt Centre at the University of Queensland has been involved in implementing a holistic methodology 'mine-to-mill optimisation' to improve the efficiency of mining operations. One of the key components of this approach is proper understanding of rock mass attributes and its variability in terms of its breakage and separation during different processes within the mine-to-mill value chain. In this chapter, the authors use the mine-to-mill approach to demonstrate the energy efficiency of mining projects by using case studies and established process models and simulations.

MINE-TO-MILL VALUE CHAIN OPTIMISATION

The traditional approach of managing mine and concentrator as separate business 'silos' offers little chance to achieve a step change in energy efficiency because in this approach, each area is optimised independently, with little understanding of the real effect on downstream processes and overall energy efficiency. A holistic approach is needed, wherein mining activities are designed to supply the mill with ore characteristics (grade, hardness and size) to increase the overall energy efficiency and productivity of the total production value chain rather than individual unit process without adversely affecting safety, slope stability and environmental risks. A schematic view of this approach is given in Figure 4.1.

Key elements of this approach are an adequate understanding of

- The influence of ore body and rock mass characteristics on the performance of key processes in the total mine-to-mill value chain (e.g. drilling, blasting, loading, hauling, blending, crushing, grinding, recovery and environmental management)
- The influence of the characteristics of the feed on the performance of key processes in the concentrator
- Interactions between the key processes and the ability to use these as leverage to optimise the performance of overall value chain

ROCK MASS CHARACTERISTICS AND ITS IMPACT ON KEY MINE-TO-MILL PROCESSES

It has been very well recognised by the mineral industry that the characteristics of the rock (both as rock mass as well as rock matrix) influence the efficiency of mineral industry processes such as drilling, blasting, slope stability, transportation, comminution and flotation. As the rock mass is a heterogeneous material, it will contain discontinuities at different scales which may range from the microscopic to kilometres long. Depending on the characterisation purpose, certain rock matrix properties and structural properties at a particular scale have greater influence than others. For example, grade, mineralogy, texture and hardness are important for crushing, grinding and separation processes, whereas the structure (+ metre scale), physical mechanical properties and hydrological properties are important for geotechnical, drilling and blasting purposes.

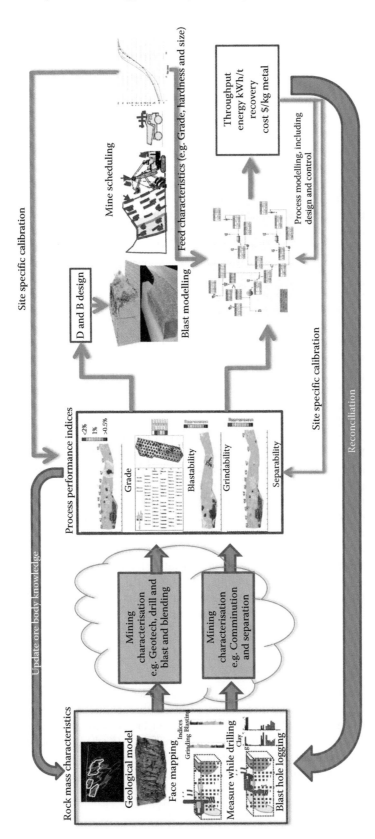

FIGURE 4.1 Holistic approach to optimise mine-to-mill production value chain.

ROCK MASS CHARACTERISATION FOR MINING

The rock mass characterisation for mining needs to consider three main purposes:

- Geotechnical
- Drill and blast
- Mine scheduling and operation

The geotechnical engineer characterises the rock mass mainly from the slope stability point of view. The boundary between rock mechanics and engineering geology relies on the definition of rock and soil in terms of its mechanical resistance and not in terms of structure, texture or weathering (Palmström, 1995). Therefore, the strength and the structural characteristics (scale is usually in metres) of the rock mass are very important for geotechnical characterisation. The strength, localisation and spatial distribution of weak planes are the most important issues for the geotechnical engineer. In many instances, a good knowledge of geological characterisation can help in geotechnical characterisation. For example, the knowledge about the genesis of ore body formation can help to estimate the faults and the structural characteristics of the ore and waste rocks. It may also help in determining the direction of major principal *in situ* stresses that will influence slope stability.

The rock mass characterisation for blasting purposes has received attention from many authors in the last decades. Despite its complexity, it has been demonstrated that certain rock mass characteristics influence the blasting results over others. Structural properties (in fraction of metres up to 10–15 m of scale) play an important role in producing the coarse end of the size distribution, while the matrix strength influences the fine end of the distribution (Scott et al., 1996). The density and the porosity of the rock mass influence the blast induced movement. The impedance (density × sonic velocity) and the hydrological properties influence the explosive selection.

The characterisation for mine scheduling to a large extent depends on grade and mineralogical characteristics of the rock mass. Clearly, the most important aspect of rock mass characterisation is whether the material is ore or waste, which is an economic definition. A rock mass is considered to be ore, if the net value of minerals extracted from that rock mass is more than the costs of extraction which includes both mining, milling and transportation. The value of minerals extracted depends on the price, throughput and recovery. In most mining operations, the mill is fed with more than one ore domain, therefore, it is necessary to understand the process performance of the blended feed.

The prediction of environmental parameters is another important consideration for all mining operations. Social license to operate is a major consideration for most operators and an understanding of rock mass in terms its environmental impacts in waste management (both overburden and tailings) is important to decrease the risk profile of many projects. Full value chain integration cannot be achieved without consideration for the long-term impacts of tailings and waste rock mined through exploitation of the ore body.

ROCK MASS CHARACTERISATION FOR MILLING

The rock mass characterisation for milling needs to consider comminution and separation processes. Comminution is the reduction of size of a particle through the application of mechanical energy during crushing and grinding processes. A number of tests have been developed to relate the energy expended to break the rock resulting in size reduction (Kick, 1883; Bond, 1952). Bond's work index is the most popularly used comminution index to quantify the resistance of the rock to crushing and grinding and is expressed by the equation

$$W = \frac{10\,W_i}{\sqrt{P}} - \frac{10\,W_i}{\sqrt{F}} \tag{4.1}$$

JKMRC has developed the 'Drop Weight Test' to determine the breakage characteristics of rocks and it relates the specific breakage energy to the percentage of the particles reduced to one-tenth of the original size. This test is used to break rocks in a range of sizes under a range of energies. The results are condensed in a relationship between the T10 and the specific breakage energy as follows:

$$T10 = A(1 - e^{-bEcs}) \tag{4.2}$$

where
　A, b = ore specific parameters
　Ecs = specific breakage energy (kWh/t)
　T10 = % of ore passing 1/10th original particle size

The T10 relationship with respect to the rest of the product size curve is expressed as a matrix of Tn values, where n = 2, 4, 25, 50, 75. By convention, the T10 is the % passing 1/10th of the original particle size, whilst the T50 is the % passing 1/50th, etc. The ore specific parameters (A and b) and the breakage matrix define the breakage characteristics of a rock at different energy levels and a detailed description of this test is given by Napier-Munn et al. (1996).

The nature of the minerals present and their association with each other control the ease with which the desired minerals can be separated from the gangue. Clearly, liberation versus particle size and the presence and influence of contaminants on the final product are key factors. Texture is being quantified to relate readily observable features of mineral specimens to their mineralogy and potential processing performance (Napier-Munn et al., 1996). Mineralogical analysis, typically undertaken via QEMSCAN particle mineral analysis or JKMRC's Mineral Liberation Analyser can be used to identify valuable mineral associations including the presence of potential deleterious and penalty minerals to assess opportunities for potential remediation. Liberation assessments will provide insight into grain size and extent of liberation at specific grind sizes to establish flotation feed and regrind sizes for optimum flotation recovery and concentrate grade, particle distribution and theoretical maximum flotation recoveries. Chemical assay, quantitative x-ray diffraction and whole rock analysis assist the calibration of the QEMSCAN measurements.

USE OF HISTORICAL AND OPERATIONAL DATA TO ESTIMATE PROCESS PERFORMANCE INDICES

Laboratory methods used to determine the process performance of a rock mass have some disadvantages. Firstly, they are biased towards stronger rock elements because they survive better in the sample preparation procedures. Secondly, the time required to obtain the data may be too long to implement any design changes if there is a change in rock properties. Thirdly, the laboratory test data do not properly represent the spatial variation of rock properties. Figure 4.2 shows the typical variation in rock mass properties in a mine bench. Depending on the geology, the variation can be in the rock matrix properties (such as strength and density) or in structure (such as massive to fractured) or both. In some situations, there may be two or more rock domains present in one blast.

An alternative is to use the operation data from the mine and mill (e.g. face mapping of exposed faces, grade control drilling, production drilling, size distribution from online systems and mill production data) to update and improve the accuracy of ore body knowledge. For example, the most detailed possible data regarding the characteristics of rock mass can be obtained from blast hole drill monitoring. The drill bit in a drill rig is a kind of a probe that can extract useful information about the rock mass properties. Drilling performance is not only associated with inherent variations of the drill rig but it can also be correlated with geology. With the technological advances in drill monitoring and data capture, it is now possible to monitor every drill hole and interpret the data to describe the rock for various purposes. For instance, based on the drill monitoring data (such as

FIGURE 4.2   Variation in the rock mass in a bench.

rate of penetration, pull down pressure and torque), it may be possible to estimate the strength. The measurement while drilling (MWD) technology offers the possibility of capturing the variations in the rock mass and interpreting them just in time so that it can be used for designing and controlling the blasting and milling operations for achieving optimum results.

Figure 4.3 shows the rock hardness as estimated by the drill monitor for each blast hole in a blast from a copper mine, where the rock in the northern end of the blast is harder than the southern end. Also, there appears to be an almost consistent occurrence of soft material at the top of most holes, which would correspond with the broken ground usually at the collars due to sub-drill damage from previous blasts.

In most cases, the blast hole is not loaded with explosives for a few hours or a few days depending on the size of the blasts. These blast holes can be further investigated with advanced measurements after drilling (MAD) and before they are loaded with explosives and subsequently blasted. The most important aspect is to convert the rock mass data from measurements before drilling, MWD and MAD to predict process performance in both mining and milling operations. These data can then be used to regularly reconcile with the online process performance measurements and update ore body knowledge. For example, on line fragmentation measurements and machine diggability data to assess blasting performance and mill performance data to assess crushing and grinding performance.

IMPACT OF FEED CHARACTERISTICS ON PROCESS PERFORMANCE

An understanding of feed characteristics and how they influence the performance of key processes in the value chain is fundamental to any optimisation. The key feed characteristics that influence the process performance are

- Grade and mineralogy
- Fragmentation or size
- Hardness

GRADE AND MINERALOGY

Grade is the most fundamental characteristic of feed and is usually estimated based on assays of valuable minerals from exploration drilling and/or from grade control drilling. Mischaracterisation of the grade boundaries both prior to, as well as a result of blasting can lead to ore loss and dilution. In standard ore control procedures, various ore zones and waste are excavated based on the pre-blast markings. During a blast, the rock mass within the blast volume is fractured and displaced. The direction and magnitude of movement is a function of ore body shape, free face conditions and blast design parameters. Excavation of ore zones without understanding the blast movement can cause significant dilution and ore loss as shown in Figure 4.4.

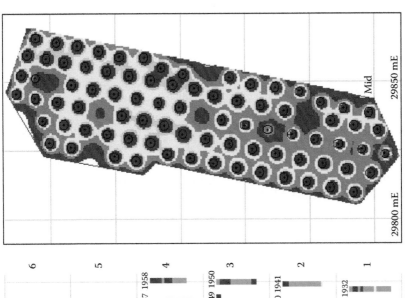

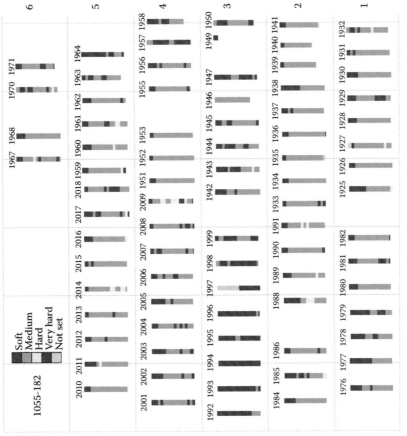

FIGURE 4.3   Rock hardness estimated from measure while drilling data.

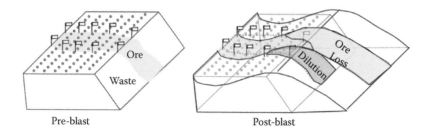

Pre-blast   Post-blast

FIGURE 4.4   Impact of blast movement on ore loss and dilution.

Ore dilution occurs when waste material is miscategorised as ore and sent for processing diluting the run of mine head grade and recovery. Ore loss takes place when valuable mineral is miscategorised as waste and sent to the waste dumps. Therefore, actual head grade in the feed to the mill can vary depending on the dilution and ore loss caused by blast movement, digging and blending processes. In operations where the dilution is a critical issue, blasts are usually designed to minimise any lateral movement of the ore. The success of such practices depends on a trade-off between production efficiency and any ore loss or dilution. However, proper blasting and grade control procedures, coupled with ore boundary adjustments to account for blast movements, can reduce ore loss and dilution (Taylor et al., 1996; Thornton et al., 2005; Thornton, 2009; Engmann et al., 2012; Rogers et al., 2012).

In addition to grade, the nature of the minerals present and their association with each other control the ease with which the desired minerals can be separated from the gangue. Identification of valuable mineral associations and the presence of deleterious and penalty minerals is important to achieve the final concentrate quality. Understanding the extent of liberation at specific grind sizes is necessary to establish optimum grind size to flotation feed and regrind strategies to maximise flotation recoveries.

In ore bodies where mineralisation is concentrated along the veins and fracture planes, increased fracturing and fragmentation along these fracture planes can concentrate the mineralised rock or grade in fines. Advanced blasting and crushing techniques can increase stress concentrations near the veins and fracture planes to promote the liberation of valuable minerals. Preliminary results from recent studies conducted by the author indicated significant grade increase in finer size fractions in certain ore types with advance blasting techniques.

FRAGMENTATION OR FEED SIZE

Fragmentation has a direct influence on the performance of mining production equipment, crushers and grinding mills. The definition of optimum particle size distribution (PSD) depends on site-specific conditions but traditionally blasting engineers are interested in +250 mm size rocks because these size particles are the ones that typically affect the efficiency of the loading and hauling operations which are a part of the mining cost centre.

The PSD from blasting must not only physically fit into the bucket of the excavator but also must do so without unduly reducing the bucket fill time. The oversize fragments not only reduce the productivity of the excavator but also increase the secondary blasting costs and equipment maintenance costs. The definition of oversize depends on the size of excavating equipment and large excavators are more tolerant to large fragments than a small backhoe or front end loader. In addition to over size, the amount of fines and the overall PSD also affect excavator productivity. Generally, the payload of trucks loaded with finely fragmented rock is significantly higher than a truck loaded with coarsely fragmented rock (Michaud and Blanchet, 1995). It is important to note that the maximum truck pay load is also determined by the specifications of the truck and mine operating practices

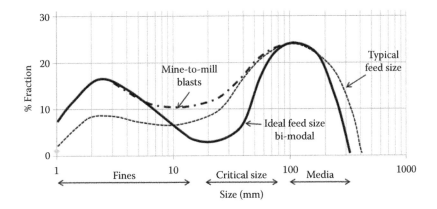

FIGURE 4.5 Feed size distribution for SAG mills. (Adapted from Kanchibotla, S. S. et al. 2011. Energy footprint reduction at KUC. JKMRC Internal Report.)

(i.e. number of passes an excavator takes to fill a truck). In addition to improved loading times, pay loads with finer PSD will also impose less stresses on load and haul equipment and can potentially reduce maintenance costs.

Primary crusher throughput and power consumption are strongly influenced by the run of mine (ROM) particle size distribution (ROMPSD) especially by the coarse particles (Eloranta, 1995; Eliiot et al., 1999). Flakey particles and clayey material coupled with moisture can bog the crusher pocket and affect its performance.

Semi-autogeneous grinding (SAG) mills operate most efficiently when they are fed with a bi-modal feed size distribution shown as the continuous line in Figure 4.5. Ore fractions smaller than the mill discharge grates and trommel screen apertures are considered as 'free grind' material as they pass through the SAG mill and are easily broken down further. Fractions of competent ore between 100 and 200 mm act as grinding media and potentially reduce the required charge of steel balls. Fractions between 20 and 80 mm are generally called 'critical size' and break slowly in SAG mills and hence should be minimised in the mill feed. Ideally, a SAG mill runs efficiently when it is fed with lots of fines (–10 mm), very little critical size and some coarse rock to work as grinding media (Napier-Munn et al., 1996). Impact of critical size on mill efficiency can be minimised by installing large holes in the grate (pebble ports) and crushing them with pebble crushers. However, removing critical size from the mill feed before it reaches the mill is even more beneficial.

SAG mill feed size distribution is influenced by the primary crusher and blasting processes. Generally, primary crushers do not produce much fines (–10 mm) and most of the fines in the SAG feed are produced during blasting (Kanchibotla et al., 1999). Primary crushers are mostly used to reduce the top size from the ROM. An inappropriate feed size distribution to the SAG mill (shown as dashed line in Figure 4.5) will reduce its throughput and make it unstable. Optimising a mineral processing plant with variable feed size is considerably more complicated and challenging compared to when it is fed with consistent feed size distribution.

HARDNESS

The effect of feed hardness is the most significant driver for SAG/ autogeneous grinding (AG) mill performance. The effect is not as pronounced for ball mill operation. In ball mills, the mass of the balls accounts for approximately 80% of the total mass of the charge and dominates both the power draw and the grinding performance of the mills (Wills, 2006). In SAG/AG mills, the ore hardness effect is intertwined with feed size. A proportion of the SAG mill grinding media (or all of it in AG mills) is derived from the feed ore. Any change in the feed size distribution will therefore result in a change in the grinding media size distribution. Any change in the feed ore hardness will affect breakage of the ore, and result in a change in the grinding media size distribution as well. The grinding media size distribution in turn will affect the breakage characteristics of the mill. Associated with the change of the breakage characteristics, the mill charge level will change, which affects the mill power draw. The mill throughput will consequently need to be adjusted to account for the change in ore hardness and maintain a stable power draw.

Laboratory experiments conducted by several researchers suggest that micro-cracks reduce the crushing and grinding resistance of post blast fragments (Revnivstev, 1988; Kojovic and Wedmair, 1995; Nielsen and Kristiansen, 1995; Chi et al., 1996). Preliminary results from recent studies conducted by the author indicated significant ore softening from high intensity advanced blasting techniques (Kanchibotla et al., 2015). Microscopic examination of some of these blast fragments showed that the blast induced micro-cracks occurred mainly along the grain boundaries. This preferred formation of micro-cracks along the grain boundaries can improve the liberation characteristics.

UNDERSTANDING THE INTERACTION BETWEEN KEY PROCESSES

Understanding the mechanisms and interaction between different processes in the mine mill chain, modelling key processes and integrating them is of vital importance to optimise the mine-to-mill value chain. Some of the key processes in the mine-to-mill value chain are drill and blast, crushing and grinding and flotation.

Drilling and blasting is the first step in the breakage and separation process and plays an important role in optimising the mine-to-mill value chain. The outcomes of drilling and blasting impact all the three feed characteristics that impact the performance of key processes and they can be predicted by using a variety of modelling approaches ranging from purely empirical to rigorous numerical models. The blasting research group at the Sustainable Minerals Institute of the University of Queensland has been working for the past three decades and has developed a variety of models to predict blasting outcomes such as fragmentation, blast induced ore loss, dilution and damage.

Most concentrators use crushing, grinding and classification as a part of comminution process followed by flotation to recover the concentrate. The JKMRC has been conducting considerable research in this area for the past 50 years and has developed a number of models to simulate different stages of comminution and flotation (Napier-Munn et al., 1996). All these models are encapsulated in simulators called JKSimMet and JKSimFloat which are used widely by mineral processors.

In this approach, site specific process models (blasting, crushing, grinding and flotation) are developed from controlled blast trials followed by comminution and flotation plant surveys. The site specific models are integrated as shown in Figure 4.1 to simulate the key mine-to-mill processes and quantify the leverage of each process on the total value chain. The majority of research at the JKMRC has so far concentrated on integrating blast fragmentation with the crushing and milling models. New research is focussed on integrating blast movement models and extending the milling processes to flotation and liberation models to predict the recovery of the final product.

CONCEPTUAL STUDY USING JKMRC MODELS

In this chapter, a conceptual study was conducted to demonstrate the impact of the holistic mine-to-mill approach on productivity, energy efficiency and sustainability. Productivity was assessed using throughput, energy efficiency using specific energy consumption and sustainability using greenhouse gas (GHG) emissions.

DRILL AND BLAST

Three blast designs were used to demonstrate the impact of this approach. Details of the three designs are given in Table 4.1. The baseline design represents the standard drill and blast parameters used in hard rock metal mine blasts. Option 1 uses tighter patterns and higher energy typically used in mine-to-mill blasts. Option 2 uses radically different blast designs with much smaller patterns, hole diameter and a novel explosive system with very high velocity of detonations and density to increase the shock energy of the explosion. As expected, the energy, costs and GHG emissions of new designs are higher than standard blast patterns. The rock mass was assumed to be hard and blocky with density of 2.7 t/m³, uniaxial compressive strength of 200 MPa and rock quality designation of 100.

ROMPSD for the three designs estimated by the JKMRC blast fragmentation model shown in Table 4.2 is fed to the crusher and SAG mill models to estimate the crushing and grinding circuit performance. The main objective of the mine-to-mill and next-generation mine-to-mill blast is to produce a much finer feed to the concentrator.

TABLE 4.1
Blast Designs Used in the Conceptual Study

Blast Design	Baseline	Option 1: Mine-to-Mill Blast	Option 2: Next-Generation Mine-to-Mill Blast
Bench height (m)	15	15	15
Burden (m)	6.8	5.5	4.0
Spacing (m)	7.7	6.5	4.5
Hole diameter (mm)	311	311	269
Stemming length (m)	8.0	7.0	4.0
Sub-drill (m)	1.0	1.5	1.5
Explosive			
Weight (kg)	523	902	959
Type	ANFO	Current Emulsion	High-Shock Explosives
Density (g/cm³)	0.86	1.25	1.35
Velocity of detonation (VOD) (m/s)	4500	5500	7000
Relative weight strength (RWS)	100	90	95
Powder factor (kg/t) Costs	0.25	0.62	1.32
Drilling and blasting cost ($/t)	$0.34	$0.93	$1.96
GHG factor (CO₂kg/t)	0.04	0.11	0.22

TABLE 4.2
Estimated ROMPSD for the Three Blast Designs

Feed Property	Units	Baseline	Option 1: Mine-to-Mill Blast	Option 2: Next-Generation Mine-to-Mill Blast
Top size	mm	370	228	92
F80	mm	134.9	101.0	49.3
Fines (−10 mm)	%	11	15	41

COMMINUTION PERFORMANCE

JKSimMet, a comminution simulation package developed at JKTech, was used to assess the effect of the different ROMPSDs on the performance of a hypothetical copper concentrator. High-shock explosives, advanced priming and timing techniques used in the next-generation mine-to-mill blasts are expected to increase the micro-fractures in the blast fragments and soften the ore hardness by 10%. Table 4.3 shows the assumed ore characteristics used in the simulations.

Simulations were conducted for a single SAG and ball mill (SABC) circuit (Figure 4.6) for each ROMPSD. Energy consumed per ton of ore processed includes the direct major equipment energy, ancillary equipment energy and embodied energy from grinding media. The embodied energy is defined as the quantity of energy necessary for the fabrication of material or consumable used in the process (Daniel et al., 2010). Table 4.4 shows the assumptions used to calculated energy efficiency

TABLE 4.3
Assumed Ore Characteristics

Ore Property	Units	Baseline	Option 1: Mine-to-Mill Blast	Option 2: Next-Generation Mine-to-Mill Blast
A*b		35.0	35.0	38.5
Bond work index (BWI)	kWh/t	18.0	18.0	17.1
Cu head grade	%	0.5	0.5	0.5
Cu recovery	%	90	90	90
Cu concentrate grade	%	30	30	30

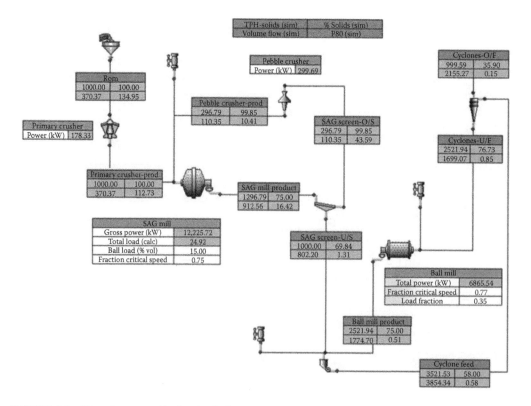

FIGURE 4.6   Flow sheets used in the simulations.

TABLE 4.4

Assumed Energy Parameters

Assumption	Units	SABC
Ancillary equipment specific energy (Vanderbeek et al., 2006)	kWh/t	2.33
Grinding media consumption (Daniel et al., 2010)	kg/t	0.9
Embodied energy in grinding media (Daniel et al., 2010)	kWh/t (steel)	6000
Electricity emission factor (EPA, 2015)	kg(CO_2)/kWh	0.690

and emissions for the plants. All the simulations were conducted with conditions to achieve a similar final grind size or cyclone over flow to maintain the recoveries. A summary of the simulation results is presented in Table 4.5.

IMPACT ON COMMINUTION CIRCUIT PERFORMANCE

Simulations indicate 10% increase in throughput (~300 tph) and 6% improvement in energy efficiency with comparable reductions in GHG emissions due to finer feed size from the mine-to-mill blast compared to conventional basting practices. The next-generation mine-to-mill blast with significantly finer feed increases the throughput of the mill by 66% or 1985 tph and reduces specific energy consumption and GHGs by almost one-third compared to the feed from conventional blasting practices (Figure 4.7). Even though the next-generation mine-to-mill blasting increases the blasting GHG emissions from 0.4 to 0.22 kg(CO_2)/t, this amount is negligible when compared to the 4.75 kg(CO_2)/t decrease that the finer blast achieves in the comminution circuit. It should be noted that the next-generation mine-to-mill blast contains no media sized rocks (+100 mm). This would have had an impact on SAG mills with low ball loads (3%–8%) which were common in the 1990s. Current SAG mills are now typically run with high ball loads of up 20% which would be sufficiently high not to be adversely affected by the low amount of media in the ore. The simulated circuit had a SAG mill ball load of 17% and performed as expected.

The simulations also indicate that to achieve an additional 2000 tph throughput with the current drill and blast practices, the mill would require another grinding line with a slightly smaller SAG mill (10.8 m dia and 14.2 MW) and a ball mill similar to the current ball mills. The additional grinding line can increase the throughput but with higher (22%) specific energy requirement and corresponding GHG emissions. An additional grinding line requires significant capital investment.

CONCLUSIONS

Future mineral deposits are expected to be deeper and of lower grade with complex mineralogy compared to the current deposits. Metal production from such deposits is likely to substantially increase OPEX, CAPEX and energy requirements as well as the carbon footprint of the operations. The traditional approach of managing mine and concentrator separately cannot achieve the energy efficiencies required to operate these future deposits. A new holistic approach is needed wherein mining activities are designed to supply the mill with ore characteristics to increase the overall energy efficiency and productivity of the total production value chain rather than individual unit process without adversely affecting safety, slope stability and environmental risks.

The value of such an approach is demonstrated thorough a conceptual study using the established JKMRC blasting, crushing and grinding models. The simulations from this study clearly demonstrate that by integrating mining and milling operations, it is possible to achieve significant energy efficiencies and reduce GHGs. Such an approach can also reduce the capital requirements for brownfield expansion and greenfield projects.

TABLE 4.5
Summary of Simulation Results

Circuit Section	Parameter	Baseline Blast	Mine-to-Mill Blast	Next-Generation Mine-to-Mill Blast	Adding Equip + Baseline Blast
New feed	Throughput (tph)	3015	3325	5000	5000
	F80 (mm)	134.9	100.9	47.5	134.9
Primary crusher	No in operation	1	1		1
	Crusher power (kW)	118	99		143
SAG mill	No in operation	1	1	1	1
	Mill diameter (m)	12.02	12.02	12.02	12.02
	Mill length (m)	6.30	6.30	6.30	6.30
	Mill power (kW)	21,957	22,257	19,565	21,950
SAG Mill2	No in operation				1
	Mill diameter (m)				10.82
	Mill length (m)				5.33
	Mill power (kW)				14,201
Pebble crusher	No in operation	2	2	2	2
	Crusher power (kW)	340	362	526	591
Ball mill	No in operation	2	2	2	2
	Mill diameter (m)	6.98	6.98	6.98	6.98
	Mill length (m)	12.5	12.5	12.5	12.5
	Mill power (kW)	13,164	13,166	13,167	13,166
Ball Mill3	No in operation				1
	Mill diameter (m)				6.25
	Mill length (m)				9.3
	Mill power (kW)				7537
Cyclones	Overflow P80 (μm)	148	149	149	150
	Major equipment total power (kW)	49,084	49,411	46,950	71,346
	Major equip. specific energy (kWh/t)	16.28	14.86	9.39	14.27
	Ancillary equip. and embodied specific energy (SE) (kWh/t)	7.73	7.73	7.73	7.73
	Total specific energy (kWh/t)	24.01	22.59	17.12	22.00
	GHG emissions (kg(CO_2)/t)	16.56	15.58	11.81	15.17

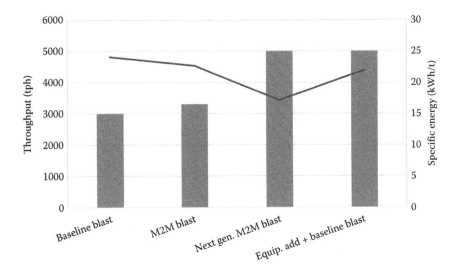

FIGURE 4.7 Simulated effect of mine-to-mill blast on SABC circuit throughput.

REFERENCES

Bond, F. C. 1952. The third theory of comminution. *Transaction of SME/AIME*, 193: 484–494.

Chi, G., M. C. Fuerstenau, R. C. Bradt, and A. Ghosh. 1996. Improved comminution efficiency through controlled blasting during mining. *International Journal of Mineral Processing*, 47: 93–101.

Daniel, M., G. Lane, and E. McLean. 2010. Efficiency, economics, energy and emissions – Emerging criteria for comminution circuit decision making. *IMPC*, Brisbane, Australia, pp. 3523–3531.

Eliiot R., R. Either, and J. Levaque. 1999. Lafarge Exshaw finer fragmentation study. *Proceedings of the 25th ISEE Conference*, Nashville, Tennessee, pp. 333–354.

Eloranta, J. 1995. Selection of powder factor in large diameter blastholes. *EXPLO 95 Conference*, AusIMM, Brisbane, September, pp. 25–28.

Engmann, E., S. Ako, B. Bisiaux, W. Rogers, and S. Kanchibotla. 2012. Measurement and Modelling of blast movement to reduce ore loss & dilution at Ahafo Gold Mine in Ghana. *In Proceedings of the 2nd UMaT Biennial International Mining & Mineral Conference – UBIMMC*, Tarkwa, Ghana.

EPA. 2015. Calculations and references, http://www.epa.gov/cleanenergy/energy-resources/refs.html (accessed 1 July 2015).

Kanchibotla, S. S., S. Tello, R. Morrisson, and B. Adair. 2011. Energy footprint reduction at KUC. JKMRC Internal Report.

Kanchibotla S. S., W. Valery, and S. Morrell. 1999. Modelling fines in blast fragmentation and its impact on crushing and grinding. *EXPLO 99*, Kalgoorlie.

Kanchibotla, S. S., T. G. Vizcarra, S. A. R. Musunuri, S. Tello, A. Hayes, and T. Moylan. 2015. Mine to mill optimisation at Paddington gold operations. *SAG 15 Conference*, Vancouver, Canada, accepted.

Kick, F. 1883. The law of proportional resistance and its application to sand and explosions. *Dinglers Journal*, 247, 1–5 (in German).

Kojovic, T. and R. Wedmair. 1995. Prediction of fragmentation and the lump/fines ratio from drill core samples at Yandicoogina. JKMRC Internal Report.

Michaud, P. R. and J. Y. Blanchet. 1995. Establishing a quantitative relation between post blast fragmentation and mine productivity: A case study. In *Proceedings of the Fragblast*, Montreal, Canada, 5, pp. 389–396.

Napier-Munn, T. J. (ed.) et al. 1996. *Mineral Comminution Circuits*, JKMRC Monograph Series in Mining and Mineral Processing 2, 2nd Edition, Chapters 3, 4 and 7.

Nielsen, K. and J. Kristiansen. 1995. Blasting and grinding – An integrated comminution system. *EXPLO 95 Conference*, AusIMM, Brisbane, Australia, September, pp. 427–436.

Palmström, A. 2005. Measurements of and correlations between block size and rock quality designation (RQD), *Tunnels and Underground Space Technology*, 20, 362–377.

Revnivstev, V. I. 1988. We really need revolution in comminution. In *Proceedings of the 16th International Mineral Processing Congress*, Stockholm, Sweden.

Rogers, W., S. Kanchibotla, A. Tordoir, S. Ako, E. Engmann, and B. Bisiaux. 2012. Solutions to reduce blast-induced ore loss and dilution at Ahafo open pit gold mine in Ghana. *2012 Transactions of the Society for Mining, Metallurgy, and Exploration*, 332.

Scott, A. et al. 1996. *Open Pit Blast Design – Analysis and Optimisation*, JKMRC Monograph No. 1, Brisbane, Australia: JKMRC.

Taylor, S., L. Gilbride, J. Daemen, and P. Mousset-Jones. 1996. The impact of blast induced movement on grade dilution in Nevada's precious metal mines. *Proceedings of the Fifth International Symposium on Rock Fragmentation by Blasting-Fragblast*, Montreal, Quebec, Canada, pp. 407–413.

Thornton, D. 2009. The application of electronic monitors to understand blast movement dynamics and improve blast designs. *Proceedings Ninth International Symposium on Rock Fragmentation by Blasting–Fragblast,* Granada, Spain, pp. 287–300.

Thornton, D., D. Sprott, and I. Brunton. 2005. Measuring blast movement to reduce ore loss and dilution. *Thirty-First Annual Conference on Explosives and Blasting Technique*, 2: 1–11.

Vanderbeek, J. L., T. B. Linde, W. S. Brack, and J. O. Marsden. 2006. HPGR implementation at Cerro Verde. *SAG Conference*, Vancouver, Canada, Vol. IV, pp. 45–61.

Wills, B. A. 2006. *Mineral Processing Technology.* 7th edn, Butterworth-Heinemann, p. 164. ISBN: 978-0-7506-4450-1.

5 Energy Use in Comminution in a Global Context

Michael J. Daniel

CONTENTS

ABSTRACT

Sustainability is discussed in the context of the electrical energy that is consumed by the mining industry for the production of everyday metal commodities. The results presented are from a study which attempts to quantify global energy consumption attributed specifically to the comminution processes, namely the crushing and grinding of metaliferous ores. It also reflects on the current understanding of eco-efficient comminution practices, a required mineral processing stage that has received a lot of attention in recent years because of the view that mining is directly and indirectly responsible for up to 45% of the global economy and that a large proportion of the consumed energy originates from the beneficiation processes in the production of mineral commodity.

The mining industry has been under pressure to review its approach to energy usage. This is further complicated as a major trend compounding energy use in the mineral extraction industry today is the necessity of having to process more massive, hard and low-grade ore bodies. As a result, a much greater amount of gangue material (waste) is being reduced in size, and to much finer sizes, due to finer mineralisation, to produce the same amount of recoverable metal. This research therefore aims to seek alternative mechanisms/processes in which the minerals can be liberated from the ore resulting in more energy efficient and economical benefits. For this, the high pressure grinding rolls (HPGR) have been considered in recent years as an alternative energy efficient comminution technology to that of the existing inefficient tumbling mills.

The goal of being able to realise a step change improvement in energy efficiency and lessened environmental impacts through reduced equivalent CO_2 greenhouse gas (GHG) emissions is being vigorously pursued. From a sustainable viewpoint, the balance between global energy consumption and economic value (currently 2%–4% of gross domestic product (GDP) in industrialised nations) is expected to change in the future. This means that as the economic cost of energy increases, the amount of energy required to do the same amount of work increases at a greater rate, and energy intensive industries such as the mining and mineral processing industry are likely to be the most affected. In particular, major processes such as comminution (crushing and grinding) that consume relatively large quantities of electrical energy generated from coal or oil are targeted.

Direct energy use in the comminution processes is reviewed. It is shown that 0.43% (87 TWh) of the global net electrical energy consumption of 20,000 TWh per annum is used to crush and grind non-ferrous ores. Of this, 33% and 53% of the energy is required to process gold and copper ores, respectively. This suggests that alternative energy efficient comminution technologies should be targeted at gold and copper mining operations in the future to be effective in reducing carbon emissions. As such, new eco-efficient flow sheets that use new devices, process less ore by pre-concentrating the feed to the plant should be considered. In addition, eco-efficient power generation technologies could tackle the problem from an energy supply perspective.

Greater eco-efficiency can also be realised by reducing the consumption of mill liners and grinding media. Though the 'dollar cost' of comminution is normally accounted for as a direct electricity expense in the process and is rarely considered for its overall energy cost or 'embodied energy' of manufacturing the steel which amounts to up to 4–6 kWh/t. Eco-efficient and sustainable development initiatives are linked to 'energy cost' and not always 'dollar cost' savings. The direct and indirect energy cost savings and the impact on the environment should be targeted.

INTRODUCTION

Comminution, or particle size reduction as it is commonly termed, remains as the only physical method by which valuable mineral components in massive rocks can be liberated for further treatment and recovery. Comminution processes are known to be very energy intensive and consume vast quantities of electrical energy. However, quantification of how much energy relative to global electrical energy consumption is not well understood.

This work attempts to quantify global energy consumption attributed to the comminution process within mining and minerals beneficiation processes. It reflects on the current understanding of efficient comminution and mineral processes within the context of emerging global phenomena such as climate change, energy resources, 'peak oil' and environmental degradation.

Comminution is broadly defined as a group of mineral processing techniques used in extractive metallurgy to manipulate the particle sizes of rocks. Comminution processes are used to break rocks into fine particle sizes for the minerals of interest to be recovered. A particle becomes liberated when the valuable mineral is physically freed from other minerals or gangue minerals through the application of mechanical energy by stressing the particles. The extent to which the particles are liberated has a direct influence on the amount of mineral of interest that may be recovered.

The devices used for comminution are generally divided into broad categories based on the size of the fragments they produce. Devices producing relatively coarse chunks (500–10 mm) are called crushers. Devices that produce finer particles are called grinders (10 mm–10 μm). This managed process of particle size reduction remains as the only commonly used and available physical method today by which valuable minerals in massive rocks are liberated for further treatment.

The energy costs are mainly associated with electrical energy and the direct and indirect 'dollar costs' and 'energy costs' coupled with increased use of grinding media. In the future, it is quite possible that a more serious situation may arise when energy costs and consumption increase rapidly resulting in the energy industry not being able to supply sufficient quantities of energy, and not being able to maintain the rate at which renewable energy is generated.

The chapter attempts to quantify and place into context the extent to which electrical energy is consumed globally during comminution processes. Comminution processes include both crushing and grinding of ores but exclude blasting processes. Annual global mining and mineral beneficiation processes are quantified along with an estimate of the energy intensity associated with the production of each commodity. Accurate data and reporting is scant, and thus the data presented here is subject to error, however, the outlying objective of the document is to

- Better understand comminution energy and sustainability, by reviewing energy efficiency drivers or 'eco-efficiency' within the context of economics 'dollar cost' and energy balances 'energy cost'
- Quantify global comminution energy consumption in respect to the production of various metaliferous and non-metaliferous commodities and matters concerning eco-efficiency and increasing 'dollar cost' of energy

GLOBAL CONTEXT

An energy balance on existing comminution practices has revealed that the 'embodied energy' of producing the grinding media in the first place is very high. Reducing or eliminating their use in processes that consume large quantities of grinding media are not eco-efficient and these processes should consider emerging sustainably attractive technologies such as HPGR, ore sorting and high voltage disintegration technologies.

An account of global comminution energy, global grinding media consumption and global mineral grades and their relation to crushing and grinding processes within the mining and minerals industry is reviewed here. The stand-alone results as presented are impressive.

Comminution processes in minerals processing account for between 50% and 70% of the electrical energy used in mining (National Research Council, 1981; DOE, 2001) and sometimes a substantial portion of the direct operational costs. The potential of significant increases in energy costs, capital costs and possible reduced revenue (as a result of lower grades and not necessarily commodity prices) is a factor that could possibly unbalance the profitability of many mining operations in the future. This unbalancing could be counteracted by researching how 'energy costs' and 'dollar costs' are utilised in comminution processes and then evaluating how the valuable mineral components are liberated from the host ore using alternative technologies and circuit designs.

Sustainable development has the following consequences. Renewable energy, climate change, environmental degradation, so-called 'peak oil', population growth and poverty are currently highly controversial and debatable global topics. Does meaningful data exist to substantiate scientific analysis to draw conclusions to these issues? Do current day research projects provide adequate data that could have a major impact in partially solving such problems?

When scientists speak of the future, what do they mean, next year, one or two decades hence, the end of the twenty-first century, the end of the third millennium or forever? When the future is examined within the context of 'sustainable development', generally the idea is aimed at trying to understand how lifestyles will have to change to meet the challenges posed by emerging global events such as climate change, peak oil, limited non-renewable energy resources and continued environmental and ecological degradation. Fragments of 'sustainable development' concepts have emerged in the form of labels such as 'sustainable society', 'sustainable energy' and 'economic sustainability' to mention a few. The general consensus is that economics principles and business always has and always will remain central to the transformation process of change and development. Will this always be so in the future? Perhaps, a paradigm shift is required here.

The author's view is that the boundaries of the roles and responsibilities of business in society have shifted. Is business fulfilling its part of its contract with society and the environment, and what exactly should the role of business be?

The current understanding of mineral processing is based on the following

- Mining and plant performance is measured on tonnage throughput and quantity of metal produced which by default is economically viable.
- Throughput in the plant is managed through various size reduction processes which indirectly result in mineral liberation.
- The material size reduction process is managed through the operation of various stages of crushers, mills, screens and classifying cyclones.

- The overall plant performance defines recovery factors which may affect the mine plan and definition of the ore reserve.
- Unless optimised (usually in economic terms), the process will result in sub-optimal exploitation of the mineral resource.
- The traditional perception of indiscriminate size reduction, as applied by most conventional comminution processes over the past 100 years often results in excessive generation of ultra-fine particles (both mineral and gangue material at the expense of large amounts of energy), which are sometimes either difficult to recover or store in downstream processes.

Focussing on effective resource consumption, reduction of waste and eco-efficiency harnesses the business concept of creating value and links it to the environment.

REAL ISSUES FACING THE MINING INDUSTRY

Recent and ongoing annual reviews by Ernst and Young are based on survey data relating to risks facing the global mining industry. Over the past 7 years, the three most significant issues facing the mining and metals industry are

1. Access to secure energy
2. Skills and social and community concerns
3. Financial and economic performance of the industry

Other ranked issues identified in the global mining initiative survey are shown in Figure 5.1 and compare views from 2008 against 2014 (Ernst and Young, 2014).

The environmental concerns are almost certainly directly linked to energy resources and the availability of (net) energy, as well as the environmental degradation caused by having to mine more massive low-grade ore deposits. These real and complex problems facing the mining industry have spurred sustainable mining interests in energy efficiency and reduced GHG emissions. In the mining process as a whole, the energy intensive process of comminution remains a focus for driving technological change and research in this area. Two very important aspects of comminution are energy consumption and grinding media consumption. Both have a significant impact on the overall 'dollar cost' of the processes, and both are strongly linked with the sustainable drive

Top 10 risks		Over 7 years	
2014		**2008 (peak of supercycle)**	
01	Productivity improvement	01	Skills shortage (now balancing
02	Captial dilemmas-allocation and access		talent needs)
		02	Industry consolidation
03	Social license to operate (SLTO)	03	Infrastructure access
04	Resource nationalism	04	Social license to operate
05	Captial projects	05	Climate change concerns
06	Price and currency volatility	06	Rising costs
07	Infrastructure access	07	Pipeline shrinkage
08	Sharing the benefits	08	Resource nationalism
09	Balancing talent needs	09	Access to secure energy
10	Access to water and energy (new to top 10)	10	Increased regulation

FIGURE 5.1 Prioritisation of the main issues of the mining industry. (Adapted from Ernst and Young. 2014. Website accessed 2014, Business risks facing mining and metals, web, http://www.businessspectator.com.au/sites/default/files/EYG_Business%20risks%20in%20mining%20and%20metals%202013-2014_June%202013.pdf.)

to reduce the overall 'energy cost' or energy consumption per tonne treated or per tonne metal produced. If the 'energy cost' of producing the grinding media is factored into the sustainable objectives, then reducing media consumption, or eliminating it altogether could have a large 'eco-efficiency' benefit to the overall process. Eco-efficiency is a term defined as being able to do the same with less.

DOLLAR COST AND ENERGY COST OF MINERAL EXTRACTION

Table 5.1 (Norgate and Rankin, 2002) shows economical ore grades required for six major metal commodities in 2000. Commodities at the time of writing are exiting a decade long mining boom period with much lower prices being obtained for most ferrous and non-ferrous metals. This shift in the cycle is reported to be mainly driven by the slowdown demand for raw materials by China and India and the dynamics of global debt and economic imbalances.

In the past 2 years, the price of many commodities has dropped by more than 30%–40%, and in the case of oil by to 50%–60% (Desjardins, 2015). These prices have the effect of making low-grade ore deposits uneconomical, and a slump in existing plant throughput expansion projects. Overall 'dollar costs' to produce each commodity are well known and controlled because the 'dollar costs' are closely linked to the financials (Figure 5.2), but the associated 'energy costs' are not very well documented and understood from a sustainability perspective.

'Dollar costs' is defined as an 'imaginary' numerical digit that is negative and is usually confined within an accounting system that governs business practices according to a country's law.

'Energy cost' is defined as 'reality' in which the laws of physics and chemical substances are used to drive physical systems contained within the natural world for the purpose of doing useful work.

Fortunately, mineral resources are abundant in Australia and provide for a large percentage of export wealth, which favours the economist's standpoint. In 2013/2014, minerals generated A$96.5 billion in exports, representing 29% of Australia's total merchandise exports (http://www.austrade. gov.au). Earnings from mineral and energy exports are expected to account for 50% of export revenue, at A$167.5 billion, in 2013/2014 (Australian Benchmarking Report, 2014). The 'dollar cost' data of producing every commodity are well documented because of the economic system under which the data are managed. Each producing mine may be ranked against the specified commodity's annual global production.

TABLE 5.1
Economic Grades, Reserves, Production Rates and Years of Supply

Metal	Economic Ore Grade (%w/w)	Reserves (Mt of Metal)	Production in 2000 (Mt/y)[a]	Years of Supply[b,c]	Years of Reserve Estimation
Iron/steel	30–60	65,000	842	77	1995
Aluminium	27–29	3910	24.2	162	1997
Copper	0.5–2	320	14.8	22	1997
Lead	5–10	65	6.6	10	1997
Zinc	10–30	142	8.9	16	1995
Nickel	1.5–3	47	1.1	43	1995

Source: Norgate, T. E. and W. J. Rankin. 2002. *Proceedings Green Processing 2002: International Conference on the Sustainable Processing of Minerals,* May, pp. 49–55. Report accessed on web at http://www.minerals.csiro.au/sd/ CSIRO_Paper_LCA_Sust.pdf.

[a] Includes primary and secondary metals.

[b] Assumes consumption rate closely balanced to total production rate.

[c] Assumes no recycling.

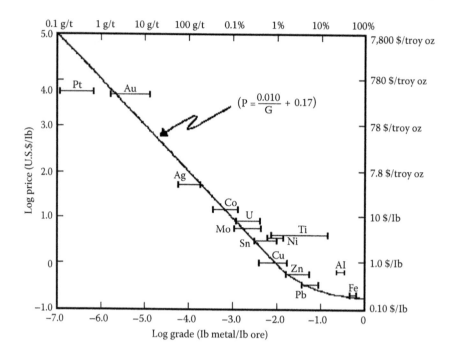

FIGURE 5.2   Relationship between the ore grade of several commodities and the 'dollar cost' of producing them.

This provides accurate information that is used mainly for investment decisions on the various commodities that are traded in a volatile price market. Unfortunately, 'energy costs' are not factored into the decision-making process.

'Energy costs' are likely to increase in terms of the quantity of energy consumed as new mineral processing plants become more energy intensive due to lower head grades and in terms of the higher 'dollar cost' of the energy. The increased 'dollar cost' of the energy is driven as a result of increased oil prices which in turn are being driven by the increased 'energy cost' required to produce the energy. Only new energy efficient processing technologies can reverse this trend.

The minerals industry is a very energy negative industry, unlike the mining of energy minerals such as coal, gas and petroleum where, although the 'dollar cost' may escalate, the 'energy output' cannot exceed the 'energy input' or 'energy cost'. In the long term, the requirement for more energy efficient and cost effective devices will need to be considered for the mining and minerals industry as well as energy resource industries.

A trend compounding the increasing 'energy cost' of producing minerals today is the necessity of having to find improved technologies to process massive, hard and low-grade ore bodies. As an example, larger open pit copper mines currently have the capacity to process more than 200,000 t/ day. This is four times the mining rate in the 1960s. Ore head grade has decreased which implies that two to three times more energy is required to produce the same metal as the mines did in the 1960s (Figure 5.3; Brown, 2004). The larger amounts of gangue material or waste are also required to be reduced in size during milling. The grind sizes are generally finer today than before as a result of finer deportment of the mineralisation zones.

This increased production owes much to the adage 'economies of scale' that have been prompted by lower grades and relatively 'cheap' and 'readily accessible energy'. The figure on the left-hand side of Figure 5.4 shows how copper ore deposits grades have decreased steadily over the past few decades where the minerals today are much more finely disseminated within the host rock.

Other factors such as increased mechanisation and increased capacities of equipment such as blast hole drills, shovels, loaders, trucks, crushers and mills have kept costs low. These trends bring

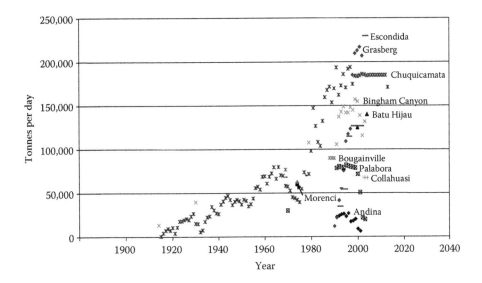

FIGURE 5.3 Evolution of increased tonnes per day being processed in several large copper mines.

with them a range of management and engineering challenges, and one area in particular which is being greatly affected is the mineral preparation stage of comminution.

The figure on the right-hand side of Figure 5.4 shows how mineral concentrations vary in abundance within the earth's crust and in the size of the ore body in which those minerals are found. Over time, the smaller rich deposits were mined out. Large massive ore bodies of relatively low grades (<0.5% in the case of copper) now remain. These mineral resources are currently mined in both open pit and underground mines in a select number of countries whose economies are often described as resource-based economies. Australia, Chile and South Africa are good examples of such countries and are expanded on briefly in the next section.

For copper ores, the size of the ore body increases exponentially as the grade decreases, as shown in Figure 5.5, which backs up Brown's comments and Figure 5.4 on the trend of increasing tonnages

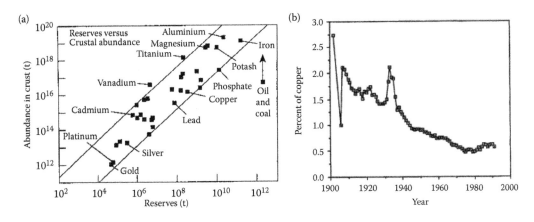

FIGURE 5.4 Left: (a) How copper resource grades have declined over the past 100 years. (After Ruth, M. 1995. Thermodynamic constraints on optimal depletion of copper and aluminum in the United States: A dynamic model of substitution and technical change. *Ecological Economics*, 15(3): 197–213.) Right: (b) Abundance of the various commodities as they exist in the earth's crust relative to estimated reserves. (Reprinted by permission from Macmillan Publishers Ltd. *Mineral Resources, Economics and the Environment*, Kesler, S. E., 391 pp, copyright 1994.)

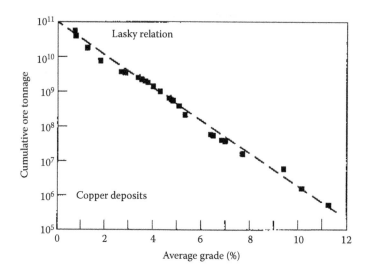

FIGURE 5.5 How the size of a copper ore body increases with decreasing copper head grades. (Reprinted by permission from Macmillan Publishers Ltd. *Mineral Resources, Economics and the Environment*, Kesler, S. E., 391 pp, copyright 1994.)

in mass mining. The typical lower-grade trend for copper is synonymous for most other mineral ore resources today. This coupled with increased tonnages, is the increased 'energy cost' of extraction as Figure 5.6 shows. Figure 5.6 expresses the total energy in MJ/kg of metal produced that increases exponentially as the ore grade is decreased. Economies of scale as previously mentioned have counteracted this trend so far, all within a long cycle of depressed commodity prices (Chapman and Roberts, 1983).

The total energy input per kg of product generally applies in the production of all commodities, including the crushing and production of bulk industrial materials such as aggregates, gypsum, limestone, potash and phosphate rock, and in the grinding of other industrial bulk commodities such as cement and coal. An analysis of the 'energy intensity' and associated global 'energy cost' in comminution of these bulk commodities as well as ferrous and non-ferrous metals is given later in the chapter.

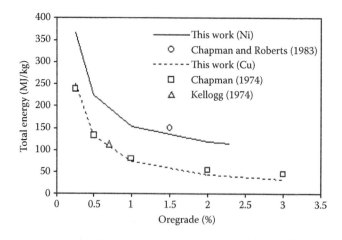

FIGURE 5.6 Energy costs of mineral extraction with respect to ore grade composition in a study by Chapman and Roberts (1983) for copper and nickel ores.

GLOBAL USE OF ENERGY IN COMMINUTION

Mining is a very energy intensive industry and comes a close second to the petroleum industry as shown in Figure 5.7. Figure 5.8 presents data that benchmarks energy efficiency as a function of GDP per capita for most of the world's industrious countries (EIA, 2001). These data highlight the importance of each nation in managing and improving energy efficiency to remain globally competitive. An example of this in relation to the minerals industry is the production of aluminium where energy requirements to produce primary aluminium have decreased by an average of 6.5% in 2006 from levels in 1990 as shown in Figure 5.7.

Energy is an important input in the mining industry, where two major energy carriers, that of electricity and diesel fuel are predominantly used. The diesel fuel energy is used for the transport of personnel, material and ore via mining production machines and the electrical energy for the ore processing equipment. In underground mining operations, electrical energy is utilised in pumping water, ventilation and refrigeration, with much of the electrical energy used in comminution processes.

A textbook view by Wills (1997) stated that grinding is the most energy intensive operation in mineral processing. Wills quotes a survey by Joe (1979) of the total plant energy consumed in a number of copper concentrators that showed that the energy consumption was on average 2.2 kWh/t for crushing, 11.6 kWh/t for grinding and 2.6 kWh/t for flotation. These energy consumptions are based on actual measured plant energy of installed tumbling mill devices. The high grinding energy intensity of 11.6 kWh/t is similar to the energy intensity for most ore types that are being processed in large copper porphyry mines today, with the exception that lower copper grades and often finer mineralisation are becoming more common. Wills (1997) further stated that all ores have an economic optimum grind size which depends on many factors, including the mineral dispersion in the gangue and the subsequent separation processes. Since grinding is the greatest single operating cost, the ore should not be ground any finer than is justified economically.

From an energy perspective, it is generally accepted that at least 50% of the energy used in the production of copper stems from crushing and grinding processes (Figure 5.9; Norgate and Haque, 2010). The 'dollar cost' of energy for comminution as a percentage of the total mining process costs

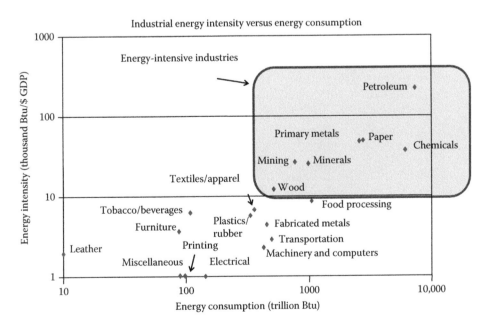

FIGURE 5.7 Energy intensity of various industries, of which mining is one of the most intensive.

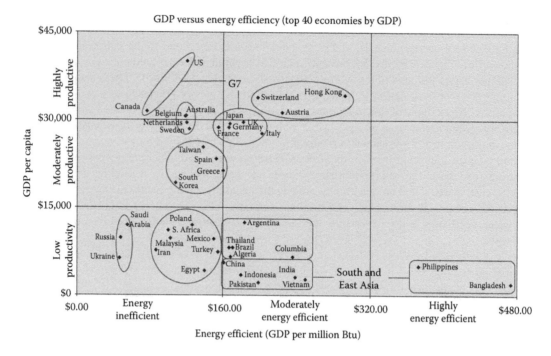

FIGURE 5.8 Benchmarks of energy efficiency as a function of GDP per capita for most of the world's industrious countries and highlights of the importance of managing and improving energy efficiency to remain globally competitive. (Adapted from EIA. 2001. 1998 Manufacturing Energy Consumption Survey. US DOE 2002, Energy and Environmental Profile of the US Mining Industry, Washington, DC.)

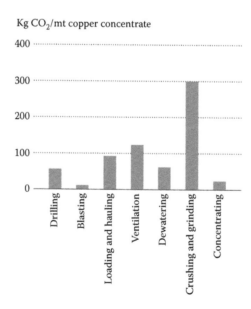

FIGURE 5.9 Depiction of the impact crushing and grinding have on the carbon footprint in the production of copper. (Adapted from Norgate, T. and N. Haque. 2010. Energy and GHG impacts of mining and mineral processing operations. *Journal of Cleaner Production*, 18: 266–274.)

is much lower. In the future, energy balances should be taken into consideration, as there is a complex link between economic and energy cycles.

In South Africa, a study by Kilani (2002) estimated that the mining industry accounted for 32 TWh of electricity which represents 18% of total national consumption. Electrical energy costs accounted for 11.4% of the working costs in the gold mining industry, and 8% in the coal mining industry. In terms of diesel fuel consumption, 515,231 kL was consumed which represents 8.7% of total national consumption. This is bearing in mind that diesel fuel costs have increased dramatically over the past 7–10 years, only to plummet in 2014/2015. Both iron ore and coal mining are adversely affected by shifting electrical energy and diesel fuel prices. The energy commodity is what drives the economies, and hence several complex economical forces are now at play creating much uncertainty and havoc within the mining and resources sectors.

In Australia, the mining industry accounts for 14% (424 PJ or 117 TW) of the national net energy consumption (Australian Bureau of Statistics, 2015). The electricity is consumed in predominantly truck-and-haul and mineral processing activities. These high quantities of energy are typical of resource-based economies, but later it will be shown that the crushing and grinding component of global energy for the processing of non-ferrous metals is relatively low.

These quoted numbers are useful because they provide some perspective of where energy is used in mining and what percentage mining energy contributes to national energy use for two major resource-based countries. Even though the data do provide a perspective of energy consumption, the data is not detailed enough to provide a comparison on what type of impact new comminution devices will have on energy.

Eco-efficiency and sustainable resource processing are important emerging criteria that are driving new comminution technologies, but it should not be confused with the still relatively low percentage that comminution energy in 'dollar costs' have in respect to the overall mining and recovery process costs. The new forms of energy accounting or 'energy footprints' and equivalent GHG emissions are being factored into mine management and new plant designs as exemplified in Figures 5.10 and 5.11.

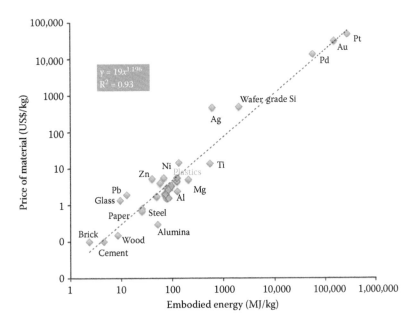

FIGURE 5.10 Example of the energy footprint and price of materials. (http://rsta.royalsocietypublishing. org/content/371/1986/20120003).

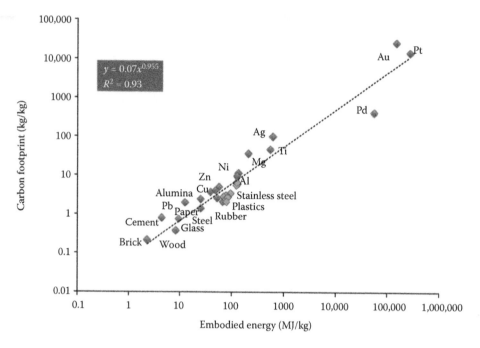

FIGURE 5.11   Example of the energy footprint and embodied energy. (http://rsta.royalsocietypublishing.org/content/371/1986/20120003).

   This form of 'sustainable mineral processing' is important even though there are not immediate large 'dollar cost' benefits (capital and operational) in choosing a more energy efficient comminution circuit that uses HPGR. An HPGR circuit when analysed on the basis that little or no grinding media is used looks much more attractive when the so-called 'unaccounted energy cost' is included into the equation. The 'unaccounted energy costs' are the total energy input that accounts for the production of the grinding media itself, both in mining the ore and producing the iron. These 'unaccounted energy costs' and not the 'dollar costs' should be factored into the decision-making criteria in new plant designs and comminution processes.

   Many major mining multinationals today, as part of their sustainability objectives, are aiming to reduce the total energy input per tonne or per ounce of metal produced as exemplified in the aluminium industry (Figure 5.12). Interestingly, the aluminium industry has several energy intensive processes other than comminution which has led to the aluminium industry being the almost environmentally unfriendly. Emerging energy efficient comminution technologies should rather target comminution intensive processes where fine grinding and high grinding media consumption is concerned. This represents specifically the hard, low-grade ores, and if 'unaccounted energy cost' and not 'dollar cost' be considered, then the case for new technology and innovation could be much more convincing.

   In the next two sections, global comminution energy consumption is analysed and presented. Major mineral and industrial commodities that are responsible for mainly crushing and grinding processes are identified, and in particular, the type of ore resource that should be studied with respect to HPGR possibilities. This analysis is based on data detailing global electrical energy generation, global comminution energy and global commodity production. This data are then compared with a similar energy-based analysis that hinges on the estimated global grinding media consumption and the associated ore tonnages that are treated in the various types of grinding media consuming comminution devices such as semi autogenous grinding (SAG) mills, rod and ball mills.

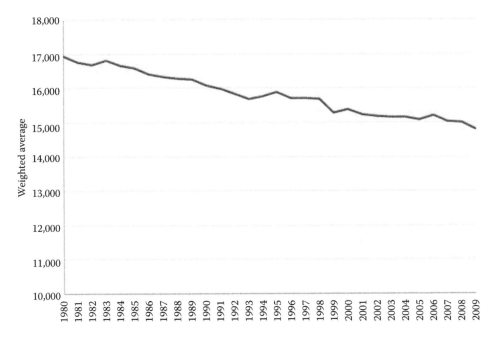

FIGURE 5.12 Energy required in the production of aluminium, where improved energy efficiency has reduced energy requirements by an average of 13% in 2009 from levels in 1990. (http://www.thealuminium-dialog.com/en/economic-impacts/energy-efficiency-for-improved-productivity.)

FUTURE DIRECTIONS

Comminution is known to be extremely energy intensive, but placing it in the context of global energy use has been very difficult due to the lack of coordinated or regulated data capture. Several researchers (Fuerstenau, 1983, 1992; Fuerstenau and Kapur, 1995; Fuerstenau and Abouzeid, 2002) have previously stated that the electrical energy consumed globally for comminution processes was of the order of 3%–4%. These cited figures have been commonly based on a report published by the committee on comminution and energy consumption in 1981 (National Research Council, 1981). At about this time, it was estimated that in the United States alone, the energy consumption for crushing and grinding of solids was in excess of 29 TWh, a figure that was re-quoted for comminution energy consumption in 1997 (DOE, 2001).

In an edition of 'State of the World – Progress Towards a Sustainable Society, 2003' (Sampat, 2003), it is claimed that electrical energy consumption in the mining industry accounts for 7%–10% of the global electrical energy of 4900–6600 TWh, although Figure 5.13 shows that the global net electrical energy consumption in 2012/2013 was much higher at 20,000 TWh or 20,000 billion kWh. The global comminution energy should however be estimated using the 4900–6600 TWh consumption figure which may exclude electrical energy transmission losses. If it is assumed that 50%–70% of the electrical energy consumption within the mining industry is attributable to mineral preparation or more specifically comminution processes (DOE, 2001) then globally, about 540–775 TWh of electrical energy is consumed in comminution processes. This then represents 3.5%–5% of the total global electrical consumption based on a consumption of 20,000 TWh and represents all comminution processes. The increase in global energy consumption in comminution processes from 3%–4% (1981) to 3.5%–5% (DOE, 2001) may be directly attributed to the mining of larger, more massive and often lower-grade ore bodies and the fact that the 3%–4% quoted in previous years was based on data from before 1981. The 3.5%–5%, however, does not reflect any associated increase in global electrical energy which has risen from 7500 TWh in 1980 to 20,000 TWh in 2013/2014. As such the mining industry is now using far more energy in 2013/2014 than it did before 1981.

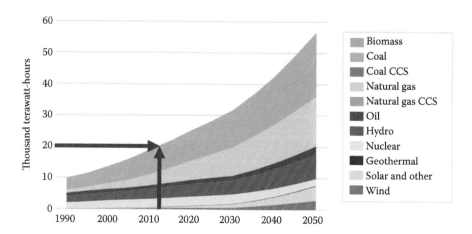

FIGURE 5.13 How the world 'net' electrical energy consumption has doubled since 1990. Nilsson, M. et al. Research Report-Energy for a Shared Development Agenda-2012, p. 30 (Figure 2.9). Stockholm environment institute. ISBN 978-91-86125-40-0. http://sei-international.org/mediamanager/documents/Publications/SEI-ResearchReport-EnergyForASharedDevelopmentAgenda-2012.pdf.)

Nowhere in the literature is the split between crushing and grinding energy consumption ever mentioned. This leads to the assumption that grinding processes are responsible for a large proportion of the global electrical energy consumption in comminution because grinding is known to be far more energy intensive than crushing. In the next section, an attempt is made to provide an analysis of energy split between crushing and grinding energy consumption based on steel ball grinding media consumption.

GLOBAL COMMINUTION ENERGY BASED ON GRINDING MEDIA CONSUMPTION

The potential to reduce energy consumption and improve mineral recovery seem very attractive from an economical and sustainable energy perspective. However, part of the cost analysis involves the quantification of the consumption of grinding media in conventional circuits as against that of HPGR circuits which do not require grinding media, other than the roll liners.

So, if the quantity of global grinding media production is known, and if a grinding media consumption rate for SAG and ball mills is known, then it could be possible to estimate the quantity of material that is subjected to grinding/milling processes. If the tonnage of each processed commodity ore is known, and an estimate of the respective energy intensity is made, then it is possible to estimate the amount of energy that will be consumed globally for each commodity in crushing and grinding.

Searching the web, estimates of global grinding media production were sourced and are given in Figure 5.14 and based on data presented by Arrium (2014), a manufacturer of steel ball grinding media. An analysis of the data presented in Figure 5.14 shows that global steel grinding medial production according to Arrium is equivalent to 4.5 million tonnes of steel per annum (Arrium websites and presentations, 2014).

The estimated global steel production in 2012 was approximately 1500 million tonnes which is illustrated in Figure 5.15 (World Steel Association). Based on the data shown in Figure 5.14, the 4.5 million tonnes (4,500 kT) of steel ball grinding media production and consumption thus represents 0.3% of global steel production. This is negligible in relation to the total amount of steel used in industry. Later in this chapter, it is shown that grinding energy consumption by comparison to global comminution energy consumption is much lower than the quantity of energy used in crushing which includes the crushing of non-ferrous and non-metallic commodities such as cement, coal and building material aggregates.

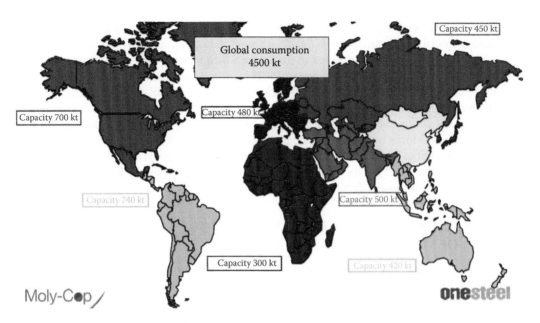

FIGURE 5.14 Cumulative production of world steel ball grinding media, and the companies that represent the global production market. (Adapted from Arrium. 2014. Arrium. Website accessed 2014, http://www.arrium.com/investor-centre/reports-presentations.)

Grinding media costs are largely dictated by global steel prices, which in the case of the mining industry are governed largely by the fixed steel price contracts that the mining companies have with steel producing blast furnaces in various parts of the world. The production of grinding media of different sizes and strength specifications requires an added cost factor when considering the entire manufacturing process of the balls. This is estimated to be of the order of 1.4–1.8 times the price of the steel on the global market.

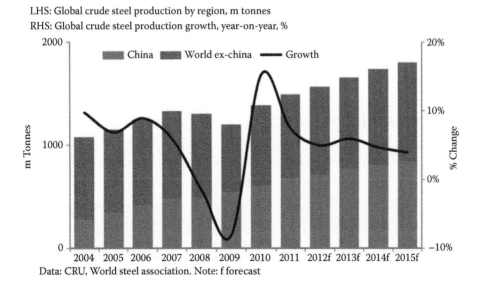

FIGURE 5.15 Quantification of global steel production and the steady increase in production.

Grinding media price depends largely on the country in which the media is manufactured and the specifications (size and hardness) and quality (chrome and carbon content) of the media. Rapidly increasing steel prices could be expected in the future which is being driven mainly by increased demand for steel in China as shown in Figure 5.15. This overall knock on effect will eventually lure large low-grade iron ore, copper and gold operations to consider HPGR as an alternative to large media consuming tumbling mills in the future.

Other mineral commodities may consider new comminution technologies from a media cost perspective, but will focus mainly on potential recovery benefits that are realised, especially with high value commodities such as the platinum group metals and base metals. In the next section, the potential reduction in steel media consumption is evaluated from a sustainable and eco-efficient perspective as an option to consider in the future.

ECO-EFFICIENCY AND THE 'ENERGY COST' (EMERGY) OF STEEL GRINDING MEDIA

The wear of mill liners and grinding media add to the 'dollar cost' and 'energy cost', sometimes referred to as emergy. Emergy defined as the 'embodied energy', that is, the quantity of energy necessary for the fabrication of a specific material. When measuring embodied energy, all energy inputs, from raw material extraction, to transport, manufacturing, assembly, installation and others are considered. Embodied energy as a concept seeks to measure the true energy cost of an item. The need for continual cost reduction and increasing regulatory and social pressure relating to GHG emissions creates a strong need for improvements in comminution energy efficiency and associated sustainable development initiatives.

New eco-efficient and sustainable development initiatives are not necessarily linked to 'dollar cost' savings, but rather the indirect 'energy cost' savings through a targeted significant reduction in the consumption of grinding media. This is attractive because of the high energy intensity required to produce the steel in the first place. Typical consumption figures for coal and coke are 700 kg/t of finished steel in the United States which is based on an ore grade of 62.5% iron. This coal consumption rate is equivalent to an approximate energy usage of 5320 kWh/t of finished steel. The consumption of oil, natural gas and electrical power in addition to coal usage is 710 kWh/t finished product. This results in a total 'energy cost' of 6030 kWh/t finished steel (Figure 5.16).

In terms of eco-efficiency, this unaccounted 'energy cost' implies the use of 6 kWh/kg of steel grinding media. In a typical steel ball milling grinding circuit, the average grinding media consumption rate is 0.7 kg of steel per tonne of ore treated which is equivalent to an additional 4.2 kWh/t of indirect and unaccounted energy. The 4.2 kWh/t represents about 33% of equivalent additional energy effort in a comminution circuit that has an operational work index of approximately 12.6 kWh/t (SvensktStal, 2006).

SHARE OF COMMINUTION ENERGY IN CRUSHING AND GRINDING

Outside of increasing 'dollar costs' and 'energy cost' of grinding media, it is interesting to back-calculate the amount of ore being ground in milling processes globally, and thus estimate the

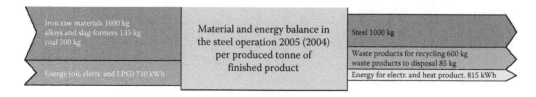

FIGURE 5.16 Energy balance over the production of each tonne of the steel grinding medial.

corresponding quantity of electricity that is consumed in milling processes. As a rule, media consumption at copper and gold processing circuits using carbon alloy steel balls is of the order of 0.5 kg of steel per tonne of ore milled in SAG mills and 0.8 kg of steel per tonne of ore milled in ball mills. If it is assumed that on average 1.3 kg of steel is consumed per tonne of ore milled in a conventional SAG-based grinding circuit, then approximately 3460 million tonnes of ore would be processed each year based on the 4.5 million tonnes of grinding media produced.

If the total comminution energy intensity is 23 kWh/t, then the total specific energy to mill the ore would be 80 TWh. Assuming a 6.5% no-load power inefficiency (large mills) then the total specific power required would be 86 TWh that is used in non-ferrous metal comminution processes. Total global electrical energy consumption is 20,000 TWh (Figure 5.13), hence the global electrical energy consumed in grinding media mills is of the order of 0.43% which is significantly lower that the energy consumption percentage of 3.5%–5% quoted in the literature and websites.

In an attempt to clarify the low milling share of comminution energy, a summary of major global mineral commodities has been assembled as shown in Tables 5.2 and 5.3 (updated data 2015) from Daniel (2007). In these tables, the energy intensity of the comminution operation is estimated and the total comminution energy is estimated based on estimated volumes and grade of each commodity.

Not surprisingly, it is shown in Table 5.2 that even though all non-ferrous metal production process are assumed to use tumbling mills and grinding media, the total electrical energy is estimated to total 87 TWh for all commodities listed. The 87 TWh corresponds closely to the 86 TWh based on grinding media consumption. From Table 5.3, it is clear that other bulk commodities such as steel, cement, aggregates and coal are produced in massive quantities with much lower energy intensity due to crushing requirements only. The large tonnages amount to a very large comminution energy consumption of 225 TWh. The total comminution energy of both bulk materials and metals production is 312 TWh which is equivalent to 1.56% of global net electrical energy consumption at 20,000 TWh. Tables 5.2 and 5.3 confirm that milling processes, although much more energy intensive, consume much less energy than do crushing processes overall.

The data in Table 5.2 confirm that approximately 0.43% (87 TWh) of the global net electrical energy consumption of 20,000 TWh is used to crush and grind non-ferrous ores to produce most major metal commodities. This comes from crushing and grinding an estimated 3360 million tonnes of ore per annum based on the total annual production of the metals and is comparable to the 3460 million tonnes of ore processed per annum based on the global steel grinding media consumption.

In the bulk industrial mineral industries of coal, iron ore, cement and aggregates, comminution energy intensities, with the exception of cement, are very much lower than those for metal commodities. The tonnages are however much larger and account for more than 2.5 times as much electrical energy (225 TWh) as is consumed in comminution processes within the metal commodity industries. This represents 1.13% of the global electrical energy consumption. The analysis of estimated comminution energy consumption of all major non-ferrous ores is useful so that potential HPGR applications might be placed into perspective where the impact to global energy consumption is concerned.

The data assembled in Tables 5.2 and 5.3 (Daniel 2007, updated by Daniel, 2015) were sourced mainly from the US Geological Survey (USGS, 2006) http://minerals.usgs.gov/minerals/pubs/commodity/).

Data representing the platinum/chrome ores overlap which depends on whether chrome or platinum is seen as the product. The same logic applies to copper and gold ores where some copper mines produce gold as a by-product. These multi-mineral ores are assumed not to affect the content of Table 5.2 with respect to estimating ore tonnages processed and corresponding comminution energy consumptions. The same applies to comminution energy intensity where a lack of coordinated data capture has led to best estimates of the energy used in the respective comminution processes.

TABLE 5.2

Estimated Global Comminution Energy Consumed in the Production of Major Metal Commodities

Major Metal Commodity	Grinding Media Used (Yes/No)	Crushing Circuit Energy Intensity (kWh/t)	Grinding Circuit Energy Intensity (kWh/t)	Bulk Ore Processed (Millions of Tonnes per Year)	Average Estimated Comminution Energy Intensity (kWh/t) for Process	Estimated Comminution Energy @ Ave Energy Intensity (TWh)	Mineral Ore Grade/ Bulk Material	Annual Global Mined Commodity Production Tonnes
Tin	Yes	0.5–2	10–25	9	20	0.2	0.8%–1%	90,000
Nickel	Yes	0.5–7	10–25	65	20	1.3	1%–3%	1,500,000
Manganese	Yes	0.5–3	10–25	20	20	0.4	45%–60%	9,790,000
Lead	Yes	0.5–5	10–25	85	20	1.7	7.50%	3,280,000
Zinc	Yes	0.5–6	10–25	70	20	1.4	15%	10,100,000
Aluminium	Yes	0.5–3	10–25	165	20	3.3	25%–30%	31,200,000
Platinum/chrome	Yes	0.5–8	20–35	90	30	2.7	3%–4 g/t	218
Chrome/platinum	Yes	0.5–8	10–25	36	20	0.7	40%–60%	18,000,000
Gold	Yes, major consumer	0.5–8	20–35	960	30	28.8	3–4 g/t	2,450
Copper	Yes, major consumer	0.5–4	20–30	1,860	25	46.5	0.5%–1%	14,900,000
Total major commodity				3,360		87.0		

TABLE 5.3

Estimated Global Comminution Energy Consumed in the Production of Major Bulk Materials

Major Mined Industrial Mineral Commodity	Grinding Media Used (Yes/No)	Crushing Circuit Energy Intensity (kWh/t)	Grinding Circuit Energy Intensity (kWh/t)	Bulk Mined Ore Processed (Millions of Tonnes per Year)	Average Estimated Energy Intensity (kWh/t) for Process	Estimated Comminution Energy @ Ave Energy Intensity (TWh)	Mineral Ore Grade/Bulk Material	Annual Global Commodity Production Tonnes
Diamonds	No	0.5–7	5–15	240	8	1.9	50 cts/100 t	24
Potash	No	0.5–4	N/A	31	3	0.1	100%	31,000,000
Phosphate	No	0.5–4	N/A	148	3	0.4	100%	148,000,000
Gypsum	No	0.5–2	5–10	110	4	0.4	100%	110,000,000
Limestone	No	0.5–2	5–10	128	4	0.5	100%	128,000,000
Cement clinker grinding	Yes	0.5–4	30–50	2200	57	125.4	100%	2,200,000,000
Iron ore/steel	Yes	0.5–2	5–16	1300	9	11.7	40%–62.5%	1,000,000,000
Aggregate crushing (est)	No	1–6	N/A	6000	4	24.0	100%	6,000,000,000
Coal grinding power station and blast furnace (est)	Yes	N/A	10–18	5500	11	60.5	100%	5,500,000,000
Total all materials				15657		225.0		

CONCLUSIONS

Mines in the future may be required to produce products within a new framework of sustainable and eco-efficient development that includes aspects of global energy and environmental restriction policies. These formidable changes and the ever increasing cost of energy (particularly if generated from non-renewable energy resources), global warming and climate change, will eventually force the mining industry to seek alternative sources of electrical energy coupled with greatly improved energy and eco-efficient comminution strategies.

Greater eco-efficiency can be realised in the reduced wear of mill liners and grinding media. Though the 'dollar cost' of comminution is accounted for as a cost consumable in the process, it is rarely considered for its 'energy cost' or the 'embodied energy' of having to manufacture the steel in the first place. Eco-efficient and sustainable development initiatives are not always required to be linked to 'dollar cost' savings. Rather the direct and indirect 'energy cost' savings and the impact on the environment should be targeted. Reducing grinding media consumption is attractive from a sustainable viewpoint because the high energy requirements to produce the steel are eliminated. These aspects are reviewed and it is shown that 0.43% (87 TWh) of the global net electrical energy consumption of 20,000 TWh per annum is used to crush and grind non-ferrous ores, and of this 33% and 53% of the energy is required to process gold and copper ores, respectively. This suggests that the new energy efficient comminution technologies such as HPGR should be targeted at gold and copper mining operations in the future to be effective in reducing carbon emissions.

The step change improvement in the overall eco-comminution process efficiency offers a real potential of playing an important role in reducing the rate of current energy resource consumption and depletion, whilst new renewable energy resources are exploited and developed. This could be achieved through a renewed approach to comminution, through improved energy efficiency, improved mineral liberation and through overall energy conservation. The study has the potential to reduce environmental impact by reduced water consumption and also by changing the properties (particle size) of the waste products produced in the mining and minerals processing industry, but these aspects are beyond the scope of this work.

The energy costs are mainly associated with electrical energy and the direct and indirect 'dollar costs' and 'energy costs' coupled with increased use of grinding media. In the future, it is quite possible that a more serious situation may arise when energy costs and consumption increase rapidly resulting in the energy industry not being able to supply sufficient quantities of energy, and not being able to maintain the rate at which renewable energy is generated. A futuristic approach to mineral beneficiation could be based on the following

- The mine managed under 'optimal exploitation of the resource', and corresponding profit/revenue ratios within the context of 'sustainable mining'.
- Mineral liberation processes are managed through optimal energy (cost/sustainability) and recovery processes (revenue).
- Optimisation of the measured liberation and energy parameters will ensure optimal exploitation of the ore resource from a comminution perspective.
- The energy efficient liberation approach emphasises the importance of value tracking the mineral in the liberation process. Any improvement in the efficiency of the comminution and liberation process will impact directly on the control of environmental damage, conservation of energy and the economics of the operation.
- This approach to comminution will ensure that the comminution energy is utilised efficiently for the sole purpose of mineral liberation, and will provide mines with data to substantiate sustainable process routes that may be economically tailored to the mineralisation of the ore body.
- Therefore, the challenge is for the mining industry to continue to meet the demand for minerals in a climate of declining head grade and price, whilst ensuring the industry's commitment to sustainability is not compromised.

Finally, the concept of sustainable mineral processing and reduced energy consumption should be investigated by considering alternative crushing and breakage principles such as high voltage electric pulsed disintegration which applies high voltage to excite electrons and produce plasma within the rocks that expand to generate shear forces from within the rocks, rather that applying an external force to create differential shear forces that ultimately lead to the rock failing and breaking. The results of such a project should include an analysis of the processing benefits with respect to the economics of the process concept and with respect to current emerging global phenomena such as global warming, carbon dioxide emissions and energy resource depletion.

REFERENCES

Arrium. 2014. Arrium, website accessed 2014, http://www.arrium.com/investor-centre/reports-presentations.

Australian Benchmarking Report. 2014. Why Australia: Benchmark Report 2016, p. 39. http://www.austrade.gov.au/International/Invest/Resources/Benchmark-Report

Australian Bureau of Statistics. 2015. Website accessed March 2015, http://www.abs.gov.au/ausstats/abs@.nsf/mf/4.604.0.

Brown, E. T. 2004. Keynote address. *JKMRC International Student Conference*, Brisbane, Australia, 6–7 September.

Chapman, P. F. and F. Roberts. 1983. *Metal Resources and Energy*. Borough Green, England: Butterworth Scientific.

Daniel, M. J. 2007. Energy efficient mineral liberation using HPGR technology. PhD thesis, University of Queensland, JKMRC, Brisbane, Australia.

Desjardins, J. 2015. The collapse of commodities in one simple chart. The Visual Capitalist, website accessed, http://www.visualcapitalist.com/the-collapse-of-commodities-in-one-simple-chart/.

DOE. 2001. Mining industry of the future. Office of Industrial Technologies (OIT), Office of Energy Efficiency and Renewable Energy, Washington, DC, DOE/GO-102001-1157, February, www.oit.doe.gov/mining.

EIA. 2001. 1998 Manufacturing Energy Consumption Survey. US DOE 2002, Energy and Environmental Profile of the US Mining Industry, Washington, DC.

Ernst and Young. 2014. Website accessed, Business risks facing mining and metals, web, http://www.businessspectator.com.au/sites/default/files/EYG_Business%20risks%20in%20mining%20and%20metals%202013-2014_June%202013.pdf.

Fuerstenau, D. W. 1983. *Comminution and Simulation Using Rod and Ball Mills*. Bureau of Mines, U.S. Department of the Interior, United States.

Fuerstenau, D. W. 1992. Comminution: Past developments, current innovation and future challenges. *Proceedings of the International Conference on Extractive Metallurgy of Gold and Base Metals*, October 26–28. *AusIMM Extractive Metallurgy Conference*. Published by Australasian Institute of Mining & Metallurgy, Parkville, Australia, Kalgoorlie, Australia, pp. 15–21.

Fuerstenau, D. W. and A.-Z. M. Abouzeid. 2002. The energy efficiency of ball milling in comminution. *International Journal of Mineral Processing*, 67: 161–185.

Fuerstenau, D. W. and P. C. Kapur. 1994. New approach to assessing the grindability of solids and the energy efficiency of grinding mills. *Minerals & Metallurgical Processing*, 11(4): 210–216.

Fuerstenau, D. W. and P. C. Kapur. 1995. Newer energy-efficient approach to particle production by comminution. *Powder Technology*, 82(1): 51–57.

Joe, E. G. 1979. Energy consumption in Canadian mills. *CIM Bulletin*, 72, 147, June.

Kesler, S. E. 1994. *Mineral Resources, Economics and the Environment*. Macmillan, New York, 391 pp.

Kilani, J. 2002. The business sense of energy management in mining. *Global Mining Initiative Conference*, http://www.icmm.com/gmi_conference/439JohnKilani.pdf.

National Research Council. 1981. Report of the committee on Comminution and Energy Consumption, document NMAB-364, National Academy Press, Washington, DC.

Nilsson, M., C. Heaps, Å. Persson, M. Carson, S. Pachauri, M. Kok, M. Olsson et al. Research Report-Energy for a Shared Development Agenda-2012, p. 30 (Figure 2.9). Stockholm environment institute. ISBN 978-91-86125-40-0. http://sei-international.org/mediamanager/documents/Publications/SEI-ResearchReport-EnergyForASharedDevelopmentAgenda-2012.pdf.

Norgate, T. and N. Haque. 2010. Energy and greenhouse gas impacts of mining and mineral processing operations. *Journal of Cleaner Production*, 18: 266–274.

Norgate, T. E. and W. J. Rankin. 2002. *Proceedings Green Processing 2002*: *International Conference on the Sustainable Processing of Minerals*, May, pp. 49–55. Report accessed on web at http://www.minerals.csiro.au/sd/CSIRO_Paper_LCA_Sust.pdf.

Ruth, M. 1995. Thermodynamic constraints on optimal depletion of copper and aluminum in the United States: A dynamic model of substitution and technical change. *Ecological Economics*, 15(3): 197–213.

Sampat, P. 2003. Scrapping mining dependence. In *State of the World* 2013: "Impossible" Environmental Revolution Is Already Happening, Chapter 6, pp. 110–129. Worldwatch Institute, Vision for a sustainable world. Web site accessed: 2016. http://www.worldwatch.org/state-world-2003.

SvensktStal. 2006. Energy and material balance, http://www.ssab.com.

USGS. 2006. Commodity statistics and information, http://minerals.usgs.gov/minerals/pubs/commodity/.

Wills, B. A. 1997. *Mineral Processing Technology: An Introduction to the Practical Aspects of Ore Treatment and Mineral Recovery.* Butterworth and Heinemann, 486 pp.

6 Mill-to-Melt Energy Efficiency Opportunities

Sheila Devasahayam

CONTENTS

ABSTRACT

This chapter aims to provide an overview of energy saving options in metal production processes. It is hoped that the information will benefit industrialists, students, researchers, metallurgists, and policy makers. Mining companies serious about both cost control and improving productivity globally need to focus on energy efficiency measures. The mineral sector consumes a lot of energy and resources. The cost of energy, especially electricity is rising sharply, in addition to the uncertainties surrounding supply. Many large well-established mining and minerals operations are doing brilliant work to reduce energy consumption and reliance on fossil fuels in their operations, with a greater focus on the efficiency of their equipment and process design. Energy savings are identified through process optimization of different plant systems using a holistic approach that considers all the major plant processes and systems, that is, mill-to-melt processes, and treats all these elements as part of one interlinked system that needs to function optimally.

Comminution (crushing and grinding), separation (froth flotation), and concentrate drying consume the most energy and are responsible for most gas emissions in the mining and mineral processing sectors. This chapter discusses the energy saving opportunities for mill-to-melt processes in terms of innovation, step change technologies, recycling, better waste management practices, and the use of renewable and alternate energy sources. Energy efficiency measures using geometallurgy three-dimensional (3D) models, smart blast technologies, eco-efficient grinding devices, classification systems, novel dewatering technologies, in primary and secondary metal production is discussed. Optimizing energy efficiency across mineral processing operations, and downstream metal production, for example, smelting and refining can directly reduce the overall energy consumption per unit mass of metal produced and thus reduce their greenhouse gas (GHG) footprint.

INTRODUCTION

The mining industry is very capital intensive and has to contend with significant volatility in the prices and demand for mineral commodities. Considerable sunk costs as well as expensive and long-lived capital infrastructure are involved in discovering and proving the viability of a deposit. This capital intensity is necessary to improve productivity, and to take up new innovations before the end of the life of the current equipment (Catchpole, 2015). Across the globe, mining productivity has declined by 20% over the past 7 years. Productivity in Australia's mineral sector has reportedly fallen by 30% since 2003 (Port Jackson Partners, Minerals Council of Australia). Figure 6.1 shows a stark decline in Australian productivity in terms of labor, capital, and multi-factor measures (Catchpole, 2015). South Africa's productivity had dropped to a 50-year low by February 2013. Chile's productivity slumped on falling ore grades, energy shortages, and industrial disputes. Canada has also experienced a sharp decline. Canada's productivity rate in the mining, oil, and gas sectors has declined 37% over the past decade.

Increasing fuel prices, ever-narrowing operating margins due to volatile commodity prices as well as increased opposition from communities to new conventional energy sources have led mining

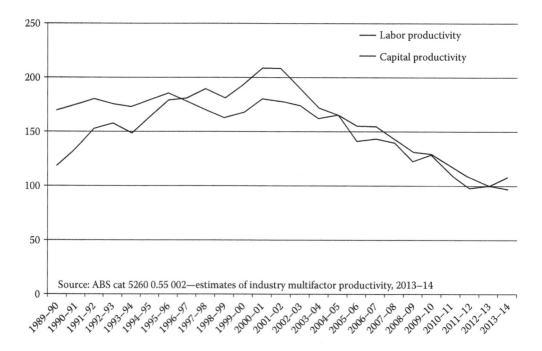

FIGURE 6.1   Australian Bureau of Statistics' measure of Australian mining productivity.

industries to embed the concept of sustainability into strategic decision-making processes and operations, especially in relation to extracting nonrenewable resources.

Phil Hopwood, the global mining leader, Deloitte has advised companies to focus on productivity, sustainable cost management, capital efficiency, and better waste management practices to enhance reuse and recycling (*A Guide to Leading Practice Sustainable Development in Mining*, 2011).

ENERGY DEMAND

Strategic priority for the global mining sector is the access to reliable and cost-effective forms of energy with focus on innovative designs, technologies, policies, and marketing methods related to conventional energy sources and renewable energy sources to maintain future energy supply. Energy efficient measures assume greater importance especially when considering future energy services (Lovins, 2004). To improve multi-factor productivity, energy productivity needs to improve through energy efficiency, system optimization, and business transformation. Multi-factor productivity measures the joint effects of many factors including new technologies, economies of scale, managerial skill, and changes in the organization of production. Energy productivity looks to outputs, quality, and quantity; not only to savings or a reduction in consumption but also ways to improve the value of output created from energy use.

Australia is behind developed economies in measures of both energy productivity and rate of improvement. The energy costs consumers more than A\$110 billion per year, about 8% of gross domestic product (GDP) (CEEC—The future, 2015). Global energy demand is expected to rise by about 35% from 2010 to 2040, less than half the growth rate seen during the period from 1980 to 2010. Most of this growth will come from developing countries, where energy consumption is increasing by nearly 70%. Of the projected growth in global energy demand, China, India, and the key growth countries (Brazil and Mexico in the Americas; South Africa and Nigeria in Africa; Egypt and Turkey in North Africa/Mediterranean; Saudi Arabia and Iran in the Middle East; as well as Thailand and Indonesia in Asia) are expected to account for 30%, 20%, and 30%, respectively (Figure 6.2; *The Outlook for Energy: A View to 2040*, 2015).

Industry accounts for 30% of primary energy usage and 50% of electricity demand. The two largest sub-sectors—heavy industry and chemicals—together account for about 85% of this growth (Figure 6.3). The industrial processes, for example, agriculture, steel, cement, asphalt, plastics, glass, textiles, aluminum, and electronics, extraction and processing of the oil, gas, and coal (Figure 6.3) consume about half of the world's energy.

RENEWABLE AND NONRENEWABLE ENERGY SOURCES

The current energy source for the largest industrial subsector, heavy industry is coal (>35%), followed by electricity (derived from renewables) and natural gas (Figure 6.4). In future, however, electricity and natural gas will become the fuels of choice for manufacturing. The combined shares of natural gas and electricity in global heavy industry are expected to climb from about 40% to about 60% by 2040. Use of coal in heavy industry is expected to fall below 2010 levels, mostly because of changes in China, the world's largest user of coal (*The Outlook for Energy: A View to 2040*, 2015).

The trend toward cleaner energy by most nations is likely to see coal's share as an important source of electricity generation drop from 40% to about 25%. According to Luke Sussams, senior analyst and coauthor of the Carbon Tracker report, the rise of cheap natural gas and increasingly stringent environmental regulations have crushed coal's share of power generation in the United States (Mathiesen, 2015, http://www.theguardian.com/environment/2015/jul/14/gassurges-ahead-of-coal-in-us-power-generation). The new climate change regulations in United States, the Clean Power Plan will mean gas generation will permanently surpass coal by 2020.

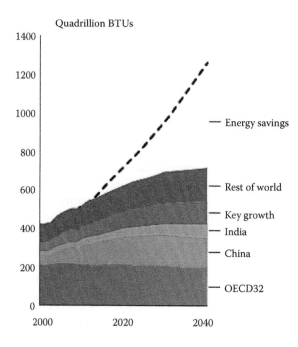

FIGURE 6.2 Energy demand in 2040. (From *The Outlook for Energy: A View to 2040*. 2015, http://cdn. exxonmobil.com/~/media/global/Reports/Outlook%20For%20Energy/2015/2015-Outlook-for-Energy_print-resolution. With permission.)

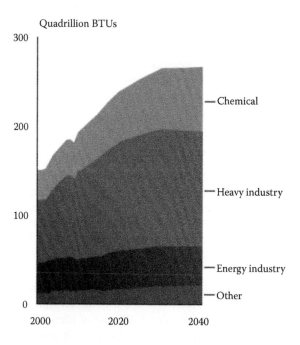

FIGURE 6.3 Industrial energy demand by sector. (From *The Outlook for Energy: A View to 2040*. 2015, http://cdn.exxonmobil.com/~/media/global/Reports/Outlook%20For%20Energy/2015/2015-Outlook-for-Energy_print-resolution. With permission.)

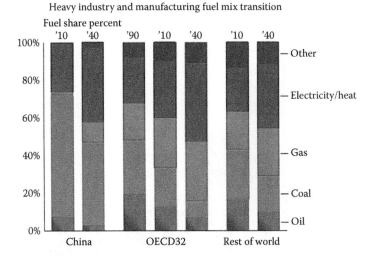

FIGURE 6.4 Energy source for heavy industry. (From *The Outlook for Energy: A View to 2040*. 2015, http://cdn.exxonmobil.com/~/media/global/Reports/Outlook%20For%20Energy/2015/2015-Outlook-for-Energy_print-resolution. With permission.)

By 2040, natural gas, nuclear, and renewables are expected to deliver more than 70% of the world's electricity (Figures 6.5 and 6.6). It is projected by 2040 global demand for energy will be met by other fuels sources such as oil, biomass, biofuels, hydroelectric, and geothermal as depicted in Figure 6.6. The United States will invest US$6 trillion in renewable energy infrastructure by 2035, compared to just US$1 trillion in nuclear and US$2.75 trillion in conventional energy (World Energy Outlook, 2013).

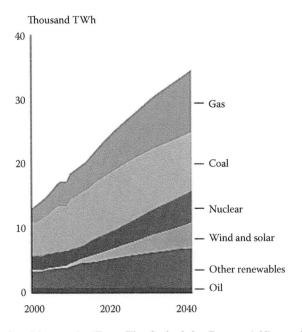

FIGURE 6.5 Global electricity supply. (From *The Outlook for Energy: A View to 2040*. 2015, http://cdn.exxonmobil.com/~/media/global/Reports/Outlook%20For%20Energy/2015/2015-Outlook-for-Energy_print-resolution. With permission.)

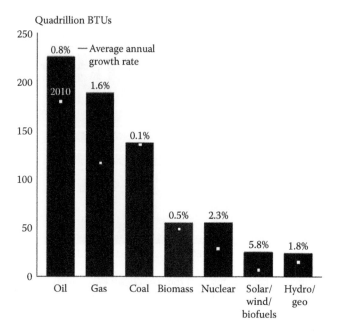

FIGURE 6.6 2040 global demand by fuel. (From *The Outlook for Energy: A View to 2040.* 2015, http://cdn. exxonmobil.com/~/media/global/Reports/Outlook%20For%20Energy/2015/2015-Outlook-for-Energy_print-resolution. With permission.)

The mining industry has traditionally relied on conventional fossil fuel sources such as diesel, oil, coal, and natural gas to meet its growing energy demand. The energy use manifests as GHG emissions as illustrated in the case of Rio Tinto operations in Figure 6.7. The industry, especially in remote locations need to gain full understanding of local renewable energy capacity—from geothermal and hydroelectric to solar and wind and biomass (EY, 2014).

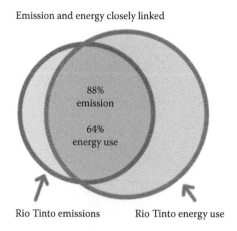

FIGURE 6.7 Link between energy and emissions for Rio Tinto's. (From Energy Efficiency Opportunities. 2008. How Rio Tinto Addresses Energy Management. Energy Efficiency Opportunities, May 22, Melbourne, Australia, http://www.schneider-electric.com/solutions/nz/en/med/333390763/application/pdf/2288_identify-opportunities-to-reduce-the-consumpt.pdf. With permission.)

ENERGY CONSUMPTION IN MINERAL PROCESSING

The energy use of the mining industry represents about 40%–60% of a mine's operating costs currently. While energy costs constitute 15% of total mining and mineral processing input costs (Syed et al., 2010), the energy consumption and intensity in mining and mineral processing keeps rising at around 6% per annum (Figure 6.8; Australian Bureau of Agricultural and Resource Economics, 2010). Comminution (crushing and grinding), separation (froth flotation), and concentrate drying consume most energy and are responsible for most gas emissions in the mining and mineral processing sectors (http://eex.gov.au/technologies/mineral-processing/). Grinding constitutes the largest among the top 10 energy-intensive processes in coal, minerals, and metals (Figure 6.9). Energy is

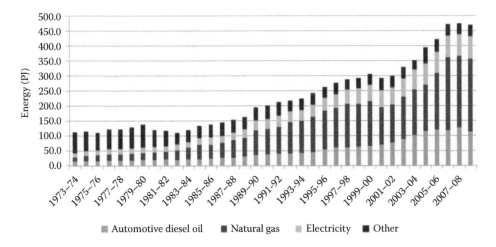

FIGURE 6.8 Energy consumption by mining in Australia. (Adapted from ABARES. 2011. *Australian Energy Statistics – Australian Energy Update 2011*. Canberra, Australia: Australian Bureau of Agricultural and Resource Economics and Sciences.)

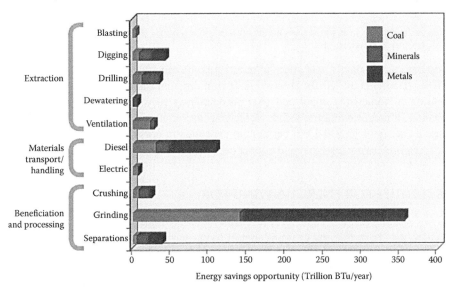

FIGURE 6.9 Energy savings opportunity in US Mining Industry for top 10 energy-intensive processes. (Adapted from Energy Efficiency Opportunities. 2008. How Rio Tinto Addresses Energy Management. Energy Efficiency Opportunities, May 22, Melbourne, Australia, http://www.schneider-electric.com/solutions/nz/en/med/333390763/application/pdf/2288_identify-opportunities-to-reduce-the-consumpt.pdf.)

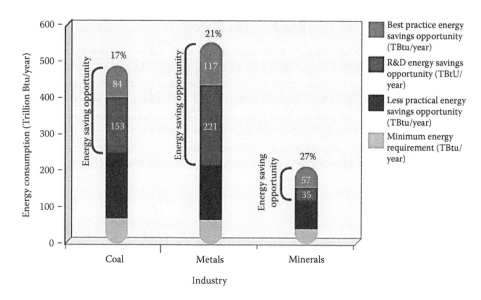

FIGURE 6.10 US Mining Industry energy bandwidth for coal, metal, and mineral mining. (Adapted from Mining Industry Energy Bandwidth Study, June 2007 by the US Department of Energy; Energy Efficiency Opportunities. 2008. How Rio Tinto Addresses Energy Management. Energy Efficiency Opportunities, May 22, Melbourne, Australia, http://www.schneider-electric.com/solutions/nz/en/med/333390763/application/pdf/2288_identify-opportunities-to-reduce-the-consumpt.pdf.)

also used in blasting, drilling, dewatering, and transporting of mineral ores. Since the average grade of mined ore bodies in Australia has halved over the last 30 years, the waste removed to access the minerals has more than doubled (Bye, 2011), with a 70% increase in energy consumption across mining operations (Figure 6.10) and fall in the multi-factored productivity by 24% (Daniel, 2007; Sandu and Syed, 2008).

ENERGY-INTENSIVE COMMODITIES

Diamonds, rare earth metals, and the platinum group of minerals are the most energy-intensive commodities (Daniel et al., 2010). Processing of iron ore, and the copper and gold ores consume most comminution energy. Copper and gold production is very energy intensive because of the small amounts of metals being produced. Nonferrous ores consume 0.56% (87 TWh) of the global net electrical energy of 15,500 TWh per annum expended in crushing and grinding. Of this, 33% and 53% is required to process gold and copper ores, respectively (Daniel and Lewis-Gray, 2011).

OPPORTUNITIES FOR ENERGY SAVING FOR MINERAL PROCESSING TECHNOLOGIES

Figure 6.10 shows that while the coal and metals have the potential to reduce their energy use by 17% and 21% respectively using best practices, minerals have the potential to reduce it by 27%. A 2005 report by the Natural Resources of Canada reported the potential to reduce the mining costs (based on energy savings) by 36% and milling costs by 47% in iron ore in an open cut mine. For gold mining, this estimate was 53% for milling.

Opportunities to reduce the energy and production costs exist through an integrated process control system which makes use of process measurements, drive, and power distribution information; to increase plant-wide efficiency, productivity, and reliability. An integrated approach to mine design and planning can synchronize energy supply and demand from the outset (CEEC—The future, 2015). Automated mine processes at the design phase reduce reliance on fossil fuels. Recent Australian studies suggest that energy savings per tonne up to 50% are practically feasible in the design of new mining and mineral processing developments (Pokrajcic et al., 2009). Applying energy efficiency strategies begins with the resource characterization stage to comminution using more efficient equipment (Johnson, 2013).

Holistic (whole-of-system) approach will facilitate utilizing full potential of energy efficiency opportunities, including selecting appropriate comminution technologies, crushing, and grinding mill configurations. This will depend on the feed-size distribution, desired product size, physical properties of the material, resource characterization, preconcentration of ore grades, fit-for-purpose efficient technologies, flexible comminution circuits, and the efficiency of motor systems indirect energy use (http://eex.gov.au/technologies/mineral-processing/opportunities/#implement-a-holistic-comminution-energy-efficiency-strategy).

Some of the key factors contributing to poor energy efficiency in mineral processing operations are (Napier-Munn et al., 2012):

- Difficult lower grade orebodies with complex mineralogy
- Access difficulties and high infrastructure costs
- Reluctance in industry to adopt new technologies, coupled with unwillingness to adopt new processes perceived as being high risk, or engage in or support risky research
- Energy efficient strategies lack support from senior management

EFFECT OF ORE CHARACTERISTICS ON THE ENERGY USE

The ore bodies determine the overall energy use in mining and mineral processing operations. Low-grade ores and a harder or more competent ore increases the energy use by over 10% for the same throughput within a mill. Similarly, the need to halve the grind size to maintain recovery rates during flotation (due to liberation issues with an ore), will increase the energy in excess of 10%, more likely by a factor of four (Energy Efficiency Opportunities, 2008). This also increases the embodied energy of primary metal production and the associated GHG emissions especially for metals like copper, gold, and nickel (Norgate and Jahanshahi, 2006).

Geometallurgical mapping and geometallurgical 3D models assist in understanding the mineralogy and geological variability within an ore body to make the right grinding choices (Bye, 2006).

Potential benefits of geometallurgy techniques include (Lund and Lamberg, 2014): better,

- Utilization of the ore resources—ore boundaries are defined in order to forecast the metallurgical performance
- Metallurgical performance—by tuning the process according to the plant feed information beforehand
- Controlled mining—due to more comprehensive knowledge of the ore body
- Changes in plant optimization—the variation in the plant feed is low, or at least better controlled
- Changes for new technological solutions—ore-derived problems are identified well ahead toward solving them

Ore-sorting technologies/pretreatment separation processes help to reject most of the gangue from the outset ahead of surface rock breakage resulting in improved energy efficiency and an increase in ore grade ahead of crushing. A better understanding of ore texture is critical to the selection of an optimum ore-separation technology (Powell and Bye, 2009). Automated ore-sorting technologies apply optical sensors coupled with electrical conductivity and magnetic susceptibility sensors to control the mechanical separation of ore into two or more categories. The CRC ORE

TABLE 6.1

Sensor-Based Sorting Applications

Sensor/Technology	Material Property	Mineral Application
RM (radiometric)	Natural gamma radiation	Uranium, precious metals
XRT (X-ray transmission)	Atomic density	Base/precious metals coal, diamonds
XRF (X-ray fluorescence)	Visible fluorescence under x-rays	Diamonds
COL (charge-coupled device [CCD] color camera)	Reflection, brightness, transparency	Base/precious metals industrial minerals, diamonds
PM (photometric)	Monochromatic reflection/absorption	Industrial minerals, diamonds
NIR (near-infrared spectrometry)	Reflection, absorption	Base metals industrial minerals
IR (infrared cam)	Heat conductivity, heat dissipation	Base metals industrial minerals
MW-IR (heating in conjunction with IR)	Sulfides and metals heat faster than other minerals	Base/precious metals

Source: Adapted from Seerane, K. and G. Rech. 2011. Investigation of sorting technology to remove hard pebbles and recover copper bearing rocks from an autogenous milling circuit. *6th Southern African Base Metals Conference 2011,* The Southern African Institute of Mining and Metallurgy, Phalaborwa, South Africa, pp. 123–135. http://saimm.co.za/Conferences/BM2011/123-Rech.pdf.

(2010) at the University of Queensland reported a 2.5 fold in average ore concentration feed to the grinding mill when using smart blasting, ore sorting, and gangue rejection at actual mines (SME Mining Finance, 2015).

Depending on the type of ore, different sensor types are used in sorting technologies (Table 6.1; Seerane and Rech, 2011). Pretreatment separation can be achieved by electric pulse fragmentation, microwave, or ultrasound (http://eex.gov.au/technologies/mineral-processing/opportunities/#fn-16865-1). Zuo et al. (2015) report on a preconcentration technique using high voltage pulses. This technique uses metalliferous grain-induced selective breakage, under a controlled pulse energy loading, and size-based screening to separate the feed ore into body breakage and surface breakage products. The combined advantages of preweakening and preconcentration, and multigrade comminution and recovery circuits offer new opportunities for barren pebbles rejection from the autogenous grinding (AG)/semi-autogenous grinding (SAG) mill pebble stream, coarse waste rejection at mine site to reduce the run of mine haulage, and pretreatment of AG/SAG mill feed.

SELECTIVE/SMART BLASTING TECHNOLOGIES

Geometallurgy facilitates the "smart blasting" approach to target the sections of the ore body with the highest ore-grade concentration (Bye, 2011) and reduces energy use per tonne of metal by 10%–50% (CRC ORE, 2011). The selective/smart blast design technology raises the grade of ore being fed to the crusher and grinding mill by improving resource characterization and reducing the net total energy consumed at the crushing and grinding stages compared to conventional blasting (Powell and Bye, 2009). The benefits include:

- Less energy is required to crush the ore to the same product size if the feed-size to the primary crusher is decreased
- Additional macro- and micro-fracturing within individual fragments from the blasting makes fragments easier to fracture further, using less energy in the crushing and grinding phase
- An increased percentage of relatively small mineral particles can bypass stages of crushing
- Software packages assist in designing effective blasting techniques, including analyzing and evaluating energy, scatter, vibration, damage, and cost

Eco Efficiencies for Size Reduction Devices

The choice of comminution equipment and design and optimization of circuits which simulate the processes of crushing, grinding, and classification has a significant impact on energy use. Run of mill size distributions due to blasting, an important input parameter to the simulation models, have a great influence on the performance of a SAG mill (Bobo et al., 2015). The eco-efficiencies of the size reduction equipment in the mineral industry are compared on the basis of: (1) capital cost, (2) installation cost, (3) operating cost, and (4) maintenance cost. An ideal crusher would (1) have a large capacity; (2) require a small power input per unit of product; and (3) yield a product of the single size distribution desired. From the viewpoint of eco-efficiency, compact size reduction devices (lowered energy consumption in manufacture) having high throughputs and lowered energy usage per unit of feed provide improved size reduction systems.

Comminution energy includes the energy directly consumed in size reduction, the energy consumed in the manufacture of comminution consumables, especially the steel grinding media and the liners (Napier-Munn et al., 2012). The overall energy cost or "embodied energy" of manufacturing the steel in the liners and grinding media, amounts to up to 4–6 kWh/t. A reduced consumption of mill liners and grinding media can lead to a greater eco-efficiency (Daniel et al., 2010).

High-Pressure Grinding Rolls

High-pressure grinding rolls (HPGR), stirred grinding mills, and vertical shaft impactors (VSI) are more energy-efficient than SAG technology. SAG mills are essentially autogenous mills, but utilize grinding balls to aid in grinding like in a ball mill. A SAG mill is generally used as a primary or first-stage grinding solution. SAG mills use a ball charge of 8%–21% SAG mills are essentially autogenous mills, but utilize grinding balls to aid in grinding like in a ball mill. A SAG mill is generally used as a primary or first-stage grinding solution. SAG mills use a ball charge of 8%–21% (Mular et al., 2002). HPGR are compact devices that require a small footprint; they are flexible adjustable machine settings such as pressure and roll speed to react to ore variations whilst maintaining a high throughput (Daniel et al., 2010). HPGR and VSI weaken the feed particles allowing a finer crushed feed ahead of the energy-intensive grinding circuits to reduce overall energy intensity. They can be used as a replacement for SAG mill for coarser grinding, primarily to reduce energy consumption and costs within crushing and grinding circuits.

The energy saving is attributed to the factors that the HPGR treatment generated the weakened particles for the downstream ball milling process, while the particle weakening effect by HPGR was more significant at coarse particle size (Shi et al., 2006). In most cases, HPGR treated materials had a better flotation response. Under various conditions, an HPGR circuit can achieve a 15%–20% reduction in direct energy (kWh/t or energy cost) compared to traditional SAG mills (Morley, 2005). HPGR circuits reduce the need for high embodied-energy grinding media, and therefore help reduce energy use over the entire lifecycle and can also provide around 25% reduction in operating costs (Daniel et al., 2010). (Embodied energy is the energy consumed by all of the processes associated with the production.) Schwarz and Seebach (2000) reported the technology required 30%–50% of the energy of conventional ball mills for the same size reduction task on cement clinker.

Stirred Mills

Small stirred mills are more eco-efficient, even for comparisons where the kWh/t values are similar for a given ore. Stirred mills allow size reduction to occur with acceptably low energy inputs in the fine and very fine grinding regions using the attrition mechanism. Inert media provide a "clean" grinding environment. Stirred mills can be operated with a classification system (usually in open circuit). The IsaMill is the highest speed version of stirred mills and Vertimill is the lowest speed version (closely related to the tower mill).

Until now the tower mill, with the power intensity of the biggest model at only 1.1 MW, was considered the most efficient technology for regrinding (Figure 6.11). The IsaMill offers a step

3 MW IsaMill

Media system under floor

Open circuit operation

FIGURE 6.11 Step change in grinding technology. (From Pease, J. 2007. Increasing the energy efficiency of processing. Presented at: Crushing and Grinding, Brisbane, Australia, September, http://www.isamill.com/EN/Downloads/Downloaded%20Technical%20Papers/Joe_Pease.pdf. With permission.)

change in the conventional grinding mill and is much more efficient than ball or tower mills for fine grinding. A 3 MW IsaMill has three times more power than the much bigger tower mill. It also can be easily installed in the open circuit in the middle of a flotation circuit (Pease, 2007). Other advantages include applications in difficult and fine-grained materials, improved liberation, and flotation selectivity (Pease, 2007).

The IsaMill (Pease et al., 2006) is more eco-efficient as it does not require an external classification system, eliminating the need for a separate classification system. It can supply sufficient energy levels to provide a degree of mechanical activation in downstream processing. This is demonstrated for the Hicom Nutating Mill having higher energy input per unit volume. Tavares (2005) and Maxton et al. (2003) have indicated that in Argyle diamonds, they could achieve the requisite reduction ratios whilst maintaining a relatively wide operating gap using the IsaMill. The use of ceramic and inert media can expand this technology to produce clean surfaces for flotation, sharp size distributions, coarser grind sizes and is now more economically viable to implement (Curry and Clemont, 2005).

Classification, Cyclone versus Stack Sizer

A classifier in the closing of a grinding circuit removes liberated mineral particles from the mill as coarse as possible in order for them to be recovered via a mineral concentrating method, such as flotation, spirals, etc. A classification system needs to achieve size separation efficiently with the objective of getting liberated particles on their way for recovery. Separation efficiencies for cyclones are in the range of 45%–65% and for the Stack Sizer, they are in the range of 85%–99%. The difference between the two technologies does not end at size separation efficiencies. Cyclones make their separations based on differences in settling rate velocities, particle size, and mass, while screening machines ignore all other variables with the exception of particle size.

In the grinding circuits which use cyclones, the heavier fine, liberated minerals report to the underflow and are fed back into the grinding mill while the lighter, middling particles report to the overflow without the liberation of the desired mineral. This is attributed to the large specific gravity differentials between the desired mineral and the gangue material. Becker et al. (2015) have reported that UG2 ore in South Africa in the platinum group minerals are typically very fine grained, with the potential for poor liberation high, especially with conventional cyclones providing a poor split of the coarse silicates and fine chromites to the over and underflows respectively. The screen on the other hand recognizes the particle size only and allows for accurate determination of liberation (Barkhuysen, 2009). However, in sieves with square mesh coarser and unliberated fraction report to finer fraction due to variation in shapes from the spherical.

The main economic advantages of screens versus cyclones in grinding circuits are

- Typically require less water for operation
- Reduction in slimes generation allows for reduced reagent consumption in flotation
- Reduced energy consumption due to reduced circulating loads
- Increased production rates (typically between 10% and 30%)
- Reduced grinding media consumption
- Reduced mill maintenance costs
- Increased efficiency in downstream classification devices such as spirals, flotation cells, and jigs as they process a narrower size range of particles more efficiently
- Increased mineral recovery rates
- Reduced concentrate dewatering costs due to coarser liberation and reduction of slimes

Strategies to improve eco-efficiency in communition can be summarized as (Napier-Munn et al., 2012):

- Size reduction effort pushed toward crushing stages rather than to expensive final grinding stages
- Applying geometallurgical information
- Developing new comminution machines, for example, new generation crushers, variations on HPGR, variations on the IsaMill, and grinding roller mill
- Eco-efficient comminution based on compact design
- Using flexible comminution circuits based on a mathematical model of the energy profile of the complete process system and for process optimization and to evaluate performance

Other actions or areas for improvement are the separation processes involving mixers and agitators. Significant energy reductions (30%–50%) through the use of new blade and tank designs and cell efficiencies offer opportunities for further energy savings (Australian Greenhouse Office and the Minerals Council of Australia, 2002). Significant opportunities for energy savings also exist in the areas of froth flotation separation, materials movement, and ventilation.

ENERGY EFFICIENCY IN FLOTATION

Flotation is an important mineral concentrating unit operation. It is efficient and complicated, at the same time challenging and less understood. Over 95% of all base metals, for example, copper, lead, zinc, and silver are recovered and concentrated in the froth of flotation machines. In coal preparation plants using flotation, approximately 5%–15% of the raw coal feed is processed through the flotation unit for recovery of fine coal particles. As the grade (the assay of valuable constituent) of mineral deposits decreases and coal and metal prices rise, the importance of flotation increases. In turn, the necessity of improving selective recoveries becomes even more critical (Kennedy, 2008).

In froth flotation, large economic gains can be made through optimization of many present processes (Shean and Cilliers, 2011). The process control of froth flotation comprises a hierarchy of three to four inter-connected tiers (Laurila et al., 2002; Figure 6.12). The upper-most tier of process control in flotation, the optimizing flotation circuits, aims to maximize the financial feasibility of

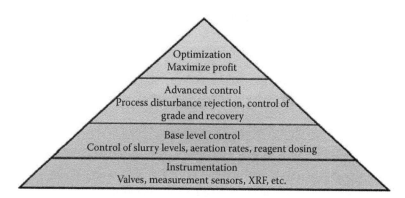

FIGURE 6.12 Process control system level hierarchy for flotation processes. (Reprinted from *Mineral Processing Plant Design, Practise and Control*, 1, Laurila, H., J. Karesvuori, and O. Tiili, Strategies for instrumentation and control of flotation circuits, 2174–2195, Copyright 2002, with permission from Elsevier.)

the process. This is achieved by determining the theoretical grade–recovery curve most profitable to operate and, subsequently, shifting the operating point orthogonally to further maximize profit (Figure 6.13). The optimizing flotation controllers locate the optimal operating point on the grade–recovery curve; and then present recovery and grade set points to lower control systems and/or plant operators/management.

Major advances seen in instrumentation and technology, and simplifications and modifications of new flotation plant designs in the last few decades have allowed for significant developments in process control, in particular, the development of base level process control (control of pulp levels, air flowrates, reagent dosing, etc.). For example, a data analysis methodology, the matrix method, optimizes flotation performance, based on mineral liberation and particle size. It detects the difference between how the mineral was prepared for flotation in grinding (feed quality), and how it was then actually recovered in flotation (flotation performance), in a single flotation cell or in a whole flotation circuit, as long as there is no further size reduction (such as regrinding) in the circuit. Cumulative sum analysis is then applied to measure and trend the specific role of both feed quality and flotation performance on overall recovery, to determine their effect on the recovery (http://www.fei.com/natural-resources/mining/ floatation/).

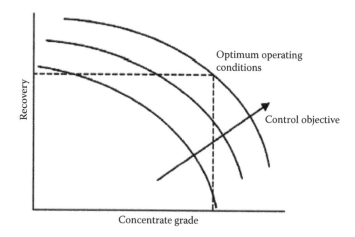

FIGURE 6.13 Grade–recovery curve illustrating optimizing control objective. (Reprinted from *Wills' Mineral Processing Technology*, 7th edn, Chapter 12, Wills, B. A. and T. J. Napier-Munn, 267–344, Copyright 2006, with permission from Elsevier.)

A long-term, automated advanced, and optimizing flotation control strategies have been more difficult to implement. This is because a computer-based control system controls the process in a fundamentally different way from a human operator. When trying to automate the actions of even the best of human operators, there is the possibility to miss an opportunity to make a full step improvement in the control of the process. This is attributed to failing to properly leverage the speed, memory, and computational power and reliability available with computer-based control (Forbes and Gough, 2003). It is hoped that this will change as a result of the development of new technologies such as machine vision and the measurement of new control variables, such as air recovery (Shean and Cilliers, 2011).

In some mineral processing sites investing in more energy-efficient froth flotation cells, such as the Jameson Cell can increase productivity. The Jameson Cell consistently produces smaller bubbles compared to earlier cell types. The mixing and adhesion occur more quickly in a smaller space, and a higher percentage of mineral is recovered. It can treat a large amount of material with a small footprint. The Jameson Cell requires a pump but has no requirement for a motor, air-compressor, or moving parts (Glencore Technology, 2012). Larger-volume flotation cells consume less energy per cubic meter of material processed, allowing for a simpler plant layout with a smaller overall physical footprint according to Jarmo Lohilahti, technology manager, flotation, Outotec. For example, it is claimed that the slower rotation speed Outotec FloatForce reduces power consumption and the carbon footprint while enabling better recovery levels (http://www.outotec.com/en/Sustainability/Sustainable-offering-for-customers/Flotations-sustainable-future/). The SuperCells, the largest flotation cells in the world, with a capacity ranging from 300 to 350 m³ created by FLSmidth Minerals—reportedly offers lower energy consumption, improved recovery, and increased operational efficiency, outperforming its predecessor (FLSmidth, 2009).

The conventional flotation technology, for example, in coal processing, allows some ultrafine impurities (e.g., clay) to be hydraulically carried into the clean coal product in the process water that reports to the froth, which is its major shortcoming. This contamination substantially reduces the overall quality of the clean coal froth product. The entrainment problem has led many coal operations to adopt column flotation cells in place of conventional flotation machines (Davis et al., 1995). Flotation columns work on the same basic principle as mechanical flotation equipment. In column flotation, there is no mechanical agitation and separation takes place in a vessel of high aspect ratio (the cell height is much greater than the cross-section of the cell.) Column flotation offers many advantages including

- Improved metallurgical performance
- Improved control
- Less energy consumption
- Reduced footprint
- Reduced capital requirement

Columns allow mineral processing plants to achieve higher profits for their concentrate by purifying concentrates, lowering shipping costs, decreasing plant footprint, and lowering smelter penalties. The column achieves efficient separation and high upgrading mainly by two methods. Firstly, by using wash water, entrained minerals can be washed from the froth before reporting to the concentrate. Secondly, columns achieve high froth levels that encourage draining and dampening of entrainment caused by high gas rates. (http://www.sgs.com.au/~/media/Global/Documents/Flyers%20and%20Leaflets/SGS-MIN-WA340-Column-Flotation-EN-11.pdf). Column cells are able to reduce entrainment by the addition of a counter-current flow of wash water to the top of the froth. The water rinses ultrafine impurities such as clay from the froth back into the flotation pulp, with less than 1% of the feed slurry reporting to the froth in a properly operated column.

IMPROVING THE EFFICIENCY OF DEWATERING AND DRYING

DRYING

At mine sites where concentrate is processed, drying is a significant energy-using activity. Conventional dryers used in the mineral processing industry are classified as hearth, shaft, and grate type. Other types of dryers used less commonly are the spray, fluid bed, pneumatic or flash conveyor, drum, stationary- and rotating-tray type, infrared type, and others. The commonly used drying mediums are hot air, combustion gas, and steam (Wu et al., 2010). The emerging drying methods include, superheated steam drying, which eliminates the fire hazard associated with conventional drying methods. The Wirbelshicht-Trocknungmit interner Abwarmenutzung dryer (Mujumdar et al., 1994) is claimed to consume 80% less energy; have 80% less dust emissions; and have a smaller capital cost compared to the rotary steam-tube dryer.

Pulse combustion drying offers the advantages of short drying time, high energy efficiency, improved product quality, and environmentally friendly operation. In pulse combustion spray driers, very hot drying gases are accelerated to about 300 mph in a resonating pulse combustion engine. The liquid is pumped into this gas stream at low pressure (about 1 psi) and very low velocity. The high-velocity pulse waves instantly atomize the liquid and drying is completed in less than a second. The rapid drying is caused by extreme temperature differentials and intense mixing in the highly turbulent environment downstream of the combustor tailpipe (http://www.pulsedry.com/tech.php).

Microwave drying has potential in minerals drying; for example, drying of ceramics, fine coal, pretreatment of refectory gold ore and concentrate, etc. It offers district advantages such as high process speed, uniform heating, high energy efficiency, better and rapid process control, and less space requirement. Simultaneous grinding and drying operations, for example, in the KDS grinder chamber achieves air temperature between 70°C and 90°C, where grinding action is accompanied by drying. Drying with heat pump or waste heat where the waste heat is recovered heat from exhaust gas at high temperature is useful for drying foods and biomaterials, minerals, coal, and clay minerals, for example, bentonite (Fath, 1991). The mining sites could also use waste heat to reduce the energy required for drying and also for solar drying of concentrate (http://eex.gov.au/technologies/mineral-processing/opportunities/#fn-16994-1).

DEWATERING

Dewatering can save on the energy required in thermal drying, to improve the handling properties of the concentrates, and to reduce the transportation costs involved in shipping. Since the 1920s, various dewatering and upgrading processes have been developed in a number of industries (Le Roux et al., 2005; Lee et al., 2006). Conventional dewatering techniques utilize thickeners, dewatering screens, vacuum filters, centrifuges, and pressure (hyperbaric) filters to reduce product moisture to a lower value (Wu et al., 2010).

New developments in dewatering technology include electro-dewatering, which can dewater adsorbed water like thermal treatment and unlike the mechanical dewatering methods; the cross-flow microfiltration technique for very fine materials; dewatering using surfactants and flocculants to enhance filtration rates by improving the wetting characteristics of the mineral surfaces; and the use of ceramic filter media in vacuum disc filters, with fine pores preventing the air breakthrough, thus permitting retention of higher vacuum levels and production of drier cakes.

Dewatering processes are grouped into evaporative (thermal) and nonevaporative (nonthermal) drying processes. Non-evaporative drying processes are attractive due to their enhanced energy effectiveness, since water is separated in liquid form and the latent heat of vaporization is not expended. Some of the non-evaporative drying of brown coal includes the Fleissner process (Fleissner, 1927) which incorporates high steam treatment at temperatures above 200°C and high operating pressures. The Evans–Siemon process is the updated Fleissner process, where pressurized

TABLE 6.2
Energy Consumption for Various Coal Dryers' Technologies

Type of Dryer	Energy Consumption per Mass of Water Removed (kJ/kg H_2O)
Rotary dryer	3700
Rotary tube dryer	2950–3100
Chamber dryer	3150
Pneumatic dryer	3100
Fluid-bed dryer	3100–3400
Fleissner process (superheated steam)	130–1750
Fluid-bed superheated steam with heat recovery	450

hot water is used to improve the thermal efficiency by avoiding the pressure cycle (Higgins and Allardice, 1973). The Koppelman (K-fuel) process (Murray, 1979) involves high pressure with low operating temperature, however, this process was not demonstrated commercially. Mechanical thermal expression dewatering process (Janine, 2005) involves the use of high temperatures, up to 250°C and pressure <12.7 MPa, resulting in noteworthy reductions in residual moisture content which may be largely attributed to the destruction of porosity in lignite. Electro-dewatering Reuter et al. (1981) does not show encouraging results. Table 6.2 shows energy consumption for various coal dryers' technologies that are used for dewatering brown coals (Wilson et al., 1992).

Dzinomwa et al. (1997) used a nonevaporative drying technique to dewater fine coal, where super absorbent polymers (SAP) were used to absorb moisture from coal. Within 4 h of polymer and coal contact, they achieved good reduction in moisture content, paving the way to cost-effective production of fine coal, removing the need for constructing large plants and dams. The moisture absorption capacity of the SAP was such that the coal product moisture was found to be comparable to that of thermal drying (Dzinomwa et al., 1997).

When coal is in contact with the SAP, the water molecules get diffused in the polymer network, the chains uncoil, and hydrogen bonds hold the water between the chains as shown in Figure 6.14 (Zohuriaan-Mehr and Kabiri, 2008). The second mechanism that affects and controls the swelling ability of an SAP is cross-linking. Increase in the cross-link density decreases the swelling capacity of the SAP, and increase in the gel strength and resistance to shear. SAP with greater resistance to shear can retain moisture within the network for greater periods of time. Thus, cross-linking can

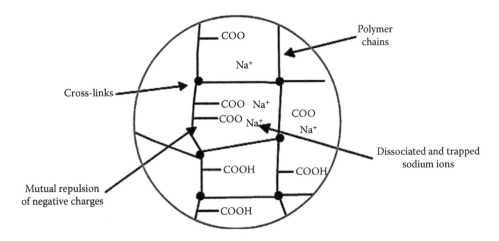

FIGURE 6.14   Mechanism of water absorption in SAP.

play a pivotal role in the way SAPs behave during swelling, thereby controlling the swelling capacity while maintaining the structural stability of the SAP.

Peer and Venter (2003) used SAP to dewater black coal fines, which enabled an effective decrease in the moisture content of coal. But the separation of fine coal from the SAP was found to be problematic. The authors found that SAP encapsulated in sachet form proved an effective means for separating the polymer from the coal. The SAP being pH sensitive makes it an interesting proposition to regenerate this moisture loaded SAP to dry SAP for further use. The SAP can also be regenerated using low temperature thermal methods (Dzinomwa et al., 1997).

At the Federation University Australia the dewatering of fine brown coal studies showed the highly oxidized Victorian brown coal that exhibits a very high affinity to moisture due to highly electronegative carbonyl groups which readily form hydrogen bonding with the moisture. Dry SAP has effectively removed approximately 44% of the moisture within an hour of contact at room temperature (Devasahayam et al., 2015, 2016).

The advantages of SAP dewatering, compared to the current technologies used are:

- Low energy consumption
- Many current nonevaporative technologies still require further drying thus creating a niche for such a product
- Fine coal particles can be dewatered which is difficult by other processes
- Low cost of operation
- Energy efficient process
- Comparable moisture removal
- Minimal safety issues

MILL-TO-MELT METHODOLOGY

Traditional grinding and flotation models seek to improve the energy efficiency of these mineral processing operations, without considering the energy used by downstream metal production, for example, smelting and refining. However, the comminution and smelting of metal-bearing ores are both highly energy-intensive processes. Optimizing the energy efficiency across these two processing stages can directly reduce the overall energy consumption per unit mass of metal produced and thus reduce their GHG footprint (Evans et al., 2009).

But even when the concentrators and smelters individually are running efficiently, it is not certain that the energy consumption of the overall system is optimized. Whether it takes less energy to remove an impurity in the concentrator or the smelter needs to be ascertained. According to Pease (2007), there is no point in reducing grinding energy if this increases smelting energy. Analysis of the complete concentrator—smelter process chain may show that there are times when the concentrator should use more energy to reduce the overall energy requirements across the mill and smelter. In order to double the energy efficiency of processing, there are tools already existing such as:

- Laboratory grinding and laboratory flotation tests
- Quantitative mineralogy
- Basic thermodynamics of smelting and refining
- To grind the right mineral in the right place to the right size, with the need to be combined in the right way

The mill-to-melt methodology allows to model energy consumption from mill to smelter with a view to reducing the energy-intensity and GHG footprint of their processing and smelting operations. It aims to integrate mass and energy models of both mineral concentration and smelting stages to provide a tool for optimizing overall energy consumption across these two energy-intensive processes. The mill-to-melt methodology includes:

- Determining what grade and recovery positions can be achieved in the concentrator by using different combinations and amounts of grinding and regrinding
- Thermodynamic calculations of the energy used to process the different concentrate grades to metal

Integrating the concentrating and smelting models to find the lowest total energy for the system. For example, in nickel metallurgy, the presence of white MgO in pentlandite causes a problem for the smelter by producing a highly viscous slag. The smelter has to add silica and lime fluxes, and heat everything up to 1350°C to counter the effect of MgO. This consumes over 10 times the energy it would take to grind the MgO out. The high temperatures also make smelting difficult and reduce refractory campaign life. So to improve energy efficiency, the concentrator could regrind this particle to recover the pentlandite (Pease, 2007). Regrinding allows gangue minerals to be removed from the intermediate concentrate, generating a higher grade final concentrate at the same recovery point (Figure 6.15). The mill-to-melt energy (E_{MM}) dividend, the ratio of the net increase in regrinding energy to the resulting net reduction in smelter energy is given by

$$E_{MM} = \frac{Nes\ smelter\ energy\ saving\ with\ regrinding}{Energy\ consumed\ in\ regrinding}$$

Although regrinding is an energy-intensive process, increased regrinding is shown to reduce the overall energy consumption in mineral processing chain. For example, regrinding reduces the overall energy in a concentrator-smelter process chain by 11% per tonne of metal produced for copper–nickel sulfide ore. Use of 1 kWh of energy for regrinding was shown to reduce the overall energy consumption by 12 kWh (Figure 6.16; Evans et al., 2009).

The mill-to-melt process requires optimization of the flotation performance of ore particles after regrinding. There is potential for greater rejection of gangue after regrinding with the potential to

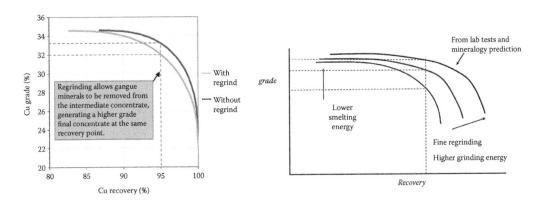

FIGURE 6.15 Effect of grinding energy on separation performance (a) and smelting energy (b) of the ore. (From Evans, L. et al. 2009. Improving energy efficiency across mineral processing and smelting operations— A new approach. In *Proceedings SDIMI 2009—Sustainable Development Indicators in the Minerals Industry.* The Australasian Institute of Mining and Metallurgy, Melbourne, Australia, pp. 9–14. With permission; Pease, J. 2007. Increasing the energy efficiency of processing. Presented at: Crushing and Grinding, Brisbane, Australia, September 2007, http://www.isamill.com/EN/Downloads/Downloaded%20Technical%20Papers/ Joe_Pease.pdf. With permission.)

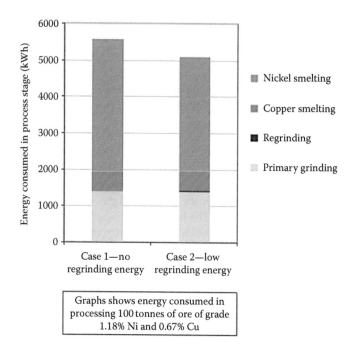

FIGURE 6.16 Relative contributions to the total energy consumed in the concentrator to smelter process chain: energy consumed in copper and nickel production with and without regrinding. (From Evans, L. et al. 2009. Improving energy efficiency across mineral processing and smelting operations—A new approach. In *Proceedings SDIMI 2009—Sustainable Development Indicators in the Minerals Industry*. The Australasian Institute of Mining and Metallurgy, Melbourne, Australia, pp. 9–14. With permission.)

further reduce the total energy requirements. When integrating the concentrator and smelter models in the mill-to-melt methodology toward energy and GHG footprint minimization studies, inclusion of mineral-dependent operating temperature calculations enhances the smelter energy modeling. This will demonstrate the disproportionate energy savings that are achievable by selective rejection of problematic gangue species at the concentrator stage. Where the smelter and concentrator are geographically separated, the mill-to-melt approach could be expanded to include the logistics chain between the concentrator and smelter, because the impact of transport energy, per tonne of metal, is also dependent on concentrate grade.

PRIMARY AND SECONDARY METALS PRODUCTION

The on-going demand for metals well into the future has necessitated a combination of primary and secondary metal production, from newly mined ores and recycled metals, respectively. It is expected the amount of metal recycled will continue to increase (Norgate et al., 2007). Primary production, in particular the minerals processing and metals production stages, requires large amounts of energy and chemicals and produces large volume wastes and potentially harmful emissions of minor elements and toxics.

The embodied energy, defined as the sum of the direct and indirect energies of the individual stages along the value-adding chain, of the common metals varies widely, from typically around 20 MJ/kg for lead and steel to over 200 MJ/kg for aluminum (Rankin, 2012). The stability of the minerals from which the metal is extracted (related to free energy of formation), the ore grade, the degree of beneficiation required, and the overall recovery determine the embodied-energy content of the primary metals.

TABLE 6.3
Embodied Energy for Primary and Secondary Production

Commodity		Embodied Energy (GJ/t)	
		Secondary Production	Primary Production
Aluminum	Alloy	17.5	212
	Cans	10.1	
Copper	No. 1 scrap	4.4	33
	No. 2 scrap	20.1	
	Low-grade scrap	49.3	
Steel	Billets	9.7	23
Lead-soft	Batteries	9.4	20
Lead-hard	Batteries	11.2	
Nickel	Alloy scrap	12.9	114
Zinc	New scrap	3.8	48
	Slab	22	

Source: Adapted from Rankin, J. 2012. Energy use in metal production. In *High Temperature Processing Symposium.* Swinburne University of Technology, Melbourne, Australia, February 3–4, pp. 7–9.

The quantity of GHG produced frequently closely follows the trends in embodied energy. Steel production contributes the greatest quantity of GHG, about 7% of global CO_2 produced from fossil fuels. Aluminum production is responsible for around 3% of global CO_2 (Rankin, 2012). Steel has a lower embodied energy compared to aluminum, attributed to iron's lower affinity to oxygen compared to aluminum. However, it is produced in large quantities resulting in higher emissions.

Step changes in technologies reducing the embodied-energy content of the metal are required to achieve energy efficiencies in primary metal production. Thermodynamic constraints impose limit to the energy reduction, most importantly, the stability of the compound from which the metal is produced. While there is a thermodynamic limit below which it is not possible to go, the technology used at present is far removed from that limit meaning there is a great potential for further improvement (Rankin, 2012).

Secondary metals production on the other hand has multiple environmental benefits over primary metals production. Firstly, the metal value is preserved and mining of ore is avoided. On current ore grades, ore to primary metal ratios for selected metals are: 6.5 t bauxite/tonnes aluminum, 115,000 t ore/tonnes gold, 140 t ore/tonnes copper, 1.6 t iron ore/tonnes iron, 4.5 t mineral sand/tonnes titanium, and 7.2 t ore/ton zinc (Wright et al., 2002; Rene van Berkel, 2007). The energy required to recycle metals is a relatively small fraction of the energy required to produce metals from their ores since energy is required largely only for melting and not mineral processing. However, when the energy required for collection and separation of scrap is included, the embodied energy of recycled metals increases as the fraction of scrap collected increases since transportation and separation costs progressively increase (Rankin, 2012). Typical energy benefits for secondary metal production are: 94% for secondary aluminum; 75% for secondary copper, 40% for secondary steel, and 70% for secondary lead (Wright et al., 2002). Secondary zinc production is, however, estimated to be about 10% more energy intensive than primary zinc production (Wright et al., 2002; Table 6.3).

ENERGY EFFICIENCY IN PRIMARY/SECONDARY METAL PRODUCTION

Alumina production is responsible for the largest energy consumption (225 PJ) followed by iron smelting and steel production (173 PJ) followed by aluminum smelting (133 PJ) in 2010–2011, accounting for 90% of metal sector's energy use (Figure 6.17; Climate Works Australia, 2013).

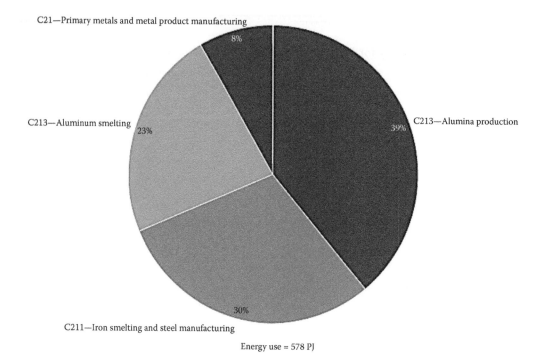

C21—Primary metals and metal product manufacturing

C213—Aluminum smelting

C213—Alumina production

C211—Iron smelting and steel manufacturing

Energy use = 578 PJ

FIGURE 6.17 Metals sub-sector energy use. (Adapted from Climate Works Australia. 2013. Industrial energy efficiency data analysis report: Metals manufacturing, http://www.climateworksaustralia.org/sites/default/files/documents/publications/climateworks_dret_ieeda_factsheet_metals_20130502.pdf.)

The largest energy savings can be found in the alumina production sector (17,000 TJ) and iron and steel manufacturing (13,500 TJ). Primary metal and metal product manufacturing sub-sector produces metals such as nickel, zinc, and copper. The primary metal and metal product manufacturing sector have the energy saving opportunities up to 12% of energy used. The energy efficient opportunities for these sectors exist in improvements to furnaces and kilns due to high usage of thermal energy (Figure 6.18). Improvements to various industrial systems represent 40% of the total energy saving potential, while improvements to process heating, thermal electricity generation, and furnaces and kilns also represent large opportunities. Savings are also identified in the use of direct fuels gas, coal, oil, biomass, and bagasse.

Aluminum Industry

Currently, the average energy consumption per ton of alumina in Australia is 12,269 MJ/t, compared to world's best practice of 9000 MJ/t achieved by the German company AOS38 (Martin et al., 1999; Bushnell, 2000; The Natural Edge Project, 2007) although this would depend on the mineralogy of the feed and cannot be directly compared. Alumina production is the intermediary step between bauxite mining and aluminum smelting involving electrolysis and thermal process. Bauxite is refined into alumina using the Bayer process. In the Bayer process, bauxite is digested by sodium hydroxide at 175°C, under pressure converting the aluminum oxide in the ore to soluble sodium aluminate (http://alteo-alumina.com/en/alumina-refining). The process began to gain importance in metallurgy together with the invention of the Hall–Héroult electrolytic aluminum process invented just 1 year earlier in 1886. The majority of energy consumed in alumina refineries is in the form of steam produced by burning gas, coal, or fuel oil used in the main refining process. The theoretical energy requirements of alumina refining at 1.7 GJ/t are significantly lower than the energy now used. The world's best practice is around 3 GJ and the Australian average is 3.9 GJ.

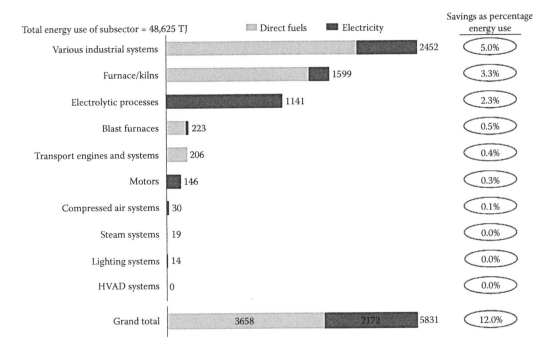

FIGURE 6.18 Energy saving in primary metal and metal product manufacturing by technology and fuel category. (Adapted from Climate Works Australia. 2013. Industrial energy efficiency data analysis report: Metals manufacturing, http://www.climateworksaustralia.org/sites/default/files/documents/publications/climateworks_dret_ieeda_factsheet_metals_20130502.pdf.)

Sixty-seven percent of the opportunities to improve energy efficiency in alumina production are achieved through improving the boiler systems and other process heating (Climate Works Australia, 2013).

A key opportunity for energy efficiency in this industry is cogeneration. Most of the energy for alumina refining requiring heat could be provided by cogeneration. Alcoa's WA refineries have negotiated an arrangement with Alinta to supply excess electricity from cogeneration to the grid. This electricity has low greenhouse intensity compared with conventional power generation, so its economics improve when a price is levied on CO_2 emissions.

Alumina production is also well-suited to use of renewable energy, through options such as solar, reforming and use of natural gas, biogas, biomass, and solar thermal. In the medium term, efficiency improvements and increased cogeneration offer substantial opportunity to cut primary energy use in the alumina industry by at least 30%. In the longer term, much more aggressive efforts to minimize heat loss, heat recovery techniques, and renewable energy sources will be important for ongoing improvement.

The energy saving opportunities by technology type and fuel category for alumina production is shown in Figure 6.19. Improvements to boiler systems represent 40% of total potential, while improvements to process heating, thermal electricity generation, and furnaces and kilns also represent large opportunities. Savings are also identified in the use of direct fuels (Climate Works Australia, 2013).

Electricity intensity of the Hall–Heroult electrowinning process at present in Australia is 13–15 MWh/t of primary aluminum produced, which is over twice the absolute theoretical limit of 6.34 MWh/t. A realistic eco-technical potential is 10–11 MWh/t, which will require the use of new materials and improved pot design (Bushnell, 2000). Ninety-two percent of the opportunities to improve energy efficiency in aluminum smelting are associated with the efficiency improvements to the process (Climate Works Australia, 2013).

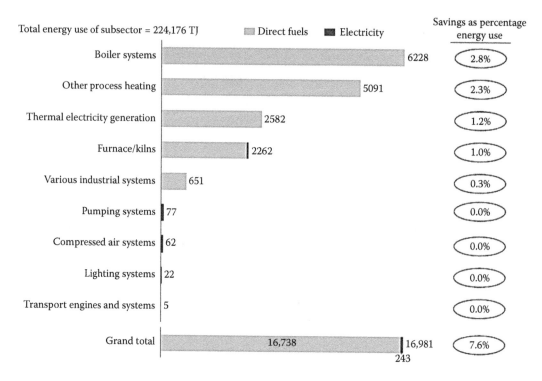

FIGURE 6.19   Energy saving in alumina production by technology and fuel category. (Adapted from Climate Works Australia. 2013. Industrial energy efficiency data analysis report: Metals manufacturing, http://www. climateworksaustralia.org/sites/default/files/documents/publications/climateworks_dret_ieeda_factsheet_ metals_20130502.pdf.)

Within smelters, a number of developments offer significant potential for energy savings, including improved smelter fume systems such as the filter bag area and the pressure drop across the bags as well as fan and motor efficiency and flow resistance. Opportunities to improve energy efficiency exist in the smelting sub-sector where the inert anodes replace carbon anodes: inert anodes reduce the significant amount of energy used to bake carbon anodes and may offer some process control benefits. Since pot lines run at close to 1000°C, there is a large amount of heat available, and the US Department of Energy has provided funding for Alcoa to develop heat exchangers that can cope with the fumes in the waste heat stream toward heat recovery. Overall, savings of 30%–40% seem feasible over a period of time within smelters, but dematerialization, recycling, and switching to lower energy materials at the point of consumption will drive much larger savings in the long term. Figure 6.20 shows a potential for 60% energy efficiency by improving various industrial systems and electrolytic processes and furnaces and kilns.

According to John Marlay, Australian CEO of Alumina, in 2007, of the world's total consumption of 36 million tonnes of aluminum, one-third consumption is actually from recycled aluminum requiring 95% less energy than the primary metal production. Over 60% of the aluminum ever produced in the world over the last 80 years is still being used. Construction materials are recycled close to 100%. In the automotive sector, it is 90%, and in packaging, depending on the particular economy, it is between 50% and 60%.

STEEL INDUSTRY

The basic process of making steel has several phases to it, and at each stage of the process, there are energy efficiency opportunities (The Steel University—Online Course at http://steeluniversity.org/ content/html/eng/default.asp?catid=&pageid=1016899460; Martin et al., 1999).

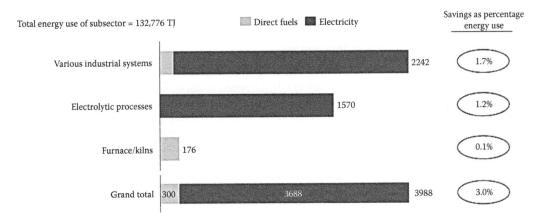

FIGURE 6.20 Energy savings in aluminum smelting by technology and by fuel category. (Adapted from Climate Works Australia. 2013. Industrial energy efficiency data analysis report: Metals manufacturing, http://www.climateworksaustralia.org/sites/default/files/documents/publications/climateworks_dret_ieeda_factsheet_metals_20130502.pdf.)

The theoretical energy intensity for making steel from hematite is around 7 GJ/t Fe. But modern iron and steelmaking practice requires around 20 GJ/t. Hence the steel industry has the potential to increase its energy efficiency by 35% through adoption of new technologies (Tables 6.4 and 6.5). Energy consumption is directly related to the environmental impact of the industry. Cost of energy accounts for 15%–20% of the total cost of steel production. Energy efficiency has always been one of the steel industry's key priorities. Over the last 40 years, the steel industry has reduced its energy consumption per tonne of steel by 50% (World Steel Association, 2015).

TABLE 6.4
Overview of a Selection of Energy Efficiency Opportunities in the Steel Sector

Current Production Technologies

Co-generation	Incorporating specially adapted turbines in place of conventional systems such as steam boilers and steam turbines can burn low-calorific-value off-gases such as coke oven gas, blast furnace gas, and basic oxygen furnace (BOF) gas, which are produced in significant quantities in integrated steel plants
	Co-location of an EAF with other industries that require heat means a shared cogeneration facility can be used
Advanced and high efficiency electricity generation technology	Combined cycle gas turbine can be used to self-generate some of the electricity used in the steel and iron industries
Coke dry quenching	To recover the heat losses in coke-making to generate electricity
Pulverized coal injection in blast furnaces	This technology allows the coke to be directly injected into the blast furnace, thereby reducing the amount of coal used for steel production by a ratio of 1.4:1
Top gas recovery turbines	To recover the top gas from blast furnaces to generate electricity
BOF gas/stream recovery systems	To recover the gas and steam from BOFs and to realize a net negative energy use in BOF steel making processes

Source: Adapted from The Natural Edge Project. 2007, http://www.naturaledgeproject.net/documents/ESSP/
EnergyTransformed/TNEP_Energy_Transformed_Lecture_5.1.pdf.

TABLE 6.5
Emerging Enabling Production Technologies

Emerging Enabling Production Technologies

Smelting reduction iron making	The smelting reduction process no longer requires the coke oven or the sinter plant, and can even run on cheap noncoking coals. These innovations reduce both the operational and up-front capital costs of iron production
Direct reduction iron-making process	Directly reduced iron production involves directly reducing iron ores to metallic iron without the need for smelting of raw materials in a blast furnace (which is the most energy-intensive process in iron production). In this process, reformed natural gas is used to convert iron ore into partially metallized iron granules
Thin slab and strip casting	Continuous casting in the integrated steelmaking process provides improved yields, with significant savings in energy. It involves the casting of molten steel directly into slabs, blooms, or billets, eliminating the need for ingot manufacture. Continuous casting replaces the conventional hot rolling mill, thereby bypassing the reheating and roughing steps in normal hot rolling mill production sequences. This produces a thin slab at lower cost with maximal use of the thermal energy of molten iron, while also minimizing additional fuel and electricity use downstream.

Source: Adapted from The Natural Edge Project. 2007, http://www.naturaledgeproject.net/documents/ESSP/ EnergyTransformed/TNEP_Energy_Transformed_Lecture_5.1.pdf.

The main opportunities for energy savings for steel industry in the future will come from the optimal selection of production processes and raw materials, increased use of economically available scrap, transfer of best practices, waste heat recovery, and reduction of yield losses (Figure 6.21). Figure 6.22 illustrates the energy saving opportunities where the blast furnace is identified to have the potential of 50%. Use of direct fuels has large potential to save energy (Figure 6.22).

The HiSmelt process developed by Rio Tinto and CSIRO is around 20% more efficient than conventional steel blast furnaces and can use iron ore fines that were previously considered to be

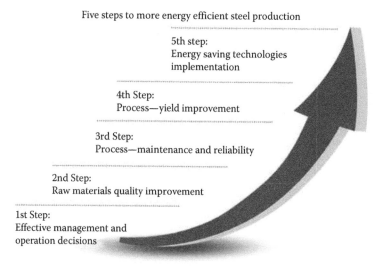

Five steps to more energy efficient steel production

5th step:
Energy saving technologies implementation

4th Step:
Process—yield improvement

3rd Step:
Process—maintenance and reliability

2nd Step:
Raw materials quality improvement

1st Step:
Effective management and operation decisions

FIGURE 6.21   Toward energy efficient steel production. (From World Steel Association. 2015. Energy use in the industry, https://www.worldsteel.org/media-centre/press-releases/2015/Energy-use-in-the-steel-industry-report-available-now.html. With permission.)

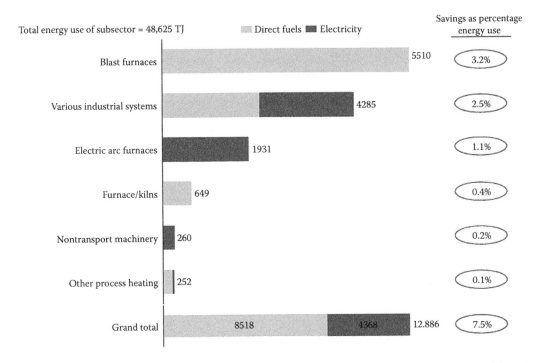

FIGURE 6.22 Energy savings in iron and steel manufacturing by technology and by fuel category. (Adapted from Climate Works Australia. 2013. Industrial energy efficiency data analysis report: Metals manufacturing, http://www.climateworksaustralia.org/sites/default/files/documents/publications/climateworks_dret_ieeda_factsheet_metals_20130502.pdf.)

waste. Electric arc furnaces (EAFs) use mainly electricity for steel production: a typical mid 1990s energy requirement was 550 kWh/t of product, but best practice is now considered to be around 300 kWh, with "ideal" performance around 150 kWh (Worrell, 2004a). There is opportunity for the Australian steel industry to make greater use of cogeneration or, alternatively, an EAF could be co-located with another industry that requires heat, so they could share a cogeneration facility. The increased recycling of steel and iron can offer further efficiency gains. Recycling takes at least 60% less energy to produce steel from scrap than it does from iron ore. Furthermore, waste disposal problems are lessened because used steel can be recycled over and over.

CONCLUSION

Energy saving opportunities for mill-to-melt processes are discussed. A number of actions are recommended that relate to comminution performance, intelligent blasting, ore sorting, classification or preconcentration, reprocessing and product recovery process design, and optimization using advanced mathematical modeling. The use of alternate energies, step change technologies, and secondary metal production plays a crucial role in reducing emissions and energy consumptions.

REFERENCES

A Guide to Leading Practice Sustainable Development in Mining. 2011, A report by Department of Resources, Energy and Tourism based on the Continuing Opportunities Energy Efficiency Opportunities (EEO) Program—2010 Report, A look at results for the Energy Efficiency Opportunities Program 2006–2010, Taken from public reports of assessments undertaken during the period July 2006–June 2010. http://www.industry.gov.au/resource/Documents/LPSDP/guideLPSD.pdf.

Australian Bureau of Agricultural and Resource Economics. 2010, http://eex.gov.au/industry-sectors/mining/#fn-166-1.

Australian Greenhouse Office and the Minerals Council of Australia. 2002. *Energy Efficiency and Greenhouse Gas Reduction Best Practice Manual—Mining*. ACT, Australia. http://eex.gov.au/technologies/mineral-processing/opportunities/#fn-16994-1.

Barkhuysen, N. J. 2009. Implementing strategies to improve mill capacity and efficiency through classification by particle size only, with case studies. *The South African Institute of Mining and Metallurgy Base Metals Conference*, pp. 101–104, July 27–31, Botswana. http://www.saimm.co.za/Conferences/BM2009/101-114_Barkhuysen.pdf.

Becker, M., A. N. Mainza, M. S. Powell, D. J. Bradshaw, and B. Knopjes. 2015. Quantifying the influence of classification with the 3 product cyclone on liberation and recovery of PGMs in UG2 ore. *Minerals Engineering*, 21(7): 549–558.

Bobo, T., J. Kemeny, F. Fernandez Romero, and M. Rocha Gill. 2015. Impact of ROM PSD on the crushing and grinding circuit throughput, CEEC—The future, http://www.ceecthefuture.org/productivity/impact-of-rom-psd-on-the-crushing-and-grinding-circuit-throughput/

Bushnell, H. 2000. Redding energy management. ACIL Consulting and Alumination Consulting Energy Efficiency Best Practice in the Australian Aluminium Industry—Sector Study Report for Energy Efficiency Best Practice Program. Department of Industry Tourism and Resources, Canberra, Australia. www.industry.gov.au/assets/documents/itrinternet/aluminiumsummaryreport20040206151753.pdf (Accessed September 4, 2007).

Bye, A. R. 2006. The application of multi-parametric block models to the mining process. *International Platinum Conference 'Platinum Surges Ahead'*, The South African Institute of Mining and Metallurgy, Sun City, South Africa, pp. 259–266.

Bye, A. R. 2011. Case studies demonstrating value from geometallurgy initiatives. *1st International Geometallurgy Conference (GeoMet 2011)*. Brisbane, Australia, pp. 9–30.

Catchpole, M. 2015. The productivity Challenge. *The AusIMM Bulletin*, https://www.ausimmbulletin.com/feature/the-productivity-challenge/.

CEEC—The future. 2015. Briefing: The energy productivity imperative, http://www.ceecthefuture.org/innovation-2/briefing-the-energy-productivity-imperative/

Climate Works Australia. 2013. Industrial energy efficiency data analysis report: Metals manufacturing, http://www.climateworksaustralia.org/sites/default/files/documents/publications/climateworks_dret_ieeda_factsheet_metals_20130502.pdf.

CRC ORE. 2010–2011. Annual Report: Transforming Resource Extraction, Queensland, Australia.

Curry, D. and B. Clermont. 2005. Improving the efficiency of fine grinding—Developments in ceramic media technology. *Proceedings of the Randol Innovative Metallurgy Conference*. http://www.isamill.com/EN/Downloads/Downloaded%20Technical%20Papers/Improving_The_Efficiency_Of_Fine_Grinding-_Developments_.pdf.

Daniel, M., G. Lane, and E. McLean. 2010. Efficiency, economics, energy and emissions—Emerging criteria for comminution circuit decision making. In *XXV International Mineral Processing Congress (IMPC) 2010 Proceedings*, Brisbane, Australia, September 6–10, pp. 3523–3531.

Daniel, M. and E. Lewis-Gray. 2011. Comminution efficiency attracts attention. *AusIMM Bulletin*, Issue 5: 20–22, 24, 30, October.

Daniel, M. J. 2007. Energy efficient mineral liberation using HPGR technology, PhD thesis, JKMRC, School of Engineering, University of Queensland, Australia.

Davis, V. L., P. J. Bethell, F. L. Stanley, and G. H. Luttrell. 1995. Plant practices in fine coal column flotation. In *High Efficiency Coal Preparation: An International Symposium, Chapter 21*, edited by S. K. Kawatra. Littleton, Colorado: SME, Inc., pp. 237–246.

Devasahayam, S., M. A. Ameen, V. Verheyen, and S. Bandyopadhyay. 2015. Brown coal dewatering using poly(acrylamide-co-potassium acrylic) based super absorbent polymers. *Minerals*, 5: 623–636.

Devasahayam, S., S. Bandyopadhyay, and D. J. T. Hill. 2016. Study of Victorian brown coal dewatering by Super Absorbent Polymers using Attenuated total reflection Fourier transform infrared spectroscopy. *Mineral Processing and Extractive Metallurgy Review*—Accepted author version posted online: March 23, 2016. http://www.tandfonline.com/doi/abs/10.1080/08827508.2016.1168417.

Dzinomwa, G. P. T, C. J. Wood, and D. J. T. Hill. 1997. Fine coal dewatering using pH-sensitive and temperature sensitive super absorbent polymers. *Polymers and Advanced Technologies*, 8: 767–772.

Energy Efficiency Opportunities. 2008. How Rio Tinto Addresses Energy Management. Energy Efficiency Opportunities, May 22, Melbourne, Australia, http://www.schneider-electric.com/solutions/nz/en/med/333390763/application/pdf/2288_identify-opportunities-to-reduce-the-consumpt.pdf.

Evans, L., B. L. Coulter, E. Wightman, and A. S. Burrows. 2009. Improving energy efficiency across mineral processing and smelting operations—A new approach. In *Proceedings SDIMI 2009—Sustainable Development Indicators in the Minerals Industry*. The Australasian Institute of Mining and Metallurgy, Melbourne, Australia, pp. 9–14.

EY. 2014. Mining: The growing role of renewable energy. Global Cleantech Center, http://www.ey.com/Publication/vwLUAssets/EY_-_Mining:_the_growing_role_of_renewable_energy/$FILE/EY-mining-the-growing-role-of-renewable-energy.pdf).

Fath, H. S. E. 1991. Diesel engine waste heat recovery for drying of clay minerals. *Heat Recovery System and CHP*, 11(6): 573–579.

Fleissner, H. 1927. The drying of fuels and the Australian coal industry. Nos. 10 and 11. Sonderdruck Spartwirtchaft,

Forbes, M. G. and B. Gough. 2003. Model predictive control of sag mills and flotation circuits, pp. 1–12, http://www.andritz.com/de/aa-automation-mpc-sag-mills-flotation-circuits.pdf. www.andritz.com.

Higgins, R. S. and D. J. Allardice. 1973. *Development of Brown Coal Dewatering*. Reserach and Deveopment department Report, State Electricity Commission of Victoria (SECV): Victoria, Australia.

Hulston, J., F. George, and C. Alan. 2005. Physico-chemical properties of Loy Yang lignite dewatered by mechanical thermal expression. *Fuel*, 84: 14–15.

Glencore Technology. 2012. Jameson Cell Rising to the Challenge, Robust, efficient, high intensity flotation technology. Brochure. http://jamesoncell.com/EN/Downloads/Documents/brochure_en.pdf.

Johnson, N. W. 2013. Review of existing eco-efficient comminution devices, http://www.ceecthefuture.org/wp-content/uploads/2013/02/Bills-review-of-Eco-eff-devices.pdf?dl=1.

Kennedy, D. L. 2008. Redesign of industrial column flotation circuits based on a simple residence time distribution model redesign of industrial column flotation circuits based on a simple residence time distribution model. Master's thesis, Virginia Polytechnic Institute and State University, Virginia. October, http://scholar.lib.vt.edu/theses/available/etd-10272008-172159/unrestricted/Thesis_Dennis_Kennedy.pdf.

Laurila, H., J. Karesvuori, and O. Tiili. 2002. Strategies for instrumentation and control of flotation circuits. *Mineral Processing Plant Design, Practise and Control*, 1, 2174–2195. ISBN-10:0873352238.

Lee, D. J., J. Y. Lai, and A. S. Mujumdar. 2006. Moisture distribution and dewatering efficiency for wet materials. *Drying Technology*, 24(10): 1201–1208.

Le Roux, M., Q. P. Campbell, M. S. Watermeyer, and S. de Oliveria. 2005. The optimization of an improved method of fine coal dewatering. *Journal of Minerals Engineering*, 18: 931–934.

Lovins, A. B. 2004. Energy efficiency, taxonomic overview. *Encyclopedia of Energy*, 2: 383–401; 6 Vols, San Diego, California and Oxford, UK: Elsevier, www.elsevier.com/locate/encycofenergy; www.sciencedirect.com/science/referenceworks/012176480X.

Lund, C. and P. Lamberg. 2014. Geometallurgy: A tool for better resource efficiency. *Topical – Metallic Minerals, European Geologist*, 37: 39–43.

Martin, N., E. Worrell, and L. Price. 1999. *Energy Efficiency and Carbon Dioxide Emissions Reduction Opportunities in the US Iron and Steel Sector Environmental Energy Technologies*. Ernest Orlando Lawrence Berkeley National Laboratory, California. http://www.energystar.gov/ia/business/industry/41724.pdf. See Industrial Energy Analysis—Iron and Steel Related Publications at http://industrial-energy.lbl.gov/node/326. Accessed September 4, 2007.

Mathiesen, K. 2015. Gas surges ahead of coal in US power generation, http://www.theguardian.com/environment/2015/jul/14/gas-surges-ahead-of-coal-in-us-power-generation.

Maxton, D., C. Morley, and R. Bearman. 2003. A quantification of the benefits of high pressure rolls crushing in an operating environment. *Minerals Engineering*, 16: 827–838.

Morley, C. 2005. The case for high pressure grinding rolls. In *Randol Innovative Metallurgy Forum*, Perth, Australia, August 21–24.

Mujumdar, A. S., H. Husain, and B. Woods. 1994. Techno-economic assessment of potential superheated steam drying applications in Canada. Technical Report of Canadian Electrical Association 9138U888. Montreal, Quebec, Canada.

Mular, A. L., D. N. Halbe, and D. J. Barratt. 2002. Mineral processing plant design, practice, and control. In *Proceedings SME*, p. 2369, California. ISBN 978-0-87335-223-9.

Murray, R. G. 1979. Stable high energy solid fuel from lignite. *Coal Process Tech*, 5: 211–214.

Napier-Munn, T., D. Drinkwater, and G. Ballantyne. 2012. The CEEC Roadmap for Eco-Efficient Comminution, Compiled by Tim Napier-Munn, Diana Drinkwater and Grant Ballantyne, from material produced at the CEEC/JKTech workshop on eco-efficient comminution, Noosa, Australia, June 12–13, http://www.ceecthefuture.org/wp-content/uploads/2013/02/2012CEECRoadmap3.pdf?dl=1.

Norgate, T. E. and S. Jahanshahi. 2006. Energy and greenhouse gas implications of deteriorating quality ore reserves *5th Australian Conference on Life Cycle Assessment Achieving Business Benefits from Managing Life Cycle Impacts*, Melbourne, November 22–24, pp. 1–10.

Norgate, T. E., S. Jahanshahi, and W. J. Rankin. 2007. Assessing the environmental impact of metal production processes. *Journal of Cleaner Production*, 15: 838–848.

Pease, J. 2007. Increasing the energy efficiency of processing. Presented at: Crushing and Grinding, Brisbane, Australia, September, http://www.isamill.com/EN/Downloads/Downloaded%20Technical%20Papers/Joe_Pease.pdf.

Pease, J. D., D. C. Curry, K. E. Barnes, M. F. Young, and C. Rule. 2006. Transforming flow sheet design with inert grinding—The IsaMill. In *Proceedings of the 38th Annual Meeting of the Canadian Mineral Processors*, Canadian Institute of Mining Metallurgy and Petroleum, Ottawa, Canada, pp. 231–249.

Peer, F. and T. Venter. 2003. Dewatering of fines using a super absorbent polymer. *The Journal of the South African Institute of Mining and Metallurgy*, 103(6): 403–410.

Pokrajcic, Z., R. D. Morrison, and N. W. Johnson. 2009. Designing for a reduced carbon footprint at greenfield and operating comminution plants. In *Proceedings of Mineral Processing Plant Design 2009—An Update Conference, Society for Mining, Metallurgy and Exploration*, edited by D. Malhotra, P. R. Taylor, E. Spiller, and M. LeVier, Tucson, Arizona, September 30 to October 3, pp. 560–570.

Powell, M. S. and Bye, A. R. 2009. Beyond mine-to-mill: Circuit design for energy efficient resource utilisation. In *Proceedings of 10th Mill Operators Conference 2009*. The Australasian Institute of Mining and Metallurgy, Adelaide, Australia, October 12–14, pp. 357–364, ISBN 978-1-921522-12-3.

Rankin, J. 2012. Energy use in metal production. In *High Temperature Processing Symposium*. Swinburne University of Technology, Melbourne, Australia, February 3–4, pp. 7–9.

Reuter, F., H. Molek, W. Kockert, and W. Lange. 1981. Laboratory desalination of brown coal by electroosmosis. *Neue Bergbautechnik*, 11, 186–187.

Sandu, S. and A. Syed. 2008. *Trends in Energy Intensity in Australian Industry*. Canberra, Australia: ABARE.

Seerane, K. and G. Rech. 2011. Investigation of sorting technology to remove hard pebbles and recover copper bearing rocks from an autogenous milling circuit. *6th Southern African Base Metals Conference 2011*, The Southern African Institute of Mining and Metallurgy, Phalaborwa, South Africa, pp. 123–135. http://saimm.co.za/Conferences/BM2011/123-Rech.pdf.

Schwarz, S. and M. Seebach. 2000. Optimization of grinding in high pressure grinding rolls and downstream ball mills. *7th European Symposium on Comminution*, Ljubljana, Vol. 2, pp. 777–788.

Shean, B. J. and Cilliers, J. J. 2011. A review of froth flotation control. *International Journal of Mineral Processing*, 100:57–71.

Shi, F., S. Lambert, and M. Daniel. 2006. A study of the effects of HPGR treating platinum ores. *SAG 2006*. Department of Mining Engineering, University of British Columbia, Vancouver, Canada, pp. 1–12.

Shi, F., W. Zuo, and E. Manlapig. 2015. Pre-concentration of copper ores by high voltage pulses. Part 2: Opportunities and challenges. *Minerals Engineering*, 79, 315–323.

SME Mining Finance. 2015. Mogalakwena selective blasting trial, CRC ORE, http://www.crcore.org.au/selectiveblasttrial.html. Pre-treatment separation can be achieved by electric pulse fragmentation, microwave or ultrasound. (http://eex.gov.au/technologies/mineral-processing/opportunities/#fn-16865-1).

Smidth, F. L. 2009. The world's largest and most efficient flotation cells, http://www.flsmidth.com/en-us/eHighlights/Archive/Minerals/2009/December/The+worlds+largest+and+most+efficient+flotation+cells#sthash.quRnFdOc.dpuf.

Syed, A., J. Melanie, S. Thorpe, and K. Penney. 2010. Australian energy projection to 2029–30. ABARE Research Report, Canberra, Australia.

Tavares, L. M. 2005. Particle weakening in high-pressure roll grinding. *Minerals Engineering*, 18: 651–657.

The Natural Edge Project. 2007, http://www.naturaledgeproject.net/documents/ESSP/EnergyTransformed/TNEP_Energy_Transformed_Lecture_5.1.pdf.

The Outlook for Energy: A View to 2040. 2015, http://cdn.exxonmobil.com/~/media/global/Reports/Outlook%20For%20Energy/2015/2015-Outlook-for-Energy_print-resolution.

van Berkel, R. 2007. Eco-efficiency in primary metals production: Context, perspectives and methods. *Resources, Conservation and Recycling*, 51: 511–540.

Wills, B. A. and T. J. Napier-Munn. 2006. *Wills' Mineral Processing Technology*. 7th edn, Chapter 12, Elsevier, Oxford, UK, pp. 267–344.

Wilson, W., B. Young, and W. Irwin. 1992. Low rank coal drying advances. *Coal*, 24–27.

World Energy Outlook. 2013. Power sector cumulative investment by type and region in the New Policies Scenario, 2013–2035.

World Steel Association. 2015. Energy use in the industry, https://www.worldsteel.org/media-centre/press-releases/2015/Energy-use-in-the-steel-industry-report-available-now.html.

Worrell, E. 2004a. Industrial energy use—Past developments and future potential. *Presentation at EEWP Expert Workshop International Energy Agency*, Paris, April 26–27, http://www.iea.org/Textbase/work/2004/eewp/worrell.pdf (accessed September 4, 2007).

Wright, S., S. Jahanshahi, F. Jorgensen, and D. Brennan. 2002. Is metal recycling sustainable? *Proceedings of the International Conference on Sustainable Processing of Minerals, Green Processing 2002*, Australian Institute of Mining and Metallurgy, Cairns, Queensland, Australia, pp. 335–341.

Wu, Z. H., Y. J. Hu, D. J. Lee, A. S. Mujumdar, and Z. Y. Li. 2010. Dewatering and drying in mineral processing industry: Potential for innovation. *Drying Technology: An International Journal*, 28: 7, 834–842.

Zohuriaan-Mehr, M. and K. Kabiri. 2008. Superabsorbent polymer materials: A review, *Iranian Polymer Journal*, 17(6): 451–477, http://journal.ippi.ac.ir.

Section II

Metallurgy/Recycling

7 Carbothermic Processing of Copper–Cobalt Mineral Sulphide Concentrates and Slag Waste for the Extraction of Metallic Values

Yotamu R. S. Hara and Animesh Jha

CONTENTS

ABSTRACT

Engineering mineral processing steps that may have the least impact on land, air and water remain a holy grail for the mining and mineral processing industry. With rising demand for cobalt and copper in the energy sector, the mineral processing industry is under pressure for producing these metals for industrial need; consequently, the environmental impact of mineral processing has shifted geographically nearer to the mining and mineral beneficiation areas of the world. In this chapter, an attempt has been made to capture the essential features of lime-based carbothermic reduction of cobalt and copper-based sulphide minerals and silicate slag derived from flash smelting and converting processes. The slag is considered as a waste product from industry, which often accumulates a high percentage of copper and cobalt, ranging between 1 and 2 wt%. The results discussed herein focus on Zambian concentrates as a sub-Saharan case study for mineral processing using lime-based reduction and extraction of metallic values.

INTRODUCTION

Industrialisation and information technology have contributed to a huge growth in consumer electronics, energy devices and energy efficient structural materials. In this context, both the cobalt and copper as metals remain indispensable for several engineering applications. Cobalt and its alloys have important applications in the areas of energy storage in lithium cobalt oxide batteries, high temperature alloys (Fleitman and Herchenroeder, 1971), cemented carbide tools (Zhai et al., 2011) and cobalt-alloy based magnetic materials (Hawkins, 1998; Zhai et al., 2011) for transduction devices (Anon, 1960; Betteridge, 1982). Copper is widely used in construction and electric cables (Malfliet et al., 2014). The detailed applications of copper and cobalt and their alloys are discussed in Anon (1960), Betteridge (1982) and Davis and A.S.M. International (2001), respectively.

Chalcopyrite ($CuFeS_2$), bornite (Cu_5FeS_4), carrolite ($CuCo_2S_4$) and cobalt-based pyrites $(Fe,Co)S_2$ (Claig and Vaughan, 1979) are commonly occurring sulphide minerals for copper and cobalt metal production, and in this respect, the concentrates from sub-Saharan countries namely, Zambian and Botswana concentrates are of particular interest for cobalt and copper metal extraction, where the copper refineries are geographically located. In this part of Africa, there is also a lack of supply of energy, on which the metal refineries and mining and mineral processing industry depend. The concentration of these minerals in the Zambian ore bodies is below 1 wt%, due to the presence of impurities or gangue minerals. The extraction of copper and cobalt sulphides from the ore bodies containing high gangue content is uneconomical because of the slag waste handling costs and the consequential energy requirement for handling slag during the metal refining stage. As the tonnage of slag per unit tonnage of metals produced increases, the metal processing is likely to be less profitable because the industry currently expends much energy in melting of the gangue minerals during pyrometallurgical extraction (West, 1982; Bodsworth, 1994). As explained above, since the concentrates of sulphide minerals are also not sufficiently rich, these minerals are unsuitable for hydrometallurgical leaching. As a result, the mineral sulphide of copper and cobalt is beneficiated prior to the pyrometallurgical extraction process (West, 1982; Bodsworth, 1994; Davenport et al., 2002; Norgate and Jahanshahi, 2010) by separating mineral concentrates using a range of physical and physico-chemical techniques from sulphides complex silicates. Amongst the physical and physico-chemical separation techniques, the selective froth flotation is quite commonly used for non-ferrous sulphide-based minerals (Davenport et al., 2002). Using selective flotation, the concentrates richer in iron sulphides (FeS_2 and FeS) are separated from the concentrates richer in copper and cobalt mineral sulphides. Since using selective froth flotation of mineral sulphide concentrates, only a part of the gangue minerals can be reduced, the remaining gangue minerals are then further beneficiated together with iron during pyrometallurgical steps, namely smelting and converting (Whyte et al., 2001). Before embarking on the main part of the text on carbothermic reduction of mineral sulphides in the presence of lime, we briefly discuss the state-of-the-art for cobalt and copper metal extraction.

CONVENTIONAL TREATMENT OF CU–CO MINERAL SULPHIDE CONCENTRATES

Metallic copper and cobalt may be recovered from their respective mineral sulphides via a combination of pyrometallurgical and hydrometallurgical operations (Davenport et al., 2002), briefly described below.

Pyrometallurgical Process

The majority of non-ferrous mineral sulphide concentrates are smelted under controlled oxidising conditions in a temperature range of 1373–1673 K (Habashi, 1969; Alcock, 1976) for removing iron and sulphur from smelted mineral sulphides (Cu_2S, $CuFeS_2$, Cu_5FeS_4, etc.) via selective oxidation. The two commonly used pyrometallurgical steps in copper and cobalt metal extraction are the smelting and converting processes during which chiefly iron and sulphur are removed (Maweja et al., 2009). For example, the copper matte smelting process involves partial oxidation of iron and sulphur, as shown in Equation 7.1, in which a copper-rich sulphide or matte phase ($Cu_2S \cdot FeS$) is formed. In the presence of an oxidising gas, the residual ferrous oxide (FeO) present in the matte and slag may be oxidised to solid Fe_3O_4, which, as reported earlier (Popil'skii, 1976; Rosenqvist, 1983; Bale et al., 2002; Maweja et al., 2009), may then act as a sink for copper and cobalt oxides which are lost as a complex spinel phase $(Cu,Co)Fe_2O_4$. It is for this reason that the partial pressure of the O_2 gas is carefully controlled during both the smelting and converting processes. For minimising the metal loss in slag, it is a common practice in industry to treat magnetite-bearing slag with SiO_2 by controlling the oxygen potential such that the loss of cobalt and copper is minimised by forming, for example, a predominantly fayalite $FeSiO_3$ rich liquid, as shown in Equation 7.2. The gangue mineral also dissolves and produces iron oxide bearing slag, which is then separated from sulphide at elevated temperatures due to the presence of miscibility gap. Often limestone is also added during slag melting to control the basicity and viscosity so that the metal loss into the slag is minimised. However, all of these steps for controlling slag properties only reduce the metal loss, and fail to eliminate or even minimise the loss. Since the sulphide-matte laden with copper, cobalt and residual iron is heavier that than the silicate slag, the liquid matte tends to sinks into the bottom of the flash smelting furnace. The matte and slag formation reactions during flash smelting are as follows:

$$2CuFeS_2(l) + 2.5O_2(g) = Cu_2S \cdot FeS(l) + FeO(l) + 2SO_2(g) \tag{7.1}$$

$$FeO(l,s) + SiO_2(s) = FeSiO_3(l) \tag{7.2}$$

With this brief background knowledge on copper and cobalt mineral processing in controlled partial pressures of oxygen, two additional important conclusions from the literature are

1. The oxidised transition metal oxides (CoO, Cu_2O and CuO) may be partially lost into the slag phase due to the high solubility of these oxide in the magnetite spinel.
2. The matte containing sulphide and oxide mixtures of copper, cobalt and iron is the primary source of precious metals (gold and silver). After smelting, any residual iron and sulphur present in copper matte are removed during the converting process (Davenport et al., 2002), which is achieved via the following two-step reaction (Davenport et al., 2002) in Equations 7.3 and 7.4:

$$Cu_2S \cdot FeS(l) + O_2(g) = Cu_2S(l) + FeO(l) + SO_2(g) \tag{7.3}$$

$$Cu_2S(l) + O_2(g) = 2Cu(l) + SO_2(g) \tag{7.4}$$

Equations 7.3 and 7.4 explain the removal of residual iron and sulphur, respectively. As in the smelting step, the slag formation is enhanced by the addition of silica, as shown in Equation 7.2. Once the slag reaction terminates by reaching a critical concentration with respect to iron oxide, any residual iron that remains as sulphide in the matte can only be removed by increasing the oxygen partial pressure during converting. In the presence of lime as a flux and alumina in gangue, with increasing oxygen partial pressure, the iron solubility in the slag increases by forming calcium ferrite and iron calcium ferrites. The ferrites have a spinel structure which is further stabilised for solubilising CuO and CoO (Davenport et al., 2002). At the end of converting the copper reaches near metallic state, called the blister copper with some impurities, which is then removed in subsequent operation, called the fire refining. When comparing the smelting and converting processes, as the iron content of the matte reduces, the equilibrium oxygen potential required for removing the remaining iron and sulphur increases. Under such oxidising conditions, the loss of cobalt into the slag becomes thermodynamically inevitable, as discussed below in the phase stability section. Increasing the oxygen potential during converting of cobalt and copper containing matte also leads to high loss of metallic values into the slag phase, from which the recovery of both copper and cobalt is energetically demanding (Banda et al., 2004).

Pyro-Hydrometallurgical Processes

Lean sulphide concentrates of copper and cobalt are sometimes treated via a pyro- and hydrometallurgical process, known as the 'roast-leach route' for reducing the high loss of cobalt which occurs during the conventional copper smelting and converting processes (Sinha and Nagamori, 1991).

The purpose of roasting is to convert the mineral sulphides into acid-soluble compounds (e.g. metal sulphates) which is achieved by blowing oxygen into the feed of mineral sulphide concentrate. Individual and overall reactions occurring between the mineral sulphides and oxygen gas are shown in Equations 7.5 through 7.8. It is necessary to control both the reaction temperature and time of roasting by minimising the formation of sparingly soluble copper and cobalt ferrites for acid leaching by keeping the exothermic roast reaction temperature below 973 K (Rosenqvist, 1983). However, the unfavourable kinetic conditions below 973 K affect the lower than equilibrium concentrations of SO_2 in the off-gas, which makes it difficult for use in the manufacture of sulphuric acid or elemental sulphur (Bodsworth, 1994; Davenport et al., 2002). It is well known in the non-ferrous industry that the failure to capture SO_2 gas from off-gas leads to environmental threats not only due to the release of greenhouse gases, but also from the increased acidity of air, soil and water in the ambient eco-system of a non-ferrous metal operation.

$$CuCo_2S_4 + 5.5O_2 = CuO + 2CoO + 4SO_2 \qquad (7.5)$$

$$CuCo_2S_4 + 7O_2 = CuSO_4 + 2CoSO_4 + SO_2 \qquad (7.6)$$

$$CuFeS_2 + 3.25O_2(g) = CuO + 0.5Fe_2O_3 + 2SO_2(g) \qquad (7.7)$$

$$CuFeS_2 + 3.75O_2(g) = CuSO_4 + 0.5Fe_2O_3 + SO_2(g) \qquad (7.8)$$

The cobalt–copper minerals from the Nkana Mines in Zambia are upgraded to higher concentrations using bulk and segregation flotation processes, which yield two streams – one rich in copper sulphide and the other rich in copper–cobalt sulphide concentrates, as shown in the process flow sheet in Figure 7.1. The cobalt–copper sulphide concentrates have a typical assay of 10 wt% Cu, 2 wt% Co and 20 wt% S. Since the relative concentrations of copper and cobalt are low with respect to gangue, during conventional smelting of such concentrates, high tonnage of slag per ton of metallic phase is produced as a result. The relatively large volume of slag-to-metal results into a proportionately larger fraction of metallic value being lost via slag entrainment (Davenport et al., 2002). In a worst case scenario, nearly all the cobalt in the mineral sulphide concentrate may be oxidised and dissolve into slag during conventional smelting. As a result, the Nkana copper–cobalt sulphide concentrates are treated via roast – leach – electro-winning process, described below.

The sulphide concentrates are roasted at temperatures below 973 K, to produce copper and cobalt oxides/sulphates. The calcined roast is then leached with sulphuric acid by dissolving the copper and cobalt. The leach solution is separated from the residue and the metallic copper is recovered from the filtered leachate via electro-winning. After electro-winning of copper, the leachate is enriched with cobalt ions which is precipitated as $Co(OH)_2$ in the pH range 6.5–7 (Mututubanya,

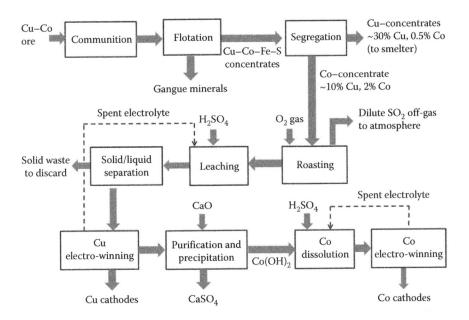

FIGURE 7.1 Simplified process flow sheet showing the treatment of Cu–Co ore at Nkana plant in Zambia. (Modified from Mututubanya, A. 2013. *Cobalt Recovery from Old Nkana Copper Slag via Solid State Carbothermic Reduction and Sulphation, in Metallurgy.* Lusaka, Zambia: University of Zambia.)

2013). The $Co(OH)_2$ precipitate is then dissolved into sulphuric acid for electro-winning metallic cobalt. The disadvantages of the 'roast-leach-electro-winning' process are

1. A dilute SO_2 off-gas is generated from the roasting process which cannot be released into the surrounding environment. Although the SO_2 sequestration seems a natural choice using limestone, it is not feasible due to the economic reasons. Consequently, there might be risk to the environment due to the presence of dilute SO_2.
2. A significant amount of lime (CaO) is consumed during the precipitation of $Co(OH)_2$ when using the spent electrolyte after the electro-winning of copper. The acid neutralisation converts CaO into $CaSO_4$ which is an industrial waste, adding more burden to the environment (Mututubanya, 2013).
3. There is high loss of cobalt into the solution during precipitation of $Co(OH)_2$ (Mututubanya, 2013).
4. Silver and precious metals that are normally present in the mineral sulphide concentrates (Bodsworth, 1994) and these metals are insoluble in sulphuric acid and, as a result, the metallic values are lost into the leach residue into the tailings.

PRINCIPLES OF METAL SULPHIDE REDUCTION

EQUILIBRIUM CONSIDERATIONS

In order to analyse the benefits of a novel process based on the reduction of mineral sulphides, it is first important to summarise by making a comparison of the temperature dependence of the Gibbs energy changes for the oxidation of elemental materials, namely copper, cobalt, iron and sulphur present in the mineral sulphides. For considering the equilibrium to compute the Gibbs energy change, the standard states are the pure metal in equilibrium with the corresponding metal oxide and pure oxygen and pure S_2 gases at the standard temperature and pressure condition. These generic equilibrium conditions define the oxidative equilibrium for smelting, converting and fire refining reactions. The temperature dependence of the values of Gibbs energy ($\Delta G°$, kJ/mol) of each oxide is plotted in Figure 7.2 from which it is evident that CoO and FeO are more stable than the SO_2 gas. By comparison, Cu_2O is less stable than SO_2, which implies that under the condition of oxidative reactions during smelting, converting and fire refining, a complete desulphurisation of matte via oxidation will lead to oxidative loss of not only iron but also of cobalt. In practice, the loss of cobalt via oxidation is greater, due to the tendency for spinel ($CoO . Fe_2O_3$) formation in the slag. It should be also noted at this point that the cobalt, iron and copper also form mutual solid solutions, which extend over a wide composition range, negating the possibility for forming purer forms of metallic copper and cobalt simultaneously. Although the overall energy balance for the oxidative process may appear to be quite favourable for the extraction of copper, it is certainly not favourable for the extraction of cobalt, especially when it is oxidised and lost into the slag phase. For this reason, an alternative to the oxidative process must be considered in view of minimising the process waste, including the generation of slag and SO_2 gas.

In view of the above-stated limitations of the oxidative process, a comparison of equilibrium conditions for the reduction process is made by considering the equilibria in Equations 7.9 through 7.11 which are already well established in the literature (Rosenqvist, 1983). The equilibrium analysis points out that the metal sulphides cannot be reduced to metal by using carbon, carbon monoxide or hydrogen, as the Gibbs energy changes for these reactions are large and positive. This is because the sulphur chemical potentials in the forward direction of reduction reactions (Equations 7.9 through 7.11), shown below, are much higher than that in the reactants, forcing the equilibrium constants to be less than unity. The chemical potential of sulphur, however, can be lowered substantially by using either lime or an alkali as an anion exchange medium.

$$2MS + C = 2M + CS_2 \tag{7.9}$$

$$MS + H_2(g) = M + H_2S(g) \tag{7.10}$$

$$MS + CO(g) = M + COS(g) \tag{7.11}$$

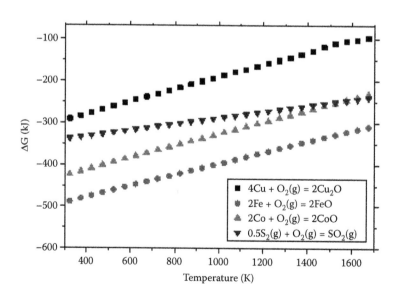

FIGURE 7.2 Comparison of the plots of the computed Gibbs energy change versus absolute temperature for the oxidation of Cu, Co, Fe and S_2 gas per mole of O_2 gas, using HSC software. (Adapted from Jha, A., P. Grieveson, and J. H. E. Jeffes. 1989. *Scandinavian Journal of Metallurgy*, 18: 31–45.)

In the presence of an anion exchange medium, such as lime, the metal sulphides present in a complex mineral sulphide converts into metal oxides, as shown below, following which the reduction may be carried out with one of the reducing agents, namely carbon (C), carbon monoxide (CO) and/or hydrogen (H_2) gas/gas mixture.

Using the examples of the reduction of a complex copper–cobalt and iron sulphide mineral with carbon in the presence of lime, it is possible to show that both pure metal and alloys may form at reasonable temperatures well below 1373 K, allowing magnetic and related physical separation of Fe/Co-rich phases from that of non-magnetic copper and its alloys. By considering calcium oxide (CaO) as anion exchange medium, the divalent metal sulphides (MS), for example, may transform via the O^{2-}–S^{2-} ion-exchange reaction (Equation 7.12) into an oxide MO and calcium sulphide (CaS) in a non-oxidising atmosphere. No sooner the anion exchange reaction occurs, the metal oxides reduce to metal via Equation 7.13 (Jha et al., 1989, 1996). The overall reaction may then be given by Equation 7.14: which can be applied for designing the reduction of metal sulphides (e.g. Cu_2S, CuS, FeS, CoS and NiS) found in naturally occurring complex sulphide minerals.

$$MS(s,l) + CaO = MO(s,l) + CaS \tag{7.12}$$

$$MO(s,l) + C = M + CO(g) \tag{7.13}$$

$$MS(s,l) + CaO(s) + C = M(s,l) + CaS(s) + CO(g) \tag{7.14}$$

For the sulphides of copper (Cu_2S), cobalt (CoS) and iron (FeS), the computed values of Gibbs energy change as a function of temperature are shown in Figure 7.2 from which it is apparent that above 925 K the values of ΔG^0 (kJ/mol) changes are all negative. The plots in Figure 7.2 show that the carbothermic reduction of CoS is thermodynamically more feasible than that for FeS and Cu_2S because of the larger negative values of ΔG^0 for CoS. The ΔG^0 versus T lines in Figure 7.2 are the univariant lines for defining the equilibrium for sulphides shown in this figure. Since the univariant lines diverge with increasing temperature, implying that the difference in ΔG^0 values between individual sulphides becomes larger, which may be appreciated by drawing a vertical line at a given temperature, for example, at 1100 K. The apparent differences in the values of ΔG^0 between Cu_2S–CoS, Cu_2S–FeS and FeS–CoS pairs in Figure 7.3 are nearly of the order of −35, −8 and −20 kJ/mol, respectively, which suggests that by utilising these modest yet significant differences in the negative values of ΔG^0, a suitable selective or preferential separation of metals may be achieved. The computed difference in the Gibbs energy also suggests that the separation of cobalt from sulphide concentrates is likely to be more feasible than that for copper and iron, for example. Such preferential separation may then be designed by optimising the process parameters. Another important conclusion from the divergence of univariant lines in Figure 7.3 may be drawn. Since with increasing temperature, the Gibbs energy difference increases between the sulphides, the condition for selective reduction of metal sulphides is much better than at lower temperatures, for example, below 1173 K.

An alternative to reducing the chemical potential of sulphur present in metal sulphide for metal production is to utilise calcium sulphate ($CaSO_4$) for desulphurising both the sulphate and sulphide (Roine, 2002; Bale et al., 2002) via the following reaction (Equation 7.15) in which the metal oxide (MO) produced may be reduced to metal via accompanying reaction (Equation 7.16) with a reducing agent. A unique feature of Equations 7.15 and 7.16 is that if a sink for the oxide (MO) is found by forming a slag; for example, $CaSiO_3$ then the partial pressure of SO_2 may increase proportionate to the reduction in the value of thermodynamic activity of the oxide.

$$MS + 3CaSO_4 = MO + 3CaO + 4SO_2 \tag{7.15}$$

$$MO(s,l) + C = M + CO(g) \tag{7.16}$$

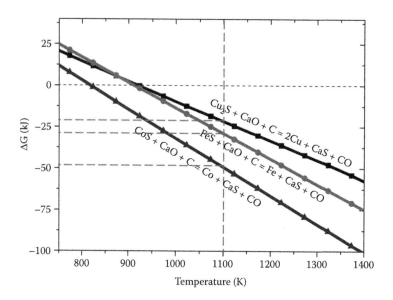

FIGURE 7.3   Plot of Gibbs energy against absolute temperature (T, K) for reduction of FeS, Cu_{2S} and CoS, per mole of CO gas (Jha et al., 1996) via MS + CaO + C = M + CaS + CO(g) reaction.

$$3CaO + 3SiO_2 = 3[CaO \cdot SiO_2] \qquad\qquad (7.17)$$

Equations 7.15 through 7.17 are analogous with the high temperature desulphurisation of iron and steel, currently practiced in industry, during which the lime is added for the removal of sulphur by forming calcium sulphide (CaS). In this case, sulphur is removed, SO_2 and lime is converted into a calcium silicate slag under inert condition, which also happens during desulphurisation of iron when silicon present in iron converts into silica and forms $CaSiO_3$ together with CaS in the slag. This principle of desulphurisation is therefore universal in its approach for other metals; for example, the equilibrium conditions defined by Equations 7.15 and 7.16, in which the chemical potential of lime is lowered in the forward direction by forming calcium silicate, $CaSiO_3$. Consequent of equilibrium with respect to $CaSiO_3$ yields higher partial pressures of SO_2 gas. The proposed approach is ideally suited for the non-ferrous operation in Zambia, where gypsum is produced in tonnage quantity from the leaching plant and the sulphide mineral concentrate contains siliceous gangue, which may be rendered into a calcium silicate slag for cement manufacture.

Based on the thermodynamic analysis, the approach based on gypsum appears more favourable than without gypsum, described Equations 7.12 through 7.14. In Figure 7.4, the univariant lines for CoS, Cu_2S and FeS are plotted for metal formation using gypsum as an ion-exchange medium by considering isotherm at 1273 K.

In the context of lime and gypsum-based chemical reaction and the corresponding equilibrium analysis for reducing conditions, the following advantages of extracting copper and cobalt from the concentrates and slags may be considered, in the presence of either gypsum or lime as a desulphurising medium for capturing both SO_2 and exchanged S^{2-} ions.

Lime may be available either from limestone or it may be formed *in situ* by using gypsum (Popil'skii, 1976; Igiehon, 1990; Bodsworth, 1994):

1. Sulphur and sulphur dioxide may be sequestrated as calcium sulphide (CaS) during the metallisation process, as shown in Equation 7.12, as a result, the risk of environmental pollution due to sulphurous gaseous species may be significantly reduced or even made negligibly small.

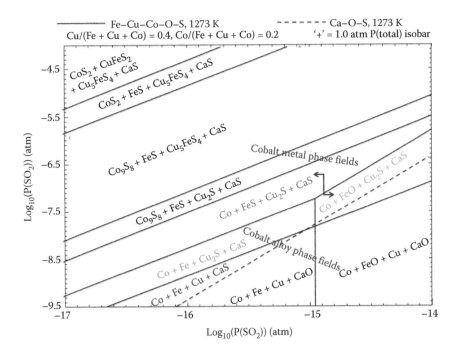

FIGURE 7.4 Cu–Co–Ca–Fe–O–S predominance area diagram at 1273 K, computed by superimposing the Fe–Cu–Co–O–S and Ca–O–S predominance area diagrams, using FactSage software 6.1. (Adapted from Roine, A. 2002. *HSC Chemistry 5.1*. Finland: Outokumpu Research Oy.)

2. Additional energy-related advantage may be derived from the processing of minerals by extracting and separating metals well below their melting points, for example, at 1273 K which avoids unnecessary melting of both the metals and especially the gangue minerals present in the mineral sulphide as a feedstock. By avoiding the need for melting, better physical separation in solid state may be achieved by introducing magnetic separation.

3. A significant reduction in overall waste volume may be potentially realisable by utilising other process waste from the leaching plant and slag from smelting and converting processes. The technique also allows reutilisation of lime which is generated from the oxidation of CaS (Hara and Jha, 2013). The carbon monoxide gas generated may be re-used to supply a part of the total energy requirement for the non-ferrous process.

As understood from the equilibrium analysis in Figures 7.2 and 7.3, the influence of the partial pressures of CO, SO_2 and O_2 gases is likely to determine the equilibrium for the metallisation of copper, cobalt and iron. For this reason, it is essential to compute and plot an isothermal Fe–Cu–Co–Ca–O–S predominance area diagram at 1273 K, which is shown in Figure 7.4, from which it is evident that the mineral sulphides, namely, $CuFeS_2$, CoS_2 and Cu_5FeS_4 are stable at high partial pressure of SO_2 gas, as a result, the metallic phases are unstable under this condition. This condition corresponds to the top left-hand corner of the phase stability diagram in Figure 7.4. However, under the reducing condition, especially in the presence of carbon and CaO, the SO_2 is unstable and it transforms into sulphides of metal (e.g. CaS and transition metal sulphides). Under the carbothermic mineral processing condition in the presence of lime or gypsum, the equilibrium combination of phases likely to exist will depend on both the partial pressures of oxygen and SO_2, yielding individual conditions for deriving metals and alloys by controlling the gas compositions. These are exemplified below.

REGIMES OF PREFERENTIAL METALLISATION OF COBALT

Preferential formation of metallic cobalt may be achieved by controlling the partial pressures of SO_2 and O_2 in the region, designated as the *cobalt metal phase fields* in Figure 7.4, where the two distinct phase fields with metallic cobalt (Co) in equilibrium are (Co + FeS + Cu_2S + CaS) and (Co + FeO + Cu_2S + CaS). These two phase fields have a unique gas composition range which may be estimated for the corresponding partial pressures of O_2 and SO_2 gases.

For the formation of cobalt-based alloys, four different partial pressure regimes may be considered and these are designated as the cobalt-alloy phase fields, namely: (Co + Fe + Cu_2S + CaS), (Co + Fe + Cu + CaS), (Co + Cu + Fe + CaO) and (Co + Cu + FeO + CaO). The main difference in the phase stabilities of alloy and metallic phase combinations is that the elemental cobalt metal is more stable at higher SO_2 and O_2 pressures, with respect to the alloy phase fields. This means by carefully selecting the gas composition during reduction, it may be possible to form purer forms of cobalt, rather than the complex solid solutions of metals as alloys. On reaching the equilibrium state for the cobalt metal bearing phase, subsequent separation of the cobalt-rich phase may be possible using the application of magnetic field. This technique may then yield a non-magnetic fraction for the recovery of copper from the conventional copper smelting process. Such an approach may therefore yield not only a high concentration of cobalt but also of copper metal for further refining. The lime, either formed *in situ* from the decomposition of gypsum or from limestone, present during smelting may be able to combine with silica and form a calcium iron silicate slag, instead of a faylite (FeO . SiO_2) slag formed usually during conventional copper smelting and converting. If CaS is present, during melting of non-magnetic sulphide-oxide mixture, containing copper, a controlled oxidation of CaS with $CaSO_4$ may yield SO_2 via Equation 7.14, shown below (Roine, 2002):

$$CaS + 3CaSO_4 = 4CaO + 4SO_2(g) \tag{7.18}$$

Based on above equilibrium analysis, a concept flow sheet for copper/cobalt production may be considered by designing a process which allows reaching the equilibrium corresponding to the (Co + Cu_2S + FeS + CaS) phase field in Figure 7.4. If this be the case, then the magnetic separation may yield a fraction rich in cobalt and FeS from the non-magnetic fraction of Cu_2S and CaS for subsequent leaching of cobalt with sulphuric acid for extracting cobalt sulphate. The metallic cobalt of purity better than 99.9% may be then won using electro-metallurgical dissociation of cobalt sulphate.

Note that the equilibrium analysis for the above two processes for cobalt separation appears to be environmentally sensitive in terms of waste generation and energy consumption, which may be better understood in the following example of the process flow sheet in Figure 7.5 for cobalt metal

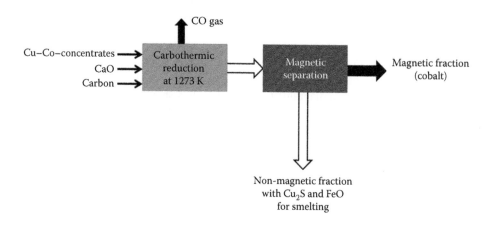

FIGURE 7.5 Proposed method for recovering cobalt via carbothermic reduction and magnetic separation.

separation. In this flow sheet, there is no need for high temperature melting of slag and alloy, nor is there any requirement for slag crushing and grinding for metal recovery.

RESULTS AND DISCUSSION

EXAMPLES OF RESULTS FOR COBALT AND COPPER SEPARATION FROM NKANA MINERAL CONCENTRATES

Preferential metallisation of cobalt was achieved when the Nkana copper–cobalt mineral sulphide concentrates with composition (wt%): Cu_2S-0.7, $CuFeS_2$-8.6, Cu_5FeS_4-5.0, $CuCo_2S_4$-2.2, FeS_2-10.1, SiO_2-19.0 and other gangue minerals −54.4 (Hara, 2013; Hara and Jha, 2013). The powdered samples of concentrates were mixed with lime and carbon and, reacted between 5 and 60 min at 1273 K by maintaining a purge of argon gas at 2 L/min for displacing the carbon monoxide gas out of the reaction vessel. The preferentially reduced metallic cobalt was found with FeS, FeO, Cu_2S and CaS, as predicted from the equilibrium between the $Co + FeO + Cu_2S + CaS$ and $Co + FeS + Cu_2S + CaS$ phase fields in Figure 7.4. These results are discussed below in details by analysing the microstructure for the phases formed.

Preferential metallisation of cobalt and iron. Preferential metallisation of iron and cobalt may be achieved since these two metals co-exist with Cu_2S in the phase fields $Co + FeO + Cu_2S + CaS$ and $Co + FeS + Cu_2S + CaS$ in Figure 7.4. Figure 7.6a and b exemplifies the conditions for selective metallisation of iron and cobalt, via the reduction FeO and CoO with carbon in the presence of lime, as described in Equations 7.12 through 7.14 and in Figure 7.3. The partially reduced sulphide minerals yield cobalt-rich metallic phase within 10 min of reduction at 1273 K, as shown in Figure 7.6a and b. In scanning electron microscopy, the particulates of metallic phases were analysed using the energy dispersive X-ray technique, from which it is evident that the metallic constituent of cobalt and iron in such particles averages above 95 wt%. Both of these metals are strongly magnetic and may be magnetically separated from the partially reduced samples, containing copper sulphide, residual iron oxides, gangue minerals and calcium sulphide.

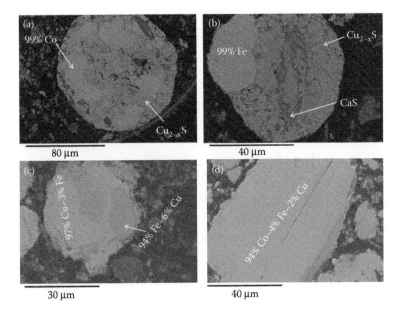

FIGURE 7.6 Carbothermic reduction of Nkana Cu–Co mineral sulphide concentrates in the presence of CaO, reacted for; (a) and (b) 10 min and (c) and (d) 60 min at 1273 K.

Complete metallisation of mineral sulphides. As explained above in Figure 7.3 and Equations 7.12 through 7.14, if the reduction reaction of Nkana sulphide concentrates is allowed to continue towards completion, the resulting metallic phase constitution is not only the cobalt-rich metallic phases, but also an aggregate of metallic mixtures of cobalt, copper and iron in alloy form in equilibrium with CaS. The reduction reaction completes in an hour at 1273 K, as shown in Figure 7.6c and d. A detailed analysis of experimental results on complete metallisation of mineral sulphides to a mixture of Cu, Co and Fe may be found in Hara and Jha (2012, 2013). Such a mixture of metallic powders may be separated using a combination of gravity and magnetic separation techniques, before embarking on to a next step for metal purification.

The above two examples demonstrate that both the partial and complete metallisation of mineral concentrates may be feasible for metallic phase separation by controlling the processing parameters, for example, time. Other factors such as the role of lime-to-mineral sulphide and lime-to-carbon ratios for partial and complete metallisation at different temperatures may be found elsewhere in the literature (Popil'skii, 1976; Hara, 2013; Hara and Jha, 2013) These results point out to a direction that there may be opportunity to readapt mineral processing for preferential separation of metallic phases directly from sulphide concentrates.

RECLAMATION OF COPPER AND COBALT IN SLAG

A copper–cobalt slag is defined as a molten mixture of silicates and metal oxides (Rosenqvist, 1983; Moore et al., 1990; Viness, 2011). During smelting and converting, the slag acts as a receptacle for the unwanted gangue minerals in metal extraction processes (Moore et al., 1990). The copper smelting and converting slag waste is an important source of copper and cobalt because the metallic constituents are often present in higher concentrations than in mineral concentrate. For example, the Nkana copper smelting slag dump in Zambia has about 2 wt% Cu and 1 wt% Co which is economically attractive for mineral beneficiation.

Copper in slag exists in metallic (Cu), sulphide (Cu_2S and Cu_5FeS_4) and oxide (Cu_2O and CuO) forms in the liquid state or after cooling copper sulphide or matte enters the slag via entrapment as shown in Figure 7.7a. Furthermore, a part of matte is poured together with slag during slag skimming. Matte that enters the slag due to entrapment and during skimming is insoluble in the slag such that it separates out during cooling. Consequently, the solidified slag contains portions of matte which may be a source of high-grade copper (>65 wt% Cu) as shown in Figure 7.7b.

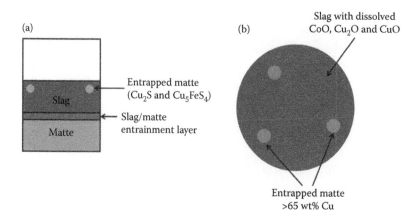

FIGURE 7.7   (a) Slag and matte separation in a furnace and matte entrapment and (b) distribution of the entrapped matte in slag.

The copper oxide in the slag is formed via oxidation during smelting and converting and it is sparingly soluble in acid, due to the high concentrations of silica. However, the solubility limits for copper oxide in slag may be dependent on the basicity of slag formed; for example, when calcium ferrite is incorporated during slag formation for copper concentrates with high silica. A specific example is the slag from direct-to-blister copper smelting process contains up to 15 wt% Cu in the form of oxide, due to highly oxidising atmosphere. The divalent state of the cobalt and copper in slag is quite important for forming stable spinel with Fe_2O_3 and/or Fe_3O_4, which then intrinsically controls the lower limit of oxygen partial pressure required. As a result, more than 90 wt% of the cobalt is present as CoO in slag in the form of cobalt–copper spinel $(Co,Cu)O . Fe_2O_3$, which is stable at high oxygen potential, required for conventional copper smelting and converting processes.

The reclamation of metallic values from slag is via: (a) recovery of matte from copper–cobalt lost into the slag, (b) froth flotation, (c) recovery of metallic copper and cobalt via leaching of oxides dissolved in the slag and (d) reduction of oxide constituents in slag in electric arc furnaces.

As explained above, a portion of copper and cobalt remain entrapped in the slag as sulphide matte and it is a mixture of sulphides and oxides of transition metals (iron, copper and cobalt). Recovery of metallic values from slag is therefore controlled by adopting different steps, discussed below for matte and oxide constituents. The entrapped matte in slag may be recovered in a number of ways as discussed below:

Settling furnace for slag-matte separation via density difference. In this method, two different properties of slag and matte are considered during the smelting of slag-matte mixture in an electric arc furnace. The density difference and phase immiscibility between the slag and matte promote separation of these two phases when the mixture is heated above 1473 K. The matte, being heavier than slag, settles down at the bottom of the electric furnace from where it is tapped off and transferred into the cradles for further processing. Since the settling technique is a simple way of recovering entrapped matte from slag, it may be applied by integrating with the overall process wherever there is slag production. Although the electric furnace is useful in reclaiming matte from integrated non-ferrous operation, it suffers from two serious drawbacks that the electric furnace operation is quite energy demanding and the oxide forms of copper and cobalt are not recovered, which implies that the metallic values, present as oxides are irretrievably lost after the matte recovery in the electric furnace settling tank.

Unlike the matte-settling electric arc furnace, *froth flotation* as a technique is implemented for the quenched slag at ambient condition. Losing high-quality heat energy from slag into the surrounding, however, is efficient, nonetheless the solidified slag is crushed and ground to particle sizes of less than 125 µm for froth flotation to work by providing a large surface area for solid (slag)-liquid (with surfactant)-gas (air) interface. The ground slag is fed into the flotation unit and suspended matte particles trapped in air bubbles are separated by floating them with the help of a flocculating reagent. The froth flotation method of matte separation is comparable with the one adopted for concentrating copper and cobalt sulphides from the ground ore. The Boro Mining Company in Chambishi, Zambia treats the Nkana copper–cobalt slag via froth flotation for recovering high-grade matte concentrates. Although the flotation is quite successful in recovering metallic values, the technique is not suitable for chemically tied metal oxide constituents in the slag. The tailings after froth flotation of the Nkana copper–cobalt slag on an average contains about 1 wt% Co and 0.5 wt% Cu. Except for the energy required for grinding slag to desired particle size, the overall energy requirement for this process is much less than that for the matte-settling electric furnace. Nonetheless, the grinding costs cannot be ignored in the overall equation for the energy requirement. The froth flotation method may not be ideal for treating slag material containing high content of dissolved copper and cobalt oxides, such as from flash smelting slag.

The *leaching* of metallic oxide values from slag is based on the dissolution principles of copper and cobalt from low-grade concentrates. After leaching, the metallic values are recovered via electro-winning (Davenport et al., 2002). For the leaching step to work for metal recovery, it is essential that copper and cobalt selectively dissolve into the acid medium, which poses significant challenge for treating copper–cobalt containing slag since it is mineralogically different from the constituents of concentrates which are explained below:

1. The oxides of copper and cobalt present in the slag are mainly dissolved in the forms of silicate, glassy and spinel ferrites (Habashi, 1969; Yang et al., 2010), when compared with the lean concentrates in which the metallic values are present in the form of sulphides.
2. The melting condition alters the pore distribution in slag, compared to natural alteration in sulphide concentrates. In a slag, therefore, the dissolved copper and cobalt oxides cannot be selectively leached out, regardless of the lixiviant used, unless the silicates and ferrites matrices are destroyed. The partially leached silica in leach solution significantly interferes with filtration due to formation of silica gel (Banza et al., 2002; Yang et al., 2010).
3. In slag, the iron oxide is relatively more abundant than the oxides of copper and cobalt, consequently the leach solution has higher concentrations of iron than either copper or cobalt which makes the electro-winning step difficult to run.
4. Since the copper and cobalt in slag are present in the forms of oxide and sulphide as in a matte (Cu_2S, FeO, CoO, Cu_5FeS_4 and CoS), the matte phase is rather less responsive to sulphuric acid attack, as a result the leaching efficiency for cobalt and copper is poor.

For above reasons, the leaching of slag has never been successful on industrial scale.

In *reductive smelting*, the copper–cobalt oxide containing slag is smelted above 1773 K in the presence of carbon via Equations 7.15 through 7.17, which yields a Cu–Co–Fe alloy. As discussed above, the reduction condition above 1273 K yields invariably a ternary alloy due to large mutual solid and liquid solubility between metallic constituents iron, cobalt and copper. With increasing iron content in the ternary Cu–Co–Fe alloy, the liquidus temperature increases reaching above 1473 K. Once the ternary alloy is formed and separated from the electric arc furnace, it is further processed via leaching and electro-winning for selective recovery of copper and cobalt.

$$MO(l,s) + C(s) = M(l,s) + CO(g) \tag{7.19}$$

At Chambishi Metals Plc., Zambia, the copper–cobalt slag is mixed with malachite ore prior to reductive smelting in order to increase the content of overall copper, which also helps in decreasing the liquidus temperature of the Cu–Co–Fe alloy. The temperature of the molten Cu–Fe–Co alloy is raised to 1873 K after which the liquid alloy is atomised by using a high pressure water jet, which yields fine metallic particles (Jones et al., 2002) for leaching. The atomised powder of ternary alloy is leached under high temperature above ambient and pressure in an autoclave using sulphuric acid. The leachate derived from the autoclave is a mixture of copper and cobalt sulphate. Under the condition of autoclave leaching, the metallic iron forms its hydroxide. Although this combination of pyrometallurgical melting, atomisation and leaching appears attractive, it is an energy demanding process and can only compete where the energy costs may be cheap. The total energy requirement for smelting of copper–cobalt slag is about 800 kJ/kg of feed material which is uncompetitive in the current energy market in the entire sub-Saharan region and this factor makes the recovery of cobalt economically unviable (Hara and Jha, 2013).

An alternative to high temperature reduction and leaching, discussed above, is low temperature carbothermic reduction of copper–cobalt slag. In this approach, the oxides of copper (Cu_2O and CuO) and cobalt (CoO) may be reduced to metallic state (Equation 7.15) at temperature as low as 1223 K (Turkdogan, 1980; Rosenqvist, 1983; Bale et al., 2002). Under optimised reducing conditions, iron present as Fe_3O_4 and $FeSiO_3$ may also be reduced either partially or completely, by forming ternary alloys of copper, cobalt and iron, which may be then magnetically separated from the copper-rich alloy. Below the results from two examples of synthetic slag reduction are compared for demonstrating the feasibility of slag reduction for the recovery of metallic values.

In order to analyse whether the copper–cobalt containing slag may be reducible at lower temperatures than 1773 K, a synthetic slag with composition: 40 wt% SiO_2, 30 wt% Fe_2O_3, 6 wt% Al_2O_3, 7 wt% CuO and 7 wt% CoO was synthesized in air at 1573 K. For investigating the low temperature carbothermic reduction of slag, the experiments were divided into three parts: (i) direct reduction of slag with carbon as a reducing agent at chosen isotherms below 1773 K, (ii) direct reduction of slag with carbon and pyrite (FeS_2) for *in situ* matte formation and (iii) direct reduction of slag with gypsum and carbon. The above investigations were undertaken with graphite, carbon black and activated charcoal in a temperature range of 1173–1323 K.

In Figure 7.8, the data for the relative kinetics of overall reduction reactions are compared for three different conditions during which the samples of synthetic slag were directly reduced with two different forms of carbon (graphitic and active charcoal) at 1273 and 1323 K. The overall chemical equilibrium may be explained by Equation 7.14. On the other hand, when the slag is treated with anhydrous gypsum ($CaSO_4$) in the presence of carbon, the overall chemical reaction is very different from that in Equation 7.14. The anion exchange between the metal oxide (CuO, CoO, FeO) and $CaSO_4$ in this reaction is governed by the reduction–calcination reaction (Banza et al., 2002) which may occur at temperatures as low as 1073 K in the presence of carbon (Jha et al., 1996).

$$CaSO_4 + 0.5C = CaO + SO_2 + 0.5CO_2(g) \tag{7.20}$$

$$MO + SO_2(g) + 1.5C = MS + 1.5CO_2(g) \tag{7.21}$$

The SO_2 gas given off from the reduction calcination of $CaSO_4$ in Equation 7.20 may react with the metallic oxides and yield metal sulphides via Equation 7.21. The summation of Equations 7.20 and 7.21 is the overall reduction–calcination reaction (Equation 7.22) for matte formation.

$$MO + CaSO_4 + 2C = MS + CaO + 2CO_2(g) \tag{7.22}$$

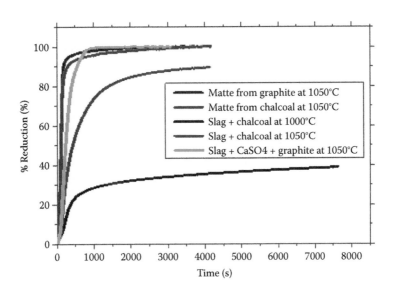

FIGURE 7.8 Comparison of the extent of overall chemical reaction completed, determined by % reduction, as a function of time for reaction of a slag with composition: 7 wt% CuO–7 wt% CoO–6 wt% Al_2O_3–30 wt% Fe_2O_3–40 wt% SiO_2 in the temperature range of 1273 K (1000°C) and 1323 K (1050°C) in argon.

However if the carbon-to-metal oxide ratio may be controlled in such a way that the calcination reaction (Equation 7.20) dominates, then it may be possible to metallise the metal oxides, such as CoO and CuO, by releasing an equilibrium gas mixture containing CO and SO_2 via Equation 7.23:

$$MO + CaSO_4 + 2C = M + CaO + 2CO(g) + SO_2(g) \qquad (7.23)$$

By comparing Equations 7.22 and 7.23, the (M + CaO) phase mixture in equilibrium with CO and SO_2 gas is more stable under reducing conditions than MS + CaO in equilibrium with CO_2. This is because above 1073 K, it is the Boudouard reaction: $C(s) + CO_2(g) = 2CO(g)$ equilibrium which helps in stabilising the CO gas, and this means the equilibrium 17 will only dominate when there is not enough carbon to support the Boudouard reaction, otherwise Equation 7.23 will dominate.

In Equation 7.22, however, if the carbon-to-metal oxide ratio is increased from MO:C = 1:2 to 1:5, the resulting reaction yields metal, rather than a matte (MS), as shown below in Equation 7.23. This reaction is inherently a carbothermic reduction of metal oxide in the presence of gypsum, in which the $CO:CO_2$ ratio may be controlled by controlling the amount of carbon used.

$$MO + CaSO_4 + 5C = M + CaS + 5CO \ (g) \qquad (7.24)$$

Treating slag with gypsum, however, is particularly advantageous because the slag contains a large fraction of silica, which during reduction calcination of calcium sulphate may be fluxed with the *in situ* formed lime via Equation 7.23 and may yield calcium silicate, as shown in Equation 7.25. The formation of calcium silicate will shift the equilibrium in the forward direction by lowering the chemical potential of lime. Under such a condition, the overall equilibrium for matte and metal may also be considered with the slag forming reaction below:

$$SiO_2 + CaSO_4 + C = CaSiO_3 + CO(g) + SO_2(g) \qquad (7.25)$$

The above discussions on equilibrium in the sulphate, calcium oxide, silicate and transition metal oxides may be understood with the help of the Fe–Co–Cu–Ca–Si–O–S predominance area diagram, which we have computed for equilibrium phase combination analysis at 1273 K and is shown in Figure 7.9. From this representation of phase stability, it is evident that oxides of copper, cobalt and iron may be sulphidised in the presence of $CaSO_4$ at low oxygen potential by forming complex ternary and binary sulphides, namely Cu_5FeS_4, CoS and FeS which may co-exist with $CaSiO_3$ and SiO_2. It is also evident from the phase stability diagram that the sulphidisation of copper to a cuprous state (Cu_2S) may also be feasible by preferentially keeping the magnetic elements in oxide forms: for example, as $CoFeO_2$, Fe_3O_4 in equilibrium with $CaSiO_3$. This is particularly advantageous for the separation of sulphide and oxide mineral values, using magnetic separation after sulphidisation of slag with gypsum, say at 1273 K. On lowering the oxygen partial pressure to a lower value than $log(P(O_2)) = -10$ atm, a majority of transition metal oxides transform into sulphide, which at and above 1273 K is likely to yield a liquid form of matte for further process. From the examples on phase combination, it is therefore apparent that for preferential separation of metallic values, it is essential to control the partial pressures of O_2 and SO_2, and this approach may be implemented in practice by varying the ratio of calcium sulphate to carbon in a given mixture.

The discussion on phase stability in the Fe–Co–Cu–Ca–Si–O–S system is verified experimentally. In Figure 7.8, the relative rates of completion, characterised by % reduction versus time (t, s) curves for matte- and metal-forming reduction reactions are compared, from which it is apparent that without the presence of $CaSO_4$ the metal oxide reduction in slag is much slower with graphite than that in the presence of gypsum and graphite. Since the graphitic form of carbon is more inert

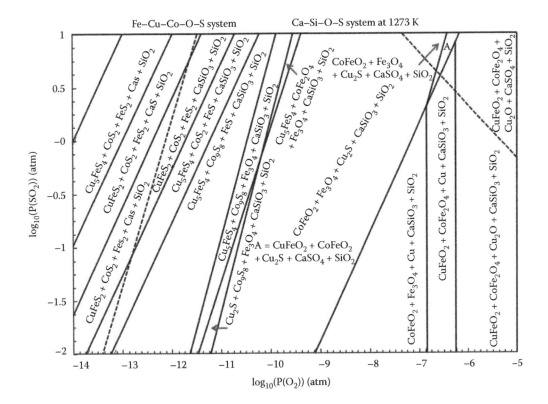

FIGURE 7.9 Fe–Co–Cu–Ca–Si–O–S predominance area diagram at 1273 K constructed by superimposing the Fe–Co–Cu–O–S (continuous line) and the Ca–Si–O–S (dotted line) predominance area diagrams, computed using Fact Sage software. (Adapted from Roine, A. 2002. *HSC Chemistry 5.1.* Finland: Outokumpu Research Oy.)

during reduction reaction than any other active forms of carbon, the relative rate of completion of overall reaction in Figure 7.8 proves that the presence of $CaSO_4$ may be necessary for matte and metal formation. These observations are consistent with previously reported data on carbothermic reduction of sulphide minerals using graphitic and active forms of carbon (Jha et al., 1996). Below we examine the phase and microstructural analysis using X-ray powder diffraction (XRD) and scanning electron microscopy (SEM), respectively.

Preferential metallisation of copper was achieved with carbon black at 1173 K as shown in Figure 7.10. However, when a more reactive activated charcoal was used, the oxides of cobalt and iron were partially metallised, as shown from the XRD pattern in Figure 7.11. The backscattered SEM images of the slag reduced with activated charcoal at 1173 and 1323 K are shown in Figure 7.12a and b, respectively. It is apparent from Figure 7.12b that even though the extent of reduction was higher than 80% (see Figure 7.8) at 1323 K, the metallic phases formed were finely distributed in the remaining unreduced matrix of slag, signifying that for a better physical liberation of the metallic constituents from the remaining slag matrix further crushing and grinding may be required. From this microstructural evidence, it is also apparent that the metallic constituents are forming at the peripheral regions of slag particles, which explains why the overall reduction rate for slag with graphite or active form of carbon slows down considerably at 1323 K. This is due to an increasing diffusion path for transition metal cations with time in the remaining slag matrix. Since the slag particles are relatively non-porous, the only way the metallic phase can nucleate is via a solid-state reduction of slag particles with carbon and carbon monoxide gas available in the surrounding inter-particle spaces between $CaSO_4$, carbon and slag.

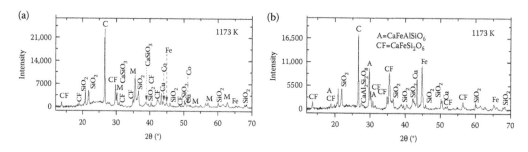

FIGURE 7.10 X-ray powder diffraction patterns after reduction of synthetic slag at molar ratio of MO:C = 1:3, with; (a) carbon black and (b) activated charcoal. Temperature T = 1173 K. Argon flow rate = 0.6 L/min.

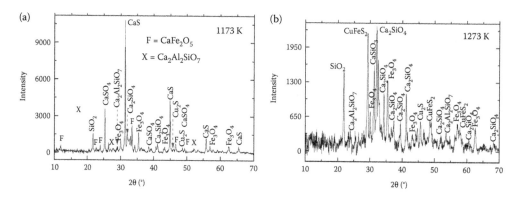

FIGURE 7.11 XRD patterns of the slag samples, sulphidised in the presence of $CaSO_4$ and graphite in the Thermogravimetric analysis (TGA) equipment at molar ratio of $MO:CaSO_4:C$ = 1:1.3:1.5. Argon flow rate = 0.6 L/min. (a) at 1173K and (b) at 1323K.

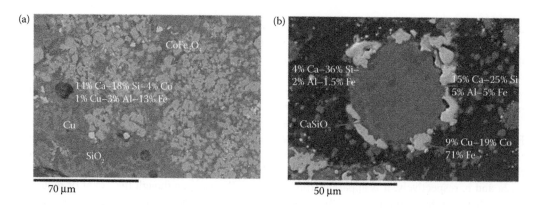

FIGURE 7.12 A comparison of back scattered SEM images of the slag samples, after the reduction of a mixture with molar ratio MO:C = 1:3 with activated charcoal. (a) at 1173K and (b) at 1323K.

Matusewicz and Mounsey (1998) investigated the sulphidisation of copper and cobalt containing slag in the presence of $CaSO_4$ and reported that the gypsum ($CaSO_4$) might not be the best ingredient for sulphidisation of transition metal oxides for transforming them into a matte phase. In the absence of details on their experimental finding, further investigation was carried out by using different types of carbon and stoichiometric ratios of $CaSO_4$ and carbon. The phase analysis of a synthetic slag (Hara and Jha, 2014) (40 wt% SiO_2, 30 wt% Fe_2O_3, 6 wt% Al_2O_3, 10 wt% CaO, 7 wt% CuO and 7 wt% CoO), discussed above in Figure 7.8, was carried out in detail in the presence of $CaSO_4$ in the presence of carbon. The phases formed at 1173 and 1273 K are compared in Figure 7.12a and b, respectively. In Figure 7.12a, the presence Cu_2S and Fe_3O_4 with the calcium ferrite ($CaFe_2O_3$) and silicate phases ($Ca_2Al_2SiO_7$) explains that the decomposition of calcium sulphate has occurred by promoting both the sulphidisation of oxides by forming matte and calcium silicate slag at 1173 K. By increasing the temperature to 1273 K, a larger proportion of iron, cobalt and copper oxides present in the slag transformed into unary and complex binary sulphides, for example, Cu_2S, $CuFeS_2$, Cu_5FeS_4 and $(Co,Fe)S$.

The experimental evidences discussed above in Figure 7.12a and b show that the slag reduction with gypsum may be a better option for copper and cobalt recovery from slag, which otherwise is considered waste.

CONCLUSIONS

We have demonstrated two different approaches for copper and cobalt metal extraction, based on the origin of the metallic values, for example, in the form of froth flotation concentrates and in the metallurgical slag waste. For sulphides present in the form of concentrates, the reduction of mineral sulphides in the presence of lime as an exchange medium was discussed, especially by highlighting the regimes of preferential metallisation which may allow magnetic phase separation. The separation of metallic values from metallurgical slag was also discussed by using carbon well below 1473 K in the presence of gypsum, as a slag forming agent for controlling the complexation of silica and alumina in the gangue mineral. This is a fundamental investigation which offers potential opportunity for process design and development.

ACKNOWLEDGEMENTS

The authors acknowledge the financial support from the Copperbelt University, Zambia, the University of Leeds, UK and the Institute of Mining, Metallurgy and Materials, in London, UK.

REFERENCES

Alcock, C. B. 1976. *Principles of Pyrometallurgy*. 2nd edn, London: Academic Press.
Anon, 1960. *Cobalt Monograph*. Brussels, Belgium: Centre d'Information du Cobalt, p. 515.
Bale, C. W. et al. 2002. FactSage thermochemical software and databases. *Calphad: Computer Coupling of Phase Diagrams and Thermochemistry*, 26(2): 189–228.
Banda, W., N. T. Beukes, and J. J. Eksteen. 2004. Factors influencing base metal recovery from waste reverberatory furnace slags in a 50 kVA laboratory DC plasma arc furnace. *Journal of the South African Institute of Mining and Metallurgy*, 104(3): 201–207.
Banza, A. N., E. Gock, and K. Kongolo. 2002. Base metals recovery from copper smelter slag by oxidising leaching and solvent extraction. *Hydrometallurgy*, 67(1–3): 63–69.
Betteridge, W. 1982. *Cobalt and Its Alloys*. 1st edn, Industrial Metals, Chichester, UK: Ellis Horwood Limited.
Bodsworth, C. 1994. The extraction and refining of metals. In *Materials Science and Technology*, edited by B. Ralph. Boca Raton, Florida: CRC Press, p. 346.
Claig, J. R. and D. J. Vaughan 1979. Cobalt-bearing sulfide assemblages from the Shinkolobwe deposit, Katanga, Zaire. *American Mineralogist*, 64: 136–139.
Davenport, W. G. L. et al. 2002. *Extractive Metallurgy of Copper*. 4th edn, Chemical, Petrochemical & Process, Pergamon, Oxford, United Kingdom, Chapter 4, pp. 57–70.
Davis, J. R. and A. S. M. International. 2001. *ASM Specialty Handbook: Copper and Copper Alloys*. Materials Park, Ohio: ASM International, vii, 652 pp.
Fleitman, A. H., R. B. Herchenroeder, and J. G. Y. Chow. 1971. Cobalt-base alloys for use in nuclear reactors. *Nuclear Engineering and Design*, 15(0): 345–362.
Habashi, F. 1969. *Principles of Extractive Metallurgy*. London: Gordon and Breach, pp. 194–196.

Hara, Y. and A. Jha. 2013. Carbothermic reduction of Zambian sulphide concentrates in presence of lime. *Mineral Processing and Extractive Metallurgy*, 122(3): 146–156.

Hara, Y. S. R. 2013. Study of reaction mechanisms for copper–cobalt–iron sulfide concentrates in the presence of lime and carbon. *JOM*, 65(11): 1597–1607.

Hara, Y. R. S. and A. Jha. 2012. A novel low energy route for the extraction of copper and cobalt metals/alloys from the Zambian sulphide concentrates. In *Characterization of Minerals, Metals, and Materials*. New York: John Wiley & Sons.

Hara, Y. S. R. and A. Jha. 2014. Kinetic and thermodynamic analysis of the reduction of oxides of Cu and Co in a SiO_2–CaO–(Al, Fe)$_2O_3$ Slag. In *Celebrating the Megascale*. New York: John Wiley & Sons, Inc., pp. 553–562.

Hawkins, M. J. 1998. Recovering cobalt from primary and secondary sources. *JOM*, 50(10): 44–45.

Igiehon, U. O. 1990. *Carbothermic Reduction of Complex Sulphides*. London: Department of Materials, Imperial College of Science.

Jha, A., P. Grieveson, and J. H. E. Jeffes. 1989. An investigation on the carbothermic reduction of copper sulfide minerals: Kinetics and thermodynamic considerations. *Scandinavian Journal of Metallurgy*, 18: 31–45.

Jha, A., S. H. Tang, and A. Chrysanthou. 1996. Phase equilibria in the metal-sulfur-oxygen system and selective reduction of metal oxides and sulfides. 1. The carbothermic reduction and calcination of complex mineral sulfides. *Metallurgical and Materials Transactions B: Process Metallurgy and Materials Processing Science*, 27(5): 829–840.

Jones, R. T. et al. 2002. Recovery of cobalt from slag in a DC arc furnace at Chambishi, Zambia. *Journal of the South African Institute of Mining and Metallurgy*, 102(Compendex): 5–9.

Malfliet, A. et al. 2014. Degradation mechanisms and use of refractory linings in copper production processes: A critical review. *Journal of the European Ceramic Society*, 34(3): 849–876.

Matusewicz, R. and E. Mounsey. 1998. Using ausmelt technology for the recovery of cobalt from smelter slags. *JOM*, 50(10): 51–52.

Maweja, K., T. Mukongo, and I. Mutombo. 2009. Cleaning of a copper matte smelting slag from a water-jacket furnace by direct reduction of heavy metals. *Journal of Hazardous Materials*, 164(2–3): 856–862.

Moore, J. J. et al. 1990. *Chemical Metallurgy*. 2 edn, London: Butterworths.

Mututubanya, A. 2013. *Cobalt Recovery from Old Nkana Copper Slag via Solid State Carbothermic Reduction and Sulphation, in Metallurgy*. Lusaka, Zambia: University of Zambia.

Norgate, T. and S. Jahanshahi, 2010. Low grade ores – Smelt, leach or concentrate? *Minerals Engineering*, 23(2): 65–73.

Popil'skii, R. Y. 1976. Superduty ceramics based on calcium oxide. *Refractories*, 17(11–12): 692–699.

Roine, A. 2002. *HSC Chemistry 5.1*. Finland: Outokumpu Research Oy.

Rosenqvist, T. 1983. *Principles of Extractive Metallurgy*. 2nd edn, Japan: McGraw-Hill, pp. 460–500.

Sinha, S. N. and M. Nagamori. 1991. Activities of CoS and FeS in copper mattes and the behavior of cobalt in copper smelting. *Journal of Electronic Materials*, 20(12): 461–470.

Turkdogan, E. T. 1980. *Physical Chemistry of High Temperature Technology*. New York: Academic Press.

Viness, A. 2011. *Extractive Metallurgy 1: Basic Thermodynamics and Kinetics*. Vol. 1, London: Wiley.

West, E. G. 1982. *Copper and Its Alloys*. Chichester, UK: Ellis Horwood.

Whyte, R., N. Schoeman, and K. Bowes. 2001. Processing of Konkola copper concentrates and Chingola refractory ore in a fully integrated hydrometallurgical pilot plant circuit. *Journal of the South African Institute of Mining and Metallurgy (South Africa)*, 101(8): 427–436.

Yang, Z. et al. 2010. Selective leaching of base metals from copper smelter slag. *Hydrometallurgy*, 103(1–4): 25–29.

Zhai, X.-J. et al. 2011. Recovery of cobalt from converter slag of Chambishi Copper Smelter using reduction smelting process. *Transactions of Nonferrous Metals Society of China*, 21(9): 2117–2121.

8 Processing of Lithium-Ion Batteries for Zero-Waste Materials Recovery

D. Marinos and Brajendra Mishra

CONTENTS

ABSTRACT

A review of current lithium-ion battery recycling processes, developed through research, as well as being practiced by industry, has been analyzed. Also, a low-temperature, low-energy method for recovering all the valuable, economically viable, components from the spent lithium-ion batteries has been developed which has allowed the determination of pathways for complete recycling of the battery. Lithium present in the electrolyte has been selectively leached from spent lithium-ion batteries. Leaching conditions were optimized including time, temperature, solid/liquid ratio, and stirring velocity. All the samples were analyzed using inductively coupled plasma–mass spectrometry (ICP-MS) for chemical composition. Leaching was conducted in a flotation machine that was able to separate plastics by flotation, creating bubbles without the use of any external chemical excess reagents. The pregnant solution was concentrated with lithium to a higher amount in order for it to be precipitated and it was shown that the solution could be concentrated by recirculation of the leach solution. The next set of research was composed of battery shredding, steel separation by magnets, leaching with distilled water, and sizing using wet sieving to account for all the components of the battery. Every fraction is sent to rare-earth rolls separation and eddy-current separation for full recovery.

INTRODUCTION

Lithium-ion batteries are the most commonly used batteries. They have found use in a wide variety of applications from electronics to automotive over the past 25 years, and their volume is expected to rise more due to their use in electric vehicle batteries (Fitzimmons, 2014; http://en.wikipedia.org/wiki/Hybrid_electric_vehicles_in_the_United_States). Lithium-ion batteries cannot be disposed into landfill due to safety, environmental reasons, and cost. Apart from this, deposits of important ores are getting lower and lower in grade. Thus, recycling is needed not only to help the environment and reduce pollution from landfilling but also as a source of valuable materials. Also, a research performed by Begum Yazicioglu and Dr. Jan Tytgat proved that it is more energy efficient to recycle the used batteries than to get nickel and cobalt from mining. They also showed that recycling of the batteries produces less CO_2 emissions than mining (Yazicioglu and Tytgat, 2011). Although recycling is beneficial, it is still worthwhile to find the best recycling route to keep energy costs as low as possible.

Thus, over the last years, there has been a lot of effort invested to find ways to recycle lithium-ion batteries. Lots of valuable materials like cobalt, lithium, aluminum, copper, nickel, and manganese are present in a lithium-ion battery making it's recycling favorable. Many attempts, including pyrometallurgical and hydrometallurgical methods, have been researched and some of them are already being used by the industry. Some of them include the following:

Umicore is the biggest recycling company for lithium-ion batteries located in Belgium. It is not a lithium-ion dedicated process; it recycles lithium-ion, lithium-polymer, and NiMH batteries. It uses both pyrometallurgy and hydrometallurgy processes (The Saft Group, 2012). Large industrial batteries, such as hybrid and full electric cars batteries, are first dismantled in a dedicated dismantling line. The battery scrap is prepared, adding slag formers, coke, limestone, etc. and is directly melted in a smelter (shaft furnace). Carbon contained in lithium-ion batteries is used as a reducing reagent in the smelting process. Hot gases produced from plastic pyrolysis are reused. A gas cleaning installation is used to ensure that no damaging dioxins or volatile organic compounds are produced during smelting, and it collects any dust produced. Fluorine from the electrolyte is also collected and can be recovered (http://www.batteryrecycling.umicore.com/UBR/). The metal fraction contains Ni, Co, and other valuable metals that are further refined and transformed into $Ni(OH)_2$ and $LiMeO_2$ (Me: Co, Ni, Mn). $Ni(OH)_2$ and $LiMeO_2$ products are used as new materials for the production of lithium-ion batteries. The slag produced from the process is utterly inert and harmless and can be used as construction material (http://www.batteryrecycling.umicore.com/UBR/). The slag also contains rare earth elements (REEs) (when processing NiMH batteries) and they are recovered as oxides by further treatment (http://www.batteryrecycling.umicore.com/UBR/). The products that are produced are $Ni(OH)_2$, $LiMeO_2$, Cu, Fe, rare earth oxides (REO), and slag for concentrate (The Saft Group, 2012). Umicore's annual tonnage capacity is 7000 tons (The Saft Group, 2012).

Accurec GmbH is a German company that was founded in 1995. This company recycles all types of batteries (Vezzini, 2014), including Ni–Cd, NiMH, Zn–Cd, and lithium-ion batteries. Accurec has a new project called "EcoBatRec" that is planned to recycle more than 55% of the lithium-ion battery materials. Accurec uses autothermal vacuum pyrolysis to condensate the electrolyte. They are planning to produce crack oil, used as oil substitute, ferromagnet and aluminum cases, steel and stainless steel, aluminum and copper foils, and lithium oxides. They are also planning to send the metal concentrate in existing processes route to recover cobalt manganese alloy (http://www.accurec.de/treatment-and-recycling/technologies/lion). Their annual tonnage comes up to 4000 tons and is expected to increase (The Saft Group, 2012).

Xstrata Nickel is a company located in Canada and Norway. Xstrata is not a lithium-ion battery dedicated recycling process. It uses both pyrometallurgy and hydrometallurgy processes. Since 2001, they recycle cobalt and nickel-containing scrap as a secondary feedstock material in their process. Their products are Ni, Cu, and Co alloys. Thus, they are mostly interested on cobalt, nickel, and copper contained in lithium-ion batteries. All the other components are removed with the slag. Xstrata has an annual capacity that comes up to 3000 tons (The Saft Group, 2012).

Toxco, Inc., which is located in Trail, B.C., was first developed just for the treatment of primary lithium-ion batteries. Today, it processes all kinds of lithium batteries (Lain, 2001; Georgi-Maschler et al., 2012) from buttons to 570-lb military batteries (Gaines and Cuenca, 2000). Toxco uses cryogenic and hydrometallurgy processes. The battery scrap is stowed in earth-covered concrete storage bunkers (Georgi-Maschler et al., 2012). Any remaining electrical energy is removed before treating them with Toxco's cryogenic patent that cools the batteries to about 200°C (Gaines et al., 2000; Lain,

2001; Georgi-Maschler et al., 2012; Al-Thyabat et al., 2013) at which lithium is almost inert and thus the risk of explosion is minimized (Lain, 2001; Georgi-Maschler et al., 2012). Then, batteries are separated to large and small to be safely shredded. The large batteries are shredded using a three-step shredding circuit. They are first put through a vertical shear, a rotary kiln, and at last through a hammer mill. Smaller batteries are directly fed into the hammer mill. During each shredding step, a caustic bath is used to neutralize any acidic components and dissolve the lithium salts. A scrubber and filter is used to treat any gas/dust (mainly hydrogen and organics) produced during shredding. The salts are dewatered and lithium carbonate is precipitated using soda ash. The sludge is further processed to recover cobalt for reuse. Any remaining large pieces are passed through a hammer mill, which recovers ferrous and nonferrous metals. Plastics and paper are removed by floating and are recovered for recycling or disposal. Carbon sludge is filtered out and collected as a cake but it has no economic value at the time to reuse or burn. Toxco tries to recycle various types of batteries and most of the materials. Thus, the recovery of the electrolyte would be valuable (Gaines et al., 2000). Their annual capacity is 3000 tons (The Saft Group, 2012).

The Sumitomo-Sony process is one of the early processes that are devoted to the recycling of lithium-ion batteries (Al-Thyabat et al., 2013). At 1992, Sony partnered with Sumitomo Metals in order to recover cobalt from secondary lithium-ion batteries (Goonan 2012). The Sumitomo-Sony process is a pyrometallurgical process (The Saft Group, 2012). The electrolyte and plastics are removed using calcination (Gaines et al., 2000; Al-Thyabat et al., 2013). The heat from the electrolyte is recovered and the cell maintains its original shape. The remaining scrap is then treated using pyrometallurgy to recover an alloy of Co–Ni–Fe. Then hydrometallurgy is used to recover cobalt out of the alloy. Cobalt is then reused for manufacturing of new batteries. Copper and stainless steel are recovered as a by-product using physical separation techniques while lithium and the rest of the components are lost in the slag (Al-Thyabat et al., 2013). Thus, their main products are cobalt and Fe, Cu, Al fractions (The Saft Group, 2012).

Akkuser is a recycling company located in Finland and it recycles NiMH, Lithium-ion, Ni–Cd, alkaline batteries and lead accumulators (http://www.akkuser.fi/en/service.htm). Their process is based on a dry technology and thus doesn't need any heating. Dry technology is developed for the recycling of lithium-ion/li-polymer and Ni–Mh batteries and mobile phones. The batteries are stored and sorted into alkaline, Ni–Mh, Li-ion, and Ni–Cd accumulators and batteries (Pudas and Karjalainen, 2011a). The sorted Ni–Mh and Li-ion accumulators are crushed (Pudas and Karjalainen, 2011a) in a two-phase crushing line (Pudas and Karjalainen, 2011b). Dusts and gases generated in a process are collected and returned into product. Magnetic separation is used to remove steel cases (Pudas and Karjalainen, 2011b). Other plastics, cardboards, and metals are further separated with other separation techniques (Pudas and Karjalainen, 2011a). Plastics and cardboards are used for energy production. Metals are delivered to smelting plants to return to production. Sorted alkaline batteries and sorted lead and Ni–Cd accumulators are delivered to appropriate processing plants (Pudas and Karjalainen, 2011a). For mobile phones, before the treatment, batteries are removed from the phones and processed separately. The crucial issue is that the metals are processed in such a manner that metal refineries can easily and economically refine the product (Pudas and Karjalainen, 2011b). Overall, the process has a very low energy consumption (0.3 kWh/kg), it has a very high recycling rate (more than 90%), it has much lower CO_2 emissions than the foundries and is a safe process. (Pudas and Karjalainen, 2011a).

The Batrec Industrie AG recycling plant is located in Wimmis, Switzerland, and it was first designed for Zn–MnO_2 batteries and was one of the first battery recycling plants (Espinosa et al., 2004). It uses the Sumitomo process from Sumitomo Heavy Industries (Georgi-Maschler et al., 2012; http://www.akkuser.fi/en/service.htm). It recycles all types of batteries including lithium-ion batteries. Lithium-ion batteries are presorted and supplied to the crushing unit in batches (http://www.batrec.ch). The batteries are crushed in a controlled CO_2 gas atmosphere (Georgi-Maschler et al., 2012; The Saft Group 2012; Al-Thyabat et al. 2013) to evaporate the electrolyte and recover it as nonusable condensate (Georgi-Maschler et al., 2012). The released lithium is neutralized to minimize atmospheric pollution (http://www.batrec.ch). In line there is a multistage separating plant that separates the individual components. The annual capacity is 1000 tons (The Saft Group 2012).

Recupyl is a company located in France and has an annual capacity of 110 tons (Zhang and Cheng, 2007). It mostly uses hydrometallurgy techniques and recycles alkaline/zinc–carbon batteries, lithium-ion, and electric vehicles batteries (http://www.recupyl.com). The battery material is crushed in an inert gas atmosphere of argon and CO_2 mixture. The off-gases are leached with water and neutralized with soda. Then they use mechanical separation to separate the steel and copper from the paper and plastics. The remaining materials, metal oxides and carbon, are sieved and they are added in stirring water, adding lithium hydroxide to alkaline the solution. This step is carried out in an atmosphere of low oxygen because there is hydrogen gas release (Vezzini, 2014). Then, after filtering, the solution of lithium salts is obtained and lithium precipitates as $LiCO_3$ using CO_2 gas or as Li_3PO_4 using H_3PO_4. The solid mixture remaining from filtration, containing solid oxide mix, inserted lithium, and carbon, is further treated with acid leaching using sulfuric acid and is filtered. Then the cobalt solution is electrolyzed to recover cobalt, and sulfuric acid and lithium salts are oxidized using sodium hypochlorite to obtain lithium sulfate and trivalent cobalt hydroxide (Vezzini, 2014).

Inmetco was originally designed to treat NiCd and NiMH (Georgi-Maschler et al., 2012; Al-Thyabat et al., 2013). Thus, a small amount of lithium-ion batteries is treated in this process (Georgi-Maschler et al., 2012). Lithium-ion batteries and NiMH batteries are fed along with iron-containing materials as a secondary charge in order to produce iron-based alloys (Al-Thyabat et al., 2013). Copper, nickel and iron are the main products recovered whereas the rest of the materials end up into the slag (Fisher et al., 2011). Carbon is used as a reducing agent in the furnace (Georgi-Maschler et al., 2012; Al-Thyabat et al., 2013). Thus, in the Inmetco's process, most of the lithium-ion battery materials are lost into the slag.

S.N.A.M. is a company located in France. It mostly uses pretreatment techniques and grinding (The Saft Group, 2012). Their main products are Co-powder and graphite (The Saft Group, 2012). Lithium-ion batteries are separated from other types of batteries using sortation techniques. Then, follows a step of pyrolysis pretreatment, crushing, and sieving. This process separates the rough crushed products from the very fine powder containing cobalt, lithium, etc. (RecLionBat project). Their statement is that they are able to recycle up to 60% of the components (from external to the electrodes) of a battery (RecLionBat project).

Although a lot of recycling processes for lithium-ion batteries are currently used industrially, they have some drawbacks: (1) not all of them are lithium-ion battery dedicated processes (Al-Thyabat et al., 2013); (2) in most of the processes, not all of the materials are recycled; (3) usually the electrolyte, plastics, and organics are evaporated or are burned by pyrometallurgy and thus are not recovered (Al-Thyabat et al., 2013); (4) hydrometallurgical processes can sometimes be questionable, from an economical point of view because of the low solid to liquid ratio (Al-Thyabat et al., 2013) and the cost and handling of the reagents (especially if they are not recovered); (5) some battery components can't be leached easily. An example is the strong bond between oxygen and cobalt, which makes leaching with commonly used reagents harder (Al-Thyabat et al., 2013). So, efforts have been focused on in recovering $LiCoO_2$ straight away.

An efficient recycling process should follow the following requirements: (1) lithium-ion batteries can be composed of different materials, which are constantly being substituted by new ones thus making it harder to find a process that would treat all the different compositions at the same time (Al-Thyabat et al., 2013); (2) the electrolyte ($LiPF_6$) should be recovered though it is very hard to recover because of its high reactivity with water, moisture, and heating (Yazami and Touzain, 1983); (3) the products produced by recycling have to be pure enough to meet the market needs and be competitive to new products.

Because of all these, a lot of research has been conducted over the last years. The main goals are to recover most of the products and be environmentally friendly. The researches mentioned below focus on flow sheets that try to recover most of the materials composing a battery and not just the cathode, which is of higher value. The sequence in which they are shown is chronological from the earliest to the latest.

Zhang et al. (2014) demonstrated a process (Figure 8.1) that included demolishing of the battery case and manual separation of the steel casing, which was followed by scraping to remove LiCoO$_2$, a small polymeric substance, and carbon powder from the aluminum foil. The anode material was leached using HCl at 80°C, for 1 hour. Cobalt was then separated from lithium using solvent extraction with PC-88A as the extractant. The next step was a single stage lithium scrubbing from the loaded solvent using a chloride solution followed by stripping of cobalt with H$_2$SO$_4$. Lithium was recovered as lithium carbonate at 100°C. The process recovered cobalt that was as pure as 99.99% or more and 80% of the lithium (Zhang et al., 1998).

Contestabile et al. (2001) developed a laboratory-scale recycling process (Figure 8.2) that recovered both aluminum and copper foils, steel cases, carbon powder, and Co(OH)$_2$. This process consists of crushing and riddling step to remove the steel casings. Then followed a hydrometallurgical treatment using N-methylpyrrolidone (NMP) at 100°C, for 1 hour. After that, filtration is performed to separate copper and aluminum foils. The remaining cathode and carbon powder is leached with HCl at 80°C, for 1 hour, then filtered to separate carbon powder, and finally Co(OH)$_2$ is precipitated and filtered (http://www.samsungsdi.com/small/p_4_1t.jsp). Although this process recovers most of the materials, the solvents were too expensive to move to scale-up operations (Jinqiu Xu et al., 2008).

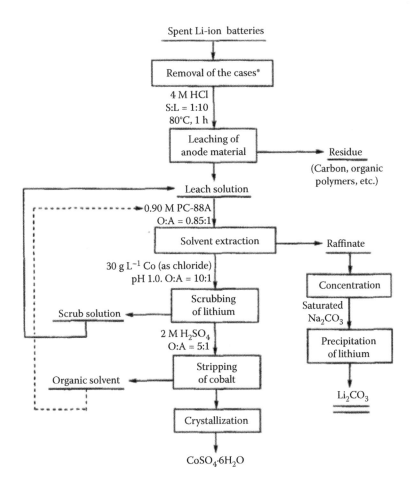

FIGURE 8.1 Flowsheet proposed by Zhang et al., where * shows that the word cases represent both the plastic and the steel cases.

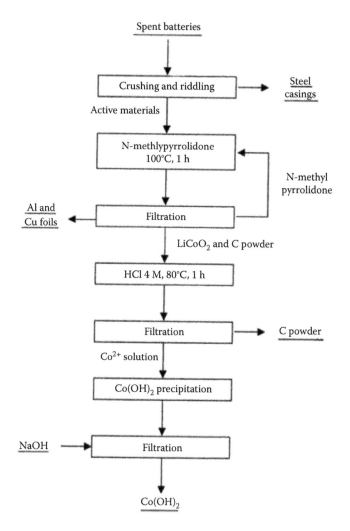

FIGURE 8.2 Flowsheet proposed by Contestabile et al.

Lain developed a process, which started with mechanical shredding in an inert, dry atmosphere to keep the electrolyte from hydrolyzing and to possibly extract the electrolyte with an appropriate solvent. The cell pieces are then treated with a solvent at around 50°C to dissolve the binder and separate the electrode particles from the residual copper, aluminum, steel, and plastic, which can be later, separated from each other using physical processing methods. The electrode particles are filtered from the binding solution that is concentrated to recover the bulk of the solvent for reuse. Electrochemical reduction is used to reduce cobalt(III) to cobalt(II), which releases lithium according to the overall reaction: $2LiCoO_2(s) + H_2O \leftrightarrow 2CoO(s) + 2LiOH(aq) + 1/2O_2(g)$. Aqueous lithium hydroxide is used as the electrolyte and graphite, which is still in solution as the collector (Xu et al., 2008).

Castillo et al. developed a flowsheet (Figure 8.3) that was able to recover steel, carbon, and $Mn(OH)_2$ as a precipitate and an alkaline solution containing sodium and lithium. The process consists of water washing followed by crushing. Then solubilization in HNO_3 (2 mol/L, at 80°C, for 2 hours) followed by filtration. The solid residue is treated pyrometallurgically at 500°C to recover steel and eliminate carbon, and the filtrate is treated hydrometallurgically using NaOH solution to recover $Mn(OH)_2$ as precipitant and sodium and lithium dissolved in solution (Castillo et al., 2002). This method uses simple equipment; it is safe, economic, and is able to recover lots of the battery materials (Jinqiu Xu et al., 2008).

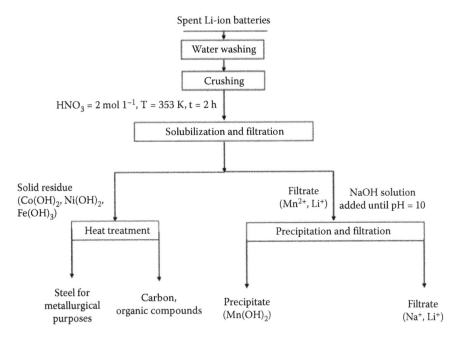

FIGURE 8.3   Flowsheet proposed by Castillo et al.

Nan et al. researched a recycling flowsheet (Figure 8.4) composed of dismantling of the batteries to remove steel crusts using a specially designed machine. Then, the solution was further processed with H_2SO_4 (to make sure no gases would form from the decomposition of the electrolyte. Also, it was shown that approximately 98% of aluminum was dissolved in this leaching step, the Al_2O_2-lixivium was filtered and rinsed. Then, most of the copper was removed from the solution by treatment of H_2SO_4. After, cobalt was recovered by oxalate deposition (~90% because of the copper

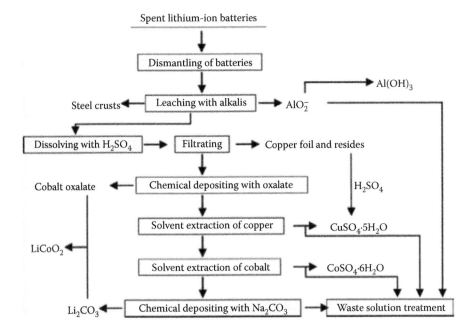

FIGURE 8.4   Flowsheet proposed by Nan et al.

impurities). Then followed a solvent extraction step to recover the remaining copper using Acorga M5640 as the extractant and H_2SO_4 for stripping. The total recovery of copper, which was recovered as $CuSO_4$, was up to 97% of the total copper in the lithium-ion battery. To extract cobalt out of the solution, Cyanex272 was used as the extractant and H_2SO_4 was used for stripping. Approximately 99% of the total cobalt in the lithium-ion battery was recovered. Then, the raffinate was concentrated and treated with a saturated sodium carbonate solution to precipitate Li_2CO_3 at 100°C, which recovered 80% of the lithium. Then lithium carbonate and cobalt oxalate was used as precursors to produce $LiCoO_2$ (Nan et al., 2005).

Shin et al. demonstrated a process (Figure 8.5) of crushing, vibration sieving, and magnetic separation to separate aluminum, copper electrodes, steel casing, and plastic packaging. Then, fine crushing and further vibration sieving were conducted to remove all the aluminum. The remaining powder is treated hydrometallurgically (leaching with hydrogen peroxide). This process could recover almost 100% of the lithium and cobalt that remained in the powder (Shin et al., 2005).

Mantuano et al. conducted a laboratory scale experiment to treat NiCd, NiMH, and lithium-ion batteries from mobile phones. The process (Figure 8.6) consisted of (1) sortation, (2) dismantling to separate iron scraps, paper, and plastic, which were sent for further treatment, (3) leaching of the remaining products with H_2SO_4 to remove gangue materials, and (4) liquid–liquid extraction using Cyanex272 in Kerosene Escaid to recover lithium and cobalt. Cyanex272 was proved to be able to separate lithium from cobalt in lithium-ion batteries ($\Delta H_{1/2} \sim 3$) (Mantuano et al., 2006).

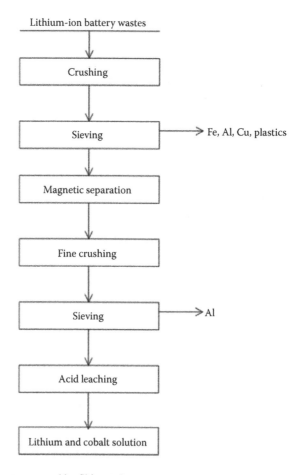

FIGURE 8.5 Flowsheet proposed by Shin et al.

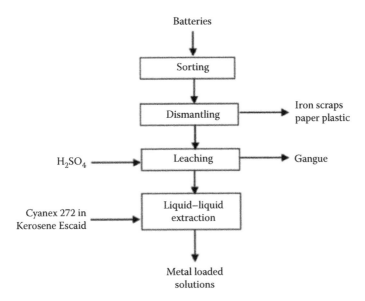

FIGURE 8.6 Flowsheet proposed by Mantuano et al. for the treatment of NiCd, NiMH, and lithium-ion batteries.

Dorella and M.B. Mansur developed a process (Figure 8.7) that included manual dismantling to separate the plastic case at first and then the metallic shell. Then followed manual separation of the anode and cathode and dried (24 hours, at 60°C). After that, battery dust (anode and cathode) was leached with H_2SO_4 and H_2O to recover metals in the solution. The solution was treated with NH_4OH to precipitate the metals; 80% of the aluminum was precipitated out of the solution

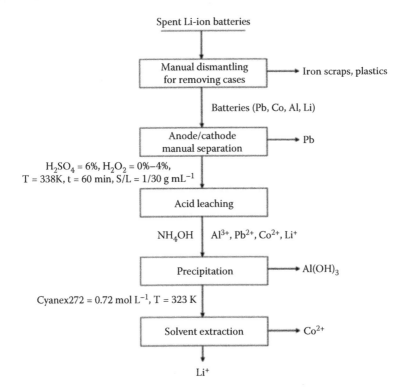

FIGURE 8.7 Flowsheet proposed by Dorella and Mansur.

as hydroxide. They concluded that at a pH of 5, most of the aluminum could precipitate without co-precipitation of other hydroxides (lithium and cobalt, less than 20% precipitation). The solution was further treated using liquid–liquid extraction with Cyanex to extract aluminum, then cobalt, and finally lithium at a $pH_{1/2}$ of 3, 4, and 6, respectively. 100% of the aluminum, 88% of the cobalt, and 33% of the lithium were extracted. These solutions would be further treated by electrowinning (Dorella and Mansur, 2007). The results showed that totally 55% of aluminum, 80% of cobalt, and 95% of lithium were recovered on total from the cathode (Jinqiu Xu et al., 2008).

F. Paulino et al. proposed a flowsheet (Figure 8.8) for the treatment of lithium batteries composed of the following: (1) manual dismantling of the battery followed by 1 hour in vacuum, which raised the temperature creating a solvent that was separated from the rest of the material by mechanical vibration and manual removal; (2) calcination of the anode, cathode, and electrolyte at 500°C, for 5 hours: (3) wet grinding producing LiF and Li_3PO_4 by evaporation; (4) leaching of the insoluble waste with H_2SO_4 and H_2O_2 at 100°C to produce $CoSO_4 \cdot 7H_2O$ or $MnSO_4 \cdot H_2O$ crystals. The lithium recovery was 90 wt%, the copper (high purity) ~94 wt%, and the manganese (high purity) was ~92 wt% having low-waste generation but high-energy consumption (Paulino et al., 2008). Of course, this process is not for lithium-ion batteries but it is mentioned to have an idea of the handling of some of the materials they have in common.

Georgi Maschler et al. developed a process (Figure 8.9) that includes the following: (1) pretreatment of lithium-ion battery packs using sortation to remove plastics, electronic fractions, and other battery types. Because of the high copper content, it is proposed to send this fraction to copper recycling processes; (2) pyrolysis in a retort furnace to evaporate the electrolyte, which was not able to recover because it contains other decomposition products as well (e.g., methyl carbonate [MC], ethyl methyl carbonate [EMC]); (3) crushing, classification, and sorting using vibrating screen, magnetic separation, and air separation, which generates an iron–nickel with aluminum fraction, an electrode foil fraction and fines, which contain the electrode material; (4) the fine fraction is pelletized or briquetted; (5) melting in a vacuum induction furnace producing a Co-based alloy and lithium concentrates; (6) leaching with H_2SO_4 and precipitation of lithium, producing lithium carbonate and sodium sulfate. In this process, 49.08% of the total battery materials can be recovered. Assuming recovery of the electrolyte condensate, they would recover 65.49% (Georgi-Maschler et al., 2012).

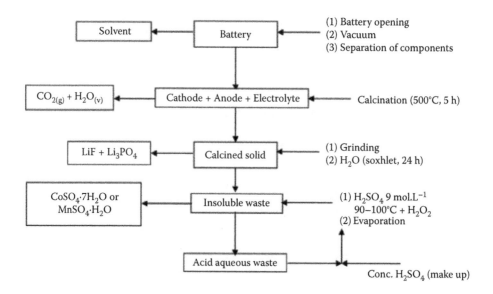

FIGURE 8.8 Flowsheet proposed by J. F. Paulino et al. for the treatment of primary lithium batteries.

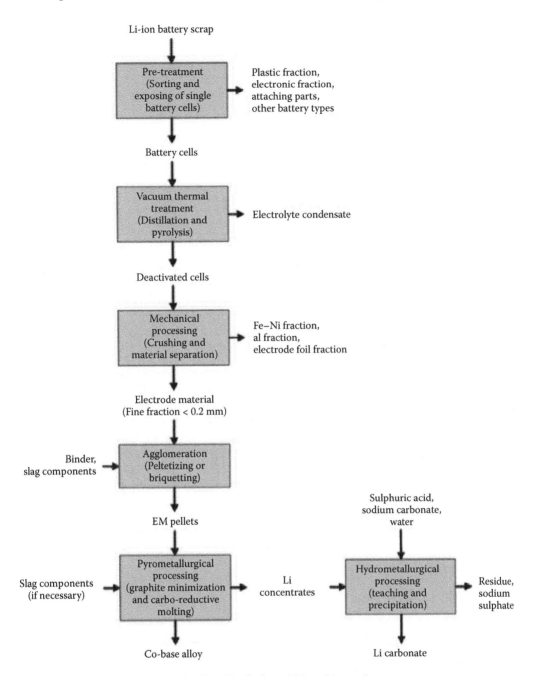

FIGURE 8.9 Recycling flowsheet developed by T. Georgi-Manschler et al.

Erick Gratz et al. proposed a flowsheet (Figure 8.10) that is able to recover most of the metals present in a lithium-ion battery. The process consists of (1) discharge and shredding; (2) removal of steel using rare-earth magnet; (3) recovery of aluminum using NaOH solution; (4) sieving to separate −250 μm fraction; (5) the +250 μm fraction is treated with heavy media separation to separate copper from plastics; (6) the −250 μm fraction is treated with hydrometallurgical processes to finally precipitate high purity $Ni_{0.33}Mn_{0.33}Co_{0.33}(OH)_2$, which can be used in manufacturing. The process has a recovery efficiency of approximately 90% of the valuable cathode metals (Ni, Co, and Mn) (Gratz et al., 2014).

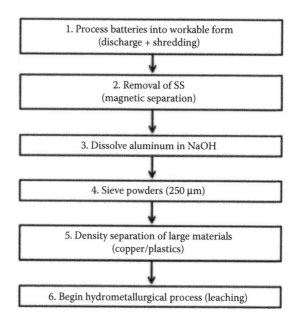

FIGURE 8.10   Process flowsheet developed by Erick Gratz et al.

Further improvements are needed to the already existing processes, to win more valuable materials, use less energy and be more environmentally benign. This experimental work is focused on finding a low-temperature, low-energy method of recovering lithium from the electrolyte and to develop pathways for complete recycling of the battery. In this research, pure $LiPF_6$ powder, which is the electrolyte material, is characterized and titrated with different solvents. Leaching conditions for leaching spent lithium-ion batteries are optimized and because leaching could be performed at room temperature, leaching is conducted in a flotation machine that is able to separate plastics by creating bubbles with no excess reagents use. Then, batteries are shredded, steel is separated using a hand magnet, the batteries are leached with distilled water and sized using wet sieving. Every fraction was sent to rare-earth rolls separation and eddy-current separation.

EXPERIMENTAL

METHODOLOGY

The following research work was conducted at CSM, funded by the government of the United States and the CR[3] group. The research aimed in finding a way to recycle the electrolyte without breaking the $LiPF_6$ compound. It was found out both from references (Jinqiu Xu et al., 2008; Al-Thyabat et al., 2013; Zhang et al., 2014) and the experiments that this was not possible due to partial decomposition of the electrolyte during shredding.

Because of the difficulty in recovering the electrolyte, the research focused in finding new recycling routes for spent lithium-ion batteries that would be able to recover most of the components and thus make the whole process more economically viable. The goal was to provide streams that would be of a smaller volume and of higher value compared to the starting material.

PROCESS DESCRIPTION

A proposed flowsheet, shown in Figure 8.11, has been developed. At first, the material was shredded in order to liberate the constituents of the lithium-ion battery. In this step, a very good exhaust

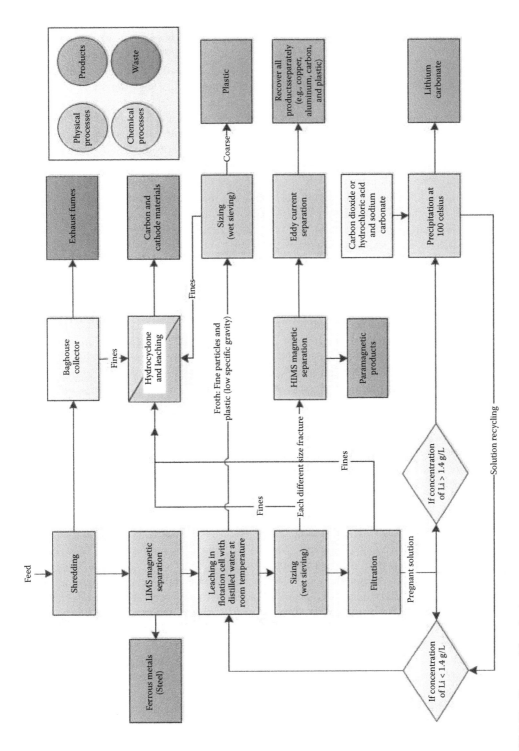

FIGURE 8.11 Proposed flowsheet.

system is needed, like a baghouse, to avoid dust and organic gas. Dust will be collected with all the fine material to be treated separately as it can be seen in the flowsheet. Then, the material is passed through a low-intensity magnetic separator to remove all the ferromagnetic material, which is primarily the steel battery shell.

The remaining material is leached in room temperature for 1 hour using distilled water as the leachant. Lithium is present in a lithium-ion battery in the cathode that is usually as $LiCoO_2$ and in the electrolyte as $LiPF_6$. $LiCoO_2$ is insoluble in water. Thus, during the leaching process, lithium present in the electrolyte, as $LiPF_6$, gets into solution. This leaching step can also be performed in a flotation cell, as there is no need of high temperature and no acid use. After leaching time has passed, the air valve can be opened, forming bubbles and thus, helping the low-specific gravity particles, which are plastic and some fine particles that are stuck on the plastic, to float. In this way, a small amount of lithium can be recovered; the battery material is washed, and plastics are separated using just one machine. The overflow is then sized, using wet sieving, into coarse and fines to wash the plastics and separate fine particles that are stuck on the plastic material. Plastics can then be recycled and the fine material will be collected to treat separately.

The remaining pulp, pregnant solution, and solids of high specific gravity, go through multiple sieves to get different size fractions. Sieving again needs to be wet in order to clean the particles from the fine carbon and cathode powder. Every different size fraction will be treated separately. The very fine will be combined with the rest of the fines and go through the hydrocyclone and/or leached to recover carbon and cathode materials. Carbon could also be recovered from fines using pine oil to float it selectively from the rest of the fines and the cathode materials could be recovered in multiple ways that have been researched.

The pulp that passes through the smallest sieve gets filtered, and the pregnant solution is recovered. Fine particles that are filtered are combined with the fines from previous steps. Lithium concentration in the solution is low and in order to precipitate lithium carbonate from the pregnant solution, according to lithium carbonate solubility, at 100°C, the concentration of lithium needs to be at least 1.4 g/L and at room temperature at least 2.4 g/L. That is based on the fact that lithium carbonate solubility decreases with increasing temperature. Thus, if the concentration of the pregnant solution is lower than that, the solution is recycled back into the leaching step to get more concentrated with lithium. If the concentration of the pregnant solution is higher than that, then lithium carbonate product is recovered using carbon dioxide or hydrochloric acid, to turn LiOH to LiCl, and sodium carbonate. This process is used industrially for the recovery of lithium.

The rest of the fractions that were obtained by sieving will be sent separately for high-intensity magnetic separation and then eddy-current separation. High-intensity magnetic separator and more specifically, rare-earth roll magnetic separator has very high magnetic field and thus could separate any paramagnetic materials from the diamagnetic ones.

After the high-intensity magnetic separation, each fraction will be sent separately to an eddy-current separator. Eddy current is a type of machine that is very attractive for this process. Theoretically, if all the particles are liberated and the size of the particles is close to each other, then a perfect separation of multiple products can be achieved by running them through the belt just once. In this case, all the material is completely liberated and the reason multiple sieves are used is to achieve almost perfect same sized products. In this step, all of the remaining products will be recovered separately (e.g., copper, any remaining carbon or plastic, and aluminum).

Not all of the flowsheet is optimized in this research. In the experiments, shredding was performed and then magnetic separation but without optimizing the conditions, because these processes have already been researched and are used successfully by industries. The leaching conditions using distilled water were optimized. A flotation machine was used for leaching to remove the plastic. Sieving was performed determining the weight and the chemical components present in each fraction. Then, an indicative experiment was conducted to see if the rare-earth rolls step and the eddy current are feasible.

EXPERIMENTAL PROCEDURE

Battery preparation: The spent batteries that were used were Japan-based and they ranged from 3.6 to 14.8 V. They were all discharged before any usage. They were provided from WPI University. The external plastic cases of the batteries were removed using two straight screwdrivers. For the first series of leaching experiments, to determine the leaching conditions, the batteries were hand shredded: The steel case was peeled off and the battery cell was cut with scissors. For the second series of leaching experiments, the batteries were shredded.

Shredding: The batteries were removed from the plastic external case and they were placed in a shredder with a motor of 2.2 kW. The shredder was fed from the top and has a sieve of 100 mm hole size. Each battery was weighted on an accuracy scale and each battery was shredded separately to make sure the sample is representative. Material balance was maintained at each step, starting with a spent battery.

Low-intensity magnetic separation: The steel from the shredded batteries was removed by hand using a hand magnet with dimensions: W*L: 2.15*1.93″, 0.70″ depth, a gap of 0.51, and an approximate pull of 25 lb pool capacity. The shredded battery material was placed on an aluminum foil and a magnet, covered with a cloth, was passed over the material several times to ensure no steel was left in the material. The magnet was removed from the cloth and the steel was separated due to gravity force.

Optimized leaching conditions: Leaching experiments were performed using a beaker placed on a heating plate and a magnetic stirrer to provide the appropriate conditions. The solvent used was distilled water. The conditions that were looked at were temperature, time, solid/liquid ratio, and stirring velocity. For every series of experiment, one parameter changed and the others remained stable. Every battery has different amount of electrolyte ($LiPF_6$), depending on their manufacturing and usage. Thus, to be able to compare the results, it was certified that for each parameter the same package of batteries was used. The temperature conditions that were looked at were room temperature and 50°C. Time conditions were 30, 60, 90, 120, and 360 minutes. The solid to liquid ratios were maintained at 25, 35, 50 g/L. The stirring conditions that were used depended on the solid to liquid ratio. For pulp densities over 35 g/L, the mixture was slurry-like and the velocity was kept below 215 rpm, which wasn't enough to create the proper turbulence. All the leaching samples were tested using ICP-OES Standard Method.

Leaching to concentrate the solution: After the appropriate leaching conditions were determined, there was a need to concentrate the solution. The solvent that was used was distilled water and the batteries were leached for 1 hour, at room temperature, using one battery in a solid to liquid ratio of 25 g/L of distilled water and 280 rpm stirring velocity. The first concentration leaching experiments were performed on a flotation machine that used a 115/230 V, 4.4/2.2 A, and 1725/1435 rpm motor. Though the flotation machine leaching worked, in order to have accurate results, the leaching experiments had to be performed in a beaker. After 1 hour, the solution was filtered and the solution was reused for next leaching. All the leaching samples were tested using ICP Standard Method. The reason, the flotation machine was not used for the experiments, is that the same liquid had to be recycled continuously to raise the lithium concentration enough, so that lithium carbonate can be precipitated. When using a flotation machine, to remove the overflow, always some of the liquid gets removed with it, which would be acceptable in an industrial scale, but for these experiments accuracy was needed. To perform the experiments following the leaching experiments, a lot of material was needed. Thus, all the battery material that was collected through all the leaching experiments was processed in the flotation machine, using a 5 L tank, to float the plastics, and then the materials were sieved into different fractions.

Sizing (wet sieving): Sizing was performed in a sieving machine, which had a motor that works at 115 V, 60 Hz, 0.65 A, and 1625 rpm. The material was passed through three different sieves, one at a time. Enough water was provided to optimize wet sieving in an industrial scale. The sieve sizes that were used were 4, 16, and 50 mesh or 4.699, 0.991, and 0.297 mm. The different fractures were collected and after multiple experiments they were sent to further processing. The very fine material (less than 50 mesh) was not treated any further. The rest of the fractions were sent separately to a rare-earth roll magnetic separator to see if anything else than steel could be separated and then to an eddy-current machine.

Jones Splitter: After drying the material at ambient temperature, a Jones Splitter was used to get representative fractions of each different fraction (+4 mesh, −4 + 16, and −16 + 50). An eighth of each fraction was kept for ICP analysis, one quarter was sent to the rare-earth rolls, a quarter was sent to the eddy-current separator and another quarter was sent to the rare-earth rolls and then to the eddy-current separator.

Rare-earth roll magnetic separation: RDI (Resource Development Inc.) was kind enough to let us use their rare-earth rolls magnetic separator. Each different fraction was processed to see if any further separation could be achieved. Half of the samples were sent to the eddy-current separation and the rest were kept for analysis.

Eddy-current separator: ERIEZ Manufacturing Co. and more specifically Jaisen Kohmuench and Zakary Ekstrom were kind enough to run some indicative eddy-current tests for us. The machine that was used has a 22-pole rare-earth rotor, 12″ in diameter and spinning at 3200 rpm. The unit that was used is used for fines separation and it is fitted with a thinner 0.125″ belt and does not have protective ceramic on the rotor cell. Two types of fraction were sent to ERIEZ: the −4 + 16 mesh and the −16 + 50 mesh.

RESULTS AND DISCUSSION

LEACHING EXPERIMENTS

Time: Time range for leaching included 0.5, 1, 1.5, 2.5, and 6 hours using 50 g of battery per liter of distilled water, 280 rpm stirring velocity, and room temperature. After the filtration, the solutions were analyzed using ICP-OES. The results are shown in Figure 8.12a, where it can be seen that the highest lithium concentration is at 1–1.5 hours.

In Figure 8.12b, the recovery of lithium is plotted assuming an 8% w/w of $LiPF_6$ in the starting battery material. The long time 6 hours experiment was performed because lithium recovery showed a small increase between 1.5 and 2.5 hours, showing a tension for increase as time increases. Though, performing the 6 hours experiments showed that the recovery decreases over time after 2.5 hours. The reason for the concentration of lithium to decrease is that as time passes, CO_2 from the atmosphere gets through the solution, forming solid Li_2CO_3 and thus decreasing the amount of lithium dissolved in the solution.

Temperature and solid-to-liquid ratio: The temperature conditions that were tested were room temperature and 50°C. Higher temperatures were not tested because of the low-decomposition temperature of $LiPF_6$. The solid/liquid ratios tested were 25, 35, and 55 g/L for each temperature using batteries from the same pack. The solid/liquid ratios tested were 25, 35, and 55 g/L for each temperature at 280 rpm of stirring velocity and an hour of leaching. The results can be seen in Figure 8.13a, where highest lithium concentration is achieved at room temperature and at a pulp density of 25 g/L decreasing gradually until it reaches 55 g/L. At 50°C, the concentration of lithium increases, showing the highest recovery at 55 g/L. The reaction is exothermic. However, the solubility depends on temperature and, in this case, as the temperature goes up, solubility increases. Solubility is also dependent on the stirring velocity. The stirring velocity was the same for all the experiments but for higher pulp densities, stirring rate could not be increased. Therefore, the recovery drops as the S/L ratio increases.

Figure 8.13b plots the recovery of lithium as a function of solid to liquid ratio at room temperature and at 50°C. At room temperature, the highest recovery of lithium can be observed at 25 g/L, again decreasing slowly until it reaches 55 g/L. At 50°C, the recovery of lithium increases, showing the highest point at 35 g/L.

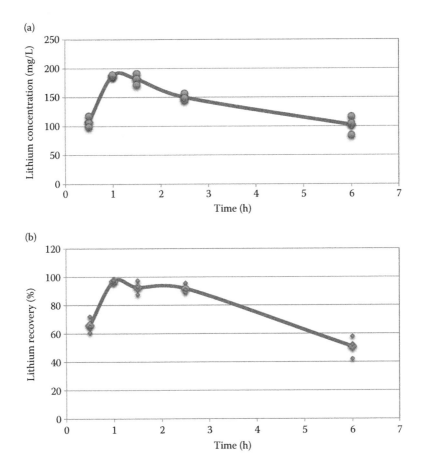

FIGURE 8.12 (a) Lithium concentration as a function of time and (b) lithium recovery (%) as a function of time (hours).

Stirring velocity: The battery material when shredded, occupies a lot of volume, mostly due to the plastics. Thus, stirring is difficult at high solid to liquid ratios. At 35–55 g/L, the stirring velocity couldn't go higher than 280 rpm. At 25 g/L, 280 rpm created a vortex and thus good transport was observed. Consequently, the optimized conditions that were chosen for the leaching are 1 hour leaching time, room temperature, 25 g/L solid to liquid ratio, and 280 rpm stirring velocity. Consequently, there was no reason of trying higher speeds that would be more energy consuming.

Leaching materials of varying size fractions: The need for this work was established by sieving shredded battery through a 50 mesh sieve and leaching the two fractions separately. The percent of lithium recovered is calculated based on the total lithium recovered in milligrams (Figure 8.14a). It is observed that most of the lithium is present in the coarse fraction. However, it should be noted that most of the cathode material is present in the fine fraction. Because the quantity of lithium recovered is already low in the range of 30–200 mg from one battery, the size fractions were not treated separately.

Leaching separate battery components: A battery was hand separated into plastic separator, the anode, and the cathode and leached separately. The solid to liquid ratio was analogous to the weight of each part. The theoretical weight of lithium in the starting material is calculated assuming 8% w/w of $LiPF_6$. The percentage of lithium that was recovered in each fraction is shown in Figure 8.14b indicating that most of the lithium is present in the cathode part. Since, lithium is present in anode and the plastics as well, no components separation is justified. Plastics occupy large

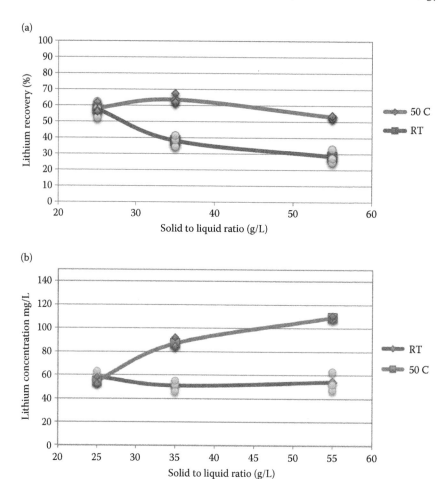

FIGURE 8.13   (a) Lithium recovery (%) as a function of solid to liquid ratio (g/L) at room temperature and at 50°C and (b) lithium concentration (mg/L) as a function of solid/liquid ratio (g/L) at room temperature and at 50°C.

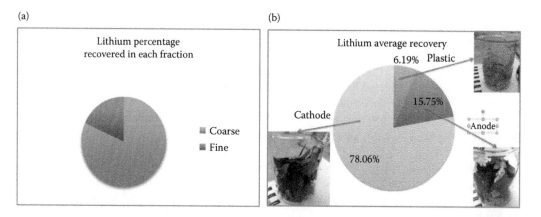

FIGURE 8.14   (a) Lithium percentage recovered in each fraction (coarse and fine) and (b) lithium percentage recovered in each fraction (plastic, anode, and cathode).

volume requiring higher material handling; it could be separated before leaching using heavy media separation if economically justified.

Leaching to concentrate the solution: Electrolysis and evaporation, although viable options for lithium recovery from leached solution, were not considered due to high energy intensities. Industrially, lithium is usually precipitated from a lithium chloride solution using sodium carbonate as lithium carbonate. Lithium carbonate solubility has the tendency to decrease with temperature. Thus at 100°C lithium carbonate solubility is 7.2 g/L and at room temperature the solubility is 13.3 g/L. Consequently, to precipitate lithium carbonate at 100°C, we had to raise the concentration of lithium at least up to 1.4 g/L. The results observed through all the experiments ranged from 30 to 200 mg/L or 0.03 to 0.2 g/L for one battery. Thus, the only way lithium concentration could be raised is if we were to use the same solution for multiple leaches.

The experiments were performed in a beaker and the solution was filtered using a vacuum pump. The solvent that was used was distilled water and the batteries were leached for 1 hour, at room temperature, using one battery in a solid to liquid ratio of 25 g/L of distilled water and 280 rpm stirring velocity. For every experiment one whole battery was used that had been previously shredded and magnetically treated to remove the steel. After 1 hour, the solution was filtered; it was analyzed through ICP and the solution was reused for the next leaching. A small amount of solution was lost during every experiment but that wouldn't impact the results as the volume was measured again before every experiment. The results obtained from the leaching experiments for lithium can be seen in Figure 8.15a. We can see that the lithium concentration is increasing. The lithium recovery is calculated according to the cumulative theoretical amount of lithium contained in $LiPF_6$, assuming 8% w/w on the battery material. We can see that around 34% of lithium is recovered. The difference seen on the fifth can be attributed to the fact that battery from a different case was used. In Figure 8.15b, we can see the lithium concentration as a function of the number of leaches. We can see that it is constantly rising while more batteries are leached but R is not close enough to 1 to represent a line. On the other hand on Figure 8.15b, where lithium concentration is represented up to the fourth leach, which was conducted using batteries from the same pack, we can see that R is very close to 1. So, if we used that line to calculate how many leaches would be needed to reach 1.4 g/L of lithium in the solution, then we can see that 44 batteries would need to be leached. Though, that is not accurate because changing battery cases also changes the slope of the graph as it can be seen in Figure 8.14b.

Figure 8.16 shows that approx. 50% of the solids in solution are near the size of 2 μm and the other half is approx. 8 μm. The use of a finer filter paper with an opening of 2.5 μm leaves half the particles in solution. Therefore, after every leach, more solids (less than ~2.5 μm) were present in the solution. A solution to this problem can be the use of fiberglass filter which would allow the solution to be used for over 40 times. It was confirmed that lithium could be precipitated either by adding hydrochloric acid to turn lithium hydroxide into lithium carbonate according to the reaction: $2LiCl + Na_2CO_3 \rightarrow Li_2CO_3 + 2NaCl$, or by using CO_2 to precipitate lithium carbonate from lithium hydroxide according to the reaction: $2LiOH + CO_2 = Li_2CO_3 + H_2O$. Precipitation was confirmed by preparing a solution with lithium concentration of 1.5 g/L.

Mineral Processing

Flotation: One whole pack of batteries (with the outside case) was shredded. Steel was not removed prior to leaching. Leaching was carried out in a floatation cell. After the leaching time passed, the air valve opened to let the bubbles float the plastic, the particles that have lower specific gravity than the solvent (de-ionized [DI] water). In Figure 8.17a, it can be seen that nothing floats while the air valve is closed, but plastic is separated when (Figure 8.17b) air valve is opened. The plastics were scooped using a spoon. Not all of the plastics were separated; the two rings that each battery has were made of plastic of higher specific gravity and thus could not float. Also, some very fine plastic were left behind because the spoon could not scoop them outside. This problem could be overcome if scooping was performed using a cloth.

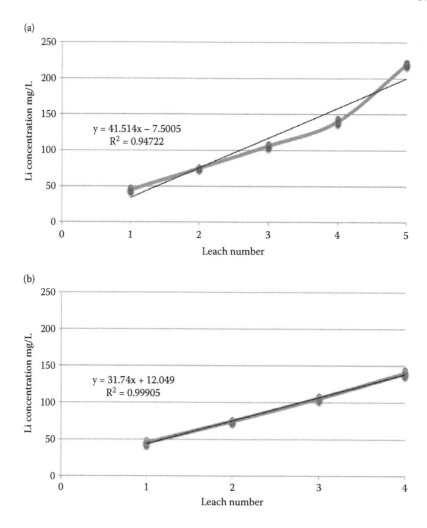

(a)

$y = 41.514x - 7.5005$
$R^2 = 0.94722$

Leach number

(b)

$y = 31.74x + 12.049$
$R^2 = 0.99905$

Leach number

FIGURE 8.15   (a) Lithium concentration as a function of the number of leaches and (b) lithium concentration as a function of the number of leaches, up to the fourth leach.

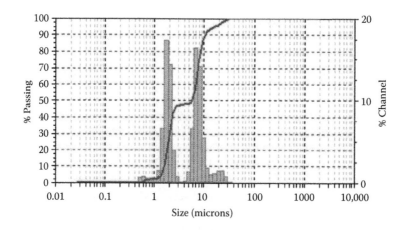

FIGURE 8.16   Size distribution: percent passing compared to size in microns.

FIGURE 8.17 Leaching in a flotation cell: (a) air valve closed and (b) air valve opened.

Size distribution: The plastics—the overflow obtained from flotation—were separated, dried in ambient temperature, and weighed. The underflow was further processed and passed through multiple sieve nos. 4, 16, 50, 200, and 400. The +4 mesh material contained some copper, aluminum, and the chips that are located inside the battery case but mostly steel. The +16 mesh material contained mostly aluminum, copper, and some wool that was present in the battery case. The +50 to −400 mesh material were similar to each other containing mostly cathode powder, carbon, and some copper. The −400 mesh, the material that was filtered, also contained some metallic traces.

The size distribution can be seen in Figure 8.18a. Around 20% w/w is the overflow, thus, it may be concluded that leaching the material in a flotation cell is feasible and removes a lot of weight and volume to allow for less use of energy in further processing steps. The +4 mesh fraction accounts for 31% of the material but it is mostly comprised of steel. The percent retained, cumulative, and passing were calculated according to the sieves. The size distribution can be seen in Figure 8.18b. Since the fine materials were similar and would be difficult to recover through mineral processing three sieve sizes, nos. 4, 16, and 50, were used.

Magnetic separation: To avoid oxidation, steel should be removed prior to leaching. The shredded material was sieved through no. 200 sieve to check if all the steel was present in the coarse fraction. The fine material, −200 mesh, was looked at for steel content using a hand magnet. Some ferromagnetic material in the fine fraction was observed. Therefore, the process step sequence included shredding, magnetic separation using low-intensity magnet, and then leaching.

Processing of a Larger Mass of Batteries

In order to test the rare-earth rolls and the eddy-current separation to see if we could achieve further separation, there was a need to process more material. Thus, all the material that had been leached, after all the experiments, was collected; some steel that was thrown away with the material was removed; and the material was floated in four batches to remove plastics using the 25 g/L solid to liquid ration. The steel that was removed can be seen in Figure 8.19a. The plastics that were removed at the flotation cell as the overflow can be seen in Figure 8.19b. The plastics were not treated any further. They could be washed to separate the very fine material stuck on them and see if they are reusable or they should be considered as waste. The underflow was sized using wet sieving in three sieves: nos. 4, 16, and 50. The −50 mesh materials were filtered out of solution and were not treated any further. They were just analyzed on x-ray fluorescence (XRF) to have an indicative composition. An accurate analysis would not be possible for this fraction since a lot of the powder was stuck in the filtrate papers. The +4 mesh fraction can be seen in Figure 8.19c. We can see the plastic rings that are of high specific gravity and thus, cannot float. Also there are some green chips and small pieces of the external case, which should be not considered, as the leaching tests were performed without the external case. Apart from this, the material is mostly composed

(a)

Sieve no	Diameter (mm)	Mass retained in each sieve (g)	Mass percentage in each fraction	Percent retained (%)	Cumulative retained (%)
4	4.75	15.54	31%	54%	54%
16	1.18	4.905	10%	17%	72%
50	0.3	0.711	1%	2%	74%
200	0.075	2.161	4%	8%	82%
400	0.038	0.708	1%	2%	84%
Pan		4.497	9%	16%	100%
Sum		28.522		100%	
Overflow		9.46	19%		
Impeller		10.486	21%		
Cleaning		2.012	4%		
Total sum		50.48	100%		

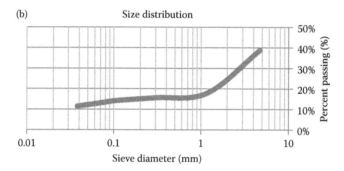

(b)   Size distribution

FIGURE 8.18   (a) Size distribution table and (b) plotted size distribution of the sieved material.

of copper and aluminum. The −4 + 16 mesh fraction can be seen in Figure 8.19d. The material is mostly composed of copper and aluminum. Some cathode powder is stuck on the aluminum pieces and some carbon in the copper ones. The −16 + 50 mesh fraction can be seen in Figure 8.19e. This fraction is mostly comprised of fine carbon and cathode powder. It also has some small quantities of copper and aluminum. The three fractions were further processed. The fractions were split using

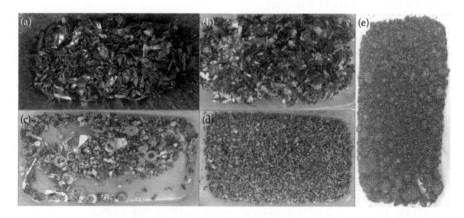

FIGURE 8.19   (a) Steel removed before floating, (b) plastic, flotation overflow, (c) +4 mesh fraction, (d) −4 + 16 mesh fraction, and (e) −16 + 50 mesh fraction.

FIGURE 8.20   +4 mesh: (a) nonmagnetic fraction and (b) magnetic fraction; 4 + 16 mesh: (c) nonmagnetic fraction and (d) magnetic fraction; and −16 + 50 mesh: (e) nonmagnetic fraction and (f) magnetic fraction.

a Jones Splitter. Firstly, an eighth of the −4 + 16 and the −16 + 50 mesh were obtained for analysis. Two quarters of the −4 + 16 and the −16 + 50 mesh and half of the +4 mesh were sent to the rare-earth rolls for separation. The other half of the +4 mesh was kept for analysis. Another quarter of the −4 + 16 mesh and the −16 + 50 mesh was send to the eddy current.

Rare-earth rolls separation: For the +4 mesh fraction, the conditions used were 115 rpm and slow feed vibration; the left splitter was at no. 0.5 position and the right was at no. 2 position. For the −4 + 16 mesh fraction, the conditions used were 50 rpm and slow feed vibration; the left splitter was at no. 3 position and the right was at no 2 position. For the −16 + 50 mesh fraction, the conditions used were 85 rpm and slow feed vibration; the left splitter was at no. 2 position and the right at no. 2 position. The nonmagnetic material and the magnetic material, for each fraction, can be seen in Figure 8.20a and Figure 8.20b, respectively. Both the magnetic and the nonmagnetic fractions were dissolved in aqua regia and analyzed using ICP analysis. The results obtained from the ICP analysis can be seen in Table 8.1. In the +4 fraction, we can notice that in the nonmagnetic fraction, copper is a lot higher. In Table 8.2, we can see the ICP results obtained in weight percentage, making it easier to compare the difference in concentration after magnetic separation. So, we can say that most of the copper ended up on the nonmagnetic fraction. The magnetic fraction is higher in aluminum content and shows some iron and nickel, which is probably steel that could have been left from imperfect hand magnetic separation. Therefore, we can say that a separation takes place, but not a very favorable separation. That could be attributed to the fact that a very small amount of the +4 mesh fraction was obtained or because of the large particle size. Maybe a larger quantity of this material would show better results.

Two quarters of the −4 + 16 mesh fraction were processed through the rare-earth rolls. The conditions used were 50 rpm and slow feed vibration; the left splitter was at no. 3 position and the right was at no. 2 position. The nonmagnetic fraction and magnetic fraction can be seen in Figure 8.20c and Figure 8.20d, respectively. Just by looking at the fractions, we can see that a separation of the products is occurring showing most of the aluminum in the magnetic fraction. This could have been expected as copper is nonmagnetic and aluminum is slightly paramagnetic. One quarter of the fraction was processed and the nonmagnetic fraction was sent to the eddy current to see if any further separation could be achieved. The other quarter was processed and both the magnetic and the nonmagnetic fractions were dissolved in aqua regia and analyzed using ICP analysis. The results obtained from ICP can be seen in Table 8.1. From the results, we can see that in the nonmagnetic fraction copper is a lot higher and in the magnetic fraction aluminum is a lot higher. Also, we can see that most of the cobalt is gathered up in the magnetic fraction which would have been expected since aluminum is part of the cathode. In Table 8.2, we can see the ICP results obtained for the −4 + 16 mesh fraction in weight percentage, making it easier to compare the difference in concentration after magnetic separation. It can be seen that aluminum was concentrated in the magnetic

TABLE 8.1

ICP Results in g/L for +4, −4 + 16, and −16 + 50 Mesh Fractions before and after Rare-Earth Rolls Magnetic Separation

Element	+4 Mesh Fraction before Separation	+4 Mesh Fraction Nonmagnetic	+4 Mesh Fraction Magnetic	−4 + 16 Mesh Fraction before Separation	−4 + 16 Mesh Fraction Nonmagnetic	−4 + 16 Mesh Fraction Magnetic	−16 + 50 Mesh Fraction before Separation	−16 + 50 Mesh Fraction before Separation	−16 + 50 Mesh Fraction Magnetic
Al	2.1	2.8	0.4	11.9	4.2	7.0	0.5	0.3	0.3
Co	0.2	0.4	0.2	5.1	1.4	3.3	11.6	9.6	3.6
Cu	8.2	5.5	0.3	25.8	23.5	1.0	0.8	1.0	0.2
Fe	0.1	0.2	0.3	BDL	BDL	BDL	BDL	BDL	BDL
Li	0.0	0.1	0.0	0.6	0.2	0.3	1.6	1.2	0.5
Ni	0.0	0.1	0.1	0.0	0.0	0.0	0.0	0.0	0.0
P	0.4	0.2	0.0	1.3	1.1	0.0	0.1	0.1	BDL
Carbon and plastic	3.2	4.2	0.4	3.7	3.1	1.9	34.5	30.6	6.6

TABLE 8.2

ICP Results in wt. pct. for +4, −4 + 16, and −16 + 50 Mesh Fractions before and after Rare-Earth Rolls Magnetic Separation

	Concentration (g/L)						Percentage (%)					
Element	−16 + 50 Mesh Fraction Analysis	−16 + 50 Mesh Electromagnetic Fraction	−16 + 50 Mesh Fraction Nonmagnetic	−16 + 50 Mesh Fraction Electromagnetic (Nonmagnetic)	−16 + 50 Mesh Fraction Magnetic	−16 + 50 Mesh Fraction Electromagnetic (Magnetic)	−16 + 50 Mesh Fraction Analysis	−16 + 50 Mesh Electromagnetic Fraction	−16 + 50 Mesh Fraction Nonmagnetic	−16 + 50 Mesh Fraction Electromagnetic (Nonmagnetic)	−16 + 50 Mesh Fraction Magnetic	−16 + 50 Mesh Fraction Electromagnetic (Magnetic)
Al	1.1	3.2	0.5	2.3	2.2	5.2	0.990%	21.695%	0.603%	12.199%	2.321%	36.803%
Co	24.1	0.8	18.4	0.8	28.5	1.5	21.690%	5.575%	21.821%	4.450%	30.348%	10.808%
Cu	1.7	2.3	1.8	9.4	1.4	1.3	1.530%	15.717%	2.165%	49.705%	1.500%	9.088%
Fe	BDL	BDL	BDL	BDL	BDL	BDL	BDL	BDL	BDL	BDL	BDL	2.662%
Li	3.4	0.1	2.3	0.1	3.8	0.2	3.060%	1.008%	2.766%	0.651%	4.062%	1.698%
Ni	0.1	BDL	0.0	BDL	0.1	0.1	0.050%	BDL	0.028%	BDL	0.113%	0.910%
P	0.1	0.3	0.1	0.5	BDL	0.2	0.130%	1.746%	0.128%	2.468%	BDL	1.703%
Carbon and plastic									69.341%	21.491%	55.896%	23.478%

Note: Magnetic and nonmagnetic refer to RE rolls separation, and electromagnetic refers to eddy-current separation.

fraction from 23 to 50 wt%, and copper was concentrated in the nonmagnetic fraction from 50 to 66.7 wt%. Also, cobalt that usually follows aluminum was concentrated in the magnetic fraction from 10% to 23%. In this fraction, we see a very good separation that could be treated further. If running more scavengers, the concentrations might be even higher.

Two quarters of the −16 + 50 mesh fraction were processed through the rare-earth rolls. The conditions used were 85 rpm and slow feed vibration; the left splitter was at no. 2 position and the right was at no. 2 position. The nonmagnetic material and the magnetic material can be seen in Figure 8.20e and Figure 8.20f, respectively. One quarter of the fraction was processed and both the nonmagnetic and the magnetic fractions were sent to the eddy current to see if any further separation could be achieved. The other quarter was processed and both the magnetic and the nonmagnetic fractions were dissolved in aqua regia and analyzed using ICP. The results obtained from ICP can be seen in Table 8.1. From the results, we can see that in the magnetic fraction aluminum and cobalt are higher. In Table 8.2, we can see the ICP results obtained for the −16 + 50 mesh fraction in weight percentage, making it easier to compare the difference in concentration after magnetic separation. It can be seen that aluminum was concentrated in the magnetic fraction from 0.603 to 2.321 wt%, and copper was concentrated in the nonmagnetic fraction from 1.526 to 2.165 wt%. Also, cobalt was concentrated in the magnetic fraction from 21.687% to 30.348%. In this fraction, we see that a small concentration after the magnetic separation, which would be a lot higher if carbon that is almost the same in both the magnetic and the nonmagnetic fraction had been separated before any treatment.

Eddy-current separation: Because only a large eddy-current machine was available at the time and the samples were too little for the machine, to avoid sample loss the separation was provided like that: For the −16 + 50 mesh fraction, the technician held the sample above the rotor contained in a larger clear bag and was able to "manually" effect a separation. For the −4 + 16 fraction samples, there was a great deal—more agitation caused by the eddy-current rotor—that they were not able to run this in the clear bag or on the large unit without significant losses. However, their statement was that for the larger particle a separation could definitely be achieved in slow feed speed; just not with the current mass. If more material was provided, they could probably create the two product streams. We could comment that since this fraction showed the best concentration after the rare-earth rolls separation, treating with eddy current would occur in a much more concentrated material.

For the −16 + 50 fraction, one quarter of the material (directly after sieving) was processed through the eddy-current machine. The electromagnetic fraction and the non-electromagnetic fraction can be seen in Figure 8.21a and b, respectively. The electromagnetic fraction was dissolved in aqua regia and was analyzed using ICP. The ICP results in concentration (g/L) can be seen in Table 8.3. From the results, we can see that the aluminum and the copper concentration increased while the cobalt and the lithium decreased. In Table 8.3, we can also see the ICP results obtained for the −16 + 50 mesh fraction in weight percentage, making it easier to compare the

FIGURE 8.21   −16 + 50 mesh fraction: (a) electromagnetic fraction, (b) non-electromagnetic fraction, (c) electromagnetic nonmagnetic fraction, (d) non-electromagnetic, nonmagnetic fraction, (e) electromagnetic magnetic fraction, and (f) non-electromagnetic magnetic fraction.

TABLE 8.3

ICP Results in (g/L) and in wt% of the Electromagnetic Fractions before and after Electromagnetic Separation by Eddy Current

Element	Concentration (g/L)						Wt. Fraction					
	−16 + 50 Mesh Fraction Analysis	−16 + 50 Mesh Electromagnetic Fraction	−16 + 50 Mesh Fraction (Nonmagnetic)	−16 + 50 Mesh Fraction Electromagnetic (Nonmagnetic)	−16 + 50 Mesh Fraction Magnetic	−16 + 50 Mesh Fraction Electromagnetic (Magnetic)	−16 + 50 Mesh Fraction Analysis	−16 + 50 Mesh Electromagnetic Fraction	−16 + 50 Mesh Fraction Nonmagnetic	−16 + 50 Mesh Fraction Electromagnetic (Nonmagnetic)	−16 + 50 Mesh Fraction (Magnetic)	−16 + 50 Mesh Fraction Electromagnetic (Magnetic)
Al	1.10	3.21	0.51	2.31	2.18	5.17	0.01	0.22	0.01	0.12	0.02	0.37
Co	24.11	0.82	18.36	0.84	28.49	1.52	0.22	0.06	0.22	0.04	0.30	0.11
Cu	1.70	2.32	1.82	9.39	1.41	1.28	0.02	0.16	0.02	0.50	0.02	0.09
Fe	BDL	BDL	BDL	BDL	BDL	BDL	BDL	BDL	BDL	BDL	BDL	0.03
Li	3.40	0.15	2.33	0.12	3.81	0.24	0.03	0.01	0.03	0.01	0.04	0.02
Ni	0.05	BDL	0.02	BDL	0.11	0.13	0.00	BDL	0.00	BDL	0.00	0.01
P	0.15	0.26	0.11	0.47	BDL	0.24	0.00	0.02	0.00	0.02	BDL	0.02
Carbon and plastic									0.69	0.21	0.56	0.23

Note: Magnetic and nonmagnetic refer to RE rolls separation, and electromagnetic refers to eddy-current separation.

difference in concentration after the eddy-current separation. It can be seen that aluminum was concentrated in the electromagnetic fraction from 0.987 to 21.695 wt% and copper was concentrated from 1.526 to 15.717 wt%. Also, cobalt decreased from 21.687 to 5.575 wt% and lithium from 3.057 to 1.008 wt%. But almost half of the electromagnetic fraction was composed of carbon and plastics. In general, the weight recovered was very little (0.2956 g out of 46.85 g). If the experiment was run properly, with a larger mass of sample, appropriate for the machine, we could expect a better concentration.

After treatment with rare-earth rolls, a magnetic and a nonmagnetic fraction were sent to the eddy current to see if any better separation could be achieved. From each fraction, an electromagnetic and a non-electromagnetic fraction was obtained. In Figure 8.21c, we can see the nonmagnetic, electromagnetic fraction and in Figure 8.21d the nonmagnetic, non-electromagnetic fraction. In Figure 8.21e, we can see the magnetic, electromagnetic fraction and in Figure 8.21f the magnetic, non-electromagnetic fraction. The electromagnetic fractions were dissolved in aqua regia and were analyzed using ICP analysis. The ICP results in concentration (g/L) can be seen in Table 8.3. From the results, we can see that in both the magnetic and nonmagnetic fractions the aluminum concentration increased. Also, in both the magnetic and nonmagnetic fractions the cobalt and lithium concentration increased. In Table 8.3, we can also see the ICP results obtained for the −16 + 50 mesh fraction in weight percentage, making it easier to compare the difference in concentration after the eddy-current separation. For the nonmagnetic fraction, we can see that aluminum was concentrated from 0.603 to 12.199 wt% and copper from 2.165 to 49.705 wt% while cobalt decreased from 21.821 to 4.450 wt% and lithium decreased from 2.766 to 0.651 wt%. For the magnetic fraction, it can be seen that aluminum was concentrated in the electromagnetic fraction from 2.321 to 36.803 wt% and copper was concentrated from 1.5 to 9.088 wt%, while cobalt decreased from 30.348 to 10.808 wt% and lithium decreased from 4.062 to 1.698 wt%. From the results, we can say that processing the material through the rare-earth rolls first and then through the eddy current results in more concentrated products. Also, it can be observed that the rare-earth rolls provide a separation between the anode materials (copper) and the cathode materials (aluminum, cobalt and lithium), while the eddy current seems to draw more copper and aluminum in the electromagnetic and leave cobalt and lithium in the non-electromagnetic.

CONCLUSIONS

- It was shown that lithium contained in $LiPF_6$ can be dissolved in distilled water and that the solution can be further concentrated.
- Leaching can be performed in a flotation cell to provide washing of the materials and plastic separation at the same time.
- Low-intensity magnetic separation has to be done after the shredding, so that steel doesn't get oxidized during leaching.
- Rare-earth rolls separation, proved to provide a separation between the anode (copper and carbon) and the cathode (aluminum and cathode powder) and eddy-current separation shows better results after the rare-earth rolls separation than on the starting material.
- Eddy-current separation indicative test provided a superior separation between copper and aluminum in the electromagnetic fraction and cobalt and lithium in the non-electromagnetic fraction.

Contribution to the field:
This research provided a low-temperature process that can recover a small amount of lithium contained in the electrolyte, while washing and separating plastics at the same time. It introduced new ways of pretreating the battery material that reduce the material that needs to be treated and thus, reduce the cost of further processing. Although more work needs to be done to optimize the

process, it is a good start showing that mineral processing can beneficiate the materials, and its advantage is that it could find application to a lot of already existing processes.

REFERENCES

Al-Thyabat, S., T. Nakamura, E. Shibata, and A. Lizuka. 2013. Adaptation of minerals processing operations for lithium-ion (LiBs) and nickel metal hydride (NiMH) batteries recycling: Critical review. *Minerals Engineering*, 45: 4–17, Doi:10.1016/j.mineng.2012.12.005.

Castillo, S., F. Ansart, C. Laberty-Robert, and J. Portal. 2002. Advances in the recovering of spent lithium battery compounds. *Journal of Power Sources*, 112(1): 247–254, Doi:10.1016/S0378-7753(02)00361-0.

Contestabile, M., S. Panero, and B. Scrosati. 2001. A laboratory-scale lithium-ion battery recycling process. *Journal of Power Sources*, 92(1-2): 65–69, Doi:10.1016/S0378-7753(00)00523-1.

Dorella, G. and M. B. Mansur. 2007. A study of the separation of cobalt from spent Li-ion battery residues. *Journal of Power Sources*, 170(1), 210–2155, Doi:10.1016/j.jpowsour.2007.04.025.

Espinosa, D. C. R., A. M. Bernandes, and J. A.o S. Tenório. 2004. An overview on the current processes for the recycling of batteries. *Journal of Power Sources*, 135(1–2): 311–319, Doi:10.1016/j.jpowsour.2004.03.083.

Fisher, A. S., M. B. Khalid, M. Widstorm, and P. Kofinas. 2011. Solid polymer electrolytes with sulfur based ionic liquid for lithium ion batteries. *Journal of Power Sources*, 196(22): 9767–9773, Doi:10.1016/j.jpowsour.2011.07.081.

Fitzimmons, M. 2014. *High Power Technology: Easy 30% Upsid as Lithium-Ion Battery Demand to Go Parabolic*. Seeking Alpha, Springer, Switzerland.

Gaines, L. and R. Cuenca. 2000. *Costs of Lithium-Ion Batteries for Vehicles*. United States Department of Energy. http://www.transportation.anl.gov/pdfs/TA/149.pdf.

Georgi-Maschler, T., B. Friedrich, R. Weyhe, H. Heegn, and M. Rutz. 2012. Development of a recycling process for Li-ion batteries. *Journal of Power Sources*, 207: 173–182, Doi:10.1016/j.jpowsour.2012.01.152.

Goonan, T. G. 2012. *Lithium Use in Batteries*. U.S Department of the Interior and U.S. Geological Survey. http://pubs.usgs.gov/circ/1371/pdf/circ1371_508.pdf.

Gratz, E., Q. Sa, D. Apelian, and Y. Wang. 2014. A closed loop process for recycling spent lithium ion batteries. *Journal of Power Sources*, 262: 255–262, Doi:10.1016/j.jpowsour.2014.03.126.

Lain M. J. 2001. Recycling of lithium ion cells and batteries. *Journal of Power Sources*, 97–98: 736–738, Doi:10.1016/S0378–7753(01)00600–0.

Mantuano, D. P., G. Dorella, R. C. A. Elias, and M. B. Mansur. 2006. Analysis of a hydrometallurgical route to recover base metals from spent rechargeable batteries by liquid-liquid extraction with Cyanex 272. *Journal of Power Sources*, 59(2): 1510–1518.

Nan, J., D. Han, and X. Zuo. 2005. Recovery of metal values from spent lithium-ion batteries with chemical deposition and solvent extraction. *Journal of Power Sources*, 152: 278–284, Doi:10.1016/j.jpowsour.2005.03.134.

Paulino, J. F., N. G. Busnardo, and J. C. Alfonso. 2008. Recovery of valuable elements from spent Li-batteries. *Journal of Hazardous Materials*, 150(3): 843–849, Doi:10.1016/j.jhazmat.2007.10.048.

Pudas, J. and T. Karjalainen. 2011a. The Best Battery Recycling. AKKUSER, http://www.cefic.org/Documents/ResponsibleCare/Awards2011/Winners/RCAward2011-Winner-AkkuSer-BatteryRecycling.pdf.

Pudas, J. and T. Karjalainen. 2011b. Mobile Phone and Battery Recycling Services. AKKUSER, http://www.teec.ee/docs/HazWa/AkkuSerPresentation.pdf.

RecLionBat project, Lithium-Ion battery recycling, http://www.snam.com/recyclion_fr.htm.

Shin, S. M., N. H. Kim, J. S. Sohn, D. H. Yang, and Y. H. Kim. 2005. Development of a metal recovery process from Li-ion battery wastes. *Hydrometallurgy*, 79(3-4): 172–181, Doi:10.1016/j.hydromet.2005.06.004.

The Saft Group. 2012. EV&HEV battery developments and prospectives: Closed-loop battery recycling. ANVIR. http://avnir.org/documentation/congres_avnir/telechargements2012/presentations/SiretLCAconf2012.pdf.

Vezzini A. 2014. *Manufactures, Materials ad Recycling Technologies*. In *Lithium-Ion batteries Advances and Applications*, edited by Gianfranco Pistoia, Chapter 23, pp. 529–551, Newnes, New York.

Xu, J., H. R., Thomas, R. W. Francis, K. R. Lum, J. Wang, and B. Liang. 2008. A review of processes and technologies for the recycling of lithium-ion secondary batteries. *Journal of Power Sources*, 177(2): 512–527, Doi:10.1016/j.jpowsour.2007.11.074.

Yazami, R. and P. Touzain. 1983. *Journal of Power Sources*, 9: 365, Doi:10.1016/0378–7753(83)87040-2.

Yazicioglu B. and J. Tytgat. 2011. *Life Cycle Assesments involving Umicore's Battery Recycling Process*. DG Environment – Stakeholder Meeting, Umicore, Elsevier, the Netherlands.

Zhang, W. and C. Y. Cheng. 2007. Manganese metallurgy review. Part I: Leaching of ores/secondary materials and recovery of electrolytic/chemical manganese dioxide. *Hydrometallurgy*, 89(3–4): 137–159, Doi:10.1016/j.hydromet.2007.08.010.

Zhang, T., Y. He, F. Wang, L. Ge, X. Zhu, and H. Li. 2014. Chemical and process mineralogical characterizations of spent lithium-ion batteries: An approach by multi-analytical techniques. *Waste Management*, 34(6): 1051–1058, Doi:10.1016/j.wasman.2014.01.002.

Zhang, P., T. Yokoyama, O. M. Itabashi, T. Suzuki, and K. Inoue. 1998. Hydrometallurgical process for recovery of metal values from spent lithium-ion secondary batteries. *Hydrometallurgy*, 47(2–3): 259–271, Doi:10.1016/S0304-386X(97)00050-9.

9 Selective Leaching
An Ecological Solution for Recovering Metals from Complex Minerals and Materials?

Gretchen T. Lapidus

CONTENTS

ABSTRACT

The increasing complexity of ores and recycled materials obliges research into innovative methods for metal separation and purification. Selective leaching may offer an ecological alternative or a complement to bulk smelting. The principles involved in determining the most advantageous agent to oxidize, reduce, or sequester a specific metal are discussed. Examples are given for precious and base metal, as well as actinide selective leaching.

INTRODUCTION

Our modern society requires metals for mobility, communication, and a myriad of uses. Base metals, such as copper, iron, lead, and manganese, or precious metals such as gold, silver, and platinum, are employed in the manufacture of almost all devices familiar to us and many other applications. In fact, the need for these metals is ever increasing. However, procuring metals from underground are increasingly more complicated and costly, since the metallic content of the ores are lower, mineralogically encapsulated, and in combination with materials of little commercial value (Wills and Napier-Munn, 2011; Free, 2013).

Until now, minerals have been processed in a similar manner throughout the world: crushing, fine grinding, flotation, and smelting for base metals or cyanidation for precious metals. This series of procedures have been operating efficiently as long as the metals in the ore, usually two or more, were separable at a particle size of between 10 and 150 µm (Wills and Napier-Munn, 2011). However, with the passage of time, these ores are becoming scarce; leaving only those whose liberation size is very small. This situation causes excess expenditures for ultrafine grinding or metal loss when the values do not separate.

A convenient solution to this situation is using more selective methods capable of extracting one metal while leaving the rest intact. This "residue" can then continue the traditional route or be subjected to another selective treatment to extract other values. This implies that whatever technique is used should be executed at relatively mild conditions (near ambient temperature and pressure). Aqueous-based extractive metallurgy, better known as hydrometallurgy, operates under these conditions, with its corresponding limitations. The method is based on the ability of specific ligands (or complexing agents) to form soluble species, preferably in a selective manner with the desired metal or metals, in aqueous solutions (leaching). The best example of this is the cyanidation process to extract gold and silver from their minerals, employed for over 120 years in the precious minerals industry.

Despite the existence of hydrometallurgical processes for several metals, most are toxic, nonselective, or performed at elevated temperature and pressure. Additionally, many of these processes generate large quantities of waste that must be treated prior to disposal. For that reason, the current hydrometallurgical routes generally offer little or no advantages compared to the pyrometallurgical ones (smelting and roasting). However, with today's problematic minerals, processes must be developed with a different philosophy: extract each metal, one by one, in different leaching stages, so as not to have to separate them downstream, as would be necessary in a total leach. Furthermore, the solutions employed in each of the different stages should be recycled within the same processing scheme, decreasing waste treatment and reagent costs.

The objective of this chapter is to introduce the basic concepts of selective leaching and the tools that are used to propose a chemical system and the operating conditions that are most feasible to extract each metal from a specific mineralogical or metallurgical environment. Applications for several precious, base and rare earth metal systems will be given.

In most minerals, a specific metal value is found in combination with others, in addition to worthless materials. In order to separate it from the rest of the constituents, in hydrometallurgy, the metal is dissolved in a principally aqueous solution. The specific mineral phase that contains the metal usually is associated with other elements in such a manner that it forms a thermodynamically stable solid. The degree of stability may be represented by its value of Gibbs' free energy or its solubility product (K_{sp}). For instance, silver naturally occurs in over 200 mineral phases (Gasparrini, 1984); however, the most common form is acanthite (Ag_2S). The maximum solubility of Ag may be expressed as follows (NIST, 2004):

$$K_{sp} = [Ag^+]^2[S^{2-}] \approx 10^{-50} \quad \text{at } 25°C$$

This indicates that in pure water, the silver ion solubility would be less than 10^{-16} M or approximately 5×10^{-15} g Ag/L.

So, how does silver dissolve in the cyanidation process? For this mineral phase, there are basically two routes: (1) the disappearance by reaction or physical removal of the silver ion or (2) the same for the sulfide ion. In cyanidation, both actions are made to occur: on one hand, the sulfide is oxidized by the oxygen in air to another sulfur species (sulfate or thiosulfate), liberating the silver ion (Luna and Lapidus, 2000). Simultaneously, the silver ion combines with cyanide ion (ligand) to form a soluble species.

$$Ag_2S + 4CN^- + O_2 + \frac{1}{2}H_2O \rightarrow 2Ag(CN)_2{}^{2-} + \frac{1}{2}S_2O_3{}^{2-} + OH^-$$

Complexation to form soluble entities lowers the free silver ion concentration, which is the limiting factor in the solubility product. This last phenomenon is the basis of selective leaching: finding an adequate ligand that sequesters the desired metal ion and complexes very little with other constituents in the ore.

Therefore, as hydrometallurgists, we look for both factors: modification of the oxidation state of the components in the mineralogical phase of interest to destabilize the solid and complexation of the desired metal ion to increase its solubility. In other words, we deal with the best (fastest, cheapest, and less polluting) way to chemically break up a specific solid, while leaving the rest practically intact. Sometimes, it is sufficient to apply only one of the measures, but many times both are necessary. Most of the time, the agents employed to change the oxidation state (oxidants or reductants), as with ligands, are aqueous phase chemical reagents, although in some cases, electrical current can be used. There are many other challenges involved in guaranteeing a successful leaching process, such as getting the reagents to the mineral phase so that they can react; and problems related to encapsulation and passivation (pore blockage) can arise. However, the present chapter will focus on achieving chemical selectivity and much less on kinetic aspects.

In the sections that follow, there are several examples of applications for selective leaching in which the author has personally been involved; the first and most extensive deals with gold and silver leaching. The next is base metal extraction from a complex sulfide and from recyclable materials. Finally, thorium and uranium dissolution from a rare earth mineral (monazite) is discussed.

PRECIOUS METALS

The fundamental concern originates from the almost universal use of cyanide, a toxic material employed to extract gold and silver. Although the industry generally controls the management of this substance and of its destruction in residues, its application causes concern in the population due to the eventuality of an accident. For this reason, much research has centered on the extraction and recovery of silver and gold with cyanide-alternatives, in particular thiourea solutions. Thiourea combines with gold and silver ions much the same way that cyanide does, although not as strongly. This fact is reflected in the predominance diagrams presented in Figure 9.1 for elemental silver in the cyanide or thiourea systems, where it may be observed that in cyanide solutions the metal is oxidized at lower ligand (cyanide or thiourea) concentrations and at inferior potentials.

Although cyanidation is a proven method since latter part of the nineteenth century, thiourea has several advantages. It is most stable in acidic solutions, which permits the use of oxidizing agents stronger than oxygen, which is employed in cyanidation. It is more selective for gold and silver and, because of this, the barren solution (without gold and silver) can be recycled back to the leaching stage. Derived from this, the acidothioureation process is effective for certain minerals that experience some types of refractoriness toward cyanide (those that experiment preg-robbing and complex sulfides).

However, thiourea has disadvantages that have limited its industrial implementation: it easily degrades and much higher concentrations are required as compared to those necessary with cyanide. Regardless of this, if it were possible to control the decomposition, the elevated cost due to the high concentration would be offset by recirculation of the leaching solution. In an excellent review article, Li and Miller (2006) mark four causes for thiourea loss: adsorption by mineral particles, complexation with base metals, thermal degradation, and irreversible oxidation due to the mineral itself or the very agent required to react with the gold. The first two are inherent in the mineral and may be minimized by acid pretreatments. Although the third is remedied by operating near room temperature, it is quite difficult to control because of the nature of the leaching reaction. Gold and silver are naturally found in phases that must be oxidized in order to liberate the metal ions so that they can complex with the thiourea. Thiourea also is oxidized very close to the redox potential required to leach the precious metals (Groenewald, 1976).

To diminish thiourea destruction, tests were performed on specific precious metals phases. The delicate nature of the solution was revealed, but also the adequate conditions for conservation of

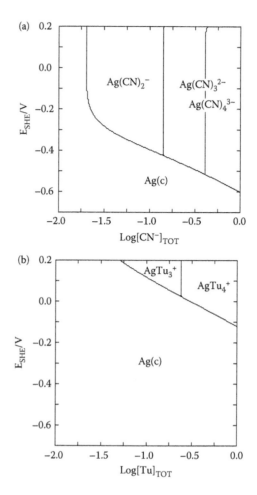

FIGURE 9.1   Predominance diagrams of silver species in (a) cyanide or (b) thiourea solutions as a function of the log [ligand]. Conditions: for the cyanide system, pH = 10 and for the thiourea system, pH = 1. $[Ag]_{TOT} = 0.01$ M.

the reagent. This knowledge and a very peculiar characteristic of thiourea (the reversibility of the first oxidation reaction) were applied to improve the efficiency of the leaching stage and diminish thiourea consumption.

It is well known that thiourea oxidation takes place in various successive steps (Gupta, 1963); the first of these, which is reversible, is the formation of formamidine disulfide ($(Tu)_2$ or FADS):

$$2CS(NH_2)_2 \leftrightarrow \underset{\text{FADS}}{(NH_2)NHCSSCNH(NH_2)} + 2H^+ + 2e^- \quad E^\circ = 0.42 \text{ V vs SHE}$$

The others lead to the irreversible destruction of the reagent:

$$(Tu)_2 \rightarrow Tu + \text{Sulfinic compounds}$$
$$\text{Sulfinic compound} \rightarrow CN - NH_2 + S^\circ$$

The oxidized product, FADS, itself has the capacity to act as an oxidant for most phases of gold and silver, which is the key point of the process described below. However, the FADS dimer is quite unstable and, over time, gradually dismutes to sulfur, cyanamide, and thiourea. It seems that various factors, including the solution redox potential and the presence of some mineral components, increase the rate of this dismutation reaction.

FADS is usually formed in situ by the action of the oxidant on the thiourea. However, it is also possible to produce FADS by direct oxidation in an electrolytic cell (Lapidus et al., 2007a). When an external oxidant is employed, the correct dosage is critical to prevent the irreversible destruction of the thiourea. However, when current is applied, it is possible to better control the reversible formation of the FADS. One drawback to the electrolytic production of FADS, as compared to the use of an oxidant such as ferric ion, is that it must be placed immediately in contact with the mineral to be leached in order to take advantage of its oxidizing power, before dismutation occurs. Another shortcoming of the direct use of FADS is that it is a rather weak oxidant, whose redox potential is very close to that of the phases to be oxidized, especially in the case of silver compounds. This situation obliges the maintenance of its concentration at a reasonably high level, without promoting the dismutation reaction. Despite the apparent sensitivity of this system, these problems are not insurmountable.

In 2008, an alternative to cyanidation, based on electro-oxidized thiourea, was tested in a bench scale operation (Lapidus et al., 2007b). The procedure consisted of the electro-formation of the leaching agent (FADS), its subsequent use in a rapid leach stage, the electro-recovery of the precious metals and recycle of the solution back to the initial stage of the process. Since the thiourea solutions are recycled within the process scheme, reagent costs are minimized. The results for a high-grade silver-lead concentrate showed that the thiourea solution can be reutilized several times without a decrease in its effectiveness (Lapidus et al., 2008).

Parallel to the thiourea process, research was performed on other systems for gold and especially silver extraction. Silver and gold are extracted in ammoniacal thiosulfate media, both occupying copper as the principal leaching agent, although each by different mechanisms. Since silver is principally present as sulfides, the silver is substituted in the solid lattice by Cu(I), as shown in the following reaction, leaving a solid chalcocite product:

$$Ag_2S + 2Cu(S_2O_3)_2^{3-} \rightarrow Cu_2S + 2Ag(S_2O_3)_2^{3-}$$

On the other hand, gold dissolution requires a soluble redox pair. In conventional practice with this system, the cupric ion is moderately stabilized by ammonia, which prevents excessive reduction.

$$Au + Cu(NH_3)_4^{2+} + 4S_2O_3^{2-} \rightarrow Au(S_2O_3)_2^{3-} + Cu(S_2O_3)_2^{3-} + 4NH_3$$

However, thiosulfate is easily oxidized by weakly complexed cupric ion, which is continuously regenerated by air from the environment. Because of this, even high concentrations of ammonia are only partially effective in maintaining a low thiosulfate consumption rate. The addition of small amounts of a stronger complexing agent, such as EDTA (ethylenediaminetetraacetic acid), in addition to the ammonia, has been proven efficient in decreasing thiosulfate consumption while increasing silver and gold dissolution (Lapidus-Lavine et al., 2001; Xia et al., 2003).

$$Au + Cu(EDTA)^{2-} + 4S_2O_3^{2-} + H_2O \rightarrow Au(S_2O_3)_2^{3-} + Cu(S_2O_3)_2^{3-} + H(EDTA)^{3-} + OH^-$$

The resulting system has a high degree of complexity, since a delicate balance exists between the different cupric and cuprous soluble species. Therefore, any additional soluble metal ions present in solution will surely influence the copper equilibria. One of these is Pb(II), which also strongly complexes with EDTA, diminishing the amount available for the stabilization of Cu(II) (NIST, 2004). However, the thiosulfate system has been shown to be very efficient in gold and silver extraction from high arsenic-containing refractory concentrates and minerals (Aazami et al., 2014; Mesa-Espitia and Lapidus, 2014).

At this point, the thiosulfate and thiourea systems are adequate for niche processes, where cyanidation is not effective. However, both have proven to be effective lixiviants and can be profitable for multiple applications.

LEAD

Another of the refractory phases that have challenged researchers in the hydrometallurgy is lead sulfide (galena). It is a constituent in many complex sulfide ores and is severely penalized in smelting operations because of lead's toxicity. The hydrometallurgical problem with this mineral is that the lead cation (Pb^{2+}) tends to form insoluble solids in the presence of sulfate or, as most other metal ions, with hydroxide. The panorama is further complicated by the fact that galena requires the oxidation of its sulfide ion to elemental sulfur (two electrons per mole of galena) to liberate the lead ion; however, at highly oxidizing conditions, the elemental sulfur converts to sulfate (six more electrons per mole of galena).

With the addition of a strong and selective complexing agent (ligand), lead is dissolved, but the soluble form of lead must have characteristics that are compatible with the recovery method and allow reuse (recirculation) of the leaching solution. This dilemma has been tormenting hydrometallurgists for over 40 years. Many options have been proposed, the most recent of which include the use of fluoborate or methanesulfonate ions (Maccagni et al., 2012; Wu et al., 2014a,b). However, these are relatively nonselective and are limited by the presence of sulfate ion.

The search for the correct ligand, using thermodynamic considerations, and parameter optimization, produced a system that not only extracted lead from minerals, but also is adequate for recycling lead from lead/acid batteries and other waste materials. The citrate ion is an effective and selective ligand for lead; it forms several soluble complexes, some of which are shown below (NIST, 2004)

$$Pb^{2+} + Cit^{3-} \rightarrow Pb(Cit)^-$$
$$Pb^{2+} + 2Cit^{3-} \rightarrow Pb(Cit)_2^{4-}$$
$$Pb^{2+} + HCit^{2-} \rightarrow Pb(HCit)$$
$$Pb^{2+} + H_2Cit^- \rightarrow Pb(H_2Cit)^+$$
$$Pb^{2+} + 2Cit^{3-} + H^+ \rightarrow PbH(Cit)_2^{3-}$$
$$2Pb^{2+} + 2Cit^{3-} \rightarrow Pb_2(Cit)_2^{2-}$$
$$2Pb^{2+} + 2Cit^{3-} + OH^- \rightarrow Pb_2(Cit)_2(OH)^{3-}$$
$$2Pb^{2+} + 2Cit^{3-} + 2OH^- \rightarrow Pb_2(Cit)_2(OH)_2^{2-}$$

where $Cit^{3-} = C_6O_7H_5^{3-}$ = citrate ion

The variety and relative concentrations of the lead complexes are better appreciated in the speciation diagram of Figure 9.2, where it can be appreciated that lead is soluble within the pH interval of approximately 5.5–8 at ambient temperatures.

The difference between using citrate to sequester the Pb^{2+} ion, as opposed to a stronger complexing agent such as EDTA, is that the lead citrate complexes are relatively weak, permitting a more efficient recovery of the lead in its metallic form after the leaching stage. Figure 9.3 illustrates this principal: at pH values in the soluble range, the reduction potential is approximately −0.3 V versus

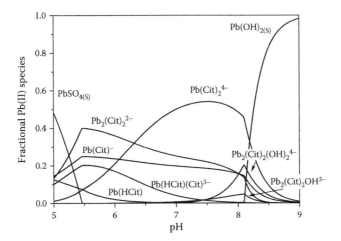

FIGURE 9.2 Speciation diagram (fraction of lead versus pH) for the Pb–citrate–sulfate system. Conditions: 0.12 M Pb$_{total}$, 0.12 M (SO$_4$)$_{total}$, and 1 M citrate$_{total}$ at 25°C.

SHE (standard hydrogen electrode) (for EDTA in the same pH interval this potential is less than −0.5 V). The aforementioned characteristics of citrate also have the advantage of their selectivity for lead, while EDTA also strongly complexes other base metals.

Another distinguishing property of lead is its affinity for oxygen-containing oxidants, such as peroxide (H$_2$O$_2$) and ozone (O$_3$). With peroxide, galena (PbS) as well as metallic lead are readily leached. As shown in Figure 9.4a and b, lead is preferentially dissolved from a complex sulfide concentrate with hydrogen peroxide in citrate solution at ambient temperature. The residue resulting from this leach maintains a similar mineralogy, except for the galena; this feature allows it to be subjected to another selective leach or to be incorporating into the traditional pyrometallurgical treatments for the remaining metals. The lead is then easily and efficiently recovered from the citrate solution.

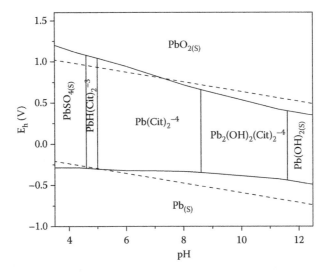

FIGURE 9.3 Predominance diagram (E$_h$ versus pH) for the Pb–citrate–sulfate system. Conditions: 0.12 M Pb$_{total}$, 0.12 M (SO$_4$)$_{total}$, and 1 M citrate$_{total}$ at 25°C.

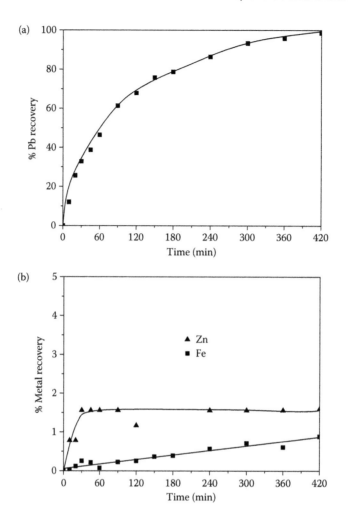

FIGURE 9.4   Effect of 0.15 M H_2O_2 concentration on (a) Pb and (b) Zn and Fe dissolution. Experimental conditions: 20 g Pb concentrate/L, pH 7, 1 M Na_3Cit, and 25°C. Concentrate: 24.9% galena, 17.1% pyrite, 21.6% sphalerite, 4.5% silver sulfides, and 1.4% chalcopyrite.

MANGANESE

Selective recovery of manganese from low-grade ores (15%–25%) and waste materials has aroused interest in recent years. Enormous reserves of low-grade pyrolusite mineral are available, but not exploited. However, the traditional techniques of acid or oxidative leaching, while effective for treating manganese carbonate (rhodochrosite), does not extract manganese from highly oxidized phases.

Manganese is similar to many others of the transition metals in that it possesses several oxidation states; its redox behavior permits its dissolution or precipitation, depending on the solution potential and pH, in a distinct manner relative to other metallic phases. As opposed to most metals, manganese is soluble when it is in a low-oxidation state Mn(II). This characteristic permits a selective separation of its ions from the others. Currently, in the industry, the reduction is performed at 800–900°C in the presence of CO, originating from coal (Zhang and Cheng, 2007). However, with low-grade minerals, this process is by no means profitable (Gutiérrez-Muñoz et al., 2012).

In recent years, various reducing agents, such as metallic iron, ferrous sulfate, peroxide, activated charcoal and mainly SO_2 have been tested (Zhang and Cheng, 2007; Bafghi et al., 2008; Biswal et al., 2011). However, only the last has been considered economically and technically viable

(Zhang and Cheng, 2007). Despite this, the use of SO_2 as the principal reagent, constitutes a health risk and would require significant infrastructure for its safe handling. In the last few years, several types of organic materials, such as glucose and corn cobs, have been proposed as reducing agents; however, this process is possible only at relatively high temperatures (80–90°C) (Furlani et al., 2006; Tian et al., 2010).

One of the most interesting proposals for extracting manganese from low-grade minerals uses electrical current to directly reduce the mineral particles, suspended in an acid solution (Elsherief, 2000; Kumari and Natarajan, 2002; Gutiérrez-Muñoz et al., 2012). Elsherief (2000) and Kumari and Natarajan (2002) successfully tested this technique; however, the leaching rate was very slow at ambient temperature with platinum cathodes.

On the other hand, several authors have reported the catalytic effect of the Fe(III)/Fe(II) couple in this system (Das et al., 1982; Mukherjee et al., 2004; Zhang and Cheng, 2007). Figure 9.5 shows the Pourbaix diagram for both the manganese and the iron species; the area highlighted in yellow denotes the conditions under which the ferrous ion can reduce manganese dioxide. However, the ferrous ion alone only can reduce one-half its quantity in moles for each manganese dioxide.

Combining the ferric/ferrous couple with that of direct galvanic reduction, Gutiérrez-Muñoz et al. (2012) electroleached a low-grade pyrolusite ore in sulfuric acid solutions, using added ferrous ion as an electric charge transfer vehicle. In Figure 9.6, the dissolved manganese

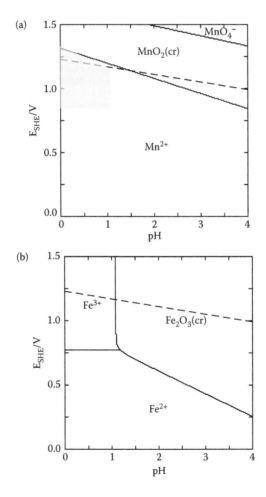

FIGURE 9.5 Predominance diagrams of (a) manganese and (b) iron as a function of solution potential and pH. Conditions: $[Mn]_{total} = [Fe]_{total}$ 0.001 M at 25°C.

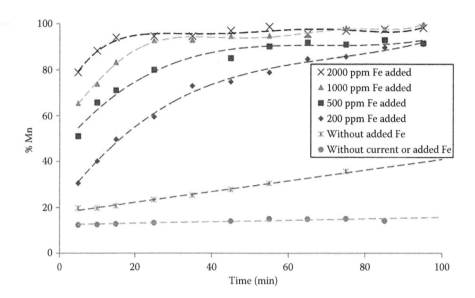

FIGURE 9.6 Comparison of manganese extractions versus time, varying the concentration of added ferrous ion. The slurry consisted of 5 g of ore (19% Mn, mostly as pyrolusite) in 250 mL of 1 M H_2SO_4. In all, except the experiment without current, 1 A was applied. (Adapted from Gutiérrez-Muñoz, M. et al. 2012. Reductive leaching of manganese ores. In *Recent Developments in Metallurgy, Materials and Environment*, edited by M. I. Pech-Canul, A. L. Leal-Cruz, J. C. Rendón-Angeles, C. A. Gutiérrez-Chavarría, J. López-Cuevas, J. L. Rodríguez-Galicia. Editorial Cinvestav IPN, pp. 13–22, 341pp. ISBN 978-607-9023-18-8.)

concentration is plotted as a function of time for different concentrations of initial ferrous ion. As may be observed, the spontaneous dissolution of manganese increased with the elevation in the ferrous ion concentration. Furthermore, the supply of electric current maintains the leaching rate. In this manner, it can be inferred that it is the Fe(II) ions, and not the direct contact of the mineral with the electrode, that foments the reduction of the manganese dioxide. According to this hypothesis, the mechanism should include the following reactions (Lapidus and Doyle, 2008; Gutiérrez-Muñoz et al., 2012):

$$S-H^{\bullet} + Fe^{3+} \rightarrow Fe^{2+} + H^+ + S$$

where S–H$^{\bullet}$ represents monatomic hydrogen formed on the surface of the cathode, which is a strong reducing agent.

The ferrous ion reacts with the pyrolusite, producing Mn^{2+} and ferric ion; the latter is then regenerated at the cathode.

$$MnO_{2(S)} + 2Fe^{2+} + 4H^+ \rightarrow Mn^{2+} + 2H_2O + 2Fe^{3+}$$

Once in solution, the manganese may be selectively recovered as a solid by oxidative precipitation, electrolytically or with a strong oxidant (Biswal et al. 2015; Cruz-Díaz et al., 2015).

RARE EARTH MINERALS AND ACTINIDES

The current methods for procurement of actinides (thorium and uranium) from their minerals are hydrometallurgical, but at severe pH and temperatures, which destroy the mineral matrix, generating large quantities of toxic residues. Research has centered on finding additives, clean oxidants,

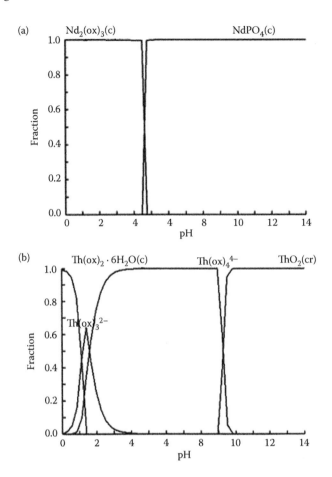

FIGURE 9.7 Speciation diagrams as a function of pH for (a) neodimium and (b) thorium in solutions containing phosphate and oxalate ions. Concentrations: $[Nd]_{total} = 0.012$ M, $[Th]_{total} = 0.0085$ M, $[PO_4{}^{3-}]_{total} = 0.079$ M, and $[Oxalate]_{total} = 0.5$ M.

and solution parameters to surmount the stability barrier of the mineral phases. At this moment, there have been efforts to selectively extract the rare earth elements (REE) and actinides from concentrates of different ore deposits from diverse locations in the world, the most difficult being monazite (Kim and Osseo-Asare, 2012; Panda et al., 2014).

However, alternative lixiviants for the actinides have been scarcely investigated in the literature. Carbonate and other carboxylate ions have been shown to complex with the rare earth and actinide ions (NIST, 2004). In particular, oxalate has been used to precipitate the rare earths from neutral and acidic solutions, while thorium and uranium remain soluble. In Figure 9.7, the thermodynamic speciation of neodymium(III) (as a representative element of the rare earths) and thorium(IV) are shown as a function of pH in a typical solution produced by leaching monazite (complex rare earth phosphate); while thorium is soluble, largely as $Th(ox)_4{}^{4-}$, between pH 2 and 8, neodymium is completely insoluble.

Thermodynamic predictions provide promising indications of a successful separation. However, on many occasions, there are kinetic limitations that hinder or prevent the dissolution process. The rare earths and actinides in monazite are already in their most soluble oxidation state (+3 for rare earths, +4 for thorium, and +6 for uranium), eliminating the need for oxidants or reductants. The leaching tests themselves are therefore the best indicators of the dominant phenomenon in the process. The insolubility of the rare earth ions in oxalate/phosphate solutions is confirmed in Figure 9.8

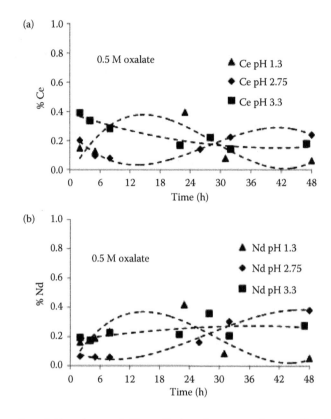

FIGURE 9.8 Extraction of (a) cerium and (b) neodimium from monazite as a function of time in 0.5 M oxalate solutions at 40°C, at different values of pH. Concentrations: $[Ce]_{total} = 0.025$ M, $[Nd]_{total} = 0.012$ M, $[PO_4^{3-}]_{total} = 0.079$ M, and $[Oxalate]_{total} = 0.5$ M.

for cerium and neodymium at different acidic values of pH. As may be observed, the percentages extracted are well below 1% with an oscillating tendency, characteristic of solubility limitations.

On the other hand, at the same conditions, the thorium and uranium solubilities are much higher as predicted by the speciation diagram and this is reflected in the leaching tests (Figure 9.9). However, their respective behaviors with pH are quite different. On one hand, thorium extraction is highest at pH 3.3; while in more acid conditions, the leaching rate and extent are lessened by the low solubility of the $Th(ox)_2 \cdot 6H_2O(c)$, probably formed at the reaction interface. It is also interesting to note that the thorium extraction reaches a maximum at relatively short leaching times, after which the dissolved thorium re-precipitates, perhaps as phosphate or oxalate. On the other hand, uranium does not present a solid oxalate compound at low pH values and, thus, becomes more soluble in strongly acid solutions.

Despite the fact that thermodynamics correctly predicts the relative solubilities of the different metal ions, the leaching kinetics are painfully slow, probably due to the dense structure of the monazite. For this process to be economical, more research is necessary to find methods that disrupt the extreme stability of this mineral.

RECYCLE

The reuse of metals contained in waste materials from consumer products or those that constitute a danger to the environment, has gained importance in our modern society. Furthermore, normally these metals are usually found in the waste materials in concentrations that are far superior to those

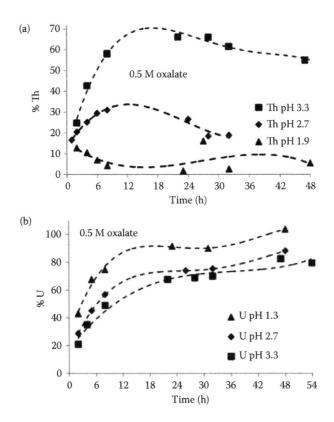

FIGURE 9.9 Extraction of (a) thorium and (b) uranium from monazite as a function of time in 0.5 M oxalate solutions at 40°C, at different values of pH. Concentrations: $[Th]_{total} = 0.0085$ M, $[U]_{total} = 0.00016$ M, $[PO_4^{3-}]_{total} = 0.079$ M, and $[Oxalate]_{total} = 0.5$ M.

found in nature. However, many times they are in combination with other elements that hinder their dissolution and/or are covered with plastic or glassy ceramics that are inert to the leaching agents. For this reason, their recovery creates different challenges than those encountered with minerals.

For example, *"e-waste"* or electronic garbage contains high levels of base and precious metals, especially copper. The sheer number of cellular phones, tablets and computers sold, and eventually discarded, as well as the batteries that power them, is much greater than anyone would have suspected a few years ago, and is increasing every day. Currently, the recycle industry mostly separates the plastics and smelts the rest, leaving for later the purification of each metal. It would be much more efficient to design a processing train which includes selective leaching of each metal, followed by its recovery in different stages (Alonso et al., 2014).

Research on selective dissolution and recovery of metals, such as copper, nickel, cobalt, manganese, zinc, gold, silver, and REE, from different types of e-waste and batteries, represents an important topic that could represent a large step toward sustainability. The REEs are of special importance, since China currently produces 99% of the worldwide demand and the possibility of recycling them opens an enormous international market.

PERSPECTIVES

It is important to point out that an enormous reserve of metals is available in those minerals classified as complex or inadequate for traditional processing methods. Although hydrometallurgical techniques have existed for many years, these have not been commercially employed for these materials because they are not selective. However, alternative methods are being investigated that are applicable to such ores. Relatively novel approaches have been tested to selectively extract the distinct mineralogical phases, using the study of their chemical nature and the election of specific media to dissolve each of the values contained within a specific ore at relatively moderate conditions. This practice would facilitate the subsequent processing of residues and solutions by traditional or innovative routes. An essential part of this methodology is the inclusion of recovery steps and reuse of all solutions within the process through recirculation.

Despite the advances achieved, there still is a need to systemize and integrate the knowledge, applying it to treat multimetallic, multiphase materials. With this generalized comprehension, it would then be possible to design environmentally friendly process schemes to treat ores and wastes that are presently not being exploited and in some cases, are considered toxic residues. Finally, to undertake these challenges, thermodynamic considerations and mathematical modeling techniques must be employed to understand, predict, and design efficient and sustainable processes.

REFERENCES

Aazami, M., G. T. Lapidus, and A. Azadeh. 2014. The effect of solution parameters on the thiosulfate leaching of Zarshouran refractory gold ore. *International Journal of Mineral Processing*, 131: 43–50.

Alonso, A. R., E. A. Pérez, G. T. Lapidus, and R. M. Luna-Sánchez. 2014. Hydrometallurgical process for rare earth elements recovery from spent Ni-HM batteries. In *Proceedings of Hydrometallurgy 2014*, Vol. I, edited by E. Asselin, D. G. Dixon, F. M. Doyle, D. B. Dreisinger, M. I. Jeffrey, M. S. Moats. Vancouver Island, British Columbia, Canada: Canadian Institute of Mining, Metallurgy and Petroleum (MetSoc), ISBN 978-1-926872-22-3, pp. 277–289.

Bafghi, M. Sh., A. Zakeri, Z. Ghasemi, and M. Adeli. 2008. Reductive dissolution of manganese ore in sulfuric acid in the presence of iron metal. *Hydrometallurgy*, 90(2–4): 207–212.

Biswal, A., B. Nayak, B. Dash, K. Sanjay, T. Subbaiah, and B. K. Mishra. 2011. Preparation of electrolytic manganese dioxide (EMD) from low grade manganese ores/residues. In *Proceedings of the European Metallurgical Conference*, 2011, Dusseldorf, Germany, pp. 1323–1333.

Biswal, A., B. C. Tripathy, T. Subbaiah, D. Meyrick, and M. Minakshi. 2015. Dual effect of anionic surfactants in the electrodeposited MnO_2 trafficking redox ions for energy storage. *Journal of the Electrochemical Society*, 162(1): A30–A38.

Cruz-Díaz, M. R., Y. Arauz-Torres, F. Caballero, G. T. Lapidus, and I. González. 2015. Recovery of MnO_2 from a spent alkaline battery leach solution via ozone treatment. *Journal of Power Sources*, 274: 839–845.

Das, S. C., R. N. Sahoo, and P. K. Rao. 1982. Extraction of manganese from low-grade manganese ores by ferrous sulfate leaching. *Hydrometallurgy*, 8(1): 35–47.

Elsherief, A. E. 2000. A study of the electroleaching of manganese ore. *Hydrometallurgy*, 55(3): 311–326.

Free, M. L. 2013. *Hydrometallurgy: Fundamentals and Applications*, Chapter 1. Hoboken, New Jersey, USA: John Wiley & Sons, ISBN: 1118732588.

Furlani, G., F. Pagnanelli, and L. Toro. 2006. Reductive acid leaching of manganese dioxide with glucose: Identification of oxidation derivatives of glucose. *Hydrometallurgy*, 81(3–4): 234–240.

Gasparrini, C. 1984. The mineralogy of silver and its significance in metal extraction. *CIM Bulletin*, 77(866): 99–110.

Groenewald, T. 1976. The dissolution of gold in acidic solutions of thiourea. *Hydrometallurgy*, 1: 277–290.

Gupta, P. C. 1963. Analytical chemistry of thiocarbamides, I. Quantitative determination of thiourea. *Zeitschrift fur Analytische Chemie*, 196: 412–431.

Gutiérrez-Muñoz, M., V. H. Brito-Ramos, E. Pérez-Anacleto, I. Pérez-Cruz, and G. T. Lapidus-Lavine. 2012. Reductive leaching of manganese ores. In *Recent Developments in Metallurgy, Materials and Environment*, edited by M. I. Pech-Canul, A. L. Leal-Cruz, J. C. Rendón-Angeles, C. A. Gutiérrez-Chavarría, J. López-Cuevas, J. L. Rodríguez-Galicia. Saltillo, Coahuila, México: Editorial Cinvestav IPN, pp. 13–22, 341pp. ISBN 978-607-9023-18-8.

Kim, E.-Y. and K. Osseo-Asare. 2012. Aqueous stability of thorium and rare earth metals in monazite hydrometallurgy: E_h–pH diagrams for the systems Th–, Ce–, La–, Nd– (PO_4)–(SO_4)–H_2O at 25°C. *Hydrometallurgy*, 113–114: 67–78.

Kumari, A. and K. A. Natarajan. 2002. Cathodic reductive dissolution and surface adsorption behavior of ocean manganese nodules. *Hydrometallurgy*, 64(3): 247–255.

Lapidus, G. T. and F. M. Doyle. 2008. Process for recovery of metal-containing values from minerals and ores. Application PCT/US08/54661 (WO/2008/103873), 22 de Febrero 2008. Mexican Patente MX 323976, granted 10 September 2014.

Lapidus, G.T., I. González, J.L. Nava, R. Benavides, and C. Lara-Valenzuela. 2008, Integrated process for precious metal extraction and recovery based on electro-oxidized thiourea. In *Hydrometallurgy 2008*, edited by C.A. Young, P.R. Taylor, C.G. Anderson and Y. Choi. Littleton, Colorado: Society for Mining, Metallurgy and Exploration, Inc. (SME), pp. 837–842. ISBN: 978-0-87335-266-6.

Lapidus-Lavine, G. T., A. R. Alonso-Gómez, J. A. Cervantes-Escamilla, P. Mendoza-Muñóz, and M. F. Ortiz-García. 2001. Mejora al Proceso de Lixiviación de Plata de Soluciones de Tiosulfato de Cobre (Improvement on the Silver Leaching Process with Copper Thiosulfate Solutions.). Mexican Patent MX 257,151, granted 26 February 2008 (in Spanish).

Lapidus-Lavine, G. T., M. C. López-Escutia, and M. T. Oropeza-Guzman. 2007a. Mejora al Proceso de Lixiviación de Plata y Oro con Soluciones de Tiourea (Improvement on the Silver and Gold Leaching Process with Thiourea Solutions.). Mexican patent MX 250,894 (in Spanish).

Lapidus-Lavine, G. T., M. C. López-Escutia, M. T. Oropeza-Guzman, I. González-Martínez, F. Rodríguez-Hernández, M. E. Poisot-Diaz, and I. Girón-Bautista. 2007b. Proceso de Lixiviación y Recuperación de Plata y Oro con Soluciones de Tiourea Electro-oxidada (Process for Leaching and Recovery of Silver and Gold with Electro-oxidized Thiourea Solutions.). Mexican patent MX 251,047 (in Spanish).

Li, J. and J. D. Miller. 2006. A review of gold leaching in acid thiourea solutions. *Mineral Processing and Extractive Metallurgy Review*, 27(3): 177–214.

Luna, R. M. and G. T. Lapidus. 2000. Cyanidation kinetics of silver sulfide. *Hydrometallurgy*, 56: 171–188.

Maccagni, M. G., J. H. Nielson, W. L. Lane, and D. M. Olkkonen. 2012, Process for the recovery of lead from lead-bearing materials. US patent application US 13/444,706, 11 April 2012.

Mesa-Espitia, S. L. and G. T. Lapidus. 2014. Improved gold extraction from a refractory arsenopyritic ore by chemical pretreatments. In *Proceedings of Hydrometallurgy 2014*, Vol. I, edited by E. Asselin, D. G. Dixon, F. M. Doyle, D. B. Dreisinger, M. I. Jeffrey, M. S. Moats. Vancouver Island, British Columbia, Canada: Canadian Institute of Mining, Metallurgy and Petroleum (MetSoc), pp. 573-580. ISBN 978-1-926872-22-3.

Mukherjee, A., A. M. Raichur, K. A. Natarajan, and J. M. Modak. 2004. Recent developments in processing ocean manganese nodules—A critical review. *Mineral Processing and Extractive Metallurgy Review*, 25(2): 91–127.

NIST. 2004. National Institute of Science and Technology, NIST critically selected stability constants of metal complexes. Database 46, version 8, compiled by R. M. Smith, A. E. Martell and R. J. Motekaitis, May 2004.

Panda, R., A. Kumari, M. K. Jha, J. Hait, V. Kumar, J. R. Kumar, and J. Y. Lee. 2014. Leaching of rare earth metals (REMs) from Korean monazite concentrate. *Journal of Industrial and Engineering Chemistry*, 20: 2035–2042.

Tian, X., X. Wen, C. Yang, Y. Liang, Z. Pi, and Y. Wang. 2010. Reductive leaching of manganese from low-grade manganese dioxide ores using corncob as reductant in sulfuric acid solution. *Hydrometallurgy*, 100(3–4): 157–160.

Wills, B. A. and T. J. Napier-Munn. 2011. *Wills' Mineral Processing Technology: An Introduction to the Practical Aspects of Ore Treatment and Mineral Recovery*. 7th ed, Oxford, UK: Butterworth-Heinemann. ISBN: 978-0-080-47947-7.

Wu, Z. H., D. B. Dreisinger, H. Urch, and S. Fassbender. 2014a. Fundamental study of lead recovery from cerussite concentrate with methanesulfonic acid (MSA). *Hydrometallurgy*, 142: 23–35.

Wu, Z. H., D. B. Dreisinger, H. Urch, and S. Fassbender. 2014b. The kinetics of leaching galena concentrates with ferric methanesulfonate solution. *Hydrometallurgy*, 142: 121–130.

Xia, C., W. T. Yen, and G. Deschenes. 2003, Improvement of thiosulfate stability in gold leaching. *Minerals and Metallurgical Processing*, 20(2): 68–72.

Zhang, W. and C. Y. Cheng. 2007. Manganese metallurgy review. Part I: Leaching of ores/secondary materials and recovery of electrolytic/chemical manganese dioxide. *Hydrometallurgy*, 89(3–4): 137–159.

10 Reductive Leaching of Metal Oxides in Hydrometallurgical Processing of Nickel Laterite Ores, Deep Sea Manganese Nodules, and Recycling of Spent Batteries/Catalysts

Gamini Senanayake, Alexander Senaputra,
L. C. S. Dharmasiri, Dong-Jin Kim, S. M. Shin,
Jeong-Soo Sohn, Kyung-Ho Park, and Jim Avraamides

CONTENTS

ABSTRACT

Reducing agents can be applied to leach metal values from natural and secondary resources such as laterite ores and deep sea manganese nodules and recycled materials such as spent batteries/ catalysts, which contain high-valent metal oxides of manganese, cobalt, and iron. Selective leaching of some metals can be achieved by selecting conditions to control the saturated solubility and dissolution rates of metals from oxides or mixed metal/oxide matrices. The leaching results are rationalized on the basis of chemical species, Eh–pH diagrams and heterogeneous kinetic models. The role of reducing agents, such as hydrogen peroxide, sulfur dioxide, and iron(II) sulfate, in acid, alkali, and in the presence of complexing ligands such as ammonia for the improvement of the leaching of metal oxides and metals is discussed. Previous studies on acid leaching of metals from nickel laterites, manganese nodules, and spent catalysts/batteries in the absence or presence of reducing agents show that the rate controlling step involves the diffusion of protons through a thickening product layer on the particle being leached. It is shown that the leaching of nickel and cobalt from manganese nodules and zinc–carbon batteries in ammoniacal sulfur dioxide solutions also obey shrinking core kinetics.

INTRODUCTION

Natural mineral resources of base metals such as laterite ores and deep sea manganese nodules and the waste materials such as spent catalysts, zinc carbon batteries (ZCBs), and lithium-ion batteries (LIBs) suitable for urban mining contain multivalent oxides as well as metals (Table 10.1). The first step of hydrometallurgical treatment involves acid or alkaline leaching. The efficiency and selectivity of leaching of metals from high-valent oxides depends on several factors: (i) mineralogy or composition of oxides and valency of metals, (ii) reactivity of the leaching agent toward the minerals in ores, or oxides and metals in catalysts, (iii) saturated solubility of oxides/hydroxides in the selected medium, complex formation of metal ions with ligands and the redox potentials in the case of multivalent oxides, (iv) relative rates of leaching of metals/oxides in the selected medium, and (v) other physicochemical factors such as temperature, particle size, and solid/liquid ratio. The presence of oxides of Fe(III), Mn(III), or Mn(IV) and Co(III) listed in Table 10.1 makes it essential to have an effective reducing agent to facilitate the leaching reaction, while the metals such as Ni and Zn need oxidants.

Table 10.2 lists selected reactions and equilibrium constants for the dissociation of water as well as the hydrolysis of CO_2, SO_2, and NH_3 to highlight the effect of temperature (HSC 6.1 database, Roine, 2012). Nickel and zinc metals can be leached in acid with H^+/H_2 couple as the oxidant whilst zinc metal is also leached in strong alkali with OH^-/H_2 couple as the oxidant. Oxides of zinc and nickel can be leached in acid or alkaline media in the presence of ligands which are capable of forming stable metal ion–ligand complexes with these cations. The equilibrium constants for nickel and zinc leaching reactions are generally lower in alkaline media than in acid media (Table 10.2). When nickel oxide is associated with the oxide phases of high-valent iron or manganese the reductive leaching becomes essential in order to leach both metals. The equilibrium constants for the reductive leaching of high-valent oxides are relatively large (Table 10.2) and reductive leaching is also generally rapid (Avraamides et al., 2006; Senanayake et al., 2011, 2015). The leaching of manganese, cobalt, and iron is facilitated by the reducing agents such as Fe(II), SO_2, and H_2O_2 while Mn(II) can also act as a reducing agent for Co(III) oxides as shown by the favorable equilibrium constants listed in Table 10.2. Large differences in solubility and dissolution kinetics of metal ions can lead to selective leaching of some metals. For example, despite the thermodynamic favorability of leaching both Ni and Al_2O_3 from nickel–alumina catalysts in acids (Table 10.2), the slow leaching kinetics of alumina allows the selective leaching of nickel with very low aluminum leaching (Parhi et al., 2013). The equilibrium constants for iron dissolution from Fe(III) oxides to produce Fe^{3+} are low even in acid solutions, but the complexation with sulfate ions makes it more favorable in acid media (Table 10.2). The selectivity of leaching zinc or nickel over iron is higher in alkaline media than in acid media in the presence of strong complexing agents for nickel and zinc. Unlike in the case of nickel and zinc, the leaching of molybdenum from MoO_2 in alkaline media has a larger equilibrium constant than in acid media (Table 10.2). However, the leachability of molybdenum will also depend on other cations as evident from the dependence of the solubility of various Mo(VI) compounds in

TABLE 10.1
Types of Oxides in Feed Material of Interest

Material	Valuable Metals/Oxides	Material	Valuable Metals/Oxides
Spent LIBs	CoOOLi	Nickel laterites	MO, MOOH, M_2O_3, MO_2, M_3O_4 (M = Ni,Co, Mn, Fe)
Spent ZCBs	Zn/ZnO, Fe_2O_3, MnO_2, Mn_3O_4	Spent catalyst 1	NiO–V_2O_5–MoO_3–Al_2O_3
Manganese nodules	MnO, Mn_3O_4, MnO_2, Ni–Co–oxides	Spent catalyst 2	Ni–Al_2O_3

TABLE 10.2
Equilibrium Constants Relevant for Oxide Leaching

Type	Reaction	Log K (25 °C)	Log K (95 °C)
	Acid–Base Equilibria of Reagents		
1	$H_2O = OH^- + H^+$	−14.0	−12.3
2	$NH_4^+ = NH_3(a) + H^+$	−9.24	−7.48
3	$SO_2(a) + H_2O = HSO_3^- + H^+$	−1.76	−2.79
4	$HSO_3^- = SO_3^{2-} + H^+$	−7.19	−7.77
5	$CO_2(a) + H_2O = HCO_3^- + H^+$	−6.47	−6.53
6	$HCO_3^- = CO_3^{2-} + H^+$	−10.3	−9.99
7	$NH_3(a) + CO_2(a) + H_2O = NH_4^+ + HCO_3^-$	2.76	0.95
8	$NH_3(a) + HSO_3^- = NH_4^+ + SO_3^{2-}$	2.04	−0.29
9	$NH_3(a) + SO_2(a) + H_2O = NH_4^+ + HSO_3^-$	7.47	4.69
10	$2NH_3(a) + S_2O_5^{2-} + H_2O = 2NH_4^+ + 2SO_3^{2-}$	8.85	4.43
	Metals Dissolution with Acids or Bases		
11	$Ni + HCO_3^- + H^+ = NiCO_3 + H_2(g)$	8.71	8.13
12	$Ni + 2H^+ = Ni^{2+} + H_2(g)$	8.01	6.16
13	$Ni + 2SO_2(a) + 2H_2O = Ni^{2+} + 2HSO_3^- + H_2(g)$	4.47	0.58
14	$Ni + CO_2(a) + H_2O = NiCO_3 + H_2(g)$	2.24	1.16
15	$Ni + 2NH_4^+ + 4NH_3(a) = Ni(NH_3)_6^{2+} + H_2(g)$	−1.68	−3.49
16	$Ni + 2HCO_3^- + 6NH_3(a) = Ni(NH_3)_6^{2+} + 2CO_3^{2-} + H_2(g)$	−3.81	−8.51
17	$Ni + 2H_2O + 6NH_3(a) = Ni(NH_3)_6^{2+} + 2OH^- + H_2(g)$	−11.1	13.2
18	$Zn + 2H^+ = Zn^{2+} + H_2(g)$	25.8	20.7
19	$Zn + HCO_3^- + H^+ = ZnCO_3 + H_2(g)$	25.4	21.4
20	$Zn + SO_2(a) + H_2O = Zn^{2+} + SO_3^{2-} + H_2(g)$	16.8	10.1
21	$Zn + 2NH_4^+ + 2NH_3(a) = Zn(NH_3)_4^{2+} + H_2(g)$	15.7	11.9
22	$Zn + 2OH^- + 2H_2O = Zn(OH)_4^{2-} + H_2(g)$	15.3	11.9
	Oxide Dissolution with Acids, Alkali and/or Complexing Agents		
23	$NiO + 2H^+ = Ni^{2+} + H_2O$	12.5	9.14
24	$NiO + 2NH_4^+ + 4NH_3(a) = Ni(NH_3)_6^{2+} + H_2O$	2.79	−0.52
25	$ZnO + 2H^+ = Zn^{2+} + H_2O$	11.6	8.59
26	$ZnO + 2NH_4^+ + 2NH_3(a) = Zn(NH_3)_4^{2+} + H_2O$	1.52	−0.14
27	$ZnO + 2OH^- + H_2O = Zn(OH)_4^{2-}$	1.14	−0.18
28	$MoO_3 + 2H^+ = MoO_2^{2+} + H_2O$	−3.35	−3.20
29	$MoO_3 + 2OH^- = MoO_4^{2-} + H_2O$	16.3	13.7

(Continued)

TABLE 10.2 (*Continued*)

Equilibrium Constants Relevant for Oxide Leaching

Type	Reaction	Log K (25 °C)	Log K (95 °C)
30	$MoO_3 + OH^- = HMoO_4^-$	7.68	7.06
31	$V_2O_5 + H_2O + 2OH^- = 2H_2VO_4^-$	12.3	10.9
32	$V_2O_5 + 4OH^- = 2HVO_4^{2-} + H_2O$	24.1	20.7
33	$V_2O_5 + 2H^+ = 2VO_2^+ + H_2O$	−1.55	−2.85
34	$V_2O_5 + 2HSO_4^- = 2VO_2SO_4^- + H_2O$		
35	$Al_2O_3 + 6H^+ = 2Al^{3+} + 3H_2O$	19.8	11.1
36	$Al_2O_3 + 2OH^- + 3H_2O = 2Al(OH)_4^-$	0.57	0.33
37	$Fe_2O_3 + 6H^+ + 2SO_4^{2-} = 2FeSO_4^+ + 3H_2O$	4.15	2.78
38	$Fe_2O_3 + 6H^+ = 2Fe^{3+} + 3H_2O$	0.27	−2.09
39	$FeOOH + 3H^+ = Fe^{3+} + 2H_2O$	0.17	−1.82
40	$FeOOH + 3H^+ + SO_4^{2-} = FeSO_4^+ + 2H_2O$	2.10	1.62
41	$Fe_3O_4 + 2H^+ = Fe^{2+} + Fe_2O_3 + H_2O$	10.6	7.65
42	$Fe_3O_4 + 2H^+ = Fe^{2+} + 2FeOOH$	10.6	7.19
	High-Valent Oxide Dissolution with Acid and Reducing Agents		
43	$MnO_2 + SO_2(a) = Mn^{2+} + SO_4^{2-}$	36.6	26.3
44	$MnO_2 + 4H^+ + 2Fe^{2+} = Mn^{2+} + 2Fe^{3+} + 2H_2O$	15.8	9.62
45	$MnO_2 + 4H^+ + 2Cl^- = Mn^{2+} + Cl_2(a) + 2H_2O$	−5.31	−3.64
46	$2MnO_2 + 4H^+ = 2Mn^{2+} + O_2(a) + 2H_2O$	−2.17	−1.23
47	$Mn_2O_3 + 6H^+ + 2Cl^- = 2Mn^{2+} + Cl_2(a) + 3H_2O$	3.67	3.10
48	$2MnOOH + 6H^+ + 2Cl^- = 2Mn^{2+} + Cl_2(a) + 4H_2O$	0.84	1.37
49	$Mn_3O_4 + 8H^+ + 2Cl^- = 3Mn^{2+} + Cl_2(a) + 4H_2O$	15.4	12.1
50	$Co_3O_4 + SO_2(a) + 4H^+ = 3Co^{2+} + SO_4^{2-} + 2H_2O$	50.8	36.9
51	$Co_3O_4 + 4H_2O_2(a) = 3Co(OH)_2 + 2.5O_2(a) + H_2O$	40.0	32.4
52	$Co_3O_4 + 2Mn^{2+} + 2H^+ = 3Co^{2+} + 2MnOOH$	8.13	4.77
53	$Fe_3O_4 + SO_2(a) + 4H^+ = 3Fe^{2+} + SO_4^{2-} + 2H_2O$	31.6	20.2
54	$2FeOOH + SO_2(a) + 2H^+ = 2Fe^{2+} + SO_4^{2-} + 2H_2O$	21.0	13.0
55	$Fe_2O_3 + SO_2(a) + 2H^+ = 2Fe^{2+} + SO_4^{2-} + H_2O$	20.9	12.6
56	$2HgO + SO_2(a) = Hg_2SO_4$	36.4	28.2
57	$2HgO + SO_2(a) + H_2O = Hg_2O + 2H^+ + SO_4^{2-}$	25.2	17.5
58	$V_2O_5 + SO_2(a) + H_2O = V_2O_4 + 2H^+ + SO_4^{2-}$	18.5	11.9
59	$V_2O_5 + 2SO_2(a) + 2H_2O = V_2O_3 + 4H^+ + 2SO_4^{2-}$	23.2	12.8
60	$MoO_3 + SO_2(a) + H_2O = MoO_2 + 2H^+ + SO_4^{2-}$	12.5	6.97

TABLE 10.3
Solubility of Molybdenum Compounds

Solid	Solubility (g/L)	Solid	Solubility (g/L)
MoO_2	Insoluble[a]	$MgMoO_4$	83
$CaMoO_4$	0.02	$(NH_4)_2MoO_4$	191
MoO_3	1	$(NH_4)_2Mo_2O_7$	169
$ZnMoO_4$	2.1	Na_2MoO_4	330

[a] Fe(II), Fe(III), and Ni(II) molybdates are also insoluble.

TABLE 10.4
Stability Constants (log β_n) for Metal Ion Ammine Complexes

n	Zn(II)	Mn(II)	Fe(II)	Cu(I)	Cu(II)	Ni(II)	Hg(II)	Co(II)	Co(III)	Pb(II)
1	2.38	1.00	1.4	5.93	4.16	2.80	8.80	2.11	7.30	1.60
2	4.88	1.54	2.51	10.9	7.47	5.04	17.5	3.74	14.0	–
3	7.43	1.70	2.68	–	10.8	6.77	18.5	4.79	20.1	–
4	9.65	1.30	3.26	–	13.1	7.96	19.3	5.55	25.7	–
5	–	–	3.25	–	–	8.71	–	5.73	30.7	–
6	–	–	–	–	–	8.74	–	5.11	35.2	–

Note: β_n defined as equilibrium constants for $M^{z+} + nNH_3 = M(NH_3)_n^{z+}$, data at 25 °C and ionic strength 0–2 from Senanayake et al. (2010) and references therein.

water listed in Table 10.3. Ammonia forms stable complex species of a range of metal ions relevant to the leaching of different types of metals/oxides discussed in this study (Table 10.4). The use of ammonia/ammonium/carbonate media for leaching offers the advantage of recycling NH_3 and CO_2 as in the Caron process for roasted nickel laterite (Senanayake et al., 2010). The use of aqueous SO_2 in reductive leaching offers sulfate anion to stabilize the metal ions in solution (Table 10.2). The use of NH_3/SO_2 and NH_3/CO_2 can provide a range of concentrations of the anions, cations, ligands, and pH at different temperatures, due to the temperature dependence of the equilibrium constants for these reactions, which can be useful for selective leaching or precipitation. The purpose of this paper is to review some of the previous work for LIBs and ZCBs and present preliminary results for manganese nodules, laterite ores, and spent catalysts.

EXPERIMENTAL

The samples of manganese nodules and spent ZCBs, LIBs, and catalysts used in this study and previous studies were supplied by the Korea Institute of Geoscience and Mineral Resources. The spent ZCBs and LIBs were subjected to physical treatment which involved discharging, dehydration, drying, and crushing steps (Kang et al., 2010a,b). After homogenizing and screening, the crushed ZCBs and LIBs were separated to powders of three different size fractions: +8 mesh, −8/ + 16 mesh, and −16 mesh. The fraction of −16 mesh (−1.19 mm) was selected for leaching. The laterite ores were received from a mining company in Western Australia. The leaching studies were conducted in glass reactors described previously (Avraamides et al., 2006; Shin et al., 2009). Preliminary fundamental leaching studies were conducted at low solid/liquid ratio of 1–2.5 g/L, as the main purpose of this work was to compare the effect of different reagents on the leaching behavior of oxides or metals. In some cases, higher solid/liquid ratios were used. In the case of laterite and manganese

TABLE 10.5

Elemental Assays of Feed Material for Leaching

Element %	Manganese Nodules	Nickel Laterite 1	Nickel Laterite 2	Spent LIBs	Spent ZCBs	Spent Catalyst 1	Spent Catalyst 2
Mn	21.8	0.22	0.33	0.11	22.7	–	0.039
Zn	0.10	–	–	–	20.8	0.1	0.21
Fe	6.86	12.4	22.5	1.80	2.65	0.3	0.68
Ni	0.96	1.80	1.27	1.40	0.03	2.0	12.75
Cu	0.79	–	–	12.2	0.03	–	–
Co	0.22	0.03	0.06	23.3	0.01	<0.03	0.012
Mo	–	–	–	–	–	1.4	–
V	–	–	–	–	–	9.0	–
Li	–	–	–	2.70	–	–	–
Al	2.80	0.60	–	–	0.28	19.5	39.2
S	–	–	–	–	–	5.0	–
Si	0.36			–	–	–	0.1
Size (μm)	90–105	90–105	90–105				

nodules, the particle size range was 90–125 μm and the mineralogy was examined using x-ray diffraction (XRD). Samples of starting material were digested and assayed for elemental composition using atomic absorption spectroscopy (AAS) or inductively coupled plasma (ICP) spectroscopy. A weighed solid sample was added to a reactor which contained 800 mL of solution of the concentrations of required reagents and preheated to the required temperature and stirred at 1000 rpm. In some cases, gases such as SO_2 and CO_2 were delivered to the reactor at the required flow rates. Samples of slurry were withdrawn at regular intervals and filtered using sintered glass crucibles. The filtered liquor was analyzed using AAS or ICP. Table 10.5 lists the elemental assays of the solids which indicate that the major components are (i) Mn, Zn, and Fe in spent ZCBs, (ii) Co, Li, and Cu in spent LIBs, (iii) Al, V, Ni, and Mo in spent catalyst 1, (iv) Ni and Al in spent catalyst 2, (v) Mn, Fe, Al, Ni, Cu, and Co in manganese nodules, and (vi) Fe, Ni, Al, and Mn in nickel laterites. Table 10.6 summarizes the mineralogy of laterite and manganese nodules.

TABLE 10.6

Mineralogical Composition of Laterite and Manganese Nodules

Material	Chemical Formula	Mineral Name	JCPDS No.
Nickel laterite	SiO_2	Quartz	00-005-0490
	$(Ni,Mg)_3Si_4O_{10}(OH)_2$	Willemseit	00-022-0711
	$Mg_3Si_2O_5(OH)_4$	Chrysotile-2Mcl	00-043-0662
	$3MgO \cdot 2SiO_2 \cdot 2H_2O$	Antigorite	00-002-0092
	$Fe_2O_3 \cdot H_2O$	Goethite	00-008-0097
	$Fe_{21.16}O_{31.92}$	Maghemite	01-089-5892
	$Mg_3Si_2O_5(OH)_4$	Lizardite-1T	00-010-0382
Manganese nodules	MnO_2	Pyrolusite	00-004-0779
	KMn_8O_{16}	Cryptomelane-M	00-004-0778
	$NiMn_2O_3(OH)_4 \, H_2O$	Asbolane	00-042-1319
	$(Na_{0.7}Ca_{0.3})Mn_7O_{14} \cdot 2.8H_2O$	Birnessite	00-013-0105
	$Na_{1.88}Al_2Si_{2.22}O_{9.48}$		00-048-0730

RESULTS AND DISCUSSION

SPENT BATTERIES

Table 10.7 lists the leaching results for cobalt from LIBs to show the importance of a reducing agent such as hydrogen peroxide to enhance the cobalt leaching. Acid alone can leach only 62% cobalt in 2 M H_2SO_4, but 6% H_2O_2 increases the leaching efficiency to 98% at 60 °C. Flowsheets have been developed for the reductive leaching of spent LIBs with hydrogen peroxide and to recover cobalt as $CoSO_4$ (Kang et al., 2010a) or Co_3O_4 (Kang et al., 2010b). Further work is in progress to selectively extract, separate, and recover lithium and other metals (Joo et al., 2016).

In general, the increase in temperature can facilitate kinetics of chemically controlled reactions, but the exothermic reactions are thermodynamically less favored at high temperatures. For example, the solubility of ZnO in NaOH solutions increases with increasing alkali concentrations as expected from the stoichiometry of the dissolution reaction in Table 10.2. However, the decrease in equilibrium constant for ZnO dissolution in alkali at higher temperatures causes a decrease in predicted solubility as shown in Figure 10.1a. The beneficial effect of high alkali on leaching zinc from ZCBs is evident from the leaching results (Shin et al., 2009; Figure 10.1b). Results in Table 10.8 and Figure 10.2a show that the leaching of zinc (major), nickel, and cobalt (minor) in spent ZCBs do not need a reducing agent to achieve 100% leaching efficiency. Sulfuric acid alone is capable of achieving 100% leaching efficiency of these three metals, as expected from the Eh–pH diagrams in Figures 10.1c and d. However, a reducing agent is essential to produce Mn(II) ions from high-valent Mn oxides as evident from the Eh–pH diagram in Figures 10.1d and 10.1e in order to achieve 100% leaching efficiency of manganese. Figure 10.1d shows that the Eh of Fe(III)/Fe(II) is lower than that of MnO_2/Mn^{2+}, $MnOOH/Mn^{2+}$, and Mn_3O_4/Mn^{2+}. Therefore, direct reaction with SO_2 or Fe(II) produced from the reduction of iron(III) oxides is responsible for enhanced leaching of manganese, as shown by the reactions in Table 10.2 and Figure 10.2a. Moreover, SO_2 has a detrimental effect on the leaching efficiency of mercury (Figure 10.2b and Table 10.8) which may be a result of reductive precipitation of Hg_2SO_4 or Hg_2O as evident from the large equilibrium constants in Table 10.2 (Senanayake et al., 2010).

Even in acid solutions under reducing conditions in the presence of SO_2, only 25% of iron is leached from spent ZCBs (Table 10.8). Figure 10.2c shows the solubility of Fe(II) and Fe(III) as a function of pH adjusted using NaOH in solutions which contained different concentrations of Na_2SO_4 (0–1 M). The precipitation of Fe(II) is unaffected by the different concentrations of Na_2SO_4 in solution. However, the pH for the precipitation of Fe(III) increases with the increase in concentration of Na_2SO_4 indicating the stabilization of Fe(III) at high sulfate content and warrants further studies. Hydrogen peroxide can also be used as a reducing agent and the leach liquor can be further processed to produce Mn–Zn–ferrite nanoparticles (Kim et al., 2009).

The effect of temperature on the speciation in NH_3/SO_2 solutions is shown in Figure 10.2d. The Eh–pH diagrams in Figure 10.1c and e indicate that the dissolution of zinc and manganese can be facilitated in alkaline ammonia media by producing $Zn(NH_3)_4^{2+}$ and $Mn(NH_3)_3^{2+}$, respectively, the latter under reducing conditions. However, the solubility limitations also need to be considered as

TABLE 10.7

Effect of Acid and Reducing Agent on Cobalt Leaching from Spent LIBs

H_2SO_4 (mol/L)	H_2O_2 (%)	S/L (g/L)	T (°C)	% Co Leached (1 h)
1	0	50	75	40
2	0	50	80	62
2	15	50	75	83
2	20	50	75	100
2	6	100	60	98

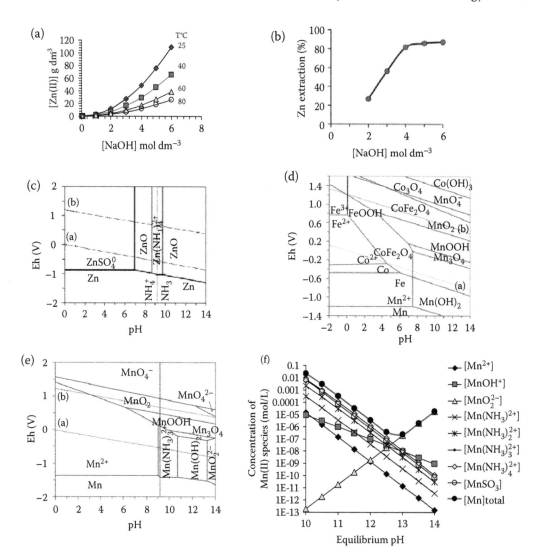

FIGURE 10.1 (a) Predicted solubility of Zn(II) in alkali based on HSC 6.1 database, (b) effect of NaOH on leaching efficiency of zinc from spent ZCBs at 80 °C, (c)–(e) predicted Eh–pH diagrams at 25 °C, and (f) predicted solubility of Mn(II) in ammoniacal solutions at 25 °C.

TABLE 10.8

Effect of Acid, Ammonia, and Reducing Agent on Metal Leaching from Zinc–Carbon Batteries

System	Leaching Efficiency of Metals (%)							
	Zn	Mn	Fe	Co	Hg	Ni	Cu	Pb
1 M H$_2$SO$_4$/30 °C	100	35	<1	44	29	100	100	<0.1
SO$_2$/1 M H$_2$SO$_4$/30 °C	100	100	25	89	0.7	100	100	<0.1
SO$_2$/2.5 M NH$_3$/60 °C	100	<1	<1	63	50	75	100	<0.1

Source: Adapted from Senanayake, G. et al. 2010. *Hydrometallurgy*, 105: 36–41.
Note: After 90 min leaching, SO$_2$ (50 mL/min).

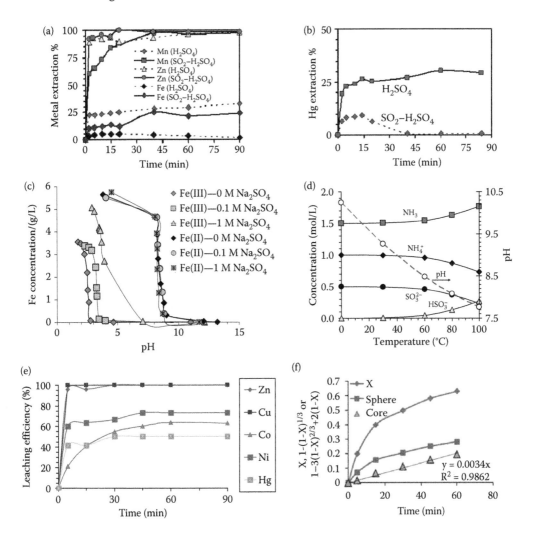

FIGURE 10.2 (a) Leaching efficiency of metals in zinc–carbon batteries waste in SO_2/H_2SO_4 (Table 10.8), (b) leaching efficiency of mercury (same conditions as, (a)), (c) pH dependence of solubility of Fe(III) and Fe(II) in the presence or absence of Na_2SO_4, (d) predicted effect of temperature on speciation and pH of 2.5 M NH_3 and 0.5 M SO_2 based on HSC 7.1 database, (e) leaching efficiency of metals in zinc–carbon batteries waste in SO_2/NH_3 (Table 10.8), and (f) applicability of shrinking core kinetic model for cobalt leaching (discussed in text).

shown in Figure 10.1f as the stability constants of Mn(II) complexes are relatively low compared to those of Zn(II) complexes (Table 10.4). Thus, the solubility of Mn(II) is largely controlled by hydroxide species and the total solubility is low and decreases with increasing pH (Figure 10.1f). As a result, the leaching efficiency of manganese in SO_2/NH_3 is <1% (Table 10.8). The reductive leaching of zinc, nickel, copper, and mercury in SO_2/NH_3 is very fast as shown in Figure 10.2e. The slow leaching of cobalt shown in Figure 10.2e can be modeled according to the shrinking core kinetics as shown in Figure 10.2f and discussed later.

MANGANESE NODULES AND LATERITE ORES

Successful leaching of nickel, cobalt, and manganese from nodules requires the presence of reducing agents such as SO_2 or $FeSO_4$, where 47%–94% Fe is also leached in acidic/SO_2 media, as shown

TABLE 10.9

Effect of Acid, Ammonia, and Reducing Agent on Metal Leaching from Manganese Nodules

Lixviant	Conditions					Leaching Efficiency			
	Additive	T (°C)	Time (h)	pH		Mn	Fe	Co	Ni
1 M H_2SO_4	None	60	0.25	<0		2	47	2	54
	SO_2	60				100	94	100	97
	$FeSO_4$	60				93	48	100	55
1 M NH_3	SO_2	42	3	9		4	3	70	50
1 M NH_4HCO_3	SO_2			9		2	1	71	63
1.5 M NH_4HCO_3	SO_2			9		1	2	91	80

Note: 25 g/L nodules, 200 mL/min SO_2.

in Table 10.9. This can be related to the coexistence of Mn and Ni in nodules, as shown by the linear relationship between the compositions of % Ni as a function of % Mn for manganese nodules of different origin in Figure 10.3f. Moreover, the Eh–pH diagram in Figure 10.1d shows a lower Eh for $MnOOH/Mn^{2+}$ and MnO_2/Mn^{2+} compared to the relevant redox couples of high-valent cobalt. This behavior indicates the possibility that Mn^{2+} can act as a reducing agent for high-valent cobalt oxides, which is also supported by the large equilibrium constant for the relevant reactions in Table 10.2. However, in ammoniacal SO_2 solutions, the leaching efficiency of manganese and iron is low (<4%, Table 10.9). The leaching efficiency of nickel and cobalt increases from 50% to 60% in ammonia solutions without carbonate to 80%–90% in the presence of carbonate, compared to only 1%–2% leaching of manganese and iron. These results warrant further studies on the ion-pair association between the cationic ammine complexes of nickel(II) and cobalt(II) with carbonate anions. It is also worth noting that the measured pH of 9 in SO_2/NH_3 solutions listed in Table 10.9 is close to the predicted pH at 40 °C in Figure 10.2d.

Table 10.10 compares the leaching of metals by the three acids HNO_3, H_2SO_4, and HCl. Lowest leaching of 2% Mn, 47% Fe, 2% Co, and 54% Ni after 20 min was observed with 1 M HNO_3, due to its oxidizing nature. When the acid was changed to 1 M HCl, these leaching efficiencies were increased to 39% Mn, 84% Fe, 50% Co, and 83% Ni which indicates the reducing role of Cl^- ions which react with the oxides such as Mn_2O_3 and Mn_3O_4 to produce Mn(II) (Table 10.2). The presence of Mn(III) oxides in manganese nodules is evident from XRD scans (Table 10.6). A similar trend is also shown for the laterite ore 1 in Table 10.10. However, as shown in Figure 10.3a and 10.3b, the beneficial effect of HCl compared to H_2SO_4 is not observed at a higher solid content of 25 g/L. Figures 10.3c, d, and e show the need for a reducing agent such as $FeSO_4$ to rapidly leach manganese, nickel, and cobalt from manganese nodules. Further tests carried out with laterite ore 2 with 0.5 M HCl at a low temperature of 40 °C and 10% solids (w/v) show very low leaching efficiency of 0.5%–6% for all metals (Table 10.10). The addition of increasing amounts of $FeCl_2$ (0.004–0.3 M) increased the leaching efficiency of manganese to 82% and cobalt to 75% due to the reductive leaching indicating the coexistence of these metals in the same mineral phase. However, iron and nickel leaching remains low due to their association with goethite and the slow leaching due to low temperature and low acid strength (Senanayake and Das, 2004).

SPENT CATALYSTS

The published Eh–pH diagrams show the pH dependence of vanadium(V) speciation at low concentrations (10^{-6} M) over the pH range 0–14: VO_2^+ (pH < 1) and $VO_2SO_4^-$ (1 < pH < 2), HVO_4^{2-} (12 < pH < 13) and VO_4^{3-} (pH > 13), but polymeric species are formed at higher concentrations (10^{-2} M) (Zhou et al., 2011). However, the equilibrium constants, Eh–pH/solubility diagrams predict

TABLE 10.10

Effect of Acid on Metal Leaching from Manganese Nodules or Laterite

Material		Conditions				Leaching Efficiency (%)					
Nodules[a]	Lixviant	Additives	T (°C)	Time (min)	pH[c]	Mn	Fe	Co	Ni	Mg	Al
	1 M HNO$_3$	None	90	20	<0	2	47	2	54	78	37
	1 M H$_2$SO$_4$				<0	3	70	9	70	100	51
	1 M HCl				<0	39	84	50	83	91	50
Laterite 1[a]	1 M HNO$_3$		95	20	<0	18	7	18	22	7	10
	1 M H$_2$SO$_4$			10		47	49	52	85	62	33
	1 M H$_2$SO$_4$			20		60	69	63	92	70	39
	1 M HCl			20		60	69	63	90	62	34
Laterite 2[b]	0.5 M HCl	None	40	120	0.3–0.8	3	0.5	6	2	–	–
		0.004 M FeCl$_2$			0.3–0.7	27	0.3	30	5		18
		0.01 M FeCl$_2$			0.3–1.1	61	0.3	64	13		
		0.3 M FeCl$_2$			0.3–1.1	82	0	75	11	8	–

[a] 2.5 g/L solids.

[b] 10% (w/v) solids.

[c] pH remained <0 or changed in the range indicated during leaching.

the low solubility of MoO$_3$ in acid solutions (Tables 10.2 and 10.3) and the existence of HMoO$_4^-$ and MoO$_4^{2-}$ ions in alkaline solutions (Park et al., 2010; Nam et al., 2011). This information on speciation is consistent with the leaching behavior of spent catalysts. In the case of spent catalyst 1 that consists of oxides of Ni–Mo–V–Al, the acid leaching with 6 M H$_2$SO$_4$ dissolves 96% vanadium (Table 10.11). The low leaching efficiency of 30% Mo is due to the unfavorable acid dissolution reaction in Table 10.2, which needs much stronger acid and drastic conditions. The slow kinetics of dissolution of Al$_2$O$_3$ in acid (Parhi et al., 2013) may also be responsible for incomplete leaching of metals from the oxide matrix in acid. Alkaline leaching of V–Mo–Al–oxides is more favorable as revealed by the larger equilibrium constants in Table 10.2. Thus, alkaline leaching with 2 M NaOH is more selective and leaches 100% V, Mo, and Al with 0% Ni (Table 10.11). Despite the comparable pH of 13, the addition of 1 M Na$_2$CO$_3$ is detrimental in alkaline leaching.

Table 10.11 shows that a solution of 2 M NH$_3$ at a low pH of 10.4 selectively leaches over 90% V and Mo, with minimal leaching of Ni and Al. The decrease in aluminum leaching from >74% at pH 13 to ≤11 at pH 10.4 is a result of the stabilization of Al(OH)$_4^-$ at high pH (Table 10.2). In acid solutions (pH 1–5), the reductive leaching of V$_2$O$_5$ is enhanced by the addition of Na$_2$SO$_3$ due to the high solubility of V(IV) compared to V(V) (Tavakoli et al., 2014). However, the addition of SO$_2$, Na$_2$SO$_3$, or Na$_2$S$_2$O$_5$ in ammoniacal media retards the leaching of vanadium and molybdenum and enhances the leaching of nickel (Table 10.11). The detrimental effect of reductive conditions in the present study may be due to the reduction of V$_2$O$_5$ to V$_2$O$_4$ or MoO$_3$ to MoO$_2$ as revealed by the published Eh–pH diagrams (Park et al., 2010; Zhou et al., 2011) and the favorable equilibrium constants for the formation of these oxides in Table 10.2, but further studies are warranted. In contrast 47%–68% nickel leaching in the presence of these reagents can be related to the reaction of NiO with NH$_4^+$/NH$_3$ and formation of Ni(NH$_3$)$_6^{2+}$ (Table 10.2). The presence of NH$_4$HCO$_3$ improves the Ni leaching further to above 80%. A solution of 1 M HCl alone can leach 90% Ni from Ni–Al$_2$O$_3$ catalyst as shown in Table 10.12. Aqueous SO$_2$ or NH$_3$ can also leach nickel due to the acidity of the solution. However, aqueous CO$_2$ leaches only 2% Ni due to the possibility of precipitation of NiCO$_3$. All reasons mentioned above are supported by the reactions of large equilibrium constants shown in Table 10.2.

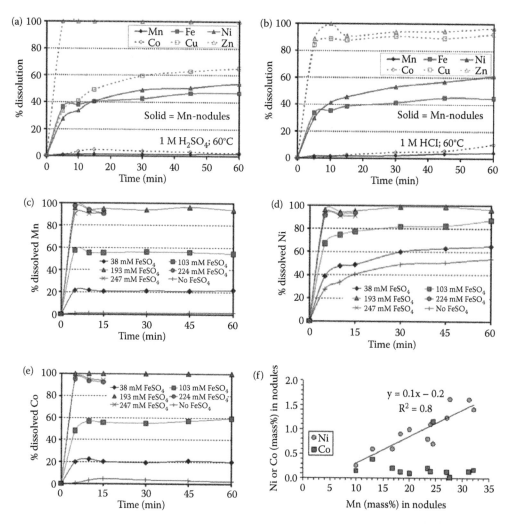

FIGURE 10.3 Leaching efficiency of metals from manganese nodules in 1 M H_2SO_4 (a) and 1 M HCl (b) at 60 °C, 25 g/L solids and the effect of $FeSO_4$ on manganese (c), nickel (d), and cobalt (e) leaching from manganese nodules in 1 M H_2SO_4 (other conditions: 60 °C, 25 g/L), correlation between Ni and Co content as a function of Mn content in nodules (f).

KINETIC MODELS

The acid leaching rate of nickel from laterite and manganese nodules is controlled by the diffusion of H^+ through a porous insoluble product layer (Senanayake and Das, 2004; Senanayake, 2011; Senanayake et al., 2011, 2015). This was indicated by the applicability of a shrinking core kinetic model for the leaching of nickel and iron from laterite ores in H_2SO_4 expressed by the equation: $1 - 3(1 - X)^{2/3} + 2(1 - X) = [6bDC/(\rho_M(1 - \varepsilon)r^2)] \cdot t = k_{app} \cdot t$ where X is the fraction of metal leached after time t, ρ_M is the molar density of the dissolving metal ion in the initial particle (mol/cm³), and ε is the particle porosity. The apparent rate constant k_{app} is directly proportional to the H^+ ion concentration (C) for acid leaching of both laterite and manganese nodules. The same kinetic model is applicable for the dissolution of nickel from Ni–Alumina catalyst in HCl (Parhi et al., 2013), and iron dissolution from spent ZCBs in H_2SO_4 (Avraamides et al., 2006). As a result, a logarithmic plot of the apparent rate constant (k_{app}) for the acid leaching of nickel and iron from laterite or manganese nodules as a function of acid concentration corresponds to a linear relationship

TABLE 10.11

Effect of Acid and Alkali on Metal Leaching from Ni–V–Mo–Al-Oxide Catalyst

| Lixviant | Additive | Other Conditions | | | Leaching Efficiency (%) | | | |
		T (°C)	Time (h)	pH	V	Mo	Al	Ni
6 M H$_2$SO$_4$	None	80	6	<0	96	30	80	67
2 M NaOH	None	80	3	13	100	100	100	0
	1 M Na$_2$CO$_3$	80	3	13	86	81	74	0
2 M NH$_3$	None	80	3	10.4	96	100	1	0
	1 M Na$_2$SO$_3$	80	3	10.4	29	21	11	47
	1 M Na$_2$S$_2$O$_5$			8.4	36	52	6	68
	SO$_2$			3.24	38	48	16	65
2 M NH$_3$	1 M NH$_4$HCO$_3$			10.4	71	48	2	93
	2 M NH$_4$HCO$_3$			10.4	77	76	0	83

TABLE 10.12

Effect of Acid and Alkali on Metal Leaching from Ni–Al$_2$O$_3$ Catalyst

Reagent	T °C	pH(initial)	pH(final)	Ni% (2.5 h)
50 mL/min CO$_2$	80	4.3	5.1	2
2.5 M NH$_3$	80	11.1	11.0	63
2.5 M NH$_4$HCO$_3$	80	8.5	8.7	76
50 mL/min SO$_2$	80	0.8	0.8	82
1 M HCl	80	<0	<0	95

of slope 1 (Senanayake, 2011; Senanayake et al., 2011). The dissolution of cobalt from spent ZCBs in SO$_2$/NH$_3$ solutions also follows a shrinking core kinetic model as shown in Figure 10.2f. Likewise, the dissolution of nickel and cobalt from manganese nodules in 1 M NH$_4$HCO$_3$ pre-saturated with SO$_2$ and maintained at a continued flow of 50 mL/min SO$_2$ at 42 °C and 25 g/L solids also obey this model as shown in Figure 10.4. Other kinetic models such as a shrinking sphere model based on a surface reaction controlled mechanism: $1 - (1 - X)^{1/3} = k_{app} \cdot t$, or film diffusion controlled mechanism: $1 - (1 - X)^{1/2} = k_{app} \cdot t$ (Levenspiel, 1972) are not applicable as they do not give linear relationships through origin in Figure 10.4. The slopes of linear relationships corresponding to a

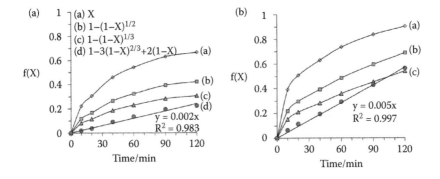

FIGURE 10.4 Shrinking core model for nickel (a) and cobalt (b) leaching from manganese nodules in 1 M NH$_4$HCO$_3$ pre-saturated with SO$_2$ and maintained at a continued flow of 50 mL/min SO$_2$ at 42 °C and 25 g/L solids.

shrinking core model in Figures 10.2f and 10.4 are of the same order 2×10^{-3} to 5×10^{-3} min^{-1}. This warrants further studies on the effect of concentrations of reagents on kinetics in order to shed more light on rate controlling step in alkaline leaching.

CONCLUSIONS

Nickel from Ni/Al_2O_3 catalysts can be selectively leached with HCl without leaching aluminum. In the absence of high-valent Mn/Co oxides, nickel oxide material can be leached in acid or NH_3/NH_4^+ without a reducing agent. Carbonate is beneficial for nickel leaching indicating complexation of the type $Ni(NH_3)_5CO_3$. When NiO exists with V–Mo–Al-oxides, NaOH selectively leaches 100% V, Mo, and Al. However, NH_3 selectively leaches only V and Mo (96%–100%). Ammoniacal NH_4HCO_3 leaches 70%–80% Ni, V, and Mo and 0% Al. Reducing agents such as H_2O_2, SO_2, or $FeSO_4$ are essential for leaching high-valent Co/Mn oxides and to improve the leaching of any associated nickel. Zinc leaching from Zn–Fe–Mn-oxides in ZCBs is selective with NaOH alone or SO_2/NH_3, but reducing agents such as SO_2 or H_2O_2 leaches all three oxides. Reducing agents (SO_2 or $FeSO_4$) are essential to acid leach Mn, Fe, Co, and Ni from manganese nodules but $NH_4HCO_3/$ SO_2 selectively leaches Ni and Co from manganese nodules.

ACKNOWLEDGMENT

The authors acknowledge the financial support from the Parker Centre for Hydrometallurgy funded by the Commonwealth Government of Australia and Leading Foreign Research Institute Recruitment Program through the National Research Foundation of Korea (NRF) funded by the Ministry of Science, ICT, and Future Planning (2013).

REFERENCES

Avraamides, J., G. Senanayake, and R. Clegg. 2006. Sulfur dioxide leaching of spent zinc–carbon-battery scrap. *Journal of Power Sources*, 159: 1488–1493.

Kang, J. G., G. Senanayake, J. S. Sohn, S. M. Shin. 2010a. Recovery of cobalt sulfate from spent lithium ion batteries by reductive leaching and solvent extraction with Cyanex 272. *Hydrometallurgy*, 100: 168–171.

Kang, J. G., J. S. Sohn, H. K. Chang, G. Senanayake, and S. M. Shin. 2010b. Preparation of cobalt oxide from concentrated cathode material of lithium ion batteries by hydrometallurgical methods. *Advanced Powder Technology*, 21: 175–179.

Kim, T. H., G. Senanayake, J. G. Kang, J. S. Sohn, K. I. Rhee, S. W. Lee, and S. M. Shin. 2009. Reductive acid leaching of spent zinc–carbon batteries and oxidative precipitation on Mn–Fe ferrite nanoparticles. *Hydrometallurgy*, 96: 154–158.

Joo, S. H., D. J. Shin, C. H. Oh, J. P. Wang, G. Senanayake, and S. M. Shin. 2016. Selective extraction and separation of nickel from cobalt, manganese and lithium in pre-treated leach liquors of ternary cathode material of spent lithium-ion batteries using synergism caused by Versatic 10 acid and LIX 84-I. *Hydrometallurgy*, 159: 65–74.

Levenspiel, O. 1972. *Chemical Reactions Engineering*. Wiley, New York.

Nam, Y. I., S. Y. Seo, Y. C. Kang, M. J. Kim, G. Senanayake, and T. Tran. 2011. Purification of molybdenum trioxide calcine by selective leaching of copper with HCl–NH_4Cl. *Hydrometallurgy*, 109: 9–17.

Park, Y. Y., T. Tran, Y. H. Lee, Y. I. Nam, G. Senanayake, and M. J. Kim. 2010. Selective removal of arsenic(V) from a molybdate plant liquor by precipitation of magnesium arsenate. *Hydrometallurgy*, 104: 290–297.

Parhi, P. K., K. H. Park, and G. Senanayake. 2013. A kinetic study on hydrochloric acid leaching of nickel from Ni–Al_2O_3 spent catalyst. *Journal of Industrial and Engineering Chemistry*, 19: 589–594.

Roine, A. 2012. *Outokumpu HSC Chemistry Thermochemical Database, ver 7.0.* Finland: Outokumpu Research Oy.

Senanayake, G. 2011. Acid leaching of metals from deep-sea manganese nodules – A critical review of fundamentals and applications. *Minerals Engineering*, 24: 1379–1396.

Senanayake, G., S. M. Shin, A. Senaputra, A. Winn, D. Pugaev, J. Avraamides, J. S. Sohn, and D. J. Kim. 2010. Comparative leaching of spent zinc–carbon batteries using sulfur dioxide in ammoniacal and sulfuric acid solutions. *Hydrometallurgy*, 105: 36–41.

Senanayake, G., and G. K. Das. 2004. A comparative study of leaching kinetics of limonitic laterite and synthetic iron oxides in sulphuric acid containing sulphur dioxide. *Hydrometallurgy*, 72: 59–72.

Senanayake, G., G. K. Das, A. Lange, J. de, Li, D. J. Robinson. 2015. Reductive atmospheric acid leaching of lateritic smectite/nontronite ores in $H_2SO_4/Cu(II)/SO_2$ solutions. *Hydrometallurgy*, 152: 44–54.

Senanayake, G., J. Childs, B. D. Akerstrom, and D. Pugaev. 2011. Reductive acid leaching of laterite and metal oxides—A review with new data for Fe(Ni,Co)OOH and a limonitic ore. *Hydrometallurgy*, 110: 13–32.

Senanayake, G., A. Senaputra, and M. J. Nicol. 2010. Effect of thiosulfate, sulfide, copper(II), cobalt(II)/(III) and iron oxides on the ammoniacal carbonate leaching of nickel and ferronickel in the Caron process. *Hydrometallurgy*, 105: 60–68.

Shin, S. M., G. Senanayake, J. S. Sohn, J. G. Kang, D. H. Yang, and T. H. Kim. 2009. Separation of zinc from spent zinc–carbon batteries by selective leaching with sodium hydroxide. *Hydrometallurgy*, 96: 349–353.

Tavakoli, M. R., S. Dornian, and D. B. Dreisinger. 2014. The leaching of vanadium pentoxide using sulphuric acid and sulfite as a reducing agent. *Hydrometallurgy*, 141: 59–66.

Zhou, X., C. Wei, M. Li, S. Qiu, and X. Li. 2011. Thermodynamics of vanadium–sulfur–water systems at 298 K. *Hydrometallurgy*, 106: 104–112.

11 Sustainable Recycling Technology for Electronic Waste

Sandip Chatterjee, Archana Kumari, and Manis Kumar Jha

CONTENTS

ABSTRACT

Electronic waste (e-waste) generation, accumulation and affordable technology to recycle e-waste are of global concern and India is facing similar challenges. The e-waste Rules, 2011 have been enacted on 12 May 2012 in India to address the issue. In India, about 95% of e-waste is, however, being recycled in the unorganised sector by crude means to primarily extract precious metals (gold, silver and copper) from printed circuit boards (PCBs), thereby harming the environment and affecting the health of the workers. This practice needs to be discouraged by providing affordable recycling technology. Authorised recyclers in India are broadly engaged in the segregation and disassembly processes. PCB recycling, however, is predominantly carried out abroad. Recovery of valuable resources from waste is essential to conserve the depletion of natural resources. E-wastes contain a number of precious metals and are alternative resources of metals, which are available in very limited quantities in nature.

The Department of Electronics and Information Technology (DeitY) has developed several affordable technologies to recycle e-waste. Processing technology was developed at the pilot scale level for the recycling of e-waste at the National Metallurgical Laboratory (NML), Jamshedpur, India. Recovery processes for precious metals from PCBs have been developed jointly by the Centre for Materials for Electronics Technology (C-MET), Hyderabad and E-Parisaraa, Bangalore. A masterbatch and value-added products were successfully developed from e-waste plastics at the Central

Institute of Plastics Engineering and Technology (CIPET), Bhubaneswar as part of technology transfer. This chapter aims to apprise the reader of DeitY's effort in developing cost-effective technology to recycle e-waste in an environmentally friendly manner.

INTRODUCTION

E-wastes generation and their accumulation is a major concern for modern society. E-waste is considered to be the fastest growing waste stream in the urbanised world. Affordable technology to recycle and achieve zero waste is a significant challenge. The ever-increasing demand for electronic products, rapidly changing technology and features in the devices and availability of improved products drive consumers to dispose of electronics products rapidly, generating e-waste at an alarming rate. The major source of e-waste in India is through the disposal of the hardware and electronic items from government offices, public and private sectors and academic and research institutes. Household consumers are also contributing a significant volume of end-of-life electronic products.

Environmental sensitivity, and the social and economic aspects associated with e-waste are the key challenges in India. The e-wastes (Management and Handling) Rules, 2011 have been enacted in India on 12 May 2012 to address the safe and environment-friendly handling, transportation, storage and recycling of e-waste, and also to reduce the use of hazardous substances during the manufacturing of electrical and electronic equipment (http://envfor.nic.in/downloads/rules-and-regulations/1035e_eng.pdf). This legislation has mandated the stakeholders including producers, consumers, transporters, recyclers and the general public to ensure that certain guidelines are met so that the environment can be protected. The rules are to be considered for amendment in 2015 to enact more stringent provisions for the stakeholders. The implementation however creates challenges for the legislators and regulators. Though the legislative measure is operational, major amounts of the e-waste are finding their way into the informal recycling sector from both manufacturers and bulk consumers. Significant amounts of materials including exported materials are being processed by this sector in a non-scientific manner.

E-waste is hazardous in nature due to the presence of toxic substances such as Pb, Cr^{6+}, Hg, Cd and flame retardants (polybrominated biphenyls and polybrominated diphenylethers). E-waste when mixed with solid municipal waste is therefore posing a major threat for environmental degradation. Informal operators are extracting precious metals especially gold, copper and silver through crude means such as dipping PCB in acidic/alkaline solutions and heating/burning of PCBs. These processes are very inefficient and extremely harmful to the workers and to the environment. This practice needs to be discouraged by providing affordable and safe technology.

The situation is not so grim in developed countries, as the laws are sufficiently stringent to address the appropriate stocking, processing, disposal and land-filling of e-waste. Moreover, the availability of appropriate technologies from earlier days in those countries had made e-waste recycling a profitable business. However, recycling is becoming economically non-viable in developed countries due to the increase in manpower and operational costs. These countries are, therefore, compelled to find out alternative destinations for disposal, where the labour cost is comparatively low and the environmental laws are not enforced so strictly.

Recovery of valuable resources from waste is essential to conserve the depletion of natural resources. E-waste contains a number of precious and valuable metals which are available in very limited quantities in nature. The availability of the affordable technology to recover these valuable resources is a major challenge in India. Attempts are also needed to bridge the formal and informal recycling sectors to develop a profitable business model. Formal recyclers are engaged in the segregation and disassembly processes in the country and predominantly dependant on businesses abroad for the recycling of PCBs.

The DeitY, the nodal Department of Electronics, is engaged in developing several affordable technologies to recycle e-waste. Development of e-waste recycling technology to recover precious metals, plastics and other valuable resources and translating these technologies to demonstration stage and thereafter commercial exploitation are the prime objectives of the programme. Only a few successful projects have been completed and demonstrated at the pilot plant level. This chapter will describe the DeitY's efforts and share some interesting results.

INVENTORY OF E-WASTES

An actual inventory of e-waste is not available in India. Studies have been conducted by various agencies to determine the inventory of e-waste in the country, where a model for the obsolescence of electronic products was considered as the basis. The Manufacturers Associations of Information Technology (MAIT), India and Deutsche Gesellschaft für Internationale Zusammenarbeit, India, jointly conducted a field study during 2007 to primarily estimate the inventory of three products, namely, televisions (TVs), personal computers (PCs) and refrigerators (MAIT-GTZ Study. 2007) (http://weeerecycle.in/publications/reports/GTZ_MAIT_E-waste_Assessment_Report.pdf). The study concluded that the total quantity of generated, recyclable and recycled e-waste was 332,979 Mt (1 Mt = 1000 kg). The inventory of PCs, mobile phones and TVs estimated were 56,324, 1655 and 275,000 Mt, respectively. The study found that the e-waste processed during 2007 consisted of 12,000 kg of PCs and 7000 kg of TVs. A total of 2.2 million PCs became obsolete in 2007. India had about 20 million PCs in 2007, which grew to 75 million by 2010. Around 14 million mobile handsets were replaced in 2007 as per another study. The Central Pollution Control Board, India, had carried out a survey during 2005, which estimated that 134,700 Mt of e-waste was generated in the country during that year. The e-waste generation reached 800,000 Mt in 2012. Considering a conservative growth rate of 10%, the yearly inventory generation that is projected for the year 2020 is shown in Figure 11.1.

RESOURCE MATERIALS IN E-WASTE

The waste electrical and electronic equipment (WEEE) contains 1000 different varieties of substances of 'hazardous' and 'non-hazardous' nature. WEEE consists of ferrous and non-ferrous metals, plastics, glass, PCBs and many other useful resources. One study by the United Nations Environment Programme describes the percentages of various valuable items available in assorted e-waste (Schluepa et al., 2009). The non-ferrous metals consist of metals like copper, aluminium and precious metals including silver, gold, platinum, palladium, etc. (Figure 11.2). In another study, the potential for precious metals recovery from various motherboards was shown to be good if the

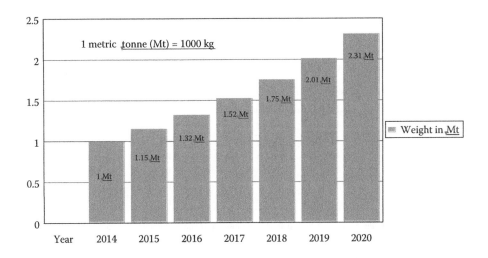

FIGURE 11.1 Projected growth of e-waste in India.

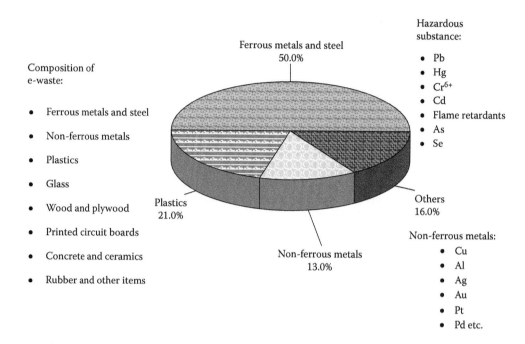

FIGURE 11.2 Composition of assorted e-waste.

appropriate technology is used (Buchert et al., 2012). Table 11.1 shows indicative quantities of gold, silver, copper and palladium that could be recovered from PCBs of various devices including cell phone, hard disc, display, etc.

STATUS OF RECYCLING CAPABILITY IN INDIA

The segregation and disassembly of WEEE recovered major volumes of small and large structural metal parts, heat sinks, ferrous metals, ferrite and ceramic components, non-ferrous metal scrap mainly Cu and Al as well as cables and wires. Appropriate segregation followed by smelting would be the best possible recycling option in India to transform these waste materials into usable raw materials. India has adequate smelter capacity to carry out this part of the recycling. The recovered

TABLE 11.1

Precious Metal Potential in Various PCBs

Items	Weight	Copper	Silver	Gold	Palladium
Cell phone	1 Mt	130,000 g	3500 g	340 g	140 g
Motherboard (310 g)	1 kg	350 g	800 mg	180 mg	80 mg
Memory cards (20 g)	1 kg	N/A	1650 mg	750 mg	180 mg
Small PCBs (28 g)	1 kg	–	800 mg	180 mg	80 mg
H/disk drive PCB (12 g)	1 kg	N/A	2600 mg	400 mg	280 mg
PCB optical drive (25 g)	1 kg	N/A	2200 mg	200 mg	70 mg
Display PCB (37 g)	1 kg	N/A	1300 mg	490 mg	99 mg

Source: Adapted from Schluep, M. et al. 2009. Sustainable innovation and technology transfer, industrial sector studies. Recycling – From E-Waste to Resources, United Nations Environment Programme and United Nations University, http://www.unep.org/pdf/Recycling_From_e-waste_to_resources.pdf.

small and large structural plastic parts from e-waste also can be recycled through globally acceptable method like chemical recovery (plastics-derived fuels), materials recovery (recycled plastics)-polycarbonate (PC), acrylonitrile butadiene styrene (ABS), PC or energy recovery (H_2, steam). In India, waste plastics are being used as filler for construction work or as alternative fuel.

Glass smelters, available in India, are able to transform the waste glass recovered from e-waste to usable raw materials. Lead extraction from cathode ray tube (CRT) glass would be the key challenge here. E-waste also contains hazardous materials including wastes like chlorofluorocarbons (CFCs), mercury (Hg) switches, CRT, batteries and capacitors, flame retardants plastic, all of which require environment friendly technology and special safety measures to recover the valuable resources presence in these components. Recovery of value materials from hazardous mixed substances is challenging and some of the research groups in India are carrying out safe separation of hazardous components at the laboratory level. Upscaling of these technologies will take further efforts. Table 11.2 indicates the potential recovery options for the resource materials from the e-waste.

BEST POSSIBLE RECYCLING OPTIONS

COLLECTION, SEGREGATION AND DISASSEMBLY

Since the implementation of the E-waste Rule 2011, collection of the e-waste is being streamlined for the formal recycling sector. The said rule mandates the bulk consumers and manufactures to hand over end-of-life electronics products only to the formal sector. The informal recycling sector is now facing problems collecting the waste materials and is thus registering as authorised collectors.

TABLE 11.2
Potential Materials Recovered from E-Waste

Recovered Items	Nature of Materials	Recycling Option	Used
1. Small and large structural metal parts, heat sinks, ferrous metal	Metals	Smelting	Secondary raw material
2. Ferrite and ceramic components, non-ferrous metal scrap mainly Cu and Al			
3. Cables and wires	Metals	Stripping + smelting	
4. Precious metal scrap, PCBs with IC chips, electronic components and connectors	Precious metals	Smelting + hydrometallurgy + electro-chemical process Mechanical shredding + screening + electrostatic separation + Falcon centrifugal separation + hydrometallurgy	Secondary raw material Copper, silver, gold, palladium, etc.
5. Small and large structural plastic parts	Plastics	Chemical recovery (plastics-derived fuels) Materials recovery (recycled plastics)-PC, ABS, PC Energy recovery (H2, steam)	Secondary raw material or alternative energy
6. Glass components	Glass	Smelting	Secondary raw material
7. Hazardous wastes such as CFC, mercury (Hg) switches, CRT, batteries and capacitors, flame retardant plastics	Materials containing hazardous substances	Recycling is possible with appropriate environmental care	Secondary raw material

Hence, a substantial amount of e-waste (nearly 80%) is still being diverted to the informal sector illegally for recycling.

The formal sector in India possesses adequate, methods and tools required to segregate and dissemble collected e-waste. Such techniques are well established. Various categories of components, modules, metals, glass and plastics are segregated from diverse sources of e-waste. The disassembly methods used are non-destructive and destructive. Non-destructive methods recover disassembled parts for reuse while the destructive disassembly separates each material type for recycling processes. The composition of various electronic components has changed significantly which makes the majority of de-soldered components now obsolete for reuse. Non-destructive methods are thus becoming unfeasible. The disassembly of e-waste yields various items such as small and large structural metal parts, heat sinks, small and large structural plastic parts, PCBs with integrated circuit (IC) chips, electronic components and connectors, ferrite and ceramic components, cables and wires and glass components.

The disassembly of CRTs also recovers items like enclosures made of plastics, cables/wires (power cords, CRT high voltage cable, etc.), iron/steel fittings and screws, lead-rich glass present in the colour picture tube and computer monitors, however, the glasses have to be separated according to lead content or as having a lead-free composition. The deflection yoke is further dismantled into its ferrite parts. The small populated PCB attached to the CRT and present in tuners requires specialised processing technology to recover precious metals. Similarly, disassembly of PCs recovers items like the metal/plastic enclosure, switch mode power supply unit, cooling fan, main mother board, auxiliary cards, hard disc drives, CD/DVD read/write devices, main power cable, printer cable, monitor, modems, scanner, etc.

RECYCLING OPTIONS FOR STRUCTURAL METAL, PLASTIC AND GLASS PARTS

The major volume of the assorted e-waste recovered includes small and large structural metal parts, heat sinks, ferrous metal, ferrite and ceramic components and non-ferrous metal scrap mainly Cu and Al. The metal smelters in India have adequate expertise for recycling the metals to produce new raw materials. The plastic waste is however being used as filler in road construction in India and in some cases as alternative fuel source. Nearly 27% of the plastic parts are recovered from e-waste. Glass smelters are also effectively using glasses recovered from WEEE to convert to usable raw materials. Lead extraction from CRT glass is the key challenge. Glass cullet is being used to prepare fresh glass parts (panels, funnels) and TV units.

RECYCLING OPTIONS FOR TECHNOLOGY-DRIVEN WASTE PARTS

Being the nodal agency of electronics, DeitY is engaged in nurturing and evolving several affordable technologies to recycle e-waste. DeitY concluded that consolidation of the recycling technologies available in various research organisations and the promotion of indigenous affordable technologies to recycle e-waste to recover resources and conserve natural resources was essential. It identified areas such as PCB recycling, converting plastics to value-added products, recycling Li-ion batteries (LIBs), liquid crystal display (LCD) screen processing, recovering copper from wires, recycling components containing hazardous substance, etc. would require special attention.

Recycling of PCBs

DeitY has prioritised the development of indigenous recycling technology of PCBs to recover precious metals. Populated PCBs constitute 3%–5% by weight of total e-waste and have a rich array of metals such as copper, silver, gold, palladium, platinum and tantalum with other metals in traces amounts. The recovery of all metals requires appropriate technology, professional skill and expensive equipment. A lack of knowledge with the substantial price of gold motivates the informal recycling sector to employ unhygienic and unscientific methods for the recovery of valuable metals such as Cu, Ag, Au, etc. Affordable technology solutions and stringent implementation of legislative measures are imperative to restrain informal operators from this practice and find environment-friendly solutions to the problem. DeitY has developed, established and matured two technologies at pilot plant scale to recover precious metals. A few projects were also successfully completed through academia–industry collaboration. The following sections describe the salient features of these technologies.

PCB Processing Technology Developed at CSIR-NML, Jamshedpur, India

Initially, processing technology for the recycling of PCBs was developed at CSIR-NML, Jamshedpur, India. PCBs are rich in metals and as a high value waste material constitute about 6% of the total weight of e-waste. Although, the actual composition of PCBs varies according to their specific class, these typically contain 28% metallic, 23% plastic and 49% ceramic material. Encapsulation of metallic components with plastics, ceramics and reinforced resin in PCBs hinders the extraction of metals via direct leaching processes. Physical beneficiation is required to achieve higher efficacy using hydrometallurgical routes. The combination of techniques used during the physical beneficiation of PCBs, viz. magnetic, electrostatic and gravity separation, is relatively inexpensive, scientifically sustainable and simplistically articulated (Kumar et al., 2012, 2015a, b). Thus, the PCBs were initially pulverised and thereafter subjected to a series of physical methods to separate metal and epoxy components. The PCBs are ground to the desired size (–1.0 mm) using various mechanical processes including physical impaction, shredding/fragmentation and granulation. Shredding breaks the PCBs into pieces via ripping or tearing for sorting into material streams having dissimilar subsequent processing demands. The granulation process is used to convert PCB scrap into fine particles. Precious metal particles are further concentrated by means of various separation techniques. Magnetic separation is used to separate magnetic materials (iron, nickel and cobalt) from the PCB powder and the aluminium-containing particles are removed by eddy current separation. The metal-rich powder is then separated from plastic-rich particles by electrostatic separation. A feed of pulverised PCB containing approximately 23% metal has been successfully enriched to more than 78% using this method. The rejected plastic stream contains less than 1% metal. Re-circulating intermediate streams during continuous operation could reasonably enhance the recovery of metals to around 90% (Das et al., 2009; Das, 2011). These beneficiated methodologies provide a novel perspective as it is devoid of any chemical, making the process eco-friendly. Single-stage air separation for metal recovery after shredding and sieving can contribute to energy conservation and, moreover, may provide optimum yields of metal.

The process requires conventional mineral processing equipment such as a shear/shredding machine, ball mill, classifier, gravity concentrator, flotation cell and high tension separator. The natural hydrophobicity of the non-metallic constituents is effectively exploited by a wet flotation process. Continuous plant operation minimises the loss of ultrafine metal values to negligible levels. Harmful effluents are avoided as physical processes are used. The operation is cost-effective and has been tested at a scale that utilised 1.0 Mt of PCBs. However to date such investigations have encountered significant constraints.

The metal-rich feed from physical upgrade is roasted and then chemically leached for precious metal recovery. After roasting at elevated temperatures, the plastic-free material is leached by sulphuric acid under atmospheric conditions at suitable temperature using air and hydrogen peroxide as the oxidant. Leaching conditions were optimised to recover approximately 93%–98% of the constituent base metals including Cu, Ni, Co, Zn, Al and Fe. The maximum recovery of metals in the shortest time with low acid consumption was achieved in an autoclave under oxygen pressure leaching conditions. More than 98% recovery of copper and other base metals was achieved under optimised conditions using 2 M sulphuric acid at elevated temperatures for 2 h with about 120 psi oxygen pressure.

After pressure dissolution, the residue was treated for gold and silver recovery. The nontoxic reagent thiourea was used to recover gold and silver from the enriched leached residue. Various

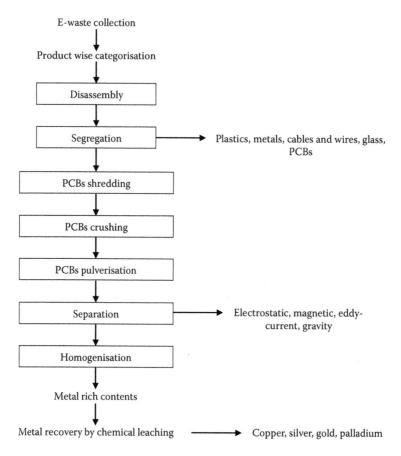

FIGURE 11.3 Process flowchart for the separation and beneficiation of precious metals from PCBs at CSIR-NML.

parameters including temperature, sulphuric acid and additive concentrations were optimised to obtain maximum recovery of silver and gold. Maximum values of 97% silver and 92.5% gold were recovered. The lead remaining in the residue was dissolved in brine solution at 80°C and 96.8% recovery was achieved. Subsequently, about 96% tin was recovered by hydrochloric acid leaching. The process was tested at 1 kg scale for the recovery of different valuable metals. An indicative process flowchart for recovery of metals/precious metals from the populated PCBs is shown in Figure 11.3.

E-Waste Processing Technologies Developed at CSIR-NML, Jamshedpur, India

CSIR-NML is actively engaged in developing various recycling processes to recover metals and valuables in an eco-friendly manner. In this respect, NML has established strong collaboration with various international agencies viz. Korea Institute of Geosciences and Mineral Resources (KIGAM), South Korea; Russian Academy of Sciences (RAS), Russia and several Government of India organisations such as DST and DeitY, as well as various Indian industries viz. M/s Eco Recycling, Mumbai; M/s ADV Metal Combine Pvt. Ltd., New Delhi and M/s. E-Parisaraa, Bangalore.

During processing, the electronic equipment is initially dismantled to remove outer casings made up of iron/ plastic and glass materials. PCBs contain the majority of valuable precious, non-ferrous, and hazardous metals and materials present in e-waste.

PCBs are initially crushed and mechanically beneficiated to produce enriched metallic components which are then leached and are subsequently processed using solvent extraction/ion exchange and electro-winning to produce pure metal sheets.

Mechanical pre-treatment processing was found to be neither feasible nor economic due to the lack of an appropriate mill/device. To solve this problem, the Indo-Korean research team has developed an alternative novel pre-treatment technique of organic swelling, which readily liberates the encapsulated thin layer of copper metal sheet and separates the outer layer of epoxy resin which mostly contains a solder material. Copper, lead and tin metals are thus easily separated and the metal-free epoxy resin is used as a filling material. The maximum recovery of metals in the shortest time with low acid consumption was achieved under oxygen pressure leaching conditions (Jha et al., 2008, 2010, 2011b, c, 2012a, b).

A joint venture of M/s Eco Recycling Company, Mumbai and CSIR-NML, Jamshedpur, has developed a hybrid process consisting of pyrolysis, beneficiation and hydrometallurgy. From this process, low-density oil and copper metal sheets have been produced (Kumari et al., 2013; Kumar et al., 2014).

A process has been developed to recycle LIBs of mobile phones to recover lithium and cobalt under a joint venture between CSIR-NML and RAS, Russia. Experiments were carried out at bench scale in which the cathodic material of LIBs was leached in acid at elevated temperature for 1 h. Leach liquor obtained was further put to solvent extraction studies to generate a pure solution of lithium and cobalt. The pure solution obtained was concentrated and evaporated to genarate cobalt sulphate and lithium sulphate salt as value-added products (Jha et al., 2011a, 2013b, c). On the basis of the above studies, an e-waste recycling company, M/s ADV Combine Pvt. Ltd. New Delhi, India, has jointly established a technology on 500 kg scale. Under this technology, the batteries were crushed and subsequently beneficiated using wet scrubbing to separate black powder, plastic and casings along with separator sheets. The beneficiated cathodic black powder is leached in acid and was found to contain valuable cobalt along with other metals viz. lithium, manganese, copper, iron, etc. The leach liquor obtained was treated to remove manganese and iron by precipitation. The remaining filtrate was used in solvent extraction to remove copper. Thereafter, cobalt is extracted using Cyanex 272 and the concentrated pure solution of stripped cobalt in sulphate medium was evaporated to generate cobalt sulphate heptahydrate salt. Cobalt metal was prepared from this solution by electrowinning. Lithium left in the raffinate from solvent extraction is recovered as a salt (Figure 11.4).

After establishing feasible technologies for the recovery of valuable metals, a process for the recovery of gold from waste mobile phones and the scrap of various equipment has been developed. Dissolution of gold present on the outer layer of PCBs of waste mobile phones and small parts of various equipment using chemical leaching followed by adsorption/cementation with subsequent heat treatment has been developed (Ha et al., 2010; Anand et al., 2013). The complete flow sheet for gold recovery from waste PCBs or the scrap from other items of equipment is shown in Figure 11.5.

The unique properties of rare earth minerals have attracted significant attention at NML and particularly their recovery from e-wastes. A process for the recovery of neodymium as a value-added product from waste hard disk of PCs was developed. In this process, the magnets are initially demagnetised and crushed to reduce the size for dissolution in an acidic medium at an optimised temperature. A neodymium salt was obtained by precipitation from the leach liquor obtained. The salt formed was further leached in hydrogen fluoride for few minutes to form the value-added product neodymium fluoride (Lee et al., 2011; Jha et al., 2013a).

Fluorescent tubes are another source of rare earth metals containing lanthanum, yttrium, terbium, cerium, europium, etc. that can be recycled. This uses a process where the fluorescent lamps are initially broken in a closed system fitted with a vacuum device and the phosphor powder lining is collected. This powder which contains rare earth metals is leached in acid at an elevated temperature. The leach liquor containing the mixture of rare earth elements (REEs) is processed using solvent extraction to generate pure solutions of individual REEs (Jha et al., 2014).

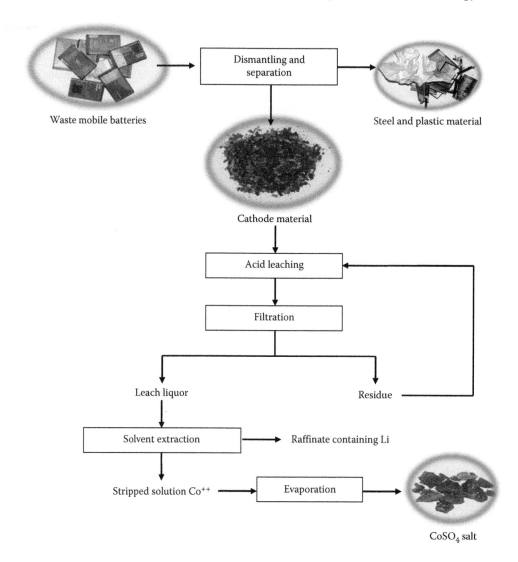

FIGURE 11.4 Process flowsheet for the extraction of cobalt and lithium from waste mobile phone batteries.

PCBs Processing Technology Developed at C-MET, Hyderabad and E-Parisaraa, Bangalore

During the development of physical separation methods as described previously, it was learnt that effective recovery of other precious metals (like platinum, tantalum, ruthenium, etc.) present in the ppm level would be a challenge. The idea of initial component segregation before the processing of PCBs was thought to be more effective for the recovery of all the precious metals. Another project on the processing of PCBs has been implemented successfully using this concept at C-MET, Hyderabad, in collaboration with the formal recycler, M/s E-Parisaraa, Bangalore.

In this project, the PCBs were depopulated using a vibratory centrifuge to separate the components. Nearly 70% of the PCBs consists of components including small outline transistors, ball grid arrays, ICs, crystals, crystal oscillators, chip capacitors, chip resistors and connectors, which were depopulated and segregated. The bare PCBs were pyrolysed through a specially designed system of 500 kg batch size capacity. Pyrolysis was found to easily shred and pulverise multi-layer circuit boards liberating the metals to enable chemical leaching. The pyrolysed metals were then calcined in air producing fragile materials to facilitate leaching and causing a 40%–50% reduction

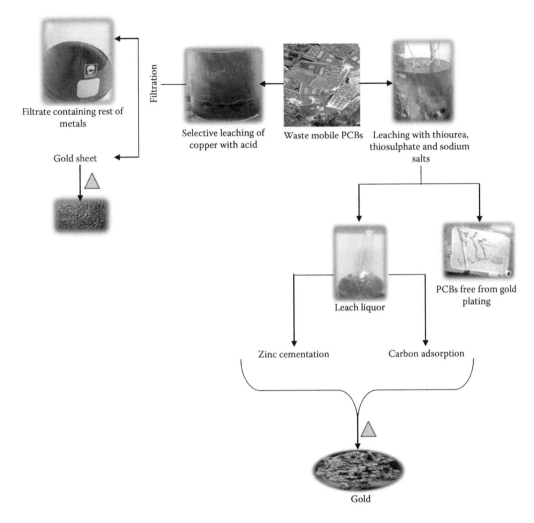

FIGURE 11.5   Process flowsheet for the recovery of gold from the PCBs of waste mobile phones and the scrap of various equipment.

of mass. Hydrometallurgical methods were used to recover metals from the calcined metal residue employing various leaching agents including nitric acid, hydrochloric acid, aqua regia and sulphuric acid. Electro-refining was then used for the separation of copper from other precious metals and for purification. Selective recovery of pure metal products directly from the residual waste stream was applied to the precious metals. Leaching of depopulated and pyrolised components and PCBs satisfactorily recovered four metals: copper, gold, silver and palladium (Reddy and Parthasarathy, 2010). An indicative process flowchart for the recovery of metals/precious metals from the populated PCBs is shown in Figure 11.6.

Recycling of Other Materials

Plastics Processing Technology Developed at CIPET, Bhubaneswar

Plastics are a significant constituent of e-waste and constitute approximately 33% by weight. In electrical and electronics products, plastics are used for insulation, flame retardation, noise reduction, sealing, housing, interior structural parts, electronic components, etc. This study has shown that e-waste contains a sizable amount (>70%) of four homogeneous polymers namely ABS,

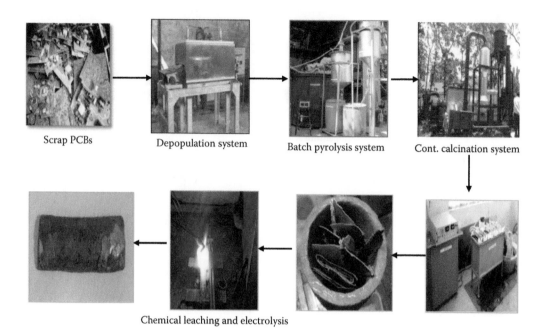

FIGURE 11.6 Process flowchart for the recovery of precious metals from PCBs developed by C-MET and E-Parisaraa.

polypropylene, polystyrene and polyurethane. The epoxy mainly used in PCBs and PVC used in wires can be considered as minor components in e-waste. Technological solutions have been achieved for the recycling of homogeneous plastics including PC, ABS, high impact polystyrene, PC/ABS blends, ABS/PS blends, etc., collected from major structural components such as the printer housing, monitor stand, monitor cover, computer casing, keyboard bottom plate, TV housing, central processor unit casing, speaker box, telephone casing, etc.

Various recycled plastic formulations were developed in this study using the above categories of plastics based on their flow parameters, processability, mechanical performance and cost-effectiveness. The processing for two of these formulations was optimised and subsequently prototype products were developed for testing and certification by relevant authorities. The processing broadly includes cleaning, drying and grinding of plastics products, optimisation of mixed plastics within virgin general purpose polycarbonate, improvement of the performance via the addition of plasticiser and evaluation of the product properties. The transfer this technology to a material manufacturer capable of moulding master batches to produce various thin-walled components is currently being undertaken. The formulation developed has the potential to promote indigenisation of the innovation in material technology which is presently being borrowed from neighbouring countries such as China and other Asian/American countries at a higher cost. This material formulation can suitably be used in products wherein PC/ABS blends dominate (Mohanty, 2011).

Processing Technology Developed for Other Items at C-MET,
Hyderabad and E-Parisaraa, Bangalore

The pyrolysis process developed at C-MET, Hyderabad and E-Parisaraa, Bangalore, for PCB recycling (discussed in section 'PCBs Processing Technology Developed at C-MET, Hyderabad and E-Parisaraa, Bangalore') could also be used for processing other complicated ingredients of e-waste such as CRT yokes, mixed wires and cables, low-value transistors and capacitors. Similarly, after pyrolysis, calcination would be used prior to leaching of the precious metals.

PILOT SCALE DEMONSTRATION PLANT FOR PCB RECYCLING

DeitY along with the Government of Karnataka has recently commenced the demonstration of continuous processing for PCBs as the second phase of the project that has been jointly implemented by the C-MET, Hyderabad and E-Parisaraa, Bangalore. The scope and objectives of the project are to

a. Establish indigenous capability in environmentally sound e-waste processing technology in order to halt the import of PCBs from foreign countries for processing and conserving natural resources
b. Provide a demonstration unit for the processing of 30 Mt/a PCBs to recover the valuable metals
c. Establish a production scale depopulation system and prototype systems for smelting (100–200 kg batch), electro-refining for copper (50–100 kg), electro-winning and solvent extraction
d. Set up pilot plant processing facilities for gold, palladium and silver recovery from PCBs
e. Service the informal recycling sector by extending the processing facility for the recovery of metals with direct economic and ecological benefits

Should the demonstration plant be successful, DeitY plans to initiate several similar facilities in other states involving partner recyclers.

RESEARCH AND DEVELOPMENT RECYCLING ROAD MAP FOR OTHER ITEMS

E-wastes are rich in specialised materials that impart special properties critical to its functionality. Various recycling technologies are required to recover the valuable materials and other resources from the structural metal, plastic parts and other electronic components generated from disassembled e-waste. Recycling of the rare element indium from LCD displays is another area that requires immediate attention. LCD displays, widely used in monitors of desktop computers, mobile phones and laptops, use indium-tin oxide (ITO) as a transparent conducting surface. A soft and silvery metal, indium is generally found as an impurity in galena and sphalerite ores. Around 70% of indium is consumed in photovoltaic cells and displays. Extraction of indium from LCD and photovoltaic devices, however, requires specialised technology.

Recycling of phosphors from the displays and fluorescent tubes to recover high-value REEs viz. lanthanum, cerium, terbium, yttrium, europium, etc. is another challenging research area. Table 11.3 indicates some of the important e-waste items and their indicative recycling options, which will be undertaken by DeitY in near future to develop appropriate recycling technologies.

CONCLUSION

E-waste recycling is becoming a non-viable business in developed countries due to the high costs of labour, transportation, electric power, etc. Furthermore, the content of precious metals is decreasing in the miniaturised electronics devices. Thus, the viability of the recycling business is of great concern. The volume of e-waste is, however, increasing worldwide at an alarmingly rate, especially in the emerging economies such as China, India and Brazil. The presence of toxic elements discourages stockpiling this waste without having appropriate disposal mechanisms. On the other hand, as e-waste contains valuable materials, the extraction of precious metals and other resources will generate raw materials that can be used by future generations. It is therefore appropriate to devise a holistic approach to manage and recycle e-waste in a sustainable manner to preserve the environment and human health. The developed countries have technology and infrastructure, whereas, the cost of labour, transporting, processing, etc. are cheaper in the developing countries. It is, therefore, proposed in this chapter that the best approach to solve these issues is to manage e-waste by involving

TABLE 11.3
Recycling Options for Other E-Waste Items

Items	Recycling Option
1. LIB	LIBs contain 15%–25% of Al, 0.1%–1% of carbon amorphous powder, 5%–15% of Cu foil, 1%–10% of diethyl carbonate, ethylene carbonate and methyl ethyl carbonate, 1%–5% of $LiPF_6$, 10%–30% of graphite powder, 25%–45% of $LiCoO_2$ and 0.5%–2% of polyvinylidene fluoride, steel, Ni and inert polymer. Cobalt and lithium could be targeted for recovery. Research and Development can be undertaken to develop environment friendly methods for safe dismantling, crushing of LIBs and leaching of valuable metals (Petrániková, 2011)
2. LCD screen	Recycling process involves acid leaching of ITO from the surface of crushed LCD glass. Leaching uses various acids (HNO_3, HCl and H_2SO_4) of different concentrations under suitable temperature. Leaching is followed by solvent extraction to separate indium from other metals. Indium can be separated into a relatively pure fraction with di-(2-ethylhexyl)phosphoric acid. Higher purity can be achieved by optimisation of the solvent extraction and through additional processes like electrolysis of the pure metal solution obtained (Yang, 2012)
3. Phosphors in fluorescent lamps and colour picture tubes	Fluorescent lamps as well as colour picture tubes are rich sources of heavy rare earth metals which can either be directly reused in new lamps or hydrometallurgically treated using different acids, similar to the processes used for extraction of rare earth metals from ores. Lamps are initially crushed and phosphor powder is collected. Different acids were used for the dissolution of rare earth metals from the phosphor powder followed by application of solvent extraction methods for the separation of individual rare earth metals from the leach liquor

the informal recycling sectors in developing countries and formal recycling sectors in the developed countries. The current chapter has described how such an approach is being developed in India.

The major volume (nearly 95%) of the segregated e-waste that is generated in, and imported into, India can be managed with the existing technologies using available facilities. The remaining 5% of the e-waste actually consists of PCBs/connectors, which are the most valuable waste components and need sophisticated technology to recycle. Providing a sustainable solution for PCB recycling is an immediate requirement for India to address the present e-waste problem. PCB recycling is the most challenging, technology-intensive process and the options used by developed countries are expensive.

DeitY through its continued Research and Development efforts has successfully developed indigenous technology on a laboratory scale and established 'proof-of-concept' for the recycling of PCBs at a pilot plant scale. The department is committed to the development of this technology at a scale of 30 Mt/a PCBs collected from both the formal and informal recycling sectors. Upon successful complementation of this operation, DeitY intends to explore transfer of the technology and commercial exploitation of the process. This will provide a sustainable solution for PCB recyclers in India. The precious metals recovered will remain in the country and Indian recyclers will not need to depend on foreign smelters for PCB recycling to recover precious metals or suffer the accompanying loss of business.

The proposed demonstration plant will not only process PCBs, but also would take care of other complicated components such as CRT yokes, mixed wires and cables, low-value transistor and capacitors, etc. DeitY's in the future plans to provide viable recycling technologies for LIBs, the extraction of REEs from phosphors, the extraction of indium from LCD screen and for other complex e-waste items.

DISCLAIMER

The views expressed in this chapter are those of the authors and do not represent the official views of the DeitY, Ministry of Communication and Information Technology or the Government of India. The facts and figures, and data used in this chapter are indicative, informative and have been extracted from various research articles and project reports submitted by various research organisations. The validation of these data would be challenging and not within the scope of the present topic. The use of the data and the recommendations presented are at the discretion of the users of the chapter.

ACKNOWLEDGEMENTS

The author acknowledges Dr. Avimanyu Das, NML, Jamshedpur; Dr. M. Raviprakasa Reddy, C-MET, Hyderabad; Dr. P. Parthasarathy, Managing Director, E-Parisaraa Pvt. Ltd., Bangalore; Dr. Smita Mohanty, Central Institute of Plastics Engineering and Technology (CIPET), Bhubaneswar; Mr. B. K. Soni, Eco-Recycling Company, Mumbai; Mr. Vinayachal Jha, ADV Metal Combine Pvt. Ltd, New Delhi, India, as well as the Korea Institute of Geosciences and Mineral Resources (KIGAM), South Korea, for their valuable suggestions on the subject and for sharing their inputs and important data from the electronic waste recycling operations.

REFERENCES

Anand, A., M. K. Jha, V. Kumar and R. Sahu. 2013. Recycling of precious metal gold from waste electrical and electronic equipments (WEEE): A review. In *Proceedings of the XIII International Seminar on Mineral Processing Technology (MPT-2013)*, CSIR-IMMT, Bhubaneswar, India, pp. 916–923.

Buchert M., A. Manhart, D. Bleher and D. Pingel. 2012. *Recycling Critical Raw Materials from Waste Electronic Equipment*. North Rhine-Westphalia State Agency for Nature, Environment and Consumer Protection, Darmstadt, Germany: Öko-Institut e.V., 24 February, www.oeko.de/oekodoc/1375/2012–010-en.pdf.

Das, A. 2011. Development of processing technology for recycling and reuse of electronic waste. DeitY, OM No. 1(7)/2007/M&C, 17 September 2007 (unpublished).

Das, A., A. Vidyadhar and S. P. Mehrotra. 2009. A novel flow sheet for the recovery of metal values from waste printed circuit boards. *Resources, Conservation and Recycling*, 53: 464–469.

E-Wastes (Management and Handling) Rules – Ministry of Environment and Forest. 2008. http://envfor.nic.in/downloads/rules-and-regulations/1035e_eng.pdf.

Ha, V. H., J. C. Lee, J. Jeong, T. H. Huyunh and M. K. Jha. 2010. Thiosulfate leaching of gold from waste mobile phones. *Journal of Hazardous Materials*, 178: 1115–1119.

Jha, A. K., M. K. Jha, V. Kumar and R. Sahu. 2013a. Recovery of neodymium as value added product NdF_3 from rare earth magnet. In *Proceedings of the XIII International Seminar on Mineral Processing Technology (MPT-2013)*, CSIR-IMMT, Bhubaneswar, India, pp. 841–846.

Jha, A. K., M. K. Jha, A. Kumari, S. K. Sahu, V. Kumar and B. D. Pandey. 2013b. Selective separation and recovery of cobalt from leach liquor of discarded Li-ion batteries using thiophosphinic extractant. *Separation and Purification Technology*, 104: 160–166.

Jha, A. K., A. Kumari, P. K. Choubey, V. Kumar and M. K. Jha. 2011a. Waste lithium ion batteries – An alternative resource to recover lithium and cobalt. *Journal of Metallurgy and Materials Science*, 53(3): 225–231.

Jha, M. K., P. Choubey, A. Kumari, R. Kumar, V. Kumar and J. C. Lee. 2011b. Leaching of lead from solder material used in electrical and electronic equipment. In *Recycling of Electronic Waste II: Proceedings of the Second Symposium, TMS Annual Meeting and Exhibition*, San Diego Convention Center, San Diego, California, pp. 25–31.

Jha, M. K., P. K. Choubey, A. K. Jha, A. Kumari, J. C. Lee, V. Kumar and J. Jeong. 2012a. Leaching studies for tin recovery from waste e-scrap. *Waste Management*, 32: 1919–1925.

Jha, M. K., J. Jeong, J. C. Lee, M. S. Kim and B. S. Kim. 2008. Recovery of copper from printed circuit boards (PCBs) scraps by mechanical pre-treatment and hydrometallurgical routes. In *Proceedings of Symposium Electronic Recycling, TMS – Annual Meeting and Exhibition*, New Orleans, Louisiana, 9–13 March.

Jha, M. K., A. K. Jha, J. Hait and V. Kumar. 2014. A process for the recovery of rare earth metals from scrap fluorescent tubes. Patent No. -0280DEL2014.

Jha, M. K., A. Kumari, P. K. Choubey, J. C. Lee, V. Kumar and J. Jeong. 2012b. Leaching of lead from solder material of waste printed circuit boards (PCBs). *Hydrometallurgy*, 121–124: 28–34.

Jha, M. K., A. Kumari, A. K. Jha, V. Kumar, J. Hait and B. D. Pandey. 2013c. Recovery of lithium and cobalt from waste lithium ion batteries of mobile phone. *Waste Management*, 33(9): 1890–1897.

Jha, M. K., J. C. Lee, A. Kumari, P. K. Choubey, V. Kumar and J. Jeong. 2011c. Pressure leaching of metals from waste printed circuit boards using sulphuric acid. *Journal of Metals*, 63(8): 29–32.

Jha, M. K., Shivendra, V. Kumar, B. D. Pandey, R. Kumar and J. C. Lee. 2010. Leaching studies for the recovery of metals from the waste printed circuit boards (PCBs). *TMS-2010, 139th Annual Meeting and Exhibition, In Proceedings of Sustainable Materials Processing and Production*, Washington State Convention Center, Seattle, Washington, pp. 945–952.

Kumar, V., A. Kumari, M. K. Jha, A. Vidyadhar and B. K. Soni 2014. Recycling of valuable metals from poly cracker ash of printed circuit boards (PCBs) by physical beneficiation and hydrometallurgical treatment. In *EPD Congress 2014*, San Diego Convention Center, San Diego, California, pp. 545–551.

Kumar, V., J. C. Lee, J. Jeong, M. K. Jha, B. S. Kim and R. Singh, 2015a. Novel physical separation process for eco-friendly recycling of rare and valuable metals from end-of-life DVD-PCBs. *Separation and Purification Technology*, 111: 145–154.

Kumar, V., J. C. Lee, J. Jeong, M. K. Jha, B. S. Kim and R. Singh 2015b. Recycling of printed circuit boards (PCBs) to generate enriched rare metal concentrate. *Journal of Industrial and Engineering Chemistry*, 21: 805–813.

Kumar, V., J. C. Lee, J. Jeong, D. Shin, M. K. Jha and V. Kumar. 2012. Enrichment of valuable metals from end-of-life printed circuit boards using mechanical recycling techniques. In *XXVI International Mineral Processing Congress (IMPC) 2012 Proceedings*, New Delhi, India, September.

Kumari, A., V. Kumar, M. K. Jha and B. K. Soni. 2013. Processing of poly-cracked ash of printed circuit boards for recovery of valuable and rare metals. In *Proceedings of the XIII International Seminar on Mineral Processing Technology (MPT-2013)*, CSIR-IMMT, Bhubaneswar, India, pp. 929–933.

Lee, J. Y., A. K. Jha, A. Kumari, J. R. Kumar, M. K. Jha and V. Kumar. 2011. Neodymium recovery by precipitation from synthetic leach liquor of concentrated rare earth mineral. *Journal of Metallurgy and Materials Science*, 53(4): 349–354.

MAIT-GTZ Study. 2007. Report on 'E-waste Inventorisation in India', http://weeerecycle.in/publications/reports/GTZ_MAIT_E-waste_Assessment_Report.pdf.

Mohanty, S. 2011. Novel recovery and conversion of plastics from WEEE to value added products. Central Institute of Plastics Engineering and Technology (CIPET), Bhubaneswar, India. DeitY, OM No. 1(8)/2010/M&C, 1 March (unpublished).

Petrániková, M., A. Miškufová, T. Havlík, O. Forsén and A. Pehkonen. 2011. Cobalt recovery from spent portable lithium accumulators after thermal treatment. *Acta Metallurgica Slovaca*, 17: 106–115.

Reddy, M. R. P. and P. Parthasarathy. 2010. Environmentally sound methods for recovery of metals from printed circuit boards. DeitY, OM No. 1(2)/2010/M&C, 24 May (unpublished).

Schluep, M., C. Hagelueken, R. Kuehr, F. Magalini, C. Maurer, C. Meskers, E. Mueller and F. Wang. 2009. Sustainable innovation and technology transfer, industrial sector studies. Recycling – From E-Waste to Resources, United Nations Environment Programme and United Nations University, http://www.unep.org/pdf/Recycling_From_e-waste_to_resources.pdf.

Yang, J. 2012. Recovery of indium from end-of-life liquid crystal displays. Thesis for the Degree of Licentiate of Engineering, Industrial Materials Recycling, Department of Chemical and Biological Engineering, Chalmers University of Technology, Gothenburg, Sweden, publications.lib.chalmers.se/records/fulltext/165702/165702.pdf.

12 Recovery of Rare Earth Elements from e-Wastes (Nd-Fe-B Spent Magnets) Using Magnesium Chloride Salts
A Preliminary Study

Komal Babu Addagatla, Sheila Devasahayam, and M. Akbar Rhamdhani

CONTENTS

ABSTRACT

Rare earth elements (REEs) are used in many applications such as in wind turbines, batteries, as catalysts, strong magnets, and in electric cars. The availability of rare earths is in partial decline. The price of REEs has escalated due to their high demand surpassing the supply. As we have only a few exploitable natural resources for mining rare earths around the world, we will have to rely on recycling the REEs from pre- and postconsumer scraps, mainly from the end-of-life products.

The current research focuses on eco-friendly processes to recover REEs from spent magnets in e-wastes. One such method involves using molten salts at higher temperatures for extracting REEs from the end-of-life products.

This work aims to recover the rare earth elements Nd, Dy, and Pr from Nd–Fe–B magnets found in e-wastes, using magnesium chloride salts heated to temperatures 800°C, 1000°C, and 1200°C for a time period of 12 hours under normal atmospheric conditions.

The reacted products are analyzed using scanning electron microscope (SEM), electron-dispersive spectroscopy (EDS), and ICP-AES (inductively coupled plasma atomic emission spectroscopy). This work sheds light on the influence of temperature on the recovery rate. The results obtained clearly show that REEs are successfully recovered from the magnets in smaller percentages, the highest recovery is observed at 1200°C (~37%). The low recovery rates are due to the influence of oxygen in the process, forming rare earth oxides, which are highly stable and cannot be processed at these temperatures.

INTRODUCTION

There are 17 rare earth elements, for example, yttrium (Y), lanthanum (La), cerium (Ce), praseodymium (Pr), neodymium (Nd), promethium (Pm), samarium (Sm), europium (Eu), gadolinium (Gd), terbium (Tb), dysprosium (Dy), holmium (Ho), erbium (Er), thulium (Tm), ytterbium (Yb), lutetium (Lu), and scandium (Sc) (Oko-Institut e.V., 2011). Rare earths are moderately abundant in the earth's crust, some more abundant than copper, gold, lead, and platinum, but not concentrated enough to be easily and economically exploitable. The rare earths are subdivided into heavy rare earth elements (HREEs) and light rare earth elements (LREEs). The elements that come under these subdivisions are represented in the periodic table as shown in Figure 12.1. REEs play an important role in most of the modern industrial activities. At present, REEs are mainly used as catalytic convertors in automobiles, alloying materials in metallurgy, and phosphors in display equipment such as cell phones, laptops, computers, fluorescent lamps, and permanent magnets (Yamamura et al., 2003). REEs are used in space applications such as space-based satellites and communication systems (Tabuki, 2010). The applications of rare earths are shown in Figure 12.2.

The REEs are extracted from the mineral ores as oxides; they are iron-gray to silvery lustrous metals which are typically soft, malleable, ductile, and usually reactive at elevated temperatures. The unique properties of these elements make them useful in various modern applications. For example, magnets made of these elements are very powerful, weigh less, and can be made smaller than the conventional magnets. The chemical behavior of REEs are similar, but some variations can be attributed to their atomic number and ionic radii, making these elements well suitable to the processes such as sorption, precipitation, and formation of colloids (Merten and Büchel, 2004).

The United States was the world leader in the production of rare earth oxides from 1960 to 1980, supplying 100% of U.S. demand and 30% of the world demand. China in the late 1970s started the production of rare earths as shown in Figure 12.3, which made it the world's dominant producer of rare earths. Decline in the production of REEs in the United States made China the leading producer of rare earths contributing up to 95% of the global production (IMCOA, 2011).

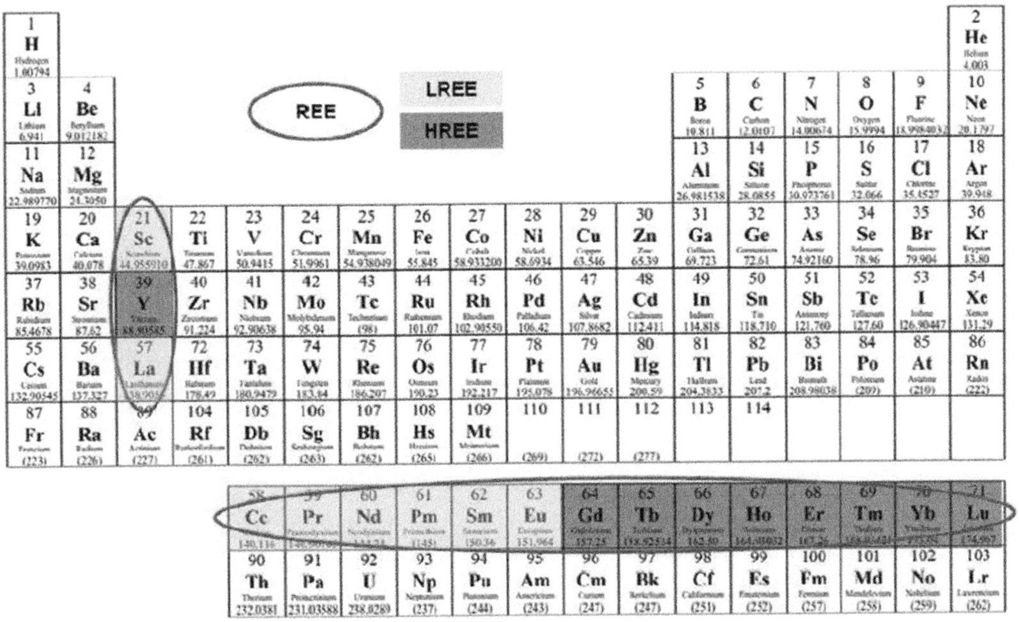

FIGURE 12.1 Periodic table showing the division between HREE and LREE. (Adapted from Schüler, D. et al. 2011. Study on rare earths and their recycling. Final Report for the Greens/EFA Group in the European Parliament, Öko-Institut e.V., Darmstadt, January 2011. http://www.resourcefever.Org/publications/reports.)

In the United States, occurrence of rare earth ores is estimated by the USGS (United States Geological Survey) as follows:

- Thorium–rare earth vein deposits (Armbrustmacher et al., 1995) of ~0.2 million tons.
- Low titanium–iron ore deposits generating 3.5–450 million tons, with a standard quantity of 40 million tons (Fosse and Graunch, 1995).
- Minable carbonatite ore deposits generating 6.6–331 million tons (Modreski and Armbrustmacher et al., 1995) of REE.

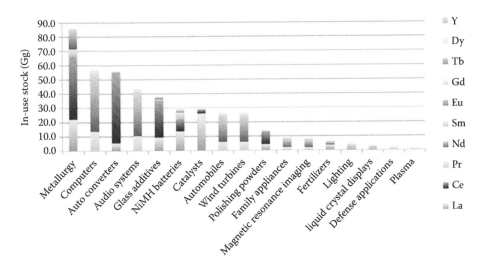

FIGURE 12.2 Use of selected REE in specific applications or industry. (Adapted from Du, X. and T. E. Graedel. 2011. *Environ. Sci. Technol.*, 45: 4096–4101.)

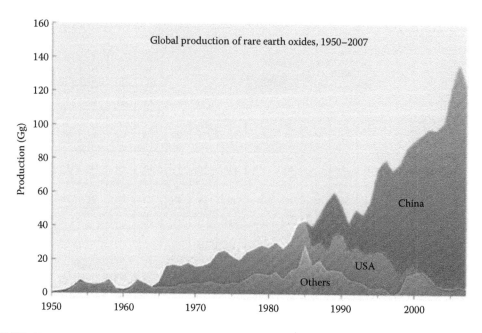

FIGURE 12.3 Global production of rare earth oxides. (Adapted from Du, X. and T. E. Graedel. 2011. *Environ. Sci. Technol.*, 45: 4096–4101.)

The world demand for REEs is estimated at 136,000 tons/year, with a global production of 133,600 tons (IMCOA, 2011). This difference is covered by previously mined and available aboveground stocks. By 2015, it is estimated that the demand will rise to 185,000 tons annually (IMCOA, 2011). Others estimate the demand for REE to globally rise to 210,000 tons (Bloomberg News, 2010). The primary force for this situation is due to the increased demand of REEs, increasing cost of these elements, supply restrictions, and policies implemented by organization for recycling selected items for the recovery of REE. According to the report given by Oko-institute (Yamamura et al., 2003), two main scenarios of REE usage are envisaged. The first scenario predicts an average annual growth of 4.5% (left bar in Figure 12.4) which represents the moderate scenario. The second scenario (right bar in Figure 12.4) predicts an annual growth of 9% which is considered as an upper limit. It considers the ongoing market for new REE applications. This scenario is supported by the growth rate with respect to specific applications (Bloomberg News, 2010).

Mining operations of REE produce a variety of solid material that can cause contamination in the environment. Extraction of rare earths after the mining process releases dangerous materials, for example, HF, HCl, SO_2, and radioactive nuclides in the environment (Oko-Institut e.V., 2011). However, the major environmental risk in mining and extraction of these REEs is associated with the treatment and disposal of the tailings (Oko-Institut e.V., 2011).

Processing and Recycling of REEs

A recent study by a Japanese research group states that around 300,000 tons of REE are present in e-waste (Tabuki, 2010).

The process of recycling the end-of-life and postconsumer products mainly involves four main steps, as given in Figure 12.5:

1. Collection
2. Dismantling
3. Separation (preprocessing)
4. Processing (recovery)

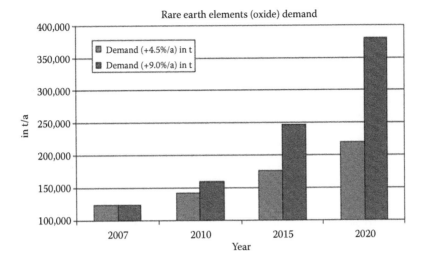

FIGURE 12.4 Demand of rare earth oxides based on the calculation of Oko-Institute. (Adapted from Panayotova, M. and V. Panayotov. 2012. Review of methods for the rare earth metals recycling. *Annual of the University of Mining and Geology "ST. IVAN RILSKI," Vol. 55: Part II, Mining and Mineral Processing,* 2012.)

Collection

It is the first step in the recycling process, where all the postconsumer products and the end-of-life products have to be collected and stored at a particular place as discussed in the recent report by UNEP (United Nations Environment Programme) (Meyer and Bras, 2011). Higher efficiency is obtained when an infrastructure is established for the collection.

Dismantling and Preprocessing

This is a critical step where a high value material is separated from lesser value materials. High value materials in this context mean REEs and some precious elements such as gold, silver, platinum, and copper which are used mostly in all the electronic equipment (Meyer and Bras, 2011). During the separation process some nonmetals get mixed with the metals in the metallic scrap making the recycling process for the mixed scrap more difficult compared with the segregated metals.

Typical steps for dismantling and preprocessing include manual or mechanical separation, mechanical shredding, screening, and manual or mechanical disassembly (Schluep et al., 2009). Figure 12.6 shows the images of the components that have been obtained after the disassembly process, where magnets are recovered which have a good amount of REE.

In the process of dismantling and preprocessing, all the hazardous and unwanted substances have to be removed, while restoring the valuable materials for recycling. Some devices such as refrigerators and air conditioners contain ozone depleting substances such as chlorofluorocarbon or hydrochloroflurocarbons, which may affect the environment. These materials require initial degassing in the preprocessing stage. LCDs and monitors contain toxic materials such as mercury, which may cause some serious health problems to the workers, so particular care must be taken during the separation of these elements (Schluep et al., 2009).

Processing

Since REEs are thermodynamically stable, a large amount of energy is required to obtain the metal or alloy. Rare earth metals are available in newer nickel-hydride and lithium systems used in batteries; recycling is still in the early stages and not yet put to practice to recover the REE from the hydride alloy. REEs such as yttrium, europium, lanthanum, and cerium are found in discharge lamps, fluorescent lamps, and also in TV tubes (Rabah, 2008).

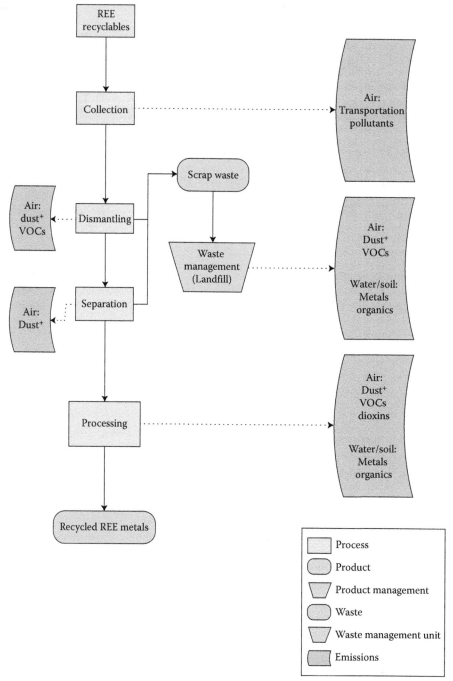

FIGURE 12.5   Recycling steps of REE and waste emission.

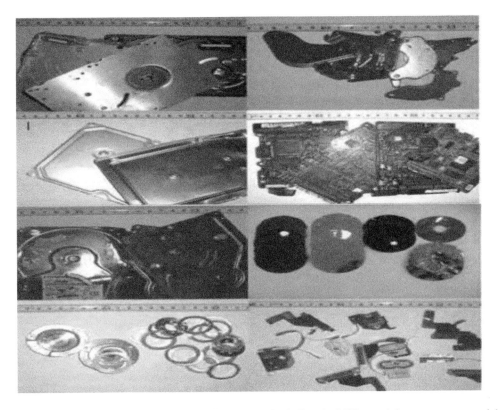

FIGURE 12.6 Components after the dismantling process including the REE containing magnets at top right corner. (Adapted from Schluep, M., C. Hagelueken et al. 2009. Recycling- from e waste to resources. United Nations Environment Programme, United Nations University, July. Retrieved from http://www.unep.org/PDF/PressReleases/E-waste_publication_screen_FINALVERSION-sml.pdf.)

There are several processing methods which can be carried out to extract the REE metals present in the material obtained from the preprocessed step. The methods used in processing are as follows (Schluep et al., 2009; ICF International., 2011; Meyer and Bras, 2011; Walton and Williams, 2011):

- Electrometallurgy process, such as electrowinning where current is passed from an inert anode through a liquid leach containing the required metal; the metal is extracted by an electroplating process where the material gets deposited on the cathode.
- Hydrometallurgy process, where strong acids or basic solutions are used to dissolve and precipitate metal of interest from a powder form. Processes such as selective leaching, solvent extraction, and selective precipitation are followed for recovering the metals. Hydrometallurgy process has been developed for the recovery of cobalt, nickel, and rare earth metals from Ni-MH batteries (Kuzuya et al., 2003). Rare earths are dissolved using sulfuric acid followed by precipitation using sodium hydroxide (Bertuol et al., 2009). Higher recovery rates are expected for the hydrometallurgy process, compared with the other processes, that is, 90%–95% of rare earth elements (RREs) can be recovered, but with considerable impact on the environment (Ellis et al., 1994). This process generates large volumes of hazardous substances that do not decompose under normal environment.
- Dry process which is still under investigation uses hydrogen gas at atmospheric pressure to turn the Nd-containing magnets to a powder that can be reformed into new magnets under heat and pressure. This research is being conducted at the University of Birmingham in the United Kingdom, where the newly produced magnets do not have the same quality as the originals, but they can be used in monitors (Davies, 2011).

- Pyrometallurgy process, which uses high temperature to convert the feed materials chemically and separate them so that required metals can be recovered. During this process REEs form rare earth oxides which make the recovery difficult. During this process some of the volatile organic compounds and other dioxins are generated and particular care is needed to manage these products.
- Tailing recycling is a process where reprocessing of the tailings are carried out to extract the remaining amounts of REEs they contain. This generally occurs at the mining stage using the same equipment that are used to process the original ore. This process is economically beneficial (University of Leeds, 2009; Xiaozhi, 2011), having good power savings, and significant environmental benefits.

PROCESS TECHNOLOGIES FOR RECYCLING REE MAGNETS

Since its discovery in 1984 (Croat et al., 1984a, b; Sagawa et al., 1984), rare earth magnets are widely used in hard disc drives (HDDs), wind turbines, and cell phones because of their excellent magnetic properties (Machida et al., 2001; Yamamura et al., 2003). REE magnets are mainly Nd–Fe–B alloys comprising the $Nd_2Fe_{14}B$ matrix, with small quantities of Pr, Gd, Tb, and Dy as well as trace elements such as vanadium, cobalt, niobium, zirconium, titanium, and molybdenum (Croat et al., 1984a, b; Yu and Chen, 1995; Gutfleisch et al., 2011). Rare earth magnets usually contain high amounts of neodymium and smaller amounts of dysprosium and praseodymium. Magnets made of rare earths are usually fragile and get fractured easily but afford the magnets better material properties (Akai, 2008).

In comparison with other permanent magnets, Nd–Fe–B magnets exhibit superior magnetic properties which include magnetic intensity, coercive force, energy densities, and other mechanical properties. Nd–Fe–B material is mostly found in electronic goods, with the highest REE content, which are the main source for REE scrap (Binnemans et al., 2013). Nd–Fe–B magnets are mainly used in HDDs with an annual turnover of 600 million (Walton and Williams, 2011). In recent years, technological innovations have seen about 20%–30% of sintered Nd–Fe–B magnets being wasted as scrap powders in the process of cutting and grinding into the required shape (Buschow, 1994; Binnemans et al., 2013). The REE magnets in the HDD are not easily identified, although they are often removed by waste electronic and electrical Equipment (WEEE).

There are three categories of waste materials generated during the production of rare earth magnets as given below:

- The left out material after vacuum melting, atomizing, or rapid quenching.
- Magnets that gets rejected during the finishing process.
- Magnetic scraps generated during the grinding operation.

There are many processes for extracting rare earth metals from magnets which are present in the e-wastes. Nd magnets easily get oxidized at high temperatures making them difficult to process. Out of the many processes available, hydrometallurgical routes are mainly used to recover rare earths from magnets. These wet processes consume huge amounts of water and chemicals, which bring unavoidable environmental issues. Thus, an economical and convenient method for recycling is in demand. Researchers have proposed a new dry recycling method for magnetic scrap, where a metallic reagent is used to recover the REE. A higher recovery rate has been observed for these processes (Zhang et al., 2010).

The recovery of rare earth magnets are classified into four processes (Schüler et al., 2011).

- Remelting the scrap to obtain the material in an unoxidized state. This is an economical process but it shows a low yield (Kara et al., 2010).
- Recovering materials in its oxide form, although this method is not economical but handling process of these materials are easy.

- Recycling the material in such a way that it can be used as recombination in other magnets.
- Selective recovery of REEs.

Methods used for the recovery of rare earth magnets are as follows (Yamamura et al., 2003):

1. Recovery using molten salts
2. Extraction using liquid metals
3. Re-sintering
4. Melt spinning
5. Formation of slags
6. Hydrometallurgy process

1. Molten slag method

In recent years chloride salts such as $MgCl_2$, KCl, $FeCl_2$, and NaCl are used for separating REEs from magnetic scrap. The rare earths are chlorinated and dissolved in molten salts such as $MgCl_2$, NaCl, and KCl. It has been concluded by studies that rare earths can be recovered efficiently using this method (Yamamura et al., 2003). The binary chlorides are reduced and vacuum distilled to separate rare earths. This method is more efficient than the solvent extraction process (Uda and Hirasawa, 2003).

Compared with all other salts $MgCl_2$ is widely used because of the high affinity of magnesium toward REEs. In Japan, molten $MgCl_2$ was used as a selective extracting agent at 1000°C (Shirayama and Okabe, 2009). Approximately 80% of Nd and Dy could be efficiently extracted as chloride after 12 h of reaction. The process was found to be eco-friendly with a significant yield. Use of LiCl-KCl and NaCl using $AlCl_3$ as flux to recover iron free Nd from Nd-Fe-B magnets is reported by Abbasalizadeh et al. (2015).

2. Liquid metal treatment

Nd is selectively extracted using liquid metals such as Mg, leaving Fe behind. The magnesium is then distilled to obtain Nd (Okabe et al., 2003; Takeda et al., 2006). Research has also been carried out (Takeda, 2004) using molten silver for selectively extracting Nd by oxidation, forming Nd_2O_3.

3. Re-sintering

Nd from hard disks has been successfully recycled by milling and re-sintering and by adding new Nd (Zakotnik et al., 2009). A similar process (Kawasaki et al., 2003), where Nd-rich alloy powder was added to ground magnet scrap before sintering has been reported.

4. Melt spinning

Magnetic powders containing isotropic Nd–Fe–B was recovered from Nickel-coated waste sintered magnets by melt spinning (Itoh et al., 2004). The recovered magnets have good magnetic properties.

5. Glass slag method

REEs are extracted as $RE\text{-}BO_3$ by treating them with boric acid.

6. Hydrometallurgy processes

Mostly 99% (Itoh et al., 2004) of rare earths are precipitated using HCl, HNO_3 forming rare earth chlorides or oxides. High-quality materials are processed through aqueous recycling method but the removal of dissolved species can cause problem.

OBJECTIVES OF THE STUDY

On the basis of the discussion in the preceding sections the current research objectives are proposed. There are limited reports on extracting REEs from Nd–Fe–B magnets using the molten slag method. The present research aims to recover rare earths from the spent magnets using eco-friendly materials such as $MgCl_2$.

- This work focuses on the recovery of REEs in the form of rare earth chloride hydrates under normal atmospheric conditions, which are stable and can be easily stored (Ellis et al., 1994).
- The recovery efficiency of this process will be analyzed.

EXPERIMENTAL DETAILS

Figure 12.7, shows the workflow of the various processes and experiments undertaken during the study.

In this study we have used the FactSage software to determine the reaction phenomena of the $MgCl_2$ salt with the NdFeB magnet, and to investigate the phases which will form during the recycling process at temperatures ranging from 500°C to 1200°C. The phases obtained are $NdCl_X$, $DyCl_X$, $PrCl_X$, $MgCl_X$, Nd_2O_3, Dy_2O_3, Pr_2O_3, and MgOH. From the results it is clear that REEs can selectively react with magnesium chloride thermodynamically above 700°C, while the reaction between other elements such as Fe, B, and Al with $MgCl_2$ will not appear. This indicates that using the chemical reaction it is possible to extract REEs from the magnet scrap.

SAMPLE PREPARATION

Magnetic samples initially had a larger dimensions of 4×4 cm, and had to be cut to smaller sizes of 1×0.5 cm to fit into the alumina crucible using a cutting machine.

For this study, a total number of 10 samples were used, of which 3 samples were subject to 800°C for 12 h; 3 samples were subjected to 1000°C for 12 h; 3 samples were subjected to 1200°C for 12 h;

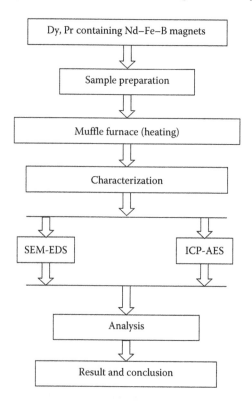

FIGURE 12.7   Flowchart representing the methodology of the project.

and 1 sample was used for characterization. The cut samples were preheated in a muffle furnace at 400°C to demagnetize the samples.

All the experiments were carried out using a muffle furnace (Nabertherm, Germany), which can reach a temperature of 1300°C. The initial and final weights of the samples were measured.

EXPERIMENTAL PROCEDURE

In this process anhydrous magnesium chloride salt (Sigma-Aldrich, 98% pure) is placed in an alumina crucible, such that it entirely covers the magnetic sample as molten salt at higher temperatures. The samples are covered by larger alumina crucible before placing it in the muffle furnace. The samples are then heated to 800°C, 1000°C and 1200°C for 12 h. The reaction scheme (Zhongsheng et al., 2014) is as follows:

$$2Nd(s) + 3MgCl_2(l) = 2NdCl_3(l) + 3Mg(l)$$

$$2Dy(s) + 3MgCl_2(l) = 2DyCl_3(l) + 3Mg(l)$$

$$2Pr(s) + 3MgCl_2(l) = 2PrCl_3(l) + 3Mg(l)$$

The magnetic sample and the salt sample are obtained by crushing the alumina crucible after it cools down. The obtained samples after the experiment and as-received samples are characterized using SEM, EDS, and ICP-AES. In the next section the results of magnetic samples and the salt samples obtained after heating to 800°C, 1000°C and 1200°C for 12 h are presented.

SAMPLE PREPARATION AND CHARACTERIZATION

After every experiment the alumina crucibles are removed carefully from the furnace. The magnetic samples and salt samples are obtained by crushing the crucibles using a hammer. The magnet and salt samples are separated and stored in different polythene covers. The magnetic samples require polishing using 320-1 μm emery papers to obtain satisfactory results using SEM-EDS.

SCANNING ELECTRON MICROSCOPE-ELECTRON DISPERSIVE SPECTROSCOPY

SEM (Gemini, Germany) studies have been carried out on the magnetic samples; Backscattered electron mode has been used to capture the SEM-EDS images to characterize the samples. SEM-EDS spot analysis has been carried to determine the elemental composition present in the magnetic samples before and after the experiments.

INDUCTIVELY COUPLED PLASMA ATOMIC EMISSION SPECTROSCOPY

ICP-AES (Spectro, Germany) analysis has been carried out on the initial magnetic sample and the salt samples to get information about the amount of REEs extracted into the magnesium salt after the reaction.

RESULTS AND DISCUSSIONS

SEM-EDS ANALYSIS

SEM Analysis of the Initial Magnetic Sample

SEM analyses of the untreated magnetic sample were carried out to determine the initial structure and the elemental composition of the sample. In Figure 12.8, the dark phase represents mainly the

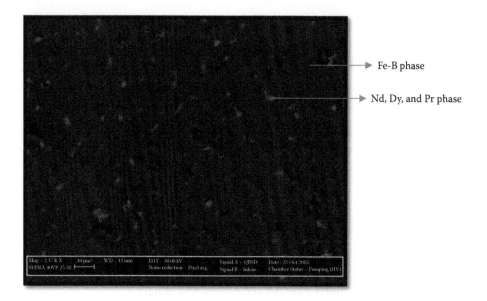

FIGURE 12.8 SEM image of the initial magnet showing the two phases.

Fe phase and the bright phase represents the REEs such as Nd (neodymium), Dy (dysprosium), and Pr (praseodymium).

EDS Analysis

In the EDS analysis, eight spot analyses were carried out to determine the elemental composition of the sample and the obtained results are presented in Table 12.1.

The SEM-EDS analysis was carried out on the magnetic samples, processed at 800°C, 1000°C, and 1200°C for 12 h and the samples were polished using emery papers ranging from 320 to 1 µm before the SEM-EDS analysis.

Samples Processed at 800°C for 12 h (SEM)

The SEM-EDS analysis was carried out on the magnetic samples, processed at 800°C for 12 h. In Figure 12.9, the dark phase represents mainly the Fe phase and the bright phase represents the REEs such as Nd, Dy, and Pr.

TABLE 12.1
Elemental Values Obtained from Intital Magnetic Samples by SEM-EDS Analysis

Spectrum	Fe	Pr	Nd	Dy	Total
Spectrum 1	40.32	14.16	41.73	3.79	100.00
Spectrum 2	15.25	20.13	63.05	1.57	100.00
Spectrum 3	16.81	23.00	57.66	2.53	100.00
Spectrum 4	7.05	23.48	68.84	0.63	100.00
Spectrum 5	56.12	12.46	28.39	3.03	100.00
Spectrum 6	66.03	7.61	23.47	2.89	100.00
Spectrum 7	69.69	6.30	19.94	4.06	100.00
Spectrum 8	67.43	5.84	20.21	3.97	100.00

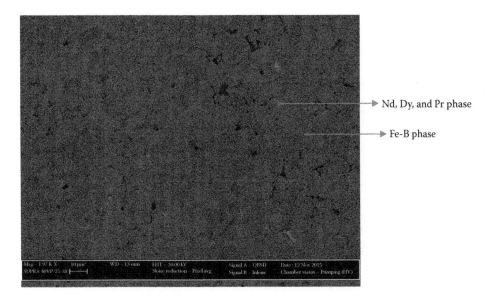

Nd, Dy, and Pr phase

Fe-B phase

FIGURE 12.9 SEM image of the magnetic sample processed at 800°C for 12 h.

EDS Analysis of the Sample

Eight random spot analyses are carried out on the sample and the obtained metal composition are tabulated in Table 12.2.

Samples Processed at 1000°C for 12 h (SEM)

The SEM-EDS analysis was carried out on the magnetic samples, processed at 1000°C for 12 h as shown in Figure 12.10.

There are two phases present in the sample after the reaction, the brighter one represents the phase enriched with REEs and the dark phase represents the phase enriched with Fe.

EDS Analysis

Eight random spot analyses give us the information about the elemental composition of the sample as shown in Table 12.3.

TABLE 12.2
Elemental Values Obtained at 800°C for 12 h through EDS Analysis

Spectrum	Cl	Fe	Pr	Nd	Dy	Total
Spectrum 1	19.58	14.64	17.55	47.12	1.11	100.00
Spectrum 2	20.50	15.34	16.26	47.18	0.72	100.00
Spectrum 3	17.63	21.71	16.12	43.99	0.55	100.00
Spectrum 4	4.68	83.92	2.06	6.63	2.71	100.00
Spectrum 5	1.82	92.17	1.05	3.34	1.62	100.00
Spectrum 6	2.95	91.02	1.37	3.56	1.11	100.00
Spectrum 7	1.78	91.82	0.94	4.39	1.07	100.00
Spectrum 8	3.17	88.41	1.76	4.36	2.29	100.00

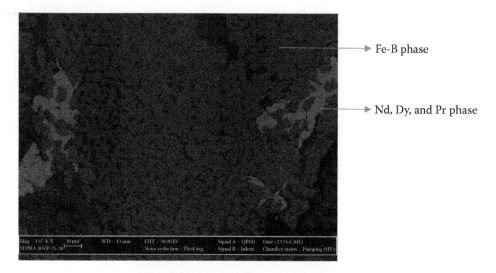

FIGURE 12.10　SEM image of the sample obtained after 12 h of heating at 1000°C.

TABLE 12.3

Elemental Values Obtained at 1000°C for 12 h through EDS Analysis

Spectrum	Cl	Fe	Pr	Nd	Dy	Total
Spectrum 1	20.11	2.75	16.78	59.25	1.12	100.00
Spectrum 2	22.79	1.17	15.60	60.75	0	100.00
Spectrum 3	21.58	1.29	17.56	59.43	0.14	100.00
Spectrum 4	19.30	6.16	17.28	55.86	1.40	100.00
Spectrum 5	5.66	75.11	2.52	12.81	3.90	100.00
Spectrum 6	5.36	84.37	2.27	4.06	3.94	100.00
Spectrum 7	14.27	80.50	0.43	0.96	3.84	100.00
Spectrum 8	6.66	88.38	0	0.64	4.91	100.00

Samples Processed at 1200°C for 12 h (SEM)

The SEM-EDS analyses were carried out on the magnetic samples, processed at 1200°C for 12 h, are shown in Figure 12.11. From the images we can clearly see that there are two different phases present in the sample after the treatment. The brighter phase mainly contains the REEs and the darker phase contains high amounts of Fe compared with REEs.

EDS Analysis

Eight random spot analyses are carried out on the obtained sample and the results are presented in Table 12.4.

From the SEM-EDS results it can be concluded that the REE (Nd, Dy, and Pr) content in the spent magnet samples has decreased at 800°C and 1200°C, whereas there is less change at 1000°C.

ICP–AES Analysis

ICP-AES analysis is carried out on the initial magnetic sample and the salt samples which are obtained after the experiment, to get information the elemental compositions of the samples. Using this analysis we can configure the recovery rate for each experiment. The result of the initial magnetic sample and the obtained salt samples are shown in Table 12.5.

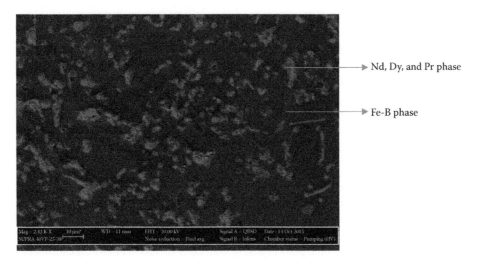

FIGURE 12.11   SEM image obtained after 12 h of heating at 1200°C.

TABLE 12.4
Elemental Values Obtained at 1200⁰ C for 12 h through EDS Analysis

Spectrum	Cl	Fe	Pr	Nd	Dy	Total
Spectrum 1	29.99	10.39	16.53	42.13	0.96	100.00
Spectrum 2	15.09	80.97	0.13	0.69	3.13	100.00
Spectrum 3	26.68	2.92	15.84	53.80	0.77	100.00
Spectrum 4	13.06	81.66	0.09	0	2.41	100.00
Spectrum 5	24.27	2.59	16.24	55.98	0.91	100.00
Spectrum 6	27.38	2.03	15.64	54.00	0.95	100.00
Spectrum 7	14.14	80.05	0.03	1.34	2.44	100.00
Spectrum 8	3.65	80.34	2.82	8.57	4.62	100.00

TABLE 12.5
ICP-AES Analysis of the Samples

Sample	Nd (wt.%)	Dy (wt.%)	Pr (wt.%)
Initial magnet	22.4	1.2	6.7
Salt (800°C, 12 h)	6.68	0.34	2.04
Salt (1000°C, 12 h)	3.99	0.18	1.26
Salt (1200°C, 12 h)	7.94	0.39	2.51

The REE extraction ratio was determined using the following expression.
REE extraction ratio (Shirayama and Okabe, 2009)

$$R_i^{\text{salt/alloy}} = \left[\frac{W_i(\text{salt})}{W_i(\text{alloy})} \right] \times 100 \, (\%)$$

$$W_i \, (\text{alloy}) = \text{Initial mass of element in alloy}$$

$$W_i \, (\text{salt}) = \text{Mass of element in salt after experiment}$$

$$i = \text{Nd}, \text{Dy}, \text{Pr}$$

TABLE 12.6
Extraction Ratios

Sample	Nd (%) (Extraction Ratio)	Dy (%) (Extraction Ratio)	Pr (%) (Extraction Ratio)
Salt (800°C, 12 h)	30	28	30
Salt (1000°C, 12 h)	18	15	19
Salt (1200°C, 12 h)	36	33	37

The results obtained through this method are presented in Table 12.6.

The above results clearly indicate that more than 30% of REE have been recovered from the spent magnet at 1200°C. The recovery rate at 800°C > 1000°C with recovery percentages of 30% and 19%, respectively. The lower recovery rate obtained are due to the reaction being carried out in air and not in an inert atmosphere. At the melting point of magnesium chloride around 714°C, the mixed chlorides lose mass continuously. The Gibbs free energy of reaction between REEs and magnesium chloride increases with increasing reaction temperatures, which implies that a high temperature is not beneficial for the extraction thermodynamically (Zhongsheng et al., 2014). However, at high temperatures the viscosity of molten chloride decreases with increase in temperature favoring the diffusion of the reactant and the product in the melt, which would enhance the recovery of REE. Hua et al. (2014), suggested the mechanism of extraction can follow (1) transport of $MgCl_2$ from bulk melt to the exterior surface of the ash layer through a melt boundary layer, (2) diffusion of molten reactant ($MgCl_2$) through the ash layer to the reaction surface, (3) chemical reaction of REEs with $MgCl_2$, (4) diffusion of molten product ($RECl_3$) outward through the ash layer, and (5) transport of the molten product. Based on shrinking core kinetic model, they concluded that the rate-controlling step was diffusion. At high temperature perfect wettability is achieved which will accelerate the reaction between the molten chlorides and the REE magnetic scraps (Zhongsheng et al., 2014). Above 415°C anhydrous magnesium chloride deliquesces to form $MgCl_2 \cdot nH_2O$, which will turn into MgO with a high melting point of 2852°C, following thermal decomposition reactions of $MgCl_2 \cdot nH_2O$, according to:

$$MgCl_2 \cdot nH_2O = Mg(OH)Cl \cdot 0.3H_2O + (n - 1.3(H_2O)) \ (1 \leq n \leq 2)$$

$$Mg(OH)Cl \cdot 0.3H_2O = MgOHCl + 0.3HCl$$

$$MgOHCl = MgO + HCl$$

This could affect the wettability lowering the extraction efficiency of REEs (Hua et al., 2014). These conflicting phenomena that occur during the extraction process affects the extraction efficiency of REE. The results obtained in this study clearly indicate that the extraction needs to be carried out in an inert atmosphere or vacuum due to the high affinity of the REE and the magnesium for oxygen to form oxides having very high melting temperatures. This will of course reduce the amount of the reactants available in molten form.

$$MgOHCl = MgO + HCl$$

CONCLUSIONS

Extraction of REEs from the magnet scraps using molten $MgCl_2$ has been investigated in this study. The reaction sequence of the samples at 800°C, 1000°C and 1200°C for 12 h have been proposed as follows:

- Images obtained from the SEM clearly shows that after the reaction there is the formation of a separate phase containing a higher percentage of REEs.
- Electron dispersive spectroscopy (EDS) analysis clearly shows that at 1200°C the amount of the Nd, Pr, and Dy content in the magnet is less compared with the samples which are processed at 1000°C and 800°C. A higher recovery rate 30% have been realized at 1200°C.
- The results obtained from the ICP-AES analysis clearly shows that more than 30% of REEs have been extracted from the magnetic sample. The recovery range obtained from the results is in the following manner, that is, 1200°C > 800°C > 1000°C.

All the above points clearly explain that it is possible to extract REEs into molten $MgCl_2$ selectively from spent Nd–Fe–B magnet alloy. The current experiments were carried out under normal atmospheric conditions where partial amounts of $MgCl_2$ gets converted to magnesium oxide perhaps limited the recovery rates. The results clearly indicate that REE extraction need to be carried out in an inert atmosphere or vacuum to increase the extraction efficiency.

FUTURE SCOPE OF WORK

- Investigating the extraction process under inert atmosphere, that is, using argon or nitrogen, which reduces the formation of rare earth oxides and magnesium oxide helping in obtaining higher recovery rates.
- Additional analysis using salts such as $AlCl_3$, KCl, and NaCl, mixed with $MgCl_2$ can be used for the process which helps in extracting the rare earths at lower temperatures.
- Recovering the magnesium which gets evaporated during the process at higher temperatures.

REFERENCES

Abbasalizadeh A., L. Teng , S. Sridhar, and S. Seetharaman. 2015. Neodymium extraction using salt extraction process. *Mineral Processing and Extractive Metallurgy*, 124(4): 191–198.
Akai, T. 2008. Recycling rare earth elements. *AIST Today*, 29: 8–9.
Armbrustmacher, T. J., P. J. Modreski, D. B. Hoover, and D. P. Klein. 1995. Mineral deposits models: The rare earth element vein deposits. United States Geological Survey. Retrieved from http://pubs.usgs.gov/of/1995/ofr-95–0831/CHAP7.pdf.
Bertuol, D. et al. 2009. Spent Ni-MH batteries—The role of selective precipitation in the recovery of valuable metals. *Journal of Power Sources*, 193: 914–923.
Binnemans, K., P. T. Jones, B. Blanpain, T. Van Gerven, Y. Yang, A. Walton, and M. Buchert. 2013. Recycling of rare earths: A critical review. *Journal of Cleaner Production*, 51: 1–22.
Bloomberg News, 2010. Global rare earth demand to rise to 210,000 metric tons by 2015. Estimate provided by Wang Caifeng, Secretary General of the Chinese Rare Earth Industry Association, October 18, 2010.
Buschow, K. H. J. 1994. Trends in rare-earth permanent-magnets. *IEEE Transactions on Magnetics*, 30: 565–570.
Croat, J. J., J. F. Herbst, R. W. Lee, and F. E. Pinkerton. 1984a. Pr-Fe and Nd-Fe-based materials: A new class of high-performance permanent magnets. *Journal of Applied Physics*, 55: 2078.
Croat, J. J., J. F. Herbst, R. W. Lee, and F. E. Pinkerton. 1984b. High-energy product Nd-Fe-B permanent-magnets. *Applied Physics Letters*, 44: 148–149.
Davies, E. 2011. Critical thinking. *Chemistry World,* January, pp. 50–54.
Du, X. and T. E. Graedel. 2011. Global in-use stocks of the rare earth elements: A first estimate. American Chemical Society. *Environmental Science & Technology*, 45: 4096–4101.
Ellis T. W., F. A. Schmidt, and L. L. Jones. 1994. Methods and opportunities in recycling of rare earth based materials. In *Metals and Materials Waste Reduction, Recovery, and Remediation*, edited by K.C. Liddell. Warren dale, Pennsylvania: TMS, pp. 199–208.
Fosse, M. P. and V. J. S. Graunch. 1995. Low-Ti iron oxide Cu-U-Au-REE deposits (Models 25i and 29b; Cox, 1986a, b): Summary of relevant geologic, geoenvironmental and geophysical information. United States Geological Survey Open-file Report No. 95-0831 (Chapter 22). Retrieved from http://pubs.usgs.gov/of/1995/ofr-95-0831/CHAP22.pdf.

Gutfleisch, O., M. A. Willard, E. Bruck, C. H. Chen, S. G. Sankar, and J. P. Liu. 2011. Magnetic materials and devices for the 21st century: Stronger, lighter, and more energy efficient. *Advanced Materials*, 23: 821–842.

Hua, Z., J. Wang, L. Wang, Z. Zhao, X. Li, Y. Xiao, and Y. Yang. 2014. Selective extraction of rare earth elements from NdFeB scrap by molten chlorides, *ACS Sustainable Chemistry and Engineering*, 2(11): 2536–2543.

ICF International. 2011. Literature review of urban mining activities and E-waste flows. Prepared for the U.S. Department of Energy, Office of Intelligence and Counterintelligence, January 21.

IMCOA. 2011. Meeting rare earth demand in the next decade. Reports of Industrial Minerals Company of Australia, March 2011.

Itoh, M. et al. 2004. Recycling of rare earth sintered magnets as isotropic bonded magnets by melt spinning. *Journal of Alloys and Compounds*, 374: 393–396.

Kara, H., A. Chapman, T. Crichton, P. Willis, and N. Morley, 2010. Lanthanide resources and alternatives. A report for Department for Transport and Department for Business, Innovation and Skills, Oakdene Hollins Research & Consulting, 26 May 2010.

Kawasaki, T., M. Itoh, and K. Machida. 2003. Reproduction of Nd-Fe-B sintered magnet scraps using a binary alloy blending technique, Japan. *Materials Transaction*, 44(9): 1682–1685.

Kuzuya T. et al. 2003. Hydrometallurgical process for recycle of spent nickel-metal hydride battery. In *Yazawa International Symposium*, edited by F. Kongoli et al. TMS 2003, Sandiego, CA, USA. 365–372.

Merten, D. and G. Büchel. 2004. Determination of rare earth elements in acid mine drainage by inductively coupled plasma mass spectrometry. *Microchimica Acta*, 148: 163–170.

Machida, K., Suzuki, S. and N. Ishigaki. 2001. Resources and recycle for rare earth magnets. *Kidorui* 39: 27.

Meyer, L. and B. Bras. 2011. Rare earth metal recycling. In *Sustainable Systems and Technology, 2011 IEEE International Symposium*, Chicago, Illinois, 16–18 May, pp. 1–6.

Modreski, P. J., T. J. Armbrustmacher et al. 1995. Mineral deposits models: Carbonatite deposits. United States Geological Survey. Retrieved from http://pubs.usgs.gov/of/1995/ofr-95-0831/CHAP6.pdf.

Okabe T. H. et al. 2003. Direct extraction and recovery of neodymium from magnet scrap. *Materials Transactions*, 44: 798–801.

Oko-Institut e.V. 2011. Environmental aspects of rare earth mining and processing. In *Study on Rare Earths and Their Recycling*. Öko-Institut e.V., Germany, p. 162.

Panayotova, M. and V. Panayotov. 2012. Review of methods for the rare earth metals recycling. *Annual of the University of Mining and Geology "ST. IVAN RILSKI," Vol.* 55: *Part II, Mining and Mineral Processing*, 2012. St. Ivan Rilski, Sofia, pp. 142–157.

Rabah, M. 2008. Recyclables recovery of europium and yttrium metals and some salts from spent fluorescent lamps. *Waste Management*, 28: 318–325.

Sagawa, M., S. Fujimura, N. Togawa, H. Yamamoto, and Y. Matsuura. 1984. New material for permanent magnets on a base of Nd and Fe. *Journal of Applied Physics* 55: 2083.

Schluep, M., C. Hagelueken et al. 2009. Recycling- from e waste to resources. United Nations Environment Programme, United Nations University, July. Retrieved from http://www.unep.org/PDF/PressReleases/E-waste_publication_screen_FINALVERSION-sml.pdf.

Schüler, D., M. Buchert, R. Liu, S. Dittrich, and C. Merz. 2011. Study on rare earths and their recycling. Final report for the Greens/EFA Group in the European Parliament, Öko-Institut e.V., Darmstadt, January 2011, p162, Öko-Institut e.V. Freiburg, Germany. http://www.resourcefever. Org/publications/reports.

Shirayama, S. and T. Okabe, 2009. Selective extraction of Nd and Dy from rare earth magnet scrap into molten salt. In *Processing Materials for Properties*. The Minerals, and Metals & Materials Society, 3rd, 2009 Meeting, San Francisco, CA, USA, February 15–19, 2009, pp. 469–474, PMP III Wiley, Hoboken, NJ.

Tabuki, H. 2010. Japan recycles minerals from used electronics. *The New York Times*, October 4. Retrieved from http://www.nytimes.com/2010/10/05/business/global/05recycle.html.

Takeda O., T. H. Okabe, and Y. Umetsu. 2004. Phase equilibrium of the system Ag-Fe-Nd and Nd extraction from magnet scraps using molten silver. *Journal of Alloys and Compounds*, 379: 305–313.

Takeda, O., T. H. Okabe, and Y. Umetsu. 2006. Recovery of neodymium from a mixture of magnet scrap and other scrap. *Journal of Alloys and Compounds*, 408–412: 387–390.

Uda T. and M. Hirasawa. 2003. Rare earth separation and recycling process using rare earth chloride In *Yazawa International Symposium*, edited by F. Kongoli et al. TMS 2003, Sandiego, CA, USA, pp. 373–385.

University of Leeds. 2009. Valuable, rare, raw earth materials extracted from industrial waste stream. *Science Daily*, December 19. Retrieved from http://www.sciencedaily.com/releases/2009/12/091215101708.htm.

Walton, A. and A. Williams. 2011. Rare earth recovery. *Materials World*, 19: 24–26.

Xiaozhi, Z. 2011. The carbon reduction effect of the tailings recycling. School of Resources & Environment Engineering, Shandong University of Technology, Zibo, Shandong, People's Republic of China, April.

Yamamura T. et al. 2003. Mechanism of electrolysis of rare earth chlorides in molten alkali chloride baths. In *Yazawa International Symposium*, edited by F. Kongoli et al. TMS 2003, Sandiego, CA, USA, pp. 327–336.

Yu, Z. S. and M. B. Chen. 1995. *Rare Earth Elements and Their Applications*. Beijing, People's Republic of China: Metallurgical Industry Press.

Zakotnik, M., I. R. Harris, and A. J. Williams. 2009. Multiple recycling of NdFeB-type sintered magnets. *Journal of Alloys and Compounds*, Institution University of Birmingham, 469(1–2): 314–321.

Zhang, X., D. Yu, and L. Guo. 2010. Test study new process on recovering rare earth by electrical reduction— P507, extraction separation method. *Journal of Copper Engineering*, 1, 1009-3842-0066-04.

Section III

Environment

13 Building Suitable Restoration Approaches in the Brownfields

Singarayer K. Florentine, Patrick Graz, Augustine Doronila,
Rachael Martin, Kim Dowling, and Nimesha Fernando

CONTENTS

ABSTRACT

Human activity has, in the recent past, resulted in substantial changes in land cover, ecosystem health, and the ability of affected ecosystems to return to their original state. This necessitates further human intervention to recreate the systems functions than the present. Earlier restoration activities have not been documented extensively. This hinders our efforts to identify approaches that might support further work. To a large extent this is attributable to a lack of clearly stated aims that would facilitate targeted management approaches. In this chapter, we propose a theoretical foundation for the development of restoration activities and programs using a "state and transition" model approach. In this context, "states" are a combination of components and processes that define the system's existence. "Transitions," on the other hand, are those activities necessary to move a system between states. Our approach recognizes the possibility of alternative states to which a system might be changed. This is particularly the case of brownfields, which are industrial or commercial sites that are inactive or underused because of real or perceived environmental pollution. Moreover, this approach also recognizes that the activities that cause the transition from one state to another are likely to differ. We discuss phytoremediation as an important restoration tool, to show plants may be used in the restoration of even severely contaminated sites. Nevertheless, it must be recognized that not all plants are suitable in all situations and it is necessary to select potential species and planting methods with care. We then provide a list of desirable plant characteristics to be considered when selecting plants for such applications.

INTRODUCTION

Intensive human activities in the last 50 years, such as clearing native vegetation for agriculture, infrastructure, and industrial development, have altered the health of ecosystems to a greater and more rapid extent than at any other comparable period of history. Land degradation may increase still further during the first half of this century (Millennium Ecosystem Assessment, 2005), accelerated by increases in resource use, population growth, and a rapidly changing climate. Compounding this concern is that activities such as mining and storage of quantities of chemical and mineral processing waste can cause negative impacts on the land for millennia (Kempton et al., 2010). Worldwide, many contaminated and subsequently abandoned sites have been left undeveloped, due to potential environmental and social threats (Albering et al., 1999; Gallagher et al., 2008; Luo et al., 2012). There are approximately 294,000 and 300,000 ha of contaminated sites in the United States of America (USEPA, 2004) and the United Kingdom (DEFRA, 2006) respectively, which are unavailable for either public or commercial usage, but which are now coming under increased pressure for reuse. At present, there are no figures available for Australia.

In view of such high rates of environmental change and the increasing amount of land taken from the available landstock, direct human intervention is essential to implement appropriate restoration techniques on these contaminated lands. This urgency was already highlighted two decades ago by Wilson (1992), and is now becoming even more pressing. The rehabilitation of these so-called *brownfields*, which is our focus here, is of particular interest, as they not only represent an important class of degraded land but also a long-term potential hazard to human wellbeing (Gallagher et al., 2008). The studies reported on by CABERNET (2005), Qian et al. (2012), and Luo et al. (2012) have shown, for instance, that many brownfield sites represent environmental and social risks because of contamination by a range of organic and inorganic pollutants.

A compounding factor, which makes concerted action on brownfield sites somewhat complicated, is that there is considerable discussion regarding the definition of a *brownfield*. Dowling et al. (this volume) have illustrated the complexities and ramifications of this debate by providing a comprehensive and scholarly description of this issue. In this chapter, we have pragmatically adopted the definition of brownfields as "…any land, which has been previously developed, including derelict and vacant land, which may or may not be contaminated" (Dixon, 2006). This has been chosen instead of that advanced by Tedd et al. (2001), which suggests that brownfields are "…lands which have been abandoned, contain or potential presence of a hazardous substance, pollutant, underutilized for industrial and commercial activities where redevelopment is problematical by real or perceived contamination." For our purposes, the latter definition is too narrowly focused on industrial and commercial activities—we strongly suggest that appropriate rehabilitation and development of brownfields should also allow options for social and environmental uses.

Over the past decade or more, brownfield redevelopment across the world has been a key item on the environmental agenda. Yet, as Bardos et al. (2000) pointed out, brownfield revitalization can only be of economical and sustainable benefit to the broader community if a sustainable and environmentally appropriate approach is involved. It is important to emphasize that while the conversion of brownfield into sustainable land is widely encouraged and supported, it presupposes that issues regarding the restoration approach used to revitalize such areas have been carefully planned and managed.

We present a theoretical framework to underpin the development of site-specific restoration work in order to ensure that best-practice approaches can be implemented. While we keep our initial focus fairly general, we later provide two examples of the implementation of phytoremediation in the restoration of brownfields. Finally, we provide a list of desirable plant characteristics to be considered when selecting plants for such applications.

STATE AND TRANSITION MODELS IN BROWNFIELDS RESTORATION

Within the broad context of environmental management, there is much support for State and Transition Models, which typically describe multiple, stable communities of varying composition (states) that are separated in space or time by discrete changes (transitions) (Westoby et al., 1989).

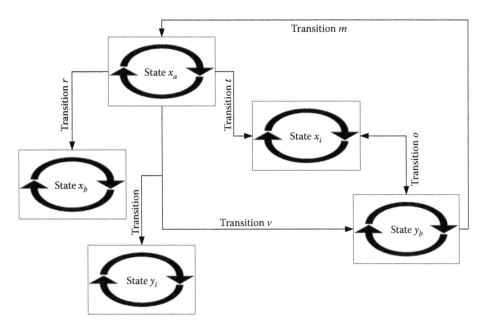

FIGURE 13.1 Possible states of a system. Each state is defined by its components and the processes that govern it. Transitions (black lines) occur when conditions within a state change beyond a given threshold particular to that condition.

We suggest that such an approach is eminently suitable for the restoration of brownfields, since these sites typically suffer from considerable, often multiple, disruptions to their original state, and thus will require sequential and long-term management strategies that are linked to separate transitions during the overall restoration process.

Consider Figure 13.1 showing five possible states of a given system. Here we consider a "state" to comprise of a set of biotic and abiotic components, together with the processes that govern the relationships between these components. The components and processes are not necessarily unique to a particular state but their relative importance may change. For instance, different savanna rangeland different states will have tree and grass components. In the different states, however, the components will differ in their presence and in the effect that they have on the development of the other component. A system remains in a given state, as long as the conditions and processes governing it will remain within particular boundaries.

However, a "transition" from one state to another occurs when one or more conditions or processes defining the state are changed beyond a critical threshold. In other words, a change in a state indicates that one or more conditions and/or ecological processes that contribute to the existence of the state have been disturbed in such a way that they cannot return spontaneously to their previous condition.

What makes the restoration of a system back to its previous state most difficult, is that within a complex system such as an environmental community, reinstatement of the initially altered processes alone might not cause the complex system to return to its overall original state. This is because when one process has been disturbed past a critical threshold, other related and integrated processes are also disturbed past their critical threshold. As a result, restoration efforts may require considerable and unexpected additional external input to address both the initial and other flow-on disturbances of the system in order that overall stability of the original state is restored.

In this chapter, we have used the state-and-transition model approach to provide guidance to our understanding of the restoration process for brownfields. These brownfields, whose initial states have been significantly disturbed, are now such complex systems that detailed knowledge of the particular example's components and their interactions are necessary, and it is essential to appreciate how intended management actions designed to address a particular process may affect the final environmental balance. Whilst instituting any restorative action, each of the contributing processes of a site will be open to modification, and thus must be carefully monitored to ensure that their disturbance does not compromise the intended outcome. At the same time, it is important to appreciate that a state itself is a dynamic system that not only consists of various physical components such as soil, water, and plant life, but necessarily includes processes that will ensure that the overall system will remain in a given state. These processes can include the natural regeneration of plants and soil, a regular inflow of water, or the managed removal of material from the site.

Restoration strategies, in general, represent those actions that need to be implemented to reinstate or assist the processes that will move the site to a desired state. To appreciate the use of the state and transition model in the context of restoration, and to illustrate the complexity of the task of restoring brownfield areas, Table 13.1 has been constructed to show three hypothetically possible alternative restoration states of the same brownfield site where there are several desired end usages.

TABLE 13.1
Alternative States to Which a Particular Brownfield Might Be Restored

Original State	Alternative States
	Parkland for Recreation
	Characteristics:
	An open area of lawn with scattered trees and flower-beds.
	A playground suitable for children aged 3–12, together with benches for supervising parents
	A toilet block
	Processes:
	Maintenance of garden beds
	Maintenance of lawn
	Maintenance of facilities
	Maintenance of individual trees
	Wetland for Biodiversity Conservation
	Characteristics:
Brownfield	A wetland with three ponds fed by runoff from surrounding urban areas.
	A bird-hide to facilitate bird observation
	Processes:
	Monitoring and maintenance of water supply
	Monitoring and maintenance of wetland flora
	Monitoring and maintenance of wetland fauna
	Weed and problem animal control
	Maintenance of facilities
	Sports Facility
	Characteristics:
	A 400 m running track surrounding a soccer field
	A building providing ablution facilities and changing rooms
	A grandstand for 3000 people
	Processes:
	Maintenance of infrastructure and facilities
	Management of infrastructure and its users

This table also shows a number of processes that are required to maintain the individual states, in order to ensure their continuity. This list of possible alternative states or the maintaining processes is not complete, but is merely provided for discussion.

It is important to note here that the processes relevant to the restoration of each of the states differ, although we may perceive that a certain overlap occurs, such as the maintenance of required facilities. However, because of the specific parameters defining the three end uses, there are considerable differences at this final level between maintenance of playground equipment versus maintenance of a bird-hide versus maintenance of a running track. Successful restoration of severely disturbed lands is not a quick or simple process, and each restoration attempt involves unique understandings of the type of initial disturbance, the concomitant disturbance of related processes, the phased introduction of mitigation strategies, and, as indicated above, the ways in which maintenance of the desired end use can be ensured.

The complexity of these systems means that as the characteristics of the alternative states change, so will the underlying equilibrium processes. Let us, for instance, change the physical characteristics of the parkland to a cycle track leading through a healthy, natural looking tree population. The supporting processes would then be required to change to the maintenance of the cycle track, and away from arboriculture toward silviculture. For optimal end-state outcomes, therefore, restoration efforts should not only focus on aspects such as the reconstruction of vegetation structure, composition, and/or diversity, but also on the restoration of functions and processes that maintain the state (Hobbs and Norton, 1996).

The important notion of "sustainability" has been incorporated as key outcome of restoration efforts (Society for Ecological Restoration International Science & Policy Working Group, 2004). This approach has a particular emphasis on ecological outcomes, intimating that at some point external restoration activities should become unnecessary. In this respect, Dixon (2006) defines sustainable development in relation to brownfields restoration on the basis of the respected Brundtland commission report (Brundtland, 1987), suggesting sustainable development is "Development that meets the needs of the present without compromising the ability of future generations to meet their own needs." However, in the context of restored brownfields, which are often lands with considerable potential economic and social value, this issue also requires a statement of *what is to be sustained?* Are they environmental and/or economic and/or social values? Development and subsequent management of brownfield sites therefore needs to develop and maintain the relevant processes that sustain the desired state, condition, and productivity of the restored system, which again adds an additional level of complexity to the task.

Notwithstanding the desire to develop a sustainable system, current natural processes may not always be sufficient to sustain a restored area as necessary, and technological and other means may need to be continued to be implemented to ensure that a critical process is provided. For instance, an irrigation system rather than natural drainage may be needed to augment water to a wetland to ensure minimum water quality or quantity. Of course, as the quality of drainage water entering the wetlands improves, this may partially or totally replace the need for artificial water input. Similarly, natural regeneration of biotic material may initially be limited, necessitating artificial regeneration or supplementation of species. Both these examples show not only the potential need for selected technological or community-based restoration inputs into a disturbed areas, they highlight the possibility that processes may be needed to maintain a system over an uncertain self-establishment period.

MANAGEMENT OF THE TRANSITION PROCESS

SELECTING THE DESIRABLE STATE—SETTING OBJECTIVES

Over the past decades, governmental, nongovernmental, and private institutions have allocated substantial resources in support of efforts aiming to restore degraded land (Wilkins et al., 2003).

Though it is generally agreed that actions must be taken to "fix" degraded lands, it is not clear at this stage, whether or not many of these past efforts have successfully achieved their intended outcomes. This uncertainty is, in part, due to a dearth of quality monitoring activity, often because the intentions or targets identified by the initiators were not sufficiently well articulated to allow meaningful evaluation (Wortley et al., 2013).

The desirability of well-defined targets for management planning has received considerable attention in a multitude of disciplines in the past (Steiner, 2008). Similarly, the articulation of clearly defined desired target states is a quintessential component of management and management planning for restoration. Without a clearly defined end-state, inconsistent or mutually opposing management strategies may be implemented or may incur expenses with little to show in the end.

Transition: Ecological Restoration

While the potential outcomes of the restoration of brownfields represent states within the STM, the transitions are the processes of restoration that are necessary to shift the system from one state to another. We must assume that transitions require active intervention to initiate the processes required to push a system from an undesirable state to a desirable one.

A key step in restoration is to distinguish between the symptoms and the underlying causes of a degraded state because of the complex interrelationships between processes in these brownfields situations. This understanding will allow a clearer identification of the specific factors that may be preventing successful restoration of a site to a desired state. Relevant here, is the work of Hobbs and Harris (2001) who emphasized the careful sequencing of restoration strategies in terms of abiotic and biotic processes. In the context of brownfield restoration, site contamination by one or more chemical residues is probably a key (abiotic) issue that may limit the possibility for reestablishment of a particular plant cover or species (biotic). A first step in restoration would therefore be to rehabilitate the soil using approaches to mitigate the abiotic factors in order to bring the site to a condition that would then permit the introduction of desired plants species and give them a chance to survive. Without the removal of the contaminating substances, the introduced species are not likely to reproduce and move the site to the desired state.

Whereas Hobbs and Harris (2001) discussed this important insight, we have modified their approach as shown in Figure 13.2. This modification considers the pragmatics of a transition from any given state to another. Specifically we need to highlight the fact that social and economic considerations may represent additional thresholds or limits to restoration if we are to take cognizance of the wider definition of sustainability mentioned earlier. This notion will now be further examined.

Emphasis of restoration in general has, in the past, been on the ecological processes affecting restored systems (Society for Ecological Restoration International Science & Policy Working Group, 2004; Hallett et al., 2013), yet social aspects clearly play an integral role in restoration (Wortley et al., 2013). In this respect, environmental degradation is often the result of human activity (Aronson et al., 2010) and remediation might require (possibly unpopular) social behavioral changes. Alternatively, restoration processes may have significant economic implications, for instance in the cost of mediating strategies and the subsequent influence on the productivity of the land (Aronson et al., 2010). Indeed, the implementation of any form of restoration activity itself is based on conscious decisions by members of society, both in terms of initial and ongoing costs, and in terms of the change of social practices occasioned by the anticipated end state.

Understanding the complexity of this restoration task, Dixon (2006) took a triple bottom line approach in the context of brownfields restoration, one that balances a consideration of ecological, social, and economic aspects. Dixon advised that these three components, together with their supporting processes, should be part of the definition as well as the selection of the desired (or alternative) end states. In this work, Dixon provides a variety of examples of potential social (such as improved quality of life), economic (such as job creation) and environmental (restoration of environmental quality) goals relevant to the context of the restored site.

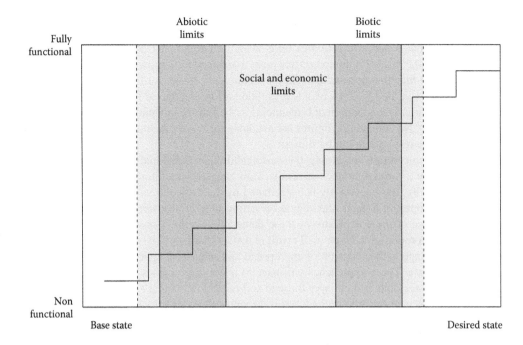

FIGURE 13.2 Conceptual model of system transition between states of varying levels of function with three thresholds, based on the model of Hobbs and Harris (2001).

It follows from this observation that transition from one state to another will therefore not only be dependent on the management of abiotic and biotic system components, but also on the social and economic settings in which they function. In most instances, these components will be social aspects such as community support and involvement for a desired end state, or economic consideration which will allow the availability of funds to affect a given transition.

At a technical level, a great deal of the focus of past restoration activities has been on the establishment of vegetation components such as trees or shrubs. Although we consider reintroduction of plant material to be considered in the restoration of brownfields, the danger in this biotic focus lies, however, in the emergence of a "one size fits all" approach to restoration. As presaged earlier, brownfield contaminants range from solvents and hydrocarbon to heavy metals (Gallagher et al., 2008; Qian et al., 2012; Sun et al., 2013). On most brownfield sites contaminated by heavy metals, the concentration are often well above the tolerance threshold of most plants, and the metals will affect metabolic process and eventually kill the plants (Gallagher et al., 2008). Hence the removal of such contaminants is an essential prerequisite step to any successful introduction of biotic material. However, contaminants such as heavy metals are expensive and difficult to remove from the soil (Qian et al., 2012), and this has immediate economic ramifications. There are a number of techniques available to remove heavy metals from contaminated soils and wastewater systems, but these are all expensive, time consuming, and require somewhat specialized procedures such as oxidation, precipitation, filtration using suitable membranes, ion exchange, and biological treatments (Tordoff et al., 2000). On the other hand, there are a range of plant species that can withstand high levels of various contaminants by exclusion and detoxification (Arthur et al., 2005; Rai, 2008) which might provide convenient vegetation strategies. In addition to chemical and physical approaches for removal of contaminants, there are an alternative suite of possibilities including capping, electrical resistance heating, steam injection, phytoremediation, or phytostabilization approaches, all or some of which can be considered in an effort to provide cost effective, sustainable, and environmentally friendly alternatives (Desouki and Feng, 2012; Qian et al., 2014).

PLANNING THE TRANSITION PROCESS

It is well known that restoration activities have swallowed significant amounts of money in the past (De Sousa, 2002), but it is anticipated that costs associated with brownfield restoration will be significantly higher than that of general habitat restoration efforts. According to the USEPA (2004) and NRTEE (2003), in the U.S. brownfields restoration efforts increased from U.S. $100 billion to over 650 billon. Similarly, restoration of brownfields in the European Union cost an estimated €100 billion (EEA 2000). No comparable figures are available for Australia, but it is likely that costs for each brownfield site restoration will be similar.

There is some uncertainty regarding the sustainability of individual restoration efforts with brownfields as only a limited number of results from long-term monitoring have been published (Wortley et al., 2013). The uncertainty is exacerbated by the difficulty with which techniques and technologies developed in a particular region or setting may be transferred to another (Wortley et al., 2013) because of the complexity of these damaged systems. Consequently, if we consider that successful restoration procedures will result in a state that supports sustainable plant growth, then the key challenge is the selection of appropriate and sequential approaches, which will meet this desired end state. These approaches will have to relate to processes that are associated with (i) developing a sustainable ground layer through soil rehabilitation (e.g., through the neutralizing or removal of toxins, stabilization of the top soil through colonization by cryptograms, facilitating water infiltration, and ensuring that nutrient levels are compatible with indigenous plants) and (ii) selection of appropriate vegetation (which will require plants which regenerate naturally, and estimating how long different structural components will take to develop so that external intervention can be attenuated).

MONITORING THE TRANSITION PROCESS

As indicated earlier, there is only limited evidence available to assess the degree of "success" of many—if not most—restoration activities undertaken across Australia and other parts of the world (Wortley et al., 2013). This lack of reflective evaluation is hampering the development of restoration efforts in general, and in particular, it makes it difficult to extend the application of successful techniques to similar areas. Given the overall costs involved in any sort of large-scale restoration, it is thus economically prudent to revisit earlier restoration attempts to develop more informed, evidence-based prescriptions for future restoration work (Hobbs and Mooney, 1993; Hobbs and Norton, 1996; Wilkins et al., 2003; Hobbs and Lindenmayer, 2007; Lake et al., 2007; Lindenmayer et al., 2008), and to adopt an adaptive, reflective management approach (see Kingsford et al., 2011; Resilience Alliance, 2015).

In the light of this suggested approach, restoration efforts could be planned as "experimental" sites, an approach that will naturally facilitate an objective evaluation of outcomes as part of a research effort. Michener (1997) has stressed that, without proper ecological audit of existing restoration sites, future restoration projects will not benefit from past attempts, a notion that has been termed in other disciplines as "double-loop learning." Michener (1997) had previously identified the need for empirical evidence to address the following generic questions: (i) Have restoration activities succeeded or failed? (ii) How and what was the science behind the success or failure? (iii) What should be done in future attempts? (ii) Will these tested procedures be successful in another environment? (iv) What will the procedures cost [both in financial and human terms]? (iv) Can they be carried out more economically without prejudicing outcomes? These questions are still relevant and pressing today, and further delay in addressing them risks a continuation of practices that do not provide optimal ecological return for the financial and human investment.

As a codicil here, it should be noted that, given the equal importance of restored components and their supporting processes, it is necessary to monitor both. For instance, if we consider the outcome of a restoration initiative to be the reinstatement of a self-sustaining ecosystem, it is important to monitor not only vegetation structure or species diversity, but also the concomitant ecosystem processes (Elmqvist et al., 2003; Dorren et al., 2004) such as successful regeneration.

PHYTOREMEDIATION IN BROWNFIELD SITES

Some recent progress in brownfields redevelopment in the United States has been documented in a number of reports, including USEPA (2004). One particular project report of interest here relates to "Brownfields Sustainability Pilots," which are an Environmental Protection Agency (EPA) effort to promote environmental sustainability within local brownfields projects (USEPA, 2004). The EPA provides communities with advice and technical assistance to help them achieve greener assessment, cleanup, and redevelopment of their brownfields, together with practical assistance support activities such as the reuse and recycling of construction and demolition materials, green building and infrastructure design, energy efficiency, water conservation, renewable energy development, and native landscaping (Schädler et al., 2011). These pilot projects represent a demonstration of best practices that can be used as exemplars by other communities across the country.

Closely related to these pilot projects is the increasing interest, over the past 20 years, in the use of phytoremediation techniques to address the restoration of polluted areas. By harnessing the natural capabilities of a range of plants, remediation can be implemented on toxic soils, contaminated ground and surface water, and unwanted sediments. Phytoremediation is a low-cost alternative to traditional brownfield remediation by physical means. Instead of removing tons of toxic soil and filling the site with new clean soil, plants can be used to remove contaminants from the soil and store it within their plant tissues (McCutcheon and Schnoor, 2003). There are several advantages to phytoremediation compared to other cleanup technologies, including the way that plants can be effectively and economically used to remove, degrade, and contain contaminants in an aesthetic, natural, and passive way. Ever since Raskin et al. (1994, 1997) performed the first phytoremediation trial, the U.S. EPA has consistently and significantly funded research into the use of plants to clean up severely degraded environments.

Spinoffs from applying phytoremediation rather than physical remediation or removal of soil mirror the benefits of a generally green environment for those people who live, work, learn, and play near a disturbed site. Using a more gentle natural remediation approach can contribute to the creation of well knit, resilient neighborhoods, and can contribute to the healing of traumas caused by past ecological and environmental degradation. These potential additional benefits of phytoremediation need further consideration and research in the context of the greening of brownfields. In the next section, a more detailed look at the development and potential use of phytoremediation will be presented.

EARLY DEVELOPMENT OF PHYTOREMEDIATION

It is of interest that the concept of using plants to clean up contaminated environments is not new. Approximately 300 years ago, plants were proposed for use in the treatment of wastewater in Berlin, Germany (Hartman, 1975). More recently, the idea of using plants to extract metals from contaminated soil was subsequently revived about 30 years ago and developed by Chaney (1997) with the first formal field trials on zinc and cadmium phytoextraction being conducted by Baker et al. (1991).

The phytoremediation of contaminants present in built environments can occur through several mechanisms. Some plants destroy organic pollutants by degrading them directly through the production and secretion of acids and enzymes into the surrounding soil, which react with these compounds and produce harmless residues. Other plants enhance degradation of organic pollutants by providing nutrients to the microbial communities in their rhizosphere (the soil surrounding the roots). There are also other plants called hyperaccumulators that take up inorganic contaminants, such as heavy metals, from soil or water and concentrate them in the plant tissue or root (Brooks et al., 1977).

The technology is attractive because the cost of phytoremediation techniques is estimated to be from 20% to 50% less than the highly engineered physical, chemical, or thermal techniques, an important consideration when there are limited funds available for environmental cleanup. This approach is particularly relevant to less economically developed countries, which have often suffered from a legacy of chemical pollution and are now unable to provide substantial funding to remove the pollutant source.

Whilst phytoremediation research is currently active in every continent, much of the work has been done in the United States (van der Lelie et al., 2001; McCutcheon and Schnoor, 2003). Sadly, it has only been in the past 10 years that phytoremediation studies have been undertaken in tropical regions, since these are often emerging markets that are experiencing major pollution problems due to rapid industrialization. Notwithstanding this relatively short research record, a number of countries are now taking the lead in harnessing this green technology, and these include China, Thailand, Brazil, Chile, and India.

Now, the science of phytoremediation has shown promising results as an innovative cleanup technology, but it is still in relatively developmental stage and more research is needed to increase the understanding and knowledge of this remediation technology. For example, one of the major problems encountered with the phytoremediation of heavy metals is their decreased bioavailabilty to plants. Generally, soil amendments are added to the soil to increase the bioavailability of heavy metals to enhance the uptake by plants, but a number of environmental concerns pertaining to the use of soil amendments have arisen and this issue will be addressed in a later section. Ongoing bench-scale studies and field demonstrations are being conducted throughout the United States in order to better understand and implement phytoremediation, and as this work progresses it is expected that it will increase its share in the environmental cleanup market. Some early work by D. Glass Associates, Inc. (Glass, 1999), which estimated markets for the field of phytoremediation, indicated that for 1998, the projected market was $16.5–$29.5 million, in the year 2000 the market was estimated at $55–$103 million, and by the year 2005, it has been estimated to reach $214–$370 million (Glass, 1999).

The European Union, through its COST Action 837 program, presented its major findings in October 2009 in Ascona, Switzerland, and showed that there was still a significant need to pursue both fundamental and applied research to provide low-cost, low-impact, visually benign, and environmentally sound remediation strategies. These strategies are well suited for use: (i) on large sites where other methods of remediation are not cost effective or practicable; (ii) at sites with low concentrations of contaminants where only a "polishing treatment" is required over long periods of time; and (iii) in conjunction with other technologies where vegetation is used as a final cap and closure of the site. Using different plants, phytoremediation can be applied as a containment measure for decomposition of the pollutant or as a removal or extraction technique. Through the development and evaluation of new, soft, appropriate, and efficient biological processes, it is possible to remove, contain, or render harmless environmental contaminants (toxic metals and difficult-to-destroy organic pollutants) in wastewaters and soils from brownfield sites.

The definition of brownfields adopted by the Canadian National Round Table on the Environment and the Economy is as follows: "brownfields are abandoned, vacant, derelict or underutilized commercial and industrial properties where past actions have resulted in actual or perceived contamination; brownfields differ from other contaminated sites in that they hold active potential for redevelopment" (NRTEE, 2003).

In 2010, Hanson received approval from the EPA to construct a 2.5 ha full-scale trial of the phytocap it had been researching for the previous 5 years. A phytocap is specially designed vegetated soil that covers a landfill, it helps to manage water and reduce the release of gases, such as methane. The phytocap used at Wollert uses a 1.5 m thick monolayer of waste rock scalps from the adjacent quarry and has been vegetated with native plants to control the percolation of rainwater into the underlying waste. This is the first large-scale phytocap approved in Australia.

Several years in development, the guidelines for using phytocaps as final cover for landfills has been released at a conference in South Australia. Produced by the Waste Management Association of Australia (WMAA) as part of the A-ACAP (Australian Alternative Covers Assessment Program) research program the guidelines are designed to pave the way for landfill operators to consider vegetated soil caps rather than the conventional compacted clay caps, which based on the research can achieve better results from an environmental and financial perspective.

The Australian research principally comprises A-ACAP and Hanson's Wollert (Victoria) trial. These programs have included the collection and analysis of empirical data from field-scale research facilities at landfill sites in five different environments in five states, including comparative data on the hydraulic performance of phytocaps and compacted clay caps. A-ACAP has also drawn upon the findings and methods of the U.S. Alternative Covers Assessment Program (ACAP), other research in the United States and elsewhere.

Phytocovers have been proposed as an alternative to traditional barrier covers, with aims to improve long-term environmental performance and reduce the financial burden of closure costs, particularly for rural landfills. This research is exploring the potential for a phytocover to be implemented at a landfill in south-east Australia. Opportunities exist at this site to use basalt quarry scalpings typical of the Victorian extractive industry for purposes of the cover substrate, while vegetation selection has focused on indigenous plant species. Although the agronomic properties of the scalpings are not ideal, preliminary planting trials have shown that the scalpings have significant potential for sustaining a careful selection of native trees and grasses. The application of mulch has not resulted in any overall plant growth advantage. A compaction trial conducted with full-scale machinery has enabled the method of placing the scalpings to be related to the in situ dry bulk density achieved. Performance of the designed phytocover profile will be monitored with four large test sections. Two of the test sections employed lysimetry for the direct collection of percolation, while the other two sections remained open to the normal fluxes of landfill gas, heat and water vapor present in the landfill environment in order to observe their effects on the establishing vegetation cover (Michael et al., 2007).

In the United States, the U.S. EPA initiated the ACAP in 1998 to review available literature; design, construct, and monitor test covers; and develop guidance on their use in the United States (Albright et al., 2004). It was the world's first large-scale quantitative trial of the comparative performance of conventional caps and phytocaps. Twelve trial sites in eight States were established representing a variety of climates from arid to humid, soil depths from 0.8 to 2.9 m, and vegetation—including grass only; grasses and shrubs; and grass, shrubs and poplar trees. The study included some side-by-side comparisons of phytocap and conventional caps using field-scale lysimeters, and data were collected over 4–5 years for most sites. ACAP used large (10 × 20 m) drainage lysimeters to determine the water balance of the test sections. Following these trials, in 2003 WMAA and PhytoLink Australia organized a National seminar tour by representatives of the U.S. ACAP. The positive response to this tour from industry, regulators, and other stakeholders led WMAA later that year to initiate A-ACAP. In 2004, a lysimeter-based trial, similar to the U.S. ACAP trials, was established at the Hanson Landfill Services' Wollert landfill (approximately 25 km north of Melbourne, Victoria). The Hanson trial was the first field-scale lysimeter-based phytocap trial in Australia. It commenced in 2004 and was developed to investigate the potential for using a waste rock substrate and native grasses, shrubs, and trees as a phytocap. The trial included an initial plant selection process and a two-year study of the proposed phytocap.

The Victorian phytocap is on the Wollert Landfill, and follows extensive trial work undertaken on the site, and water balance modeling to extrapolate results to wetter and drier climatic years. Hanson (the site operator), placed 1.5 m of <40 mm quarry scalps with a high soil content using an excavator to dump and place material. In some areas, however, trucks were used to dump material, then placed by a dozer, ripped and graded (Bateman, pers. comm., 2011). Material was placed as it became available from the quarry over a one-year period and placement was confirmed by adherence to a method, rather than by testing.

A significant issue with the Wollert phytocap was the establishment of native vegetation (ibid.). Native trees, shrubs and grasses were sown over 2 years on the site using direct seeding and tubestock. Initially, the tubestock established faster than the direct seeding, but this is expected to be reversed in the second spring when the direct seeding will grow rapidly. Hanson expects that the vegetation will be completely established in approximately 5 years, and self-sustaining with minimal maintenance in 10 years from planting.

Competition from weeds and damage from birds and hares has proved a high maintenance issue. Though Hanson selected material with low fertility to reduce weed establishment, because of the generally very wet conditions over the last 12 months, weeds have invaded some areas. The native vegetation contractor for the site has also noted that the native vegetation will need slashing in the future to reduce soil nitrogen accumulation, which can lead to invasion of weed species. Other vegetation issues for the site were the lack of contractors experienced in the broad-scale establishment of native grasses and the availability of seed, particularly native grasses. This scale of direct seeding of native grasses is not common, but proving viable with careful control.

Two other important issues for the full-scale projects are the need for access roads and the location and connection of instrumentation. Compared to the full-scale cap, trial areas are of relatively small area size and duration. Hence, access is not required. Access roads, however, may be required in the future for monitoring or access to other areas. At the Wollert Landfill, access roads will be kept unvegetated. Hence, these roads were lined with an impermeable barrier to prevent excessive drainage through the cap. The need for this level of access should be considered during the design process. In addition, if monitoring equipment is required, the distance between monitoring stations can be large and the risk of cables being damaged or signals becoming attenuated (resulting in poor or missing data) over long distances. Therefore, Hanson has decided to use radio-linked stations at Wollert (Salt et al., 2011).

The monolithic phytocap includes one vegetated soil layer over the interim cover (Figure 13.3). In practice, this may include more than one layer, depending on the materials available. As an alternative,

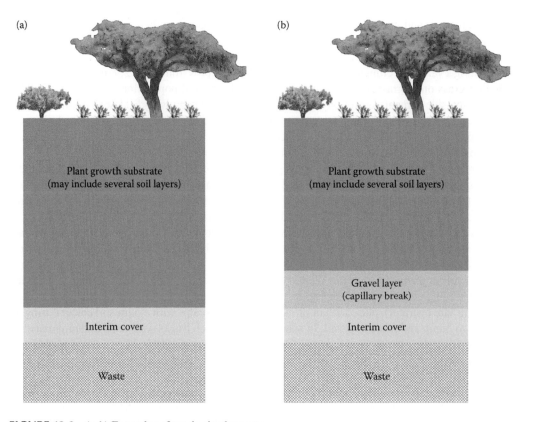

FIGURE 13.3   (a, b) Examples of two basic phytocaps.

a coarse layer may be placed at the bottom of the soil layer(s) to act as a capillary break. The advantage of this layer is that it can increase the moisture holding capacity of the overlying soil, thus, reduce the potential for drainage to occur. Capillary breaks are sometimes used in conventional cover designs.

PLANT SPECIES FOR BROWNFIELD RESTORATION EFFORTS

The term phytoremediation was coined 20 years ago and refers to the use of higher plants to remediate contaminated land by removing toxins (especially heavy metals), through their roots and subsequent transport to aerial plant organs (Salt et al., 1998). Phytoremediation programs have been successfully applied to meet a range of different objectives and environmental outcomes. One of the key successes of brownfield restoration efforts using this approach depends on the suitability of the species used in the restoration programs. Given the variable site characteristics, possible states and general environmental variability we consider various characteristics that species should have for, rather than attempting to compile an exhaustible plant list.

The ability to cope with the environment in general, that is, with general soil condition, temperature, and rainfall. This might include species indigenous to the area but also foreign plants. In such case, it is important to include suitable management practices in management and monitoring if the species should take on weedy characteristics.

The following section describes the criteria, which are generally accepted as being crucial for the success of phytoremediation efforts. These criteria may not be applicable to every contaminated site because of the variability observed among plant species (Fischerová et al., 2006; it is intended merely as an overarching guide.

Plants that can hyperaccumulate phytotoxic contaminants (Chaney et al., 1997). Hyperaccumulators have an unusual ability to grow in metal-rich sites without any adverse effects of metal phytotoxicity (Baker et al., 2000). Hyperaccumulators are capable of extracting potentially phytotoxic elements to concentrations up to 100 times higher to those found in nonaccumulator plants (Merrington, 1995).

Plants that can hypertolerate phytotoxic contaminants (Chaney et al., 1997). These plants are characterized by a high tolerance for contaminant accumulation and it is thought that such tolerance is correlated with levels of certain endogenous compounds such as cysteine and glutathione (Freeman et al., 2004). It is suggested that the development of physiological tolerances may be a survival strategy against competition from other species (Baker and Brooks, 1989).

An ability to efficiently desorb insoluble metals from the soil matrix is an important functional characteristic of plants used in phytoremediation of metal-contaminated sites (Fischerova et al., 2006). Most metals are found in insoluble forms in the soil and are therefore not available for phytoextraction. Plants use two methods to desorb metals from the soil: (i) acidification of the rhizosphere, and (ii) secretion of ligands capable of chelating the metal (Mukhopadhyaya and Maiti, 2010). Both essential and toxic metals may be released into solution using these processes (Lasat, 2000).

Plant biomass considerations. An ideal species should have one of the following characteristic combinations: (i) a low biomass plant with very high metal accumulation capacity OR (ii) a high biomass plant with enhanced metal uptake potential (Varun et al., 2015). An ability to tolerate and at the same time accumulate multiple metal contaminants and/or metal–organic mixtures are highly desirable (Varun et al., 2015).

Root physiology will determine what type of plant is suitable for a particular site. Tree roots grow deeper and are therefore suitable for deep contamination. Small plants like ferns and grasses have been used where contamination is shallow (USEPA, 2012). Regardless of contamination, root biomass should also be a consideration as an increased surface area would lead to higher levels of contaminant uptake. Furthermore, a dense root system will stabilize the soil and prevent soil erosion. Dense root systems may also reduce the amount of water (and potentially hazardous leachate) percolating through the soil matrix (Gosh and Singh, 2005). This is particularly important for phytostabilization programs.

Longer plant growing seasons will increase the time in which contaminants are extracted by the plant each year (USEPA, 2012).

Ability to tolerate high planting densities. Phytoremediation often requires higher planting densities (compared with standard agronomic rates) to overcome decreased germination rates due to contaminated soils. Higher planting densities are also needed to maximize biomass production across a contaminated site (USEPA, 2001).

An ability to grow in low nutrient conditions. Many contaminated sites are devoid of nutrient rich soils (Favas et al., 2014).

Plants that are visually appealing and that are accepted by the public. Plants that may be integrated into greenspace plans are useful for community engagement programs (USEPA, 2001). This is not often mentioned in the literature but may be an important consideration when selection plant species in urban or metropolitan environments.

Low maintenance species. Depending on available funding for phytoremediation maintenance, there may be requirements for mowing, fertilizing, replanting, pruning, harvesting, and monitoring (USEPA, 2001). Plants that require a lower frequency of maintenance activities would be desirable under low-funding conditions.

CONCLUSIONS AND RECOMMENDATIONS

The four pillars for sustainable development are (i) economic prosperity, (ii) environmental protection, (iii) social and community well-being, and (iv) governance. If the key objective of brownfield reclamation is to achieve "sustainability," land remediation practitioners should aim to address each of these four pillars during remediation planning and implementation. In this chapter, we have discussed some of the key challenges facing not only remediators of brownfields, but also the community and other key stakeholders. The State and Transition model presented in the previous sections describes a theoretical method for land restoration that acknowledges the different "states" and periods of "transition" by which reclaimed brownfield sites may undergo during their shift toward the desired outcome. The various states and transitional stages that are attained during the restoration process present real opportunities for jobs creation and community engagement. Table 13.1, for example, lists a number of different job opportunities for maintenance and monitoring activities. Community engagement through "citizen science" programs would also be encouraged. The implementation of long-term monitoring of reclaimed brownfield sites is vital for protecting the surrounding environment and human health, and should be incorporated into any restoration plan. Objectives that are clear and demonstrate mutually beneficial outcomes for regulators, land owners, and the community are vital for the success of restoration efforts. Development of a transparent list of restoration objectives early on will assist in measuring success.

REFERENCES

Albering, H. J., S. M. van Leusen, E. J. C. Moonen, J. A. Hoogewerff, and J. C. S. Kleinjans. 1999. Human health risk assessment: A case study involving heavy metal soil contamination after the flooding of the River Meuse during the winter of 1993–1994. *Environmental Health Perspectives*, 107(1): 37–43. doi:10.2307/3434287.

Albright, W. H., C. H. Benson, G. W. Gee, A. C. Roesler, T. Abichou, P. Apiwantragoon, B. F. Lyles, and S. A. Rock. 2004. Field water balance of landfill final covers. *Journal of Environmental Quality*, 33: 2317–2332.

Aronson, J., J. N. Blignaut, S. J. Milton, D. Le Maitre, K. J. Esler, A. Limouzin, C. Fontaine et al. 2010. Are socioeconomic benefits of restoration adequately quantified? A meta-analysis of recent papers (2000–2008) in restoration ecology and 12 other scientific journals. *Restoration Ecology*, 18(2): 143–154. doi:10.1111/j.1526-100X.2009.00638.x.

Arthur, E. L., P. J. Rice, P. J. Rice, T. A. Anderson, S. M. Baladi, K. L. D. Henderson, and J. R. Coats. 2005. Phytoremediation—An overview. *Critical Reviews in Plant Sciences*, 24(2): 109–122. doi:10.1080/07352680590952496.

Baker, A. J. M. and R. R. Brooks. 1989. Terrestrial higher plants which hyperaccumulate metallic elements—A review of their distribution, ecology and phytochemistry. *Biorecovery*, 1(1): 81–126.

Baker, A. J. M., S. P. McGrath, R. D. Reeves, and J. A. C. Smith. 2000. Metal hyperaccumulator plants: A review of the ecology and physiology of a biological resource for phytoremediation of metal-polluted soils. In *Phytoremediation of Contaminated Soil and Water*, edited by N. Terry and G. Banuelos. Boca Raton: Lewis Publishers; pp. 85–108.

Baker, A. J. M., R. D. Reeves, and S. P. McGrath. 1991. In situ decontamination of heavy metal polluted soils using crops of metal-accumulating plants—a feasibility study. In *In Situ Bioreclamation*, edited by R. E. Hinchee and R. F. Olfenbuttel. Boston: Butterworth-Heinemann, pp. 539–544.

Bardos, P. R., C. P. Nathanail, and A. Weenk. 2000. *Assessing the Wider Value of Remediating Land Contamination: A Review*. R&D P238. UK Environmantal Agency.

Brooks, R. R., J. Lee, R. D. Reeves, and T. Jaffre. 1977. Detection of nickeliferous rocks by analysis of herbarium specimens of indicator plants. *Journal of Geochemical Exploration*, 7: 49–57. doi:10.1016/0375-6742(77)90074-7.

Brundtland, G. H. 1987. *Our Common Future: Report of the 1987 World Commission on Environment and Development*. Oslo: United Nations.

CABERNET. 2005. Brownfield Definition. http://www.cabernet.org.uk.

Chaney, R. L., M. Malik, Y. M. Li, S. L. Brown, E. P. Brewer, J. S. Angle, and A. J. M. Baker. 1997. Phytoremediation of soil metals. *Current Opinion in Biotechnology*, 8(3): 279–284. doi:10.1016/S0958-1669(97)80004-3.

DEFRA, 2006. Defra Circular 01/2006, Environmental Protection Act 1990: Part 2A, Contaminated Land. London, UK: Defra—Department for Environment, Food and Rural Affairs.

D. Glass Associates, Inc. 1999. U.S. and International Markets for Phytoremediation, 1999–2000. USEPA 2008.

Desouki, S. H. and H. Feng. 2012. Metal contaminant source, transport and fate in the environment and phytoremediation methods. In *Metal Contamination, Sources, Detection and Environmental Impact*, edited by H. B. Shao. Hauppauge, New York: Nova Science Publishers, Inc. pp. 81–94.

De Sousa, C. A. 2002. Measuring the public costs and benefits of brownfield versus greenfield development in the Greater Toronto Area. *Environment and Planning B: Planning and Design*, 29(2): 251–280. doi:10.1068/b1283.

Dixon, T. 2006. Integrating sustainability into Brownfield regeneration: Rhetoric or reality? – An analysis of the UK development industry. *Journal of Property Research*, 23(3): 237–267. doi:10.1080/09599910600933889.

Dorren, L. K. A., F. Berger, A. C. Imeson, B. Maier, and F. Rey. 2004. Integrity, stability and management of protection forests in the European Alps. *Forest Ecology and Management*, 195(1–2): 165–176. doi:10.1016/j.foreco.2004.02.057.

Elmqvist, T., C. Folke, M. Nyström, G. Peterson, J. Bengtsson, B. Walker, and J. Norberg. 2003. Response diversity, ecosystem change, and resilience. *Frontiers in Ecology and the Environment*, 1(9): 488–494. doi:10.1890/1540-9295(2003)001[0488:RDECAR]2.0.CO;2.

Favas, P. J., J. Pratas, M. Varun, R. D'Souza, and M. S. Paul. 2014. Phytoremediation of soils contaminated with metals and metalloids at mining areas: Potential of native flora. Environmental risk assessment of soil contamination. In *Environmental Risk Assessment of Soil Contamination*, edited by M. C. Hernandez-Soriano. inTech., pp. 485–517. http://dx.doi.org/10.5772/57469.

Fischerová, Z., P. Tlustoš, J. Száková, and K. Šichorová. 2006. A comparison of phytoremediation capability of selected plant species for given trace elements. *Environmental Pollution, Soil and Sediment Remediation (SSR) Soil and Sediment Remediation (SSR)*, 144(1): 93–100. doi:10.1016/j.envpol.2006.01.005.

Freeman, J. L., M. W. Persans, K. Nieman, C. Albrecht, W. Peer, I. J. Pickering, and D. E. Salt. 2004. Increased glutathione biosynthesis plays a role in nickel tolerance in *Thlaspi* nickel hyperaccumulators. *The Plant Cell*, 16(8): 2176–2191. doi:10.1105/tpc.104.023036.

Gallagher, F. J., I. Pechmann, J. D. Bogden, J. Grabosky, and P. Weis. 2008. Soil metal concentrations and vegetative assemblage structure in an urban brownfield. *Environmental Pollution*, 153(2): 351–361. doi:10.1016/j.envpol.2007.08.011.

Glass, D. J. 1999. Current market trends in phytoremediation. *International Journal of Phytoremediation*, 1: 1–8.

Gosh, M. and S. P. Singh. 2005. A review on phytoremediation of heavy metals and utilization of it's by products. *Asian Journal on Energy and Environment*, 6(4): 214–231.

Hallett, L. M., S. Diver, M. V. Eitzel, J. J. Olson, B. S. Ramage, H. Sardinas, Z. Statman-Weil, and K. N. Suding. 2013. Do we practice what we preach? Goal setting for ecological restoration. *Restoration Ecology*, 21(3): 312–319. doi:10.1111/rec.12007.

Hartman, W. J. 1975. *An Evaluation of Land Treatment of Municipal Wastewater and Physical Siting of Facility Installations*. Office of the Chief of Engineers, Army.

Hobbs, R. J. and J. A. Harris. 2001. Restoration ecology: Repairing the earth's ecosystems in the new millennium. *Restoration Ecology*, 9(2): 239–246. doi:10.1046/j.1526-100x.2001.009002239.x.

Hobbs, R. J., and D. Lindenmayer. 2007. From perspectives to principals: Where to from here? In *Managing and Designing Landscape for Conservation: Moving from Perspectives to Principles*, edited by D. Lindenmayer and R. J. Hobbs. Oxford: Blackwell.

Hobbs, R. J. and H. A. Mooney. 1993. Restoration ecology and invasions. In *Nature Conservation 3: Reconstruction of Fragmented Ecosystems: Global and Regional Perspectives*, edited by D. A. Saunders, R. J. Hobbs, and P. R. Ehrlich. Chipping Norton: Surrey Beatty & Sons.

Hobbs, R. J. and D. A. Norton. 1996. Towards a conceptual framework for restoration ecology. *Restoration Ecology*, 4(2): 93–110. doi:10.1111/j.1526-100X.1996.tb00112.x.

Kempton, H., T. A. Bloomfield, J. L. Hanson, and P. Limerick. 2010. Policy guidance for identifying and effectively managing perpetual environmental impacts from new hardrock mines. *Environmental Science & Policy*, 13(6): 558–566. doi:10.1016/j.envsci.2010.06.001.

Kingsford, R. T., H. C. Biggs, and S. R. Pollard. 2011. Strategic adaptive management in freshwater protected areas and their rivers. *Biological Conservation*, 144(4): 1194–1203. doi:10.1016/j.biocon.2010.09.022.

Lake, P. S., N. Bond, and P. Reich. 2007. Linking ecological theory with stream restoration. *Freshwater Biology*, 52(4): 597–615. doi:10.1111/j.1365-2427.2006.01709.x.

Lasat, M. M. 2000. Phytoextraction of metals from contaminated soil: A review of plant/soil/metal interaction and assessment of pertinent agronomic issues. *Journal of Hazardous Substance Research*, 2(5): 1–25.

Lindenmayer, D., R. J. Hobbs, R. Montague-Drake, J. Alexandra, A. Bennett, M. Burgman, P. Cale et al. 2008. A checklist for ecological management of landscapes for conservation. *Ecology Letters*, 11(1): 78–91. doi:10.1111/j.1461-0248.2007.01114.x.

Luo, X., S. Yu, and Xiang-dong Li. 2012. The mobility, bioavailability, and human bioaccessibility of trace metals in urban soils of Hong Kong. *Applied Geochemistry, Recent Progress in Environmental Geochemistry—A Tribute to Iain Thornton*, 27(5): 995–1004. doi:10.1016/j.apgeochem.2011.07.001.

McCutcheon, S. C. and J. L. Schnoor. 2003. *Phytoremediation: Transformation and Control of Contaminants*. John Wiley & Sons, Inc. Hoboken, New Jersey. 1024p.

Merrington, G. 1995. Historic metalliferous mine sites: A major source of heavy metal contamination? *Land Contamination Reclamation*, 3: 173–179.

Michael, R. N., S. T. S. Yuen, A. J. M. Baker, W. S. Laidlaw, and C. S. Bateman. 2007. A sustainable approach for hydraulic control of landfills using quarry scalpings and native plants. *Australian Journal of Multi-Disciplinary Engineering*, 5: 39–48.

Michener, W. K. 1997. Quantitatively evaluating restoration experiments: Research design, statistical analysis, and data management considerations. *Restoration Ecology*, 5(4): 324–337. doi:10.1046/j.1526-100X.1997.00546.x.

Millenium Ecosystem Assessment. 2005. *Ecosystems and Human Well-Being: Biodiversity Synthesis*. Washington, DC: World Resources Institute.

Mukhopadhyay, S. and S. K. Maiti. 2010. Phytoremediation of metal mine waste. *Applied Ecology and Environmental Research*, 8(3): 207–222.

NRTEE. 2003. Cleaning up the Past, Building the Future—A National Brownfields Redevelopment Strategy for Canada Report prepared by The National Roundtable on the Environment and the Economy http://www.nrtee-trnee.ca/

Qian, Y., F. J. Gallagher, H. Feng, and M. Wu. 2012. A geochemical study of toxic metal translocation in an urban brownfield wetland. *Environ Pollut.*, 166: 23–30.

Qian, Y., F. J. Gallagher, H. Feng, M. Wu, and Q. Zhu. 2014. Vanadium uptake and translocation in dominant plant species on an urban coastal brownfield site. *Science of the Total Environment*, 476–477: 696–704. doi:10.1016/j.scitotenv.2014.01.049.

Rai, P. K. 2008. Heavy metal pollution in aquatic ecosystems and its phytoremediation using wetland plants: An ecosustainable approach. *International Journal of Phytoremediation*, 10(2): 131–158.

Raskin, I., P. B. A. N. Kumar, S. Dushenkov, and D. E. Salt. 1994. Bioconcentration of heavy metals by plants. *Current Opinion in Biotechnology*, 5(3): 285–290. doi:10.1016/0958-1669(94)90030-2.

Raskin, I., R. D. Smith, and D. E. Salt. 1997. Phytoremediation of metals: Using plants to remove pollutants from the environment. *Current Opinion in Biotechnology*, 8(2): 221–226. doi:10.1016/S0958-1669(97)80106-1.

Resilience Alliance. 2015. Resilience Alliance – HOME. http://www.resalliance.org/.

Salt, D. E., R. D. Smith, and I. Raskin. 1998. Phytoremediation. *Annual Review of Plant Physiology and Plant Molecular Biology*, 49(1): 643–668. doi:10.1146/annurev.arplant.49.1.643.

Salt, M., P. Lightbody, R. Stuart, W. H. Albright, and R. Yeates. 2011. *Guidelines for the Assessment, Design, Construction and Maintenance of Phytocaps as Final Covers for Landfills*. 20100260RA3F. Waste Management Association of Australia.

Schädler, S., M. Morio, S. Bartke, R. Rohr-Zänker, and M. Finkel. 2011. Designing sustainable and economically attractive Brownfield revitalization options using an integrated assessment model. *Journal of Environmental Management*, 92(3): 827–837. doi: 10.1016/j.jenvman.2010.10.026.

Society for Ecological Restoration International Science & Policy Working Group. 2004. *The SER International Primer on Ecological Restoration*. 2nd ed. Society for Ecological Restoration International. www.ser.org.

Steiner, F. R. 2008. *The Living Landscape: An Ecological Approach to Landscape Planning*. 2nd ed. Island Press.

Sun, L., Y. Geng, J. Sarkis, M. Yang, F. Xi, Y. Zhang, B. Xue, Q. Luo, W. Ren, and T. Bao. 2013. Measurement of polycyclic aromatic hydrocarbons (PAHs) in a Chinese Brownfield redevelopment site: The case of Shenyang. *Ecological Engineering*, 53: 115–119. doi:10.1016/j.ecoleng.2012.12.023.

Tedd, P., J. A. Charles, and R. Driscoll. 2001. Sustainable brownfield re-development—risk management. *Engineering Geology* 60.1: 333–339.

Tordoff, G. M., A. J. Baker, and A. J. Willis. 2000. Current approaches to the revegetation and reclamation of metalliferous mine wastes. *Chemosphere*, 41(1–2): 219–228.

USEPA. 2001. *Brownfields Technology Primer: Selecting and Using Phytoremediation for Site Cleanup*. EPA-542-R-01-006. United States Environmental Protection Agency.

USEPA. 2004. *Cleaning Up the Nation's Waste Sites: Markets and Technology Trends*. 2004 ed. Washington, DC: United States Environmental Protection Agency.

USEPA. 2012. *A Citizen's Guide to Phytoremediation*. USEPA Fact sheet EPA 542-F-1216. United States Environmental Protection Agency. www.epa.gov/superfund/sites www.cluin.org.

van der Lelie, D., J. P. Schwitzguebel, D. J. Glass, J. Vangronsveld, and A. Baker. 2001. Assessing phytoremediation's progress in the United States and Europe. *Environmental Science and Technology*, 35(21): 447–452.

Varun, M., R. D'Souza, P. J. Favas, J. Pratas, and M. S. Paul. 2015. Utilization and supplementation of phytoextraction potential of some terrestrial plants in metal-contaminated soils. In *Phytoremediation: Management of Environmental Contaminants*, edited by A. A. Ansari, S. S. Gill, R. Gill, G. R. Lanza, and L. Newman. Vol. 1. New York: Springer.

Westoby, M., B. H. Walker, and I. Noy-Meir. 1989. Opportunistic management for rangeland not at equilibrium. *Journal of Range Management*, 42(2): 266–274.

Wilkins, S., D. A. Keith, and P. Adam. 2003. Measuring success: Evaluating the restoration of a grassy eucalypt woodland on the cumberland plain, Sydney, Australia. *Restoration Ecology*, 11(4): 489. doi:10.1046/j.1526-100X.2003.rec0244.x.

Wilson, E. O. 1992. *The Diversity of Life*. London: Penguin.

Wortley, L., J. Hero, and M. Howes. 2013. Evaluating ecological restoration success: A review of the literature. *Restoration Ecology*, 21(5): 537–543. doi:10.1111/rec.12028.

14 Mining and the Environment

Greg You and Dakshith Ruvin Wijesinghe

CONTENTS

ABSTRACT

While mining contributes to the economy, it has some adverse impacts on the environment. However, with the advancements in mining technologies, the industry has identified the importance of maintaining environmental sustainability during mining operations which combine various components such as resource efficiency, economics, safety and environment. The reduction of environmental impacts of mining relies on the balance between these factors. Australia is one of the world's leading nations in mineral supplies, and has been working upon maintaining environmental sustainability from the beginning of mining operation until the mine closure. Ensuring environmental sustainability in mining needs not only the dedication of the industry in compliance with the laws and regulations but also requires consultations between industries and communities. This chapter summarises some major environmental problems associated with mining, such as acid mine drainage (AMD), land subsidence, mine waste, mine fire and issues related to blasting. These issues are discussed with reference to various laws and regulations that are implemented in Australia.

INTRODUCTION

From the dawn of the civilisation, mining has been evolved as a major source of generating numerous resources for human beings. Mining industry records a history of centuries as such; the oldest mine identified has a radiocarbon age of 43,000 years, which was located at Lion Cave in Switzerland (Sengupta, 1993). However, with the nature of the various activities associated with mining, the environmental impact it creates has become inevitable. Without adequate regulations to govern the industry, it may create conflicts and confrontations among nature, community and miners. In the earlier years of mining industry, environmental impact has not been identified as a major concern. But the present industry has gained more awareness which has led to the development of rules and regulations specifically related to the environmental aspects of mining activities. Therefore, it is increasingly important to identify the relationships between mining and environment, to maintain global consistency and sustainability.

Mining can be explained as the extraction of minerals, in solid, liquid or gaseous form, from the earth's crust (Bell and Donnelly, 2006). In any type of mining activity, the occurrence of environmental degradation can befall at the very first stages during the exploration. The first and foremost impact is the visual distress aroused by the destruction of land. Once a mine is established, noise, dust and vibration can be generated from the mining operations. In the case of underground mining, ground subsidence may occur. Furthermore, degradation of watercourses, mine flooding, AMD, air pollution, mine fire, sedimentation and erosion are among the major environmental issues that may be associated with mining. Finally, inadequate mine waste disposal can cause environmental degradation (Sengupta, 1993; Mitchell et al., 1994). Therefore, reducing the environmental impacts of mining should be treated as one of the prime focuses in ensuring sustainability of mining and its environment.

Reducing the environmental impacts of mining does not solely ensure its sustainability. Sustainability is a combination of resource efficiency, communities, economics, safety and environment. Therefore, when reducing the environmental impacts of mining, it is always important to keep a balance among these factors. Being a leading supplier of minerals to the world market, Australia has benchmarked environmental mining sustainability in various terms. The country has implemented a set of laws which enforce environmental accountability, social responsibility and commercial success. Similarly, the mining industry around the world has led to a paradigm shift in achieving sustainability as a part of success in mining.

Different concepts have been introduced in the mining industry to combine economic growth and environmental sustainability. The environmental sustainability in mining needs an understanding of the historical trends in production as well as the relationship between production and its intensity (Mudd, 2010). At present, environmental issues are combined with economic and social factors which are also termed as triple bottom line criteria. The relationships among them can create a number of scenarios on how sustainability can be achieved in mining. Optimising all the scenarios to achieve an environmentally viable, economically feasible and socially sound mining practice can only result in sustainable mining. Within these interrelated parameters, this chapter is dedicated to discuss some environmental impacts associated with mining and how they can be overcome while ensuring sustainability. The mining-related environmental laws and regulations play an important role in sustainable planning and operation of mining. Thus, they will be discussed by considering Australia as an example.

ENVIRONMENTAL MANAGEMENT SYSTEMS IN MINING

Environmental management systems (EMS) provide a systematic framework for organisations to assess the significant environmental impacts that may occur due to the organisations' activities. All environmental issues that are related to mining operations are managed under EMS. There are five basic components as follows:

- Environmental policy
- Planning
- Implementation and operation
- Checking and corrective action
- Management review (Environmntal Management, 2015)

Most of the mining and mineral organisations operate their EMS in compliance with AS/NZS ISO 14001:2004. However, Australia has identified the importance of strengthening the environmental policies in mining and mineral industry. The current approach in environmental management in Australia can be described as a partnership approach which is a combination of regulatory, economic, volunteer measures, agreement between governments, industry and community groups. There is a significant potential in improving the EMS in Australia. Even though the reporting of the state of the environment is strong in the Australian context, environmental monitoring and environmental data are often inadequate in terms of consistency and coverage (OECD, 1997). Therefore, this creates barriers in implementing proper EMS in mining operation systems.

MAJOR ACTS IN MINING AND ENVIRONMENTAL PROTECTION

Mining and mineral exploration in Australia are generally governed under the Environmental Protection Act 1994. The major objective of introducing this act was to promote environmental protection while allowing the development which improves the total quality of life of present and future generations. This can be simply explained by the term ecologically sustainable development (Environmental Protection Act 1994, 2015). However, since the establishment of the act, it has been a complex task for the government and business to follow up with its requirements due to the emerging environmental issues and changing community expectations. In response to the issue raised by business and the government on the complexity and difficulties in navigating the act, a green tape reduction project was established in 2012 as an amendment of Environmental Protection Act 1994 (Queensland, 2012). The key objective of the Greentape Reduction Act is to introduce an integrated approval process for all environmentally relevant activities (Department of Environment and Heritage Protection, 2013). Some of the other objectives of introducing the Greentape Reduction Act are to develop a licencing model proportionate to environmental risks, to provide flexible operational approvals, to streamline the approval processes and to clarify information requirements (Robertson, 2012).

MAJOR ENVIRONMENTAL ISSUES IN MINING

Mining and subsequent mineral processing can produce various environmental impacts. The impacts may vary with the mineral that is mined. Some of them can directly impact the surrounding eco-systems while others may be hazardous for the human lives in the short and long term. Most of the effects are dependent on the time as impacts of the present mining operations can last for years.

MINE WATER

Water is used in mining for various mineral processing activities, dust suppression, hydrometallurgical extraction and ore washing. The water from aquifers and/or surface water bodies is used in mining. Moreover, water can be obtained as a by-product from mine dewatering. Larger volumes of water are generally produced in mining activities which can sometimes be higher than the amount of total solid waste generated. The mine water consists of dissolved minerals and can be highly toxic when it is directly discharged to the environment. At present, the mine water is disposed of into the settling ponds or subsequently released from the ponds in compliance with the regulations of the environment protection authority (EPA).

ACID MINE DRAINAGE

AMD is one of the major environmental problems in mining. This phenomenon occurs from the natural oxidation of sulphide minerals in rock or waste that are exposed to air and water (Bell and Donnelly, 2006). AMD has a high contribution for the water pollution problems of coal and metal mining operations. Metallic ore deposits, phosphate ores, coal seams, oil shales and mineral sands have the highest potential to expose for the sulphide oxidation to produce AMD (Lottermoser, 2010). There exist a number of processes that contribute to the generation of AMD such as

1. Uncontrolled discharge of the process water generated by tailings, flotation or leaching
2. Rainfall runoff generated from the mining areas, such as open pits and tailing dams
3. Groundwater interactions with underground or surface excavation processes
4. Surface water contacts with the ore/rock stock piles and tailings, etc.

The major chemical parameters that are contributing to the AMD can be identified as the pH value, temperature, concentration of oxygen, degree of saturation, chemical activity with Fe^{3+}, surface area of exposed metal sulphide and the chemical activation energy required to initiate acid generation. In addition, the process of 'biotic microorganism' can accelerate the reaction rate (Bell and Donnelly, 2006). Although AMD is supposed to be generated from mining activities associated with sulphide minerals, it will not produce when the sulphide material is non-reactive or when the rock contains alkaline materials to neutralise the acidity (Bell and Donnelly, 2006).

AMD generated in coal strip mines is one of the most hazardous waters generated in mining. The major characteristics of AMD in coal mines are low pH, high electrical conductivity, total dissolved solids, sulphate, nitrate, iron, aluminium, sodium, calcium and magnesium values (Zielinski et al., 2001; Vermeulen and Usher, 2006). The trace metal contents of coal mining can be high due to components such as Cd, Co, Cr, Cu, Li, Ni, Pb, Sr, Zn and metalloids (Lussier et al., 2003; Wu et al., 2009). These elements can create a number of environmental impacts for the receiving waters. The mine water generated in coal mines may not be necessarily classified as acid. The majority of mine water around the world has got neutral pH values. It is the high electrical conductivity and dissolved solids that create the characteristics of acidity in AMD. In some instances, there can be significant percentages of radioactive minerals in the coal mine water such as uranium, thorium and radium (Lottermoser, 2010).

Prevention, Control and Treatment of AMD

Generally, at any mine site, mine waters need be controlled and treated before discharge, to the required standards in compliance with environmental regulations. First of all, it is vitally important to identify the potential of AMD generation from the mining activities, which can be done by collecting the available water quality data and conducting static and kinetic tests. One of the popular static tests available for the acid generation prediction is 'acid-based accounting'. This method can be used to get an estimation of the amount of acid generating and neutralising minerals present in the water. However, most of the AMD control strategies are site specific which are governed by external factors such as the characteristics of the mineral, nature of the mine waste, site geology and the climatic conditions of the mine site.

Once the constituents of the water are identified, the next step is to continue with the AMD management measures. There are three major strategies in controlling the AMD generation, namely control of acid generation, control of acid migration and collection and treatment of AMD (Connelly et al., 1994). The best approach is the source control by preventing the possible oxidation, generation of acids or leaching of the contaminants. Acid generation can be controlled by removing or isolating the sulphide material and exclusion of water or air (Bell and Donnelly, 2006). The exclusion of water or air can be done by placing a cover over acid generating materials. Another successful method was found to be covering with capillary barrier effects (Bussière et al., 2004). The control of acid generation can also be done by reducing the volumes of water generated from the operation. Some of the techniques that are used to reduce mine water are interception and diversion of surface waters through upstream dams, diversion of runoff from undisturbed catchments, maximisation of recycling or reuse of water, segregation of water types of different quality, controlled release into nearby waters, sprinkling of water over dedicated parts of the mine sites, use of evaporative ponds and installation of covers over sulphidic wastes (Lottermoser, 2010).

The treatment of AMD is typically site specific, depending on the characteristics of the site and the mineral mined. AMD treatment can be achieved by developing treatment systems that incorporates a number of processes such as evaporation, neutralisation, wetland processes and releasing and dilution by using natural water. The AMD treatment systems can be classified as active and passive treatment ones. The active treatment can be defined as methods that encourage the neutralisation of acids such as adding different chemical reagents. Active methods require continuous monitoring and active engagement to maintain the reagent mixing procedure to neutralise the acids. Passive systems are more popular than the active ones. Passive systems use the natural water flow and devices which incorporate a number of chemical and biological processes to neutralise the acidic water (Walton-Day, 2003; Hallberg and Johnson, 2005); refer to Figure 14.1 for an illustration of the passive AMD treatment.

Principles of Sustainable Mine Water Management, Laws and Regulations

The environmental management aspects related to mining and mineral processing operations are required to be in compliance with both state and commonwealth government laws. Currently, most of them are subjected to state government laws only. The legislations related to mining industry are enforced by the relevant state environmental authority, state department of the mines or commonwealth department of mines. Table 14.1 presents some of the state and commonwealth laws and acts related to mining water management in Australia (*Minesite Water Management Handbook*, 1997).

In earlier days, the discharge quality of mine water vastly varied among states. However, recently a more uniform approach is taken across the nation towards the mining water discharge by introducing Australian and New Zealand Environment and Conservation Council (ANZECC) water quality guidelines which imposes discharge quality levels for receiving water (*Minesite Water Management Handbook*, 1997). The sustainable mine water management should be further achieved according to the principles consistent with the ISO14001 standards. The general principles can be categorised into different stages as management, planning, implementation, measurement, evaluation, review and improvement. The management must be achieved with leadership and commitment of employees at all levels.

The importance of water should be considered in line with the operational policies and goals where possible. In the planning stage, the decisions will be taken on factors such as the opportunities of water management and threats, stakeholder participation, the context of water (dry/wet, arid/tropic, urban/remote, etc.) and legal considerations are complied. In the implementation, the efficiency of water is considered for all stages from the design to ongoing operations. Other factors that are given special attention in the implementation stage are the sustainability and the adequate resources availability. In the measurement, evaluation, review and improvement stages, monitoring programmes must be continuously carried out in each and every stage; reports should be generated and the water management plans should be reviewed to assure sustainability in mine water management (O'Brien et al., 2008).

When managing the water quality, relevant standards and acts play a critical role. They may be highly variable in different geographical areas. However, within this chapter, some of the Australian acts and standards are discussed as a reference. The water quality issues faced in Australian mining

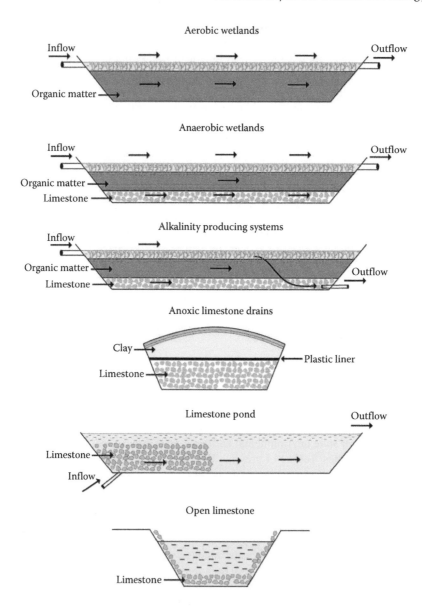

FIGURE 14.1 Scheme of passive AMD treatment.

are directly addressed by three major acts, namely, the Mining Act 1978, the Mines and Safety Inspection Act 1994 and the Environmental Protection Act 1986. Another act is considered if the water is involved with the rivers, which is known as Water and Rivers Commission Act 1995. In the water quality management, there are principles and guidelines that have been jointly developed by the Agriculture and Resource Management Council of Australia and New Zealand, and the ANZECC with inputs from the National Health and Medical and Research Council (Water and Rivers Commission, 1999).

LAND SUBSIDENCE

Land subsidence, also called ground subsidence, may occur when ore, petroleum, gas or groundwater is extracted from the underground. In mining, there is a tendency of land subsidence if the old

TABLE 14.1

Australia Laws and Acts for Mine Water Management

State Legislation	Commonwealth Legislation
• Mining Act	• Environment Protection (Sea Dumping) Act 1981
• Environmental Protection Act	• Great Barrier Reef Marine Park Act 1975
• Local Government Act	• Petroleum (Submerged Lands) Act 1967
• Clean Waters Act	• Protection of the Sea (Prevention of Pollution from Ships) Act 1983
• Groundwater Act	• Seas and Submerged Lands Act 1973
• Pollution of Waters by Oil Act	• National Parks and Wildlife Conservation Act 1975
• Environmental Protection/Marine (Sea Dumping) Act	• Environment Protection (Alligator Rivers Region) Act 1978
• Marine and Harbours Act	• Environment Protection (Impact of Proposals) Act 1974
• Petroleum (Submerge Lands) Act	• Industrial Chemicals (Notification and Assessment) Act 1989
• Coastal Protection Act	• World Heritage Properties Conservation Act 1983
• Soil Conservation Act	
• Dangerous Goods Act	
• Radiation Control Act	

workings (stopes) are not backfilled, where a part of the ore body is extracted and the rest is left to support the roof of the workings. Therefore, land subsidence occurs where pillars or support systems deteriorate or fail over time. The degree of land subsidence may be affected by the thickness of the ore body, size, shape and distribution of the support, extraction ratio, thickness and geological and geotechnical characteristics of the overburden and the mining method used (Sengupta, 1993).

Land subsidence involves in vertical and lateral movements of the ground. It can be categorised into three main types as pit subsidence, trough subsidence and shaft subsidence. The major difference between them is the way of movement of the strata during the process. For example, pit subsidence generally occurs from the failure of the roof. Pit subsidence can generate collapses of 2.0–2.5 m deep and 0.6–12 m in diameter. Typical pit subsidence is generated at mines of depths below 30 m. When the mines are deeper than 50 m, the frequency of this type of subsidence reduces (Sengupta, 1993).

Issues from Land Subsidence

Land subsidence may cause damage to structures, infrastructures and water resources. Ground movements in subsidence can occur in either horizontal or vertical direction; in particular, the differential movements create most of the damage. For example, vertical subsidence can affect the surrounding utilities such as drainage systems, railways, roads and structures, due to the formation of cracks and undulations. Especially, the structures for high-speed activities such as airport runways or motorways are susceptible to damages from subsidence (Darling and Nieto, 2011). The damage to the buildings can be further divided into functional and cosmetic damages. Functional damage is also known as structural damage, including the misplacement of building doors, windows and walls, the leakage from roofs and the damage for the utility lines. The cosmetic damage is less hazardous than the structural damage, but it may be further developed into structural damage. Cracks developed along the building exterior or interior walls, cracks on the plaster and separation of floors are considered as cosmetic damages (Sengupta, 1993). Other cases include the damage to the underground pipelines and drainage infrastructure. The pipelines that are constructed in the ground can be deformed and failed due to mine subsidence. If the material that the pipelines are constructed is unable to handle the rotations, stresses and strains occurred by the ground subsidence, material failure occurs. This type of failure may create huge financial losses for the local governments to restore the infrastructure. Hence, preventive measures need be implemented in order to reduce the impacts.

Similarly, the land subsidence may also alter and create impacts on the surface water. Owing to ground subsidence, fissures can be generated. This can lead to the loss of water from the water body. When the water body is considerably large such as a lake or a river, mine floods may be triggered, which can be catastrophic. Therefore, there are more strict rules and regulations for the mining activities under the surface water bodies. Similar to the surface water bodies, groundwater can also be affected. The fissures can alter the hydrogeological regime in the region (Wang et al., 2008). The local geology, especially the presence of such discontinuities as faults, folds or joints can influence the subsidence to various extents. The loss of water can directly affect the surface water eco-systems in the long term. Subsidence also increases the chances to generate 'sink holes'. These sink holes can be further developed and cause structural damages to buildings (Sengupta, 1993; Darling and Nieto, 2011).

Control and Prevention of Land Subsidence

The impacts from land subsidence due to mining can be controlled or prevented through a number of measures, such as adoption of comprehensive planning, preventive architectural and structural design, adequate mining techniques (partial mining, backfilling, post mining stabilisation, etc.) and strata consolidation (Sengupta, 1993; Darling and Nieto, 2011).

The planning stage of mining operation can create a significant effect on reducing the impacts of ground subsidence. Through planning, the extraction rate of mining can be decided according to subsidence reduction. High extraction rates for a fractured rock can increase ground subsidence. Post-mining stabilisation and backfilling of the voids are widely used for the reduction of subsidence. The typical backfilling materials are waste rocks and tailings. In strata consolidation, binding materials are used to bind the shallow strata beneath the structure. By binding the structure and strata into a single unit, the whole structure will move as a unit, which can significantly reduce the damage (Sengupta, 1993).

Land Subsidence Related Policies, Standards and Regulations

The regulations and guidelines related to mine subsidence vary with the mining operations. As an example, when starting an underground coal mine, consent should be obtained in accordance with the Environmental Planning and Assessment Act 1979. The environmental impacts associated with mining subsidence are assessed under Part 4 of the Act. One of the major documents that need to be prepared to obtain the consent for the mining operation under this Act is to prepare an extraction plan which describes how the environmental impacts from the subsidence would be managed throughout the operation (NSW Government, 2015).

Mine subsidence may also present a hazard to the health and safety along with its environmental impacts. Therefore, the major act or regulation associated with mine subsidence in Australia is the code of practice required for underground mining operations, under Section 274 of Work Health and Safety Act. The objective of the code is to ensure that the consideration given for the ground control or geotechnical aspects of the design, operation and closure of the underground mine is adequate for the health and safety of the workplace (Safe Work Australia, 2011).

MINE WASTE

Mining produces some unwanted gangue as a by-product of different processes throughout its production life cycle. As stated earlier, mine waste can create environmental impacts of different aspects. When comparing different minerals, coal mining generates the highest amount of mine waste followed by nonferrous, ferrous and industrial minerals (Lottermoser, 2010). Mine wastes generated from different mining activities largely vary in their properties and therefore require specific attention on control, prevention and remediation measures. Sulphidic waste, coal mine waste, radioactive waste and tailings and their relevant regulations are discussed in this session.

Sulphidic Mine Wastes

Sulphidic mine wastes are associated with deposits such as metal ores, phosphorous ores, coal seams and mineral sands. Large amounts of sulphidic mine waste can be found in heap leach piles, coal spoil heaps, tailing piles, mine runoff and low grade ore stock piles. Polymineralic aggregates are the major components of sulphidic mine wastes. Chemical weathering of these polymineralic aggregates can be classified into acid producing, acid buffering or non-acid generating or consuming reactions. Therefore, the rate and the type of reactions occurred during a particular mineral extraction are responsible for the generation of acidic wastes which can later be turned into AMD.

Coal Mine Wastes and Relevant Regulations

As stated earlier, coal extraction produces the highest amount of mine wastes. Coal is a combustible sedimentary rock that consists of carbon, sulphur and significant amounts of hydrogen. The sulphur in coal can be classified into three major categories as pyritic, sulphate and organic sulphur. The majority of sulphur is bound organically within solid carbonaceous materials. Thus, this form of sulphur does not produce acid or mine waste. Oxidation of pyrite is responsible for the generation of sulphate sulphur which can be occurred from the weathering of coal before or after the mining operation. Oxidation of the sulphides, the organic matter, leaching of the trace materials, weathering of silicate and carbonate from coal mine wastes can particularly generate harmful impacts for the receiving waters.

Another problem that is associated with coal mine waste is the tendency for spontaneous combustion. Coal usually contains very fine grained micrometre-sized pyrite at which spontaneous combustion can be easily initiated by exposure to atmospheric oxygen or groundwater that contains oxygen.

Spoil heaps are created from the overburden which is removed to gain access to the underlying coal seam. Coal spoil heaps can contain different physical, chemical and biological characteristics which can vary from the deposit that creates a heterogeneous nature. Generally, coal spoil heaps can diminish the environmental quality by reducing the aesthetics, and may gradually develop more disastrous outcomes such as mine fires.

The spoil heaps can contain varying amounts of coal that are not completely removed from the separation process. Having a look at the chemical composition of the spoil heaps, silica can be 80% or more. Other constituents that can be found in small amounts are oxides of aluminium, calcium, magnesium, titanium, iron, potassium and sodium (Kirby et al., 2010).

The major cause for spoil heap fires can be explained as the oxidation of pyrite which accelerates the combustion of carbonaceous material in coal. Spoil heap combustion can produce harmful gases such as carbon dioxide, carbon monoxide, hydrogen sulphide and sulphur dioxide. The generation of these types of gases can be further accelerated by activities such as excavating the burning spoils (Kirby et al., 2010).

Spontaneous combustion of coal spoil heaps can be controlled if the coal is located in oxygen-deficient humid environments . Some of the recommended methods in the literature are covering the burning spoil heaps with discard which is compacted and blanketing the heaps. Methods such as injection with combustion prevention or non-combustible materials, and water spraying can also be used to reduce combustion. Bell and Donnelly (2006) suggested injecting a curtain wall of pulverised fuel ash specifically to control combustion in hot spots.

Coal mining is one of the industries in Australia which contributes significantly to the economy. Therefore, the regulatory bodies have identified the importance of sustainable environmental management in coal mining. The relevant legislations govern coal mining are

- Environmental Planning and Assessment Act 1979 – assessment and approval of new mines and extension of existing mines
- Mining Act 1992 (and the Mining Amendment Act 2008) – mineral exploration licences and mining leases, including provisions for environmental management and rehabilitation

- Protection of the Environment Operations Act 1997 – environment protection licences that regulate environmental impacts during operations (NSW Government, 2010)

Radioactive Mine Wastes and Relevant Regulations

The radioactive mine wastes are mostly associated with uranium mine wastes, but also occur during the mining of beach and placer sands. The significant level of radioactivity makes them different from other wastes. Each year, 45 m³ of radioactive waste is raised from uranium mining which are stored in more than 100 sites across Australia (Australia's Uranium, 2015). These radioactive wastes are classified into two categories as high level and low level wastes, and can be further identified as long- and short-lived wastes. However, unlike the chemicals in most of the other mine wastes, radioactivity cannot be destroyed using chemical reactions, physical methods or other engineered solutions. Therefore, special care should be taken when handling radioactive mine waste to ensure that the waste is not unsafely disposed to the receiving environments.

Exposure to the radiation levels of these minerals can lead to hazardous outcomes for human health and the surrounding environment. Therefore, methods acceptable by law and approved by EPA for the safe handling and disposal of these minerals need to be implemented within radioactive mining operations.

Environmental Impacts and Control of Radiation Hazards

Low levels of radioactivity are always present in the natural environment. Radioactive elements are also available in the human body, air, soil and plants. However, the harmful levels of radiation exposure can occur through inhalation or ingestion of radioactive particulates and exposing to the radioactive emitters. The radioactivity can be concentrated in the food chains. The plants or animals can accumulate the radioactivity through contaminated soil, air or water.

Therefore, precautions should be taken to reduce the potential radioactivity exposure levels, for example, install covers for the tailing dams. In uranium mine sites, a major source of radon exposure is the release of Rn-222 to the environment from the tailings. The exposure levels can be reduced by covering the tailing dumps. Keeping the tailings wet can also decrease the addition of airborne particles to the atmosphere. Dust suppression is another control measure to reduce the impacts of radioactivity. Appropriate ventilation can also be provided to reduce the radon gas levels. In an open cut mine, natural ventilation can be incorporated while for the underground mines engineered ventilation systems are required to reduce the hazardous outcome generated from radiation wastes. Regular monitoring programmes need to be implemented to measure the radiation dose and these must be followed by strict health regulations and protective wear and equipment handling procedures for mine workers. Some of the generic environmental impacts of radioactive mine wastes are summarised in the literature (Davy and Levins, 1984; OECD, 1999), include excessive radioactivity levels, failure of tailings dams, water contamination, AMD and atmospheric emission.

Laws Related to Radioactive Mine Wastes

The management of radioactive wastes in Australia is handled by the Radiation Protection and Control Act 1982. The disposal of radioactive wastes should be kept within the EPA-defined ranges to maintain the environmental sustainability in air, water and the surroundings. The major legislations addressing radioactive waste disposal are

- Radiation Protection and Control (Ionising Radiation) Regulations 2000
- Radiation Protection and Control (Transport of Radioactive Substances) Regulations 2003
- Radiation Protection and Control (Non-Ionising Radiation) Regulations 2013 (EPA South Australia, 2015b)

Prior to starting any mining or milling operations that produce radioactive mine wastes, a waste management plan has to be developed in accordance with the guidelines in the above acts and approved by the regulatory authorities. The plan should cover all important aspects such as management of waste, working rules, recordkeeping, segregation and other controls (EPA South Australia, 2015a).

Tailing, Its Management and Governing Laws

In mineral processing, the residue after processing is known as tailings. Tailings may contain some minerals that are unable to be recovered. Tailings consist of solids and solution, which are stored in a tailing dam where the solution is drained. The solution may consist of chemicals from mineral processing. The acidity of the tailing solution is dependent on the nature of the processing technique used. Sometimes it may be strongly acidic, containing iron, aluminium, manganese, trace metals, chloride, fluoride and sulphate. The characteristics of tailings vary in different deposits according to their physical and chemical characteristics.

The tailings in tailing dams need to be monitored continuously to reduce environmental pollution and the problems that can be occurred due to the failure of dams. The monitoring programmes should be designed to address the dam performance, impoundment stability and environmental aspects (Lottermoser, 2010), and such parameters as filling rate, consolidation, grain size distribution, water balance and chemical considerations should be measured. The impoundment stability monitoring programmes are designed to measure slope stability and the pore water pressure of the tailings. The environmental monitoring includes parameters such as the radioactivity levels, meteorological measurements, information on tailings chemistry and mineralogy, AMD generation and groundwater quality.

The uncovered tailing dams can be subjected to water and wind erosion over time due to their grain sizes. These eroded tailing particles can be distributed in the atmosphere and pollute air, soil and water resources. Therefore, tailing covering methodologies have been introduced to minimise these impacts. Tailings can be covered by either wet or dry covers. As wind erosion is one of the major issues occurred in tailing dams, wet covers are generally applied by covering the tailing dams with water. Dry capping of the tailing dam is another rehabilitation methodology used to control the environmental impacts occurred. Furthermore, high-density tailings slurries or pastes as backfilling materials are used in underground mines as part of the mining method to backfill the voids, in the meantime, also as strategies to support the ground to minimise land subsidence, and to contain mine tailings to reduce environmental impact.

A sustainable tailing management is mainly concerned with tailing storage and there are codes of practices used in sustainable tailing storage and management. The concept of sustainable tailing management is generally achieved by a risk-based approach starting from mine planning, design, construction, operation, closure and rehabilitation of mine tailing in storage facilities (Williams et al., 2007). Within this procedure, the risks are identified in two major phases, namely operational risks and post-mine closure risks.

A number of different risk assessment methods are available. Risk analysis provides opportunities to identify and quantify the likelihood, cost and the outcomes of a problem. One of the risk assessment processes proposed by the AS/NZS 4360:2004 is as follows:

Step 1. Establishing the context and the parameters environmentally, socially and geographically to decide the design criteria.
Step 2. Identification of the hazards, the occurrences and potential reasons.
Step 3. Analysing the risks to determine the likelihood of risks.
Step 4. Evaluating the risks by comparing them against the design criteria, conducting sensitivity analysis and making decisions on addressing the risks.
Step 5. Addressing selected risks by implementing different remedial plans (Williams et al., 2007).

Tailing storage facility (TSF) is another important factor that needs to be considered sustainable in tailings management. A TSF may include pits, dams, ponds, integrated waste landforms, erosion protection bunds, levee banks, diversion channels, spillways and seepage collection trenches associated with the storage of tailings. TSF should be planned and designed according to the guidelines in the Mining Act 1978 and as per the requirements of the Mines Safety and Inspection Act 1994 (Department of Mines and Petroleum, 2013).

Mine Fire

Coal mine fires are caused by coal oxidation when the seams or the spoil heaps are exposed to ambient air. Fires are caused by spontaneous combustion, which is one of the processes in coal oxidation. When coal seams or spoil heaps interact with atmospheric oxygen at ambient temperature, rate of oxidation could be accelerated which leads to spontaneous combustion. These spontaneous combustion can be further lead to hazardous mine fires that can damage properties, usable mineral deposits and also cause threat to mine workers and the surrounding ecosystems. However, the mechanisms causing the generation of coal mine fires are not yet fully understood, due to the varying chemical nature of coal (Singh et al., 2007).

Sustainability and Control of Coal Mine Fires

One of the major gases responsible for coal mine fires is methane. Reducing or controlling the methane emissions can contribute to reduce coal mine related fires and combustions. In some mining operations, the emitted methane is diverted and used as fuel instead of directly releasing to the atmosphere. In different countries, these degasification systems are being installed in coal mining operations and the emission estimations have been calculated to identify the amount of methane that is being released to the environment and the gases that are used as fuel sources. This can provide valuable insights in managing and controlling coal mine fires.

The emissions from deep coal mine workings are the most explosive kinds of methane gas when they interact with air. The following actions can be taken to control and manage methane emissions from deep coal mining: pre-mining degasification, enhanced gob well recovery, deep mine ventilation and developing methane recovery systems (Hester and Harrison, 1994). In pre-mining degasification, methane is recovered from unmined coal seams before coal is mined. As a result, methane is removed from the coal seam, preventing it from releasing and interacting with the air in the workface or other space. In the gob well recovery method, methane is recovered from mining gob areas. Gob area is a highly fractured area of rock and coal which is created after the caving of mine roof when the coal is removed. The gob areas have a high potential of releasing methane gas into the mine, and reducing these gases in advance can reduce the ventilation requirements. Surface wells can be applied for methane recovery. The recovered gas can be used for power generation and industrial heating application within mining areas. Ventilation is compulsory for the underground coal mining. This is generally done by using fans and the methane concentrations in the ventilated air can be diluted. New technologies can be applied to recover methane from the air, which increases the safety of the mine site while producing a useful energy source.

Blasting

Blasting is widely used in mining to obtain fragmentation of rock and ore. Blasting is one of the activities encountered in mining operations which may create damage to property, humans, infrastructure and the surrounding environments. The major displacement effects of blasting can be categorised into two major types as permanent and transient displacements. In permanent displacements, fly rocks are generated from the top of the blast hole. Fly rocks can be transported to long distances of up to 100–1000 m. Fly rocks can damage the surrounding properties, mining machinery, vehicles and mine workers. Blasting mats are one of the technologies used for the control of fly rocks. The transient displacements are occurred from the vibrations that propagate outwards from the blast. Some of the side effects of transient displacements are structural distortion, faulted or displaced cracks, falling objects, cosmetic cracking of wall coverings, excessive instrument and machinery response and micro-disturbances (Darling and Nieto, 2011).

Blasting can also increase the tendency of cracking with different household activities in the long term. The impacts from blasting can be reduced by carefully planning of the blasting design and using proper explosives and blasting technologies. The environmental considerations, public safety and well-being must be taken into account from the beginning of the blasting plan. Many countries have developed guidelines to reduce impacts induced by the blasting operations.

Environmentally Sustainable Blasting, Laws and Regulations

Blasting is an activity carried out in all mining operations that use drilling and blasting methods. Two main impacts from blasting are overpressure and ground vibration, which can create numerous environmental impacts. Hence, most of the regulations related to blasting sustainability have provided measures to address problems related to these two factors. The blasting operations in Australia are regulated by the following acts and guidelines:

- Australian and New Zealand Environment Council, technical basis for guidelines to minimise annoyance due to blasting overpressure and ground vibration
- Department of Environment, Climate Change and Water (DECCW), assessing vibration: a technical guideline
- Environmental Planning and Assessment Act 1979, administered by the Department of Planning
- Protection of the Environment Operations Act 1997, administered by DECCW (Newsouth Wales Minerals Council, 2009)

The impacts of mine blasting can create structural damage and nuisance to the surrounding communities. Therefore, threshold limits have been defined based on field studies and researches for overpressure and ground vibration. The limits are as follows:

- Ground vibration – 5 mm/s with 5% allowable to 10 mm/s or less in an annual period
- Overpressure – 115 dB (Lin peak) with 5% allowable to 120 dB (Lin peak) or less in an annual period (Newsouth Wales Minerals Council, 2009)

MINE CLOSURE, ENVIRONMENTAL SUSTAINABILITY, LAWS AND REGULATIONS

In starting any mining operation, a plan for mine closure is one of the important aspects that need to be addressed. An abandoned or poorly closed mine is a liability to the governments, communities and the mineral companies (Bell et al., 2006). Adequate mine closure is vital in achieving mine sustainability. If the closure is not properly planned or not implemented according to the rules and guidelines, there can be pollution and hazards which can be continued for many more years or generations. Therefore, the overall objective of mine closure planning is to reduce the long-term environmental, social and economic impacts after termination of mining.

Mine closure planning can have a number of objectives. According to ANZMEC/MCA (2000), some of the objectives are

- Engage all the stakeholders and their interests for the mine closure planning
- Ensure the closure is happening in a timely manner
- Ensure the company has sufficient funds to support the closure and does not leave any liabilities
- Establish a set of criteria and guidelines for successful completion of the closure (Bell et al., 2006)

However, it has been identified that Australia had recorded numerous incidents of unplanned mine closures, unsafe workings, hazardous mine sites and unclaimed lands. This implies that there still exists an inadequacy in mine closure in contrast with best practice standards. Furthermore, the concepts and standards related to mine closure and mine rehabilitation are becoming more demanding than they were before and reflect the changes in public priorities and environmental awareness (Smith, 2007).

There are different mine closure laws available for different jurisdiction. However, most jurisdictions are using Western Australia's mine closure laws which are the Amended Mining Act of 1978. The mine closure planning is further governed by the Environmental Protection Act, 1986. Risk assessment is another important aspect that is carried out in mine closure. In each of the jurisdictions, there are several legal obligations that must be complied throughout the mine's life. From the past few years, the requirement for upfront planning of mine closure has been increased. Proper planning for mine closure can benefit the mining companies in several ways such as early planning can reduce the high costs incurred during mine closure and the rehabilitation success, reducing the company's financial risk in respect of closure, gaining easy access from stakeholders and avoiding ongoing liability due to the poor mine closure practices (Merwyk van Tony, 2013).

Once the mine closure is completed, post-closure management is conducted to ensure whether the closure has been sustainable. This requires continuous monitoring and ongoing management. The continuous monitoring includes stakeholder participation in different areas such as mining managers and general public to assess the post-closure related issues. Generally, post-mine closure activities are planned according to the requirements of the sites, land owners and the type of land use. Some of the ongoing post-closure activities can be

- Noxious weed control
- Exclusion or control of grazing animals
- Control of public access
- Fire management
- Maintenance of safety signs and fences (Bell et al., 2006)

Therefore, it is evident that environmental sustainability and management is an ongoing process that needs to be achieved and maintained throughout the beginning of mining life cycle to mine closure and even after years of mine closure.

SUMMARY

Mining is one of the major industries in the world which is required for the production of essential minerals for the economy. However, mining industry also has disastrous consequences associated with it for the natural environment which cannot be avoided. Mining industry which is dependent on the extraction of minerals from the earth has also environmental, economic and social aspects associated with it. Finding the right balance between these factors is essential to maintain sustainability in mining. Environmental issues associated with mining such as AMD, mine subsidence, mine wastes, mine fires and blasting-related problems can be mitigated through proper mine planning and management practices. Mining-related environmental management should be taken into consideration from the land acquisition for the mine to years after the closure of the mine. Furthermore, nations have developed specific governing laws and regulations for each of the different environmental problems to control the environmental impact of mining to ensure sustainability. Developing environmental sustainability in mining is not only an economical responsibility of the mining industry, but also a social responsibility from the start to post-closure of the mine.

REFERENCES

ANZMEC/MCA. 2000. Strategic framework for mine closure, Australian and New Zealand Minerals and Energy Council, Canberra and Minerals Council of Australia, Canberra.

Australia's Uranium. 2015, http://www.world-nuclear.org/info/Country-Profiles/Countries-A-F/Australia/# Radioactive_wastes (Retrieved 21 July 2015).

Bell, C., K. Lawrence, B. Biggs, E. Bingham, E. Bouwhuis, N. Currey and A.-S. Deleflie. 2006. *Mine Closure and Completion: Leading Practice Sustainable Development Program for the Mining Industry.* Canberra: Commonwealth of Australia.

Bell, F. G. and L. J. Donnelly. 2006. *Mining and Its Impact on the Environment.* Geological Society of America: CRC Press.

Bussière, B., M. Benzaazoua, M. Aubertin and M. Mbonimpa. 2004. A laboratory study of covers made of low-sulphide tailings to prevent acid mine drainage. *Environmental Geology*, 45(5): 609–622.

Connelly, R. J., K. J. Harcourt, J. Chapmen and D. Williams. 1994. Approach to remediation of ferruginous discharge in the South Wales coalfield and its application of closure planning. *Paper Presented at the 5th International Mine Water Congress*, Nottingham UK, pp. 521–531.

Darling, P. and A. Nieto. 2011. *SME Mining Engineering Handbook.* 3rd edn, 11 pp. Englewood, Colo: Society for Mining, Metallurgy, and Exploration.

Davy, D. and D. Levins. 1984. Air and water borne pollutants from uranium mines and mills. *Mining Magazine March*, 1984: 283–291.

Department of Environment and Heritage Protection. 2013. A quick guide to the Greentape Reduction Act, https://www.ehp.qld.gov.au/factsheets/pdf/environment/greentape-reduction-act.pdf.

Environmntal Management. 2015. Alcoa in Ausralia, http://www.alcoa.com/australia/en/info_page/mining_environ_man.asp (Retrieved 15 June 2015).

Environmental Protection Act 1994. 2015, https://www.legislation.qld.gov.au/legisltn/current/e/envprota94.pdf.

EPA South Australia. 2015a. *Guidelines for Waste Management Application/Plan to Dispose of Unsealed Radioactive Substances* [Online]. Available: www.epa.sa.gov.au/files/4771321_guidelines_wmp.pdf (Accessed 23 January 2015).

EPA South Australia. 2015b. Radiation Legislation and Regulations, http://www.epa.sa.gov.au/environmental_info/radiation/legislation_and_regulations (Retrieved 20 February 2015).

Hallberg, K. B. and D. B. Johnson. 2005. Microbiology of a wetland ecosystem constructed to remediate mine drainage from a heavy metal mine. *Science of the Total Environment*, 338: 53–66.

Hester, R. E. and R. M. Harrison. 1994. *Mining and Its Environmental Impact.* Cambridge, UK: Royal Society of Chemistry.

Kirby, B., C. Vengadajellum, S. Burton and D. Cowan. 2010. Coal, coal mines and spoil heaps. In Timmis, K. N. (ed.) *Handbook of Hydrocarbon and Lipid Microbiology.* Berlin Heidelberg: Springer-Verlag, pp. 2277–2292.

Lottermoser, B. 2010. *Mine Wastes: Characterization, Treatment and Environmental Impacts.* Berlin Heidelberg: Springer-Verlag.

Lussier, C., V. Veiga and S. Baldwin 2003. The geochemistry of selenium associated with coal waste in the Elk River Valley, Canada. *Environmental Geology,* 44(8): 905–913.

Merwyk van Tony. 2013. The increasing importance of mine closure planning (Retrieved 12 July 2015).

Minesite Water Management Handbook. 1997. Minerals Council of Australia. http://www.minerals.org.au/file_upload/files/resources/environment/Minesite_Water_Management_Handbook.pdf (Accessed 10 February 2015).

Mitchell, P. B., C. P. Waller and K. Atkinson. 1994. Prediction of water contamination arising from disposal of solid wastes. *Hydrometallurgy '94: Papers presented at the International Symposium 'Hydrometallurgy '94'* organized by the Institution of Mining and Metallurgy and the Society of Chemical Industry, Cambridge, England, 11–15 July, Dordrecht: Springer, Netherlands, pp. 1011–1024.

Mudd, G. M. 2010. The environmental sustainability of mining in Australia: Key mega-trends and looming constraints. *Resources Policy*, 35: 98–115. doi: 10.1016/j.resourpol.2009.12.001.

Newsouth Wales Minerals Council. 2009. *Blasting and the NSW Minerals Industry.* Available: http://www.nswmining.com.au/NSWMining/media/NSW-Mining/Publications/Fact%20Sheets/Fact-Sheet-Blasting.pdf (Accessed 15 January 2015).

NSW Government. 2010. Environmental compliance and performance report. *Management of Dust from Coal Mines* [Online]. Available: http://www.epa.nsw.gov.au/resources/licensing/10994coalminedust.pdf (Accessed 30 January 2015).

NSW Government. 2015. Subsidence management, http://www.resourcesandenergy.nsw.gov.au/miners-and-explorers/applications-and-approvals/environmental-assessment/subsidence-management   (Retrieved 20 February 2015).

O'Brien, D., P. Eaglen, G. Brassington, B. Smith, D. Way, A. Bones, M. Stutsel, T. Jeremy, A. Barger and K. Brown. 2008. Leading practice sustainable development program for the mining industry. *A Guide to Leading Practice Sustainable Development in Mining* [Online]. Available: http://www.industry.gov.au/resource/Documents/LPSDP/guideLPSD.pdf.

OECD. 1997. Environmental Performance Reviews: Australia. *Conclusions and Recommendations* [Online]. Available: http://www.environment.gov.au/system/files/resources/6d83bc2d-62ce-47c1-95b9-7d289d2a8285/files/oecd.pdf (Accessed 9 December 2014).

OECD. 1999. Environmental Activities in Uranium Mining and Milling: A Joint Report. Available: https://www.oecd-nea.org/ndd/pubs/1999/766-environmental-activities.pdf (Accessed 16 December 2014).

Queensland. 2012. Environmental Protection (Greentape Reduction) and Other Legislation Amendment Bill 2012, https://www.legislation.qld.gov.au/Bills/54PDF/2012/EnvProGROLAB12Exp.pdf.

Robertson, M. 2012. Environmental Protection (Greentape Reduction) and Other Legislation Act 2012 (Qld), http://www.mccullough.com.au/icms_docs/133738_Resources_-_17_August_2012.pdf.

Safe Work Australia. 2011. *Code of Practice – Ground Control for Underground Mines* [Online]. Available: https://submissions.swa.gov.au/SWAforms/Archive/mining/documents/GroundControlforUnderground Mines.PDF (Accessed 2 January 2015).

Sengupta, M. 1993. *Environmental Impacts of Mining Monitoring, Restoration and Control.* Boca Raton, Florida: CRC Press.

Singh, A. K., R. Singh, M. P. Singh, H. Chandra and N. Shukla. 2007. Mine fire gas indices and their application to Indian underground coal mine fires. *International Journal of Coal Geology*, 69(3): 192–204.

Smith, B. 2007. *Mining for Closure Planning, Integrated Mine Closure.* University of New South Wales [Online]. Available: https://www.be.unsw.edu.au/sites/default/files/upload/pdf/schools_and_engagement/resources/_notes/5A3_16.pdf.

Vermeulen, P. and B. Usher. 2006. Sulphate generation in South African underground and opencast collieries. *Environmental Geology*, 49(4): 552–569.

Walton-Day, K. 2003. Passive and active treatment of mine drainage. *Raeside R. Short Course Series Volume*, 31: 335–359.

Wang, G. Y., G. You and Y. L. Xu. 2008. Investigation on the Nanjing gypsum mine flooding, in: H. Liu, A. Deng and J. Chu (eds), *Proceedings of the 2nd International Conference on Geotechnical Engineering for Disaster Mitigation and Rehabilitation*, 30 May–2 June 2008, Nanjing, China, Science Press Beijing (ISBN 978-7-03-021634-2) and Springer (ISBN 978-3-540-79845-3), pp. 920–930.

Water and Rivers Commission. 1999. *Water Quality Protection Guidelines. Nos. 1–11, Mining and Mineral Processing* [Online]. Available: https://www.water.wa.gov.au/__data/assets/pdf_file/0004/4855/44631.pdf (Accessed 20 December 2014).

Williams, D., A. Minns, R. Yardi, G. Bentel, B. Brown and M. Gowan. 2007. Tailings management. *Leading Practice Sustainable Development Program for the Mining Industry* [Online].

Wu, P., C. Tang, C. Liu, L. Zhu, T. Pei and L. Feng 2009. Geochemical distribution and removal of As, Fe, Mn and Al in a surface water system affected by acid mine drainage at a coalfield in Southwestern China. *Environmental Geology*, 57(7): 1457–1467.

Zielinski, R. A., J. K. Otton and C. A. Johnson. 2001. Sources of salinity near a coal mine spoil pile, north-central Colorado. *Journal of Environmental Quality*, 30(4): 1237–1248.

15 Sustainable Management of Mine Induced Water

Muhammad Muhitur Rahman, Dharmappa Hagare,
Muttucumaru Sivakumar, and Raghu N. Singh

CONTENTS

ABSTRACT

Water is an essential part of mining industries. Fresh water is needed in different stages of mining activities. Mine induced water (MIW), water generated by different mining activities, contains harmful ingredients which cannot be reused or disposed without proper treatment. While the amount of water used by different mining industries varies depending on the type of mining product, so do the characteristics and amount of MIW. For example, the amount and characteristics of MIW generated from coal mining is very different than the characteristics of MIW generated from metal mining operations. The characterisation of MIW is also essential to devise a proper management, treatment and reuse scheme. In this chapter, characterisation and quantification of MIW from different activities in coal and metal mining are reviewed. Then impacts of MIW on the environment are identified. Later, different available options are discussed to sustainably manage the quality of MIW from different mining activities.

INTRODUCTION

Water is an integral part of the mining industry. Water is considered as a favourable medium in the mining industry, which is used to process coal and ore, to manage waste and to suppress the dust created by mining activities. More specifically, water is used by mineral industries for operational activities such as (Prosser, 2011):

- To transport ore and waste in slurries and suspension
- For separation of minerals through chemical processes
- For physical separation of material such as in centrifugal separation
- In cooling systems around power generation
- For suppression of dust, both during mineral processing and around conveyors and roads
- For washing equipment

Safe and sustainable management of water in a mine is important for the safety of workers as well as the surrounding environment. For sustainable management of water in a mining industry, sources and quality of water to be used as well as produced MIW from different mining activities is crucial. Especially mines situated in arid and semi-arid regions, where water is scarce, only proper management of water may guarantee smooth operation of mining activities. Such management includes characterisation of influent and MIW from mining activities, efficient treatment and supplement to water intake through effluent reuse schemes, and sustainable disposal of effluents to the environment.

Australia is one of the largest coal and mineral producing countries in the world. Australia is rich in reserves of a number of precious metals as well as different types of coals. At a glance, mineral resources of Australia include (ABS, 1996; Britt et al., 2014):

- Base metals: zinc, lead, copper, nickel, cobalt, antimony, cadmium, tin and tungsten
- Precious metal: gold, silver, platinum and diamond
- Energy minerals: black coal, brown coal and uranium
- Metallic minerals: iron ore, bauxite, magnesite, manganese and vanadium
- Mineral sands: ilmenite, rutile and zircon
- Other minerals: lithium, tantalum, phosphate and rare earth oxides

Australia possesses world's largest reserve of diamond, gold, iron ore, lead, nickel, rutile, tantalum, uranium, zinc and zircon. For other minerals namely, antimony, bauxite, black coal, brown coal, cobalt, copper, ilmenite, lithium, manganese, silver, tin, tungsten and vanadium rank in the top six worldwide (Britt et al., 2014). Globally, Australia is ranked fifth (behind China, the United States, India and Indonesia) as a coal producer (Britt et al., 2014). Most of Australia's black coal is located in Queensland (QLD) and New South Wales (NSW). The Bowen Basin in QLD and the Sydney Basin in NSW dominate black coal production in Australia, which contain about 60% of the nation's black coal reserve. Black coal is also available in the Surat, Clarence-Moreton and Galilee Basins in QLD and in the Gunnedah Basin in NSW. In terms of brown coal, most of the reserves are located in Victoria (VIC) with approximately 93% in the Latrobe Valley. For another important mineral, that is iron ore, Australia possesses the largest reserve in the world and ranked second as an iron ore producer. Most of the iron ore mines of Australia are located in the Pilbara region of Western Australia (WA), which is about 89% of the total reserve of Australia (Britt et al., 2014).

INFLUENT AND MIW ASSOCIATED WITH MINING INDUSTRIES

UNDERSTANDING WATER BALANCE IN MINING INDUSTRIES IN AUSTRALIA

Nowadays due to strict environmental pollution prevention regulations imposed by government environmental agencies and increased environmental awareness of residents, mining industries in advanced countries like Australia explore conventional and non-conventional sources of water for mining activities. Australia is one of the advanced countries in the world, which follows a complex water account system (Figure 15.1) for calculating water consumption by its mining industries. In Australia, water in mining industry is sourced mainly from surface and groundwater. State-based water authorities extract water from the environment including rivers, lakes, dams and groundwater. Water extracted from the environment is termed as *self-extracted water* (Figure 15.1). Then, the water authorities supply (in exchange of economic benefit) the water to mining industry through a non-natural network, such as pipe or open channels (termed as *distributed water*). The self-extracted water may or may not be treated by the water authorities before supply to the industries. On top of the water received from the water authorities, many mining facilities extract groundwater by themselves to use in the mining activities. Water authorities also supply *reuse water* to meet the mining industry needs. The reuse water is the treated water (to primary, secondary or tertiary levels) which is collected from different sources including waste water (from sewerage systems), drainage water or other water providers (ABS, 2014). In addition, stormwater collected by water authorities using infrastructure separate to sewerage systems and, depending upon its intended use, may or may not be treated before being supplied as reuse water.

Water discharged to the environment by mining industry may be *regulated* or *unregulated*. The regulated discharge refers to the water that is used *in-stream* of the mining facilities (Figure 15.1). For example, mine dewatering is categorised as self-extracted in-stream (non-consumptive) use by the mining industry. The in-stream water is usually used on-site or subsequently discharged to the environment. According to ABS (2004), water consumption by mining industries in Australia is calculated as

$$\text{Water Consumption} = \text{Water use}_{\text{Distributed}+\text{Self-extracted}+\text{Reuse}}$$
$$- \text{Water supplied to other users} - \text{In stream water use} \quad (15.1)$$

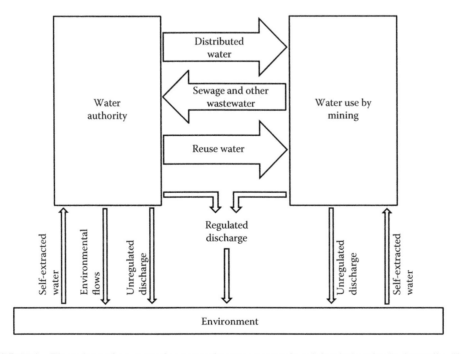

FIGURE 15.1 Flow chart of sources of water and water account in mining industries in Australia. (Based on ABS. 2014. *Water Account, Australia 2012–2013*. Canberra, Australia: Australian Bureau of Statistics.)

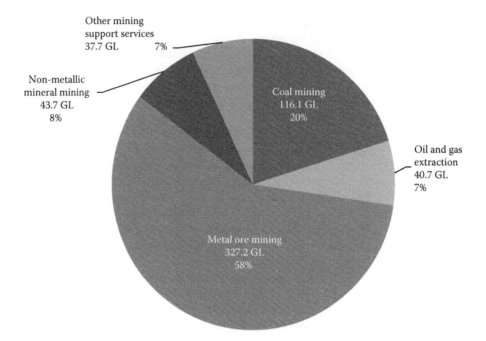

FIGURE 15.2 Yearly average water consumption by different mining sectors in Australia between 2008 and 2013. (Adapted from ABS. 2006. *Water Account, Australia 2004–2005*. Canberra, Australia: Australian Bureau of Statistics; ABS. 2014. *Water Account, Australia 2012–2013*. Canberra, Australia: Australian Bureau of Statistics.)

Amount of water used by mining industries depends on the amount of mineral produced. Usage of water by mining industries increases with the increased production of minerals. Types of mineral industry are also important in quantifying the amount of water usage.

Figure 15.2 shows yearly average water consumption by different mining industries in Australia. As seen from the figure that metal ore and coal mining in Australia are the highest consumers of the water compared to other mining industries. As such, it is obvious that effluent from metal ore and coal mining needs more attention than others. Therefore, characterisation and management of effluent from metal ore and coal mining are elaborated in the subsequent sections of this chapter. It should be mentioned that amount of water consumed by different mining industries is calculated based on Equation 15.1. State wide water consumption by mining industries in Australia is shown in Figure 15.3. As shown in the figure, WA is the highest consumer of water, followed by QLD and NSW. On average between 2000 and 2013, WA, QLD and NSW consumed 47% (241 GL), 24% (123 GL) and 14% (75 GL) of the total consumed water, respectively. It should be noted that the above-mentioned amount of water was consumed by industries including coal, metal ore, non-metallic minerals, oil and gas and other mining support services.

CHARACTERISTICS OF MIW

Characteristics of MIW are controlled by mineral deposit type, mining and metallurgical processing methods, the design of waste storage facilities and site-specific conditions (McLemore, 2008). Therefore, before discussing characteristics of MIW, it is essential to understand the types of mining methods and steps involved in the processing of that particular mineral.

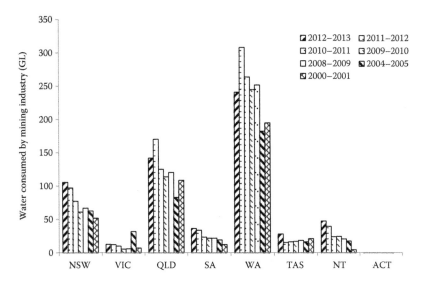

FIGURE 15.3 State wide water consumption by mining industries in Australia between 2000 and 2013. (Adapted from ABS. 2004. *Water Account, Australia 2000–2001.* Canberra, Australia: Australian Bureau of Statistics; ABS. 2006. *Water Account, Australia 2004–2005.* Canberra, Australia: Australian Bureau of Statistics; ABS. 2014. *Water Account, Australia 2012–2013.* Canberra, Australia: Australian Bureau of Statistics.) NSW = New South Wales; VIC = Victoria; QLD = Queensland; SA = South Australia; WA = Western Australia; TAS = Tasmania; NT = Northern Territory; ACT = Australian Capital Territory.

Mining Methods

Generally, there are two mining methods practiced by different mining industries:

- Surface
- Underground

Surface Mining

Surface mining is carried out when mineral deposits are located close to the surface (typically 50–60 m below the surface). Surface mining can be conducted by *strip mining* and by *open pit mining* (McLemore, 2008). Strip mining is practiced when the overlying layer of waste rock (layer consisting of soil and rock) on the mineral deposit is too thin. Typically, in a strip mining operation, a single strip of overlying waste rock is removed and placed aside, which expose the underlying mineral layer for mining. Then, a second strip of overlying layer is removed and placed in the channel left by the first strip. The process continues till the entire layer is mined. In the case of open pit mining, successive layers of minerals from the surface are removed to some depth, which eventually form an open bowl or pit. The depth of the pit is determined by the depth of the mineral deposit. In NSW, Australia, open cut mining is mainly practiced for coal mining (such as the Mount Arthur coal mine), however, it is also used for gold and copper mining (NSW-MC, 2015).

Underground Mining

Underground mining is carried out when the mineral deposits are located several hundred metres from the surface and involves creating tunnels from the surface to the mineral deposits. Underground mining can be conducted as follows (McLemore, 2008; NSW-MC, 2015):

- *Room-and-pillar mining.* A series of rooms are excavated into the mineral deposit, leaving pillars of minerals to support the mine. The method is used for coal as well as metal mining. In advanced countries like Australia, the method was prominent till the 1960s, however, with the availability of better technology now it is used only in a small number of coal mines (i.e. Yancoal's Tasman Mine near Newcastle, NSW).

- *Longwall mining*. Mechanical shearers plough minerals (i.e. coal), which is taken away by a conveyor in the reverse direction of the end of a cut. While cutting of minerals goes on, hydraulic-powered jack supports the roof of the mine. The hydraulic jack support is moved forward as mining continues. Occasionally, the roof is blasted to reduce the overburden pressure on the coalbed being mined. The longwall method is more efficient than room-and-pillar method as 80% of the entire coal face can be removed with this method. In Australia, longwall mining exists in Centennial Coal's Angus Place mine, near Lithgow, NSW.
- *Block caving mining*. Minerals (i.e. gold and copper) are extracted by collapsing the mineral deposit under gravitational force (own weight of the mineral deposit). Australia's first block cave mine opened in 1997 near Parkes, in Central West NSW.

How Mining Processes Affects MIW?

Steps involved in the processing of coal are different from extracting metals from ores. This is because the mineralogy (structural and physical characteristics) of metallic deposits is generally more complex than coal deposits. Compared to coal deposits, metallic deposits produce more than one commodity. Coals and metals also differ in the process of geological formation. While coal deposits always occur in stratigraphic layers (or beds), metallic deposits can occur in such layers but also in strips and in stockworks (in geology, a stockwork is a complex system of structurally controlled or randomly oriented veins). Due to this reason, in an aboveground metal mine, most of the mined material is waste, and only a small portion of the collected material is economically viable. According to McLemore (2008), a profitable lead or zinc mine may result in 90% or more of the total rock mined being placed in waste storage facilities; for copper mines, the wastage is 95% and for gold mines, almost 99% is waste material. On the other hand, coal produces much less waste rock from underground as well as open pit mines. The diversity and amounts of recovered minerals and chemical reagents involved in extracting metals are far greater than those used in coal extraction. Therefore, the potential for impacting the environment by the diverse range of chemicals used in metal extraction is higher in metals processing than in coal. Steps involved in coal as well as metal processing are shown in Figures 15.4 and 15.5, respectively. As shown in Figure 15.4, coal processing typically requires crushing, grinding and sizing followed by physical separation of pyrite and shale by gravity or flotation. Impurities such as rocks, soil, iron compounds and carbonates are usually removed by utilising their differences in specific gravity through the process of sedimentation. In the washing process, approximately 20%–30% by weight of the coal is lost and rejected. Coals are unwashed when supplied for the cement and power generation industry, and washed for the supply to steel works and export market. The washing process may be carried out at the colliery and at the steel works site.

In the case of metal processing (Figure 15.5), chemical extraction, hydrometallurgy, beneficiation, smelting and refining are involved. Normally, iron ores undergo crushing and partial grinding plus cobbing and dressing of a purely mechanical nature without using chemicals. Base metal ores, on the other hand, are ground to finer than 0.1 mm and concentrated by flotation after adsorption of organic reagents in dosages of the order of 100 g/t. Gold and uranium ores, are ground and leached with cyanide (dosage ~100 g/t) and sulphuric acid (dosage ~50 kg/t), respectively. Therefore, as far as environmental pollution is concerned, iron ores are harmless compared to ores that need to undergo floatation and leaching processes (Kihlstedt, 1972). In addition, the rejected portion of the metal ore after the beneficiation process may include coarse and fine particulate in the wash water and form slurry known as wet tailings or slime. According to Kihlstedt (1972), tailings produced from ore grounded to 90% by weight finer than 0.1 mm can be expected to contain at least about 2% by weight of material finer than 1 μm as a form of fine slime. This fine slime has extremely low settling velocity and a considerable adsorptive capacity, which may create different environmental problems.

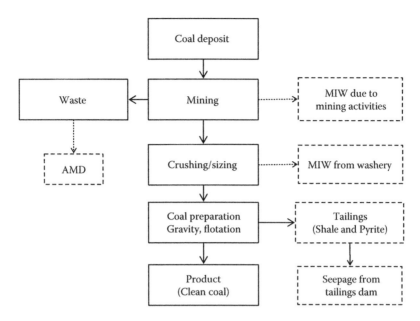

FIGURE 15.4 Steps involved in coal processing. (After McLemore, V. T. 2008. *Basics of Metal Mining Influenced Water.* Society for Mining Metallurgy and Exploration (SME).)

Coal MIW

MIW from an underground coal mine can be categorised as (Dharmappa et al., 2000):

- Mine water
- Process water
- Stormwater
- Domestic wastewater

Mine Water

Mine water is the most probable source in the case of underground mining. The source is seepage from the excavated area of the mine. The mine water is collected in underground sumps with a nominal retention time. There is an opportunity for recycling of mine water in the areas of firefighting and underground dust suppression within the mining complex. The quantity of the mine water greatly depends on the level of groundwater table and the soil conditions. The quality of the mine water widely varies from mine to mine depending upon the local conditions. Typical characteristics of underground mine water are shown in Table 15.1. As shown in the table, the mine water consists of a considerable amount of dissolved minerals which give high hardness to the water. In the case of acid main drainage, the pH can be as low as 2–3. The MIW has low biochemical oxygen demand (BOD) and chemical oxygen demand (COD), high conductivity and moderate concentrations of trace elements.

Process Water

The process wastewater is generated by the wash down facilities in the workshop, wash down bay, clean down of the coal conveyor, the dust suppression facility, oil storage and diesel filling area, and on-site coal processing facility. The likely contaminants are oil, coal and other solid materials. Here again, the quantity and the characteristics of the process wastewater varies widely. Typical characteristics of process wastewater generated from coal mine washery are shown in Table 15.2. The process wastewater contains near neutral pH, moderate dissolved solid and high filterable residues.

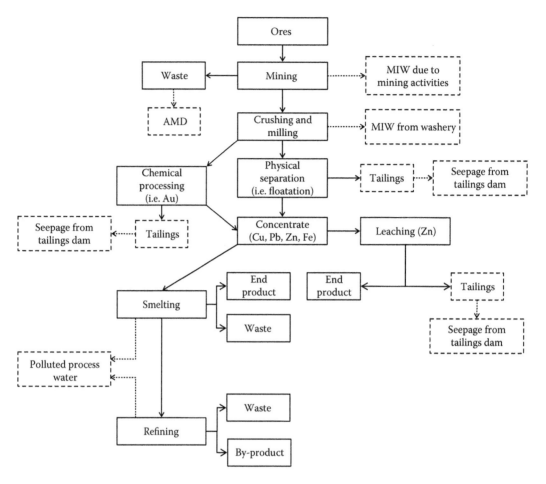

FIGURE 15.5 Steps involved in metal ore processing. (After McLemore, V. T. 2008. *Basics of Metal Mining Influenced Water*. Society for Mining Metallurgy and Exploration (SME).)

Stormwater

During moderate to heavy storms, the mining operations at the site produce a large volume of stormwater or surface runoff. Surface runoff can be computed by knowing the land area exposed to direct rain, and the rainfall rate. The characteristics of the surface runoff depend largely on the type of the mining operation as well as on the land use of the exposed area and the amount of rainfall. Table 15.3 presents the characteristics of surface runoff from one of the collieries in NSW, Australia (Sivakumar et al., 1994). As seen in Table 15.3, the suspended solids content in the MIW is considerably high, whereas the dissolved solids are moderate.

Domestic Wastewater

Domestic wastewater in a colliery may generate from *bathhouse* and *kitchen and toilet*. The bathhouse MIW is predominantly contaminated by coal fines from the mine workers and soaps used in their showering. Detergents and disinfectants are also used to clean the bathhouse. Generally, wastewater from kitchen and toilet are piped to septic tanks. The effluents from these tanks are discharged into evaporation ponds. Solids in all the septic tanks are periodically pumped out and transported off-site (Wingrove, 1996). Characteristics of wastewater from the bathhouse are shown in Table 15.4.

TABLE 15.1

Characteristics of Mine Water from a Colliery

Parameter	Unit	Mean/Range
Conductivity, EC[a]	dS/m	0.6–10
Total dissolved solids (TDSs)[a]	mg/L	500–2000
Suspended solids[a]	mg/L	10–100
Hardness (as CaCO3)[a]	mg/L	500–2000
pH[a,b]	–	3.6–9.5
Sulphate[b]	mg/L	60–165
Biochemical oxygen demand (BOD$_5$)[a]	mg/L	5
Chemical oxygen demand (COD)[a]	mg/L	10–100
Trace elements[b]		
Aluminium (Al)	mg/L	0.26–0.49
Boron (B)	mg/L	0.009–0.02
Barium (Ba)	mg/L	0.045–0.052
Beryllium (Be)	mg/L	0.004–0.005
Calcium (Ca)	mg/L	8.59–12.07
Cadmium (Cd)	mg/L	<0.01
Cobalt (Co)	mg/L	0.34–0.46
Chromium (Cr)	mg/L	<0.005
Copper (Cu)	mg/L	0.006–0.007
Iron (Fe)	mg/L	1.22–4.40
Potassium (K)	mg/L	3.18–3.92
Lithium (Li)	mg/L	0.019–0.025
Magnesium (Mg)	mg/L	6.54–8.0
Manganese (Mn)	mg/L	1.92–2.52
Molybdenum (Mo)	mg/L	<0.01
Sodium (Na)	mg/L	2.93–4.02
Nickel (Ni)	mg/L	0.78–1.02
Phosphorus (P)	mg/L	<0.05
Lead (Pb)	mg/L	<0.2
Sulphur (S)	mg/L	26.94–30.96
Silicon (Si)	mg/L	1.87–4.26
Strontium (Sr)	mg/L	0.033–0.040
Titanium (Ti)	mg/L	<0.005
Vanadium (V)	mg/L	<0.005
Zinc (Zn)	mg/L	2.38–3.06

[a] Dharmappa et al. (1995).
[b] Cohen (2002).

Metal MIW

As discussed in section 'Process Water', the quality of effluent from metal mining is diverse in range and depends on the chemical characteristics of the mineral to be extracted. Typical characteristics of effluent from metal mining are shown in Table 15.5, which may be used as an indication of the quality of the generated effluent from different metal mining including iron ore, pyrite, lead, gold and copper. As seen from the table, some of the parameters, such as COD and SO$_4$ were found in considerable amount in discharges from gold mine washery and iron ore (Pyrite). Some total dissolved solid (TDS) and total suspended solid (TSS) values were found high in gold mine washery water, among which the TSS subsided by 99.9% (from 143,000 to 75 mg/L) when the MIW was

TABLE 15.2

Characteristics of Process Water from a Colliery

Parameter	Unit	Range
pH[a,b]	–	8.24–8.87
Electrical conductivity (EC)[a,b]	dS/m	0.51–0.58
Total solids (TS)[a,b]	mg/L	608–1200
Total dissolved solids (TDSs)[a,b]	mg/L	140–801
Settleable solids (SS)[a,b]	mL/L	0.5–1.2
Net filterable residues (NFR)[a,b,c]	mg/L	400–13,650
Oil & gas (O&P)[a,b]	mg/L	301–939
Barium (Ba)[a,b]	mg/L	1.0–1.2

[a] Sivakumar et al. (1994).
[b] Wingrove (1996).
[c] Dharmappa et al. (2000).

TABLE 15.3

Characteristics of Surface Runoff after Storm Events

Parameter	Concentration
Total suspended solids (mg/L)	400–600
Total dissolved solids (mg/L)	200–300
Conductivity (μs/cm)	300–700
Total iron (mg/L)	8–11
Colour	Black

Source: Adapted from Sivakumar, M., S. Morton and R. N. Singh. 1994. Case history analysis of mine water pollution in New South Wales. *Fifth International Mine Water Congress.* Nottingham, 18–23 September, pp. 679–689.

TABLE 15.4

Characteristics of Bathhouse Wastewater from a Colliery

Parameter	Unit	Range
pH [a]	–	8.24–8.77
Electrical conductivity (EC)[a,b,c]	dS/m	0.55–0.58
Total solid (TS)[a,b,c]	mg/L	358–812
Total dissolved solid (TDS)[a,b,c]	mg/L	201–756
Settleable solid (SS)[a,b,c]	mL/L	≤0.1
Net filterable residues (NFR)[a,d]	mg/L	4–157
Oil & gas (O&P)[a,b,c]	mg/L	13–123
Barium (Ba)[a,b,c]	mg/L	0.8–1.0

[a] Wingrove (1996).
[b] Mishra et al. (2008).
[c] Laus et al. (2007).
[d] Dharmappa et al. (2000).

TABLE 15.5
Characteristics of Mine Water from Metal Mining

Parameter	Unit	Mean/Range					
		Discharge from Iron Ore (Fe_2O_3)[a]	Discharge from Iron Ore (FeS_2)[b]	Discharge from Metal Mine (Pb, Cu, Zn)[b]	Discharge from Lead Mine through Soughs[b]	Discharge from Copper Mine[c,d,e,f,g]	Discharge from Gold Mine[h,i]
pH	–	6.4–7.4	2.5	2.8–3.0	7.0–7.2	2.1–6.7	9.0–9.8
EC	dS/m				0.42–0.58	1.62	4.56
TSS	mg/L	5.7–15.2				218	1160–143,000
TDS	mg/L	41–51					2280
Cl⁻		10–18					
Hardness	mg/L	8–12					
SO_4^{2-}	mg/L		5110	441–846	27–57		644
COD	mg/L	8–15					1240
As						0.1	0.03–7.2
Al	mg/L	0.04–0.11	84.21	13.9–20.1			
Ba	mg/L				0.07–0.336		
Ca	mg/L				86–98		
Cd	mg/L				0.0005–0.0016		
Co	mg/L	0.08–0.24					
Cr	mg/L	1.76–5.14					
Cu	mg/L	0.88–2.34	0.16	0.03–0.068	0.0004–0.0009	10.3–128	5.4–7.8
Fe	mg/L	1.04–3.04	1460		0.0005–0.01	0.4–120	3.8
K	mg/L	0.28–0.96					
Mn	mg/L	0.012–0.35	3.05				
Mo	mg/L	0.07–2.8					
Na	mg/L	0.7–1.8					
Ni	mg/L	0.35–0.95					
Pb	mg/L				0.006–0.017		0.2–2.1
Zn	mg/L	0.57–1.4	0.94	38–72	0.007–0.338		0.1

[a] Wash water discharge, Ghose and Sen (2001).
[b] Banks et al. (1997).
[c] Khan et al. (2010).
[d] Mahiroglu et al. (2009).
[e] Rozon-Ramilo et al. (2011).
[f] Korać and Kamberović (2007).
[g] Jordanov et al. (2007).
[h] Wash water discharge, Acheampong et al. (2013).
[i] Benavente et al. (2011).

stored in a dam; however, the TDS remained almost same (Acheampong et al., 2013). Aluminium, ferrous and sulphate was found higher in the discharge from pyrite iron ore.

ENVIRONMENTAL IMPACTS OF MINING ACTIVITIES ON THE WATER RESOURCES

Mining activities may affect water resources including surface as well as groundwater through the generation of non-process water. The non-process water in this chapter refers to the water that is not treated or reused within the mine and reaches to surface water and seeps to groundwater in the presence of rainfall. Seepage and runoff of non-process water may occur from (Street, 1993):

- Underground mine workings
- Abandoned or inactive mine workings
- Waste rock, overburden and tailing piles
- Dump piles and tailings
- Chemical storage areas
- Slag dumps areas
- Contaminated soil areas near smelter operations
- Flue dust pile

As stated in section 'Characteristics of Mine Induced Water', specific types of mining methods are practiced for specific types of mineral ore that affect the quality of non-process water in different ways. In surface mining, composition of non-process water is changed (i.e. polluted) due to the interaction between water and mineral ore. Pollution of water may occur by the mineral ore itself or by excavated waste rock when transported to storage facilities during mining. In the case of underground mining, groundwater flow may be altered during underground excavation. Moreover, both surface and underground mining can expose previously unexposed and unweathered pyritic (iron sulphide, FeS_2) surfaces through blasting, and that exposure can result in acid mine drainage (AMD) containing sulphuric acid. The sulphuric acid and ferric ion may react with minerals they contact and release metals and other constituents to the water body. In addition, leaching locally can introduce oxidising fluids, which may enhance acid generation. When underground and open pit mines that are below the established water table are flooded with water, acid production ceases due to lack of oxygen (McLemore, 2008; Pondja et al., 2014).

ACID MINE DRAINAGE

AMD is a widely occurred environmental problem in mining areas. Generation of AMD depends on the environmental condition and composition of the water, mine wastes, ore and host rocks (McLemore, 2008). Generally, AMD contains a high concentration of iron and sulphate due to the dissolution of iron sulphides and formation of sulphuric acid and iron ions. Minerals containing sulphide in many ore deposits is pyrite (FeS_2), which is exposed to air and water due to mining activities and plays a key role in the generation of AMD. Other sulphide minerals such as pyrrhotite (Fe_xS_x, where x = 0.0–0.2), marcasite (FeS_2), chalcocite (Cu_2S), millerite (NiS) and galena (PbS) may also cause oxidation in water with a low pH and generate AMD (Lottermoser, 2010). According to McLemore (2008), AMD is formed with five elements, namely, air (oxygen), iron (as ferrous ion), sulphur (as sulphide or sulphate ion), water and microorganisms. The process of formation of AMD from pyrite (FeS_2) can be shown by the following oxidation and reduction Equations 15.1 through 15.4 (Pondja et al., 2014):

$$2FeS_2 + 7O_2 + 2H_2O \rightarrow 2FeSO_4 + 2H_2SO_4 \tag{15.2}$$

$$2Fe^{2+} + \frac{1}{2}O_2 + 2H^+ \rightarrow 2Fe^{3+} + H_2O \tag{15.3}$$

$$Fe^{3+} + 3H_2O \rightarrow Fe(OH)_3 + 3H^+ \tag{15.4}$$

$$FeS_2 + \frac{15}{4}O_2 + \frac{7}{2}H_2O \rightarrow 4H^+ + 2SO_4^- + Fe(OH)_3 \tag{15.5}$$

Pyrite reacts directly with oxygen and water in the presence or absence of microorganisms forming an acid solution (Equation 15.2). The ferrous ion in the solution continues to react with oxygen to produce ferric ion in the solution (15.3); the reaction takes place in the pH range of 2.5–3.5. Ferric ion is a powerful oxidising agent for most sulphides, and as long as this ion is present in the solution, pyrite will continue to oxidise. Oxidation and hydrolysis of ferric ion under slightly acidic to alkaline conditions lead to the formation of insoluble ferric hydroxide (Equation 15.4). The produced ferric hydroxide is the reason why AMD is red or orange in colour. It also affects streams by blocking sunlight and covering the stream bed with a thick red blanket. The precipitates can be harmful to aquatic life. As shown in Equation 15.5, by producing hydrogen ions, AMD affects the acidity of a stream, which is commonly measured by pH values. When the produced sulphuric acid lowers the pH of the streams and rivers, it causes the death of water habitats including fish and other aquatic species that cannot live in stronger acidity levels (Pondja et al., 2014).

AMD also affects the natural buffering ability of water in a stream or river. The hydrogen ion in the AMD neutralises the carbonate and bicarbonate ions to form carbonic acid (H_2CO_3), which weakens the natural buffering system of the water (COTF, 2015).

$$H^+ + CO_3^{2-} \rightarrow HCO_3^- \tag{15.6}$$

$$H^+ + HCO_3^- \rightarrow H_2CO_3 \tag{15.7}$$

$$H_2CO_3 \rightarrow H_2O + CO_2 \tag{15.8}$$

The affected carbonic buffering system slowly becomes unable to maintain suitable pH to water habitat and is completely destroyed below a pH of 4.2, where all carbonate and bicarbonate ions are converted to carbonic acid. The carbonic acid readily breaks down into water and carbon dioxide and makes the water acidic.

IMPACT ON GROUNDWATER

Groundwater pollution can occur both directly and indirectly as a result of surface mining (Rauch, 1980). Direct degradation of groundwater occurs when mine drainage from pits and ponds reaches to groundwater situated down gradient from a surface mine. Mine drainage may infiltrate to groundwater due to rainfall and degrade groundwater directly. Indirect degradation of groundwater may occur by blasting, which causes a temporary shaking of the rock and results in new rock fractures near working areas and often up to several hundred feet away from the surface mines. Mine drainage from ponds may reach to underlying aquifers using these fractures and degrade groundwater. Another factor, the distance of a groundwater supply source (i.e. wells) from a mine, is also crucial to determine groundwater contamination due to mining activities. For example, in northern West Virginia, severe contamination of groundwater occurs within about 60 m horizontally of a surface coal mine drainage source (Rauch, 1980). Especially, if the groundwater source is located near a rock fracture zone within the vicinity of a mine, contaminated water from the mine will move toward the groundwater more rapidly.

Chemical properties of groundwater are influenced by geochemical reactions between rock types and water composition. Reactions between groundwater and aquifer minerals have a significant role on groundwater quality (Cederstrom, 1946; Gupta et al., 2008; Subramani et al., 2010). Typically, groundwater contaminated by coal mine drainage has lower pH, hardness and acidity, total iron, manganese, aluminium and suspended solids than untreated surface mine drainage (Rauch, 1980). The reason is that polluted groundwater normally undergoes a higher degree of natural neutralisation compared to polluted surface water, because of its greater contact with carbonate minerals and slower rates of movement. However, the longer water-rock interaction may not be suitable for a karst aquifer system (an aquifer formed by the dissolution of the rock), which may cause an increased amount of TDSs in the groundwater (Qiao et al., 2011). Sulphate in the mine contaminated groundwater is not naturally precipitated and mostly remains in the water. Often, this sulphate is a good tracer or indicator of present and past mine drainage pollution (Rauch, 1980). A rule of thumb is that most wells and springs with more than 100 mg/L of sulphate are probably being contaminated by coal mine drainage sources located within a hundred metres uphill of these water supplies. However, wells and springs with less than 100 mg/L sulphate are either not affected or are not significantly contaminated by mine drainage (Rauch, 1980).

SUSTAINABLE MANAGEMENT OF MIW

Managing MIW from mining activities is not straight forward. Sustainable management of MIW depends on the type of mining, characteristics of generated MIW and economic viability of the adopted management options. There is no one-solution-fits-all approach of managing MIW from mining activities. Management options for MIW arising on-site (and entrapped within on-site) is different from that of MIW that occurs naturally or post closure of mine (i.e. AMD, seepage from tailing dam). Therefore, the management of MIW can be discussed as

- Management of on-site MIW
- Management of off-site MIW

MANAGEMENT OF ON-SITE MIW

In this section, management option of MIW within the mining area (i.e. mine water, process water, stormwater and domestic wastewater) will be discussed. Typically, management of on-site MIW involves five stages (Dharmappa et al., 1995):

- Identification of sources and characterisation of MIW
- Volume and strength reduction using cleaner production methodologies
- Segregation of different qualities of MIWs
- Reuse within the premises
- Final disposal

The decision process involved in the overall management of on-site MIW is shown in Figure 15.6. As shown in the figure after characterisation of the MIW, the possibility to reduce the volume and strength of the MIW is investigated. The goal of volume and strength reduction is to minimise or eliminate the MIW being generated from different mining activities. The steps involved in the volume and strength reduction are: (1) conserving the use of water; (2) changing (mining) production methods to decrease waste; (3) recycle of MIW within the premises (i.e. use of mine water for dust suppression); and (4) elimination of batch or slug discharges of process wastes. In most cases, there will be some MIW generated even after applying volume and strength reduction. This MIW needs to be segregated based on the characteristics of the individual waste streams. As mentioned in section 'Coal Mine Induced Water', MIW generated by a typical underground coal mining activity generates mine water, process water, sewage and surface runoff. Out of these four sources, sewage is organic waste and the rest are essentially inorganic wastes. Moreover, the mine water consists of more dissolved solids and less suspended solids compared to process water and surface runoff. The process water and surface runoff mainly contains suspended solids with relatively low

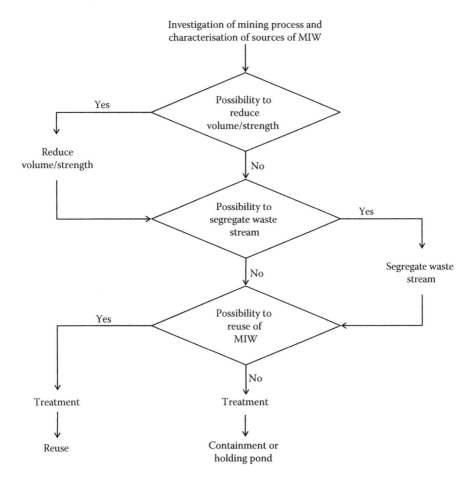

FIGURE 15.6 Strategy of the management of MIW in a mining industry. (After Dharmappa, H. B., M. Sivakumar and R. N. Singh. 1995. Wastewater minimization and reuse in mining industry in Illawarra region. In *Proceedings of Water Resources at Risk, International Mine Water Association*, Denver, Colorado, May 14–18, pp. 11–22.)

to moderate dissolved solids. Therefore, the varied characteristics of different streams necessitate adopting separate treatment systems appropriate to each type of stream before reuse. The next step includes investigation of the reuse options for each of the individual waste streams after a certain level of treatment. Finally, if the reuse option as well as the prevention and control strategies are not feasible for any of the streams, appropriate treatment may be desirable. While volume and characteristics of different on-site mining MIW are discussed in sections 'Influent and Mine Induced Water Associated with Mining Industries', the treatment and management options will be emphasised in this section.

Management and Treatment Alternatives of Coal MIW

Treatment alternatives of coal MIW depend on the types of end use after treatment. If the treated water is intended for use for drinking, the final effluent must meet the water quality standard set by WHO (2011) or by NHMRC–NRMMC (2011). Otherwise, recycling water standard set by NRMMC–EPHC–AMC (2006) may be followed for the re-use of the effluent as process water or for gardening.

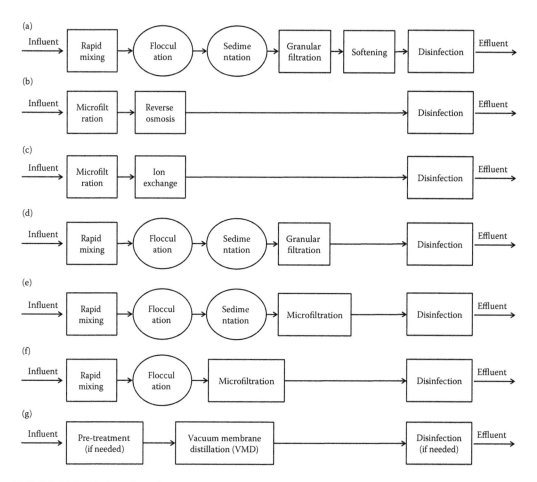

FIGURE 15.7 Typical flow diagram of different treatment processes (a–f) for treating coal mine water for reuse purposes. (After Dharmappa, H. B., M. Sivakumar and R. N. Singh. 1995. Wastewater minimization and reuse in mining industry in Illawarra region. In *Proceedings of Water Resources at Risk, International Mine Water Association*, Denver, Colorado, May 14–18, pp. 11–22.)

For the treatment of mine water for the purpose of reuse, instead of availing one single treatment process, a combination of treatment processes results in better quality effluent (Singh, 1994; Dharmappa et al., 1995). For example, a combination of different conventional treatment processes including coagulation, flocculation and granular filtration can be used for solid–liquid separation, and a softening process can be used to remove hardness from mine water (Figure 15.7a). Recently, use of membrane processes (i.e. microfiltration, ultrafiltration and reverse osmosis (RO)) are finding increasing acceptance in the water and wastewater industries instead of using conventional treatment processes like coagulation, flocculation, sedimentation and ion exchange. Membrane filtration processes are able to produce high quality effluent with better reliability and footprint. However, their economic viability on a pilot scale should be worked out on the case by case basis. Membrane filtration processes can also be used in combination with conventional treatment processes. As shown in Figure 15.7b and c, the microfiltration process can be used to remove suspended contaminant and RO or ion exchange processes can be used to remove dissolved contaminants. Besides treating mine water, these system configurations can also be used for treating process water as well as surface runoff.

For the treatment and reuse of MIW having more of solid particle (suspended solid) and less dissolved impurities (i.e. process water and surface runoff), it is possible to use the treatment combination shown in Figure 15.7d. However, membrane filtration (microfiltration) can be introduced instead of granular filtration to improve the effluent quality (Figure 15.7e). This could be further improved by replacing both sedimentation and granular filtration by microfiltration (Figure 15.7f). Sivakumar et al. (2013) and Ramezanianpour and Sivakumar (2014) have successfully shown that a vacuum membrane distillation technique (Figure 15.7g) using a hydrophobic membrane is able to treat mine water and produce high quality effluent at a laboratory scale for all purpose reuse and remove 99.9% of the dissolved salts.

For the final disposal of effluent (if effluent is not reused), some additional treatment is required to meet the effluent guidelines proposed by NRMMC–EPHC–AMC (2006). For example, process water from the washery contains high suspended solids and a significant amount of oil and some dissolved contaminants. For the disposal of this process water, a coarse solid separator, oil separator (by flotation) and a sorption process (by activated carbon) are required. Another way to remove oil from process water is to first separate the oil by a corrugated plate interceptor, and then the effluent flows to a settling pond, where the oil is further removed by a skimmer. The clarified water from the settling pond is drawn off to a filter lagoon, where the effluent is polished to meet desired effluent disposal standard (Sivakumar et al., 1994).

In practice, management, treatment, reuse and disposal scheme of MIW within a mine follows an integrated approach, which is shown in Figure 15.8. The main components of the MIW management system comprise a tailings dam, a filter dam, an intermediate dam, a settlement dam and the main dam. Wastewater from the surface amenities (i.e. bathhouse) first goes to a stabilisation pond before being discharged into the main dam. A number of sediment traps are built in the wash down bays and the stormwater systems before the water enters the settlement dam. As shown in the figure, effluent is being reused for colliery operations. The effluent is also used for firefighting and stockpile sprays. Other reuse of MIW includes using the aquifer inflow water for extensive revegetation of large areas of land.

Management of storm runoff is an important issue to deal with during the mining operation. Inefficient stormwater management may create two problems (Singh et al., 1998):

- Hydraulic overloading of the process water treatment dams at the time of storm events
- Allowing uncontaminated runoff to become contaminated

The main aim of managing clean water runoff is to ensure it remains uncontaminated. It is also desirable to remove the coal fines load and control the release of the runoff off-site to prevent downstream siltation and flooding. This can be achieved through the use of catch drains, interceptor dykes, berms, open channels and pipelines (Street, 1993). Pollution of runoff can be prevented by

- Increasing the vegetative ground cover to reduce quantity of soil particles to be picked up by runoff
- Installing straw bale barriers and check dams in diversion channels to decrease the channel flow velocity and thereby allow sediments to settle out of the flow
- Preventing drips, overflows, leaks or other material releases from vehicles, workshop areas, the washery and the conveyor belt
- Enclosing material storage areas (i.e. washery, workshop) with curbing barriers to divert runoff around the polluted areas
- Ensuring trucks are well positioned to minimise spillage of materials during loading and unloading operations, and are covered in windy condition when operated within the site.

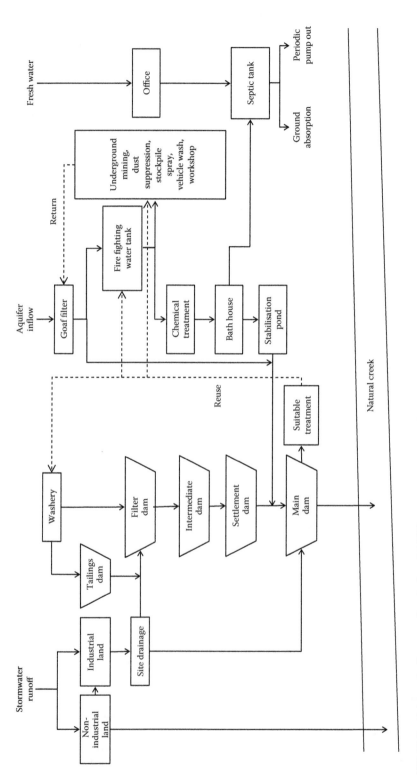

FIGURE 15.8 Schematic diagram of typical coal MIW management system in Australia. (After Dharmappa, H. B. et al., *Journal of Cleaner Production*, 8(1), 23–34, 2000.)

In addition to implementing the above-mentioned pollution prevention strategy, management options should be devised to tackle the hydraulic overloading and contamination of storm runoff. Some management options were proposed by Dharmappa et al. (2000):

Option 1

Diversion channels can be used to collect clean and dirty stormwater runoff and convey it directly to the natural creek system. The clean and dirty water diversion channels may or may not be combined.

Option 2

Diversion channels can be used to collect clean and dirty stormwater runoff and convey it to the existing process water sedimentation dams (e.g. the intermediate, settlement or main dams in Figure 15.8).

Option 3

Diversion channels can be used to collect clean and dirty stormwater runoff and convey it to the process water sedimentation dams, where these dams have been modified to increase their maximum capacity and thus increase their freeboard volume.

Option 4

Separate diversion channels can be used to collect clean and dirty stormwater runoff and convey it to purpose-built clean and dirty stormwater detention basins. To implement this option, suitable land and heavy capital expenditure are required.

Option 5

Separate diversion channels can be used to collect clean and dirty stormwater runoff and convey it to purpose-built detention basins. The stormwater will be slowly released into holding tanks or dams to store the clarified water for future use.

The options for the stormwater management discussed above can be used as a guideline only and should be modified according to the local climate conditions.

Treatment Alternatives of Metal MIW

Selection and design of an appropriate on-site treatment scheme for metal MIW depends on the type of product as well as the chemistry of the produced water. Several treatment technologies are available to treat metal MIW, such as (Gazea et al., 1996; Gusek and Figueroa, 2009):

- Neutralisation – using lime, hydroxides and limestone
- Oxidation – by aeration, chlorination, ozonation and using permanganate and peroxides
- Reduction – using sulphur dioxide and sulphites
- Catalysis – using copper sulphate
- Chemical precipitation – using hydroxides, oxides, phosphates, sulphides and polymers
- Coagulation followed by clarification– using ferric sulphate or chloride, alum and sodium aluminate
- Adsorption – using activated carbon, activated alumina and biosorbent
- Membrane separation – by RO, nanofiltration, ultrafiltration and microfiltration
- Ion exchange – using synthetic resins and zeolite
- Biological process – for example, bioreduction of metals, denitrification and sulphate reduction
- Electrochemical process – for example, electrocoagulation
- Distillation

The main goal of the treatment strategy of metal MIW is to control pH and decrease metal and anion concentration to meet the disposal standards of local regulatory organisations. Other important factors such as sulphate concentration, dissolved metals and oxygen levels should

TABLE 15.6

Treatment Alternatives of Different Metal MIW

Treatment Option	pH	Sulphate (mg/L)	Dissolved Metals	Oxygen Level	Relative Capital Cost	Estimated Operating Cost (US$/m³)
Acid neutralisation	Acidic	>10,000	Low or high	High	High	0.79–1.85
Ion exchange	Acidic to neutral	500–10,000	Low		Moderate	0.79–2.11
Reverse osmosis	Acidic to alkaline	500–10,000	Low		High	2.64–12.20
Biological sulphate reduction	Near neutral	500–10,000	High	Anoxic	Low	Highly variable

Source: Adapted from Gusek, J. J. and L. A. Figueroa. 2009. *Mitigation of Metal Mining Influenced Water.* Littleton, Colorado: Society for Mining, Metallurgy and Exploration.

be considered to determine which treatment system will be the best to treat the metal MIW. Table 15.6 shows some suggested treatment alternatives for different combinations of water chemistry.

As discussed in section 'Management and Treatment Alternatives of Coal Mine Induced Water', in practice only a single treatment process may not be able to treat the MIW to meet the disposal standard, however, a combination of several treatment processes may achieve the goal. If a combination of treatment processes is required, costs associated to the treatment processes should be carefully considered. Operational costs of treatment systems vary depending on the components of the system, which include chemicals and materials required, sludge handling and disposal and transportation costs (Gusek and Figueroa, 2009). Operating costs can be reduced if by-products from the metal MIW treatment process (i.e. elemental sulphur, pyrite or a specific metal) can be recovered from the waste stream and sold. For example, a lone system ion exchange treatment is a moderately expensive system (Table 15.6) because the ion exchange media needs to be regenerated frequently to maintain good effluent quality. However, in certain situations, ion exchange systems may be cost effective if the MIW in question contains high levels of metals that are difficult to remove by neutralisation alone. In this case, an ion exchange process implemented prior to neutralisation can reduce the quantity of neutralisation chemicals. Also, economically valuable metals can be retrieved from the ion exchange process and will make the system economically efficient. RO can be used as a single system to treat metal MIW producing effluent to produce clean water. However, the system incurs high capital cost (Table 15.6), high energy consumption and high operating cost due to the need to dispose of brine wastes. On the other hand, biological treatments of metal MIW are the least expensive (Table 15.6), but the system is very complicated to operate. For true sustainability assessment, full life cycle costing of the treatment processes including cost of ecosystem services as well as cost of reduction of greenhouse gas emission strategies should be considered.

MANAGEMENT OF OFF-SITE MIW

Off-site MIW mainly includes AMD and seepage from tailing dam reaching groundwater. A number of treatment alternatives are available for the treatment of off-site MIW. Similar to on-site MIW, after successful treatment, the effluent is possible to reuse depending on the quality of the effluent. However, most of the time the effluent is discharged to natural streams, which helps to improve the stream health by way of supporting the required environmental flows.

TREATMENT OF AMD

Treatment of AMD may be direct (also called *active* process) or indirect (also called *passive* process). The direct treatment processes include conventional aeration, neutralisation and precipitation processes, some of which is discussed in sections 'Management and Treatment Alternatives of Coal Mine Induced Water' and 'Treatment Alternatives of Metal Mine Induced Water'. In a direct treatment process, AMD is pumped or diverted from nearby polluted lakes or streams to the treatment plant. A direct treatment plant consists of mixing/reactor tanks, collection/disposal of treatment sludge and disposal system of treated water. The main goal of the treatment of AMD is to increase the pH and to remove metals before reaching the AMD to natural streams. Alkaline reagents (bases) are used to increase the pH of AMD, which include (Taylor et al., 2005):

- Hydrated lime [$Ca(OH)_2$]
- Quicklime (CaO)
- Caustic soda (NaOH)
- Ammonia (NH_3)
- Magnesium oxide or hydroxide [MgO or $Mg(OH)_2$]
- Mineral carbonates such as limestone, dolomite, magnesite and witherite

Quicklime or hydrated limes are used to neutralise AMD because of their availability and for being strong bases. During the neutralisation process, metals such as Fe^{2+}, Fe^{3+}, Al, Cu, Zn and Pb are precipitated in the form of metal hydroxides. As shown in Equation 15.9, the sludge resulting from this process contains metal hydroxide and gypsum ($CaSO_4$).

$$Ca(OH)_2 + metal^{2+} \text{ or } metal^{3+} + H_2SO_4 \rightarrow metal(OH)_2 \text{ or } metal(OH)_3 + CaSO_4 + H_2O \qquad (15.9)$$

Management of sludge in a neutralisation treatment process is a concern in terms of sludge handling. The cost and management of sludge handling depends on the mass and volume of sludge produced, sludge density and the chemical composition and stability of precipitates in the sludge. Sludge containing Fe^{3+} is more stable than sludge containing Fe^{2+} (Pondja et al., 2014). Based on the sludge density, AMD neutralisation treatment plants may be of two types:

- Low density sludge (LDS) plants
- High density sludge (HDS) plants

Figure 15.9 shows basic configuration of a LDS plants, which involve mainly three treatment stages (1) reagent mixing and dosing stage, where alkaline reagents are mixed with AMD; (2) reaction stage, where the solution is mixed using a mechanical stirrer and aerated to oxidise iron from Fe^{2+} to Fe^{3+} at a certain pH and (3) flocculation and clarification stage, where polymer is added to enhance flocculation, and the solids are separated from the liquid in a thickener or clarifier. While supernatant from the clarifier is discharged as effluent, sludge from the clarifier base is removed and generally disposed to a drying bed. The main disadvantage of LDS is that produced sludge has less than 5% (by weight) solids, therefore, substantial storage is required. Moreover, the sludge needs dewatering to increase the sludge density. These disadvantages are partially omitted when HDS plants are used. The main difference in HDS and LDS system is that in an HDS plant a proportion of alkaline treatment sludge from the thickener underflow is recycled back through the plant (shown by dotted line in Figure 15.9) to complete the first phase of neutralisation (Taylor et al., 2005).

Indirect or passive treatment processes are based on natural chemical processes to precipitate dissolved metals, sulphate and other chemicals from AMD. Physically and chemically these systems simulate natural systems, however in a confined, monitored and controlled setting. In the indirect

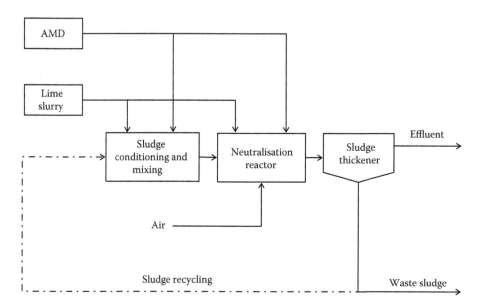

FIGURE 15.9 Flow diagram of a typical LDS or HDS for the direct treatment of AMD. (After Taylor, J., S. Pape and N. Murphy. 2005. A summary of passive and active treatment technologies for acid and metalliferous drainage (AMD). *5th Australian Workshop on Acid Mine Drainage*. Fremantle, Australia.)

treatment system the continuous input of chemicals are not required and geochemical reactions are typically assisted by microbes and plants (Gusek and Figueroa, 2009). However, the downside of this process is that generally it is not used for mine sites which are in operating condition but post closure of the mine. Moreover, not all metals (i.e. manganese) are completely removed from AMD due to maximum pH limitations (Taylor et al., 2005). One of the most commonly used indirect treatment systems is constructed wetlands, which may be aerobic or anaerobic in nature. Aerobic wetlands (Figure 15.10a) are shallow ponds where mine water is aerated while it flows slowly through the vegetation, and the dissolved iron of AMD is oxidised and precipitated within the wetland. Aerobic wetlands must receive alkaline water (often diverted from a pretreatment passive system) and solely provide residence time and aeration to allow certain metals whose solubility is dependent on the redox state of the water (i.e. iron, manganese, chromium and arsenic) to precipitate. For a successful treatment of AMD by this system, the influent AMD should have average acidity <500 mg CaCO$_3$/L and pH >6. Moreover, the typical residence time will be 1–5 days and dissolved oxygen concentrations need to have reached saturation with respect to the atmosphere early within the residence time of the water in the wetland (Taylor et al., 2005).

In the case of anaerobic wetlands (Figure 15.10b), a bed of limestone and a layer of biodegradable organic matter are added where AMD passes through organic rich material that strips oxygen from the water resulting in anaerobic conditions. Alkalinity in the wetlands can be generated by sulphate reducing bacterial activity, which uses the organic matter as a carbon source and sulphate as an electron acceptor for growth. In the bacterial conversion of sulphate to hydrogen sulphide, bicarbonate alkalinity is produced (Equation 15.10).

$$2H_2O + SO_4^{2-} + C_{(organic\ matter)} \rightarrow H_2S + 2HCO_3^- \tag{15.10}$$

For a successful treatment of AMD by an anaerobic system, the influent AMD should have average acidity <500 mg CaCO$_3$/L and pH >2.5. The typical residence time needed is 1–5 days. The ambient oxygen concentration is acceptable near the surface, however, at subsurface the oxygen concentration should be less than 1 mg/L (Taylor et al., 2005).

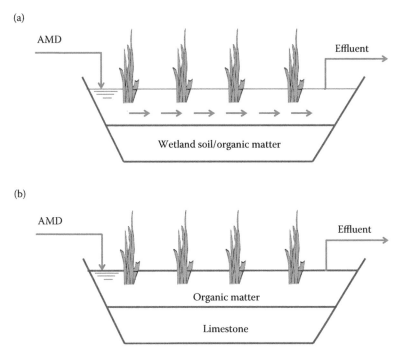

FIGURE 15.10 Cross section of typical (a) aerobic and (b) anaerobic treatment system. (After Pondja E.A. Jr, Persson, K.M. and Matsinhe, N.P., *Journal of Water Resource and Protection*, 6(18), 1646, 2014.)

Other indirect systems of AMD treatment include (Johnson and Hallberg, 2005; Gusek and Figueroa, 2009):

- Anoxic limestone drains – where limestone is used to intercept AMD in anoxic condition and suitable to treat moderately acidic AMD.
- Successive alkalinity producing system (SAPS) and reducing and alkalinity producing system (RAPS) – a vertical flow system, which remove dissolved oxygen and reduce Fe^{3+} to Fe^{2+} to enable increase in alkalinity to be achieved. Usually, water draining a RAPS flows into a sedimentation pond, and/or an aerobic wetland, to precipitate and retain iron hydroxides.

Groundwater Remediation

Contamination of groundwater is possible to avoid using containment such as vertical barriers (Mulligan et al., 2001), which may be suitable to avoid contamination of groundwater by MIW. Vertical barriers reduce the movement of contaminated groundwater. The barrier may of slurry walls, grout or geomembrane curtains and sheet pile. To prevent the transport of contaminants past the barrier, the barrier should be extended to a clay or bedrock layer of low permeability. However, a treatment option is also available, which is an indirect biological system, namely, a permeable reactive barrier (PRB). PRBs are essentially buried layers of reactive materials to intercept groundwater plumes of AMD (Figure 15.11). PRBs are constructed by digging of a trench or pit in the flow path of contaminated groundwater, and by filling the void with reactive materials that are sufficiently permeable to allow uninterrupted flow of the groundwater.

The reactive materials contain organic matter and limestone (Johnson and Hallberg, 2005; Taylor et al., 2005). Reductive microbiological processes within the PRB generate alkalinity (which is further enhanced by the dissolution of limestone and/or other basic minerals) and remove metals (such as As, Cd, Cu, Fe, Ni, Pb and Zn) as sulphides, hydroxides and carbonates (similar to

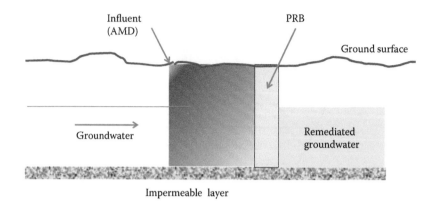

FIGURE 15.11 Cross-section of typical PRB to remediate groundwater contaminated with MIW. (After Mulligan, C., Yong, R. and Gibbs, B., *Engineering Geology*, 60(1), 193–207, 2001.)

Equation 15.10). For the effective treatment, the influent AMD should have average acidity <500 mg $CaCO_3/L$, pH >3, and average flowrate less than 1 L/s. Key issues which may affect the performance of PRB are the mass of available reactive material and the available volume of pore spaces (and permeability) of the barrier. Metal precipitation and substrate compaction can result in a decrease in porosity and permeability of the barrier. Moreover, to keep increased bacterial activity, dissolved oxygen should be low and temperature less than 5°C.

CONCLUSION

MIW irrespective of whether it is generated from coal or metal ore mines can be sustainably managed. As described in this chapter, there is ample scope for the sustainable management of on- and off-site MIW, including reuse within the mining industries. A number of treatment options are available, however, the choice of which option to use to remediate MIW should be considered based on economical and environmental factors. Legislation is a key factor in determining the remedial options, because the effluent needs to meet the standard set by regulatory agencies. It is important that the legislation should be up-to-date as treatment technologies are continuously upgraded. For example, strict regulation may ascertain the recovery and recycling of metals instead of disposing it as sludge (which results from treatment processes). Intentions of mining companies to increase reuse and recycling of MIW are also important. In advanced countries like Australia, mining companies use company policies as a framework to develop corporate pollution minimisation strategies (Driussi and Jansz, 2006). Most of the mining companies in Australia are committed to improving the sustainability of their operations and minimising their environmental impacts. Increasing the synergies in mining and mineral processing companies with local governments, community support groups, farmers and environmental researchers may help implementing suitable treatment options and that will lead to various recycling options for MIW. There is a need for considerable research in the development of such symbiosis.

REFERENCES

ABS. 1996. Australian Bureau of Statistics, Mineral Account, Canberra, Australia, http://www.abs.gov.au/ausstats/abs@.nsf/mf/4608.0#.
ABS. 2004. *Water Account, Australia 2000–2001*. Canberra, Australia: Australian Bureau of Statistics.
ABS. 2006. *Water Account, Australia 2004–2005*. Canberra, Australia: Australian Bureau of Statistics.
ABS. 2014. *Water Account, Australia 2012–2013*. Canberra, Australia: Australian Bureau of Statistics.
Acheampong, M. A., K. Paksirajan and P. N. Lens. 2013. Assessment of the effluent quality from a gold mining industry in Ghana. *Environmental Science and Pollution Research*, 20(6): 3799–3811.

Banks, D., P. L. Younger, R.-T. Arnesen, E. R. Iversen and S. B. Banks. 1997. Mine-water chemistry: The good, the bad and the ugly. *Environmental Geology*, 32(3): 157–174.

Benavente, M., L. Moreno and J. Martinez. 2011. Sorption of heavy metals from gold mining wastewater using chitosan. *Journal of the Taiwan Institute of Chemical Engineers*, 42(6): 976–988.

Britt, A. F., A. Whitaker, S. Cadman, D. Summerfield, P. Kay, D. C. Champion, A. McKay et al. 2014. *Australia's Identified Mineral Resources 2014*. Canberra, Australia: Geosciences Australia.

Cederstrom, D. J. 1946. Genesis of ground waters in the Coastal Plain of Virginia. *Economic Geology*, 41(3): 218–245.

Cohen, D. 2002. Best practice mine water management at a coal mining operation in the Blue Mountains. MS thesis, University of Western Sydney, Sydney, Australia.

COTF. 2015. Acid Mine Drainage: Chemistry [Online], http://www.cotf.edu/ete/modules/waterq/wqchemistry.html (accessed September 11, 2015).

Dharmappa, H. B., M. Sivakumar and R. N. Singh. 1995. Wastewater minimization and reuse in mining industry in lllawarra region. In *Proceedings of Water Resources at Risk, International Mine Water Association*, Denver, Colorado, May 14–18, pp. 11–22.

Dharmappa, H. B., K. Wingrove, M. Sivakumar and R. N. Singh. 2000. Wastewater and stormwater minimisation in a coal mine. *Journal of Cleaner Production*, 8(1): 23–34.

Driussi, C. and J. Jansz. 2006. Pollution minimisation practices in the Australian mining and mineral processing industries. *Journal of Cleaner Production*, 14(8): 673–681.

Gazea, B., K. Adam and A. Kontopoulos. 1996. A review of passive systems for the treatment of acid mine drainage. *Minerals Engineering*, 9(1): 23–42.

Ghose, M. and P. Sen. 2001. Characteristics of iron ore tailing slime in India and its test for required pond size. *Environmental Monitoring and Assessment*, 68(1): 51–61.

Gupta, S., A. Mahato, P. Roy, J. Datta and R. Saha. 2008. Geochemistry of groundwater, Burdwan District, West Bengal, India. *Environmental Geology*, 53(6): 1271–1282.

Gusek, J. J. and L. A. Figueroa. 2009. *Mitigation of Metal Mining Influenced Water*. Littleton, Colorado: Society for Mining, Metallurgy and Exploration.

Johnson, D. B. and K. B. Hallberg. 2005. Acid mine drainage remediation options: A review. *Science of the Total Environment*, 338(1): 3–14.

Jordanov, S. H., M. Maletić, A. Dimitrov, D. Slavkov and P. Paunović. 2007. Waste waters from copper ores mining/flotation in 'Bučbim' mine: Characterization and remediation. *Desalination*, 213(1): 65–71.

Khan, S., A. E.-L. Hesham, A. Ahmad, S. Houbo and C. Daqiang. 2010. Physio-chemical characteristics and bacterial diversity in copper mining wastewater based on 16S rRNA gene analysis. *African Journal of Biotechnology*, 9(46): 7891.

Kihlstedt, P. 1972. Waste water in metal mining industries. *Pure and Applied Chemistry*, 29(1–3): 323–332.

Korać, M. and Kamberović, Z. 2007. Characterization of wastewater streams from Bor Site. *Metalurgija*, 13(1): 41–51.

Laus, R., R. Geremias, H. L. Vasconcelos, M. C. Laranjeira and V. T. Fávere. 2007. Reduction of acidity and removal of metal ions from coal mining effluents using chitosan microspheres. *Journal of Hazardous Materials*, 149(2): 471–474.

Lottermoser, B. G. 2010. *Mine Wates: Characterization, Treatment and Environmental Impacts*. 3rd edn, Berlin Heidelberg: Springer.

Mahiroglu, A., E. Tarlan-Yel and M. F. Sevimli. 2009. Treatment of combined acid mine drainage (AMD) – Flotation circuit effluents from copper mine via Fenton's process. *Journal of Hazardous Materials*, 166(2): 782–787.

McLemore, V. T. 2008. *Basics of Metal Mining Influenced Water*. Littleton, Colorado: Society for Mining Metallurgy and Exploration (SME).

Mishra, V. K., A. R. Upadhyaya, S. K. Pandey and B. Tripathi. 2008. Heavy metal pollution induced due to coal mining effluent on surrounding aquatic ecosystem and its management through naturally occurring aquatic macrophytes. *Bioresource Technology*, 99(5): 930–936.

Mulligan, C., R. Yong and B. Gibbs. 2001. Remediation technologies for metal-contaminated soils and groundwater: An evaluation. *Engineering Geology*, 60(1): 193–207.

NHMRC–NRMMC. 2011. *Australian Drinking Water Guidelines 6*, Canberra, Australia. https://www.nhmrc. gov.au/guidelines-publications/eh52.

NRMMC–EPHC–AMC. 2006. *Australian Guidelines for Water Recycling: Managing Health and Environmental Risks*. Canberra, Australia: NRMMC–EPHC–AMC.

NSW-MC. 2015. Mining methods – NSW mineral council [Online]. Available: http://www.nswmining.com. au/industry/mining-methods Accessed September 11, 2015.

Pondja, E. A., Jr, K. M. Persson and N. P. Matsinhe. 2014. A survey of experience gained from the treatment of coal mine wastewater. *Journal of Water Resource and Protection*, 6(18): 1646.

Prosser, I. P. 2011. *Water: Science and Solutions for Australia*, Melbourne, Australia: CSIRO.

Qiao, X., G. Li, M. Li, J. Zhou, J. Du C. Du and Z. Sun. 2011. Influence of coal mining on regional karst groundwater system: A case study in West Mountain area of Taiyuan City, northern China. *Environmental Earth Sciences*, 64(6): 1525–1535.

Ramezanianpour, M. and M. Sivakumar. 2014. Mine drainage treatment for reuse. In *Responsible Mining: Sustainable Practices in the Mining Industry*, edited by D. Serafin. Chapter 16, Colorado, USA: Society for Mining, Metallurgy and Exploration, pp. 461–488.

Rauch, H. 1980. Effects of coal mining on ground-water quality in West Virginia. In *Proceedings of the Symposium on Surface Mining for Water Quality*. Morgantown, West Virginia, March 14, http://wvmd-taskforce.com/proceedings/80/80rau/80rau.htm

Rozon-Ramilo, L. D., M. G. Dubé, C. J. Rickwood and S. Niyogi. 2011. Examining the effects of metal mining mixtures on fathead minnow (*Pimephales promelas*) using field-based multi-trophic artificial streams. *Ecotoxicology and Environmental Safety*, 74(6): 1536–1547.

Singh, G. 1994. Augmentation of underground pumped out water for potable purpose from coal mines of Jharia coalfield. In *Proceedings of 5th International Mine Water Congress*, Nottingham, pp. 679–689.

Singh, R. N., H. Dharmappa and M. Sivakumar. 1998. Study of waste water quality, management in Illawarra coal mines. In Aziz, N. (ed.), *Coal 1998: Coal Operators' Conference*. University of Wollongong & the Australasian Institute of Mining and Metallurgy, Wollongong, Australia, pp. 456–473.

Sivakumar, M., S. Morton and R. N. Singh. 1994. Case history analysis of mine water pollution in New South Wales. *Fifth International Mine Water Congress*. Nottingham, 18–23 September, pp. 679–689.

Sivakumar, M., M. Ramezanianpour and G. O'Halloran. 2013. Mine water treatment using a vacuum membrane distillation system. *APCBEE Procedia, 4th International Conference on Environmental Science and Development (ICESD)*, Dubai, UAE, Vol. 5, 157–162.

Street, M. 1993. Water runon/runoff management. Report. US Environmental Protection Agency, Office of Solid Waste, Washington, D.C.

Subramani, T., N. Rajmohan and L. Elango. 2010. Groundwater geochemistry and identification of hydrogeochemical processes in a hard rock region, Southern India. *Environmental Monitoring and Assessment*, 162(1–4): 123–137.

Taylor, J., S. Pape and N. Murphy. 2005. A summary of passive and active treatment technologies for acid and metalliferous drainage (AMD). *5th Australian Workshop on Acid Mine Drainage*. Fremantle, Australia.

WHO. 2011. *Guidelines for Drinking Water Quality*. 4th edn, WHO, Geneva. http://www.who.int/water_sanitation_health/publications/2011/dwq_guidelines/en/.

Wingrove, K. 1996. Wastewater management in Illawarra coal mines. BE thesis, University of Wollongong, Wollongong, Australia.

16 Bioflocculants Relevant for Water Treatment and Remediation

K. A. Natarajan and K. Karthiga Devi

CONTENTS

ABSTRACT

Flocculation is extensively employed for clarification through aggregation of colloids and other suspended particles in a suspension to form flocs. Flocculation finds usefulness in many fields such as wastewater treatment; downstream processing, food and fermentation processes. Flocculating agents could be inorganic, synthetic, or natural. Typical examples of natural flocculating agents are bioflocculants produced by microorganisms. Bioflocculants are the metabolic products of microorganisms produced during their growth. They are usually high molecular weight biopolymers, and are synthesized and released outside the cell. Application of eco-friendly bioflocculants in wastewater treatment has attracted significant attention lately with high removal capability in terms of suspended solids, turbidity, color, dyes, and heavy metals. Natural polysaccharides derived from microorganisms are chemically modified by inclusion of synthetic, nonbiodegradable monomers onto their backbone to produce bioflocculants. This chapter is aimed to provide an overview of the development and flocculating efficiencies of microbes-based bioflocculants. Furthermore, the processing methods, characterization strategies, flocculation mechanism, and the current challenges are discussed. Commercial applicability of bioflocculants in the treatment of waste water for removal of heavy metals, suspended solids, color, and turbidity is illustrated with examples. Hence, the possibility to apply natural bioflocculants in food and beverage, mineral industries is discussed and evaluated.

INTRODUCTION

Rapid industrialization in developing countries has been accepted as an inevitable choice for faster economic growth. However, it has significantly raised the rate of water pollution including heavy metal discharge especially from industrial sources, and this has become a major environmental concern (Yim et al., 2007). The disposal of effluents and even domestic sewage without appropriate treatment could result in long-term undesirable negative impacts, especially on the environment and human health, thus necessitating the need for adequate treatment of wastewater before discharge into the environment. Earlier studies have shown that chemical coagulation is a viable process for the treatment of wastewater. Nowadays, clarification (coagulation–flocculation and sedimentation) is employed worldwide in the water/wastewater treatment industry since it is a convenient and efficient technique for removing suspended solids, colloids, and cell debris. This process involves the use of chemical coagulants such as ferric salts, aluminum sulfate, aluminum chloride, and polyacrylamide. However, many coagulants and flocculants have been implicated in undesirable health conditions (Deng et al., 2003). For example, polyacrylamide contains acrylamide monomers known to be both neurotoxic and carcinogenic to humans. On the other hand, a link between aluminum in drinking water (from alum) and human neurological disorders, such as dialysis-associated encephalopathy, was established with excess aluminum in the dialysate fluid shown to be harmful to dialysis patients (Master et al., 1985; Kowall et al., 1989). This health implication of chemical coagulant/flocculant has necessitated the need for safer alternatives.

Removal of heavy metals from wastewater and industrial effluents is usually achieved by physicochemical processes, including chemical precipitation, ion exchange, electrochemical treatment, membrane technologies, and adsorption (Eccles, 1999). Each of these methods has its advantages and disadvantages. Chemical precipitation and electrochemical treatments are not always effective, especially when metal ion concentration in an aqueous solution is lower than 50 mg/L. Moreover, such treatments produce large amounts of sludge which need to be safely disposed. Ion exchange and membrane technologies are complex and expensive. Adsorption, on the other hand, is generally recognized as an effective method for the removal of heavy metals from wastewater. Nevertheless, the high cost of activated carbon, which is usually the adsorbent of choice, limits its use in adsorption. A search for a cost-effective and easily available adsorbent has therefore led to the use of the technique of biosorption, where biosorbents are preferred (Aksu, 2005).

Biosorption is a process that utilizes inexpensive dead and waste biomass to adsorb heavy metals. It is a process of rapid and reversible binding of ions from aqueous solutions onto functional groups that are present on the surface of biomass. Biosorption offers several advantages over conventional treatment methods, such as cost-effectiveness, efficiency, minimization of chemical/biological sludge, limited requirement of additional nutrients, and regeneration of biosorbent with the possibility of metal recovery.

Bioflocculants are biodegradable polymers, which are harmless to the ecosystem. Bioflocculation is a dynamic process involving the synthesis of extracellular polymers by living cells, such as microorganisms, during their growth. Bioflocculants can be produced economically on a large scale and can easily be recovered from fermentation broth. Besides the biosorbents, several types of extracellular bacterial products containing proteins and polysaccharides also bring about metal removal from solutions. Biosorbents and bioflocculants derived from microorganisms have recently been identified as potential agents for heavy metal and turbidity removal from water.

This status review focuses on the published research results on the use of bioflocculants produced by microorganisms for water treatment and remediation. The coverage includes methods of production of bioflocculants, their characterization and applications in metals, and turbidity removal from waste waters.

BIOFLOCCULANTS

Bioflocculants are defined as extracellular/biodegradable polymers synthesized by living cells/ microorganisms during their growth (Auhim and Odaa, 2013). Bioflocculants promote flocculation by the formation of bridges between them and other particles resulting in the aggregation and precipitation of suspended particles, colloids, and debris by living cells. Generally, when the suspended particles are flocculated into large flocs, they settle down, resulting in a clarified solution that can be easily removed. They are also capable of adsorption and entrapment of metal ions. Bioflocculants are mostly composed of proteins, polysaccharides, glycoproteins, nucleic acids, and some other macromolecular compounds. The basic carbohydrate structure of most exopolysaccharides does not change with the bacterial growth conditions, but the functional groups attached to this basic carbohydrate structure can vary widely and may have a dramatic effect on the properties of the polymer and hence their effectiveness in various applications (Lopez et al., 2003).

Polysaccharides that constitute the outermost layer of bacterial cells are thought to mediate interaction between cells and their adhesion to surfaces. Some polysaccharides are liberated outside the cell and some are bound to the cell envelope, while under certain physiological conditions, extracellular polysaccharide molecules may remain bound and associated with the cell surface without detectable anchoring (Prasertsan et al., 2006; Fang et al., 2010).

Microbial extracellular polysaccharides present in the bioflocculants are either associated with and often covalently bound to the cell surface in the form of capsules or secreted into the environment in the form of slime. They are referred to as capsular or slime exopolysaccharides, respectively. L-glutamate, sucrose, fructose, galactose, as well as glucose are usually required as the main substrates for the production of biopolymer flocculants (Figure 16.1). Bioflocculants extracted from different microorganisms are listed in Table 16.1 along with their compositions and possible uses.

MECHANISMS OF BIOFLOCCULATION

Extensive research has been conducted to understand the mechanisms of bioflocculation and solid/ liquid separation processes. According to Salehizadeh and Shojaosadati (2001), bioflocculants cause aggregation of particles and cells by bridging and charge neutralization.

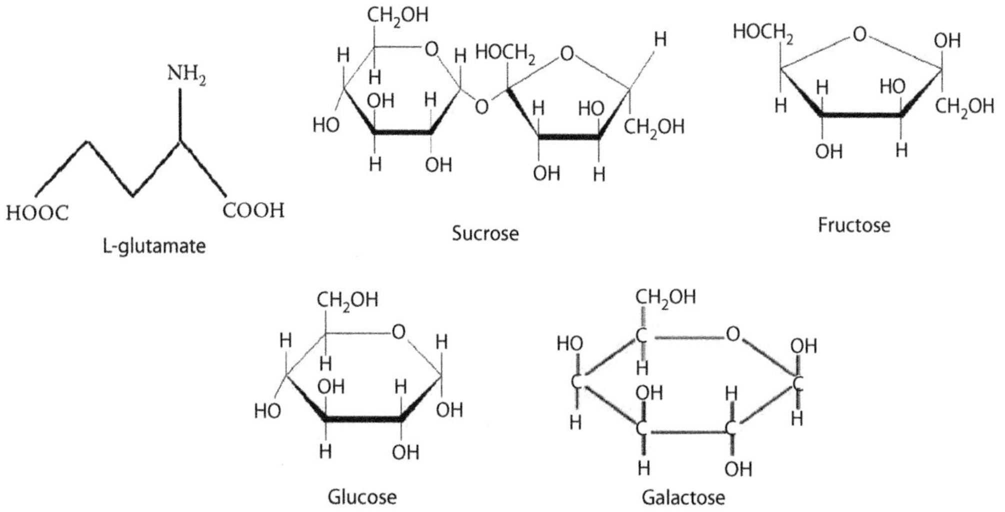

FIGURE 16.1 A pictorial representation of L-glutamate, sucrose, fructose, glucose, and galactose.

TABLE 16.1

Extracted Bioflocculants from Different Organisms

Strains	Extracted Bioflocculant Composition	Uses	References
Bacillus firmus	Glucose, fructose, mannose, and galactose	Flocculation of kaolin	Salehizadeh and Shojaosadati (2002)
Sorangium cellulosum	Glucose, mannose, guluronic acid, and proteins	Flocculation of kaolin	Zhang et al. (2002)
Bacillus mucilaginosus	Neutral sugar, uronic acid, and amino sugar	Starch waste water treatment	Deng et al. (2003)
Enterobacter cloacae WD7	Neutral sugar, amino sugar, and uronic acid	Flocculation of kaolin	Prasertsan et al. (2006)
Klebsiella mobilis	Neutral sugar	Efficient decolorization agent for disperse dyes	Wang et al. (2007)
S. ficaria	Neutral sugar	Chemical oxygen demand (COD) and turbidity removal in waste water treatment	Gong et al. (2008)
Multiple-microorganism consortia	Protein and polysaccharides	Treatment of indigotin printing and dyeing in waste water treatment	Zhang et al. (2007)
Bacillus mucilaginosus	Rhamnose, xylose, sorbose, glucose, and galactose	Removal of COD, biological oxygen demand (BOD), and suspended solids (SS) in waste water	Lian et al. (2008)
Bacillus sp. strain F19	Neutral sugar, uronic acid, amino sugars, and protein	Effective flocculation of kaolin, activated carbon, and fly coal suspension	Zheng et al. (2008)
Proteus mirabilis TJ-1	Neutral sugar, glucuronic acid, and amino sugar	Environmental bioremediation and other biotechnological processes	Xia et al. (2008)
S. aureus (A22), *P. plecoglossicida* (A14), *P. pseudoalcaligenes* (A17), *Exiguobacterium acetylicum* (D1), *Bacillus subtilis* (E1), and *Klebsiella terrigena* (R2)	Protein with a high concentration of uronic acid	Waste water treatment	Buthelezi et al. (2010)

(Continued)

TABLE 16.1 (Continued)
Extracted Bioflocculants from Different Organisms

Strains	Extracted Bioflocculant Composition	Uses	References
B. licheniformis	Neutral sugar, amino sugar, and uronic acid	Suspended solids flocculation in sugar refinery process	Xiong et al. (2010)
C. daeguense W6	Protein and polysaccharides	Mineral flocculation	Liu et al. (2010)
Bacillus subtilis and Pseudomonas spp	Glyco proteins	Flocculation of suspended solids	Zaki et al. (2011)
Herbaspirillum sp. CH7, Paenibacillus sp. CH11, Bacillus sp. CH15, and Halomonas sp.	-	Heavy metal removal in industrial effluents	Lin and Harichund (2011)
Klebsiella pneumoniae strain NY1	Quinovase-containing protein polysaccharide	Precipitation of cyanobacteria and municipal waste water treatment	Nie et al. (2011)
Indigenous Bacterial Isolates Bacillus subtilis (E1), Exiguobacterium acetylicum (D1), Klebsiella terrigena (R2), S. aureus (A22), and P. pseudoalcaligenes (A17)	Protein with high concentration of uronic acid	Dye removal in industrial waste water effluents	Buthelezi et al. (2012)
Azotobacter chrococcum	Alginate polysaccharide	Turbidity removal, river waste water treatment	Auhim and Odaa (2013)
Bacillus and Streptomyces	Non-ketose polysaccharides	Swine waste water treatment	Zhang et al. (2013)
Enterobacter sp. ETH-2	Mannose, glucose, and galactose	Flocculation of kaolin	Tang et al. (2014)
Ocenobacillus sp. and Halobacillus sp.	Uronic acid, neutral sugar, and protein	Industrial wastewater and river water treatment	Cosa and Anthony (2014)
B. licheniformis (NCIM2472) and B. firmus (NCIM2264)	Polysaccharide-conjugated proteins	Mineral flocculation, dye decolorization, heavy metal removal	Karthiga Devi and Natarajan (2015b).
Bacillus amyloliquefaciens ABL19	Polysaccharide and protein	Mineral flocculation	Ogunsade et al. (2015)

BRIDGING

Bridging occurs when a flocculant forms threads or fibers that adsorb onto more than one particle, capturing and binding them together. The polymer molecule comes into contact with the colloidal particles, the reactive groups of the polymer attach to the surface of the particle thus leaving other portions of the molecule extending into the solution. The polymer/flocculant adsorbs onto the surface in loops (i.e., the segments extending into the solution) and tails (i.e., segments adsorbed on the surface). Thus, when another particle (with free adsorption site) comes into contact with these extended loops and tails, it attaches forming a particle–polymer–particle aggregate, thereby acting as a bridge. This attachment of segments leading to the formation of strong bonds is made more effective due to the short distances involved, which can overcome electrostatic repulsions between particles having the same net charge (Ganczarczyk, 1983). This mechanism is dependent on the molecular weight of the bioflocculant, the charge on the polymer and particle, ionic strength of the suspension and the nature of mixing (Salehizadeh and Shojaosadati, 2001; Ebeling et al., 2005). This mechanism is much prevalent in high molecular weight polyelectrolytes of large low density (Sharma et al., 2006). Hence, high molecular weight flocculants produce larger, loosely packed flocs and fragile flocs (Ebeling et al., 2005). For effective interparticle bridging, this mechanism functioning simultaneously with charge neutralization brings about efficient settling resulting in shear resistant flocs. A simplistic model illustrating the bioflocculation mechanism is given in Figure 16.2.

CHARGE NEUTRALIZATION

Charge neutralization occurs when the flocculant is oppositely charged to the particles. Since the particle surfaces are adsorbed toward the bioflocculant, they can effectively approach close to each other and attractive forces become effective (Salehizadeh and Shojaosadati, 2001). In other words, a highly cationic polymer/flocculant may be adsorbed onto the surface of a negatively charged particle in a flat conformation and flocculation is promoted through reduction of the overall negative charge on the particle and through interparticle repulsion. This kind of mechanism is linked with reduced electrophoretic mobility. Moreover, the polymer/flocculant adsorption may have a net positive charge due to the polymer having a higher charge density. In such a case, the positive regions of the bioflocculant will be adsorbed onto the negative regions on the particle and this is referred to as heterocoagulation (Sharma et al., 2006).

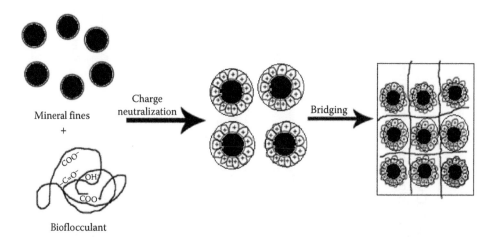

FIGURE 16.2 Simplistic model for bioflocculation. (Adapted from Karthiga Devi, K. and K. A. Natarajan. 2015b. *Minerals & Metallurgical Processing Journal*, 32: 222–229.)

METHODS OF PRODUCTION AND PURIFICATION OF BIOFLOCCULANTS

The aim of any extraction method is to generate as much of the bioflocculant as possible without damaging the integrity of the bacterial cell wall. There are several physical and chemical methods, such as high-speed centrifugation, boiling in acid or alkali, and utilization of cation exchange resins for the extraction of bioflocculants from different bacteria. As yet, no single method has been adopted as a standard procedure.

A different method for bioflocculant extraction utilizing organic solvents has been reported to give favorable results (Kurane et al., 1994; Buthelezi et al., 2012). Various methods of production and purification of bioflocculants have been well described (Yokoi et al., 1995; Gong et al., 2008; Lian et al., 2008; Xia et al., 2008; Gao et al., 2009).

A typical case study is given below. A viscous culture broth of *Bacillus mucilaginosus* and *multiple-microorganism consortia* was diluted with nine volumes of distilled water and then centrifuged to remove cells at 20,000 g for 10 min to concentrate the cultures. Two volumes of cold ethanol were added to the cold culture broth. The precipitate obtained was redissolved in deionized water and 2% cetylpyridinium chloride solution was added with stirring. After 12–18 h, the precipitate was collected and dissolved in a 0.5 N NaCl solution. Two volumes of cold ethanol were then added to obtain the precipitate, which was then washed with ethanol. The crude bioflocculant was dialyzed at 4°C overnight against deionized water and the precipitate was vacuum dried. About 700 mg pure bioflocculant was obtained from 1 L culture broth (Deng et al., 2003; Zhang et al., 2007).

Capsular extracellular polymeric substances (EPS) could be solubilized by adding 2 mL of 5.0 M NaCl and 4 mL of 0.05 M disodium ethylene di-amine tetra acetic acid (EDTA) to 100 mL of the culture. For this purpose, after adjusting to neutral pH, in an orbital shaker, the contents were centrifuged at 18,000 rpm at 15°C for 30 min to precipitate the cells. EPS in the culture supernatant fluid was precipitated by addition of 3-volumes of ice cold isopropanol and the precipitated EPS was collected on a Whatman filter paper No. 1 and dissolved in water and precipitated again by addition of 3-volumes of ice cold isopropanol, collected and dissolved in water at room temperature (Auhim and Odaa 2013; Guo et al., 2013).

CHARACTERIZATION METHODS

Bioflocculants can be characterized using Fourier transform infrared (FTIR) spectroscopy, differential scanning calorimetry (DSC), and matrix-assisted laser desorption/ionization (MALDI)-time of flight (TOF) (mass spectrum analysis). The functional groups of the bioflocculant can be determined using FTIR spectroscopy. The molecular weight distributions of the constituents of the bioflocculants can be determined using MALDI-TOF. Thermal stability of the purified bioflocculants as determined by DSC can be illustrated as heat flow–temperature curves.

FTIR Spectroscopy

The composition of the bioflocculants produced by different microorganisms differs as illustrated in Table 16.2. The flocculating activity of the purified bioflocculant solely depends on the chemical structure which is related to the functional groups in the molecule.

Thermogravimetric analysis (TGA) is a method of thermal analysis in which changes in physical and chemical properties of materials are measured as a function of increasing temperature (with constant heating rate) or as a function of time (with constant temperature and/or constant mass loss). TGA can provide information about physical phenomena, such as second-order phase transitions, including vaporization, sublimation, absorption, adsorption, and desorption. Likewise, TGA can provide information about chemical phenomena including chemisorptions, desolvation (especially dehydration), decomposition, and solid–gas reactions (e.g., oxidation or reduction).

TABLE 16.2

FTIR Characterization of the Extracted Bioflocculants from Various Bacterial Sources

Organism	Identified Functional Groups in the Extracted Bioflocculant	References
Bacillus mucilaginosus	C = O stretching vibration in COOH, C = O in the carboxylate, respectively, indicating the presence of carboxyl group in bioflocculant. C–O stretching in ether or alcohol	Deng et al. (2003)
Vagococcus sp. W31	C–H stretching band caused the absorption peak at 2980 cm⁻¹. The peak at 1115 cm⁻¹ was caused by C–O stretching, thus indicating the presence of a methoxyl group. The absorption peak of CH (840 cm⁻¹) due to sugar derivatives	Gao et al. (2006)
S. ficaria	The absorption peak at 3425 cm⁻¹ was characteristic of OH stretching from the hydroxyl group. The peak from 2926 cm⁻¹ due to aliphatic C–H group, 1600 and 1075 cm⁻¹ were characteristic of C = O and C–O groups. 1000–1200 cm⁻¹ are characteristic of sugar derivatives	Gong et al. (2008)
Proteus mirabilis TJ-1	Broad stretching intense peak at around 3400 cm⁻¹ is characteristic of hydroxyl and amino groups. Asymmetrical stretching peak suggests carboxyl groups	Xia et al. (2008)
Ruditapes philippinarum	The absorption peak at 3415 cm⁻¹ is characteristic of –OH stretching vibration. The single peak at 1660 cm⁻¹ is the absorptive band of –OH deformation vibration or C = O stretching vibration from conjugate aldehydes. The absorptive peaks at 1401 cm⁻¹ are probably caused by tertiary alcohols and phenol	Gao et al. (2009)
B. licheniformis	Hydroxyl group, caused by the vibration of OH or NH in the sugar ring. Carbohydrates indicated COH asymmetrical stretching vibration. C = O symmetrical and asymmetrical stretching of a carboxylate group. The strong absorption band at 1063 cm⁻¹ indicated asymmetrical stretching vibration of ester linkage. The bioflocculant is a type of heteropolysaccharide containing some protein	Xiong et al. (2010)
Penicillium sp.	The peaks at 1299.10 and 1112.86 cm⁻¹ due to P = O and C–O bands. Strong absorption peak observed at 535.29 cm⁻¹ is characteristic of sugar derivatives	Liu and Cheng (2010)
Bacillus subtilis and Pseudomonas sp.	The weak stretching band known to be typical of carbohydrates, indicated COH asymmetrical stretching vibration. The bending vibration of CH₃ and the scissor vibration of CH₂. CAO symmetrical and asymmetrical stretching of a carboxylate group in the bioflocculant. Asymmetrical stretching vibration of a C–O–C ester linkage. The bioflocculant contains glycoproteins	Zaki et al. (2011)
Pseudomonas aeruginosa	Broad stretching characteristic of hydroxyl and amine groups. Weak C–H stretching vibration band indicating the presence of carboxyl groups. Absorption peak at S = O stretching indicated the presence of sulfate	Gomaa (2012)
Azotobacter chrococcum	Peaks attributed to the –CH3 groups present at carboxyl groups showed the structure of the alginate	Auhim and Odaa (2013)

DIFFERENTIAL SCANNING CALORIMETRY

DSC is a thermoanalytical technique in which the difference in the amount of heat required to increase the temperature of a sample and the reference is measured as a function of temperature. Thermogravimetric property analysis of the purified bioflocculant could be used to elucidate its behavior when subjected to heat. This enables us to understand its pyrolysis property when exposed to very high temperatures. The thermal stability of the purified bioflocculants as determined by

DSC is represented as heat flow–temperature curves. DSC analysis brings out the essential difference between glass transition and thermal transitions such as crystallization and melting.

Gong et al. (2008) observed that the bioflocculant produced by Serratia ficaria could retain its flocculating activity after being heated at 100°C for 15 min, mainly due to the polysaccharide backbone. Li et al. (2010) reported that the bioflocculant produced by Agrobacterium sp. M503 retained its flocculating activity by up to 70°C and a further increase in temperature up to 121°C had no effect on the flocculating activity. High thermostability of a compound bioflocculant CBF-F26 was observed when the purified bioflocculant was heated to more than 100°C for 30 min. The residual flocculating activity of this bioflocculant was more than 90% (Okaiyeto et al., 2013; Karthiga Devi and Natarajan, 2015a,b).

MATRIX-ASSISTED LASER DESORPTION/IONIZATION

It is a soft ionization technique used in mass spectrometry, allowing the analysis of biomolecules (biopolymers) and large organic molecules (such as polymers, dendrimers, and other macromolecules), which tend to be fragile and fragment when ionized by more conventional ionization methods.

MALDI methodology is a three-step process. First, the sample is mixed with a suitable matrix material and applied to a metal plate. Second, a pulsed laser irradiates the sample, triggering ablation and desorption of the sample and the matrix material. Finally, the analyte molecules are ionized by being protonated or deprotonated in the hot plume of ablated gases, and can then be accelerated into whichever mass spectrometer is used to analyze them (Liang et al., 2013; Karthiga Devi and Natarajan, 2015a).

The challenges to polysaccharides determination are that the glycan moiety has different chemical properties than proteins or nucleic acids (Forsberg et al., 2000; Faber et al., 2001; Lattova et al., 2005; Mischnick et al., 2005; Nielsen et al., 2010) and most of them are insoluble that causes problems in chemical manipulation such as permethylation and reducing end labeling. Since tandem mass spectrometry (MS) for the analysis of polysaccharides has not been reviewed in detail, only a few reports have been described. However, the key aspects of fragmentation procedures are the same with those glycoproteomic studies on oligosaccharides and may be applied for further study on other source of polysaccharides. Tandem MS with MALDI and electro spray ionization (ESI) used for permethylated polysaccharides is a powerful tool for linkage analysis of saccharides'. The use of tandem MS is driven by the need to produce structural information on glycans (Harvey, 1999; Zaia, 2004). Tandem MS can analyze a mixture of positional isomers directly. Sequential stages of tandem MS are performed in series and the stages of MS are abbreviated MSn. The masses of the product ions of glycan substructures may be selected for dissociation in further stages. In addition, tandem mass analysis of permethylated glucan can be referred to gas chromatography (GC)/MS on their methylated acetyled alditols and in comparison with the spectra of synthetic standards for the structural analysis of polysaccharides. Most glycan tandem mass spectra are produced by collision-induced dissociation (CAD), a technique in which selected precursor ions are dissociated by collision with gas atoms in a collision cell. Typically, the weakest bonds rupture to produce the most abundant product ions. It is possible to dissociate permethylated glycans using high-energy CAD that uses a MALDI-TOF/TOF or ESI MS instrument, under which conditions bond rupture is kinetically controlled and cross-ring cleavage ions are more abundant for structural analysis.

USE OF BIOFLOCCULANTS IN THE REMOVAL OF METALS FROM WATER

Heavy metal pollution is a remarkable environmental problem. Anthropogenic activities, such as mining operations and the discharge of industrial wastes, have resulted in the accumulation of metals in the environment and the food chain, bringing about serious environmental pollution, threatening ecosystems and human health (Chisti, 2004). For example, Kozul et al. (2009) demonstrated that chronic exposure to arsenic significantly compromises the immune response to influenza. To ensure environmental pollution control, removal of toxic heavy metals from industrial wastewaters is essential (Guangyu and Thiruvenkatachari, 2003).

Biological materials (Zheng et al., 2008), algae (El-Sheekh et al., 2005), protozoans (Rehman et al., 2008), yeasts, fungi (Guangyu and Thiruvenkatachari, 2003), and plants (Heredia and Martin, 2009) have been shown to play significant roles in the removal and recovery of heavy metal. Among the bioremediants, bioflocculants have gained increasing attention since they are environment-friendly, biodegradable, and non-toxic (Shih et al., 2001). Bioflocculants contain various organic groups, such as uronic acids (containing a carbonyl and a carboxylic acid function), glutamic and aspartic acid in the protein component or galacturonic acid and glucuronic acid in the polysaccharide component which are responsible for binding metals.

Polysaccharides produced by *Bacillus firmus* MS-102 are excellent absorbents for Pb, Cu, and Zn, respectively (Salehizadeh and Shojaosadati, 2003). This can be attributed to charge density, attractive interaction, and types of conformation of polysaccharide with the adsorbed ions. Generally, charge densities of bivalent ions increase with the decrease in ionic radius charge density: Pb < Cu < Zn.

For Pb, the interaction with polysaccharides is weaker than with Cu and Zn, and the bioflocculant has fewer tails and many more loops and occupies much less surface area of ion. For Zn, the attractive interaction with the bioflocculant is stronger and results in a nearly flat conformation on the surface with few loops and tails extending into the solution. There are many more free negative functional groups on the polyelectrolyte after adsorption of ions onto the bioflocculant in the case of Pb than in the case of Cu and Zn. Therefore, the adsorbed amount of ions onto the bioflocculant is greater for Pb than for Cu and Zn after equilibrium. The metal adsorption by the biosorbent materials from *B. firmus* obeys Langmuir and Freundlich models.

Bioflocculant from the Gram-negative bacilli *Klebsiella pneumoniae* sp. (capsulated) and slime layer forming *Pseudomonas aeruginosa* showed the highest copper and mercury removal of 87% and 89%, respectively, at 20 mg/L. The optimum adsorption of lead (79%) and cadmium (79%) by the bioflocculant was recorded at 40 mg/L, whereas the highest arsenate and zinc removal of 72% and 80%, respectively, was recorded at 60 mg/L. The enhancement in metal adsorption could be due to an increase in electrostatic interactions, involving sites of progressively lower affinity for metal ions (Puranik and Pakniker, 1999). No increase in metal uptake will be observed where the binding sites are saturated by the metals. Higher efficiencies in removing heavy metals at lower bioflocculant concentrations make them very attractive in the treatment of industrial effluents/wastewaters. Heavy metal adsorption was found to be lower at low and high pH values. At lower pH values, a high concentration of protons competes for the same anionic sites on the polymer as the divalent cations. The mass of protons leads to their preferential binding and thus the divalent cation binding is low (Sahoo et al., 1992). As the pH increases to its optimum value, which differ from one metal ion to another, the adsorbing surface gets saturated with negative charges resulting in an increased efficiency to bind and adsorb metal ions of positive charges. While at a pH higher than its optimum value, hydroxo species of the metals can be formed which do not readily bind to the adsorption sites on the surface of the adsorbent/bioflocculant (Kacar et al., 2000).

USE OF BIOFLOCCULANTS IN DYE DECOLORIZATION

It is known that 90% of reactive textile dyes entering activated sludge sewage treatment plants will pass through unchanged and will be discharged to rivers. Not all dyes currently used could be degraded and/or removed with the physical and chemical processes, and sometimes the degradation products are more toxic. Traditional textile finishing industries consume about 100 L of water to process about 1 kg of textile materials. New closed-loop technologies such as the reuse of microbially or enzymatically treated dyeing effluents could help to reduce this enormous water consumption. Several hybrid anaerobic and aerobic microbial treatments have been suggested to enhance the degradation of textile dyes. Under anaerobic conditions, azo-reductases usually cleave azo dyes into the corresponding amines, many of which are mutagenic and/or carcinogenic. Furthermore, azo-reductases have been shown to be very specific enzymes, thus cleaving only azo bonds of selected dyes. By contrast, laccases act oxidatively and less specifically on aromatic rings, thus having a potential to degrade a wider range of compounds.

The dye-removal potential of bioflocculants produced by bacterial isolates indigenous to a wastewater treatment plant has been investigated. Bioflocculants from all the isolates were able to remove the different dyes tested from industrial wastewater to varying degrees, possibly by causing aggregation of particles and cells due to bridging and charge neutralization as previously reported (Salehizadeh and Shojaosadati, 2001). Bridging occurs when the flocculant extends from the particle's surface into the solution for a distance greater than the distance over which the interparticle repulsion acts. In this case, the biopolymer can adsorb onto other particles to form flocs (Hantula and Bamford, 1991; Levy et al., 1992). This demonstrates that the removal of all the dyes tested was directly proportional to the bioflocculant concentration until an optimum concentration was reached (Shubo et al., 2005). Moreover, a higher decolorization of dyes can be achieved by increasing the concentration of the bioflocculants. In order to be effective in destabilization, a polymer molecule must contain chemical groups that can interact with sites on the surface of the colloidal particles. A particle–polymer–particle complex is thus formed in which the polymer serves as a bridge. If a second particle is not available in time, the extended segments may eventually adsorb onto other sites on the original particle, so that the polymer is no longer capable of serving as a bridge (Salehizadeh and Shojaosadati, 2002). The addition of divalent cations such as Mn^{++}, Mg^{++}, and Ca^{++} enhanced both the flocculating activity and the decolorization of the dyes. Cations stimulate flocculating activity by neutralizing and stabilizing the residual negative charge of functional groups and by forming bridges between particles. Divalent and trivalent cations increase the initial adsorption of biopolymers on suspended particles by decreasing the negative charge on both the polymer and the particle. Divalent cations have been shown to bind to the bioflocculants to form complexes, thereby stimulating flocculation and decolorization. These techniques decolorize textile wastewater by partially decomposing the dye molecules and thereafter leaving the harmful residues in the effluent. Fang et al. (2010) reported that the removal of dyes (anionic azo-dyes) by flocculants may be attributed to both charge neutralization and bridging effect, with the former being the dominant mechanism. The dye structure appears to be conducive to chelation/complex formation reactions with coagulants leading to the formation of insoluble metal dye complexes that may precipitate from the solution. Color removal could probably be due to aggregation or precipitation and adsorption of coloring substances onto the coagulant species.

Decolorization efficiency using bioflocculants for methylene blue, crystal violet, and malachite green was found to be 86%, 97%, and 99%, respectively. A novel bacterium (ZHT4-13) produced a bioflocculant (MBF4-13) (Gao et al., 2009) which exhibited a strong decolorizing ability for blue and violet series of dyes, but had a relatively small decolorizing ability for red, pink, and orange series dyes. Deng et al. (2005) reported that the bioflocculant produced by *Aspergillus parasticus* was more effective for Reactive Blue 4 and Acid Yellow 25 than for Basic Blue B. The results have shown that depending on the dye used, the bioflocculant exhibited different decolorizing efficiencies (Gao et al., 2009).

Several acid, reactive, and direct dyes are used in decolorization (Deng et al., 2005). Depending on the dye used, the bioflocculant exhibited different decolorization efficiencies. The bioflocculant was more effective for anionic dyes than cationic dyes. The decolorization ability is related to the size of the dye molecule and the number of the sulfonic groups in one dye molecule since the bigger molecules adsorbed onto the bioflocculant may prevent others from being adsorbed, and more sulfonic groups in one molecule would absorb more bioflocculants. However, the decolorization of the dyes is not consistent with the size and the number of sulfonic groups. This phenomenon may be attributed to space resistance and the complex multipoint adsorption between the dye and the flocculant molecules during the flocculation process. Although the decolorization of AcidBlue45, DirectBlue1, AcidOrange8, and ReactiveOrange16 were relatively low, higher decolorization was achieved using more flocculants. The decolorization efficiencies of up to 96.1%, 98.4%, 98.3%, and 99.6% were obtained for AcidBlue45, DirectBlue1, AcidOrange8, and ReactiveOrange16, respectively, using a flocculant dosage of 150 mL/L. A larger volume of the culture broth was needed to achieve a higher decolorization efficiency (Deng et al., 2005).

Buthelezi et al. (2012) reported that whale dye was most easily decolorized, followed by medi-blue, mixed dyes, and fawn dye, respectively. One explanation for this, according to Pearce et al. (2003), could be the acidic nature of the fawn dye. Acid dyes are the most problematic and difficult to decolorize due to their inert chemical structure and the attached phenyl, methyl, methoxy, nitro, and sulfonate groups. Syafalni et al. (2012) suggested that the removal of color from dye solutions is complex and may be due to the physicochemical processes of coagulation and/or chelation–complexation-type reactions. These techniques decolorize textile wastewater by partially decomposing the dye molecules and thereafter leaving the harmful residues in the effluent. Fang et al. (2010) reported that the removal of dyes (anionic azo-dyes) by flocculants may be attributed to both charge neutralization and bridging effect, with the former being the dominant mechanism. The dye structure appears to be conducive to chelation/complex formation reactions with coagulants leading to the formation of insoluble metal dye complexes that may precipitate from the solution. Color removal could probably be due to aggregation or precipitation and adsorption of coloring substances onto the coagulant species (Karthiga Devi and Natarajan, 2015a,b).

TURBIDITY REMOVAL THROUGH BIOFLOCCULANTS

Fine particles such as ore and clay fines remain suspended in water leading to turbidity. Water turbidity due to suspended particles can be efficiently removed by the process of water clarification using bioflocculants. The addition of metal ions to kaolin suspensions during the bioflocculation process is required to induce effective flocculation by cation-dependent bioflocculants, such as bioflocculants produced by *Halomonas* sp. (He et al., 2010), *Bacillus circulans* (Li et al., 2009a,b), *Pseudoalteromonas* sp. (Li et al., 2008), *S. ficaria* (Gong et al., 2008), and *Bacillus licheniformis* (Li et al., 2009a,b). Generally, cations are added to neutralize the negative charges of cation-dependent bioflocculants and kaolin particles, thereby increasing the adsorption of the bioflocculant onto kaolin particles (Dermilim et al., 1999). A cation-independent bioflocculant IH-7 exhibited a flocculating efficiency at pH 7 of 97% without the addition of metal ions. Addition of monovalent cations such as K^+ and Na^+ and divalent cations such as Mg^{++}, Fe^{++}, and Ca^{++} could not enhance the flocculating rate of IH-7, whereas addition of Fe^{+++} inhibited flocculation. Zheng et al. (2008) showed that addition of K^+, Na^+, Mg^{++}, Fe^{++}, and Ca^{++} could not evidently enhance the flocculating rate of the bioflocculant produced from *Bacillus* sp. A schematic illustration of turbidity removal using bioflocculants is illustrated in Figure 16.3.

Bioflocculant IH-7 produced by *Aspergillus flavus* was found to be a cation-independent bioflocculant. It has the ability to flocculate activated carbon, kaolin clay, soil solids, and yeast cells without cation addition. Its flocculation performance was better than that of polyacrylamides at a wide range of flocculant concentrations, pH, and initial kaolin concentrations. IH-7 was also found to be a thermostable bioflocculant with the potential to replace the commonly used chemical flocculants used in water treatment industries, especially in cold weather regions, as well as for the treatment of turbid water with a salinity of up to 10% w/w. Flocculation of different types of suspended solids such as activated carbon, soil solids, and yeast cells were tested using IH-7 and 93% of flocculating rate in activated carbon and soil solids suspensions were achieved (Li et al., 2008). IH-7 was found to be a better bioflocculant for activated carbon removal at neutral pH than MBFF19 (bioflocculant) produced by Bacillus sp. F19 (Zheng et al., 2008).

Flocculation efficiency of bioflocculants generally increases in the presence of $CaCl_2$. It was found that addition of a cation to the reaction mixture was necessary to induce effective flocculation by forming complexes of the bioflocculant and kaolin clay mediated by a cation (Kurane and

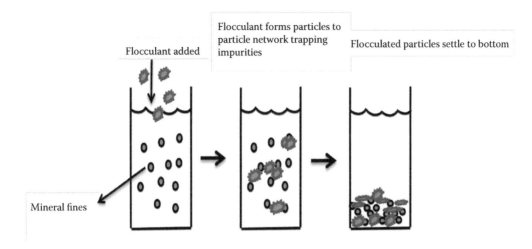

Flocculant added

Flocculant forms particles to particle network trapping impurities

Flocculated particles settle to bottom

Mineral fines

FIGURE 16.3 Removal of suspended fine particles through bioflocculants.

Matsuyama, 1994). Different cation types (such as Na^+, K^+, Mg^{++}, Al^{+++}, Zn^{+++}, and Fe^{+++}) with the same anion Cl^- were tested at a constant dosage of 10 mL 3% (w/v) for 200 mL kaolin suspension. Flocculation activity of 38A bioflocculant was stimulated in the presence of Ca^{++}, K^+, and Na^+, but inhibited by Al^{+++}, Mg^{++}, Fe^{+++}, and Zn^{+++}. Among these cations, Ca^{++} was found to be the most favorable ion exhibiting the highest flocculating activity. Addition of $MgCl_2$ resulted in the highest flocculating activity of the bioflocculant MBF-W6 derived from *Chryseobacterium daeguense*. Yokoi et al. (1995) emphasized that the cation could stimulate the flocculation by neutralization and destabilization of residual negative charges of carboxyl groups in an acidic polysaccharide, forming bridges which bind kaolin particles to each other.

APPLICATIONS OF BIOFLOCCULANTS IN VARIOUS WASTEWATER TREATMENTS

It is one of the most common liquid wastes in the food industry. During the production of starch, starch wastewater is produced in the process of bran–starch separation of corn, with most of finely suspended bran particles and almost all corn proteins retained in the wastewater. Sedimentation is commonly used to recover suspended solids, but the settling time is often very long and the separation efficiency normally low. Most of the conventional flocculants produce adverse effects on animals and the environment. Development of a bioflocculant MBFA9 and its application in starch wastewater treatment has been reported (Deng et al., 2003). MBFA9-bioflocculant was also found to be effective even at low dosages. For kaolin suspension, the relationship between flocculating rate and the dosage of MBFA9 is shown in Figure 16.4. It is apparent that the flocculating rate can reach 99% with a dosage as low as only 0.1 mL/L. The dosage of culture broths used as bioflocculants for kaolin suspensions is usually in the range of about 1.0–150 mL/L, and often Ca^{++}, Fe^{+++}, or Al^{+++} are used as coagulants to achieve high flocculating rates. Starch wastewater treatment using the bioflocculant MBFA9 could be used in drinking water processing and downstream processing in the food and fermentation industries because of its lack of toxicity. Starch wastewater was used to evaluate the flocculating activity of the bioflocculant MBFA9. Bioflocculant dosages played an important role in the flocculation rate of kaolin clay suspensions.

Flocculation efficiency was found to be high at neutral pH. The carboxyl groups present in the bioflocculant could be stretched out due to electrostatic repulsion and the stretched molecular chains provide more effective sites for particle attachment. Flocculation efficiency decreased at higher bioflocculant dosages can be attributed to oversaturation of excess polysaccharide on many mineral

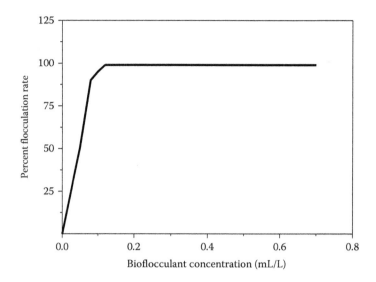

FIGURE 16.4 Relationship between bioflocculant concentration and percent flocculation rate for kaolin suspension. (Adapted from Deng, S. B. et al. 2003. *Applied Microbiology and Biotechnology*, 60: 588–593.)

surface binding sites, leading to a reduction in particle-to-particle binding. Incomplete dispersion of excess polysaccharides also would impede effective flocculation of clay fines (Yokoi et al., 1997). During the flocculation of kaolin suspensions, the flocculation rate was found to be higher in the presence of calcium ions (Kurane and Matsuyama, 1994; Zaki et al., 2011). Under acidic conditions, ionization of COOH in the bioflocculant will be blocked, restricting the bridging actions, while under an alkaline condition, COO− and OH− functional groups would increase, promoting its flocculating efficiency (Zhang et al., 2010). Bioflocculants exhibit different electrostatic states at different pH values affecting the bridging efficiency for kaolin clay (Karthiga Devi and Natarajan, 2015b).

CONCLUSIONS

Bioflocculants mainly derived from microorganisms are becoming increasingly attractive in waste water treatment due to their cost efficiency, environmental acceptability, and multiple applications. In this chapter, methods of extraction of bioflocculants from different microorganisms, their characterization, and major properties are discussed. Structural features of various bioflocculants with respect to functional groups and flocculant properties are illustrated with examples. Mechanisms of bioflocculation of fine particles in aqueous suspensions in the presence of different types of bioflocculants are analyzed and compared with those of conventional chemically derived coagulants and flocculants. Industrial applications of bioflocculants are illustrated with relevance to removal of heavy metals from water, dye decolorization, and water turbidity removal and clarification. On the basis of extensive case studies, it becomes evident that the use of bioflocculants is set to revolutionize water treatment and pollution control industries.

ACKNOWLEDGMENTS

The authors acknowledge the Department of Science and Technology (DST), Govt. of India. They also acknowledge the National Academy of Sciences, India (NASI) for the Platinum Jubilee Senior Scientist Fellowship awarded to Professor K. A. Natarajan and for the Senior Research Fellowship awarded to K. Karthiga Devi).

REFERENCES

Aksu, Z. 2005. Application of biosorption for the removal of organic pollutants: A review. *Process Biochemistry*, 40: 997–1026.

Auhim, H. S. and N. H. Odaa. 2013. Optimization of flocculation conditions of exopolysaccharide biofloccu-lant from *Azotobacter chrococcum* and its potential for river water treatment. *Journal of Microbiology and Biotechnology Research*, 3: 93–99.

Buthelezi, S. P., A. O. Olaniran, and B. Pillay. 2010. Production and characterization of bioflocculants from bacteria isolated from wastewater treatment plant in South Africa. *Biotechnology and Bioprocess Engineering*, 15: 874–881.

Buthelezi, S. P., A. O. Olaniran, and B. Pillay. 2012. Textile dye removal from wastewater effluents using bio-flocculants produced by indigenous bacterial isolates. *Molecules*, 17: 14260–14274.

Chisti, Y. 2004. Environmental impact of toxic pollutants. *Biotechnology Advances*, 6: 431–432.

Cosa, S. and O. Anthony. 2014. Bioflocculant production by a consortium of two bacterial species and its poten-tial application in industrial waste water and river water treatment. *Polish Journal of Environmental Studies*, 23: 689–696.

Deng, S. B., R. B. Bai, X. M. Hu, and Q. Luo. 2003. Characteristics of a bioflocculant produced by *Bacillus mucilaginosus* and its use in starch wastewater treatment. *Applied Microbiology and Biotechnology* 60: 588–593.

Deng, S. B., G. Yu, and Y. P. Ting. 2005. Production of a bioflocculant by *Aspergillus parasiticus* and its application in dye removal. *Colloids and Surfaces B: Biointerfaces*, 44: 179–186.

Dermilim, W., P. Prasertsan, and H. Doelle. 1999. Screening and characterization of bioflocculant produced by isolated *Klebsiella* sp. *Applied Microbiology and Biotechnology*, 52: 698–703.

Ebeling, J. M., K. M. Rishel, and P. L. Sibrell. 2005. Screening and evaluation of polymers as flocculation aids for the treatment of aqua-cultural effluents. *Aquacultural Engineering*, 33: 235–249.

Eccles, H. 1999. Treatment of metal-contaminated wastes: Why select a biological process. *Trends in Biotechnology*, 17: 462–465.

El-Sheekh, M. M., W. A. El-Shouny, M. E. H. Osman, and E. W. E. El-Gammal. 2005. Growth and heavy metals removal efficiency of *Nostoc muscorum* and *Anabaena subcylindrica* in sewage and industrial wastewater effluents. *Environmental Toxicology and Pharmacology*, 19: 357–365.

Faber, E. J., M. J. van den Haak, J. P. Kamerling, and J. F. G. Vliegenthart. 2001. Structure of the exopolysac-charide produced by *Streptococcus thermophilus* S3. *Carbohydrates Research*, 331: 173–182.

Fang, R., X. Cheng, and X. Xu. 2010. Synthesis of lignin-base cationic flocculant and its application in remov-ing anionic azo-dyes from simulated wastewater. *Bioresource Technology*, 101: 7323–7329.

Forsberg, L. S., U. R. Bhat, and R. W. Carlson. 2000. Structural characterization of the O-antigenic polysac-charide of the lipopolysaccharide from *Rhizobium etli* strain CE3. *Journal of Biological Chemistry*, 275: 18851–18863.

Ganczarczyk, J. J. 1983. Application of activated sludge process to industrial wastewater treatment. In: *Activated Sludge Process: Theory and Practice*. New York: Marcel Dekker Inc., pp. 199–227.

Gao, J., H. Bao, M. Xin, Y. Liu, Q. Li, and Y. Zhang. 2006. Characterization of a bioflocculant from a newly isolated *Vagococcus* sp. W31. *Journal of Zhejiang University SCIENCE B*, 7: 186–192.

Gao, Q., X. H. Zhu, J. Mu, Y. Zhang, and X. W. Dong. 2009. Using *Ruditapes philippinarum* conglutination mud to produce bioflocculant and its applications in wastewater treatment. *Bioresource Technology*, 100: 4996–5001.

Gomaa, E. Z. 2012. Production and characteristics of a heavy metals removing bioflocculant produced by *Pseudomonas aeruginosa*. *Polish Journal of Microbiology*, 61: 281–289.

Gong, W. X., S. G. Wang, X. F. Sun, W. L. Xian, Y. Y. Qin, and B. Y. Gao. 2008. Bioflocculant production by culture of *Serratia ficaria* and its application in wastewater treatment. *Bioresource Technology*, 99: 4668–4674.

Guangyu, Y. and V. Thiruvenkatachari. 2003. Heavy metals removal from aqueous solution by fungus *Mucor rouxii*. *Water Resources*, 37: 4486–4496.

Guo, J., C. Yang, and Y. Zeng. 2013. Treatment of swine wastewater using chemically modified zeolite and bioflocculant from activated sludge. *Bioresource Technology*, 143: 289–297.

Hantula, J. and D. H. Bamford. 1991. The efficiency of the protein dependent flocculation of *Flavobacterium* sp. *Applied Microbiology and Biotechnology*, 36: 100–104.

Harvey, D. J. 1999. Matrix-assisted laser desorption/ionization mass spectrometry of carbohydrates. *Mass Spectrometry Reviews*, 18: 349–450.

He, J., J. Zou, Z. Shao, J. Zhang, Z. Liu, and Z. Yu. 2010. Characteristics and flocculating mechanism of a novel bioflocculant HBF-3 produced by deep-sea bacterium mutant *Halomonas* sp. *World Journal of Microbiology and Biotechnology*, 26: 1135–1141.

Heredia, J. B. and J. S. Martin. 2009. Removing heavy metals from polluted surface water with a tannin-based flocculant agent. *Journal of Hazardous Materials*, 165: 1215–1218.

Kacar, Y., C. Arpa, S. Tan, A. Denizli, O. Genc, and M. Y. Arica. 2000. Biosorption of Hg(II) and Cd(II) from aqueous solution: Comparison of biosorptive capacity of alginate and immobilized live and heat inactivated *Phanerochaete chrysosporium*. *Process Biochemistry*, 37: 601–610.

Karthiga Devi, K. and K. A. Natarajan. 2015a. Isolation and characterization of a bioflocculant from *Bacillus megaterium* for turbidity and arsenic removal. *Minerals & Metallurgical Processing Journal*, 32: 222–229.

Karthiga Devi, K. and K. A. Natarajan. 2015b. Production and characterization of bioflocculants for mineral processing applications. *International Journal of Mineral Processing*, 137: 15–25.

Kowall, N. W., W. W. Pendlebum, J. B. Kessler, D. P. Perl, and M. F. Beal. 1989. Aluminium-induced neuro-fibrillary degeneration affects a subset of neurons in rabbit cerebral cortex, basal forebrain and upper brainstem. *Neuroscience*, 29: 329–337.

Kozul, C. D., K. H. Ely, R. I. Enelow, and J. W. Hamilton. 2009. Low-dose arsenic compromises the immune response to influenza A infection *in vivo*. *Environmental Health Perspectives*, 117: 1441–1447.

Kurane, R., K. Hatamochi, T. Kakuno, M. Kiyohara, M. Hirano, and Y. Taniguchi. 1994. Production of a bioflocculant by *Rhodococcus erythropolis* S-1 grown on alcohols. *Bioscience, Biotechnology and Biochemistry*, 58: 428–429.

Kurane, R. and H. Matsuyama. 1994. Production of a bioflocculant by mixed culture. *Bioscience, Biotechnology and Biochemistry*, 58: 1589–1594.

Lattova, E., S. Snovida, H. Perreault, and O. Krokhin. 2005. Influence of the labeling group on ionization and fragmentation of carbohydrates in mass spectrometry. *Journal of American Society* for Mass Spectrometry, 16: 683–696.

Levy, N., S. Magdasi, and Y. Bar-Or. 1992. Physico-chemical aspects in flocculation of bentonite suspensions by a cyanobacterial bioflocculant. *Water Resources*, 26: 249–254.

Li, Q., H. L. Liu, Q. Qi, F. Wang, and Y. Zhang. 2010. Isolation and characterization of temperature and alkaline stable bioflocculant from *Agrobacterium*. sp. M503. *New Biotechnology*, 27: 789–794.

Li, W. W., W. Z. Zhou, W. Z. Zhang, J. Wang, and X. B. Zhu. 2008. Flocculation behaviour and mechanism of an exopolysaccharide from the deep-sea psychrophilic bacterium *Pseudoalteromonas* sp. SM9913. *Bioresource Technology*, 99: 6893–6899.

Li, Z., R. Y. Chen, H. Y. Lei, Z. Shan, T. Bai, Q. Yu, and H. L. Li. 2009a. Characterization and flocculating properties of a novel bioflocculant produced by *Bacillus circulans*. *World Journal of Microbiology and Biotechnology*, 25: 745–752.

Li, Z., S. Zhong, H. Y. Lei, R. W. Chen, Q. Yu, and H. L. Li. 2009b. Production of a novel bioflocculant by *Bacillus licheniformis* X14 and its application to low temperature drinking water treatment. *Bioresource Technology*, 100: 3650–3656.

Lian, B., Y. Chen, J. Zhao, H. H. Teng, L. Zhu, and S. Yuan. 2008. Microbial flocculation by *Bacillus mucilaginosus*: Applications and mechanisms. *Bioresource Technology*, 99: 4825–4831.

Liang, J. R., X. X. Ai, Y. H. Gao, and C. P. Chen. 2013. MALDI-TOF MS analysis of the extracellular polysaccharides released by the diatom *Thalassiosira pseudonana*. *Journal of Applied Phycology*, 25: 477–484.

Lin, J. and C. Harichund. 2011. Industrial effluent treatments using heavy-metal removing bacterial bioflocculants. *Water SA*, 37: 599–607.

Liu, L. F. and W. Cheng. 2010. Characteristics and culture conditions of a bioflocculant produced by *Penicillium* sp. *Biomedical and Environmental Sciences*, 23: 213–218.

Liu, W. J., K. Wang, B. Z. Li, H. L. Yuan, and J. S. Yang. 2010. Production and characterization of an intracellular bioflocculant by *Chryseobacterium daeguense* W6 cultured in low nutrition medium. *Bioresource Technology*, 101: 1044–1048.

Lopez, E., I. Ramos, and M. A. Sanroman. 2003. Extracellular polysaccharides production by *Arthrobacter viscosus*. *Journal Food Engineering*, 60: 463–467.

Master, C. L., G. Multhaup, G. Simms, J. Pottgiesser, R. N. Martins, and K. Beyreuther. 1985. Neuronal origin of a cerebral amyloid: Neurofibrillary tangles of Alzheimer's disease contain the same protein as the amyloid of plaque cores and blood vessels. *EMBO Journal*, 4: 2757–2763.

Mischnick, P., W. Niedner, and R. Adden. 2005. Possibilities of mass spectrometry and tandem-mass spectrometry in the analysis of cellulose ethers. *Macromolecular Symposia*, 223: 67–77.

Nie, M., X. Yin, J. Jia, Y. Wang, S. Liu, Q. Shen, P. Li, and Z. Wang. 2011. Production of a novel bioflocculant MNXY1 by *Klebsiella pneumoniae* strain NY1 and application in precipitation of cyanobacteria and municipal wastewater treatment. *Journal of Applied Microbiology*, 111: 547–558.

Nielsen, T. C., T. Rozek, J. J. Hopwood, and M. Fuller, 2010. Determination of urinary oligosaccharides by high-performance liquid chromatography/electrospray ionization–tandem mass spectrometry: Application to Hunter syndrome. *Analytical Biochemistry*, 402: 113–120.

Ogunsade, O. O., M. K. Bakare, and I. O. Adewale. 2015. Purification and characterization of bioflocculant produced by *Bacillus amyloliquefaciens* ABL 19 isolated from Adeti Stream, Ilesa, Osun State, Nigeria. *Nature and Science*, 13: 55–64.

Okaiyeto, L., U. U. Nwodo, L. V. Mabinya, and A. I. Okoh. 2013. Characterization of a bioflocculant produced by a consortium of *Halomonas* sp. Okoh and *Micrococcus* sp. Leo. *International Journal of Environmental Research and Public Health*, 10: 5097–5110.

Pearce, C. I., J. R. Lloyd, and J. T. Guthrie. 2003. The removal of colour from textile wastewater using whole bacterial cells. *Dyes and Pigments*, 58: 179–196.

Prasertsan, P., W. Dermlim, H. Doelle, and J. F. Kennedy. 2006. Screening, characterization and flocculating property of carbohydrate polymer from newly isolated *Enterobacter cloacae* WD7. *Carbohydrate Polymers*, 66: 289–297.

Puranik, P. R. and K. M. Pakniker. 1999. Biosorption of lead, cadmium and zinc by Citrobacter strain MCM B-181: Characterization studies. *Biotechnology Progress*, 15: 228–237.

Rehman, A., F. R. Shakoori, and A. R. Shakoori, 2008. Heavy metal resistant freshwater ciliate, *Euplotes mutabilis*, isolated from industrial effluents has potential to decontaminate wastewater of toxic metals. *Bioresource Technology*, 99: 3890–3895.

Sahoo, D. K., R. N. Kar, and R. P. Das. 1992. Bioaccumulation of heavy metal ions by *Bacillus circulans*. *Bioresource Technology*, 41: 177–179.

Salehizadeh, H. and S. A. Shojaosadati. 2001. Extracellular biopolymeric flocculants: Recent trends and biotechnological importance. *Biotechnology Advances*, 19: 371–385.

Salehizadeh, H. and S. A. Shojaosadati. 2002. Isolation and characterisation of a bioflocculant produced by *Bacillus firmus*. *Biotechnology Letters*, 24: 35–40.

Salehizadeh, H. and S. A. Shojaosadati. 2003. Removal of metal ions from aqueous solution by polysaccharide produced from *Bacillus firmus*. *Water Resources*, 37: 4231–4235.

Sharma, B. R., N. C. Dhuldhoya, and U. C. Merchant. 2006. Flocculants—Ecofriendly approach. *Journal of Polymers and Environment*, 14: 195–202.

Shih, I. L., Y. T. Van, L. C. Yeh, H. G. Lin, and Y. N. Chang. 2001. Production of a biopolymer flocculant from *Bacillus licheniformis* and its flocculation properties. *Bioresource Technology*, 78: 267–272.

Shubo, D., G. Yu, and T. Yen Peng. 2005. Production of a bioflocculant by *Aspergillus parasiticus* and its application in dye removal. *Colloids and Surfaces B: Biointerfaces*, 44: 179–186.

Syafalni, S., I. Abustan, I. Norli, and T. S. Kwan. 2012. Production of bioflocculant by *Chryseomonas Luteola* and its application in dye wastewater treatment. *Modern Applied Science*, 6: 13–20.

Tang, W., L. Song, Y. Li, J. Qiao, T. Zhao, and H. Zhao. 2014. Production, characterization, and flocculation mechanism of cation independent, pH tolerant, and thermally stable bioflocculant from *Enterobacter* sp. ETH-2. *PLoS One*, 9: 1–19.

Wang, S. G., X. A. Gong, X. W. Liu, L. Tian, Y. E. Yue, and B. Y. Gao. 2007. Production of a novel bioflocculant by culture of *Klebsiella mobilis* using dairy wastewater. *Biochemical Engineering Journal*, 36: 81–86.

Xia, S., Z. Zhang, X. Wang, A. Yang, L. Chen, J. Zhao, D. Leonard, and N. J. Renault. 2008. Production and characterization of a bioflocculant by *Proteus mirabilis* TJ-1. *Bioresource Technology*, 99: 6520–6527.

Xiong, Y., Y. Wang, Y. Yu, Q. Li, H. Wang, R. Chen, and N. He. 2010. Production and characterization of a novel bioflocculant from *Bacillus licheniformis*. *Applied and Environmental Microbiology*, 76: 2778–2782.

Yim, J. H., S. J. Kim, S. H. Ahn, and H. K. Lee. 2007. Characterization of a novel bioflocculant p-KG03 from a marine dinoflagellate *Gyrodinium impudicum* KG03. *Bioresource Technology*, 98: 361–367.

Yokoi, H., O. Natsuda, J. Hirose, S. Hayashi, and Y. Takasaka. 1995. Characteristics of a biopolymer flocculant produced by *Bacillus* sp. PY-90. *Journal of Fermentation and Bioengineering*, 79: 378–380.

Yokoi, H., T. Yoshida, S. Mori, J. Hirose, S. Hayashi, and Y. Takasaki. 1997. Biopolymer flocculant produced by an *Enterobacter* sp. *Biotechnology Letters*, 19: 569–573.

Zaia, J. 2004. Mass spectrometry of oligosaccharides. *Mass Spectrometry Reviews*, 23: 161–227.

Zaki, S., S. Farag, G. Abu Elrees, M. Elkady, M. Nosier, and D. Abd El Haleem. 2011. Characterization of bioflocculants produced by bacteria isolated from crude petroleum oil. *International Journal of Environment Science and Technology*, 8: 831–840.

Zhang, C. L., Y. N. Cui, and Y. Wang. 2013. Bioflocculants produced by gram-positive *Bacillus* xn12 and *Streptomyces* xn17 for swine wastewater application. *Chemical and Biochemical Engineering Quarterly*, 27: 245–250.

Zhang, J., Z. Liu, and S. Wang. 2002. Characterization of a bioflocculant produced by the marine *Myxobacterium Nannocystis* sp. NU-2. *Journal of Applied Microbiology and Biotechnology*, 59: 517–522.

Zhang, Z., S. Xia, J. Zhao, and J. Zhang. 2010. Characterization and flocculation mechanism of high efficiency microbial flocculant TJ-F1 from Proteus mirabilis. *Colloids Surfaces B: Biointerfaces*, 75: 247–251.

Zhang, Z. Q., B. Lin, S. Xia, X. Wang, and A. Yang. 2007. Production and application of a novel bioflocculant by *Multiple-microorganism consortia* using brewery wastewater as carbon source. *Journal of Environmental Sciences*, 19: 667–673.

Zheng, Y., Z. L. Ye, X. L. Fang, Y. H. Li, and W. M. Cai. 2008. Production and characteristics of a bioflocculant produced by *Bacillus* sp. F19. *Bioresource Technology*, 99: 7686–7691.

17 Sustainability and Regional Development
When Brownfields Become Playing Fields

Kim Dowling, Singarayer K. Florentine, Rachael Martin, and Dora C. Pearce

CONTENTS

ABSTRACT

Brownfields present a continuing and evolving problem for land use planners, developers, public health authorities and for the public. Legislative frameworks and fears over liability combined with an inconsistent and sometimes ad hoc approach across national and international borders are key contributors to poor outcomes. In a world hungry for more development, brownfields present an alluring opportunity to improve land value, improve environmental and human health outcomes and to limit the expansion of the anthropogenic footprint by reusing this 'poor value' land. The absence of a systematic approach to the understanding of brownfields contamination, their remediation and the associated risks will significantly impact on whole-of-population health outcomes in the next century.

INTRODUCTION

Fertile soil, potable water, sustainable energy, clean air and the abundant mineral resources provided by our planet are all essential for the sustainability of human civilisation. However, the importance of soil in the public debate is often overlooked with much political and social emphasis being placed on water, air and energy and on the financial returns of mining. Given the extensive footprint of human development, it is unsurprising that much of the 'hidden' soil profile has already been affected by human activities and, in an increasingly space hungry world, attention is turning to the use of so-called brownfield sites.

Brownfields are defined by the USEPA as 'property, the expansion, redevelopment, or reuse of which may be complicated by the presence or potential presence of a hazardous substance, pollutant, or contaminant'. Human activities and operations such as mining, intensive agriculture, manufacturing and poor waste management systems have created large areas of contaminated land systems throughout the world. As early as 2004, it was estimated that 294,000 ha of contaminated sites were recorded in the United States (USEPA, 2004) while over 300,000 ha of potentially contaminated lands were identified in the United Kingdom (DEFRA, 2011). In 2014, reports suggest that China has identified 19.4% of its agricultural land as contaminated, with Greenpeace positing that treating 1 million hectares of contaminated soil will cost at least 140 billion yuan (22.6 billion USD) in 2015.

Brownfields are scattered across both urban and rural landscapes and affect all nations irrespective of economic development. The global reach of the issue has refocused the attentions of policy makers and planners to consider redevelopment and rehabilitation of these sites (De Sousa, 2003). However, soil contamination is often complex and heterogeneous and may include a range of toxic organic, inorganic, anthropogenic or geogenic compounds (e.g. Sultan and Dowling, 2006; Pearce et al., 2010; Mackay et al., 2013; Martin et al., 2013; Arnold, 2014; Méndez-Fernández et al., 2015; Taylor et al., 2015; Zierold and Sears, 2015), and have the potential to remain in the soil for poorly defined and significantly long periods of time. Contaminants may be present in very high concentrations in both the soil and sediment profiles (e.g. Chen et al., 2005; Feng and Qiu, 2008) and pose a serious threat to human health (Csavina et al., 2012). Hence remediation and repurposing of brownfield sites is desirable but also complex, expensive and potentially hazardous.

As communities expand into fringe areas land once used for industry, intense agriculture or manufacturing is now required for other commercial or residential development and with this expansion comes the need to build schools and play grounds and facilities for other, potentially vulnerable, sections of our community. It is essential that we should build 'sustainability' and public safety into redevelopment but little consensus exists among the policy makers, governments and local planning authorities. Often these hidden threats are poorly evaluated: these landscapes are under increasing pressure to be redeveloped and made 'productive' and time and sufficient funds are often constraining factors. We must always ask ourselves: Is there equal pressure to make them safe?

Of key concern in this discussion is how we should assess and redevelop brownfield sites, and what sustainable principles should be considered to redevelop a brownfield site to acceptable safe levels of contamination as defined now and potentially into the future (Figure 17.1). As we acknowledge that most of the surface of our planet records the effects of human occupancy and that there is a need to 'reuse or recycle' this land, we also ask the question: In a global market, how can legislators keep up with the needs of developers?

Brownfield sites present complex risk profiles that need to be carefully assessed (Alker et al., 2000). This task is further complexed by ownership, jurisdiction and historical factors that often coexist in a tight fiscal environment. Each site needs to be characterised in terms of the threats it poses and the duration of that threat. Characterisation and assessment informs the range of suitable future land uses based on some agreed value placed on the projected use. However, agreeing on a value is in itself complex and even if the physical and chemical characteristics of the site were well understood, the social and economic drivers to repurpose the site might be in conflict and as such a hands-off strategy is unlikely to be an option.

EXAMPLE 17.1: URBAN COMPLEXITY

In what is now central Melbourne, Australia, town gas was manufactured from coal between 1861 and 1927, after which the site was used for a multitude of purposes including production

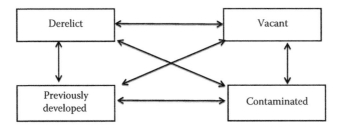

FIGURE 17.1 Terms associated with the concept of brownfield. (Adapted from Alker, S. et al. 2000. *J. Environ. Plann. Manage.*, 43(1): 49–69.)

of defence vehicles up until the 1960s. The site was identified as significantly contaminated with an array of compounds including free tar, complex hydrocarbons, heavy metals and inorganics distributed in fill material, clay and rock formations. Local groundwater was also shown to be affected. The Environment Protection Authority (EPA) Victoria, an independent statutory authority established under the Environment Protection Act 1970 determined via audit that a clean up was required in 2014. This requirement was levelled at the Crown (the Government). It is clear that the purpose of the EPA is to protect and improve the environment and prevent harm to human health but the motivation was not some systematic review of prior land use and mapping of contamination. This part of Melbourne is experiencing significant gentrification. The site is within easy walking distance of extensive urban parklands, commercial districts and desirable housing. The local council saw great potential for development and offered an Urban Design framework to attract development and commence discussions for repurposing this brownfield site. The fact that this site could be remediated is in part thanks to the pressing need for land in our urban environments. The question raised here is why remediation has taken so long to be an agenda item.

EXAMPLE 17.2: COMPETING NEEDS

Consider a brownfield site in a regional city where land pressure and competing needs suggest multiple remediation options. Historically, a solidification or stabilisation technology, perhaps using a capping cement layer might be used to limit environmental liabilities and protect human health by trapping the hazardous contaminant. Capping the material is relatively easy to cost and has proven results. However, the option of transformation into an urban park presents a more difficult costing challenge. The cost effectiveness of various remediation technologies, on-site or off, potential off-site contamination during capital works, the addition of topsoil, appropriate revegetation strategies can be calculated, but the value of the impact and contribution to the physical and aesthetic quality of the area is difficult to quantify. More recently, policymakers have broadened the concept of a park's contribution to society to include contributions to larger urban policy objectives including job opportunities, youth development, public physical and mental health, art and community building, etc.

The value of brownfield sites may be gauged using a triple bottom line approach, in which value is expressed in terms of separate financial, social and environmental benefits (Norman and MacDonald, 2004). Redeveloping (or at least remediating) these brownfields may increase community and economic opportunities at several levels, including an increase in quality of life, it may increase environmental values and could result in a reduction in urban sprawl as a result of appropriate land use assignment (CABERNET [Concerted Action on Brownfield and Economic Regeneration Network], 2006). Appropriate land use assignment requires transparency in reporting and a detailed analysis of the site conditions and both these aspects may prove difficult. Many local planning authorities and commercial investors cite commercial-in-confidence reasons for not providing complete information to our communities and while this may not be inappropriate, it further complexes the evaluation of options.

Significant factors that complicate the assessment of brownfield sites include; extensive time of use, lack of detailed records showing prior use and disposal history, time since closure, poor containment, multiple ownership (potentially including site relinquishment before assessment) and unclear lines of responsibility. Contaminated sites can be many decades or even centuries old allowing time for off-site migration of toxic components through surface run-off, dust dispersion, illicit dumping and even harvesting of soils for use off-site. This is often exacerbated by limited vegetation or surface cover. Such poorly defined and potentially diffuse contaminated sites clearly have costs associated with rehabilitation and often no clear responsible person or group to receive or accept the costs. Authorities, often on fixed term government tenure are required to provide judgement on these matters and may face a public backlash from a sector of their constituency for almost any outcome. A developer may be seen as a saviour for suggesting any solution to a potential public health issue linked to a contaminated site.

Globally there are financial and environmental drivers to redevelop brownfield sites and broad support to do so in accordance with sustainable development principles (Nijkamp et al., 2002; Lange and McNeil, 2004; Karakosta et al., 2009). However this work is hampered by the lack of appropriate frameworks, policy platforms or clear guidance in terms of best practice. To facilitate sustainable remediation the impacts of remediation activities (as well as the possible decision not to remediate) need to be balanced against the community's demand for clean environments with recognition that all development comes with both short- and long-term traditional economic and green costs.

In this chapter we discuss the human–contaminant interface (both historical and contemporary), hazard identification with specific reference to mining activities, the impacts of this exposure and finally, the ways our regulatory context might evolve to better address the needs of humans without unacceptable degradation of the larger environment.

THE INTERFACE BETWEEN HUMANS AND CONTAMINANTS

EARLY OBSERVATIONS

Environmental and health impacts associated with industrial activities are not restricted to the post-industrial world. As early as 400 BC, Xenophon noted concerns regarding the health effects of mining and metallurgy in the ancient Mediterranean. Others, including Lucretius (98–55 BC), Strabo (63 BC–24 AD) and Pliny (23–79 AD) reported links between poor human health and mining activity. Nonetheless, because of the economic importance of these activities the deleterious impacts on human health were considered acceptable (Makra and Brimblecombe, 2004). The cultural capital facilitated by mining, extraction and metallurgy is clearly seen in the art, literature and architecture of the ancient world. The health impacts are harder to quantify.

The same can be said of Europe in the late middle ages through to the industrial revolution, by which time mining and ore processing industries were recognised as primary sources of metal contamination on a global scale (Ernst, 1998). However, the resultant environmental degradation and ongoing pollution were viewed as secondary in importance to resource extraction and profit maximisation (Lottermoser, 2010). Progress was deemed most important and the cost to largely poorer peoples and the environment was ignored both in knowledge and ignorance.

Early mitigation efforts were largely ineffective and mining continued to operate with little consideration for environmental and human health until the twentieth century. And even in the twenty-first century, global practice is inconsistent and arguably open to corruption or convenient interpretation. Global corporations and mid-range firms in the extractive industries proclaim their commitment to 'corporate environmental and social responsibility' but their actual implementation, particularly in developing countries or in contexts with a limited legislative framework, is questionable. Slack (2012) cites environmental crises in Guatemala, Edwards et al. (2014) considers risks and opportunities for mining in Africa while Dashwood (2014) catalogues the patchwork of differential global practices and applauds the move of major corporations to embrace responsibility in a context of limited legislation or in the absence of reasonable regulatory tools. Clearly some corporations behave well, but without transparency and a solid legislative context this behaviour is optional. Given the geographic dimension of the issue, is this approach tenable? With increasing human populations wanting greater development and access to the artefacts of modern living, we must remain cognisant of the social and economic consequences of greater contaminant exposure.

BROWNFIELDS AS SITES OF CONTAMINANT EXPOSURE FOR HUMAN POPULATIONS

As we move into brownfields, assessment of exposure is essential for reasons encompassing risk mitigation, occupational health and safety, setting remediation goals, determining appropriate development options, costing and accountability. Human contaminant exposure pathways involve a myriad of intermediary stages and may extend well beyond the boundaries of the local population. However many definitions are limited in scope, for example, pathway of exposure is defined as the link between environmental releases and local populations that might come into contact with, or be exposed to, environmental contaminants (ATSDR, 2005). Brownfield sites need a broader definition as their contaminants may have been present and poorly contained for decades or even centuries. The temporal dimension is significant. That said, the identification and assessment of contact or 'exposure' requires a systematic approach to determine if an exposure risk and hence a public health hazard exists. Such a scheme is exemplified in the *Public Health Assessment Guidance Manual* (2005) produced by the Agency for Toxic Substances and Disease Registry (ATSDR, 2005) and although the emphasis and significance varies between jurisdictions, the five elements of exposure pathway remain relevant:

- *Source of contamination*: May include, but are not limited to, leaking drums, waste dumps, buried wastes, emission stacks (chimneys), tailings dams, etc.
- *Release mechanism and transport*: Contaminants may be transported away from primary source, through and across different environmental media including soils, water, air and biota. This process may be also change the nature of the contaminants.
- *Exposure point*: Examples include drinking water, contaminated food stuffs and dust.
- *Exposure route*: Inclusive of ingestion, inhalation, etc.
- *Potentially exposed population*: Specific to the subgroup at risk, for example, residents or the elderly.

Once the exposure elements have been identified, a conceptual site model (CSM) for human exposure can be developed and should form an integral part of a successful risk assessment procedure (ATSDR, 2005). The CSM provides a framework for the characterisation of contaminated sites and various formats of the model have been used by different regulatory agencies (NEPM, 2011; USEPA, 2011). It is also a representation of site-related information regarding potential contamination sources, receptors and exposure pathways between those sources and receptors. Furthermore, the CSM can be used as a management tool to help visualise and understand the characteristics of the site and the chemicals involved prior to implementing a recovery strategy (USEPA, 2011). A note of caution is raised by some statutory authorities, when a generic CSM is used instead of a site-specific model. The diverse nature of brownfields means that the use of a single template is a dangerous oversimplification.

Using an abandoned mine site as an example, Figure 17.2 shows how the five exposure elements may be integrated into the CSM.

PATHWAY EXAMPLES WITH A FOCUS ON MINING

Given historically poor legislative controls on mining, and the subsequent mobility and availability of potentially toxic elements, it is prudent to consider a few examples, to illustrate the impacts of past poor practice.

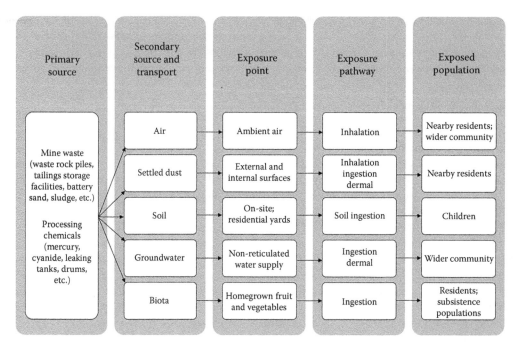

FIGURE 17.2 Example of a CSM for exposure to potentially toxic materials at an abandoned mine site. (Adapted from Australian and United States sources.)

EXAMPLE 17.3: CHILDREN'S PLAY AREAS

Reclaimed land is not a synonym for a remediated site and brownfield sites that have been converted into recreational spaces may pose an ongoing risk to the community. This is primarily due to poor definition of the contamination and inadequate containment or remediation of the primary and secondary contaminant source(s) on the site. Open spaces designated for children's play, inclusive of a child's backyard or neighbourhood, present a risk from past poor practice and potentially ongoing extractive activities.

A recent study (Taylor et al., 2014) conducted in Australia's oldest lead–zinc mining city of Broken Hill, highlighted the extent to which children may come into contact with arsenic, cadmium and lead contaminated dust from playground equipment. The innovative methods used in the study were designed to directly quantify the metal exposure hazard available to children while using playground facilities. The authors measured the metal content on hand wipes that had been wiped over the surfaces of play equipment at several playgrounds. In addition, using a 'simulated play' behaviour (Taylor et al., 2013), one of the researchers 'played' on the play equipment in a manner that mimicked the exploratory behaviour of a child. The researchers' hands were cleaned using a dust wipe before and after play to compare metal loadings on hands pre- and post-play at four playgrounds at varying distances from the smelter. A statistically significant increase in lead loadings on the post-play wipes was recorded, with a maximum mean increase of 72 times the pre-play loadings recorded at the playground closest to the smelter. It was concluded that the contamination on the playground equipment was sourced from remobilised dust from historical metal deposition as well as emissions from current mining operations. In addition to active mining operations, this study illustrates the contribution of historical contaminants on contemporary exposure risks. The Australian National Health and Medical Research Council (NHMRC) sets the childhood blood lead intervention level at 5 μg/dL and it is estimated that approximately 50% of children under 5 years old in Broken Hill have a blood lead level in excess of the intervention value.

EXAMPLE 17.4: TRACE METALS IN HOUSEHOLD DUST

Brownfield sites that host unremediated and exposed mine wastes represent permanent primary sources of contamination, particularly in arid and windy localities where trace metals become mobilised and transported through the air. Previous research has shown that abandoned mine wastes in particular are important sources of dust and are often characterised by higher trace metal concentrations than the bulk mine waste material (Kim et al., 2011; Martin et al., 2014). Our own research shows that historic mine tailings may contain up to 45% dust (i.e. particles with an aerodynamic diameter of 100 μm), and thus have the potential to be highly mobile and have multiple pathways into the human body.

When dust particles become airborne and settle in indoor environments, humans may come into contact with elevated levels of trace metals via inhalation or direct ingestion of dust particles. Once again, this is of great concern because of the unique risk factors associated with children.

To determine whether indoor dust is a significant source of arsenic exposure by residents living in an historic gold mining community, a correlation analysis was conducted on the arsenic content in settled household dust and the residents' hair and toenails (Hinwood et al., 2003). The study reported a statistically significant correlation between the arsenic content in household dust, hair (correlation coefficient, 0.82) and toenails (0.54). This study shows that residents in areas close to mining activity may be exposed to arsenic and that household dust contributes to this exposure.

EXAMPLE 17.5: URBAN VEGETABLE GARDENS

Social equity and environmental justice are two of the driving forces behind urban garden initiatives. In older industrialised cities and near peri-urban developments, urban agriculture has become a popular option for the transformation of brownfields as well as for social re-engagement within the community. These transformations are motivated by: (i) an increasing interest in affordable and locally produced foods (Hettiarachchi et al., 2010), (ii) the need of the community to restore underutilised and damaged land (Ferris et al., 2001) and (iii) a desire to provide nutrition resources regardless of socio-economic status (Campbell, 2012).

However, growing foodstuffs in unremediated or poorly remediated brownfields (either at the home or in a community garden setting) can increase the likelihood of human exposure to contaminants directly (i.e. through soil ingestion) and indirectly (i.e. via the food chain). Common urban soil garden contaminants include lead, arsenic, cadmium, zinc and polycyclic aromatic hydrocarbons (PAHs).

The most common human vectors are: (i) consumption of contaminants that have deposited from the atmosphere onto plant surfaces (Sharma et al., 2008) and (ii) consumption of contaminants that have been transferred directly from the soil into the plant (Cui et al., 2004). Other potential human vectors to urban garden contaminants include dust inhalation and direct ingestion of soil, particularly for vulnerable population subgroups such as infants (Hough et al., 2004).

Several studies have investigated the accumulation of heavy metals in vegetables grown in contaminated soils. Asteraceae vegetables (e.g. artichokes, lettuce, endive, etc.) pose one of the greatest risks to human health due to their enhanced metal uptake compared to other vegetable families (Liu et al., 2013; Ramirez-Andreotta et al., 2013). Furthermore, leafy vegetables typically have higher metal concentrations than non-leafy vegetables (Liu et al., 2013). In one greenhouse experiment, lettuce leaves and radish roots grown in mine wastes accumulated significantly more metals than beans and tomatoes (Cobb et al., 2000). Silverbeet in particular has been shown to have increased arsenic values when grown in the mining affected soils of the gold fields of Victoria, Australia (Harvey et al., 2003).

The social and health benefits of play, of engaging in outside activities linked to green space and of growing and eating fresh produce and the like are not questioned and in a space-poor world, we strive to provide more of these opportunities for an exploding human population. It is imperative that we balance the risk to the opportunity and ensure that the social good stays clearly in the sights of land use planners.

FACTORS AFFECTING RISK

AT-RISK GROUPS

Specific groups within our societies experience elevated risk for a variety of reasons including exposure intensity and duration, genetic predisposition, age, financial status, education levels and prejudice. For example, in a typical community, children, the elderly and immune compromised individuals all have heightened risk profiles. When considering the development of brownfield sites, those at risk deserve special consideration.

The age at which environmental exposure commences or occurs, including exposures prior to conception and before birth, may impact on the exposure pathway, absorption, metabolism and excretion of xenobiotic chemicals, and hence the nature and severity of their effects (Bearer, 1995). Children are particularly vulnerable to environmental toxins due to their low breathing zone, increased surface-to-volume ratio, greater intake of air, food and water compared with adults, as well as their immature metabolic pathways and rapid growth and development (Bearer, 1995; Landrigan et al., 1999). Children also have unique play and activity patterns that increase their potential for exposure to contaminants (e.g. repetitive hand-to-mouth and object-to-mouth behaviour). This leads to greater exposure through direct contact with soil and dust.

Older adults are also vulnerable to toxins as they have a reduced capacity to compensate for impairment caused by toxins and their exposure time to contamination is considerably longer (potentially their entire lifetime) and hence their body may already have elevated contaminant levels. The increased risk to older adults has received limited attention, and there is scant research covering the adverse effects of long-term, low-level exposure to environmental toxins on a physiologically stressed or impaired elderly population (Risher et al., 2010).

The recognition that exposures to toxins could produce DNA mutations represented a major landmark for risk assessment and public health management (Wogan, 1992) and the differential response to environmental toxins by individuals with specific genetic characteristics is still being explored (Baccarelli and Bollati, 2009). Links between environmental toxins and the disruption of methylation cycles, disruption to metabolism and/or to DNA repair suggest genetic risk factors which, at the current time, are poorly understood.

GEOGRAPHICAL DISADVANTAGE

Simply put, distance matters: proximity to primary or secondary toxin sources increases the risk of adverse health outcomes (Zota et al., 2011). For example, Zusman et al. (2012) examined cancer incidence rates linked to residential proximity at the Kiryat Haim industrial zone in Northern Israel and demonstrated statistically significant correlations for both lung and non-Hodgkin lymphoma patients with distancing from the industrial zone.

Geographical disadvantage has been quantified in a novel study by Currie et al. (2013). They reviewed the housing market response to the openings and closings of 1600 industrial plants that emitted toxic pollutants. Amongst other findings, they determined that housing values within 1.6 km decrease by 1.5% when plants open, and increase by 1.5% when plants close. Further, while the housing value impact was most pronounced within 800 m from plants, a statistically significant infant health impact extended up to 1.6 km away (Currie et al., 2013).

Evidence of a disease cluster is difficult to obtain unless symptoms are acute and affect numerous individuals. If the exposed community is remote or sparsely populated, prone to relocation and resettlement, a minority group of lower socio-economic status, is poorly educated, disempowered, or if disease onset has a long latency period, then the trigger to seek evidence of a causal association between environmental contaminants and disease outcome may be delayed or impossible. This is complexed further by the fact that brownfield sites are often 'hidden' by a lost history or masked by subsequent development, and suggests that sophisticated mapping tools with high-quality historical data are required for appropriate site assessment and risk evaluation.

But it is not impossible: GIS (geographic information systems) and spatial geostatistical techniques can be applied to incorporate patient level data georeferenced to residential address or locality in order to investigate disease clustering or high-risk areas. Aggregate population-level studies have been conducted to explore transmission patterns of infectious diseases such as influenza (Hu et al., 2012; Gog et al., 2014), map lung cancer mortality rates (Goovaerts, 2010) (both relevant because they may reflect an impacted immune system) and investigate rates of cancer incidence in association with soil arsenic levels (Pearce et al., 2012). The United States has developed a rapid inquiry facility to facilitate investigations of links between the environment and health outcomes (Beale et al., 2010).

Climatic and seasonal effects may add new, or exacerbate existing, environmental interfaces. For example, wind borne dispersion of airborne particulates originating from contaminated sites may increase, thus increasing the risk of exposure through inhalation or ingestion (Meza-Figueroa et al., 2009). Similarly, increased (or decreased) rainfall may cause changes in environmental interfaces. Climatic fluctuations and increases in extreme weather events may also contribute.

DIAGNOSTIC DIFFICULTIES

Even when the risk of a substance is known, identification of an issue at a brownfield site may prove difficult as a result of complexity, ignorance or errors of omission. For example, the risks associated with exposure to asbestos, arsenic, coal ash, radioactive waste or metals such as lead have been well documented (Hinwood et al., 1999; Arnold, 2014; Huang et al., 2014; Liu and Lewis, 2014; Pearce et al., 2012; Zota et al., 2011; Zierold and Sears, 2015) yet investigative journalists continue to uncover examples of exposure to workers and the surrounding environment as we repurpose and redevelop sites.

Diagnosis and attribution is often complex and conflicted. For example, long latency periods between exposure and disease outcomes, differences in exposure levels and susceptibilities due to genetics and nutritional status add to the complexity of determining the effects of arsenic (Naujokas et al., 2013).

In previous centuries, asbestos was commonly used as insulation and now many abandoned buildings, some already razed, are inundated with this contaminant (Lin et al., 2015). Elevated lead levels associated with construction, painting and historical petroleum products remain trapped in the soil profile to be rediscovered during the redevelopment phase. The fact that they have remained as a contaminant repository for many years prior to discovery has almost certainly already had undiagnosed health impacts.

The fact that there are no globally accepted and benchmarked assessment regimes for brownfield site developments means that unrehabilitated sites continue to release contaminants and that redevelopment occurs in circumstances that are less than well understood.

ATTITUDES AND PERCEPTIONS OF RISK

When confronted with news that a site is contaminated, humans react. Media reports are often emotive and public concerns about contaminated sites are heightened by a lack of confidence in environmental policy, regulations and the officials that enforce them. Public perceptions of risk combined with a fear of negative health outcomes are influencing the development of public policy regarding the environment and remediation which many see as positive. However, there is a perception by policymakers and remediators that the public are irrational (Wandersman and Hallman, 1993), yet there are excellent reasons for this 'irrationality'.

The 'stigma' associated with remediated contaminated land is tangibly demonstrated in the assessment of the negative impact on property value (Bond, 2001). Negative intangible factors such as the inability to completely remediate the site, the unseen remnants of the previous use, the risk of changes in legislation or remediation standards, or fear linked to limited understanding (Bond,

2001) play out in an environment where statistical likelihoods of effect are poorly understood and those with the most limited ability to act are often the most affected, specifically low socio-economic groups, and those with poor education.

Anthropogenic disasters such as contaminated sites caused by profit-driven motives are the subject of psychoanalytical literature aimed at aiding people to cope with this style of catastrophe (Guglielmucci et al., 2015). Real and measurable negative health impacts inclusive of mental health are recorded when we do not appropriately remediate these brownfield sites, and adequately communicate the nature and effectiveness of the remediation. Increased attention is now placed on participation by communities and individuals affected by brownfield sites in all facets of decision making (Ramirez-Andreotta et al., 2014).

NEW TECHNOLOGIES AND INDUSTRIES

It is challenging for regulation to keep up with new technologies and industries. The literature is full of examples of the development and use of a new product or technology which is later shown to have deleterious impacts on human and environmental health. Breakthrough technologies and products such as asbestos, dioxins (largely the by-products of industrial processes), the outflow from smelter stacks such as fluoride from aluminium production and mercury used in gold recovery all showed great benefit and promise when introduced. The precautionary principle inherent in much environmental legislation struggles to keep up with ever evolving and expanding databases of new compounds. The balance between development and safety is not easy to regulate and the setting of the balance point is problematic.

EXAMPLE 17.6: UNCONVENTIONAL NATURAL GAS

Unconventional natural gas development (UNGD), which includes coal seam gas (CSG), is of concern because of the potential to contaminate large tracts of land with hydrocarbon products and contaminants. While purporting to alleviate energy shortfalls and contribute to development, the long-term impacts to human and environmental health are largely untested (Werner et al., 2015). Monitoring of co-produced CSG water, which may contain residual fracking fluid constituents, has detected hazardous substances with known adverse health consequences, such as benzene, previously linked with leukaemia, and given rise to fears of contamination of surface and ground waters (Navi et al., 2015). Adverse psychological impacts have also been attributed to perceived invasive aspects of mining in agricultural areas of Australia (Hossain et al., 2013).

The underlying questions that must be asked in all such cases are: Will this production result in the development of large-scale brownfield sites reacquiring remediation? What level of assurity offsets the economic gain and how do we factor in the very expensive business of remediation noting that remediation technology may have not yet been invented?

THE RISKS OF LOW-LEVEL AND LONG-TERM EXPOSURE

Acute exposures to toxic chemicals, whatever the source or exposure pathway, may trigger the onset of symptoms that are quickly detected and therefore diagnosed and treated in a timely manner. Chronic, or low-level long-term exposures, may cause adverse health effects that are less easily identifiable. Symptoms may have a long latency period, be less specific, or appear as exacerbations of commonly experienced adverse health effects, particularly for multifactor exposures.

Symptoms of diseases associated with unanticipated environmental exposures may go unrecognised for long periods, particularly when exposure pathways are not obvious, where there are few diagnosed cases, or when the chain of notifications is unclear and regulatory procedures are absent or incomplete. Further, perceptions of environmental hazards may be misconstrued or difficult to 'prove', especially when exposed communities are disempowered and funding sources for epidemiological investigations limited.

Lack of evidence due to a deficit of monitoring or research must not, under any circumstances, be interpreted as a 'negative' association nor as the absence of a plausible causal pathway between an environmental exposure and adverse health outcomes.

BROWNFIELD SITES OF THE FUTURE

Where humans interact with the landscape, the landscape is impacted. As the human population soars, a greater landscape impact is predicted and the need to repurpose brownfield sites becomes more pressing. The mantra of reduce, reuse, recycle applies here and – even if all mining, process-ing, agricultural, urbanisation and manufacturing technologies could achieve no landscape affects – the legacy of the past and the need to make this land safe, remains.

In the simplest case, where there is a functioning and strong market, and where contamination is limited in scope and complexity and where land use remains unchanged, the private sector is most likely to initiate and implement remediation and reuse.

However, when a brownfield is redeveloped from industrial to residential use, and complexity exists (multiple ownerships, variety of contaminants and extended life of operation) greater public input and potentially public subsidies are required.

Redevelopment of a brownfield site by ether public or private capital can be considered an invest-ment, but the allocation of scarce remediation funds will still be controversial. The precautionary principle inherent in most environmental legislation requires that if a development has a risk of causing harm to the public or to the environment, the burden of proof that it is not harmful resides with the developers. Further, this implies that there is a social responsibility to protect the public from exposure to harm. In progressive legislature such as in the European Union, the precautionary principle has been made a statutory requirement and this together with the education of communi-ties serves to ensure accountability.

At a minimum, we need transparency in the development of brownfield sites and, equally, in any development that may result in the creation of further brownfield sites.

CONCLUSION

Historically, human activities including mining, manufacturing and processing have impacted on the landscape, creating the very brownfield sites that are increasingly required for safe and sustain-able development. Our needs for resources, both traditional and exotic, combined with a rapidly growing population means that not only will we be forced to redevelop increasing numbers of brownfield sites, but we will also be creating sites that will be the brownfield of the future.

The precautionary principle requires that we limit harm and, in the globally connected market in which corporations operate, the legislative context is a patchwork of differing rules and responsi-bilities. The move by some groups to integrate 'corporate environmental and social responsibility' into business practice irrespective of the country or province they are working in is clearly excellent but does not remove the need to embed good environmental and social practice at the local level.

Identification of contaminated sites and the risks they pose, strategies to promote risk awareness among potentially exposed communities and implementation of policies and practices that enable exposed communities to be proactive in risk reduction are needed. It is logical to 'reuse' land and make it productive rather than expand the human landscape footprint. This is consistent with us collectively reducing our impact. The key is a consistent legislative context that is equitable to both communities and corporations which share the cost of this past practice.

Without a deep understanding of each brownfield site, the risks to the human population and the environment in general will only be magnified. As urban environments encroach on poorly understood historical sites, and as industry itself encroaches on essential human environments, it becomes inevitable that human populations will come into contact with contaminants in the air, water, soil and food profiles.

It is how we approach this challenge that will determine whole-of-population health outcomes in the next century.

REFERENCES

Alker, S., V. Joy, P. Roberts and N. Smith. 2000. The definition of brownfield. *Journal of Environmental Planning and Management*, 43(1): 49–69.

Arnold, C. 2014. Once upon a mine: The legacy of uranium on the Navajo Nation. *Environmental Health Perspectives*, 122(2): A44.

ATSDR (Agency for Toxic Substances and Disease Registry). 2005. Public Health Assessment Guidance Manual, Chapter 6 – Exposure evaluation: Evaluating exposure pathways, http://www.atsdr.cdc.gov/hac/phamanual/ch6.html#6.1

Baccarelli, A. and V. Bollati. 2009. Epigenetics and environmental chemicals .*Current Opinion in Pediatrics*, 21(2): 243.

Beale, L., S. Hodgson, J. J. Abellan, S. LeFevre and L. Jarup. 2010. Evaluation of spatial relationships between health and the environment: The rapid inquiry facility. *Environmental Health Perspectives*, 118(9): 1306–1312.

Bearer, C. F. 1995. Environmental health hazards: How children are different from adults. *The Future of Children*, 5(2):11–26.

Bond, S. 2001. Conjoint analysis: Assessing buyer preferences for property attributes to assist with the estimation of land contamination stigma. In *The Conference of Pacific Rim Real Estate Society,* Adelaide, pp. 21–24.

CABERNET (Concerted Action on Brownfield and Economic Regeneration Network) 2006. Homepage information, www.cabernet.org.uk

Campbell D. 2012. High street outlets ignoring guidelines on providing calorie information. *The Guardian, London*. Available: http://www.guardian.co.uk/business/2012/mar/15/high-street-guidelines-calorie-information.

Chen, T. B., Y. M. Zheng, M. Lei, Z. C. Huang, H. T. Wu, H. Chen and Q. Z. Tian. 2005. Assessment of heavy metal pollution in surface soils of urban parks in Beijing, China. *Chemosphere*, 60(4): 542–551.

Cobb, G. P., K. Sands, M. Waters, B. G. Wixson and E. Dorward-King. 2000. Accumulation of heavy metals by vegetables grown in mine wastes. *Environmental Toxicology and Chemistry*, 19(3): 600–607.

Csavina, J., J. Field, M. P. Taylor, S. Gao, A. Landázuri, E. A. Betterton and A. E. Sáez. 2012. A review on the importance of metals and metalloids in atmospheric dust and aerosol from mining operations. *Science of the Total Environment*, 433: 58–73.

Cui, Y., Q. Wang, Y. Dong, H. Li and P. Christie. 2004. Enhanced uptake of soil Pb and Zn by Indian mustard and winter wheat following combined soil application of elemental sulphur and EDTA. *Plant and Soil*, 261(1–2): 181–188.

Currie, J., L. Davis, M. Greenstone and R. Walker. 2013. Do housing prices reflect environmental health risks? Evidence from more than 1600 toxic plant openings and closings, No. w18700. National Bureau of Economic Research.

Dashwood, H. S. 2014. Sustainable development and industry self-regulation developments in the global mining Sector. *Business & Society*, 53(4): 551–582.

DEFRA. 2011. Defra circular 01/2006 – contaminated land. UK Department for Environment, Food & Rural Affairs. https://www.gov.uk/government/publications/defra-circular-01-2006-contaminated-land

De Sousa, C. A. 2003. Turning brownfields into green space in the City of Toronto. *Landscape and Urban Planning*, 62(4): 181–198.

Edwards, D. P., S. Sloan, L. Weng, P. Dirks, J. Sayer and W. F. Laurance. 2014. Mining and the African environment. *Conservation Letters*, 7(3): 302–311.

Ernst, W. O. 1998. The origin and ecology of contaminated, stabilized and non-pristine soils. In *Metal-Contaminated Soils: in situ Inactivation and Phytorestoration*, Eds. J. Vangronsveld and S. D. Cunningham. New York: Springer, pp. 17–29.

Feng, X. and G. Qiu. 2008. Mercury pollution in Guizhou, Southwestern China–An overview. *Science of the Total Environment*, 400(1): 227–237.

Ferris, J., C. Norman and J. Sempik. 2001. People, land and sustainability: Community gardens and the social dimension of sustainable development. *Social Policy & Administration*, 35(5): 559–568.

Gog, J. R., S. Ballesteros, C. Viboud, L. Simonsen, O. N. Bjornstad, J. Shaman and B. T. Grenfell. 2014. Spatial transmission of 2009 pandemic influenza in the US. *PLoS Comput Biol*, 10(6): e1003635.

Goovaerts, P. 2010. Combining areal and point data in geostatistical interpolation: Applications to soil science and medical geography. *Mathematical Geosciences*, 42(5): 535–554.

Guglielmucci, F., I. G. Franzoi, M. Zuffranieri and A. Granieri. 2015. Living in contaminated sites: Which cost for psychic health? *Mediterranean Journal of Social Sciences*, 6(4): 207.

Harvey, G., K. Dowling, H. Waldron and D. Garnett, 2003. Trace element content of vegetables grown in the Victorian goldfields: Characterization of a potential backyard hazard. *Proceedings of the ANA 2003 Fifth Conference on Nuclear Science and Engineering*, Australia, pp.168–172.

Hettiarachchi, A. G. M., B. S. Martin, A. A. Rae, A. P. Defoe, A. D. Presely and A. G. M. Pierzysnki. 2010. Gardening on Brownfields Sites: Evaluating trace element transfer from soil to plants and their transformations in soils. In *19th World Congress of Soil Science, Soil Solutions for a Changing World*, 1–6 August, Brisbane, Australia. Published on DVD.

Hinwood, A. L., D. J. Jolley and M. R. Sim. 1999. Cancer incidence and high environmental arsenic concentrations in rural populations: Results of an ecological study. *International Journal of Environmental Health Research*, 9(2): 131–141.

Hinwood, A. L., M. R. Sim, D. Jolley, N. de Klerk, E. B. Bastone, J. Gerostamoulos and O. H. Drummer. 2003. Hair and toenail arsenic concentrations of residents living in areas with high environmental arsenic concentrations. *Environmental Health Perspectives*, 111(2): 187.

Hossain, D., D. Gorman, B. Chapelle, W. Mann, R. Saal and G. Penton. 2013. Impact of the mining industry on the mental health of landholders and rural communities in southwest Queensland. *Australasian Psychiatry*, 21(1): 32–37.

Hough, R. L., N. Breward, S. D. Young, N. M. Crout, A. M. Tye, A. M. Moir and I. Thornton. 2004. Assessing potential risk of heavy metal exposure from consumption of home-produced vegetables by urban populations. *Environmental Health Perspectives*, 112(2): 215.

Hu, W., G. Williams, H. Phung, F. Birrell, S. Tong, K. Mengersen and A. Clements. 2012. Did socio-ecological factors drive the spatiotemporal patterns of pandemic influenza A (H1N1). *Environment International*, 45: 39–43.

Huang, L., H. Wu and T. J. van der Kuijp. 2014. The health effects of exposure to arsenic-contaminated drinking water: A review by global geographical distribution. *International Journal of Environmental Health Research*, (ahead-of-print), 1–21.

Karakosta, C., H. Doukas and J. Psarras. 2009. Directing clean development mechanism towards developing countries' sustainable development priorities. *Energy for Sustainable Development*, 13(2): 77–84.

Kim, C. S., K. M. Wilson and J. J. Rytuba. 2011. Particle-size dependence on metal (loid) distributions in mine wastes: Implications for water contamination and human exposure. *Applied Geochemistry*, 26(4): 484–495.

Landrigan, P. J., W. A. Suk and R. W. Amler. 1999. Chemical wastes, children's health, and the superfund basic research program. *Environmental Health Perspectives*, 107(6): 423.

Lange, D. A. and S. McNeil. 2004. Brownfield development: Tools for stewardship. *Journal of Urban Planning and Development*, 130(2): 109–116.

Lin, C. K., Y. Y. Chang, J. D. Wang. and L. J. H. Lee. 2015. Increased standardised incidence ratio of malignant pleural mesothelioma in Taiwanese asbestos workers: A 29-year retrospective cohort study. *BioMed Research International*, Article ID 678598, 10 pages. doi:10.1155/2015/678598.

Liu, X., Q. Song, Y. Tang, W. Li, J. Xu, J. Wu and P. C. Brookes. 2013. Human health risk assessment of heavy metals in soil–vegetable system: A multi-medium analysis. *Science of the Total Environment*, 463: 530–540.

Liu, J. and G. Lewis. 2014. Environmental toxicity and poor cognitive outcomes in children and adults. *Journal of Environmental Health*, 76(6): 130.

Lottermoser, B. G. 2010. *Tailings*. Berlin: Springer, pp. 205–241.

Mackay, A. K., M. P. Taylor, N. C. Munksgaard, K. A. Hudson-Edwards and L. Burn-Nunes. 2013. Identification of environmental lead sources and pathways in a mining and smelting town: Mount Isa, Australia. *Environmental Pollution*, 180: 304–311.

Makra, L. and P. Brimblecombe. 2004. Selections from the history of environmental pollution, with special attention to air pollution. Part 1. *International Journal of Environment and Pollution*, 22(6): 641–656.

Martin, R., K. Dowling, D. Pearce, J. Bennett and A. Stopic. 2013. Ongoing soil arsenic exposure of children living in an historical gold mining area in regional Victoria, Australia: Identifying risk factors associated with uptake. *Journal of Asian Earth Sciences*, 77: 256–261.

Martin, R., K. Dowling, D. Pearce, J. Sillitoe and S. Florentine. 2014. Health effects associated with inhalation of airborne arsenic arising from mining operations. *Geosciences*, 4(3): 128–175.

Méndez-Fernández, L., P. Rodríguez and M. Martínez-Madrid. 2015. Sediment toxicity and bioaccumulation assessment in abandoned copper and mercury mining areas of the Nalón River basin (Spain). *Archives of Environmental Contamination and Toxicology*, 68(1): 107–123.

Meza-Figueroa, D., R. M. Maier, A. de la O-Villanueva, A. Gómez-Alvarez, A. Moreno-Zazueta, J. Rivera and J. Palafox-Reyes. 2009. The impact of unconfined mine tailings in residential areas from a mining town in a semi-arid environment: Nacozari, Sonora, Mexico. *Chemosphere*, 77(1): 140–147.

Naujokas, M. F., B. Anderson, H. Ahsan, H. V. Aposhian, J. Graziano, C. Thompson and W. A. Suk. 2013. The broad scope of health effects from chronic arsenic exposure: Update on a worldwide public health problem. *Environmental Health Perspectives*, 121(3): 295–302.

Navi, M., C. Skelly, M. Taulis and S. Nasiri. 2015. Coal seam gas water: Potential hazards and exposure pathways in Queensland. *International Journal of Environmental Health Research*, 25(2): 162–183.

NEPM. 2011. National Environment Protection (Assessment of Site Contamination) Measure. Schedule B2 Guideline on Site Characterisation, National Environment Protection Council (NEPC), Canberra.

Nijkamp, P., M. Van Der Burch and G. Vindigni. 2002. A comparative institutional evaluation of public–private partnerships in Dutch urban land-use and revitalisation projects. *Urban Studies*, 39(10): 1865–1880.

Norman, W. and C. MacDonald. 2004. Getting to the bottom of 'triple bottom line'. *Business Ethics Quarterly*, 14(02): 243–262.

Pearce, D. C., K. Dowling, A. R. Gerson, M. R. Sim, S. R. Sutton, M. Newville and G. McOrist. 2010. Arsenic microdistribution and speciation in toenail clippings of children living in a historic gold mining area. *Science of the Total Environment*, 408(12): 2590–2599.

Pearce, D. C., K. Dowling and M. R. Sim. 2012. Cancer incidence and soil arsenic exposure in a historical gold mining area in Victoria, Australia: A geospatial analysis. *Journal of Exposure Science and Environmental Epidemiology*, 22(3): 248–257.

Public Health Assessment Guidance Manual. 2005. Agency for Toxic Substances & Disease Registry. http://www.atsdr.cdc.gov/hac/PHAManual/toc.html

Ramirez-Andreotta, M. D., M. L. Brusseau, J. F. Artiola and R. M. Maier. 2013. A greenhouse and field-based study to determine the accumulation of arsenic in common homegrown vegetables grown in mining-affected soils. *Science of the Total Environment*, 443: 299–306.

Ramirez-Andreotta, M. D., M. L. Brusseau, J. F. Artiola, R. M. Maier and A. J. Gandolfi. 2014. Environmental research translation: Enhancing interactions with communities at contaminated sites. *Science of the Total Environment*, 497: 651–664.

Risher, J. F., G. D. Todd, D. Meyer and C. L. Zunker. 2010. The elderly as a sensitive population in environmental exposures: Making the case. In *Reviews of Environmental Contamination and Toxicology*, Volume 207. New York: Springer, pp. 95–157.

Sharma, R. K., M. Agrawal and F. M. Marshall. 2008. Heavy metal (Cu, Zn, Cd and Pb) contamination of vegetables in urban India: A case study in Varanasi. *Environmental Pollution*, 154(2): 254–263.

Slack, K. 2012. Mission impossible? Adopting a CSR-based business model for extractive industries in developing countries. *Resources Policy*, 37(2): 179–184.

Sultan, K. and K. Dowling. 2006. Seasonal changes in arsenic concentrations and hydrogeochemistry of Canadian Creek, Ballarat (Victoria, Australia). *Water, Air, and Soil Pollution*, 169(1–4): 355–374.

Taylor, M. P., D. Camenzuli, L. J. Kristensen, M. Forbes and S. Zahran. 2013. Environmental lead exposure risks associated with children's outdoor playgrounds. *Environmental Pollution*, 178: 447–454.

Taylor, M. P., S. A. Mould, L. J. Kristensen and M. Rouillon. 2014. Environmental arsenic, cadmium and lead dust emissions from metal mine operations: Implications for environmental management, monitoring and human health. *Environmental Research*, 135: 296–303.

Taylor, M. P., S. Zahran, L. Kristensen and M. Rouillon. 2015. Evaluating the efficacy of playground washing to reduce environmental metal exposures. *Environmental Pollution*, 202: 112–119.

USEPA. 2004. Updated report on the incidence and severity of sediment contamination in surface waters of the United States, National Sediment Quality Survey. U.S. Environmental Protection Agency, Office of Water, Washington, DC, EPA-823-R-04–007, November.

USEPA, I. 2011. Integrated risk information system (IRIS). Arsenic, inorganic. CASRN, 7440-38. Available: http://www.epa.gov/iris/ (cited 30 March 2016).

Wandersman, A. H. and W. K. Hallman. 1993. Are people acting irrationally? Understanding public concerns about environmental threats. *American Psychologist*, 48(6): 681.

Werner, A. K., S. Vink, K. Watt and P. Jagals. 2015. Environmental health impacts of unconventional natural gas development: A review of the current strength of evidence. *Science of the Total Environment*, 505: 1127–1141.

Wogan, G. N. 1992. Aflatoxins as risk factors for hepatocellular carcinoma in humans. *Cancer Research*, 52(7 Supplement), 2114s–2118s.

Zierold, K. M. and C. G. Sears. 2015. Are healthcare providers asking about environmental exposures? A community-based mixed methods study. *Journal of Environmental and Public Health*. Article ID 189526, 7 pages. http://dx.doi.org/10.1155/2015/189526.

Zota, A. R., L. A. Schaider, A. S. Ettinger, R. O. Wright, J. P. Shine and J. D. Spengler. 2011. Metal sources and exposures in the homes of young children living near a mining-impacted Superfund site. *Journal of Exposure Science and Environmental Epidemiology*, 21(5): 495–505.

Zusman, M., J. Dubnov, M. Barchana and B. A. Portnov. 2012. Residential proximity to petroleum storage tanks and associated cancer risks: Double Kernel density approach vs. zonal estimates. *Science of the Total Environment*, 441: 265–276.

18 Sustainability of Minerals Processing and Metal Production for European Economies in Transition

Vladimir Strezov, Natasa Markovska, and Meri Karanfilovska

CONTENTS

ABSTRACT

Mineral processing and metal production are some of the most important industrial activities for economic development. The countries in economic transition have introduced various incentives to promote mineral processing and metal production. These countries have often lower levels of environmental regulations enabling the mineral processing and metal production industries to operate with small investments in environmental control. This chapter reviews the current state of economic input and environmental performance for a selected range of countries in economic transition to determine the role of industrial activities in their economic and sustainable development.

AN OVERVIEW OF THE HISTORY OF SUSTAINABLE DEVELOPMENT

The economic development through industrialisation and the environmental depletion of natural resources, air, water and soil quality and their impacts on biodiversity have influenced the process of awareness and played an important role for sustainable development. While the definition of sustainable development was first proposed by the Brundtland Commission in 1987, this concept has a history dating further back. Table 18.1 displays the historical events leading to and progressing from the Brundtland Report.

The international awareness of human impact on regional environmental quality and global natural resources reached a momentum of concern in the late 1960s. The Scandinavian countries had an important leading role for raising international awareness, partly due to the severe impact from the industrialised Europe on regional acid deposition and deterioration of the quality of their lakes. This situation exacerbated with the oil crisis and fears of depletion of the global oil reserves, prompting the importance for international action to address the impact of industrial and economic

TABLE 18.1

Major Historical Events and Publication Outputs Leading to Future Shaping of the Sustainable Development Process

Year	Event/Organisation	Publication/Output
1972	UN Conference on the Human Environment, Stockholm, Sweden	Stockholm Declaration United Nations Environment Program
1972	The Club of Rome	The Limits to Growth
1980	International Union for Conservation of Nature and Natural Resources	World Conservation Strategy, Living Resource Conservation for Sustainable Development
1983–1987	World Commission on Environment and Development	Our Common Future
1992	United Nations Conference on Environment and Development	Rio Declaration on Environment and Development
	Rio Summit (Earth Summit), Rio de Janeiro, Brazil	Agenda 21
1997	Rio + 5, New York, USA	General Assembly Resolution (S-19/2)
2002	WSSD (Rio + 10), Johannesburg, South Africa	Johannesburg Declaration Plan of Implementation of the WSSD
2012	UN Conference on Sustainable Development (Rio + 20), Rio de Janeiro, Brazil	The Future We Want

development on social welfare and environmental quality. On the basis of a proposal from Sweden in the late 1960s, the United Nations held the Conference on the Human Environment, resulting in a Stockholm Declaration and establishment of the United Nations Environment Program (UNEP). The Stockholm Declaration contained 26 principles concerning environment and development, such as 'non-renewable resources must be shared and not exhausted' and 'pollution must not exceed the environment's capacity to clean itself'. The declaration further stated that 'man is both creature and moulder of his environment, which gives him physical sustenance and affords him the opportunity for intellectual, moral, social and spiritual growth'.

In the same year, the independent think tank The Club of Rome commissioned the report 'The limits of Growth' (Meadows et al., 1972). This report presented a system analysis model of the relationship between global development and its impact on resource depletion. The model indicated continuous growth in population and economy until 2030 after which a turning point was predicted due to the depletion of minerals, quality of soil and water. This report is the first to use the term 'sustainable' (Grober, 2007, p. 6) in relation to economic development and resource depletion: 'We are searching for a model output that represents a world system that is: 1. sustainable without sudden and uncontrolled collapse; and 2. capable of satisfying the basic material requirements of all of its people'.

In 1980, the International Union for Conservation of Nature and Natural Resources (IUCN, 1980) published the report 'World Conservation Strategy, Living Resource Conservation for Sustainable Development', which was commissioned with financial support and advice from the UNEP and the World Wildlife Fund. The report, for the first time, applies the term 'sustainable development' and further states that 'the object of conservation is to ensure Earth's capacity to sustain development and to support all life'. One of the major concluding remarks of this report was that 'human beings, in their quest for economic development and enjoyment of the riches of nature, must come to terms with reality of resource limitation and the carrying capacities of ecosystems, and must take account of the needs of future generations'.

In late 1983, the United Nations appointed the former Norwegian Prime Minister to lead the World Commission on Environment and Development, resulting with the report 'Our Common Future'. This report defined sustainable development as a development that 'meets the needs of the present without compromising the ability of future generations to meet their own needs' (Brundtland, 1987, p. 16), and initiated the new process, principles and direction for human development.

In 1992, the United Nations Conference on Environment and Development, known as the Rio Summit or Earth Summit, resulted in a Rio Declaration on Environment and Development and the action plan Agenda 21. The assembled leaders further signed the Convention on Climate Change, and with that human interaction with climate change became an integral part of the principles for sustainable development. The commitment to sustainable development was reviewed in the Rio + 5 summit by the Special Session of the General Assembly to Review and Appraise the Implementation of Agenda 21.

The World Summit on Sustainable Development (WSSD) or Earth Summit 2002 (UN, 2002) resulted with the Johannesburg Declaration and the Plan of Implementation of the WSSD, which reaffirmed the international commitment to sustainable development.

Most recently, the Rio + 20 summit in 2012, the UN Conference on Sustainable Development, intended to secure renewed political commitment for sustainable development. The conference focused on two themes (i) green economy in the context of sustainable development and poverty eradication and (ii) the institutional framework for sustainable development. A publication 'The Future We Want' (UN, 2012) resulted from this conference.

MINERAL PROCESSING, METAL PRODUCTION AND SUSTAINABLE DEVELOPMENT

The role of mineral processing and metal production industries to sustainable development has been one of the integral subjects of assessment in all relevant UN publications and resolutions. Mineral processing industries have a significant impact on sustainable development, as they are some of the primary industries that not only contribute to the economic well-being, employment, but also to some of the most damaging environmental implications and resource depletion. The Brundtland Report (1987) states that although use of minerals depletes these resources for future generations, they still can be used after taking into consideration 'the criticality of that resource, the availability of technologies for minimising depletion, and the likelihood of substitutes being available'. Recycling as an option is also considered as an integral part of the planning process. Agenda 21 (UN, 1992) proposes action steps to promote sustainable development through industrial development. Protection of the atmosphere and waste minimisation are the most important action steps, which can be achieved through a more efficient use of resources and materials by the industry, installing or improving pollution abatement technologies and replacing the ozone-depleting substances.

Minerals have no intrinsic value in the form of ore and when buried in the earth (Petrie, 2007). Their value is added through mineral processing and production of metals. Management of the rate of conversion of these minerals into economic capital and the way this economic capital is then invested is the primary driver for sustainability of this industry. Giurco and Cooper (2012) point out on the importance for defining the level at which it is acceptable for the natural capital to be transformed into an economic capital. This is primarily because the natural capital is exhaustible through mining; hence depletion of mineral resources is one of the most significant factors for industrial sustainability of the mineral processing industries. According to Hilson and Murck (2000), examples of sustainable mineral processing include extending the longevity of the mineral reserves through conservation and increased recycling of the minerals and metals.

The minerals industry is often divided into four sub-sectors (Azapagic, 2004): (i) energy minerals (coal, oil and gas); (ii) metallic minerals (iron, copper, zinc, etc.); (iii) construction minerals (sand, gravel and gypsum) and (iv) industrial minerals (borates, calcium carbonates, kaolin, plastic clays and talc). It is only in the case of energy minerals that these resources are non-renewable and depletable, while the other minerals can only dissipate in use, but can be, theoretically, infinitely recovered or reused. According to Auty and Warhurst (1993) and Auty and Mikesell (1998), sustainability of mineral processing includes conserving the mineral assets through stagnation of mineral development and investing part of the mineral revenue to compensate for the loss of output from the depletion of the resource. The earlier model prepared by Meadows et al. (1972) predicted the most vulnerable minerals with the highest depletion rates and identified platinum, gold, zinc and lead as minerals with insufficient quantities to meet the demands by 2030.

Governance plays an important role for managing industrial sustainability. The difference in the governance of the developed countries, comparing to the countries with economies in transition is substantial. Developing countries lack standards, quality of management of mines and emission regulations, which result in the employment of basic and low technology of mineral extraction methods and refining processes (Hilson and Murck, 2000). It is often the case when mining companies from the developed countries operate under lower environmental regulations when their operations are based in developing countries. At times, they even adopt lower standards and contribute further to deterioration of the environmental quality, rather than providing examples for sustainable industrial practices. Some examples include the 2000 cyanide spill of the tailings dams' failures at the Baia Mare goldmine in Romania caused by an Australian mining company; social conflict in 1989 due to environmental damages from copper mining on Bougainville Island in Papua New Guinea by an Australian mining company and cyanide spills in the OK Tedi river in Papua New Guinea in 1984 caused, again, by an Australian mining company (O'Faircheallaigh, 2009). The discrepancy in environmental emissions from mineral processing and metal production industries in developed and countries with economies in transition is a subject that needs to be further explored.

AIR QUALITY INDICATORS OF ENVIRONMENTAL SUSTAINABILITY

This work presents an evaluation of the environmental and socio-economic contribution of mineral processing and metal production industries to the sustainability of the countries with economies in transition and compares with the performance of the same industries in developed countries. The countries with economies in transition selected for this work were the signatories of the UN (1979) Convention on Long-Range Transboundary Air Pollution and consisted of Albania, Belarus, Bosnia and Herzegovina, Macedonia, Moldova, Montenegro, Serbia and Ukraine. The study is limited to the European countries in transition because of the comprehensive emission reporting database available for these countries as their obligation towards the UN (1979) Convention. The performance of the overall air quality indicators and, specifically, the sustainability of the mineral processing and metal production industries from these countries were compared with the EU, United States and Australia. Several databases were selected as sources of information. These were Euromonitor International (2015), Centre on Emission Inventories and Projections (CEIP, 2015), Australian National Pollutant Inventory (NPI, 2015), USA National Emissions Inventory (NEI, 2015) and the World Resources Institute's Climate Data Explorer (WRI, 2015).

Table 18.2 presents the annual per capita national emissions of each country for the priority air pollutants (NO_x, SO_2, CO, PM_{10} and $PM_{2.5}$) and greenhouse gas emissions for the period of 2011/2012. Table 18.2 shows that the per capita pollutant emissions are not based on the scale of the economy, as developed countries, specifically Australia and United States, emit the largest amounts of all pollutants (except for SO_2 where the largest emission is from Bosnia and Herzegovina). On the other hand, the average emissions from the EU countries are very low when compared with the remaining countries. Table 18.2 further gives indication of the pollutants emitted by the European countries with economies in transition, when compared with the average EU emissions and indicates the areas for improvement. While NO_x emissions in all countries are within similar magnitudes, emission of SO_2 differs significantly, with Bosnia and Herzegovina, Macedonia, Montenegro, Serbia and Ukraine all showing substantially larger emissions when compared with EU, albeit they are similar to Australian emissions, a country with one of the most stringent environmental regulations. Bosnia and Herzegovina emits significantly larger amounts of PM_{10} than the average EU citizens, while Montenegro emits substantially larger amounts of both PM_{10} and $PM_{2.5}$ when compared with EU. In case of greenhouse gas emissions, it is only Belarus that emits larger amounts,

TABLE 18.2

Annual Per Capita Atmospheric Emissions of Priority Air Pollutants in kg/Person and Annual Greenhouse Gas Emissions in t/Person for 2011/2012 for All Activities

	NO_x	SO_2	CO	PM_{10}	$PM_{2.5}$	CO_{2-e}
Australia	73	98	116	35	1.4	24
United States	40	18	210	60	18	20.5
EU	16	8	43	3.6	2.5	9
Albania	7.5	6.6	29	5.3	4.1	2.1
Belarus	18	6.7	93	6.7	5.2	10.5
Bosnia and Herzegovina	13	112	31	11	5	7.7
Macedonia	17	43	32	5.8	3	6.2
Moldova	9	1.7	27	2.3	1.5	3.5
Montenegro	21	65	53	19	8.1	5.7
Serbia	29	40	41	4.6	2.5	8.8
Ukraine	13	30	65	5.7	2.9	8.7

when compared with EU, but still more than twice less compared with Australia and United States, while the remaining countries with economies in transition emit similar or far lower amounts of greenhouse gases compared with EU.

ECONOMIC AND SOCIAL INDICATORS OF INDUSTRIAL PERFORMANCE

Table 18.3 compares a selection of generic and industry-specific economic and employment-based social parameters of the countries of interest. The main difference between the developed and the

TABLE 18.3

Economic and Employment Parameters of Manufacturing, Mineral Processing and Metal Production Industries

	Aus	USA	EU	Alb	Bel	B&H	Mac	Mol	Mon	Ser	Ukr
GDP (,000€/p)	55.5	48	32	3.7	7.2	4.4	4.7	2.0	6.6	5.2	2.6
Wage per hour (€/p)	21.8	21.3	11.1	1.7	2.6	3.7	2.8	1.5	4.8	2.9	0.7
Wage in manufacturing relative to average wage (%)	4.1	1.2	10.7	−6.5	−22	30	−24.3	−15.6	−2.1	−13.8	15.7
Employment in manufacturing (%)	8.7	9.1	15.7	–	16.6	18.8	19.9	10.4	8.8	16.7	12.9
Employment in mining and quarrying (%)	1.9	0.6	0.46	–	0.3	3.1	1.3	0.6	1.1	1.1	2.9
Production of iron and steel (Mt)	8.87	124.5	261.7	0.52	2.9	1.5	0.34	0.12	0.07	0.54	41.9
Production of iron and steel (% of GDP)	0.79	0.75	1.32	4.8	2.25	9.7	3.25	2.0	1.8	1.6	7.8
Production of aluminium (%GDP)	0.24	0.03	0.02			1.3			3.3		
Production of gold (%GDP)	0.72	0.06	0.01							0.16	
Production of silver (%GDP)	0.07	0.004	0.01				0.18			0.01	
Production of slab zinc (%GDP)	0.07	0.003	0.02							0.24	0.01
Production of copper (%GDP)	0.41	0.06	0.18							0.64	0.1
Production of refined lead (%GDP)	0.03	0.01	0.02			0.04					0.06
Income from total metal production (%GDP)	2.34	0.91	1.6	4.8	2.25	11.1	3.4	2.0	5.1	2.6	8

countries with economies in transition is in the gross domestic product (GDP) and wage per hour indicators showing a 10-fold difference between the two groups of countries. Developed countries experience higher than average wages in manufacturing, while for the countries with economies in transition, the wage in manufacturing is mostly much lower than the average wage, except for Bosnia and Herzegovina and Ukraine where employment in manufacturing attracts greater average wages.

Employment in manufacturing ranges between 9% and 20% of the total employed population in each country indicating significant social importance of this activity to sustainable development. Employment in mining and quarrying contributes to a much lower rate of employment, ranging between 0.3% and 3% of the total employed population in each specific country. Employment in mining and quarrying in the countries with economies in transition has either comparable or far greater importance than the average EU and US employment rate and is even comparable (with exception of Belarus) to the employment rate in the mineral-intensive Australia.

Production of iron and steel is the major metal producing industry contributing between 0.75% and 10% to the national GDP economies for the selected countries. While EU and United States are the major iron and steel producing regions, the relative contribution of this industry to the national economy is much smaller for these countries than for the countries with economies in transition. This is also the case for the overall economic contribution of all considered metal producing industries, which contribute significantly more to the relative national economy of countries with economies in transition, compared with the developed countries.

ENVIRONMENTAL INDICATORS OF FERROUS METAL PRODUCTION

Production of ferrous metals is the major mineral processing and metal production operation for all countries considered in this comparative analysis and the environmental performance of this industry is further explored here. Table 18.4 compares emissions of the priority air pollutants (NO_x,

TABLE 18.4

Atmospheric Emissions of Priority Air Pollutants from Ferrous Metal Production Expressed in Grams of Emitted Pollutant per Tonne of Produced Metal for 2011

	Aus	USA	EU	Belarus	Maced.	Molda	Monte	Serbia	Ukraine
NO_x	1,130	37,990	154	510			120	24	865
SO_2	755	526,170	170	240					1,450
CO	21,410	10,710	8,600	4,770			1,550	8,580	15,480
PM_{10}	530	200	110	75	580	410	30	400	400
$PM_{2.5}$	30	20,000	80	72	450	365	30	310	30
Pb	0.2	38	1.3	8	15	5	2.5	16	1.7
As	0.2	0.03	0.03	0.1	1.3	0.08	0.01	1.0	0.007
Cd	0.02	0.5	0.04	0.3	0.06	0.2	0.2	0.2	0.03
Cr	0.2	35	0.3	2.4	14.5	1.3	0.09	11	1.2
Ni	0.09	8.7	0.1	0.3	0.5	0.2	0.6	0.5	0.9
Hg	0.02	3.9	0.03	0.007	0.3	0.02	0.05	0.09	0.01
NMVOC	210	300	83		480		45	275	42
Dioxins	1.6×10^{-7}		8.7×10^{-7}	29×10^{-7}	65×10^{-7}	0.1×10^{-7}	7.3×10^{-7}	33×10^{-7}	
PAH	0.08	0.14	0.22		10			6.4	0.02
PCB		1.4×10^{-6}	0.002	0.003	0.02	8.1×10^{-6}	0.007	0.03	
NAEI[a]	2.6	9.5	1.0	1.8	6.7	1.4	1.0	5.7	1.7

[a] NAEI – normalised atmospheric emission index.

SO_2, CO, PM_{10}, $PM_{2.5}$ and Pb) and air toxics, which include priority metals (As, Cd, Cr, Ni and Hg), non-methane volatile organic compounds (NMVOC), polycyclic aromatic hydrocarbons (PAH), dioxins and polychlorinated biphenyls (PCBs). The values presented in Table 18.4 were converted to emission of grams of each pollutant per produced tonne of metal. Considering the differences in emission reporting databases, the pollutants considered here were for 'iron and steel production' and 'ferroalloy production' in the case of European countries, 'industrial processes: ferrous metals' for United States and 'basic ferrous metal manufacturing' for Australia.

The results presented in Table 18.4 show that the iron and steelmaking facilities in United States emit the largest amount of pollutants per produced metal for NO_x, SO_2, $PM_{2.5}$, Pb, Cd, Cr, Ni and Hg. The second country with most polluting iron and steelmaking industry is Macedonia which reports the highest atmospheric emission values for PM_{10}, As, NMVOC, dioxins and PAH. Australia reports the highest emission values for CO, while Serbia the highest emission values for PCBs, when compared with the amount of produced ferrous metal.

The overall environmental performance of the iron and steelmaking facilities at the national level was used to estimate a normalised atmospheric emission index (NAEI). The index was calculated according to Strezov et al. (2013) by first normalising each pollutant across the countries where the country with the lowest emissions for the specific pollutant was given a value of 0, while the country with the largest emission of each pollutant attracted a pollutant index of 1. The NAEI is a summary of each individual pollutant index. The EU countries exhibited the lowest NAEI of the iron and steelmaking facilities indicating the lowest environmental impact, while United States had the largest overall index. The NAEI for the countries with economies in transition and for the iron and steelmaking facilities ranged from very low, in the case of Montenegro with the same index value as the average EU performance, to very high, in the case of Macedonia, albeit this country being still with a lower pollutant index than the United States. Considering the European countries with economies in transition have aspirations to join the EU, they will have to consider aligning their environmental performance to the average EU emissions. This will mean that the iron and steelmaking facilities in the countries with high levels of emissions, specifically Macedonia and Serbia, will have to undergo upgrading and incorporation efficient emission control technologies to lower the overall pollutant emissions. In the case of Macedonia, two of the major steelmaking plants were under social and governance pressure to install particle capture devices (Independent, 2014), with constructions for one of the plants completed by 2015, which would significantly improve the environmental performance for this country.

CONCLUSIONS

Mineral processing and metal production industries play an important role for the environmental and socio-economic development of nations. On one hand, these industries have positive impacts on economic development and employment, which is even more profound for countries with economies in transition. On the other hand, mineral processing and metal production industries have adverse impacts on the environment and on resource depletion. The NAEI for each country, which was developed in this work, shows no clear trend between the developed and the European countries with economies in transition on the level of environmental implication of mineral processing and metal production. This is specifically the case for ferrous metal production, as the European countries in transition, although lagging behind with environmental technology adoption, tend to follow the EU leadership in environmental standards and implementation of environmental technologies. The environmental performance of the EU ferrous production industries ranks well above the other developed countries considered here, such as United States and Australia. The study also shows the importance for a wider acceptance and integration of the national reporting inventories by the other countries with economies in transition, such as China and India, which would enable broader scientific discussions and developing industrial sustainability standards of the industrial activities across different economies.

REFERENCES

Auty, R. and A. Warhurst. 1993. Sustainable development in mineral exporting economies. *Resources Policy*, 19(1): 14–29.

Auty, R. M. and R. F. Mikesell. 1998. *Sustainable Development in Mineral Economies*. Oxford, UK: Clarendon Press.

Azapagic, A. 2004. Developing a framework for sustainable development indicators for the mining and minerals industry. *Journal of Cleaner Production*, 12: 639–662.

Brundtland, G. H. 1987. *Our Common Future: Report of the 1987 World Commission on Environment and Development*. Oxford, UK: Oxford University Press.

CEIP. 2015. Centre on Emission Inventories and Projections, http://webdab.umweltbundesamt.at/official_country_year.html (accessed 13 March 2015).

Euromonitor International. 2015. Statistical database (accessed 2 March 2015).

Giurco, D. and C. Cooper. 2012. Mining and sustainability: Asking the right questions. *Minerals Engineering*, 29: 3–12.

Grober, U. 2007. *Deep Roots – A Conceptual History of 'Sustainable Development' (Nachhaltigkeit)*. Best.-Nr. P2007–002 Wissenschaftszentrum Berlin fu"r Sozialforschung (WZB), Berlin, Germany, February.

Hilson, G. and B. Murck. 2000. Sustainable development in the mining industry: Clarifying the corporate perspective. *Resources Policy*, 26: 227–238.

Independent. 2014. Jugohrom, Makstil Could Be Closed Until They Meet Environmental Standards, 16 October, http://www.independent.mk/articles/10442/Jugohrom,+Makstil+Could+Be+Closed+Until+They+Meet+Environmental+Standards#sthash.zx6aVCYn.dpuf (accessed 16 March 2015)

IUCN. 1980. *World Conservation Strategy: Living Resource Conservation for Sustainable Development*. IUCN–UNEP–WWF. Gland, Switzerland.

Meadows, D. H., D. L. Meadows, J. Randers and W. W. Behrens III. 1972. *The Limits to Growth: A Report for the Club of Rome's Project on the Predicament of Mankind*. A Potomac Associates Book, New York, USA.

NEI. 2015. National Emissions Inventory, http://www.epa.gov/ttn/chief/net/2011inventory.html (accessed 13 March 2015).

NPI. 2015. Australian National Pollutant Inventory, http://www.npi.gov.au/ (accessed 13 March 2015).

O'Faircheallaigh, C. 2009. Public policy processes and sustainability in the minerals and energy industries. In *Mining, Society, and a Sustainable World*, edited by J. R. Richards. New York: Springer, pp. 437–467.

Petrie, J. 2007. New models of sustainability for the resources sector, a focus on minerals and metals, transaction of the institution of chemical engineers, Part B. *Process Safety and Environmental Protection*, 85(B1): 88–98.

Strezov, V., Evans, A. and T. Evans. 2013. Defining sustainability indicators of iron and steel production. *Journal of Cleaner Production*, 51: 66–70.

UN. 1979. *Convention on Long-Range Transboundary Air Pollution*. Geneva, Switzerland: UN.

UN. 1992. Agenda 21. *United Nations Conference on Environment & Development*, Rio de Janerio, Brazil, 3–14 June.

UN. 2002. Plan of Implementation of the World Summit on Sustainable Development. *World Summit on Sustainable Development*, 4 September, Johannesburg, South Africa.

UN. 2012. The Future We Want. *UN Conference on Sustainable Development*, Rio de Janeiro, Brazil, 20–22 June.

WRI. 2015. World Resources Institute Climate Data Explorer, http://cait2.wri.org/wri/Country GHG Emissions (accessed 3 March 2015).

19 Toward Sustainable Design of Landfill Clay Liners
A Case Study

Samudra Jayasekera

CONTENTS

ABSTRACT

The design and maintenance of landfills for waste disposal has received great attention over the last few decades due to the increasing interest and attention paid to environment protection issues, such as ground and ground water protection. Over the years, landfill liner construction has seen innovative new practices such as addition of engineered clays, synthetic lining material and designing of more sophisticated leachate collection systems. The main objective of such practices is to enhance the landfill liner performance as a hydraulic barrier and to minimise or prevent the migration of landfill leachate into the surrounding hydro-geological system. The compacted clay liners (CCLs) are among the most widely used hydraulic barrier system in modern waste containment systems.

When a waste containment system is in operation, leachate is generated at different concentrations and characteristics over a considerably long period of time. The subsequent clay–leachate interaction processes influence the mineralogical stability and physical properties of the clay. Understanding of the possible effects of leachate over a considerably long period of time is therefore of a significant importance in effective designing of clayey barrier systems for sustainable waste containment facilities.

This chapter provides the results of a study aimed at investigating the effects of landfill leachate on the performance of a compacted basaltic clay soil, over a period of time. For this purpose, a typical Melbourne basaltic clay with varying percentages of montmorillonite clay was selected and a synthetic leachate was developed based on the composition of typical municipal waste landfill leachate reported in the literature. The clay–leachate interactions were allowed to take place under controlled anaerobic laboratory conditions. Samples were then tested for the effect of leachate over different time periods to identify the variations in grain size distribution (GSD) and hydraulic conductivity characteristics.

The analyses of test results suggest that the long-term behaviour of a CCL could be significantly affected by the clay–leachate interactions, due to the possible alterations in the physical and mineralogical properties of the clay.

INTRODUCTION

The rapid growth of global population and increasing industrialisation in many parts of the world inevitably increase the demand on well-engineered waste management practices. The escalating volume of waste being generated and the obvious need to dispose this waste in a socially and environmentally acceptable manner has led to the development of waste management and control regulations in many parts of the world. As a result, in many countries, considerable efforts are being focussed towards the minimisation of waste production. Despite the global influence towards waste minimisation, the need for land-based waste disposal in eliminating solid urban wastes continues to be an unavoidable practice in many countries. According to the findings of the study commissioned by the Australian Department of the Environment, Water, Heritage and the Arts, the report prepared by Wright Corporate Strategy Pty Limited reported that in 2010, Australia's licensed landfills have been receiving around 21 million tonnes of waste each year. The annual disposal rate has been broadly consistent over the last decade despite substantial growth in resource recovery and recycling (WCS, 2010). Over the past few decades, the United States has experienced a substantial increase in municipal solid waste (MSW) generation (i.e. more than 180% in total, and on average about 4.3 lb of MSW/person) (Reddy et al., 2015). The reports from Europe, Asia and around the world confirm the same trends in waste generation and management (e.g. Jayasekera and Mohajerani, 2001; Fauziah and Agamuthu, 2012; Scharff et al., 2007).

THE NEED FOR SUSTAINABLE LANDFILLS

Although many well-documented, established strategies exist for the treatment of MSW, waste disposal in landfills will continue to be used well into the foreseeable future in the United States, Australia and around the world (WCS, 2010; Reddy et al., 2015). In many countries, the concepts of sustainable landfilling have been introduced to make waste disposal sites more acceptable to neighbouring communities and reduce the overall cost of waste management and environmental protection (e.g. by minimising the potential for future contamination that would cost more to remediate than prevent in the first place). Despite the widely practiced and promoted efforts in waste reduction, reutilisation and recycling (the 3Rs concept) throughout the globe, a final disposal element is still necessary since there is neither technology available to prevent all unwanted materials from the waste stream nor funds available to achieve zero waste. Therefore, sustainable landfilling is a necessary part of an efficient integrated waste management system. Modern MSW landfills are designed, constructed, and maintained to minimise the adverse environmental impacts from the waste disposal over both the short and long terms (Fauziah and Agamuthu, 2012; Scharff et al., 2007).

CHALLENGES IN LANDFILLING AS A WASTE MANAGEMENT PRACTICE

A landfill is an area of land or an excavation in which wastes are placed for permanent disposal. A few decades ago, it had not been uncommon to dump refuse and waste into abandoned sand or rock quarries or in areas designated as undesirable for any development activity.

One of the primary concerns in landfilling is the potential of ground water pollution due to the perceived migration of leachate, the liquid generated as a result of degradation of waste and percolating water, into the surrounding hydro-geological system. It is reported that in the United States,

75% of the landfills have caused groundwater pollution (Bouzza and Van Impe, 1998). As a consequence, over the years, new regulations have come into effect in an attempt to ensure that the risk of migration of leachate from landfill is minimised by specifying the construction of some form of a barrier (liner) between waste body and the ground water regime. The primary objective of such a landfill liner is to perform as a hydraulic barrier and minimise permeation of landfill leachate into surrounding hydro-geological system.

As reported in the literature (e.g. Cameron and McDonald, 1982 cited by Bagchi, 1990), contamination of groundwater due to leachate migration has been observed at several instances. It has become evident that even non-hazardous waste landfills; that is, landfills accepting putrescible wastes such as household wastes, are producing highly polluting leachate (Belevi and Baccini, 1989; Robinson and Gronow, 1993). As reported by Belevi and Baccini (1989), municipal landfills will generate leachate containing significant pollutant concentrations for centuries.

The following extract from Robinson and Gronow (1993, p. 822), clearly explains the severity of long-term effects of municipal landfill leachate on groundwater contaminations.

> What is becoming clear to many landfill researchers and operators is that a large, deep, high density domestic waste landfill, operated in a typical manner as at present (in the UK), will continue to produce strong and polluting leachates well in excess of values considered acceptable for discharge to surface or groundwater (Department of the Environment, 1993), for a large number of decades and possibly over timescales in excess of a century.

Due to the increasing interest paid to environmental protection issues such as ground and ground water protection, the design and maintenance of landfills has become a challenging and sophisticated issue. In countries like Australia, the construction of new landfills has come to be under strict public scrutiny and of social, environmental and political importance. Construction of new landfills is therefore becoming subjected to strict regulatory controls. However, landfill technology is still in infancy, and a comprehensive knowledge base in landfill processes and performances is not yet in existence. It becomes apparent that records of controlled landfills are only available for about 20 years (Belevi and Baccini, 1989). The model developments and predictions of landfill behaviour too have become complicated due to complex nature of landfill liner materials, leachate characteristics and their unpredictable variation with time. Therefore, the long-term behaviour of landfills is not known from experience. As a result, carefully engineered lining systems and more sophisticated leachate collection systems are becoming the norm and acceptable practice even with the considerably high costs associated with such practices.

LANDFILL LINER DESIGN PRACTICES

For several decades, the only practical means of liner design had been the construction of a hydraulic barrier consisting of a layer of compacted clay. The use of a single layer of geomembrane, over the native foundation soil emerged as an effective lining system in early 1980s. Over the last two decades, the use of synthetic lining material as a common component of a liner system has developed to an extent where a composite liner system is becoming the norm for many landfills, especially for hazardous waste landfills, in some countries. Leachate detection and collection systems are also becoming a usual component in the lining system.

It is also apparent that in developing countries, landfill design and management policies are still in the early stages or rather non-existent. Dumping of wastes both hazardous and non-hazardous in an undesirable manner can still be in practice. In these areas, there can be a serious risk of continuous groundwater and surface water pollution.

The highest concern in the landfill design is the design of the landfill liner and leachate control and collection system. Apart from that, the landfill design would include the design of the leachate disposal system and landfill closure system. In this chapter, the attention is paid to the challenges faced in designing of landfill liners.

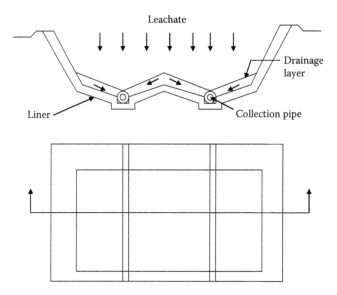

FIGURE 19.1 Typical layout of landfill liner system. (After Manassero, M., C. H. Benson, and A. Bouzza. 2000. Solid waste containment systems, *An international conference on Geotechnical and Geological Engineering, GeoEng2000*, Melbourne, Australia, 19–24 November, Vol. 1, pp. 520–642.)

A landfill lining system comprises of two major components; liner and the leachate collection network. Figure 19.1 shows a cross section and plan view of a typical landfill single liner with leachate collection and removal system. The leachate collection system helps in removing the leachate out from the landfill and maintains a low hydraulic head on the liner. The landfill liner is the barrier between the waste body and the foundation soil/rock of the waste disposal area. The main objective of the landfill liner is to serve as a hydraulic barrier and minimise migration of leachate in to surrounding hydro-geological system. A liner can consists of either a single layer of compacted clay, geomembranes, geotextiles or composites of soil and geotextile/geomembrane.

Geotextiles are fabrics made from petroleum products such as polyester, polyethylene and polypropylene and can be woven, knitted or non-woven. The primary objectives of a geotextile component in a geosynthetic clay liner (GCL) are filtration (free seepage of water from one layer to another, when placed between two soil layers), drainage (rapid movement of water from soil to various drainage outlets), separation (keep various soils layers separate when placed in between) and reinforcement (increase load bearing capacity of the soil due to tensile strength of the geotextile). Commercially available geotextiles can have mass per unit area ranging between 150 and 700 g/cm^2 and thicknesses between 0.25 and 7.5 mm.

Geomembranes consist of thin flexible sheets of rubber or plastic materials, which have extremely low permeability. The polymers used for geomembranes could be thermoplastic (e.g. polyvinyl chloride, polyethylene, polyamide, etc.) or thermoset. The most widely used geomembrane is high density polyethylene. Geomembranes can have thicknesses from 0.25 mm (as single piles) to 5 mm (when single piles are laminated together).

Geomembranes have proved to provide extremely low hydraulic conductivity values. It has been reported by Rowe (1997), that geomembranes can be expected to perform as an effective hydraulic barrier for approximately 25 years. However, the long-term fate of geomembranes due to leachate interactions over a long period of time is still unknown.

The selection of liner material and liner type (single, composite, etc.) primarily depends on the type of waste to be deposited in the landfill. The liner material must be chemically compatible with leachate for the expected active life of the landfill. The commonly used liner materials are clayey soil, synthetic lining materials such as geomembranes and geotextiles and/or soil modified by mixing with other suitable clay soils. A landfill liner could consist of one or few layers of these liner materials. Depending on the material components of the liner, landfill liners are commonly classified as GCLs and CCLs.

For a CCL, generally, most regulatory agencies have recommended a hydraulic conductivity of the clay soil not greater than 1×10^{-9} m/s. The minimum thickness varies from about 0.5 to 5 m, depending on the waste type and liner system. It has been suggested by Benson et al. (1994) that a soil with liquid limit greater than 20 and plasticity index greater than 7 can generally result a hydraulic conductivity of not greater than 1×10^{-9} m/s. The guidelines proposed by Daniel (1993) (cited by Kodikara, 2001) are percentage fines >20%–30%, plasticity index >7%–10%, percentage gravel <30% and maximum particle size <25–50 mm.

There are investigations that have been carried out to investigate the clay–leachate interactions and their possible effects on hydraulic conductivity, clay mineralogy and some soil engineering properties (e.g. Campbell et al., 1983; Belevi and Baccini, 1989; Sai and Anderson, 1991; Warith and Yong, 1991; Peters, 1993; Manning et al., 1995; Munro et al., 1997; Batchelder et al., 1998). Most of these studies have been conducted for short time periods. Only very limited studies are reported where the effects of leachate, over a considerably longer period of time (incorporating time as a variable) has been investigated while such studies conducted on the smectite family of clays are even fewer. The studies conducted by Jayasekera (2002) and Jayasekera and Mohajerani (2001a, b, 2002) are among very limited studies found in investigating the long-term effects of landfill leachate on CCL. Thus, very little has been discovered about the long-term performance of a landfill clay liner with the effect of leachate. This represents a limitation in the current knowledge required for the understanding of long-term behaviour of a CCL. Any significant variation of original soil properties with the effects of leachate may influence the effective performances of a clay liner with time.

Therefore, the aim of this research was to investigate the effects of MSW landfill leachate on GSD and hydraulic conductivity, the two most influential clay properties in CCL design of a selected smectite family clay soil, over a period of time and to evaluate the significance of any effects on the long-term satisfactory performances of a landfill CCL.

MATERIALS AND METHOD

EXPERIMENTAL SOIL

The soil used in this study was collected from a landfill liner construction site situated in Ravenhall in the Shire of Melton, a northwestern suburb of Melbourne Metropolitan. At the time of collection of the soil, the *in situ* soil had been excavated, screened through a 75 mm sieve and piled in bulk volume. According to the information gathered, the soil was in preparation for the construction of a liner of a proposed additional cell of a nearby existing landfill.

It was decided to add different percentages of commercially available sodium (Na) bentonite to the processed base soil, in order to obtain a range of soil samples with varying percentages of total montmorillonite clay minerals. By this means, the investigation program could be conducted for a range of soil samples with different percentages of total montmorillonite and analyse the effects of leachate for soil samples with varying clay contents. Accordingly, the laboratory experimental program was conducted for five soil samples, with reference numbers S1, S2, S3, S4 and S5. The soil S1 is the soil processed in the laboratory (by drying and sieving through 4 mm sieve) with no addition of Na bentonite, while soil compositions S2, S3, S4 and S5 have incremental additions of Na bentonite of 5%, 10%, 15% and 20% by weight, to the processed soil, respectively. Table 19.1 shows some engineering properties of original experimental soils used in this study.

LEACHATE

The synthetic leachate formulated in this study is comprised of cations, anions and organic compounds of a typical landfill leachate at its most reactive stage. The leachate composition is based on leachate

TABLE 19.1

Some Engineering Properties of Original Experimental Soils Used in This Study

Soil Composition	Soil Reference Number	GSD Gravel (%)	Sand (%)	Silt (%)	Clay (%)	Consistency Limits LL (%)	PL (%)	PI (%)	Maximum Dry Density (t/m³)	Optimum Water Content (%)
Soil	S1	4	24	9	63	78	41	37	1.30	32
Soil + 5% bentonite	S2	4	23	8.5	64.5	90	44	46	1.305	31
Soil + 10% bentonite	S3	4	18	11	67	103	47	56	1.32	30.5
Soil + 15% bentonite	S4	3.5	17	11.5	69	120	48	72	1.345	30
Soil + 20% bentonite	S5	3.5	14	12	71	141	50	91	1.365	30

data from several researchers (e.g. Metry and Cross, 1976; Robinson and Gronow, 1993; Bouazza and Van Impe, 1996; Didier and Comega, 1997; Batchelder et al., 1998; Gau and Chow, 1998).

As have been reported by Didier and Comega (1997), the chemical compounds that could have a greater influence on the selected clay (montmorillonite) were chosen in the formulation. It is known that Na, Ca, Mg and K have a greater influence on montmorillonite clay swelling properties (Lambe and Whitman, 1977). Similarly, among the organic substances, the acetic acid and sodium acetate (short-chain fatty acids) are known to be highly influential on the interlayer structure of montmorillonite clay mineral (Didier and Comega, 1997). Therefore, these chemical compounds were used in the formulation of the synthetic leachate. Table 19.2 presents the chemical compound composition of the synthetic leachate formulated in this study.

EXPERIMENTAL PROGRAM

Suitable quantities of experimental soils (loose state – uncompacted) were completely immersed in leachate and tested at different time intervals. By this means, it was attempted to simulate a 'worst case' scenario, where the leachate could have migrated completely through the liner and occupied all the pore spaces, which could expect to happen over a long period of time, possibly over few decades or even centuries. The results obtained are believed to permit a reasonable understanding of the possible trends in 'long-term' effects of leachate on the examined properties of experimental soils.

TABLE 19.2

Chemical Compound Composition of the Synthetic Leachate Formulated in This Study

Component Formula	Concentration (g/L)	Cations Symbol	Concentration (g/L)	Anions Symbol	Concentration (g/L)
CH_3COOH (acetic acid)	3.6	H^+		CH_3COO^-	3.54
CH_3COONa (sodium acetate)	4.92	Na^+	1.38	CH_3COO^-	3.54
$CaCl_2$ (calcium chloride)	4.98	Ca^{2+}	1.798	Cl^-	3.18
$MgSO_4$ (magnesium sulphate)	2.405	Mg^{2+}	0.484	SO_4^{2-}	1.916
NH_4Cl (ammonium chloride)	1.483	NH_4^+	0.5	Cl^-	0.983
KOH (potassium hydroxide)	1.119	K^+	0.78	OH^-	0.319

Separate test specimens from each soil were set for testing at different time intervals (e.g. 7 days, 28 days, 56 days, 132 days, 260 days, 365 days, etc.), in order to evaluate the effect of time on the soil–leachate interactions. Samples were kept under anaerobic conditions (closed) at an average laboratory temperature of 22°C.

The clay–leachate interactions were allowed to take place at these controlled conditions. At the end of the required reaction period, the excess liquid (leachate) was drained away and the soil was allowed to dry at an average temperature of 22°C, in order to prepare test specimens for various laboratory tests.

The experimental program of the parent study consisted of a range of different soil engineering tests. All experiments were conducted in accordance with AS 1289.3. (1995). This chapter reports the test results of

 i. GSD (Grain size distribution – sieve analysis in combination with hydrometer analysis) and
 ii. Hydraulic conductivity analysis, the two most important soil engineering properties affecting deign of landfill liners

RESULTS AND DISCUSSION

EFFECTS OF LEACHATE ON GSD OF EXPERIMENTAL SOIL WITH TIME

It is known that with time, clay–leachate interactions could cause some significant alterations in the original grain (particle) size distribution (GSD) of the experimental soils.

These results confirm that there are clear alterations to the GSD of the experimental soils. Figure 19.2 presents the effects of leachate on GSD of experimental soil S1 after different leaching times. Similar results were observed for other soils.

The clay contents of the original soils have gradually increased with the reaction (leaching) time and stabilised with longer periods as shown in Figure 19.3 while there is a corresponding reduction in the coarser particle fraction (Figure 19.4). This confirms that the leachate reactions on coarser particles of the soil have resulted in a physio-chemical weathering/degradation environment, which disintegrates the coarser fraction of the original soil into clay size particles. The eventual termination of the increase of clay content with longer periods of time could be due to the possible decrease of susceptible coarse particles for further physico-chemical degradation.

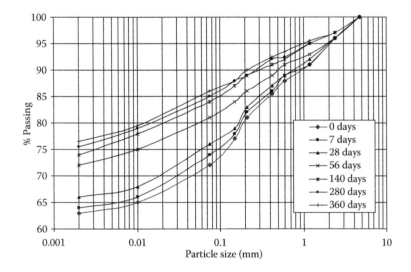

FIGURE 19.2 Effects of leachate on GSD of experimental soil S1 with time.

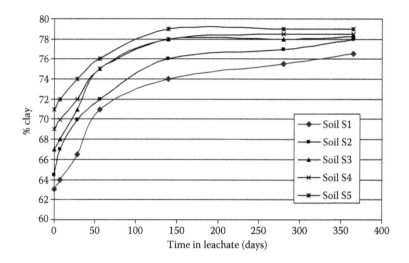

FIGURE 19.3   Effects of leachate on clay content of experimental soils with time.

From the test results, it is also apparent that the production of any new clay fractions diminishes quicker for soils containing lesser fractions of susceptible coarser particles. For example, soil S5, which has the least amount of coarser fraction, seems to cease the production of clay particles earlier than those soils containing comparatively more coarse particle fractions. Soil S1, which contains the most coarse particle fraction appears to continue the production of new clay particles even after 360 days in leachate, at a comparatively significant rate. In response to the increase in clay fraction in the soil, it is clearly visible that coarser particle fraction (particles larger than 2 μm) is gradually decreasing as shown in Figure 19.4.

The effects of leachate on various sizes of particles in the experimental soils, ranging from 4 mm to 2 μm were also analysed. It is observed that comparatively finer particles (silt, fine and medium sand) are more susceptible to degradation/weathering effects than the coarser particles fraction. This can be seen from the GSD results presented for soil S1 in Figure 19.2.

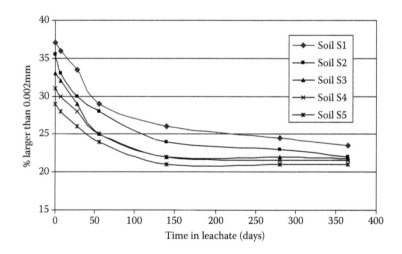

FIGURE 19.4   Effects of leachate on coarse particle fraction (>2 μm) of experimental soils with time.

EFFECTS OF LEACHATE ON HYDRAULIC CHARACTERISTICS OF EXPERIMENTAL SOILS WITH TIME

The experimental evidences so far revealed that mineralogical properties of clay particles can also undergo some changes due to leachate effects with time. Due to these physical and mineralogical changes in the soil, it is reasonable to expect changes in the micro-structure of the clay liner, in terms of possible alterations in void ratios with time. Any modification in void ratio can significantly influence the hydraulic performances of the clay.

The oedometer test based on one-dimensional consolidation theory was the major method employed in identifying compressibility properties. It has been stated that the estimation of hydraulic conductivity using consolidation results shows no discernible difference from the direct measurements (e.g. Lambe and Whitman, 1979; Bagchi, 1990; Mollins et al., 1996; Stewart et al., 1999, etc.).

The hydraulic conductivity values for original experimental soils were calculated using the consolidation test results and the widely used following equation:

$$k = C_v m_v \gamma_w \tag{19.1}$$

where

    k is hydraulic conductivity
    C_v is coefficient of consolidation
    m_v is coefficient of volume compressibility
    γ_w is unit weight of water

As presented in Figure 19.5, it can be seen that the hydraulic conductivity of experimental soils decreases with the increase of clay content due to the reduction of void ratio and thus the void space available for liquid flow.

As discussed before, the immediate clay–leachate interactions can directly influence on the clay mineral chemistry in several ways; that is, alterations in the thickness of the diffuse double layer, K^+ fixation and ion exchange.

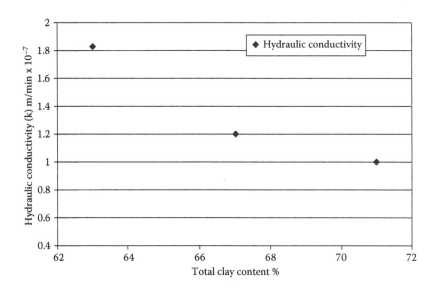

FIGURE 19.5 Hydraulic conductivity values for original experimental soils S1, S3 and S5 for a consolidation pressure of 400 kPa.

Other than these immediate consequences, with time, irreversible transformation of clay minerals from one type to another, physico-chemical degradation, changes of GSD can also take place at different degrees of intensity. These alterations, especially in the clay mineral chemistry and stability can influence the hydraulic characteristics of the soil. The performances of clay minerals in terms of natural attenuation ability as well as sorption capacity can be altered since these processes are directly influenced by clay mineral chemistry.

Studies conducted by several researchers (e.g. Anderson, 1982; Alawaji, 1999) clearly indicate that the void ratio increases due to immediate interactions between clay and leachate due to significant loss in double layer volume. The increase in the void volume increases the micro paths available for liquid flow. Consequently, the hydraulic conductivity (k) could increase unless the consolidation due to *in situ* field stresses compensates for the chemically induced changes in the micro-structure of the clay liner. The experimental evidences of the present study also show that the changes in the micro-structure of the clay liner in terms of alterations to void ratio continue to occur with time due to the long-term effects of leachate resulting in possible consequences such as physico-chemical degradation effects and mineral transformation processes.

Effects of Leachate on Void Ratio (e) with Time

The initial void ratio (e_o) (void ratio before applying any consolidation pressure) computed for original experimental soils and after different clay–leachate interaction periods were analysed to investigate the variation of initial void ratio with time with the effects of leachate. It can be clearly seen from the results presented in Figure 19.6 that the initial void ratio gradually increases with the clay–leachate interaction period for all the tested experimental soils.

The void ratios at the end of primary consolidation (e_p) for each load increment were also calculated in order to further investigate the variation of void ratio with time under different consolidation pressures. The results of variation of void ratio e_p of soil S1, with time, under different consolidation pressures are presented in Figure 19.7. Similar results were observed for other experimental soils.

From these void ratio calculations, it is clearly evident that for a given soil under a given consolidation pressure, the void ratio increases with the increase in clay–leachate interaction period. Similar results, that is, increase in void volume due to leachate effects have been observed by several researchers (e.g. Anderson, 1982; Alawaji, 1999) especially for smectites (montmorillonite) clay

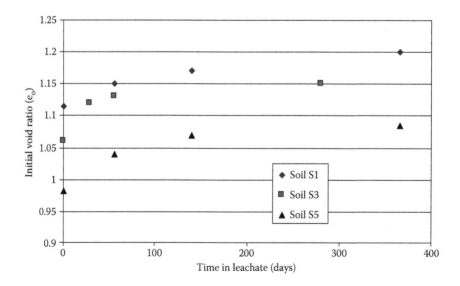

FIGURE 19.6   Effect of leachate on initial void ratio (e_o) of experimental soils S1, S3 and S5 with time.

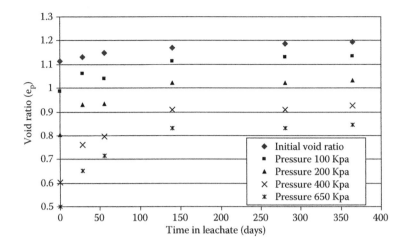

FIGURE 19.7 Effects of leachate on void ratio (e_p) with time for soil S1 under different consolidation pressures.

soils. However, most of these studies have identified only the immediate effects of leachate on clay minerals and the studies conducted to evaluate the variation of these effects over a period of time are very few.

From these experimental evidences, it can be concluded that the void ratio increase in this manner is due to the collective effect of the following consequences due to continuous leachate effects with time; (i) collapse of diffuse double layer of the clay minerals, especially in montmorillonite clay minerals, (ii) changes in the GSD due to degradation/disintegration of coarser particles and (iii) transformation of clay minerals from one type to another.

From these void ratio-time plots, it can also be observed that, for the given experimental soils, the rate of change of void ratio is more noticeable when the initial total clay content is less and the initial coarse particle fraction is high (refer Figure 19.6). The rate of change of void ratio tends to decrease with increase in the initial original clay content. The rate of change of void ratio for these experimental soils, in the order of dominancy is from S1 to S2 to S3 to S4 to S5.

It is also noticed that, the changes in the void ratio continue to occur for a long period of time, when the initial total clay content is less and the initial coarse particle fraction is high. For example, soil S1, which has the lowest clay content and the highest coarse particle fraction, shows observable higher rates of void ratio increases for the tests conducted after a considerably long period of leachate effects. The intensity of the rates of void ratio increase for other soils gradually decreases with long time leaching effects. This suggests that the void ratio increase due to leachate effects has a noticeable relationship with the GSD of the experimental soils. When there is a higher coarse particle fraction that are susceptible to degradation and disintegration due to leachate effects, the void ratio continue to increase for a longer time due to leachate interactions. This is due to the availability of coarse particles for physico-chemical weathering to take place for a longer time. However, when the coarse fraction is less, as for experimental soil S5, the degradation/disintegration process would come to a hold quicker and the effect of this degradation/disintegration on the void ratio increase would diminish earlier. The different rates of changes in the GSD were clearly evidenced from GSD test results. From the results presented in Figure 19.7, it can be seen that there is an obvious significant increase in e_p of the experimental soil with time. These changes in void ratio could eventually increase the hydraulic conductivity of the original soil. The probable variation of hydraulic conductivity with time with the effect of leachate (due to change in void ratio) was calculated using the consolidation test results as described previously in this chapter.

The estimations of variations of hydraulic conductivities with the effect of leachate with time for soils S1, S3 and S5 (under a consolidation pressure of 400 kPa) are presented in Figure 19.8 and

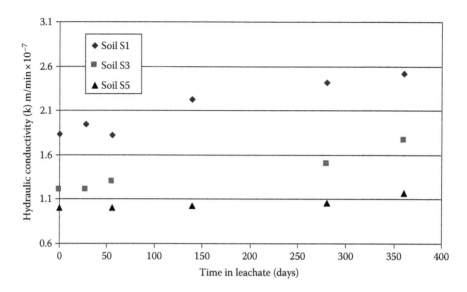

FIGURE 19.8 Effects of leachate on hydraulic conductivity of soils S1, S3 and S5 with time under a consolidation pressure of 400 kPa.

TABLE 19.3

Effects of Leachate on Hydraulic Conductivity (k) of Experimental Soils S1, S3 and S5 under a Consolidation Pressure of 400 kPa

	Hydraulic Conductivity (k) m/min × 10⁻⁷					
Soil Reference	**Time in Leachate (Days)**					
	0	**28**	**56**	**132**	**280**	**400**
S1	1.83	1.94	1.82	2.22	2.42	
S3	1.120	1.21	1.3		1.49	1.77
S5	1.01		1.03	1.038	1.06	1.173

tabulated in Table 19.3. It is quite clear that the hydraulic conductivity of the experimental soil could be significantly increased with time as a result of alterations to the void ratio due to the effect of leachate with time.

Hydraulic conductivities of the experimental soils were also predicted using some empirical relationships. The relationship proposed by Casagrande, $k \propto e^2$, which yields Equation 19.2 and the Kozeny–Carman relationship, $k \propto e^3/1 + e$, which yields Equation 19.3 were used in these predictions.

$$\frac{k_1}{k_2} = \frac{e_1^2}{e_2^2} \tag{19.2}$$

$$\frac{k_1}{k_2} = \frac{e_1^3 / 1 + e_1}{e_2^3 / 1 + e_2} \tag{19.3}$$

The predicted and estimated hydraulic conductivity values for experimental soil S1 for a consolidation pressure of 400 kPa is presented in Figure 19.9. Similar relationships could be observed for other experimental soils too for different consolidation pressures.

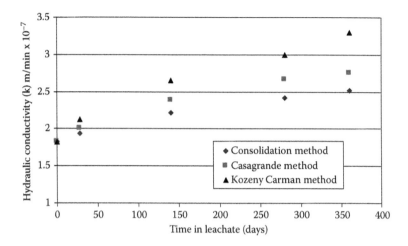

FIGURE 19.9   Estimated and predicted hydraulic conductivities for soil S1.

From these results can be seen that the hydraulic conductivity of experimental soils is a function of time that the soil has been exposed to leachate and also dependent on the consolidation pressure and the original clay content (montmorillonite clay fraction).

CONCLUSIONS

In relation to the design and construction of landfill liners, CCLs geotextile clay liners (GCLs) are becoming the preferred liner systems in many countries. The design specifications for CCLs are generally comprised of a specified maximum hydraulic conductivity of clay soils and a minimum thickness of the clay layer. It is expected that the clay liner would retain its permeability properties throughout its service life by retaining its original physical, chemical, structural and mineralogical stability.

From the test results presented in this chapter and the subsequent discussions based on those test results, the following conclusions in relation to GSD and hydraulic characteristics of the experimental soils can be made:

i. The clay–leachate interactions between the experimental soils and leachate used in this study lead to some noticeable increases in the void ratio. This void ratio increase is due to effects of several clay–leachate interaction processes such as collapse of diffuse double layer of the clay minerals, especially in montmorillonite clay minerals, changes in the GSD due to degradation/disintegration of coarser particles and transformation of clay minerals from one type to another.

ii. The effects of leachate on the increase of void ratio depend on the original GSD of the experimental soils. When the degradable coarser particle fraction is high, the rate of change of void ratio is also high and continues for a longer time period.

iii. It is observed that hydraulic conductivity gradually increases with the leachate reaction period. The rate of increase of hydraulic conductivity is high at the beginning and tends to slow down with time. It can be seen that the rate of increase of hydraulic conductivity is directly related to the rate of increase of the void ratio.

The experimental findings of this study clearly demonstrate that the long-term interactions between clay and leachate could significantly affect the GSD and hydraulic conductivity, the two most desirable engineering properties in selecting clay soil in the liner design stage.

REFERENCES

Alawaji, H. A. 1999. Swell and compressibility characteristics of sand-bentonite mixtures inundated with liquids. *Applied Clay Science*, 15: 411–430.

Anderson, D. 1982. Does landfill leachate make clay liners more permeable? *Civil Engineering ASCE*, September, 66–69.

AS 1289.3. 1995. Methods of testing soils for engineering purposes, Australian Standard, Sydney, Australia.

Bagchi, A. 1990. *Design, Construction and Monitoring of Sanitary Landfill*. John Wiley & Sons, New York, USA.

Batchelder, M., J. D. Mather and J. B. Joseph. 1998. Mineralogical and chemical changes in mineral liners in contact with landfill leachate. *Waste Management & Research*, 16(5): 411–420.

Belevi, H. and P. Baccini. 1989. Long-term behaviour of municipal solid waste landfills, *Waste Management & Research*, 7: 43–56.

Bouzza, A. and W. L. Pump. 1997. Settlement in the design of municipal solid waste landfills. In *Proceedings of 1st ANZ Conference on Environmental Geotechnics, GeoEnvironment 97*, edited by A. Bouazza, J. Kodikara and R. Parker. Rotterdam, The Netherlands: Balkema, pp. 339–344.

Bouzza, A. and W. F. Van Impe. 1998. Liner design for waste disposal sites. *Environmental Geology*, 35(1): 41–54.

Campbell, D. J. V., A. Parker, J. F. Rees and C. A. M. Ross. 1983. Attenuation of potential pollutants in landfill leachate by Lower Greensand. *Waste Management & Research*, 1: 31–52.

Didier, G. and L. Comega. 1997. Influence of initial hydration conditions on GCL leachate permeability. In *Testing and Acceptance Criteria for Geosynthetic Clay Liners, ASTM STP 1308*, edited by L. W. Well. ASTM, pp. 181–195.

Fauziah, S. H. and P. Agamuthu. 2012. Trends in sustainable landfilling in Malaysia, a developing country. *Waste Management & Research*, 30(7): 656–663.

Gau, S. H. and J. D. Chow. 1998. Landfill leachate characteristics and modeling of municipal solid wastes combined with incinerated residuals. *Journal of Hazardous Materials*, 58: 249–259.

Jayasekera, S. and A. Mohajerani. 2001a. A study of effects of municipal landfill leachate on a basaltic clay soil. *Australian Geomechanics*, 36(4): 63–73.

Jayasekera, S. and A. Mohajerani. 2001b. An investigation of long term effects of leachate on landfill clay liners. In *Proceedings of 2nd ANZ Conference on Environmental Geotechnics, GeoEnvironment 2001*, edited by D. Smith, S. Fityus and M. Allman. New Castle, Australia: The Australian Geotechnics Society, pp. 359–364.

Jayasekera, S. and A. Mohajerani. 2002. Long term strength and bearing capacity characteristics of a basaltic clay soil subjected to landfill leachate. In *Proceedings of International Workshop on Environmental Geomechanics*. Monte Verita, Ascona, Switzerland, pp. 237–244.

Kodikara, J. K. 2001. Design of compacted clayey liners for waste containment. In *Proceedings of 2nd ANZ Conference on Environmental Geotechnics, GeoEnvironment 2001*, edited by D. Smith, S. Fityus and M. Allman. New Castle, Australia The Australian Geotechnics Society, pp. 321–333.

Lambe, T. W. and R. V. Whitman. 1979. *Soil Mechanics*. New York: John Wiley & Sons.

Manassero, M., C. H. Benson, and A. Bouzza. 2000. Solid waste containment systems, *An international conference on Geotechnical and Geological Engineering, GeoEng2000*, Melbourne, Australia, 19–24 November, Vol. 1, pp. 520–642.

Manning, D. A. C., J. Owen and C. R. Hughes. 1995. Dissolved silica in landfill leachates and its possible significance for clay barrier stabilities. In *Waste Disposal by Landfill – GREEN'93*, edited by R. W. Sarsby (ed.). Rotterdam, The Netherlands: Balkema.

Metry, A. and F. L. Cross Jr. 1976. *Leachate Control and Treatment*, Vol. 7, Environmental Monograph Series. Westport, Connecticut: Technomic Publishing.

Mollins, L. H., D. I. Stewart and T. W. Cousens. 1996. Predicting the properties of bentonite-sand mixtures. *Clay Minerals*, 31: 243–252.

Munro, I. R. P., K. T. B. MacQuarrie, A. J. Valsangkar and K. T. Kan. 1997. Migration of landfill leachate into a shallow clayey till in southern New Brunswick: A field and modelling investigation. *Canadian Geotechnical Journal*, 34: 204–219.

Peters, T. 1993. Chemical and physical changes in the subsoil of three waste landfills. *Waste Management & Research*, 11: 17–25.

Reddy, K. R., R. K. Giri and S. H. Kulkarni. 2015. Two-phase modeling of leachate recirculation using drainage blankets in bioreactor landfills. Environmental Modeling and Assessment, 20: 475–490, doi: 10.1007/s10666-014-9435-1.

Robinson, H. D. and J. R. Gronow. 1993. A review of landfill leachate composition in the UK. In *Proceedings Sardinia 93, Fourth International Landfill Symposium*, CISA, Environmental Sanitary Engineering Center, Cagliari, Italy, 11–15 October, pp. 821–832.

Rowe, R. K. 1997. The design of landfill barrier systems: Should there be a choice? *Ground Engineering*, 30: 36–39.

Sai, J. O. and D. C. Anderson. 1991. Long-term effect of an aqueous landfill leachate on the permeability of a compacted clay liner. *Hazardous Waste and Hazardous Materials*, 8(4): 303–312.

Scharff H., B. Kok and A. H. Krom. 2007. The role of sustainable landfill in future waste management systems. In *Proceedings Sardinia 2007, Eleventh International Waste Management and Landfill Symposium*, CISA, S. Margherita di Pula-Cagliari, Sardinia, Italy, 1–5 October.

Stewart, D. I., T. W. Cousens, P. G. Studds and Y. Y. Tay. 1999. Design parameters for bentonite enhanced sand as a landfill liner. In *Proceedings of the Institution of Civil Engineers, Geotechnical Engineering*, Vol. 137. October, pp. 189–195.

Warith, M. A. and R. N. Yong. 1991. Landfill leachate attenuation by clay soil. *Hazardous Waste and Hazardous Materials*, 8(2): 127–141.

Wright Corporate Strategy Pty Limited. 2010. A Review of the Application of Landfill Standards, Study commissioned by Australian Department of the Environment, Water, Heritage and the Arts, March.

Section IV

Energy

20 Water and Energy Nexus
Impact of Energy Development on Water Resources

Michael Hightower

CONTENTS

ABSTRACT

Water is an essential natural resource that impacts all aspects of life—clean and abundant supplies of water are vital for supporting food production; public health; industrial, mineral, and energy development; and a healthy environment. Water is an integral part of energy and mineral extraction, production, and generation. It is used directly in hydroelectric power generation and is used extensively for thermoelectric power plant cooling and associated air emissions control, for growing and refining biofuels, for oil and natural gas (NG) refining, and for energy resource transportation. Water is also used or generated in mineral and energy resource extraction, such as hydraulic fracturing, oil and NG produced water generation, coal and uranium mining and processing, and mineral mining operations, including processing, dust control, and mine dewatering. Energy development can also impact water quality in surface water and groundwater, including thermal impacts from thermoelectric power plant cooling water thermal discharges, degradation of water quality from oil and gas development produced water leaks and disposal issues, energy refining waste water and refining by-product water impacts, and mine water drainage contamination issues.

Unfortunately, unsustainable uses of water resources, population dynamics and migration patterns, rapid economic growth in many developing countries, and climate change impacts on precipitation and the environment are all altering freshwater resources and available water supplies across the globe. These issues are exacerbating the water needs and water quality impacts of energy development. Therefore, as nations try to balance the efficient use of water to support agricultural, human health, energy, industrial, and ecological needs in the coming decades, it is clear that the water footprint, like the carbon footprint, will become a critical factor in determining sustainable and resilient energy development worldwide.

INTRODUCTION

This chapter provides an overview of emerging freshwater resource issues, growth in competing demands for freshwater, and the current and future water supply and water quality issues and challenges facing energy development. The discussions will focus on the three major segments of the energy sector and their water use, consumption, and water quality impact challenges, including

- Thermoelectric power generation—cooling and air and carbon emissions control
- Fossil and alternative transportation fuels—production, refining, and processing
- Energy resource mining and extraction—conventional oil and NG development, coal and uranium mining and processing, oil shale and oil sand mining, and hydraulic fracturing for unconventional fossil fuel development

GLOBAL WATER STRESS AND IMPACTS ON ENERGY DEVELOPMENT

The growing concerns between water resource availability and the ability to support sustainable energy growth were first highlighted in a U.S. Department of Energy Report to Congress prepared in 2007 by Sandia and Los Alamos National Laboratories in cooperation with the National Energy Technology Laboratory and the Electric Power Research Institute (DOE, 2007). That report noted a number of trends in water resource availability and the growing water demands in other sectors that could significantly challenge future energy reliability and sustainability. Since then, concerns over growing water constraints and their impact on future energy development have been recognized by energy, water, and financial and economic development leaders worldwide (WEF, 2009; WEC, 2010; IEA, 2012; DOE, 2014). These challenges include

- Many regions of the globe are already in high levels of water stress that is being exacerbated by climate change, which will limit water availability for future energy development.
- The projected water needs to support new industrial growth, public health, and additional food supplies will significantly increase competition for water resources in many developed and especially developing countries.
- Many new energy technology approaches to reduce carbon emissions and the environmental and ecological impacts of energy production and development are often currently more water use intensive than the approaches they are expected to replace.

Each of these issues will impact future energy resiliency and reliability and likely require significantly better cooperative and integrated management of water and energy resources as discussed in the following paragraphs.

One of the major challenges facing new energy production and generation globally is the already high level of water stress relative to water supply availability around the world as shown in Figure 20.1 (WRI, 2015).

The red colors in Figure 20.1 represent regions where total water withdrawals are above 80% of renewable water supplies, while the lighter colors represent lower levels of water stress. The map shows that it is common for the same country to have regional variations in water stress due to variations in water availability and regional development across a country. As can be seen in Figure 20.1, the mid-latitudes contain many of the most water-stressed regions, where renewable water supplies and precipitation are generally limited by global weather and rainfall patterns. The high water stress levels are often due to a combination of the low renewable water supply availability and unsustainable surface water and groundwater withdrawal and consumption practices for existing water demands.

Unfortunately, these mid-latitude regions are also where climate impacts on precipitation are expected to have the largest reduction in renewable water supplies. Current climate projections

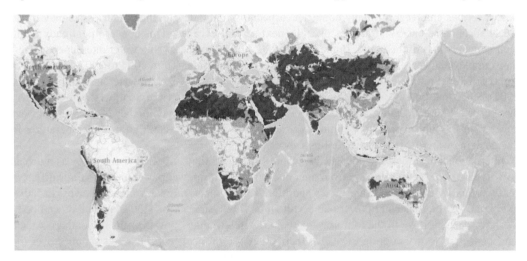

FIGURE 20.1 Global water stress. (Adapted from World Resources Institute. 2015. *Aqueduct—Monitoring and Mapping Water Risk*, Washington, DC. http://www.wri.org/our-work/project/aqueduct.)

TABLE 20.1
Global Water Use by Sector

Water Use by Sector	Developed Countries (%)	Developing Countries (%)
Agriculture	64	84
Energy and manufacturing	27	9
Domestic supplies	9	7

suggest that precipitation in the mid-latitudes will be reduced by 20%–30% by the middle of this century, and that surface water supplies could be reduced by 40%–60% due to changing climate patterns (NG, 2009; Llewellyn, 2013). This suggests a growing surface water supply stress and major impacts on groundwater recharge in many of these regions.

A second challenge is the growing water demand by major water use sectors to support population and economic growth, and the need for additional domestic water supplies to support improved public health and sanitation, especially in developing regions such as Asia, South America, and Africa. Overall, the increased water needs projected globally is expected to be about 10%–20% for the agricultural sector and 30%–40% for the domestic water supply sector by 2035. This will significantly increase the competition for limited freshwater and other water resources in many regions.

Also of importance in major developing countries and regions, such as India, China, Southeast Asia, and South America is that growth in the energy sector will require significant additional water resources to support the expected economic growth. Table 20.1 provides an overview of the relative water use by sectors for developed and developing countries (OECD, 2007; Schreir and Pang, 2012).

As highlighted in Table 20.1 and noted in a report by the World Economic Forum in 2009, developing countries could need to increase water allocated for energy development by a factor of three to four to approach the current levels of per capita energy resources available in developed countries (WEF, 2009). This is expected to put the energy sector into increased competition with other sectors for already limited water supplies in many of these countries. This was also highlighted by the World Energy Council in their 2010 Water and Energy report, which identified several regions of the globe where water supply availability is currently insufficient to meet proposed energy development (WEC, 2010).

The competition for water resources to meet growing water demands in multiple sectors could pose a serious challenge in many countries, especially many developing countries, as they try to create sustainable energy strategies over the next several decades. This concern is already being seen in many areas of the world. In the Report to Congress on energy and water interdependencies, for example, many cases were highlighted across the United States from 2004 to 2007 where power plants or biofuel refineries had been denied permits because of the lack of water availability (DOE, 2007). Since that time, we have seen instances of droughts in Texas, the Southwest, the Southeast, the Northeast United States, and California that have caused a large number of coal and nuclear power plants to reduce or stop operations because of low water flows or high water temperatures in the receiving waters that reduced cooling capacity.

International examples include France, where since 2005 they have had to curtail by as much as 20% its electric power production from their nuclear and hydropower power plants in three different years, primarily due to low water flows and high water temperatures during heat waves. In India, in February 2013, a thermal power plant with an installed capacity of 1130 MW shut down due to a severe water shortage in the Marathwada region. Already, many power plants in Southern Africa have converted to dry cooling to reduce their water use to meet reduced water supplies over the past few decades. Recurring and prolonged droughts are threatening hydropower capacity in many countries, such as Sri Lanka, China, and Brazil. These are all examples of mounting water stresses that are impacting energy development and energy supply reliability.

A final challenge is the growing water demand of new and proposed alternative energy supply development. A 2012 study by the International Energy Agency, presented in its 2012 World Energy Outlook, projects that world energy demand will grow by as much as 30% by 2035, but that water consumption for energy generation and production will likely increase by 85% (IEA, 2012). This suggests that much of the new energy generation will be based on technologies and processes that increase water use and consumption, which will only exacerbate water resource competition with other major water use sectors.

The global trends of decreasing freshwater supply availability and increasing demands for water in energy and other sectors highlight the need that future energy development should factor in the sustainable use and consumption of water to ensure future reliable, secure, and resilient energy supplies. The following sections discuss water use and consumption trends for the three major energy sectors: electric power, transportation fuels, and extraction of energy minerals, and directions and technology improvements that would help to reduce freshwater use by the energy sector.

REDUCING ELECTRIC POWER WATER NEEDS AND IMPACTS

Significant attention is often focused on the water issues in the electric power generation segment of the energy sector because it accounts for about 40% of water withdrawals in most developed countries (WEF, 2009). For example, about 50% and 60% of freshwater withdrawals in the United States and Canada, respectively, are for electric power generation. While water withdrawals are high, the water consumption is often much lower, with water consumption being the water taken out of the local watershed through evaporation or disposal. The water withdrawal and consumption of a power plant is driven by three major design and operational factors, including

- The electric power generation approach—thermoelectric steam cycle, combustion cycle, hydroelectric water turbines, wind turbines, or photoelectric
- The type of the cooling system used—open-loop or once-through, closed-loop or evaporative, dry or air-cooled, hybrid cooling, or refrigerated cooling
- The type of air emissions control required—wet or dry scrubbers for SOX, NOX, mercury, etc. and more recently and importantly CO_2 capture and sequestration

Over the past two decades, there have been several policy, environmental, ecological, and technological trends that will significantly influence the future mix of electric power generation and associated cooling and emissions control and the overall water demands and consumption. Policy changes to reduce ecological impacts of power plant cooling systems on water systems, requirements for CO_2 capture, changes in hydropower operations to reduce ecological and water quality issues, and the cost reductions seen in renewable electric generation technologies will have significant impacts on future technology selection and therefore water use and consumption. The water challenges associated with these various changes and trends are discussed below, with suggestions on ways to reduce the freshwater footprint of the electric power sector and support sustainable water resources management.

TRENDS AND CHALLENGES IN ELECTRIC POWER GENERATION WATER USE

Table 20.2 shows the range of water withdrawal and water consumption of the major electric power generation and cooling approaches, while Figures 20.2 and 20.3 show detailed data of water use and water consumption for technology-specific power plants and cooling approaches (Pate et al., 2007; Macknick and Newmark, 2012). Table 20.2 and Figures 20.2 and 20.3 are complementary, showing both general and more detailed technology specific trends for water withdrawal and water consumption. In Table 20.2, the information is presented in terms of liters per megawatt-hour of electric power (L/MWh_e). In Figures 20.2 and 20.3, the information is presented in terms of gallons

TABLE 20.2
Electric Power Plant Water Use

| | | Water Use Intensity (L/MWh$_e$) | | |
| | | Steam Condensing[a] | | Other Uses[b] |
Plant Type	Cooling Process	Withdrawal	Consumption	Consumption
Fossil/biomass steam turbine	Open-loop	80,000–200,000	800–1200	120–360
	Closed-loop	1200–2400	1200–2000	120–360
	Dry	0	0	120–360
Nuclear steam turbine	Open-loop	100,000–240,000	1600	120
	Closed-loop	2000–4400	1600–2900	120
	Dry	–	–	–
NG combined cycle	Open-loop	30,000–80,000	400	40
	Closed-loop	900	800	40
	Dry	0	0	40
Coal integrated gasification combined cycle	Closed-loop	800	750	600
	Dry	0	0	600
Geothermal steam	Closed-loop	8000	3000	20
	Dry	0	0	20
Concentrating solar	Closed-loop	3000	2800	40
	Dry	0	0	40
Wind and solar photovoltaics	N/A	0	0	10
Carbon Sequestration for Fossil Energy Generation				
Fossil or biomass	All	~80% increase in water withdrawal and consumption		

[a] Values are included for a range of plant designs, cooling water temperatures, and locations.
[b] Includes water for equipment washing, air emissions control, restrooms, and other water uses.

of water per megawatt-hour (gal/MWh$_e$) of electric power generation. To convert the data in Figures 20.2 and 20.3 to liters, multiply by 3.8.

As shown in Table 20.2, traditional approaches to electric power generation require a significant amount of water, with most of this water consumed to provide cooling water for steam condensers, though water is also consumed by other plant functions. For example, a 500-MW coal-fired thermoelectric plant with a closed-loop cooling system would consume about 24 million L of water per day, the same as a city of 50,000–60,000 people.

About 54% of current U.S. generating capacity is thermoelectric generating stations using open-loop cooling. In open-loop cooling systems, water is warmed as it passes through the cooling system, but it is not evaporated in the cooling system. Open-loop cooling requires access to a large body of water such as lakes, reservoirs, large rivers, the ocean, and estuaries for water withdrawal. Plant efficiency is highest when the condensing temperature is lowest, so the use of cold surface water and groundwater has historically been the preferred cooling fluid. During periods when drought or other circumstances lead to water levels or water temperatures outside of design limits, power plant operations can be restricted. This has been seen in a number of droughts in the United States, Europe, and Asia in the past decade. Aquatic life can be adversely affected by impingement on intake screens and entrainment in the cooling water, and by the discharge of warm water back to the source (DOE, 2007).

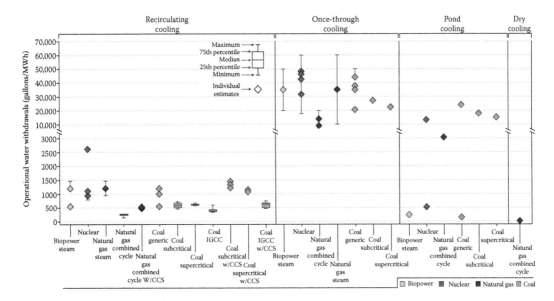

FIGURE 20.2   Electric power plant water withdrawal. (Adapted from Macknick, J. et al. 2012. *Operational Water Consumption and Withdrawal Factors for Electricity Generating Technologies: A Review of Existing Literature.* National Renewable Energy Laboratory, Golden, Colorado.)

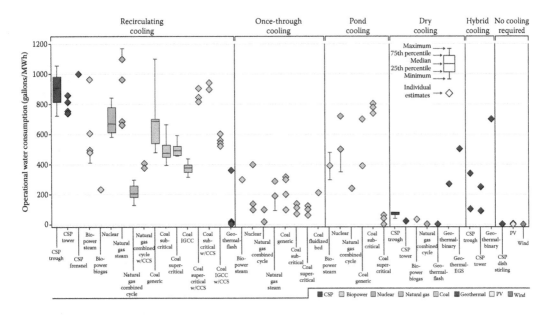

FIGURE 20.3   Electric power plant water consumption. (Adapted from Macknick, J. et al. 2012. *Operational Water Consumption and Withdrawal Factors for Electricity Generating Technologies: A Review of Existing Literature.* National Renewable Energy Laboratory, Golden, Colorado.)

Where access to sufficient surface water is not possible due to physical or other constraints, a closed-loop evaporative cooling is typically chosen. Most thermoelectric plants installed since the mid-1970s are cooled in a "closed" loop by evaporation of the cooling water in a wet-cooling tower or cooling pond. These systems withdraw less than 5% of the water withdrawn by open-loop systems, but most of the water withdrawn is consumed through evaporation. Evaporation of water concentrates contaminants, including water treatment additives, so the concentrated waters or blow-down from these systems can require treatment before release.

Another use of water in fossil-fired steam plants, especially coal plants, is for emissions control, such as flue gas desulfurization. While not all plants use wet scrubbing for flue gas removal, those that do consume about 60–70 gal/MWhe more water than those that do not (NETL, 2006). The emerging trend in reducing or sequestering carbon emissions will have a large and significant impact on fossil fuel power plant water consumption. Carbon sequestration requires energy and can decrease net plant output by as much as 25%. As a result, water requirements for steam condensing and other plant use could increase by 80% for equivalent energy output, compared to current approaches without carbon capture (Moore, 2010). This will essentially double water demands for electric power plants, probably unacceptable in many regions, suggesting that major improvements in carbon capture technology, new plant designs, and/or emissions trading and emissions management policies be developed.

The second-largest source of electric power in the United States is hydroelectric power, which produced from 6% to 10% of U.S. electricity generation between 1990 and 2003 (Hightower, 2010a). The growth in hydropower production from 1950 until about 1975 reflects the addition of large hydroelectric facilities. Variation in production after that time reflects changing water availability due to years of abundant precipitation and years of drought and changes in management to support reduced ecological impacts. Hydroelectric power water consumption is not included in Table 20.2 because estimating water consumption to hydropower is very difficult. A significant amount of water is lost by evaporation from reservoirs associated with hydroelectric plants. The United States Geological Survey (USGS) does not include these losses in its reports, but with an average loss for U.S. hydroelectric reservoirs of 16,000 L/MWhe and annual generation of approximately 300 million MWhe, total losses from hydroelectric reservoirs are estimated at 15 billion L/day. But since hydroelectric power is not the primary function of most reservoirs, decreasing hydroelectric power production would not reduce water losses. However, increasing the efficiency of hydroelectric operations would yield more power from these reservoirs without increasing water losses, and could replace power from other water-consuming power plants, such as thermoelectric plants.

Beyond conventional hydroelectric power, other renewable energy sources, including non-conventional hydropower, wind, solar, and geothermal energy systems, today account for less than 1% of the nation's electricity generation. Solar photovoltaic, solar dish engine, wind, and air-cooled geothermal hot water (binary) power systems offer a significant advantage over other electricity generation technologies—they consume no water while producing electricity. Large renewable energy power plants may use modest amounts of water for domestic use and solar plant operators may use small quantities of water for collector washing, as noted under other use for solar trough and solar power tower plants in Table 20.2, though overall that water use is extremely small.

The cost and intermittency of low-water-use renewable energy technologies have been a barrier to increased use. However, as costs decline, the use of these technologies has increased significantly since early 2000. Some of these technologies, such as wind, solar photovoltaics, and kinetic hydropower (run-of-river, wave) produce electricity only when the resource is present, which reduces the value of the electricity they produce. Connecting modest amounts of intermittent renewable sources to the grid has not been shown to undermine grid stability, but currently there are issues and concerns associated with potential large-scale deployment of renewable electric power generation on transmission grid stability and energy reliability. In some cases, renewable sources, such as solar, may have the potential to improve grid operations by providing power when it is most needed, during the hottest part of the day, and approaches are being developed, such as advanced microgrids, to try and improve the integration of these technologies into many national electric power grids to reduce both the carbon and water footprints of electric power generation.

Considerations for Reducing Electric Power Water Consumption

As noted in Table 20.2 and discussed above, the greatest reduction in water consumption in the electric power generation sector will likely come from moving toward the use of water-efficient steam-condensing technologies, such as dry and hybrid cooling. Also important is the continued use of once-through cooling technologies where they can be designed to be environmentally friendly because of their lower water consumption. If existing once-through cooling systems, especially those using seawater, are replaced with closed-loop evaporative cooling, then consumption of freshwater for steam condensing could double in coastal regions. Efforts to optimize intake and outfall structure designs and turbine technologies to minimize the environmental and ecological impacts would have great benefits not only to the ecology but also water consumption reduction and energy resiliency.

The need to better match power loads with low water use but intermittent renewable power sources, such as wind and solar, is badly needed. Evaluated individually, these technologies may appear to have unacceptable impacts on grid operations, but examined with dispatchable loads, the impact on grid operations might be minimized with the benefit of significantly reduced water consumption. Other ways to balance these intermittent power production technologies might be able to offset with the integration of low-emission hydropower, which has quick response capability.

Air emissions treatment for fossil thermoelectric power plants currently has a modest role in water use and consumption in the electric sector. However, the high parasitic power demand of carbon capture and sequestration and new air emission requirements could double water consumption by the electric power sector to implement current carbon capture technologies. This could significantly increase the water consumption costs or energy penalty costs for reducing water use cooling technologies. Therefore, options to reduce energy and water requirements for carbon capture are desperately needed.

Many technologies have been identified as having the potential to reduce water use and environmental impacts in the electric power generation sector. In general, technologies that have the potential to use and consume much less water have not yet been used widely in the United States and other countries on large power plants to date, but systems are being built and deployed on large plants in arid countries such as South Africa and in the arid western United States. The lack of data on new technology cost and performance, and in some cases, the lack of utility experience have been major obstacles to widespread use. Therefore, research and development to improve the cost and performance of these technologies, as well as technology demonstration efforts are needed to provide the operational data needed to accelerate their utilization. The major considerations to significantly reduce water consumption and the water footprint include the following.

Dry Cooling

Compared to water, air has very low heat capacity. Thus, air-cooled condensers must have a large surface area and require powerful fans to move high volumes of air through them, making them often very large as shown in Figure 20.2. Direct dry cooling via air-cooled condensers can greatly reduce power plant water use, but dry cooling also reduces plant performance and efficiency, especially during hot weather. Dry cooling relies on the temperature of the ambient air, and as the ambient temperature increases, plant efficiency decreases. Over the course of a year, the output of a plant with dry cooling will be about 2% less than that of a similar plant with evaporative closed-loop cooling, depending on the local climate. However, in the hottest weather during the summer, when power demands are often highest, dry-cooled power plant efficiency and plant output may decrease by 25% (Hightower, 2010b). Wind can also interfere with air-cooled condenser air flow, causing the hot exhaust to be recycled to the condenser air inlet, decreasing cooling system and power plant performance. Decreased plant efficiency means increased fuel use and increased emissions.

Dry cooling systems are currently sized to balance performance penalties with system cost. (As the condenser surface area is increased, the cooling system performance increases, but costs also increase.) In total, dry-cooled systems impose a cost penalty ranging from 2% to 5% over wet-cooled systems. These figures only represent the economic impact on plant operation. A utility might also incur additional costs to purchase replacement power for losses in efficiency during hot weather when replacement power is often most expensive (Hightower, 2010b).

Existing wet-cooled plants may not be candidates for retrofit with dry cooling because they may lack the additional land area required for the air-cooled condensers, and because existing steam turbines may not be compatible with the higher backpressure associated with dry cooling. An additional barrier to use of dry cooling for nuclear plants is that nuclear plants must have cooling systems which have been shown to meet their continued need for emergency cooling during a grid outage or other plant event.

As an alternative to ducting steam to air-cooled condensers, an indirect dry cooling system can be used, which essentially replaces the evaporative cooling towers with dry cooling of water. As in a conventional closed-loop system, the cooled water is circulated to a steam condenser at the turbine exit. While this eliminates the need to duct steam long distances to the air-cooled condensers, the additional heat-exchange step makes the efficiency and hot-weather performance of an indirect system even poorer than that for a direct system.

Research and development to reduce the high cost of these units requires materials that cost less but still have excellent heat transfer characteristics and for which lower-cost fabrication techniques are possible. Improved understanding of air flow through the condensers could enable higher performance and less fan capacity. Also needed is improved understanding of air flow in the vicinity of the exit of the condenser, where high winds can cause the hot air to be recycled back into the condenser. For nuclear plants, dry cooling must be shown to meet the additional safety requirements associated with emergency cooling, or additional emergency cooling water supplies would need to be available nearby.

Hybrid Cooling

Hybrid cooling systems combine dry cooling and wet cooling to reduce water use relative to wet systems while improving hot-weather performance relative to dry systems. Annual cooling water requirements would be about 20% of the requirements for a conventional evaporative cooling system (an evaporatively cooled 350 MW coal plant consumes about 8 billion L of water a year). This alternative approach is useful when extended operation is needed (e.g., a thousand hours per year), which uses an air-cooled condenser paired with a conventional evaporative cooling tower to augment cooling in the hottest weather. This alternative approach has been shown to be useful in small power plants, but Micheletti and Burns (2002) caution that the operation of parallel cooling systems can be significantly more complex because it requires controlling and coordinating variable fan speeds on both systems as well as recirculating water flow. In addition, they note "the difficulty of constructing large, low-pressure, bifurcating ductwork from the turbine exhaust to both an air-cooled condenser and a water-cooled condenser may limit this type of system to small power applications." But hybrid systems with combined cycle NG plants have gone to dry systems use in the winter and wet system support in the summer. The costs of the two systems are a trade-off in the operating simplicity. The California Energy Commission in 2002 recommended a focused research and development effort to improve hybrid cooling with improved water treatment and materials as a high energy reliability but very-water-efficient cooling approach.

Wet Surface Air Cooling

Wet surface air coolers are designed to keep the cooling surfaces wet at all times and thus reduce the opportunity for fouling and corrosion. In conventional evaporative cooling towers as well as in spray-enhanced dry cooling towers, poor quality water can foul the cooling surface, which leads to loss of performance and, potentially, corrosion. Spray enhancement of dry cooling must use highly treated water because any loss of thermal conductivity will have a significant impact on the performance during dry operation. Conventional evaporative towers use a combination of water treatment, chemical additives, and periodic blowdown (discharge of cooling water that has been through multiple cycles of evaporation and which is then replaced with freshwater) to prevent fouling. The unique design of wet surface air coolers allows them to be used with nontraditional water sources, including blowdown from conventional cooling towers, wastewater treatment plant effluent, and brackish water. Used as an auxiliary cooling technology in combination with conventional cooling towers, they have demonstrated an increase in water consumption efficiency of about 5% in combined cycle plants (Hightower, 2010b).

Wet surface air coolers have been demonstrated both in small-scale applications and as a water-conserving technology in parallel with conventional wet-cooling towers. The ability of this technology to accommodate lower-quality waters, perhaps including nontraditional waters such as wastewater treatment plant effluent and brackish water, is what makes this technology attractive. However, the use of lower-quality water requires optimization of material selection and water treatment to minimize operational costs. Many brackish groundwater sources, including produced water, contain high levels of carbonates, sulfates, and silicates not found in seawater or estuaries. Additional water treatment and materials research will be required to accommodate the use of these nontraditional water sources and reduce freshwater use.

Alternative Cooling Fluid Technology

Additional opportunities to reduce water consumption in thermoelectric cooling include the capture of cooling water condensate for reuse and use of other cooling fluids such as Brayton cycle supercritical CO_2 systems. These have been considered for many decades. EPRI (1989) recommended that economic technical studies be performed on ammonia-based phase-change cooling systems incorporated into both dry and hybrid cooling towers. Since that time, few additional base-load power generation facilities were built and the recommendations have not been pursued. Recovering heat from condenser water discharge, though studied in the past, is still lacking in economic analyses and demonstration on a commercial scale. As the United States and other countries move toward the retirement of many older coal-fired power plants and the addition of a substantial number of new power plants from 2020 to 2040, the use of cooling approaches that use alternative fluids, where the waste heat could be recovered and water demands for cooling minimized, could become economically attractive because of their wider range of applicability.

Ecologically Friendly Open-Loop Cooling

The high-volume water intake structures required for open-loop cooling can present negative environmental ramifications such as impingement and entrainment of aquatic organisms and high temperature outfalls that impact aquatic organisms. As a result, few new thermoelectric power plants are being designed to use open-loop cooling. The development and demonstration of improved intake and outfall systems that could offer reduced impacts on fish and aquatic species could help ease the permitting of and improve the ability to use available large surface water resources, such as coastal waters, for cooling. Improved intake and outfall structures and operations could reduce the trend toward the use of only closed-loop cooling systems. This could expand the use of seawater in some regions or large water bodies in other regions, in both cases helping to reduce future freshwater consumption by as much as a factor of two as highlighted in Table 20.2, which shows that water consumption in once-through or open-loop cooling systems is about half that of closed-loop cooling in similar size plants.

A number of improved intake structure designs have been developed, but few have been evaluated for ecological impact and species-protection performance for various reasons, including a recent trend toward NG electric power generation for new systems, which require significantly smaller water resources and therefore does not need large intake systems. The lack of data on new designs means that installation of improved intake structures is precluded in part because power plant designers often have older data on which to base their requests for approval of installations of new high-volume intake structures. Overcoming the concerns and performance issues of intake and outfall systems will require new modeling and technology research efforts, as well as support for pilot-scale testing and demonstrations to fully assess the cost and performance of these new devices and the ecosystem impacts in various applications and aquatic environments. Alternative heat sinks for open-loop cooling, such as underground mines (Feeley, 2005), of advanced cooling pond approaches, may have applications in some regions and could provide access to nontraditional water resources for cooling, thus minimizing the use of freshwater.

Improved Carbon Emissions Control

While current air emissions control only accounts for about 10% of the water consumption in fossil fuel power plants, current studies indicate that carbon sequestration could reduce net plant electricity output by 25%, and increase water consumption per unit of energy output by up to 80% (Moore, 2010). U.S. EPA Regulation 111D and similar regulations in developing countries suggest that future carbon capture requirements for thermoelectric power plants will impact most new coal and NG power plants. This policy alone in the United States would essentially double the water consumption of fossil fuel-based electric power plants. Therefore, developing less water-intensive air emissions and carbon capture and sequestration technologies, alternative emissions control schemes, and new plant designs will be desperately needed to reduce the significant impacts these regulations could have on freshwater consumption by power plants in water-stressed regions.

Reassess Hydropower and Innovative Hydropower Systems

Current estimates suggest that between 30,000 and 70,000 MW of additional hydroelectric generating capacity is available in the United States, with suggested nominal values of about 40,000 MW. This includes both large and small hydroelectric possibilities, and represents the potential for doubling of current hydroelectric capacity (Hall, 2005). The development of this potential could displace nonrenewable generation sources, but existing hydropower is under environmental pressure. While economies of scale have limited past applications of small hydroelectric applications, the additional benefits of zero air emissions, secure supplies, and the ability to complement intermittent electric power generation technologies such as wind and solar have increased the overall consideration of the benefits of hydroelectric power generation.

A firm estimate of the amount of hydropower affected by environmental pressures is not available and needs to be determined. Fish-friendly turbine technology has the potential to address environmental concerns at existing installations as well as to enable development of new plants. Research and development to improve the cost and performance of current and new hydropower technologies and system operations, as well as technology demonstration efforts, are needed to provide the design and performance data needed to accelerate the utilization of hydropower and reduce freshwater use and overall water consumption for electric power generation.

While traditional hydropower is often considered as having only limited regional applicability, the broader opportunities for small-head and kinetic hydropower and interest in better utilizing existing capabilities could improve water efficiency while maintaining environmental water quality at existing and new sites for both conventional and nonconventional hydropower. This includes looking at integration of fish-friendly turbines, better assess fish protection and mortality reduction options and assess overall ecological benefits, and assess designs and approaches to improve water quality.

A number of novel applications of hydropower have been proposed and are in varying stages of development, with some systems undergoing field trials in the United States and the United Kingdom. Most commonly being considered are kinetic hydropower systems, and low-head hydro systems. These systems recover energy from the natural flow of rivers, tides, or currents, or make use of small-head differences of 2–4 m to create electric power. While the amount of power per unit may be small, the number of units that can be integrated in a waterway, a water distribution system, or a large surface irrigation system can create significant local electric power capabilities. There is increasing interest in using these novel approaches in small-hydro applications and to also recover energy from man-made flows, such as irrigation systems. In addition to pilot-scale demonstrations, data are needed to develop the cost and performance data necessary to understand the technical and environmental risks of the technology and to refine designs.

A wide range of alternative systems has been proposed and developed, including horizontal and vertical axis systems, lift and flutter vanes, and venturi devices. Proof of concept demonstrations that move beyond single systems at single sites to multiple-element installations and evaluations will be needed in order to fully assess the overall cost and technical performance of these systems, the impact of the operation of these types of systems on the ecology and environment such as fish mortality and erosion, and compatibility with other competing water course issues such as shipping in large areas or irrigation practices in smaller applications.

Improve Renewable Energy Grid Integration

Excluding hydroelectric power, renewable energy today provides less than 1% of U.S. electrical energy, while, in Denmark, wind alone provides over 20% of the nation's electrical energy without consuming any water. Enhanced use of renewable technologies that do not use water during normal operation would improve the overall water use efficiency of the electric power generation system in direct proportion to market penetration, especially in water-limited regions with large renewable energy potential. Better integration into the nation's electric power grid through improved access to transmission, cost-effective storage, or the use of smart grid nodes or advanced microgrids would help accelerate implementation of these water-efficient technologies to a level that would have a significant impact on water use and consumption in all regions.

Increased use of low-water-use renewable energy technologies, such as wind, solar photovoltaics, solar dish engine, tidal, and other energy systems, can reduce water demands and water consumption in the electric power sector but face two obstacles to widespread deployment, cost, and intermittency. While costs for many of these technologies are decreasing, cost/benefits still remain an obstacle to widespread deployment. Roadmaps to address the general economic, performance, and reliability issues of renewable energy already exist for many of these technologies. A second obstacle to widespread deployment is intermittency: the intermittent nature of these sources makes the energy available from these sources less reliable than electric power generated from thermoelectric power plants that can provide electric power when needed. However, just as the power system is composed of technologies that provide power continuously 24 h a day, such as coal, nuclear, and NG, and technologies that provide power in response to demand, such as gas turbines and diesel engines, and intermittent renewable technologies that provide power only when the resource is available, such as wind and solar, our electric power grid serves loads that can operate continuously or for only part of the day. Some of these loads can be controlled to operate when power is most plentiful. The value of energy from water-efficient intermittent sources could be enhanced by better integration with controllable loads as well as with other power sources. Improved value would enable larger deployment and accompanying reduced water use.

The impacts of intermittency of renewable generation technologies on grid operations have been studied in Europe, where the fraction of power provided by intermittent renewable energy is much higher than in the United States. Accommodating the intermittency of renewable energy sources by storage of electricity currently continues to be prohibitively expensive on a large scale. However, water is more easily stored. Water storage can be incorporated within the generation system, through management of hydroelectric resources or through pumped hydroelectric storage. Storage can also occur on the load side of the system. In some cases, renewable sources may produce more power than a utility can absorb given the need to maintain adequate dispatchable capacity online to regulate the system. Such excess power could be used by dispatchable loads within the water infrastructure, for example, for water or waste water treatment. Operating the water infrastructure as a controllable load could enable expanded use of water-efficient intermittent renewable generating technologies, saving both water and energy. However, a clear understanding of opportunities is lacking, but modeling and analysis could identify the benefits and impacts to the electric power system, water infrastructure, and water supplies.

REDUCING TRANSPORTATION FUEL WATER NEEDS AND IMPACTS

Most transportation fuels currently used globally are primarily derived from petroleum. In order to reduce dependence on imported oil, many countries have developed initiatives to assess and develop alternative transportation fuels from their domestic resources. Alternative transportation fuels include renewable biofuels, fuels refined from nonrenewable crude oils produced from oil shale and oil sands, and nonrenewable synthetic liquid fuels derived from coal and NG. Also of longer-term interest is hydrogen from fossil and biomass sources and from the electrolysis of water using wind, solar, or nuclear power. Biogas (primarily methane) produced from the anaerobic digestion of agricultural wastes and the organic fraction of municipal solid wastes is also a fuel that can be used for the generation of electrical power and heat as well as a transportation fuel.

Figure 20.4 provides an illustration of many of the major energy feedstock and processing pathways being used or investigated for the production of alternative fuels (EIA, 2006; Huber et al., 2006). Included are both thermochemical pathways used for most feedstock and biochemical pathways used for ethanol and oil-based biodiesel. Processing pathway details and technology approaches are generally well documented and are the subject of significant research.

TRENDS AND CHALLENGES IN ALTERNATIVE FUELS WATER USE

Significant efforts have been underway globally to advance biofuel production from various forms of biomass and research is continuing on conversion technologies and processes to improve biofuel production efficiencies and reduce costs. Biofuels production with a range of agricultural crops such as sugar cane, starch grains, soy, and other sugar and seed oil crops are currently major biomass feedstock. The relative fuel yield per hectare of land for different biomass feedstock is presented in Table 20.3.

The successful development and commercialization of lignocellulosic and algal conversion processes and technologies is of growing interest for building a broader biorefining capacity from

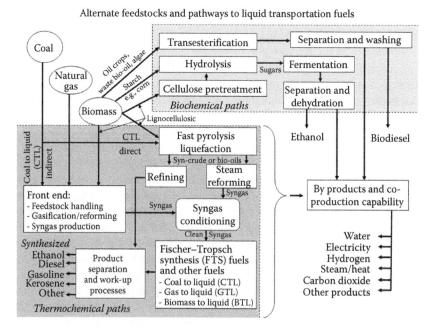

FIGURE 20.4 Feedstock and pathways for several transportation fuel alternatives. (Adapted from Pate, R. et al. 2007. *Overview of Energy—Water Interdependencies and the Emerging Demands on Water Resources.* SAND2007-1349C. Sandia National Laboratories, Albuquerque, New Mexico.)

TABLE 20.3
Biofuel Production per Hectare for Various Feedstocks

	Typical Biofuel Production (L/Hectare per Year)
Corn	40
Soybeans	100
Safflower	150
Sunflower	200
Rapeseed	250
Oil palm	1300
Micro algae	2000–12,000

various agricultural, forestry, and waste wood biomass sources to help reduce the use of crop land and food crops as a biofuel feedstock (Hightower and Pate, 2011). The research scope addressed by these efforts includes innovations in biofuel feedstock production, biochemical conversion technologies and processes, thermochemical conversion technologies and processes, by-product utilization, and technology integration and deployment noted in Figure 20.4.

In each of the steps and pathways noted in Figure 20.4, water is directly used and consumed within each of the various input feedstock and intermediary product development, conditioning, conversion, separation, and cooling processes steps noted. Washing, reforming, hydrolysis, fermentation, and refining all require water. The water requirements for each of the transportation fuels developed through these process pathways are presented in Table 20.4. Included for reference is the water consumption for refining of conventional gasoline and diesel fuels, which consumes about 1.5 L of water for every liter of fuel produced (CH2MHill, 2002). Unfortunately, as shown in

TABLE 20.4
Alternative Transportation Fuel Water Consumption

Fuel Type and Process	Water Quantity Issues	Water Quality Issues	Water Consumption (Liters of Water per Liters of Fuel)
Conventional fossil fuel Crude oil refining	Water needed to refine crude oil	Waste water from crude refining	1.5
Biofuels processing and refining Grain ethanol processing Irrigation of corn for ethanol Biodiesel processing Soy irrigation for biodiesel	Water needed for processing and irrigation where required	Waste water from processing, agriculture runoff contamination from fertilizers, herbicides, and pesticide	~4 ~1000 ~1–2 ~6000
Lignocellulose processing Biomass to liquids	Water for processing and hydrologic flow impacts	Waste water from refining, water quality changes with perennial crops	~2–6 ~2–6
Nonconventional fossil fuel Oil shale—*in situ* retort Oil Shale—*ex situ* retort Oil sands	Water needed to heat, extract, and refine	Waste water from mining and refining, *in situ* water contamination, and surface runoff	~2 ~3 ~2–6
Synthetic fuels Coal to liquids Hydrogen/electrolysis Hydrogen/NG reforming	Water needed for synthesis or for steam reforming of NG	Waste water from processing	~4–9 ~3 ~7

Table 20.4, most alternative fuels are currently from three to five times, to as much as 1000 times more water use intensive when irrigation is needed, relative to water consumption for traditional gasoline and diesel fuels. Only recently has attention been focused on this relationship between alternative fuels and water quantity, and the broader water quality, land and arable land requirements, and implications of various alternative transportation fuel production and refining technologies and options (NAS, 2007).

Water quantity and quality issues are closely tied to both energy feedstock production and the conversion processing associated with each of the alternative fuels being considered. Future alternative fuels development and production, based on current and projected approaches and processes, will likely increase demand on freshwater resources by at least a factor of three relative to current water consumption for traditional petroleum fuels refining. The regional production and refining nature of many of these alternative fuels, especially biofuels, suggests that these increased water demands will have larger impacts in some regions and could pose significant competition for available regional water supply resources in regions with already limited water supply availability. These water concerns have forced many energy and water policy analysts to highlight sustainable water resource and water quality impacts as major considerations in developing long-term transportation fuel strategies and policies to support broader overall regional and national energy security, sustainability, and reliability goals.

CONSIDERATIONS FOR REDUCING TRANSPORTATION FUEL WATER CONSUMPTION

Based on the issues highlighted in Table 20.4, presented below is a series of priorities suggested to help address the emerging water quality and quantity issues associated with any large-scale transportation fuel production. These are summarized from a wide range of findings from reports and studies conducted predominately in the United States. But the needs identified were very similar across the arid western, as well as the less arid central, and nonarid eastern regions of the United States. At first, this seemed surprising, but on closer look, the regional and local nature of many alternative transportation fuel resources will have more local water impacts and concerns, and therefore has more widespread support. It also suggests that local water issues, when it comes to local alternative fuel sources such as biofuels or even coal liquefaction or oil sands, will be driven by local and regional water availability issues, rather than national water resource data of water availability.

One of the largest water demands and water consumption for most alternative fuel processing technologies is associated with cooling. The technical needs and approaches for reducing water use for process cooling are similar to those associated with cooling applications for thermoelectric power production. Cooling water demand can be reduced not only with better cooling technologies, but also through improvements in integrated fuel processing system design and implementation that optimizes thermal energy management to minimize waste heat requiring cooling.

Another very large water demand for biofuels is irrigation for biomass feedstock production. While there are efforts to move to rainfed biomass feedstock, irrigation of even a small percentage of these crops to improve fuel supply reliability could dwarf most other biofuel water demands. And finally, there are many opportunities to reduce water use and consumption during feedstock pretreatment, processing, and conversion.

The following paragraphs offer short discussions and descriptions of the technologies or approaches that have been identified as having the potential to reduce water use and environmental impacts for alternative transportation fuels production, development, and refining. In general, approaches with the potential to use and consume less water are under significant research and evaluation. Algal fuels that use waste or brackish water, more efficient irrigation, recycling of waste water, and use of brackish water in oil sands and oil shale development, are all approaches being investigated and being tried today in many locations, especially in high-water-stress regions. But additional research and development to improve the cost and performance of these technologies and approaches, as well as technology demonstration efforts will still be needed to provide the operational data required to accelerate the widespread utilization of these approaches. The major considerations that could significantly reduce water consumption and the water footprint include the following.

Improve Alternative Fuel Process Cooling

As in thermoelectric power plants, dissipation of waste heat from nonconventional fuel feedstock pretreatment and fuel processing typically uses and consumes relatively large quantities of water for cooling. Improved materials and systems that can reduce net water demand and consumption for cooling can potentially bring about significant water savings in the production of alternative fuels. Cooling system research needs highlighted in the section "Reducing Electric Power Water Needs and Impacts" represent directions that can also address the reduction of water use for cooling in the processing of most alternative fuels, including dry and hybrid cooling, the use of non-freshwaters for cooling, etc.

Integrated Energy Use to Reduce Cooling Requirements

Reducing the amount of waste heat needing to be dissipated in alternative fuel processing will directly reduce the demand for cooling for fuel production. This involves making maximum productive use of the heat energy generated in exothermic processes to supply heat needed for other endothermic processes, and by using what would otherwise be waste heat for use in the generation of combined electric power and process heat for auxiliary processes such as moisture reduction pretreatment (drying) of biomass gasification input feedstock, and water treatment and reuse throughout the fuel processing facility. These improvements are needed to reduce cooling requirements and increase net water use efficiency and reduce makeup water demand in integrated alternative fuel processing systems. Integrated system design, implementation, and pilot scale-up demonstrations are needed to validate performance, assess cost/benefit trade-offs, and lower technical risks for commercialization.

Assess Water Impacts of Increased Irrigation for Biofuel Feedstock

There are cost/benefit trade-offs associated with the use of water and high-value agricultural land to produce irrigated crops for biofuels versus other competing products and markets. This could translate into an additional several billion liters per day of water demand for biofuels production, depending on the volume of net irrigation increase that would occur. This is expected to be less of an issue in using agricultural crop residues, other agricultural and forestry wastes, and in moving toward the future use of largely rainfed cellulosic energy crops. However, large-scale production of perennial energy crops involving millions of hectares, even when rainfed, can have water resource impacts and unintended local consequences due to alterations of hydrologic flows (Hightower and Pate, 2011). Potential impacts include soil moisture deficits, reduced subsurface aquifer recharge, and reduced surface inflows to adjacent lakes, streams, and rivers. This can represent a significant additional unplanned water demand if the agricultural sector responds in a way that creates a net increase in water use for irrigation to supply the growing demand for biofuel feedstock. The issues, trade-offs, risks, and impacts of increased water demand for biofuel feedstock irrigation is an area where additional critical analysis and quantitative assessment is needed from an energy-water and transportation fuel supply reliability perspective, especially as it pertains to fuel supply reliability in drought conditions and under future climate change considerations. Key elements of this will be better understanding and characterization of the issues and trade-offs associated with increased use of irrigation for biomass energy crops, and the issues, risks, trade-offs, and approaches to mitigate possible hydrologic flow impacts that could result from massive national expansion of perennial energy crop production, and the exploration of alternative biomass feedstock, such as oil-producing microalgae, that could also offer the potential for relatively high biofuel productivity using nontraditional waters and land that may not be suitable for agriculture.

Assess Hydrologic Flow Impacts from Energy Crops

Projections for the potential future growth of perennial lignocellulosic biomass energy crops vary by country but could range from 10 to 20 million hectares or more in support of biorefinery-based production of biofuels and other biomass-based co-products. The wide-scale planting of perennial energy crops (herbaceous grasses and fast rotation woody crops could alter hydrologic flows due to their deep, extensive root systems and dense canopies. Depending on local hydrologic conditions, this could contribute to local water-deficit impacts during drier periods. The relative water use efficiency of biomass production with herbaceous and woody perennial energy crops is higher than for many other agricultural commodity crops, but the absolute water consumption is also relatively large. As with other biofuels, lignocellulosic energy crops could see additional use of irrigation, depending on energy markets, production incentives, and concerns about ensuring the reliability and productivity of energy feedstock supplies from drier regions or during what could be abnormally dry periods in wetter regions.

Also, water quality impacts will also likely be associated with any additional agricultural runoff involving the use of pesticide, fertilizer, and herbicide inputs for increased biofuel production and will have to be assessed. A combination of modeling and field experimental work is needed to more thoroughly identify, understand, and mitigate the risks of potential adverse water quality impacts from large-scale energy crop production.

Model Climate Change Impacts on Biofuel Reliability and Irrigation Needs

Development and production of biofuels will likely be most impacted by potential climate change impacts and precipitation rates. Resource reliability could be significantly impacted. Key needs include development of climate models that can capture and predict regional precipitation and weather conditions and effects of changing climate and short- and long-term variability of weather patterns on available water supply quantity, quality, and timing, both in terms of precipitation and irrigation water availability. Fundamental questions that need to be answered include what the interrelated water, short/long-term weather variations, and climate change impacts will be on the reliability, quality, timing, and volume of biomass production, and related biofuel production, supply availability, price, and reliability over time. Also of interest is whether future climate trends will require significant additional irrigation to support long-term biomass feedstock reliability and what impact that would have on regional water sustainability and energy reliability.

Use of Low Water Use Alternative Biomass Feedstocks

Oil-producing microalgae represent one type of promising and a potentially significant renewable feedstock for biofuels that can be grown without the need for higher-quality arable land, while using nontraditional waters and CO_2 waste streams as a nutrient source. Large-scale algal growth places minimal stress on the supply and demand cycle for agricultural products utilized for human and animal consumption, and can potentially be scaled to allow for much higher overall production levels on a per hectare basis when compared to conventional terrestrial biomass plants, particularly those that produce oils. Algal-oil-based biofuels can have significantly higher energy density and are potentially more fungible within existing transportation fuel infrastructure than ethanol. The potentially high uptake of CO_2 from concentrated waste stream sources provides additional greenhouse gas abatement benefits. But considerable improvements are still needed to overcome several existing impediments to the cost-effective production and scale-up of these nontraditional biofuel resources.

Optimize Biofuels Processing and Biomass Conversion for Water Reduction

Water use, consumption, and impacts in biofuels production are predominantly associated with biomass feedstock production, and with the biorefinery utility functions that provide steam, process heat, and process cooling. All other processing and conversion water used in current biofuel processing plants and in future biorefinery system designs already appear to be moving toward achieving "zero discharge" with the intent of maximizing the capture, treatment, and reuse of water within the processing and conversion system (Aden et al., 2007). This is driven largely by the need to ensure compliance with environmental regulations, but has the added benefit of improving water use efficiency. Since many of the processes and technologies are still in development, major emphasis has been and continues to be placed on making the needed and critical advances in the efficiency and cost performance of the technologies and processes, with water impacts being of lesser concern. As the technologies and processes mature, and given the emerging issues associated with water demand and availability in the context of the expected major biofuels production volume increase, incremental improvements can provide significant cumulative water savings.

Evaluate Use of Nontraditional Waters in Alternative Fuels Production

Demands on freshwater resources for alternative fuel production can potentially be reduced by expanding the use of nontraditional waters. The use of brackish water in oil sand development, hydraulic facing for shale gas and shale oil, and algal biofuel production, and the use of high organic content waste water for methane production, for example, are already becoming a common practice. In each of these cases, the use of nontraditional water resources has displaced freshwater use and consumption. This has already significantly reduced the competition for freshwater resources in several water-stressed regions of Canada and the United States. While applications involve determining the regional and local water availability and matching nontraditional water source quality with alternative fuel development and processing, current trends suggest that many alternative fuel development approaches really can be done economically and effectively without using or consuming freshwater. This should become a major goal for all future large-scale alternative fuel development considerations and projects as a way to support sustainable water and energy development.

REDUCING ENERGY EXTRACTION AND MINING WATER NEEDS AND IMPACTS

The final component of the energy sector that uses water resources is energy extraction and refining, and mining and processing. This includes primarily conventional and nonconventional oil and gas development and coal, uranium, oil shale, and oil sand mining and processing. In this section, we provide information on the use and consumption of water in the various stages of energy resource extraction and production, including energy extraction, refining, mining, and processing. Data are presented at an average national scale based on a significant set of regional data collected within North America. While water use in mining and oil and gas development is location specific, averaging data from multiple humid and arid regions across North America helps to provide the general trends and relative impacts needed to compare water use in this energy segment to other energy development approaches. This section includes discussion of nontraditional fuels development such as biofuels, and nonconventional mineral resource development such as tight oil and tight gas hydraulic fracturing (fracking) and extraction.

To permit a relative comparison of the water needed to develop and utilize this broad range of energy resources, the water consumption information is presented in two ways. One is the volume of water used per unit energy produced (liters of water per million British thermal units), and the other metric is liters of water used per liter of oil equivalent. The first approach allows direct comparison of water use for different types of fuels, while the second approach provides an ability to compare the water use values in a format that is more easily comparable to water use and consumption by other alternative fuels and is similar of a common measurement convention (barrels of oil equivalent) typically used in evaluating various energy resource production values. For some mined and extracted energy resources, such as tight NG and uranium, the comparisons are not as relevant, but are given to provide the reader with a feel of the scale of water use per unit of energy for these different energy resources. The intent is to try and demystify with actual current operations and production data the relative water impacts of both conventional and nonconventional mining and extraction technologies and approaches. It is hoped that the discussions will provide some basic metrics to help readers better assess, compare, and identify sustainable energy and water technologies and approaches that can support improved future energy and water security, reliability, and resiliency.

TRENDS AND CHALLENGES IN ENERGY EXTRACTION AND MINING WATER USE

A short summary of the different energy mining and extraction approaches and their water use are provided in this section. The general results are summarized in Table 20.5 for the energy options discussed, including both conventional and unconventional oil and NG extraction, hydraulic fracking, and refining; coal mining and various processing and fuel use options; and uranium mining and processing. Most of the information presented in this section is taken from extensive literature reviews conducted by Sandia for the Energy Water Interdependencies Report to Congress (DOE, 2007) and available in the appendices of that report. The most recent information on the emerging shale gas, shale oil, and oil sand water use developments have been included to supplement this section with the most recent information on water use intensity in this dynamic energy development area.

Coal Mining

Water needed for coal mining varies by mining method, whether it is surface mining (~90% of current western U.S. coal) or underground mining (~65% of current U.S. Appalachian coal). Typical mining processes that require water include coal cutting in underground mines and dust suppression for mining and hauling activities. In addition, reclamation and revegetation of surface mines also require water, and requirements can be highly variable depending on a variety of factors, including

TABLE 20.5

Water Consumption in Energy Extraction and Mining

Energy Extraction or Mining Process	Water Quality Impacts	Average Water Consumption	
		L/MMBTU	L/L Oil Equivalent
Oil production	Brackish produced water spills and	8	
Conventional extraction	freshwater contamination, and waste	4–8	0.25
Hydraulic fracking	water from processing	40	0.13–0.25
Oil refining			1.25
NG development	Brackish produced and frack flowback	5	0.16
Conventional production	water spills and freshwater	8	0.25
Hydraulic fracking	contamination, waste water from	4	0.13
Gas processing	processing		
Oil shale and oil sand	Waste water from mining and refining,	96	3
Mining and processing	*in situ* water contamination, and	32	1–2
Oil shale mining	surface runoff	200	6
Oil shale *in situ* recovery		30	1–2
Oil sand mining			
Oil sand *in situ* recovery			
Coal mining and processing	Mine drainage and dewatering, waste	24	0.75
Coal mining	water from processing, tailing pile	8	0.25
Coal washing/processing	spills	240	7.5
Coal liquefaction		320	10
Coal gasification			
Uranium mining and processing	Mine drainage and dewatering, waste	16	0.50
Uranium mining	water from processing, tailing pile	32	1
Uranium processing	spills		

coal properties, mining waste disposal methods, and mine location. Estimates of water requirements for mining activities range from 40 to 400 L per metric ton of coal mined (4–24 L/MMBTU), with the lower range applicable to open pit coals with minimal revegetation activities, and the higher end applicable to underground mining of many coals. Coal can be washed to increase the heat content and partially remove sulfur. An estimated 80% of underground coal is washed (Toole, 1998). Open-surface coals typically are found in homogeneous seams with low sulfur content and often do not require washing. Water requirements for coal washing are quite variable, with estimates of roughly 80–160 L/t of coal washed (4–8 L/MMBTU).

In addition to water withdrawn to meet the needs described above, water might naturally accumulate in subsurface mines and be pumped to the surface. Western mines may also have coal seams that must be dewatered. This water, while incidental to mining activities, nevertheless represents withdrawal of groundwater. Water pumped from mines may be used to supply process needs, including cutting and washing. Excess mine water and discharged process water are contaminated and require treatment via settling ponds or other means. On occasion, water is spilled from mining operations; 1.2 billion L of coal sludge spilled in an incident in Kentucky in October 2000. Recycling of water in the underground mining process can dramatically reduce water consumption.

A rough estimate of U.S. national water consumption required for coal extraction in 2000 (mining and washing) was approximately 300 million to 1 billion L/day, or approximately 4%–12% of total freshwater withdrawals for the entire U.S. mining sector in 2000 (Hutson et al., 2004). This highlights the water demand for energy mining and the potential regional impacts on water resources that can and do occur because mining is localized in specific regions in most countries. In many locations such as the United States, China, and Australia, much of the coal resources exist in arid regions of those countries, which has significant impacts on local water resource sustainability (Figure 20.5).

FIGURE 20.5 Open-surface coal mine. (Adapted from U.S. Department of Energy. 2007. *Energy Demands on Water Resources: Report to Congress on the Interdependency of Energy and Water.*)

Uranium Mining

Estimates of water consumption in the mining uranium varies from less than 4 L/MMBTU for underground mining to 24 L/MMBTU for surface mines. Uranium is primarily mined in three states in the United States, Wyoming, Texas, and Nebraska, though there is on and off interest in opening older mines in New Mexico and Utah. This highlights the regional nature of uranium mining and how in some countries such as the United States the mines are located in arid regions where water resources are highly stressed. Depending on the type of mining and locations, uranium mines can generate from 3 to 5 million gallons of water a day from mine dewatering that must be handled and disposed. In the past, this water was simply disposed of by pumping into dry arroyos. However, under current regulations, the water must be treated to remove trace metals before disposal. After treatment, these water resources could be reused or used for other applications to improve water sustainability.

Uranium Processing

Water is also consumed in milling and enriching uranium, with total water consumption estimated at 25–30 L/MMBTU. About half of this estimated consumption is attributed to enrichment by gaseous diffusion. When enrichment by centrifuge is used, the consumption for milling and enrichment drops to about 16–20 L/MMBTU.

Conventional Oil and Gas Extraction

As shown in Table 20.5, onshore oil and gas exploration and extraction have relatively minor water requirements. Water consumption for NG extraction is negligible, and oil extraction and production requires approximately 20–20 L per barrel of oil equivalent (boe) output or 3–8 L/MMBTU or 0.25 L of water/L of oil equivalent.

Oil and Gas Produced Water

One item to note for both conventional and nonconventional oil and NG extraction is the generation of produced water. The amount of water generated varies by the age and type of well, but common produced water values vary from 6 to 10 L of water produced per L of oil. Produced water varies in quality but is often brackish to very saline in nature with total dissolved solids (TDS) (salts) varying from 2000 parts per million (ppm) to 200,000 ppm TDS. Nominal TDS values range from 30,000 to 60,000 ppm TDS. With sea water having a TDS of about 35,000 ppm, the salinity of produced water can be quite an environmental issue if spilled or if it gets into either fresh surface water or groundwater aquifers. While commonly about 70% of the produced water is recycled and used for enhanced oil recovery, another large percentage is injected in deep disposal wells where geologic formations exist. While surface water disposal and disposal at waste water treatment plants have been common in some regions, these methods of produced water disposal are being eliminated in many locations. This can require trucking to areas with deep disposal wells or on-site or regional water treatment and/or water recycling, all of which are currently relatively costly.

Oil Shales

Oil shales are shale rocks impregnated with tightly bound oil. This is very different from shale oils that are liquid and sit within the fissures of shale rocks. We will discuss shale oils and shale gases a little later. Oil shales are large geologic shale formations relatively near the surface, less than 300 m in depth. The world's largest oil shale deposit is the Green River Formation that covers parts of Colorado, Utah, and Wyoming in the arid western United States. Initial work on recovering oil from these shales focused on mining and above-ground processing (retorting). These initial studies showed that this approach consumed 2–5 L of water per L of refinery-ready oil (60–160 L/MMBTU). Most economic studies identified a proposed production rate of 500 million L/day of oil, and therefore an approximate water demand of 1–2 billion L/day, in an arid region already under water stress.

A lower-cost approach and also a lower water use approach to oil shale mining now in prototype testing is *in situ* (below ground) retorting. Electric heaters are used to heat the shale underground, and the products are collected from producer wells. About 250–300 kWh/boe of electricity is required to drive this process. Approximately two-thirds of the energy product is oil, and one-third is similar to NG. It would take the equivalent of all the produced NG to generate enough electricity for heating the shale. Water consumption with this method appears to be dominated by electricity production, but processing and decommissioning operations would also use water, although much of it might be recycled. If electricity were being provided by an evaporatively cooled combined cycle gas turbine power plant, about a liter of water would be consumed for every liter of refinery-ready oil produced (32–72 L/MMBTU). Hybrid or dry cooling could reduce water demands even more.

Water quality concerns are also an issue, with the primary concern being that the postprocessed shale residue has a high salt content that could migrate from surface impoundments to surface waters. High salinity damage is currently already a major problem in the Colorado river basin where the most of the shale resources reside. Water quality issues for *in situ* treatment are similar to those for surface retorting, except that the shale is left underground. While this reduces concerns with surface water contamination, it could cause groundwater contamination.

Oil Sands

Oil sands are a significant energy resource in Canada, but a relatively minor resource in most other countries. In northern Alberta, large sand deposits contain a hundred meters of oily sand. In most operations, the oil sands are surface mined, the oil washed out with a combination of surfactant laden water, and then reapplied to the mine pit as a sand slurry. The water consumption at a commercial oil sand plant in Canada is approximately 3–6 L of water per L of oil (80–200 L/MMBTU).

As oil sand development has expanded and the oil sand formation start to drop further below the surface, there has been movement away from surface mining to *in situ* recovery. The process used is called steam-assisted gravity drainage, where steam is injected into these deeper oil-containing sands. This liberates the oils, which flow downward where the oil and cooled steam are collected. The oil is then separated and the water is treated, heated, and then reinjected as steam. This substantially reduces the overall water demand and associated water consumption. In addition, some operations now use brackish groundwater in place of fresh surface water to further reduce freshwater use and consumption. Overall, this process has dropped freshwater use to about a third of what is normally used in oil sand surface mining operations.

Shale Oil and Shale Gas Extraction

Hydraulic fracturing of shale formation has significantly increased oil and NG production in the United States over the past decade. It has allowed the United States to develop significant oil and NG reserves in many regions of the United States and along with crude from oil sands development in Canada, has significantly reduced projected North American imports of both NG and petroleum. But shale gas and shale oil plays occur throughout the globe and have the potential to significantly change oil and NG production and trading. While this has increased the availability of oil and NG and reduced oil and NG prices, the need for water to conduct the needed hydraulic fracturing has worried water managers and environmental groups worldwide about the potential future impact on water resources and sustainable development.

The hydraulic fracturing (fracking) process, shown in Figure 20.6, uses significant volumes of water in deep horizontal wells to overpressure shale formations such that they fracture (GWPC, 2009). This liberates the trapped shale oil or shale gas such that it can be more easily extracted. The length of horizontal wells are increasing as the technology matures, which is increasing the amount of water needed per hydraulic fracture process. The July 2015 data from the United States suggest that the water volumes used in hydraulic fracturing has grown from 8 million L per full well fracture to as much as 40 million L per full well fracture. Figure 20.7 shows the locations in the United

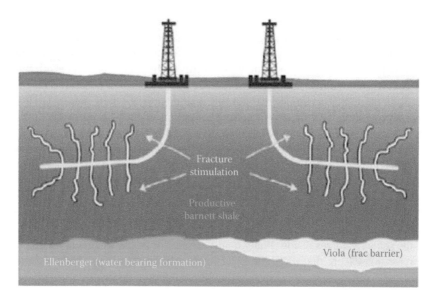

FIGURE 20.6 General hydraulic fracturing approach. (Adapted from Groundwater Protection Council. 2009. *Modern Shale Gas Development in the United States*. Groundwater Protection Council, Oklahoma City, Oklahoma.)

States with the highest average water use for hydraulic fracturing. In Figure 20.7, a cubic meter is 1000 L (Gallegos and Varela, 2015).

The data shown in Figure 20.7 represent more than 81,000 wells across the United States. The shale plays where hydraulic fracturing dominated the majority of oil and gas drilling were also the same plays where water use was highest, ranging between 2.6 and 9.7 million gallons of water per well. The study also noted that water use varied greatly across different types of oil and gas

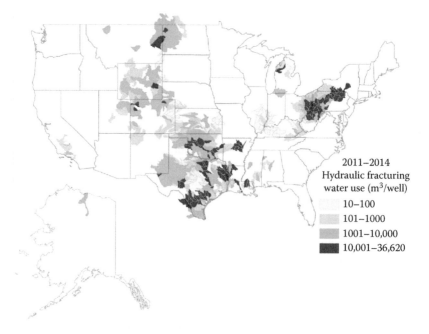

FIGURE 20.7 Water use by region for hydraulic fracturing in the United States. (Adapted from Gallegos, T. et al. 2015. *Water Resources Research*, 51(7): 5839–5845.)

reservoirs—shale reservoirs need the most water, while tight oil formations are drilled using gel-based fracking liquids instead. That is why drillers in the Bakken in North Dakota and other plays in Colorado were not among the biggest water users. The data also showed that water use in drilling has exploded in volume since 2000 coinciding with the surge in development of shale gas. A decade ago, fracking operations required less than 1 million L of water per well, but commonly between 16 million and 20 million L of water are used to frack oil and gas wells, respectively.

The advantages of the larger hydraulic fracturing efforts is that more oil and NG can be produced per well. Based on current estimates of a yield of 50–170 million cubic meters of NG per well, the actual water use per MMBTU is not much different than conventionally produced NG, about 8 L/MMBTU versus 5 L/MMBTU. While data are more widespread for shale oil development with hydraulic fracturing, the trends suggest similar results, that the water use intensity is likely to be the same or even less than for conventional oil extraction. These values are both presented in Table 20.5. Therefore, while the newer nontraditional oil and NG extraction technologies seem to use more water, the overall water use intensity has not changed. But locally, the high volume of shale gas and shale oil wells drilled in such a short period in some regions could have significant impacts on local water resource sustainability.

The biggest environmental issue documented with hydraulic fracturing is poor surface completions through shallow fresh groundwater aquifers, and surface water contamination from surface water pits. Figure 20.8 shows a typical horizontal well drilling pad in the eastern United States, showing the typical surface disposal and mud pits (GWPC, 2009). This type of drilling pad layout is also typical of conventional oil and gas well development such that the operations of conventional and nonconventional extraction are very similar. If these surface pits are not lined or maintained properly, they can leak or spill fracked or produced water into surface water systems. Disposal of the hydraulic fracturing flow back and produced water, because of the high salt content generated on contact with the shales, continues to be a major environmental and cost concern for its disposal and reuse.

Oil and Gas Processing

Refineries are large industrial complexes, with the water withdrawal rate at a typical refinery ranging from 12 to 20 million L/day. Most of the water is lost to evaporation, with about 30%–40% discharged as wastewater. Since process water may contact the petroleum product, wastewater may contain residual product as well as the water treatment chemicals and the increased dissolved solids

FIGURE 20.8 Typical horizontal well drilling pad. (Adapted from Groundwater Protection Council. 2009. *Modern Shale Gas Development in the United States.* Groundwater Protection Council, Oklahoma City, Oklahoma.)

typical of blowdown from steam systems and cooling towers. Total water consumption is about 1–2 L of water for every liter of product (1–2 L/MMBTU). This value is relatively consistent for both nominal conventional and alternative crudes from oil shales, shale oils, and oil sands.

NG requires minimal processing after extraction. Approximately 1 L/MMBTU are consumed for gas processing.

CONSIDERATIONS FOR REDUCING ENERGY EXTRACTION AND MINING WATER NEEDS AND IMPACTS

From the data presented in Table 20.5, it is clear that the water use intensity of both conventional oil and gas and nonconventional oil and NG energy development processes are similar. But it does appear that the regional drilling intensity and the sheer volume of water needs for the number of wells being drilled in different regions could become a water sustainability concern that is not yet well understood. A review of nonconventional energy resource extraction and mining across North America shows that improvements in technologies and approaches to reduce freshwater use and consumption are already underway. This includes water recycling, reuse of produced and fracking water, and use of nontraditional water supplies to replace the use of freshwater, all focused on minimizing freshwater use in water-stressed regions. In many cases, existing water stress in the production regions is driving these changes, highlighting that water issues are indeed driving energy resource development and that sustainable water management has become a key in many regions to sustainable energy development.

The bigger concern in this segment of the energy sector is on setting appropriate environmental and water management goals across transboundary oil and gas basins with different political and regulatory approaches and associated economic realities. The need to establish standard best water management practices for sustainable energy and water development across different regional basins is important to establishing common operating and regulatory frameworks to efficiently support and assure appropriate energy development. A few high-priority ideas industry and energy and water managers have pioneered include the following.

Accelerate the Use of Nontraditional Waters in Hydraulic Fracturing

As discussed, the oil and gas industry creates significant volumes of brackish produced water that in many states cannot be currently legally used for other beneficial uses. This has forced the use of freshwater in several regions, while other states encourage the use of produced water for other uses. This highlights the need for establishing a more coordinated set of regional standards and best practices that considers the regional water sustainability and water quality needs associated with energy development.

Accelerate Reuse of Hydraulic Fracking Waters

Of the water used in hydraulic fracturing, about 50%–70% is lost in the subsurface. That means that the water is essentially consumed and is not returned to the hydrologic cycle. It seems a waste to use freshwater in a way that most of it will be lost forever. Several fracking companies have developed fracking chemistries that will allow the use of waters with up to 250,000 ppm TDS as fracking fluids. Supporting more use of these types of fluid chemistries allows a reduction in freshwater use and at the same time provides environmentally friendly options for frack fluid disposal.

DRIVING TOWARD A SUSTAINABLE ENERGY FUTURE

As discussed in this chapter, the energy sector in most developed countries accounts for over 40% of freshwater withdrawals, which has started to put the energy sector into competition with other major water use sectors such as public water supplies, food and agriculture, and industry and manufacturing in many countries. This has raised questions about how to best manage and balance water use by sector to support sustainable economic development, especially in developing regions where water supplies are already stressed or where climate impacts will exacerbate water stress. It has also been pointed out that energy mining, extraction, production, and refining can also have negative impacts on surface water and groundwater quality and the environment if not managed and operated properly.

Unfortunately, the common environmental sustainability metrics used today, such as reducing the carbon footprint or reducing the thermal footprint of thermoelectric power plants, often overlook factors such as water quantity and broad water quality impacts in their overall application. Developing environmental sustainability policies optimized around a single metric or criterion can often have serious unintended consequences on system-level sustainable growth and development. As noted in this chapter, current energy sustainability drivers do not adequately consider water concerns in the face of emerging global water stress and future climate change impacts on freshwater supplies in many regions of the world. The water consumption tables provided for different energy develop approaches and technologies show that many new energy technologies to reduce carbon emissions can be anywhere from two to four times, and in some cases as much as 100 times more water use intensive than the energy approaches or technologies they are proposed to replace. On the other hand, some technologies, such as wind and solar photovoltaic electric power generation, and dry or hybrid cooling for thermoelectric power plants can significantly reduce water consumption. But many of these approaches currently have reliability and year-round availability concerns that must be addressed to solve current cost and energy reliability challenges.

Over the past decade, many agencies have studied the general issues and challenges of the energy and water nexus and how to integrate water sustainability concerns into energy planning. U.S. groups such as the Department of Energy, the National Science Foundation, the General Accountability Office, the National Research Council, and the Electric Power Research Institute, and nonprofits such as the Johnson Foundation have undertaken efforts to identify directions and innovative approaches needed to help drive a sustainable energy and water system. Internationally, many groups such as the United Nations, the World Bank, the World Council on Sustainable Development, the Canada Institute, the World Economic Forum, the World Energy Council, the Woodrow Wilson Center, the World Resource Council, and the International Energy Agency have all undertaken initiatives to improve the awareness and accelerate development and implementation of energy approaches that can both improve both energy reliability and sustainability and water sustainability. The major directions and drivers identified include the following.

REDUCE FRESHWATER USE FOR ELECTRIC POWER AND TRANSPORTATION FUELS

Many approaches exist that could help reduce freshwater consumption for electric power generation. However, technologies such as dry and hybrid cooling, and low water use renewable energy technologies have cost or intermittency issues that must be improved. Since virtually all new alternative transportation fuels will increase water consumption, major scale-up of these fuels must include approaches that use less water for growing, mining, processing, or refining.

DEVELOP MATERIALS AND WATER TREATMENT TECHNOLOGIES THAT MORE EASILY ENABLE USE OF NONTRADITIONAL WATER RESOURCES

With freshwater supplies becoming more limited, waste water reuse and nontraditional water use, including sea water, brackish groundwater, and produced water, will be needed. New water treatment technologies that can meet emerging water quality requirements at much lower energy inputs will be important. These improvements could reduce energy use for water treatment and pumping, while accelerating the use of nontraditional water resources in the energy sector, such as for cooling or hydraulic fracking.

IMPROVE WATER ASSESSMENT AND ENERGY AND WATER SYSTEMS ANALYSIS AND DECISION TOOLS

Compounding the uncertainty of available water supplies is a lack of data on water consumption. Improved water use and consumption data collection and better water monitoring are needed. Improved decision support tools and system analysis approaches are also needed to help communities and regions better understand and collaborate to sustainably develop solutions that minimize freshwater demand and consumption. Another need is the development of regional climate change models that can better predict regional or watershed level impacts of climate variability on items such as precipitation and evapotranspiration. These capabilities will help support a better understanding of water resource availability and improved water management.

IMPROVE OPPORTUNITIES TO INTEGRATE ENERGY AND WATER INFRASTRUCTURE PLANNING

Water and energy infrastructures are often currently managed independently. There are potential economies of scale by integrating or colocating energy and water infrastructure and conducting integrated planning. Waste heat from power plants or refineries and waste water from water treatment plants could be better utilized to reduce both energy and freshwater resources.

The areas noted identify the priorities necessary to enable a more balanced use of water, natural, and financial resources to support a secure, resilient, and sustainable energy future. They illustrate that sustainability in general, and energy sustainability in particular, must be considered and evaluated within a system-level, natural resource context. It is clear that all elements of the energy sector—electric power generation, fuel production, and energy extraction and mining—are dependent on water, land, air, and environmental resources to support their operations. Therefore, sustainable water, land, air, and environmental management must be integrated into all energy sector decisions and policies if we really want to create a sustainable and resilient energy portfolio that supports sustainable economic growth and development worldwide.

REFERENCES

Aden, A., S. Phillips, J. Jechura, and D. Dayton. 2007. *Thermochemical Ethanol via Indirect Gasification and Mixed Alcohol Synthesis of Lignocellulosic Biomass.* NREL/TP-510-41168.

CH2M HILL, Byers, W., G. Lindgren, C. Noling, and D. Peters. 2002. Water use in industries of the future. In *Industrial Water Management: A Systems Approach.* 2nd ed. Center for Waste Reduction Technologies, American Institute of Chemical Engineers.

Electric Power Research Institute. 1989. *Survey of Water-Conserving Heat Rejection Systems.* EPRI GS-625, Project 1260–59, Palo Alto, CA.

Feeley, T. et al. 2005. *Addressing the Critical Link between Fossil Energy and Water.* National Energy Technology Laboratory, Pittsburg, PA.

Gallegos, T. et al. 2015. Hydraulic fracturing water use variability in the United States and potential environmental implications. *Water Resources Research*, 51(7): 5839–5845.

Groundwater Protection Council. 2009. *Modern Shale Gas Development in the United States*, Oklahoma City, Oklahoma.

Hall, D. 2005. Hydropower capacity increase opportunities. *Renewable Energy Modeling Workshop on Hydroelectric Power*, National Renewable Energy Laboratory, Washington, DC.

Hightower, M. 2010a. Water and energy challenges—Emerging hydropower opportunities. *78th International Conference on Large Dams,* Hanoi, Vietnam, May 2010.

Hightower, M. 2010b. *Water Use in Electric Power*, University Council on Water Resource Conference, Seattle Washington, July 14, 2010.

Hightower, M. and R. Pate. 2011. Improving water efficiency in biofuels production, Energy supply reliability and environmental considerations. *Seventh International Starch Technology Conference*, Urbana, Illinois. June 5–8, 2011.

Hutson, S. et al. 2004. *Estimated Use of Water in the United States in 2000.* Circular 1268. U.S. Geological Survey.

International Energy Agency. 2012. *2012 World Energy Outlook.*

Llewellyn, D. and S. Valdez. 2013. *West Wide Climate Risk Assessment: Upper Rio Grande Impact Assessment.* Bureau of Reclamation, Albuquerque.

Macknick, J. et al. 2012. *Operational Water Consumption and Withdrawal Factors for Electricity Generating Technologies: A Review of Existing Literature.* National Renewable Energy Laboratory, Golden, Colorado.

Micheletti, W. and J. M. Burns. 2002. Emerging issues and needs in power plant cooling systems. In *Proceedings of the Workshop on Electric Utilities and Water—Emerging Issues and R&D Needs,* Pittsburgh, PA, July 23–24.

Moore, S. 2010. The water cost of carbon capture. *IEEE Spectrum,* Special Report, Water vs. Energy, New York, May 2010.

National Academy of Sciences. 2007. *Colloquium on the Water Implications of Biofuels Production in the United States.* Washington, DC.

National Energy Technology Laboratory. 2006. *Estimating Freshwater Needs to Meet Future Thermoelectric Generation Requirements.* DOE/NETL-2006/1235.

National Geographic, April 2009. *National Geographic Magazine,* Washington DC.

Organization for Economic Co-operation and Development. 2007. *Environmental Outlook Baseline.* Paris, France.

Pate, R. et al. 2007. *Overview of Energy—Water Interdependencies and the Emerging Demands on Water Resources.* SAND2007-1349C. Sandia National Laboratories, Albuquerque, New Mexico.

Toole-O'Neil, B. et al. 1998. Mercury concentration in coal—unraveling the puzzle, *Fuel,* 78: 47–54.

Schreir, H. and G. Pang. 2012. *Virtual Water and Global Food Security,* www.wmc.landfood.ubc.ca/webapp/VWM/.

U.S. Department of Energy. 2007. *Energy Demands on Water Resources: Report to Congress on the Interdependency of Energy and Water.*

U.S. Department of Energy. 2014. *The Water Energy Nexus—Challenges and Opportunities.*

World Economic Forum. 2009. *Energy Vision Update 2009: Thirsty Energy: Water and Energy in the 21st Century.*

World Energy Council. 2010. *Water for Energy Report,* London.

World Resources Institute. 2015. *Aqueduct—Monitoring and Mapping Water Risk,* Washington, DC. http://www.wri.org/our-work/project/aqueduct.

21 Alternative Energy Sources for the Mineral Sector
An Overview

Sheila Devasahayam and Raman Singh

CONTENTS

ABSTRACT

The cost of energy is projected to rise due to factors including infrastructure costs to transmit and distribute energy, commodity prices, exchange rates and geopolitical factors. Energy efficiency is one of the most cost-effective strategies for managing rising energy costs and growing concerns over greenhouse gas emissions. The mining industry has traditionally relied on conventional fossil-based fuel sources such as diesel, oil, coal and natural gas. This industry has a great opportunity to explore renewable sources, in order to meet its growing energy demands. Mining companies, especially those at remote locations, need to understand the nature and availability of local renewable energy – from geothermal and hydroelectric to solar and wind and biomass. The alternative energy sources often translate to reducing CO_2 emissions. In this chapter, alternative energy sources such as solar energy, microwave energy, geothermal energy, nuclear energy, blue energy, bioenergy and waste energy recovery and their application in the mineral sector are discussed.

INTRODUCTION

When considering alternative energy sources, future electricity demand cannot be undermined and it will be an increasingly important vector in the energy systems of the future (ETP, 2014). Four key points for cost-effective and practical solutions to increase energy efficiency and to reduce electricity demand as well as carbon emissions between now and 2050, according to Tracking Clean Energy Progress, 2014, are

- Rise in coal use should be accompanied by carbon capture and storage (CCS) to fundamentally comply with climate change objectives.
- Natural gas can, in the short term, play a dual role of replacing coal and supporting the integration of variable renewable energy (VRE). In the medium to longer term, it cannot be seen as a low-carbon solution.
- Balancing VRE supply and patterns of energy demand can and must be actively managed.
- Energy technology perspectives (ETP, 2014) analysis by the International Energy Agency (IEA) finds that electricity storage alone is not a necessary game changer for the future energy system.

The global demand for energy is set to increase 36% by 2035, with the industry facing greater energy price increases and volatility, and managing costs sustainably is a priority for the sector (EY, 2014). South Africa's Department of Minerals and Energy estimates that the mining industry uses 6% of all the energy consumed in South Africa. In Brazil, the largest single energy consumer is the mining giant Vale, which accounts for around 4% of all energy used in the country. In the U.S. State of Colorado, mining has been estimated to account for 18% of total industrial sector energy use, whereas the U.S. mining industry overall uses 3% of industry energy (McIvor, 2010).

Since the mining industry has traditionally relied on conventional fossil-based fuel sources such as diesel, oil, coal and natural gas to meet its growing energy demand, generating electricity as a by-product of the mining process is an attractive proposition for the mining industry where the consumption of energy is high. Many of the world's largest mining companies are evaluating greater use of renewable energy, a trend set to intensify rapidly, as part of a broader strategy to lock in long-term fixed electricity prices and availability while minimising exposure to regulatory changes, market pricing and external fuels. The mining sector is expanding into new and often remote locations as a response to increasing demand from growing and emerging markets, and they also have to deal with unreliable power supply from the grid and uncertain power prices.

In most instances, grid-connected electricity supply needs to be supplemented with on-site generation, typically large-scale diesel-based generation (that is dependent on diesel fuel supply). The more remote the mine, the more likely is the requirement for off-grid power (Richardson, 2014).

The international mining sector is making industry-wide direct investments into renewable energy infrastructure by deploying innovative energy-saving strategies (Figure 21.1). For example, Diavik Diamond Mine, Canada is looking to generate 10% of their 20–25 MW mine electrical load with renewables (EY, 2014). However, the scale and pace of renewable energy deployment across the sector is considerably slower than the business case warrants. The reasons are as follows:

- Lack of proactiveness in outsourcing to experts for developing, funding and delivering renewable energy assets
- Renewables are seen as 'non-core', with significant internal resource/opportunity cost
- Limited over-arching divisional or regional energy strategy
- Lack of clear strategic view or response to the opportunities for renewable energy developers (EY, 2014)

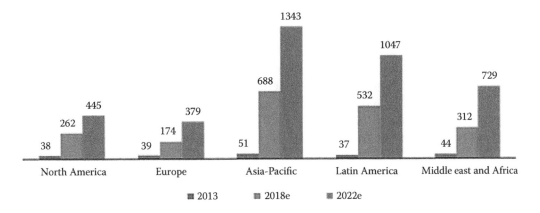

FIGURE 21.1 Renewable energy investment in the mining industry (base case, US $m), world markets: 2013–2022. (Adapted from Renewable Energy for the Mining Industry Revenue by Technology, Aggressive Investment Scenario, World Markets: 2013–2022; Renewable Energy in the Mining Industry, Navigant Consulting, Inc., 2013; EY. 2014. Mining: The growing role of renewable energy. Global Cleantech Center, http://www.ey.com/Publication/vwLUAssets/EY_-_Mining:_the_growing_role_of_renewable_energy/$FILE/EY-mining-the-growing-role-of-renewable-energy.pdf.)

ALTERNATIVE ENERGY SOURCES

Alternative energy sources are broadly defined as energy sources that do not cause or limit net emissions of carbon dioxide and thus avoid the environmental impacts associated with the combustion of fossil fuels. So, they are generally renewable sources of energy that do not require the input of fossil fuels, for example, solar energy, wind power, bioenergy, geothermal energy, wave and tidal power, hydrogen and fuel cells, and CCS as well as hydropower and energy derived from solid and liquid waste (European Commission, 2006b).

The unit cost of alternative energy is generally above that of conventional energy. In addition, high initial costs are involved in the development of alternative energy technologies and establishment of large-scale alternative energy production facilities. However, even though alternative energies have relatively low efficiency compared to conventional energy sources, they can still be a potent source of wealth and employment (WIPO, 2009) when adopted with an appropriate policy framework.

By 2040, the global demand for energy will be met by other fuel sources such as oil, biomass, biofuels, hydroelectric and geothermal (The Outlook for Energy: A View to 2040, 2015). Natural gas, nuclear power, wind and solar are set to be used more and more for electricity. By 2040, natural gas, nuclear and renewables are expected to deliver more than 70% of the world's electricity. The IEA estimates the world's remaining recoverable natural gas resources to be about 28,500 trillion cubic feet as of year-end 2013, more than 200 times the natural gas the world currently consumes annually. Natural gas will supply 135% more electricity in 2040 than in 2010, and overtake coal as the largest source of electricity.

Nuclear capacity is expected to increase by about 90%, primarily due to the growth in this sector in China and India. Solar capacity will grow by more than 20 times from 2010 to 2040. Wind capacity will expand by almost five times (the vast majority of them as onshore wind). However, even by 2040, wind and solar each will still meet far less electricity demand than nuclear power.

Interestingly, we will still continue to need many gas- and coal-fired power stations, since when there is no wind, or it is cloudy, conventional power stations will need to supply energy (Germany, Federal Ministry of Economics and Technology (2013) from 'Energy reforms on path to success'). The dramatic reduction in solar and wind energy costs in recent years has made renewable energy an increasingly attractive alternative to fossil-fuelled power, which may be highly relevant for remote mines that are not connected to the electricity grid.

SOLAR ENERGY

Solar energy productions are of two types: solar thermal energy and photovoltaic power production (WIPO, 2009). The worldwide solar power capacity has increased from 110 MW in 1992 to 178 GW, that is, 1% of global electricity demands. Worldwide deployment and installed capacity is expected to increase rapidly to greater than 500 GW between now and 2020 (http://www.solarpowereurope.org/). By 2050, solar power is anticipated to become the world's largest source of electricity, with solar photovoltaics and concentrated solar power contributing 16% and 11%, respectively.

At a price of 3–4 USD per watt, solar cells have relatively poor economic efficiency to conventional thermal power generation. However, this can be addressed by employing more efficient and cheaper materials. Developing thin-film solar cells using amorphous silicon, polycrystalline silicon, copper–indium–gallium diselenide and cadmium telluride can lead to low-cost and high-efficiency solar cells. Nanotechnologies are increasingly implemented in solar cell research, for instance, in the production of dye-sensitised solar cells or multi-junction thin-film solar cells (WIPO, 2009). Electricity from solar and wind power can cost up to 70% less than when it is generated by diesel power at mining sites. This also provides an environment-friendly solution to energy demand at remote sites as diesel can be expensive and getting it to remote and difficult locations also consumes fuel.

Pablo Burgos, Solarpack (Judd, 2013), provides a description of Solar PV for meeting the energy needs of a large mine and its integration into the energy supply. Codelco has replaced 85% of diesel demand with 51.8 GWh solar thermal energy at one facility and stands to save the expense of almost 2 months of fuel annually (Judd, 2013). It was estimated that installing and using a solar power plant at an Australian mining site could offset about 600,000 gallons in diesel fuel use (Richardson, 2014). According to the Australian Renewable Energy Agency Chairman, Greg Bourne, the technology can be used with a lot more confidence nearer a large grid, if the technology and the control systems can be proven for forecasting and intermittency.

Advantages

- Clean and unlimited source of energy
- Local energy production
- Low regional variations in availability compared to fossil fuels
- Short construction and installation period
- Low maintenance costs (solar thermal)

Disadvantages

- Power generation capacity is dependent on local weather conditions
- Large installation areas are required because of low energy density
- Large initial investment required

MICROWAVE ENERGY

Microwaves are radio waves with wavelengths ranging from as long as 1 m to as short as 1 mm. Domestic and industrial microwave ovens generally operate at a frequency of 2.45 GHz corresponding to a wavelength of 12.2 cm and energy of 1.02×10^5 eV (Rademacher and Montgomery, 1989). A distinct advantage of microwave heating over conventional heating (e.g. heating directly by thermal input in furnaces) is that the microwave energy is extremely efficient in the selective heating of materials as no energy is wasted while achieving 'bulk heating' by microwave. Unlike conventional heating sources, microwave energy is often able to heat the entire material simultaneously, taking advantage of distinctive microwave absorption properties of a given material. This enables shorter heating times, less energy consumption and reduced energy loss during the process, as a result of selective penetrating/heating capabilities (Wicks et al., 1997).

It is a commonly held view that all metals reflect microwave or even from their plasma in interaction with microwave, and hence cannot be effectively heated using a microwave field. Ishizaki et al. (2006) have successfully developed an entirely microwave-based technology for steel making with 50% reduction in carbon dioxide emission. They succeeded in producing high-purity pig irons in a multi-mode microwave reactor from iron ore powder with carbon as a reducing agent in the nitrogen atmosphere.

The emerging reduction technologies for titanium from ore produce metal powder instead of sponge. Conventional methods for sintering and melting of titanium powder are expensive and energy intensive as well as they require high vacuum (>10^{-6} Torr). Titanium reacts with oxygen at high temperatures to form a very stable oxide, which adversely affects its mechanical properties. Other melting processes such as plasma arcs require electrode consumption, and direct induction heating of the titanium powder is problematic due to high melt energy required (Technikon#1412-555 NA, 2007). On the other hand, microwave sintering or melting in argon at atmospheric pressures is potentially cost effective and energy efficient due to the possibility of direct microwave heating of the titanium powder augmented by hybrid heating in a ceramic casket (Bruce et al., 2010). Microwave heating has the following additional advantages compared with conventional heating techniques:

Advantages

- Higher heating rates
- No direct contact between the heating source and the heated material
- Selective heating may be achieved
- Easy to control the heating or drying
- Reduces the equipment size and waste

GEOTHERMAL ENERGY

Thermal energy derived from magma heat and that stored in soil, underground water or surface water can be used for heating or cooling buildings by means of a ground coupled heat pump system. Such systems operate by having a heat exchanger embedded in a borehole that supplies the energy for the evaporation and condensation of a refrigerant. Geothermal liquid can also be used to drive turbines and thus generate electricity (WIPO, 2009).

As of 2005, the largest consumers of geothermal power were the United States, with a geothermal power generation capacity of 7817 MW and Sweden with a capacity of 3840 MW (KEMCO, 2007). Geothermal energy can help reduce a mine's environmental impact and greenhouse gas (GHG) emissions, thereby improving its acceptability within communities. A local geothermal plant also results in jobs creation, and can support community development through projects such as geothermal district heating (Patsa et al., 2015).

For mines located in areas of high geothermal potential, geothermal energy can partially substitute the electrical power needs of mining operations. Geothermal fluids can be utilised in a variety of ways during the operational life cycle of a mine. Hot fluids are used directly in applications such as raffinate heating in copper production and enhanced heap leaching for the extraction of gold and silver (Patsa et al., 2015). In underground mines in areas of high geothermal potential, *in situ* geothermal power generation has the potential to reduce ventilation loads. Geothermal fluids also provide energy for space heating, typically a substantial load for northern mines. In the closure and post-closure phases of a project, irrigation by hot water (produced using geothermal energy) can enhance reclamation rates.

In Papua New Guinea, geothermal resources were used to build a power plant at the gold mine of Lihir Gold on Lihir Island, which today provides the mining operation with 'green' electricity

(Roxby Downs Sun, 2009). BHP Billiton corporate affairs manager, Richard Yeeles, in a recent interview indicated that the investigation into using geothermal heat in the Olympic Dam region was continuing, for the purpose of base-load power generation. The main factors affecting the successful integration of geothermal energy in a mining development are the presence of a proven, accessible and extractable resource; the relative price of alternate energy options; the distance to the grid; the potential for coproduction and/or minerals extraction and the availability of communities and other industries in the vicinity of the mine.

Advantages

- Highly economical due to the combined use of heating and cooling
- Stable supply due to the absence of significant fluctuations in ground temperature
- Excellent space utilisation

Disadvantages

- High initial investment (for exploration and surface development)
- Settlement of underlying supporting soil stratum may be difficult to investigate

BLUE ENERGY

Blue or marine energies that include wind, wave, tidal and ocean thermal energy have the potential to enhance the efficiency of harvesting the energy resource, minimise land-use requirements of the power sector and reduce GHG emissions by about 65 Mt CO_2 in 2020. Blue energy technologies will need investments in grid connections and transmission capacity (Blue Growth, 2012).

Wind Power

Wind currents can be used to generate electricity by using wing-shaped rotors to convert kinetic energy from the wind into mechanical energy and a generator to convert the resulting mechanical energy into electricity (WIPO, 2009). Wind power now costs a mere 1–2 cents/MW, which is half the cost of solar heat or photovoltaic power (Yoon, 2004).

In 2011, offshore wind accounted for 10% of installed capacity, employed 35,000 people directly and indirectly across Europe and represented €2.4bn in annual investments. By the end of 2011, the total offshore capacity was 3.8 GW. On the basis of Member States' National Renewable Energy Action Plans, the electricity produced from wind power in 2020 will be 494.6 TWh and of that 133.3 TWh will be generated offshore. By 2030, the annual installation of offshore capacity could exceed that onshore. Offshore wind could meet 4% of the EU electricity demand by 2020 and 14% by 2030. Continued efforts to reduce the cost of offshore wind technology are expected to accelerate this growth (Blue Growth, 2012).

Rio Tinto invested in a 9-MW wind farm at Diavik Mine in arctic conditions with the aim to generate 10% of 20–25 MW mine demand from renewables. It expects to reduce its diesel use by approximately 4 million litres and CO_2 emissions by 12,000 tonnes. Glencore Xstrata is looking to meet half of its needs from wind power at Raglan Mine (off grid). Placer Dome, acquired by Barrick Gold, has invested in wind energy in Chile and Argentina (McIvor, 2010).

Advantages

- Low maintenance cost due to minimal fuel expenses and the possibility of unmanned remote control
- Unit cost of energy production is comparable to that of fossil fuels

- Short construction and installation period
- Land on which installations are constructed can also be used for farming due to high installation height
- Installations can also be developed into tourist attractions

Disadvantages

- Wind power can only be produced economically in areas where average wind speeds exceed 4 m/s
- Installations are sensitive to changes in the natural environment, including new obstacles to wind currents
- Significant noise generated by a wind mill requires that these be placed away from residential areas

Hydropower

Electricity can be generated through the conversion of potential energy of water contained in a reservoir using a turbine and a generator (WIPO, 2009). At present, only about 10% of total water resources usable for hydropower production have been exploited (KEMCO, 2007).

In Australia, and Tasmania in particular, the electricity needs of remote mining operations that lacked access to coal prompted hydroelectric development, for example, in the remote tin mining operation at Mt Bishoff, Mt Lyell Mining (9 MW scheme) and Railway Company copper mine in Queenstown, and the Pioneer Tin Mining Co. in eastern Tasmania. The Complex Ores Company built a 49.1-MW scheme at Waddamana for their new electrolytic zinc refining process requiring a low-cost source of large amounts of electricity (Harris, 2011).

Hydropower has several benefits, compared with using compressed-air equipment, in the underground mining environment. Peter Fraser (Hydro Power Equipment (Pty.) Ltd., http://www.mining-global.com/tech/529/Hydropower-Taking-Over-Compressed-Air-Within-Drilling-Applications, 2014) says, 'Hydropower equipment is energy efficient because of the intrinsic efficiency of the positive displacement of water hydraulics and the absence of leaks. It also has lower operating costs, as localised hydropower systems feeding a local network of small bore pipes on a half-level stope are noted as the lowest-cost method to power all the operations required for conventional mining'. Peter Fraser reiterates that the scope of hydropower is applicable to all mines that use water, with further opportunities to reduce mining costs through new valves for energy recovery systems, pressure control valves and water conservation valves (http://www.saimm.co.za/Conferences/RiseOfMachines/019-Fraser).

Other applications for hydropower in underground platinum mines include the establishment of hydropowered box-hole chutes, roof bolting and stope cleaning. Implementation of hydropower at a mainly compressed-air operation could mean that mines could cherry-pick hydropower solutions most suited to their needs while maintaining the compressed air equipment. Compliance with noise and other environmental requirements is yet another driving factor behind the adoption of hydropower in the mining industry, in platinum and gold mines. It is reported that the technology enables small mines to move steadily away from using inefficient compressed air. 'For example, a new section in the mine could be developed as a stand-alone localised hydropower operation, while the compressed-air operation can be maintained for the existing operation' (Kotze, 2013).

Rio Tinto has its own hydropower generating facilities with a combined generating capacity of over 3,500 MW. The hydropower capacity that could be developed through grid-based arrangements with the mining industry in sub-Saharan Africa (excluding South Africa) is estimated at 3–5 GW by 2020 (Meike van Ginneken, 2014).

Vale is committed to developing projects to boost the use of renewable energy sources that generates a significant portion of its energy requirements through hydroelectric power plants (owning stakes in eight hydroelectric plants). In Tanzania, Placer Dome acquired by Barrick Gold, has invested in hydroelectric power (McIvor, 2010).

Advantages

- Low maintenance cost
- Relatively short design and construction period
- High energy conversion efficiency

Disadvantages

- High initial investment cost
- Water supply can be unstable

Wave Tidal Power and Ocean Thermal Energy

The energy from incoming and outgoing tides and from waves can be harnessed to generate electricity using, for instance, turbines (Renewable Northwest, 2007; WIPO, 2009). Unlike fossil-fuelled power plants, wave and tidal energy facilities generate electricity without any pollutant emissions or GHG (Renewable Northwest, 2007). Wave potential is usually measured in kilowatt per metre of wave front, and a good wave climate has an annual average of 20–40 kW for every metre of wave front, though it varies enormously from 0 kW/m in flat seas to over 1000 kW/m in major storms.

Ocean thermal energy uses the temperature difference between the surface and the deep ocean. This disparity can be used to condense and vaporise a working fluid to drive a turbine. Worldwide, ocean energy accounts for a negligible proportion of total electricity generation. Total worldwide installed wave and tidal power generation capacity is expected to grow by 47 MW between 2005 and 2009, reaching a total capacity of 227 MW by 2009 and 240 GW by 2050 (http://www.think-globalgreen.org/WAVEPOWER.html). The major portion of this capacity will be located in the open ocean, while only a small fraction (24%) will be located on the seashore (KEMCO, 2007). The share of ocean energy in world electricity generation is projected to increase by 2030, albeit only modestly. Tidal power plants currently in operation include installations in Rance in France (completed in 1967 with a capacity of 400 kW), Kislaya Guba in the Russian Federation (completed in 1968 with a capacity of 800 kW), Annapolis in the United States (completed in 1986 with a capacity of 20,000 kW) and Jiang Xia in China (completed in 1980 with a capacity of 3000 kW) (KEMCO, 2007). In order to establish effective tidal power facilities, a significant tidal change is required as well as high-capacity reservoirs. As a result, the locations where such facilities can be established are limited. In Australia, solar power and wind power are the most popular forms of off-grid renewables because they can be installed virtually anywhere (http://www.cleanenergycouncil.org.au/technologies/off-grid-renewables.html#sthash.6fDAHU47.dpuf). Australia has a significant potential ocean energy resource, especially along its western, northern and southern coastlines if both waves and tides are considered. A prime location would be off the Queensland coast, which has a differential of about 10° between surface and abyss (Hayward, 2012). CSIRO announced that energy from the ocean could supply 11% of Australia's demand by 2050 (Dopita and Williamson, 2010). That is enough to power a city the size of Melbourne. Despite its potential, there are significant constraints on the future development of ocean energy in Australia. Two limitations in particular are: technologies for the commercial conversion and utilisation of ocean energy are still immature; and capital costs, including grid connection, are high relative to other energy sources (Australian Energy Resources, 2013). In Australia, the commercialisation of wave and tidal power generation will create off-grid opportunities for island communities in the future. BioPower Systems and Worley Parsons have joined forces to deliver the final engineering and design for a 250-kW wave energy project in Victoria, which is scheduled for completion in 2015 (http://www.cleanenergycouncil.org.au/technologies/marine-energy.html#sthash.XzuQrQ8A.dpuf).

There is a potential for tide-generated electricity to contribute to the energy requirements of the mining sector in Kimberley and Pilbara areas. The 1.2-MW tide turbine was to be installed at Koolan Island (Western Australia) to meet 20% of the power needs of the mining operations when operational in 2010. However, the facility does not exist anymore due to issues with Atlantis Resources. However, the industrial loads of remote mining operations are commonly serviced by gas-fired generators. Hence the tidal or wave options, in the absence of capital grants or other subsidies such as feed-in tariffs, will need to compete with the prevailing, long-run, marginal cost of gas generation (http://arena.gov.au/files/2013/08/Chapter-11-Ocean-Energy.pdf).

Advantages

- Large-scale power generation possible
- No storage facility required
- Lightweight engineering designs
- Smooth surfaces and slow-moving operation, for minimal impact on marine species
- Limited seabed footprint with removable foundations
- Efficient energy conversion
- Reliable grid-ready power supply
- Modular and scalable, enabling large plant capacity

Disadvantages

- High initial construction cost
- Low energy density
- Impacts coastal ecosystems
- Distant from users

Hydrogen and Fuel Cells

Fuel cells are widely regarded as being the most reliable next-generation alternative energy technology, based on their level of technological maturity, lack of limitations with regard to environmental considerations and location and supply of primary energy sources. The main challenge faced in producing hydrogen lies in the need to find means of production that does not cause pollution and does not require excessive space.

Hydrogen gas can be produced from water through electrolysis or from hydrocarbons through steam reforming and can be used in combustion engines or to generate electricity in fuel cells. Hydrogen fuel cells convert the chemical energy generated by the reaction between hydrogen and oxygen into electrical energy. Electrolysis of water requires a far greater energy input than steam reforming and is therefore less economical. The energy input may be lowered using improved catalysts. Water electrolysis using electricity generated from solar heat, wind power, geothermal heat or other types of alternative energy and natural gas reforming methods can be potentially integrated into conventional hydrogen production methods.

Hydrogen fuel cells are commonly identified by the type of electrolyte used and include polymer electrolyte membrane fuel cells, solid oxide fuel cells, molten carbonate fuel cells (MCFC) and phosphoric acid fuel cells (WIPO, 2009). The integration of MCFC into a conventional large-scale power station has been shown to reduce carbon dioxide emissions by around 15%, where the MCFC is fed emissions generated from the combustion of fossil fuels in the power station (Hayase, 2005).

Hydrogen storage technologies involve low-temperature liquid hydrogen, metal hydrides and carbon nanotubes. The South African government is keen to develop a local hydrogen economy to help transform the country's mining industry into an exporter of processed minerals (McIvor, 2010). AngloPlat uses large-scale stationary fuel cell systems for power generation using coal-bed methane as the fuel source to generate electricity that is returned to the South African grid. The project involved the installation of a 200-kW demonstration stationary fuel cell power generation plant in Lephalale, Limpopo province, in South Africa (McIvor, 2010). Lack of supply of fuel in particular is a constraint as there are only a few companies manufacturing fuel cell generators of the required power, which is significantly lower than the market opportunity.

Advantages

- Water is the primary source
- Different primary materials can be used (water, natural gas, methanol, coal gas, etc.)
- No carbon emissions on combustion or reaction
- Hydrogen is easily storable and transportable as high-pressure gas, liquid or hydride
- No noise is produced by fuel cells
- No coolant is required for fuel cells
- Installation is simple and requires little space
- Very high efficiency of conversion to electrical power ideal for dispersed electrical generation
- Excellent load-following capability
- Suitable for cogeneration (heat plus electricity) (Dopita and Williamson, 2010)

Disadvantages

- High cost of manufacturing hydrogen from non-hydrocarbon sources
- Lack of hydrogen fuelling infrastructure (combustion engines)
- Low durability and dependability (fuel cell)

NUCLEAR ENERGY

The share of electricity in total in total energy consumption is only 20%, the balance 80% being derived from oil, natural gas, coal and other energy resources. Uranium makes significant contribution to the world's energy needs alongside coal (Catchpole and Robins, 2015). These two energy sources still represent at least one quarter of the world's energy needs. The current trend to reducing GHG emissions will drive a substantial reduction in demand for coal (particularly thermal coal) while accelerating the demand for uranium at an unprecedented rate. Recycling of nuclear fuel provides for greater resource utilisation, and reduced burden on deep, geological repositories (SME, 2013). KPMG (2015) predicts increasing demand and prices in the near term, with China being a key driver along with the re-start of Japan's nuclear energy sector. China's nuclear power industry could more than triple its installed capacity by 2019 with 29 nuclear reactors currently under construction (KPMG, 2015).

Large-scale nuclear hydrogen energy can replace expensive hydrocarbon fuel from the energy sector. High-temperature gas-cooled nuclear reactors (HTGRs) can provide process heat to refineries, chemical and fertiliser production facilities and this heat can reserve precious fossil fuels for feedstock uses (SME, 2013). There is an active development of atomic energy as an electricity source. Gas research institutes and general atomics in the United States developed potential pathways to implementing hydrogen production as an energy carrier and a storage medium in 1970 (Greg et al., 2013; Klimova et al., 2014).

Nuclear energy source based on HTGR in the production of reducing gases has application in direct reduction technologies in metallurgy (Klimova et al., 2014). One of the promising ways to reduce the energy consumption of metal production is the use of direct reduction technologies. The two limiting factors that limit the development of direct-reduced iron (DRI) methods are

- The need for a rich ore to maximise the productivity
- The need of natural gas or other reducing gases and a sufficient quantity of energy resources to run the process effectively

The current DRI technologies use up to 400 m^3 of natural gas per tonne of metallic pellets containing 70% of iron by weight produced, in the presence of carbon monoxide and hydrogen at high temperatures. The overall reactions are

$$Fe_XO_Y + YCO = XFe + YCO_2$$

$$Fe_XO_Y + YH_2 = XFe + YH_2O$$

Hydrogen has a lower reducing ability than CO if the reaction temperature is lower than 810°C but at temperatures greater than 810°C, it is a strong reducing agent. For a high output of iron, a H$_2$-to-CO ratio of 1 is used. The rate of reduction is increased when pure hydrogen is used as the reducing gas.

HTGR heat can produce economically efficient large-sale hydrogen and hydrogen-containing gases. The heat produced in HTGR when fed into the methane converter through a pipe system with a helium flow allows us to reduce the organic fuel consumption and to stimulate the development of hydrogen production technologies. The production of hydrogen and hydrogen containing gases through steam methane conversion follows:

$$CH_4 + H_2O = CO + 3H_2 - 206\,kJ/mol$$

$$CO + H_2O = CO_2 + H_2 + 41\,kJ/mol$$

The steam methane conversion process has better efficiency and ecological features than electrolysis. The electrolysis energy consumption is five times higher than the inherent energy capacity of the hydrogen whereas for the methane conversion process, it is 1.14 times the hydrogen energy capacity. The reducing gases for metallurgy in future will be derived from the use of organic raw material and a steam methane conversion process.

CARBON CAPTURE AND STORAGE

Carbon capture and storage (CCS) is a means for reducing GHG emissions into the atmosphere that does not require a reduction in the use of fossil fuels. Carbon dioxide generated by fossil power production and other industrial processes can be captured and stored in order to prevent it from entering the atmosphere. For this purpose, technologies must be used to separate and capture carbon dioxide from emissions and to either convert the carbon dioxide into resources such as methanol or store the carbon dioxide in a geological deposit. A concentrated pure stream of CO$_2$ is a requirement for the CCS process. In the production of synthetic natural gas and hydrogen from biomass, carbon monoxide is converted to CO$_2$ and removed to increase the energy content of the final energy product resulting in a pure stream of CO$_2$. The bio-ethanol production from fermented sugar beet results in a pure stream of CO$_2$. Gasification of black liquor in pulp and paper manufacturing allows for easy capture of CO$_2$ in addition to the production of liquid fuels (Alphen, 2011).

In the production of synthetic natural gas and hydrogen from biomass, carbon monoxide is converted to CO$_2$ and removed to increase the energy content of the final energy product resulting in a pure stream of CO$_2$. The bio-ethanol production from fermented sugar beet results in a pure stream of CO_2. Gasification of black liquor in pulp and paper manufacturing allows for easy capture of CO_2 in addition to production of liquid fuels (Alphen, 2011).

The cost of capturing the gas currently corresponds to around 70%–80% of the total cost of CCS (Wee, 2008; WIPO, 2009). Boilers, turbines, iron and steel furnaces and cement kilns require a capture step to concentrate relatively dilute streams of CO_2. A number of other processes depend upon the removal of CO_2 as part of the process itself. The CCS falls into three main categories, namely, post-combustion capture, pre-combustion capture and oxy-fuel technologies.

Carbon Capture in the Iron and Steel Sector

Overall, CCS in iron and steel production could save around 0.5–1.5 Gt CO_2 per year by 2050, which is 10%–15% of the total reduction attributable to CCS in the IEA scenarios. However, this will not only depend on technology development, but also on a global level playing field. The cost of CCS for blast furnaces is uncertain. Capture costs are estimated at €20/t CO_2 to €25/t CO_2, although changes in furnace efficiency can have a significant impact on the process economics.

In blast furnace, oxyfuelling to generate a pure CO_2 off-gas, CCS, used together with oxygen injection, could result in a reduction of 85%–95% of the CO_2 emissions attributable to the core production processes. Chemical absorption method using alkaline solvent for absorbing CO_2 has a high recovery rate of CO_2 and is applicable to a large amount of flue gas at normal pressures, but not suitable for gas containing high-level impurities (Tenaka, 2005). For high concentrations of CO_2, such as in blast furnace gas, membrane separation that uses polymeric and ceramic porous membranes to selectively separate CO_2 in gas is considered. The challenges are high cost of the membranes, low CO_2 recovery rate, high pressure requirement and the removal of the impurities before the treatment (Tenaka, 2005; Jiang, 2009).

The natural gas-based DRI production already applies carbon capture to enhance flue gas quality. CO_2 capture is by pre-combustion (gasification) and pressure swing adsorption (PSA), vacuum PSA (VPSA) or chemical absorption. The FINEX technology with some process redesign could capture all CO_2 with a reduced energy penalty. Finex eliminates the first step in the steel-making process of sintering and coking and allows the direct use of low-cost ore fines and coal, bringing overall plant installation and operational costs down as well as producing less air pollutants than traditional methods. The HIsarna smelting process produces top gas that is nitrogen free and highly concentrated CO_2 using pure oxygen instead of air. This process could capture up to 80% CO_2 in the iron-making process from iron ore and coal if equipped with CCS such as PSA or VPSA (Alphen, 2011).

Carbon Capture in Alumina Industry

The residue carbonation process adds carbon dioxide to bauxite residue (red mud), which is a mixture of minerals that are left behind when alumina is removed from bauxite (Figure 21.2). Although it is thoroughly washed, the residue retains some alkaline liquor and requires long-term storage. Mixing CO_2 into residue reduces its pH level to the levels found naturally in many alkaline soils, where it can potentially be re-used in road base, building materials or soil amendments. Carbonation's second sustainability benefit is that it locks up CO_2, which would otherwise be emitted to the atmosphere. The Kwinana carbonation plant will lock up 70,000 tonnes of CO_2 a year, the equivalent of taking over 17,500 cars off the road (http://www.alcoa.com/australia/en/info_page/pots_rd.asp).

Advantages

- Abundant carbon dioxide sources amenable for capture (coal-, oil- and natural gas-based power production, bioenergy, iron, cement, pulp production, etc.)
- Integrated gasification combined cycle (IGCC) fuelled using biomass can be combined with CCS on a much smaller scale than coal-fuelled IGCC
- Technology widely used in the chemical industry

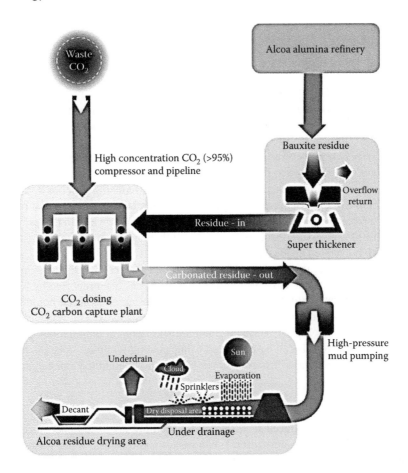

FIGURE 21.2 Carbon capture in Alcoa alumina refinery. (Adapted from http://www.alcoa.com/australia/en/info_page/pots_rd.asp, CO_2 Capture in Biomass Conversion Sector.)

Disadvantages

- Adds to costs of energy production
- Reduces energy efficiency
- Capturing carbon dioxide remains costly and difficult

Waste-to-Energy

Household and other waste can be processed into liquid or solid fuels or burnt directly to produce heat that can then be used for power generation ('mass burn'). Refuse-derived fuel is a solid fuel obtained by shredding or treating municipal waste in an autoclave, removing non-combustible elements, drying and finally shaping the product. It has high energy content and can be used as fuel for power generation or for boilers.

Advantages

- Relative short time required for commercialisation
- Low price of materials
- Eliminates waste and diminishes GHG emissions from landfills

Disadvantages

- Requires advanced technologies to prevent toxic emissions

Waste Plastic-Recycling Process

The Japanese steel industry has been developing waste plastic-recycling processes using its iron-making process. There are two major feedstock-recycling processes applied in commercial-scale operations; the blast furnace feedstock recycling and coke oven chemical feedstock recycling.

In the blast furnace process, waste plastics are used as a substitute reducing agent for coke and pulverised coal in a blast furnace to produce pig iron (Jans and Weiss, 1996; Asanuma et al., 2000, 2009). Pig iron is produced in a blast furnace by melting and reducing iron ore by the use of coke. Waste plastics composed of carbon and hydrogen act as a reducing agent in the blast furnace process to remove oxygen from iron oxide, one of the main constituents of iron ore. In this process, collected plastic waste is dechlorinated and finely pulverised, and then injected into the blast furnace through tuyeres along with pulverised coal as a reducing agent.

The use of waste plastics in coke ovens to produce 20 mass% coke, 40 mass% hydrocarbon oil (tar and light oil) and 40 mass% coke oven gas (Figure 21.3) has been reported by Nomura (2015). The coke is utilised as an iron ore-reducing agent in blast furnaces, hydrocarbon oil as raw materials for the chemical industry and the gas as a fuel for power plants. The waste plastic-recycling process using coke ovens is high performance and holistic in that a large amount of waste plastic can be treated, useful materials can be recovered using existing facilities, coke quality can be kept the same and the chlorine released from waste plastics can be fixed in the existing ammonia liquor spray system.

Bioenergy

Bioenergy refers to energy produced from biomass, including dedicated energy crops and trees, agricultural food and feed crops, agricultural crop wastes and residues, wood wastes and residues, aquatic plants, animal wastes, municipal wastes and other waste materials. Large-scale power plants based on biomass gasification have been erected by countries around the world. According to the Directive 2009/28/CE, the organic fraction of municipal solid waste (MSW) and the biogas produced and exploited from MSW are considered biomass and the energy thereof is considered biomass energy.

The use of bioenergy has contributed significantly to the reduction of GHG emissions in the United States and member states of the European Union. The European Union targets for 2020 are that 20% of the energy needs must come from renewable energy and half of it must be covered by biomass. In Finland, power plants based on biomass gasification accounts for a total output of 167 MW of electricity and 240 MW of thermal energy for use in district heating. The Australian

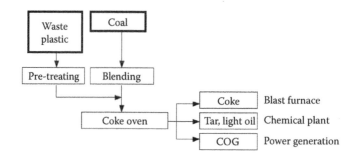

FIGURE 21.3 Outline of waste plastic-recycling process using coke ovens. (With kind permission from Springer Science+Business Media: *Journal of Sustainable Metallurgy*, Use of waste plastics in coke oven: A review, 1(1), 2015, 85–93, Nomura, S.)

Bioenergy Roadmap for stationary energy indicates that bioenergy could be expanded four-fold by 2020. Another study indicated that bioenergy could provide 29% of the electricity mix by 2040.

Biomass can be converted into solid, liquid or gas fuels or used to produce thermal energy through biochemical or physical conversion processes (WIPO, 2009). The main conversion routes to produce biomass energy include direct combustion, gasification, pyrolysis, biochemical, fermentation and biomass IGCC (Dopita and Williamson, 2010).

Gasification is a process of converting waste into synthesis gas containing carbon monoxide, hydrogen and methane. The gas has a net calorific value of 7.5–17.5 MJ/m^3 and therefore can be used to generate energy (electricity and heat) or for chemical conversion to various products. During the stages of pyrolysis and gasification, the carbon reacts with stoichiometric quantities of air and steam at temperatures between $400°C$ and $1000°C$ to form the syngas, for example, carbon monoxide, carbon dioxide and hydrogen (Table 21.1). The ratio of CO/CO_2 in the gasifier is determined by the Boudouard reaction. At temperatures lower than $700°C$, synthesis gas will contain more carbon dioxide, while at higher temperatures, most of the product will be carbon monoxide (Angelova et al., 2014). Different types of gasifiers produce syngas with different compositions and calorific values as shown in Table 21.2.

When using air as oxidant, the resulting gas has relatively high nitrogen content and a calorific value of 6–9 MJ/m^3, and is suitable for use as fuel for boilers, but is unsuitable for transportation by a pipeline. Gasification using oxygen and steam produces a gas that contains relatively high concentrations of hydrogen and carbon monoxide. Its calorific value is between 10 and 20 MJ/m^3.

TABLE 21.1

Main Reactions in Heterogeneous and Homogeneous Phase during the Solid Waste Gasification Process

Oxidation reactions

1. $C + 1/2O_2 \rightarrow CO$	-111 MJ/kmol	Carbon partial oxidation
2. $CO + 1/2O_2 \rightarrow CO_2$	-283 MJ/kmol	Carbon monoxide oxidation
3. $C + O_2 \rightarrow CO_2$	-394 MJ/kmol	Carbon oxidation
4. $H_2 + 1/2O_2 \rightarrow H_2O$	-242 MJ/kmol	Hydrogen oxidation
5. $C_nH_m + n/2O_2 \leftrightarrow nCO + m/2H_2$	Exothermic	C_nH_m partial oxidation

Gasification reactions involving steam

6. $C + H_2O \leftrightarrow CO + H_2$	$+131$ MJ/kmol	Water-gas reaction
7. $CO + H_2O \leftrightarrow CO_2 + H_2$	-41 MJ/kmol	Water-gas shift reaction
8. $CH_4 + H_2O \leftrightarrow CO + 3H_2$	$+206$ MJ/kmol	Steam methane reforming
9. $C_nH_m + nH_2O \leftrightarrow nCO + (n + (m/2))H_2$	Endothermic	Steam reforming

Gasification reactions involving hydrogen

10. $C + 2H_2 \leftrightarrow CH_4$	-75 MJ/kmol	Hydrogasification
11. $CO + 2H_2 \leftrightarrow CH_4 + H_2O$	-227 MJ/kmol	Methanation

Gasification reactions involving carbon dioxide

12. $C + CO_2 \leftrightarrow 2CO$	$+172$ MJ/kmol	Boudouard reaction
13. $C_nH_m + CO_2 \leftrightarrow 2nCO + m/2H_2$	Endothermic	Dry reforming

Decomposition reactions of tars and hydrocarbons

14. $pC_xH_y \rightarrow qC_nH_m + rH_2$	Endothermic	Dehydrogenation
15. $C_nH_m \rightarrow nC + m/2H_2$	Endothermic	Carbonisation

Source: Reprinted from *Waste Management*, 32, Arena, U., Process and technological aspects of municipal solid waste gasification. A review, 625–639, Copyright 2012, with permission from Elsevier.

TABLE 21.2

Composition and Calorific Value of the Produced Synthesis Gas According to the Type of Gasifier

Type Gasifier	Oxidising Agent	H_2 (%)	CO (%)	CO_2 (%)	H_2O (%)	CH_4 (%)	C_nH_m (%/ppm)	N_2 (%)	NH_3 (mg/Nm³)	H_2S (%/ms/Nm³)	HCl (%/ms/Nm³)	Calorific Value (MJ/m³)
Fixed-bed updraft gasifier	Air	23.4	39.1	24.4	–	5.5	4.93	–	–	0.05	–	7.5
Fixed-bed downdraft gasifier	Air	7–9	9–13	12–14	10–14	6–9	–	47–52	–	–	–	–
Bubbling fluidised bed gasifier	O_2/steam	20.0	22.0	10.5	–	5.0	–	10.0	–	–	–	13.0
	O_2	37.5	40.0	15.0	3	<1.0	–	3.0	–	–	–	11.24
	Air	6.0–12.0	14.0–15.0	16–17	–	3–4	2.9–4.1	36–58	–	–	–	9.11
Circulating fluidised bed gasifier	O_2/steam	30.1	33.1	30.6	–	5.7	770	0.4	63	0.03	–	15.00
	Air	18.0	16.0	16.0	–	5.5	2.38	42.0	2200	150	150	11.29
Plasma gasifier		42.5	45.3	4.3	0.01	–	2.56	5.2	–	0.11	0.05	17.44
		36.5	46.8	11.8	1.5	–	3.3	3.3	–	–	–	16.50

Sources: European Commission. 2006a. Integrated Pollution Prevention and Control Reference Document on the Best Available Techniques for Waste Incineration, ftp://ftp.jrc.es/pub/eippcb/doc/wi_bref_0806.pdf; Cifemo, J. and J. Marano, 2002. Benchmarking biomass gasification technologies for fuels. Chemicals and Hydrogen Production Report, US Department of Energy National Energy Technology Laboratory, USA; Reprinted with permission from Yun, Y., S. W. Chung, and Y. D. Yoo. Syngas quality in gasification of high moisture municipal solid wastes. *American Chemical Society, Division of Fuel Chemistry*, 48(2): 824. Copyright 2003 American Chemical Society.; Angelova, S. et al. 2014. Municipal waste utilization and disposal through gasification. *Journal of Chemical Technology and Metallurgy*, 49(2): 189–193, http://dl.uctm.edu/journal/node/j2014-2/12.%20S.%20Angelova_189-193.pdf.

This is suitable for gas transportation through pipelines over relatively short distances. It can be used in the synthesis of methanol and gasoline hydrocarbons.

Energy from MSW

The morphological composition and calorific value of municipal waste presented in Table 21.3 show that the percentage of the organic (combustible) components in MSW is high and depends upon the population and so it is appropriate to apply an energy recovery method for these wastes. Major barriers to the implementation of biomass energy in Australia have been the low cost of fossil fuels, specifically coal and the lack of complete understanding of bioenergy among policy makers and the general public. This results from the intrinsic complexity and variety of bioenergy production techniques (Dopita and Williamson, 2010).

Most power plants of this type are based on fluidised-bed IGCC technology (KEMCO, 2007). Large scale and high efficiency are currently necessary to achieve economic efficiency in energy production using biomass gasification. However, the quantity and stability of supply of biomass are limited, thus putting a significant damper on the adoption of this type of energy production. To overcome this problem, significant research is being invested into achieving higher efficiency in small-scale gasification power generation.

Advantages

- Raw material is abundant
- Production is carbon-neutral insofar as biomass absorbs and fixes carbon dioxide from the atmosphere during photosynthesis

TABLE 21.3

Morphological Composition and Calorific Value of Waste Generated in Bulgaria

Composition	Population				Calorific Value (MJ/m³)
	<3000 (%)	3000–25,000 (%)	25,000–50,000 (%)	>50,000 (%)	
Organic waste					
1. Food waste	4.86	12.56	20.85	28.80	4.8–6.8
2. Paper	3.87	6.55	10.45	11.10	15.7–18.6
3. Cardboard	1.30	0.70	1.63	9.70	16.3–38.0
4. Plastic	5.21	8.98	9.43	12.00	22.6–38.0
5. Textile	3.48	4.70	3.40	3.20	16.2
6. Rubber	1.15	0.45	1.10	0.60	29.9
7. Leather	1.36	1.35	2.10	0.70	29.9
8. Garden waste	14.12	14.00	5.53	6.80	4.2
9. Wood waste	2.14	2.28	1.58	1.30	16.1–19.6
Non-organic waste					
1. Glass	8.85	3.40	8.78	9.90	0
2. Metal	2.88	1.30	2.83	1.70	0
3. Slag: ashes, soil, sand, inert construction materials, including unidentified	50.78	43.73	32.35	14.20	0

Source: MOEW, National program for the management of waste 2009–2013, 2008, Bulgaria. Angelova, S. et al. 2014. Municipal waste utilization and disposal through gasification. *Journal of Chemical Technology and Metallurgy*, 49(2): 189–193, http://dl.uctm.edu/journal/node/j2014-2/12.%20S.%20Angelova_189-193.pdf.

Disadvantages

- Low energy density
- Unstable supply of raw materials

WASTE HEAT MINIMISATION AND RECOVERY

The thermal efficiency of process heating equipment, such as furnaces, ovens, melters, heaters and kilns is determined from the ratio of heat delivered to a material and heat supplied to the heating equipment. Minimising and recovering waste heat reduce energy costs and GHG emissions efficiently. Opportunities for heat recovery are based on operations that apply to many industrial sites, such as boilers, refrigeration plants, air compressors and prime movers (engines or turbines).

The use of energy for industrial processes can be broadly classed into those that are carried out in low- and high-temperature ranges. The low-temperature range covers processes operating at or discharged at temperatures up to 400–500°C found in chemicals, food and drink, paper and board and the textile and laundering sectors. The high-temperature range covers any process operating or discharging above 500°C, generally associated with furnaces or kilns, in the metals, minerals, ceramics and glass sectors. The most prevalent use of energy recovered from processes occurs in the low-temperature range (typically between ambient and 200°C).

The opportunities for heat recovery from high-temperature processes include furnaces. In any industrial furnace, the combustion products leave the furnace at a temperature higher than the stock temperature (see Figure 21.4). The higher the temperature of the excess heat, the more likely it is that the recovery will be cost effective. The energy efficiency of the process can be improved by recovering heat from the waste gas stream or the waste heat can be used elsewhere to reduce primary energy use.

The factors affecting the heat recovery include the flow rate of the waste gas stream, the temperature or 'quality' of the waste gases, the composition of the waste gases, getting the equipment up to its peak performance by reducing heat losses, improving production scheduling and closely controlling gas–air ratios (Figure 21.4). Additional significant improvements come from recapturing waste heat through direct load pre-heating, combustion air pre-heating or steam generation (Industrial Technologies Program, 2004). Waste heat recovery technologies such as cogeneration have the potential to produce electricity below the price charged by the local electricity provider. These technologies reduce dependency on the electrical grid (http://eex.gov.au/technologies/waste-heat-minimisation-and-recovery/#fn-15014-1).

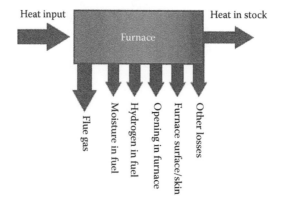

FIGURE 21.4 Typical heat flows and losses in a furnace. (Reproduced from Carbon Trust UK HQ. 2011. *Heat Recovery: A Guide to Key Systems and Applications.* Carbon Trust UK HQ, London, UK. With permission.)

Common sources of waste heat include

- High- and low-temperature industrial heating systems (furnaces, kilns, ovens, dryers and boilers) and their exhaust gases
- Industrial technologies such as compressed air systems
- Ventilation system extracts in refrigeration systems of buildings

The heat recovery opportunities include

- Pre-heating of combustion air for boilers, ovens, furnaces, kilns and other industrial process pre-heating
- Hot water generation, including pre-heating of boiler feed water
- Space heating of buildings through heat recovery in ventilation systems
- Power generation using cogeneration and trigeneration systems

Waste heat sources in relation to their level of quality are identified in Table 21.4. These areas present a number of opportunities for cost-effective heat loss minimisation and recovery. The most cost-effective heat recovery technologies tend to improve the energy efficiency of heat generating processes themselves. Table 21.5 provides a more detailed analysis of which heat exchange or cogeneration technology to use for each specific waste heat recovery opportunity. Examples of waste heat recovery strategies and enabling technologies are given in Table 21.6.

TABLE 21.4
Waste Heat Sources and Quality

Waste Heat Source	Quality of Waste Heat
1. Heat in flue and exhaust gases	The higher the temperature, the greater the potential value for heat recovery
2. Heat in vapour streams	As above, but when condensed latent heat also is recoverable
3. Convective and radiant heat lost from exterior of equipment	Lower grade if collected may be used for space heating or air pre-heats
4. Heat losses in cooling water	Lower grade, but useful energy savings if heated water is used elsewhere to transfer heat or in commercial or district residential heating; the lower grade heat can be supplemented by the use of heat pumps to raise the quality
5. Heat losses in providing chilled water or in the disposal of chilled water	High grade if it can be used to reduce demand for refrigeration; low grade if refrigeration unit used as a form of heat pump
6. Heat stored in products leaving the process may be significant	Quality depends on temperature
7. Heat in gaseous and liquid effluents leaving process	Poor if heavily contaminated and thus requiring alloy heat exchanger
8. Waste heat from air compressors	Low or high grade depending on compressor type
9. Factory/warehouse exhaust ventilation	Low grade for pre-heating replacement fresh air; efficiency can be up to 75% depending on type of heat exchanger used

Source: Carbon Trust UK HQ. 2011. *Heat Recovery: A Guide to Key Systems and Applications.* Carbon Trust UK HQ, London, UK.

TABLE 21.5
Sample of Waste Heat Recovery Strategies and Enabling Technologies

High- and Medium-Temperature Industrial and Manufacturing Processes	Temperature (Celsius)	Common Waste Heat Recovery Strategies (General)	Heat Recovery Technological Options
Furnaces (metals)	600–1650	• Steam generation for process heating or for mechanical/electrical work	Cogeneration and trigeneration (containing heat exchangers, prime movers and electric generator technologies) as well as heat exchange technologies including the following:
Cement kiln (dry process)	620–730		
Glass melting furnace	1000–1550		
Solid waste incinerators	650–1000	• Combustion air pre-heat	
Fume incinerators	650–1450	• Furnace/kiln/oven load pre-heating	
Steam boiler exhausts	230–480		
Gas turbine and exhausts	370–540	• Transfer heat to medium-/low-temperature processes	• Recuperators
Drying and baking ovens	230–600	• Pre-heating boiler feedwater	• Regenerators
		• Using waste heat boilers for electricity generation	• Heat wheels
			• Heat pipes
		• Pre-heating recycled feed materials before they are melted in furnaces	• Ebullient cooling systems
			• Forced circulation systems (cooling system)
			• Economisers
			• Deep economisers
			• Waste heat recovery boilers
Cooling water and steam production from furnaces, ovens, kilns, hot processed liquids and solids	27–260	• Pre-heating boiler feedwater	• Waste heat recovery boilers
		• Steam generation for process heating or for mechanical/electrical work	• Cogeneration and trigeneration (containing heat exchangers, prime movers and electric generator technologies)
Building ventilation systems	20–25	• Heat exchange to the incoming cool fresh air by the hot outgoing exhaust air	• Thermal wheels
			• Plate heat exchangers
			• Heat pipes
Refrigeration plant or water chillers	20–90	• Recovery of superheat and hot water to be re-used	• Refrigerant-to-water heat exchanger (de-superheater)

Source: https://eex.govspace.gov.au/technologies/waste-heat-minimisation-and-recovery/technology-background-waste-heat-minimisation-and-recovery.

Cogeneration

Cogeneration, also called combined heat and power, is a technological system that produces electrical or mechanical power and useful heat from a single fuel having capacities of 2–200 + MW (Educogen, 2001). Cogeneration systems can be designed to use any type of fuel. However, some fuels have benefits beyond the scope of the cogeneration system. Favourable fuels include organic scraps from crops (cotton and bagasse) and processing (tallow, sawdust, fabric, packaging), and biogases (methane) from manure, vegetable scrap digestion and landfill emissions. Where organic scraps and biogases are not available cost effectively, natural gas is the favourable option over other fossil fuels, such as coal and oil.

Cogeneration is used in heavy industry, oil refineries, pulp and paper manufacturing plants, food manufacturing plants and textile manufacturing plants. Cogeneration converts 60%–90% of fuel energy to useful power and heat, whereas conventional centralised electricity generation usually converts 25%–50% of fuel energy to electricity (Gans et al., 2007).

TABLE 21.6

Examples of Waste Heat Recovery Strategies and Enabling Technologies

Heat Recovery Technology	Temperature Range	Typical Sources of Waste Heat	Typical Uses	Type of Heat Exchange
Cogeneration and trigeneration	Low–high	Exhaust from gas turbines, reciprocating engines, incinerators and furnaces, cement kilns	Electricity and steam generation	Gas–liquid
Radiation recuperator	High	Soaking or annealing ovens, melting furnaces, incinerators, radiant-tube burners, reheat furnace	Combustion air pre-heat	Gas–gas
Convection recuperator	Medium–high	Soaking or annealing ovens, melting furnaces, incinerators, radiant-tube burners, reheat furnace	Combustion air pre-heat	Gas–gas
Furnace regenerators	High	Melting furnaces, reheat furnaces	Combustion air pre-heat	Gas–gas
Burner regenerators	High	Radiant-tube burners	Combustion air pre-heat	Gas–gas
Rotary regenerators	High	Exhaust ventilation, exhaust from boilers, high temperature furnaces such as aluminium	Combustion air pre-heat	Gas–gas
Metallic heat wheel	Low–medium	Boiler exhaust, curing and drying ovens	Combustion air pre-heat, space heat	Gas–gas
Hygroscopic heat wheel	Medium	Boiler exhaust, curing and drying ovens	Combustion air pre-heat, space heat	Gas–gas
Ceramic heat wheel	Medium–high	Large boilers, incinerator exhaust, melting furnaces	Combustion air pre-heat	Gas–gas
Plate-type heat exchanger	Low–medium	Exhaust from boilers, incinerators and turbines Drying, curing and baking ovens	Combustion air pre-heat, space heat	Gas–gas, gas–liquid
Heat pipe	Low–high	Waste steam, air dryers, kilns (secondary recovery), reverberatory furnaces (secondary recovery) Drying, curing and baking ovens	Combustion air pre-heat, boiler makeup water pre-heat, domestic hot water, space heat Drying, curing and baking ovens	Gas–gas, gas–liquid
Finned tube heat exchanger	Low–medium	Boiler exhaust	Boiler feedwater pre-heat	Gas–liquid
Waste heat boilers	Low–high	Exhaust from gas turbines, reciprocating engines, incinerators, furnaces	Hot water or steam generation	Gas–liquid

(Continued)

TABLE 21.6 (Continued)

Examples of Waste Heat Recovery Strategies and Enabling Technologies

Heat Recovery Technology	Temperature Range	Typical Sources of Waste Heat	Typical Uses	Type of Heat Exchange
Tube shell and tube exchanger	Low–medium	Refrigeration condensates, waste steam distillation condensates, coolants from engines, requiring heating air compressors, bearings and lubricants	Liquid feed flows requiring heating	Gas–liquid, liquid–liquid
Deep economisers	Low	Exhaust gases from gas turbines, reciprocating engines, incinerators and furnaces, cement kilns		Gas–gas
Indirect contact condensation recovery	Medium–high	Exhaust from gas turbines, reciprocating engines, incinerators and furnaces, cement kilns	Water is run across systems with hot gases which cools the gases (past condensation point) whilst turning the water into steam	Gas–liquid
Direct contact condensation recovery	Medium–high	Exhaust from gas turbines, reciprocating engines, incinerators and furnaces, cement kilns	Water is run across systems with hot gases which cools the gases (past condensation point) whilst turning the water into steam	Gas–liquid
Transport membrane condenser	High–low	Exhaust gases from gas turbines, reciprocating engines, incinerators and furnaces, cement kilns	Boiler feedwater makeup	Gas–liquid
Heat pumps (upgrading low-temperature waste heat)	Low–high	Low-temperature product streams found in process industries including chemicals, petroleum refining, pulp and paper and food processing	Upgrading waste heat to a higher temperature, or using waste heat as an energy input for driving an absorption cooling system	Liquid–gas
Thermal wheels	Low	Building ventilation exhaust streams	Heat exchange to the incoming cool fresh air by the hot outgoing exhaust air	Gas–gas
Plate heat exchangers	Low	Building ventilation exhaust streams	Heat exchange to the incoming cool fresh air by the hot outgoing exhaust air	Gas–gas
Heat pipes	Low	Building ventilation exhaust streams	Heat exchange to the incoming cool fresh air by the hot outgoing exhaust air	Gas–gas

Source: https://eex.govspace.gov.au/technologies/waste-heat-minimisation-and-recovery/technology-background-waste-heat-minimisation-and-recovery.

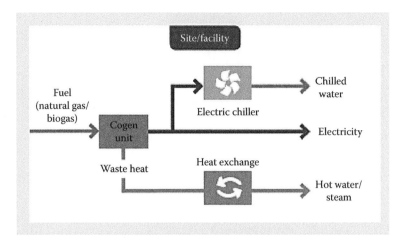

FIGURE 21.5 Cogeneration system. (Adapted from http://www.environment.nsw.gov.au/resources/business/140685-cogeneration-feasibility-guide.pdf.)

Cogeneration is a type of distributed energy generation. Distributed generation has several benefits over centralised generation. Cogeneration systems with the improved energy conversion efficiency result in lower operating costs and payback periods of 3–5 years. Major components of a cogeneration system include prime mover, electric generator, heat exchanger and control system (Figure 21.5; http://www.environment.nsw.gov.au/resources/business/140685-cogeneration-feasibility-guide.pdf).

At Alcoa alumina refineries, cogeneration power units under the partnership between Alcoa and Alinta produce both electricity and heat from the same fuel source delivering greenhouse benefits. Gas-fired cogeneration is the most thermally efficient and greenhouse friendly non-renewable energy source compared to coal-fired plants. At the Pinjarra and Wagerup alumina refineries (in Western Australia [http://www.alcoa.com/australia/en/info_page/pots_energy.asp]), it is envisaged that the use of cogeneration systems can save overall 1.8 million tonnes of greenhouse emissions as compared to the current use of coal-fired electricity.

Trigeneration

Trigeneration is an extension to cogeneration that involves the simultaneous production of electricity, heating and cooling. Trigeneration systems include an absorption chiller (or other thermally powered refrigeration device) that converts the waste heat into chilled water for cooling. A schematic presentation of the principal elements of a trigeneration system is shown in Figure 21.6.

Advantages

- Improved fuel efficiency – reducing the amount of fuel needed to provide energy means that fewer resources are required
- Reduced primary energy costs – natural gas is generally cheaper than grid-purchased electricity
- In Australia, improved National Australian Built Environment Rating System and Green Star ratings – an improved rating can have a financial benefit in rental returns and improved asset values
- Reduced network upgrade costs – with a reduced need to pay for expensive electricity network upgrades if site energy demand increases
- Reduced impact of a carbon price – the carbon price impact is greater on grid electricity than on natural gas, so purchasing less grid electricity will reduce exposure to a carbon price

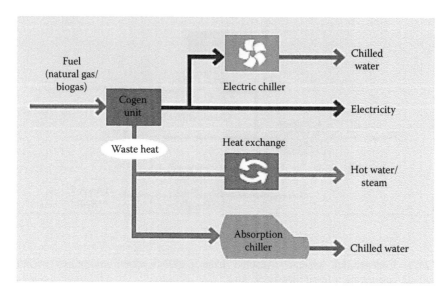

FIGURE 21.6 Trigeneration system. (Adapted from http://www.environment.nsw.gov.au/resources/business/140685-cogeneration-feasibility-guide.pdf.)

- Reduced CO_2 emissions – through reducing fuel use and replacing coal with natural gas, which has significantly lower CO_2 content
- Deferring new transmission and distribution infrastructure
- Reducing electrical losses in transmission and distribution
- Alleviating peak power demand and the potential for overload
- Providing security to local consumers by being a source of emergency or standby power
- Providing competition in generation by increasing generation diversity
- Providing opportunities to use diverse, local and favourable fuels

Disadvantages

- Cogeneration systems can be relatively complex and capital intensive and should have a plant life of 20 years

CONCLUSIONS

The mining industry has traditionally relied on conventional fossil-based fuel sources such as diesel, oil, coal and natural gas. This industry has a great opportunity to explore renewable sources in order to meet its growing energy demand. Compliance with environmental requirements is a driving factor for the mineral sector to adopt renewable energy resources for their various operations. There is a great need for mining companies, especially those at remote locations to understand the nature and availability of local renewable energy – from geothermal and hydroelectric to solar and wind and biomass. In this context, an overview of the many opportunities to reduce energy costs and greenhouse emissions in mineral sector is presented. Examples of application of the generation technologies based on renewable technologies such as solar systems, microwave energy, blue energy, heat pumps and many other technologies, where they can reduce the energy costs and improve the energy prospects for the mineral sector in many mining operations and metal productions are discussed, including the advantages and disadvantages of these energy technologies. The following is anticipated:

- The current trend to reducing GHG emissions will drive a substantial reduction in demand for coal (particularly thermal coal) while accelerating the demand for uranium at an unprecedented rate.

- By 2040, natural gas, nuclear and renewables are expected to deliver more than 70% of the world's electricity.
- By 2040, the global demand for energy will be met by other fuels sources such as oil, biomass, biofuels, hydroelectric and geothermal (The Outlook for Energy: A View to 2040, 2015).
- Natural gas, nuclear power, wind and solar are set to be used more and more for electricity.
- The mining companies will increasingly meet their energy requirements from renewable resources.

REFERENCES

Angelova, S., D. Yordanova, V. Kyoseva and I. Dombalov. 2014. Municipal waste utilization and disposal through gasification. *Journal of Chemical Technology and Metallurgy*, 49(2): 189–193, http://dl.uctm.edu/journal/node/j2014-2/12.%20S.%20Angelova_189-193.pdf

Arena, U. 2012. Process and technological aspects of municipal solid waste gasification. A review. *Waste Management*, 32: 625–639.

Asanuma, M., T. Ariyama, M. Sato, R. Murai, T. Nonaka, T. Okochi, H. Tsukiji and K. Nemoto. 2000. Development of waste plastic injection process in blast furnace. *ISIJ International*, 40: 244–251.

Asanuma, M., M. Kajioka, M. Kuwabara, Y. Fukumoto and K. Terada. 2009. Establishment of advanced recycling technology for waste plastics in blast furnace. *JFE Technical Report*, 13: 34–40.

Australian Energy Resources. 2013. Chapter 11: Ocean Energy, http://arena.gov.au/files/2013/08/Chapter-11-Ocean-Energy.pdf.

Blue Growth. 2012. *Blue Growth – Opportunities for Marine and Maritime Sustainable Growth.* Communication from the Commission to the European Parliament, the Council, the European Economic and Social Committee and the Committee of the Regions. COM/2012/0494 final, http://eur-lex.europa.eu/legal-content/EN/TXT/?uri=CELEX:52012DC0494.

Bruce, R. W., A. W. Fliflet, H. E. Huey, C. Stephenson and A. M. Imam. 2010. Microwave sintering and melting of titanium powder for low-cost processing. *Key Engineering Materials*, 436, 131.

Carbon Trust UK HQ. 2011. *Heat Recovery: A Guide to Key Systems and Applications.* Carbon Trust UK HQ, London, UK.

Catchpole, M. and W. Robins. 2015. Future global energy demand. *The AusIMM Bulletin*, August, 2015, pp. 28–31. https://www.ausimmbulletin.com/feature/future-global-energy-demand/.

Cifemo, J. and J. Marano, 2002. Benchmarking biomass gasification technologies for fuels. Chemicals and Hydrogen Production Report, U.S. Department of Energy National Energy Technology Laboratory, USA.

Dopita, M. and R. Williamson. 2010. *Australia's Renewable Energy Future*, Australian Academy of Science 2009, Canberra, Australia. http://theconversation.com/ocean-power-making-waves-in-australias-clean-energy-future-9689.

Educogen. 2001. *The European Education Tool on Cogeneration.* 2nd ed. The European Association for the Promotion of Cogeneration, Brussels, Belgium, 8pp, http://www.cogen.org/Downloadables/Projects/EDUCOGEN_Tool.pdf (accessed 17 April 2007).

ETP. 2014. Energy Technology Perspectives, http://www.iea.org/etp/etp2014/.

European Commission. 2006a. Integrated Pollution Prevention and Control Reference Document on the Best Available Techniques for Waste Incineration, Spain. ftp://ftp.jrc.es/pub/eippcb/doc/wi_bref_0806.pdf.

European Commission. 2006b. *The State and Prospects of European Energy Research: Comparison of Commission, Member and Non-Member States R&D Portfolios.* EC, Brussels. http://www.estatisticas.gpeari.mctes.pt/archive/doc/portfolios_report_en.pdf (accessed 5 March 2009).

EY. 2014. Mining: The growing role of renewable energy. Global Cleantech Center: http://www.ey.com/Publication/vwLUAssets/EY_-_Mining:_the_growing_role_of_renewable_energy/$FILE/EY-mining-the-growing-role-of-renewable-energy.pdf.

Gans, W., A. M. Shipley and R. N. Elliot. 2007. *Survey of Emissions Models for Combined Heat and Power Systems.* ACEEE, USA, p 1, http://aceee.org/pubs/ie071.pdf?CFID=1902973&CFTOKEN=31285910 (accessed April 2007.).

Harris, D. 2011. Hydroelectricity in Australia: Past, Present and Future, http://ecogeneration.com.au/news/hydroelectricity_in_australia_past_present_and_future/055974/.

Hayase, M. 2005. Development method and technology trend of fuel cell. *Transistor Technology (Japan)*, 42(8): 173–189.

Hayward, J. 2012. Ocean power making waves in Australia's clean energy future, http://theconversation.com/ocean-power-making-waves-in-australias-clean-energy-future-9689.

Industrial Technologies Program. 2004. *A Best Practices Process Heating Technical Brief, Waste Heat Reduction and Recovery for Improving Furnace Efficiency, Productivity and Emissions Performance*. U.S. Department of Energy Office of Energy Efficiency and Renewable Energy, USA. http://www1.eere.energy.gov/manufacturing/tech_assistance/pdfs/35876.pdf.

Ishizaki, K., K. Nagata and T. Hayashi. 2006. Production of pig iron from magnetite ore–coal composite pellets by microwave heating. *ISIJ International*, 46(10): 1403–1409.

Janz, J. and W. Weiss. 1996. Injection of waste plastics into the blast furnace of Stahlwerke Bremen. *3rd International Cokemaking Congress, CRM-VDEh*, Gent, Belgium, 114–119.

Jiang, K. 2009. WP2 – Future Energy Technology Perspectives. 2.2 Energy Intensive Sectors Technology Assessment, http://s3.amazonaws.com/zanran_storage/www.nzec.info/ContentPages/44532556.pdf

Judd, E. 2013. Solar PV for Codelco and Collahuasi. *Canadian Clean Energy Conferences. Renewables and Mining*, Toronto, Canada.

KEMCO (Korea Energy Management Corporation). 2007. New Renewable Energy RD&D Strategy 2030, http://www.energy.or.kr/.

Klimova, V., V. Pakhaluev and S. Shcheklein. 2014. Production of a reducing environment for metallurgy using nuclear energy. In *Energy Production and Management in the 21st Century (2 Volume Set), The Quest for Sustainable Energy*, edited by C. A. Brebbia, UK: Wessex Institute of Technology; E. R. Magaril and M. Y. Khodorovsky, WIT Press, UK.

Kotze, C. 2013. Hydropower increasingly used as opposed to compressed air in platinum, gold mines, Creamer Media's Mining Weekly, Africa/Europe edition, August 2013. http://www.miningweekly.com/article/hydropower-increasingly-used-as-opposed-to-compressed-air-in-platinum-gold-mines-2013-08-232013.

KPMG. 2015. Commodity Insights Bulletin, January 2015, Uranium (Q2, 2014 and Q3, 2014).

McIvor, A. 2010. Mining and Energy, http://www.cleantechinvestor.com/portal/fuel-cells/6422-mining-and-energy.html.

MOEW, National Program for the Management of Waste 2009–2013, 2008, Bulgaria.

Naterer, G. F. and I. D. Calin 2013. *Hydrogen Production from Nuclear Energy*. Springer, London, UK.

Nomura, S. 2015. Use of waste plastics in coke oven: A review. *Journal of Sustainable Metallurgy*, 1(1): 85–93.

Patsa, E., D. Van Zyl, S. J. Zarrouk and N. Arianpoo. 2015. Geothermal energy in mining developments: Synergies and opportunities throughout a mine's operational life cycle. *Proceedings World Geothermal Congress 2015*, Melbourne, Australia, 19–25 April.

Rademacher, S. E. and N. D. Montgomery. 1989. Base level management of radio frequency radiation protection program. AFOEHL Report 89-023RC0111DRA, USA.

Renewable Northwest. 2007. Wave & Tidal Energy Technology, http://www.rnp.org/node/wave-tidal-energy-technology.

Richardson, J. 2014. Renewables can cost 70% less than diesel power at mining sites, http://reneweconomy.com.au/2014/renewable-energy-can-cost-70-less-diesel-power-mining-sites-11278.

SME. 2013. Nuclear Power, http://www.smenet.org/SME/media/Government/201312_Nuclear-Power.pdf.

Sun, R. D. 2009. Mining sector seeing geothermal as a 'future opportunity', http://www.thinkgeoenergy.com/mining-sector-seeing-geothermal-as-a-future-opportunity/.

Technikon#1412-555 NA. 2007. Energy efficiency of microwave melting, 1412-555 NA, January, http://www.afsinc.org/files/1412-555%20energy%20public_1383851915248_7.pdf.

Tenaka, R. 2005. Carbon Capture Technologies in Japan. British Embassy Tokyo: Science and Innovation Section, https://ukinjapanstage.fco.gov.uk/resources/en/pdf/5606907/5633507/36463X.pdf.

The Outlook for Energy: A View to 2040. 2015. http://cdn.exxonmobil.com/ ~ /media/global/Reports/Outlook%20For%20Energy/2015/2015-Outlook-for-Energy_print-resolution.

Tracking Clean Energy Progress. 2014. http://www.cleanenergyministerial.org/Portals/2/pdfs/CEM5-IEA-TCEP_2014_Report.pdf.

van Alphen, K. 2011. CO_2 Capture: Industrial Sources Technology Roadmap IEAGHG Summer School, Global CCS Institute, http://www.ieaghg.org/docs/General_Docs/Summer_School/2011/Presentations/24_2011INDUSTRIALSOURCES-VANALPHEN.pdf

van Ginneken, M. 2014. A decade of sustainable hydropower development – What have we learned? http://www.hydroworld.com/articles/print/volume-23/issue-1/articles/a-decade-of-sustainable-hydropower-development-what-have-we-learned.html.

Wee, J. H. 2008. Reduction of carbon dioxide omission using CCS (carbon capture & storage). EMC (Environmental Management Corporation), www.emc.or.kr/include/board_union/union_filedown. asp?bbs_id = 1879 (accessed March 10, 2009).

Wicks, G. G., D. E. Clark and R. L. Schulz. 1997. Microwave technology for waste management applications: Treatment of discarded electronic circuitry. DOE Report, WSRC-MS-97-0299.

WIPO. 2009. Patent-Based Technology Analysis Report – Alternative Energy Technology. Geneva, Switzerland: World Intellectual Property Organization, http://www.wipo.int/patentscope/en/technology_focus/alternative_energy.htm.

Yoon, C. S. 2004. Alternative Energy. Intervision Co. LTD., Korea.

Yun, Y., S. W. Chung and Y. D. Yoo. 2003. Syngas quality in gasification of high moisture municipal solid wastes. *American Chemical Society, Division of Fuel Chemistry*, 48(2): 824.

22 Nuclear Power
Current Status and Prospects

Gail H. Marcus

CONTENTS

ABSTRACT

Nuclear power presently accounts for about 12% of the electricity production around the world and is one of the major sources of low-carbon, reliable base load electricity production. As a result, active programs are underway in several countries around the world to build more nuclear power plants. However, a history that started with ties to nuclear weapons programs and includes several serious accidents, as well as concerns about waste disposal and other factors, have raised questions about nuclear power in some countries. As a result, other countries are closing operating nuclear power plants or curtailing plans to build new ones. New reactor design concepts are under development that may help address some of the concerns. This chapter addresses the major issues, current technologies, and advanced concepts under development.

BACKGROUND

History

Nuclear power is arguably the newest major energy-producing technology. It did not begin to provide any useful energy at all until the mid-1950s (Controlled Nuclear Chain Reaction, 1992; Marcus, 2010). By contrast, the other major sources of electricity today have been in use for decades, or even centuries, and mankind has made limited use of the sun and wind since ancient times, albeit with much more primitive technology than exists today. Given the short length of time nuclear power has been a part of the world's energy-producing portfolio, it is significant that it now accounts for over 10% of the world's electricity production, and in a few countries, is the major source of electricity.

The earliest work on nuclear reactors was inextricably linked to the rush to develop nuclear weapons and nuclear propulsion during and immediately following World War II. Those efforts shaped much of nuclear technology as we know it today. In an effort to accelerate weapons development, a wide range of potential nuclear technologies was tested, and the technologies that became established in the early days were those that could be developed and built most quickly, while others were discarded, perhaps prematurely.

As a result, a small number of nuclear technologies dominate the nuclear market today. However, current needs and concerns are spurring a serious look at alternative nuclear technologies that might better meet future needs and emerging markets.

Current Status

From its early beginnings in a few countries that were engaged in weapons development activities, the use of nuclear power for electricity production spread, and today, there are more than 400 nuclear power plants in over 30 countries around the globe. Table 22.1 shows the countries with operating nuclear power plants today.

TABLE 22.1

Operational Power Reactors in the World by Country

Country	Number of Reactors (2015)	Electrical Capacity (MWe)	Percentage of Total Electricity from Nuclear Power in 2014
Argentina	3	1627	4.0
Armenia	1	375	30.7
Belgium	7	5921	47.5
Brazil	2	1884	2.9
Bulgaria	2	1926	31.8
Canada	19	13,500	16.8
China	28	24,025	2.4
Czech Republic	6	3904	35.8
Finland	4	2752	34.6
France	58	63,130	76.9
Germany	8	10,799	15.8
Hungary	4	1889	53.6
India	21	5308	3.7
Iran	1	915	1.5
Japan	43	40,290	0
Mexico	2	1330	5.6
Netherlands	1	482	4.0
Pakistan	3	690	4.3
Romania	2	1300	18.5
Russia	34	24,654	18.6
Slovakia	4	1814	56.8
Slovenia	1	688	37.2
South Africa	2	1860	6.2
South Korea	24	21,667	30.4
Spain	7	7121	20.4
Sweden	10	9651	41.5
Switzerland	5	3333	37.9
Taiwan	6	5032	18.9
Ukraine	15	13,107	49.4
United Kingdom	16	9371	17.2
United States	99	98,708	19.5
Total	438	379,055	10.9

Source: International Atomic Energy Agency (IAEA) Power Reactor Information System (PRIS), https://www.iaea.org/PRIS/WorldStatistics/OperationalReactorsByCountry.aspx (retrieved September 15, 2015); World Nuclear Association. 2015. World Nuclear Power Reactor and Uranium Requirements, http://www.world-nuclear.org/info/Facts-and-Figures/World-Nuclear-Power-Reactors-and-Uranium-Requirements/ (retrieved September 15, 2015); 2015. Nuclear News world list of nuclear power plants. *Nuclear News*, 58(3): 39–72, http://www.ans.org/pubs/magazines/nn/y_2015/m_3.

Notes: The number of reactors shown for each country includes all reactors listed as operational as of September 15, 2015. It includes reactors that are presently shut down but remain in operational status. The table does not show reactors that are under construction. The percent of total electricity from nuclear power in each country is for the calendar year 2014, the last full year for which data was available. All reactors in Japan were shut down during that entire year.

In addition, several countries with nuclear power are expanding their use of the technology with active new-build programs, and a number of countries that have not had nuclear power in the past are either building plants today, initiating new programs, or considering the use of nuclear power in the future.

On the other hand, other countries are planning to reduce, or even to phase out, their use of nuclear power. The situation in individual countries is complex and will be addressed in more detail at the end of this chapter. First, though, it may be useful to consider some of the broad issues, pro and con, that have affected, and are affecting, nuclear power development.

Nuclear power developed rapidly in the early years. Its development was spurred by several factors that varied somewhat, depending on the country. First, most of the Western world was experiencing growing demands for electricity. Therefore, countries that had engaged in weapons development during World War II turned their acquired expertise and infrastructure to peaceful applications after the war. This expertise was shared with other countries through the International Atomic Energy Agency (IAEA) and the "Atoms for Peace" program initiated by President Eisenhower (Röhrlich, 2013).

The Arab oil embargo of the early 1970s added another dimension to the interest in nuclear power. Most electricity was supplied by fossil fuels, so countries such as France, which lacked their own fossil fuel resources, made a committed and intensive effort to develop nuclear power as their primary source of electricity. Nuclear power eventually grew to supply about three quarters of France's electrical power.

Even in countries that imported only a fraction of the fossil fuel they needed, the Arab oil embargo of the early 1970s sensitized countries to their vulnerability to supply disruptions. Although the oil embargo affected the transportation sector more than the electricity-production sector in most parts of the world, many countries felt they needed to improve the security of their overall energy supply.

Indeed, the current issues of natural gas supply from Russia to Europe demonstrate that such concerns remain. Since only very small amounts of uranium are required to run reactors and they can operate for months, or even several years, on one fuel load, it is not as easy for another country to cut off supplies and create a near-term crisis. Furthermore, uranium resources are fairly wide-spread and there are substantial resources in politically stable countries.

Other benefits of nuclear power were not as apparent in the early days, but have become important in more recent years. One of these benefits is that nuclear power generates very little air pollution or carbon emissions. Another benefit is that nuclear power provides reliable base load power. Under normal operation, it is not subject to short-term fluctuations and, since refueling is infrequent, it is not subject to delivery disruptions due to the weather.

CONTROVERSIES ABOUT NUCLEAR POWER

Despite these benefits, public opinion is mixed or negative in regions of the United States and in some parts of the world (Public Attitudes to Nuclear Power, 2010). Again, there are several reasons for this, ranging from the aesthetic (current nuclear power plants are large and they are sited on shorelines, which are often scenic areas and are desirable venues for residences and recreational activities) to concerns about safety.

The safety concerns are complex. In part, these concerns arise from the link in some people's minds between nuclear power and nuclear weapons. Other concerns arise from the way individuals approach risks, in general. Studies have shown that people tend to exaggerate risks they cannot control and to minimize risks they think they can control. That is the reason that most people fear flying or the operation of nuclear power plants more than they fear driving a car, even though cars statistically cause more fatalities.

Although concerns about nuclear power pre-dated the most serious accidents at nuclear power plants, each of the accidents exacerbated public concerns about safety. The three most significant accidents were Three Mile Island in the United States in 1979, Chernobyl in the U.S.S.R. (now Ukraine) in 1985, and Fukushima in Japan in 2011.

The causes and consequences of these accidents are complex and the details are beyond the scope of this chapter. For purposes of this discussion, what is important is that, while the accidents did not result in significant losses of life, they all had very serious economic consequences for the utilities involved and for the local population. In addition, they all increased the opposition to the use of nuclear power and caused disruptions for other operating reactors and for planned reactors, both in the countries where the accidents occurred and in other countries.

On the positive side, in the case of Chernobyl, the accident was a consequence of features of the reactor that did not exist in other designs, and reactors of that design have now largely been phased out. Also, many of the measures that are now standard at nuclear power plants were instituted to address the deficiencies that led to these accidents, making nuclear power safer overall.

Public concerns are not the only reasons for some of the ups and downs nuclear power has experienced over the years. Building nuclear power plants has proven to be a long and costly undertaking in many countries, with significant cost and schedule overruns. While many attribute the delays and cost overruns to regulatory oversight, it is a fact that large projects of many types seem prone to delays and cost escalation (Flyvbjerg et al., 2003). Whatever the causes, the high costs of building new nuclear power plants have helped put the brakes on some planned construction projects and have spurred interest in new designs that can be constructed more efficiently.

BASIC CONSIDERATIONS

SIMILARITIES AND DIFFERENCES TO OTHER TECHNOLOGIES

At the most fundamental level, nuclear reactors are quite similar to most currently operating power plants that produce electricity. The fuel—uranium for nuclear power plants, and coal, oil, or natural gas for fossil fuel plants—is used to boil water to create steam, and the steam is used to drive turbine generators that produce electricity.

As in the case of fossil plants, after the steam has transferred its energy to the turbine, it feeds into a condenser. The condenser consists of a large number of tubes through which water from the sea or from a river or lake circulates. The condenser cools the steam from the turbine and condenses it back to water, which is then returned to the reactor using recirculation pumps. The cooling water in the condenser is returned to the sea or other body of water at a temperature a few degrees higher than it started. The two loops are isolated from each other so the water does not mix.

Where there is not a large body of water available to serve as the coolant for the turbine exhaust, or in cases where the power plant would cause a possibly detrimental increase in the temperature of a body of water, a cooling tower is used. Cooling towers are not unique to nuclear power plants. They are used for other types of power plants, and for other industrial facilities, where sufficient external cooling water is not available. When a cooling tower is used, the water that is fed to the condenser to cool the exhaust from the turbines is then circulated to the cooling tower where it, in turn, is cooled by spraying it inside the tower, thus transferring its heat to the air. It condenses again in the bottom of the tower and is recirculated to the condenser.

Of course, many details are different. In particular, the fuel, what it does, and how it is handled, is very different. One major difference is that the fuel for a nuclear power plant does not burn like a fossil fuel. Rather, a nuclear power plant relies on a process called fission. In this process, neutrons hitting atoms of uranium cause them to split. The splitting produces two smaller elements, and releases some energy. The resulting heat boils the water.

Because of the way the fission process must be controlled, the fuel has to be handled very differently from fossil fuel. There are very stringent requirements on the configuration of the fuel to assure that there are enough fission events to maintain the reaction, but that the number of fission events stays at a stable level. In addition to the configuration of the fuel, other measures, such as the use of control rods, allow the operators of the nuclear power plant to increase and decrease the power of the reactor as needed.

FISSION REACTION

Most nuclear reactors in use in the world today are called thermal nuclear reactors. In these reactors, the nucleus of a heavy fuel element, usually uranium, absorbs a slow-moving, or thermal, neutron. This causes the atom to become unstable, and it splits, or fissions, into two smaller atoms, known as fission products. The fission process releases two or three fast-moving neutrons plus the kinetic energy of the recoiling fission products. Generically, the reaction is

$$1\,\text{neutron} + 1\,^{235}\text{U atom} \rightarrow \text{fission fragments} + 2\,\text{or}\,3\,\text{neutrons} + \text{energy}$$

The atom can break up in different ways, yielding a range of different atoms as the fission products. Some common fission products include isotopes of zirconium, cesium, strontium, and iodine. Many of the fission products are radioactive, with half-lives ranging from a few hours to thousands of years. They in turn decay into other elements, usually by the emission of beta particles. On average, the kinetic energy of the fission products totals a little over 200 MeV (million electron volts). The kinetic energy transfers heat to the surroundings, and the neutrons emitted create a chain reaction.

Without anything to stop the reaction, each successive fission event yields more and more neutrons, and the reaction would increase exponentially. That is what happens in a bomb. However, a nuclear reactor includes materials that absorb some of the neutrons, thus allowing the reaction to continue in a self-sustaining, but controlled manner.

Natural uranium contains about 99.3% of the isotope uranium-238 (^{238}U) and 0.7% of the isotope uranium-235 (^{235}U). (It also contains trace amounts of ^{234}U.) ^{235}U and ^{238}U behave differently when hit by neutrons. Most reactors today make use of the fission reaction in ^{235}U, which has a higher "cross-section" for absorbing "slow," or "thermal" neutrons, than does ^{238}U. ^{235}U also has a higher cross-section for absorbing slow neutrons than it does for the fast neutrons emitted when the atoms are split. Therefore, the neutrons must be slowed down to make the process work. This is done by collisions with a "moderator." Moderators are usually lightweight atomic nuclei, as neutrons transfer more of their energy to moderators that have a mass close to that of the neutron. Typical moderators in current reactors are water (hydrogen), heavy water (deuterium), and graphite (carbon).

Other fission reactions are possible. To date, these have been used only to a limited extent, but they are important for some of the advanced reactor designs.

^{238}U is not directly usable as nuclear fuel. Although fast neutrons (>1 MeV) can cause some ^{238}U atoms to fission, the number of neutrons they produce are not enough to sustain a reaction. However, ^{238}U can absorb a neutron to form ^{239}Pu, which is fissionable. In fact, this process can produce more fuel than is consumed, thus creating a "breeder" reactor. The resulting plutonium can remain in the reactor and be consumed as fuel, or can be removed and separated from the uranium to fuel other reactors.

A third possible fuel is uranium-233 (^{233}U). ^{233}U does not occur naturally, but rather, can be created from thorium, namely, ^{232}Th, leading to what is often called a thorium fuel cycle. ^{232}Th itself is not fissile—that is, it cannot be split like ^{235}U. However, it is considered "fertile." That means it can absorb a neutron that will transmute the ^{232}Th to ^{233}U, which in turn is fissionable. This is conceptually similar to the transmutation of ^{238}U to ^{239}Pu. The thorium in a reactor therefore needs to be combined with a fissile material to enable the conversion to ^{233}U. This material is called a "driver." The ^{233}U that is produced can either be chemically separated and recycled into new fuel, or it may be used *in situ*, depending on the reactor concept. A variety of existing and advanced reactor concepts can conceptually be designed to use thorium.

NUCLEAR FUEL

In contrast to fossil fuels, nuclear power has a much more complex fuel cycle. First, as with fossil fuels, the ore must be mined. Following that, however, it must be refined, enriched to increase the proportion of ^{235}U, and fabricated into fuel rods. The fuel is usable in the reactor for a period that varies with the design and other features, but is usually around a year. When the fuel reaches the end of its useful life in the reactor, it is highly radioactive for a long period of time and must therefore be sequestered for a long period of time to protect the environment and the public. However, as noted, it contains useful quantities of plutonium that can be recovered by reprocessing the used fuel. There are several different processes possible for enrichment and reprocessing (Tsoulfanidis, 2014).

URANIUM AND THORIUM RESOURCES

Most reactors today use uranium. Uranium resources can be found in a number of countries around the world. Open pit and underground mining and *in situ* recovery techniques are all used to mine uranium ore, depending on the nature of the resource. One respected global source of data on uranium resources is the so-called *Red Book*, a biennial global report on uranium resources, production, and demand produced jointly by the Organization of Economic Cooperation and Development Nuclear Energy Agency and the IAEA (2014).

According to this report, which incorporates resource estimates from 45 countries, the total identified resources (reasonably assured and inferred) as of January 1, 2013 were 1,956,700 tonnes of uranium metal (tU) in the $USD 80/kgU ($USD 30.80/lb U3O8) category and 5,902,900 tU in the $USD 130/kgU ($USD 50/lb U3O8) category. In the highest cost category ($USD 260/kgU or $USD 100/lb U3O8), total identified resources were 7,635,200 tU. Total undiscovered resources (prognosticated resources and speculative resources) as of January 1, 2013 were 7,697,700 tU. However, these numbers include only data provided by the various countries, and the 2014 issue of the report did not include data from the United States, as the United States did not provide an update. The total undiscovered resources in the previous (2012) report, which did include the United States, were 10,429,100 tU.

Current global usage is about 61,600 tU/year. At that rate of usage, the world's present resources of uranium are enough to last for about 100 years. If demand increases, of course, the present resources would be depleted sooner. However, the cost of fuel is a small fraction of the cost of operating a nuclear power plant, so it may be feasible to tap more costly resources. Also, resource discoveries, advanced reactor technologies that recycle uranium, and other developments are likely to increase this estimate.

For example, there is a considerable amount of usable fuel left in spent fuel that is removed from reactors. This includes unused ^{235}U and the ^{239}Pu that is created from the ^{238}U. This fuel can be recovered through a process known as reprocessing or recycling. There are several reprocessing technologies available, which will be described later. Reprocessing can potentially reduce the need for fresh uranium by as much as 25% or 30%. Reprocessing also has the advantage of reducing the volume and half-life of the spent fuel, thus making disposal of the resulting waste somewhat easier. Since waste disposal has been a thorny issue for nuclear power (as discussed below), this is a potentially important consideration.

In addition, there are potential unconventional sources of uranium. For example, seawater contains very small amounts of uranium (0.003 parts per million). (By contrast, the very highest-grade ores contain 200,000 ppm uranium.) The ability to extract uranium from seawater would yield a huge increase in the amount of uranium available for reactors, making it a virtually inexhaustible source of fuel. However, the uranium in seawater exists in very low concentrations and it has not yet been proven whether it can be extracted in an energy- and cost-efficient manner.

Finally, although most reactors today use uranium as fuel, thorium can be used as an alternative fuel cycle, and is actively being considered for some of the advanced reactor concepts. Less data are available for thorium than for uranium, but the *Red Book* estimates that there are about 6.212

million tonnes of total known and estimated resources of thorium and about 6.355 million tonnes of reasonably assured and inferred resources recoverable at a cost of $80/kg Th or less.

FUEL CYCLE

Uranium ores first undergo an initial refining process in which they are crushed to produce a powder, processed to leach out and concentrate uranium compounds, and dried. The process is called milling and the resulting product is called yellowcake and is approximately 85% U_3O_8 (Figure 22.1).

As noted earlier, ^{235}U makes up only a small fraction of natural uranium. Therefore, for most reactors today, the fraction of ^{235}U in the fuel must be increased from 0.7% to a little under 5% in order to permit a self-sustaining reaction in a water moderated and cooled reactor. (This level of enrichment is below the levels required for a nuclear weapon.) The process of increasing the percentage of ^{235}U is known as enrichment. The yellowcake is chemically converted into uranium hexafluoride (UF_6), a compound that, when heated, becomes a gas that can be fed into enrichment plants. There are several enrichment processes, described below.

After being enriched, the resulting uranium is made into a form suitable for use in the reactor. This is a much more complex step than is necessary for fossil fuels. It requires both that the form enables the sustained fission reaction to take place, and that it maintains its integrity under the temperatures and radiation levels in a reactor. The latter is necessary to prevent the release of radioactive fission products over the life of the fuel.

Most reactors today use fuel fabricated in the form of ceramic pellets of uranium dioxide (UO_2). The pellets are loaded into long tubes, called fuel rods. The fuel rods are coated with an element designed to contain the fission products. Zirconium is usually used because it has a good combination of characteristics for this purpose. First, it has a low cross-section for absorbing thermal neutrons, so it does not interfere with the fission reaction process. Second, it has a high corrosion resistance and a high level of ductility, so it is resistant to corrosion from the cooling water and to cracking under stress.

However, at high temperatures, such as in an accident, zirconium is oxidized by the surrounding water and releases hydrogen gas. This phenomenon occurred in the Fukushima accident. Therefore, research is being done on alternative cladding materials.

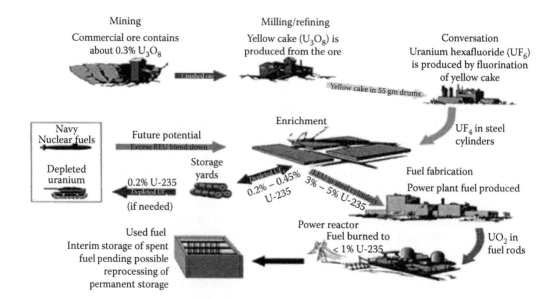

FIGURE 22.1 Nuclear fuel cycle. (Adapted from U.S. Department of Energy, http://www.energy.gov/ne/nuclear-fuel-cycle.)

The resulting fuel rods are then loaded into a reactor where the fission reaction takes place. The assembly of fuel rods, together with structural materials and control systems, is known as the reactor core.

The fission process produces radioactive fission products, some of which have very long decay periods (half-lives). These materials must be sequestered from the environment for many thousands of years. Currently, most "spent fuel," as this material is usually called, is being stored in temporary facilities, as discussed below. There are plans in a number of countries for "permanent" storage facilities, and construction of such facilities is underway in a couple of countries, but to date, no such storage facility has been completed and is in operation. Critics consider this one of the shortcomings of nuclear power.

Others note that the spent fuel actually still contains a significant amount of unused ^{235}U, and that reprocessing the spent fuel (or "used fuel," as some call it) can yield additional fuel—as well as reducing the amount of long-lived radioisotopes. There are several technologies that can be used for reprocessing, and these will be discussed later as well.

ENRICHMENT PROCESSES

Gaseous diffusion enrichment. Historically, the major process by which the enrichment was accomplished was the gaseous diffusion enrichment process (Figure 22.2). In this process, the yellowcake is first converted to uranium hexafluoride (UF_6), a gas that is pumped through porous membranes with holes that are so small that it is difficult for the UF_6 molecules to pass through. The isotope enrichment occurs when the lighter UF_6 containing the ^{235}U (and small amounts of ^{234}U) atoms pass through the barriers (diffuse) faster than the heavier UF_6 molecules containing ^{238}U. It takes hundreds of successive barriers before the UF_6 contains enough ^{235}U to be used in reactors. At the end of the process, the enriched UF_6 gas is condensed and converted to fuel.

Gas centrifuge enrichment. The gaseous diffusion process is very energy intensive, so in recent years most countries have turned to a newer process called the *gas centrifuge enrichment process* (Figure 22.3). The gas centrifuge enrichment process uses a large number of interconnected centrifuges, or rotating cylinders. The UF_6 gas is placed in a cylinder and rotated at high speed. The rotation creates a strong centrifugal force that pushes the heavier gas molecules (containing ^{238}U) toward the outside of the cylinder and the lighter molecules (containing ^{235}U) toward the center. Thus, the gas at the center is enriched very slightly. This gas is withdrawn from the center and fed into the next higher stage. Even though it takes hundreds of centrifuges to enrich the uranium enough to be used in reactors, the process requires far less energy and far fewer stages to accomplish a given level of enrichment than is the case for the gaseous diffusion process.

Laser isotope enrichment. Extensive research has been done on the use of lasers to enrich uranium. So far, none of these has reached the stage of commercial application. There are several variants, but all involve the use of specially tuned lasers to separate isotopes of uranium by taking

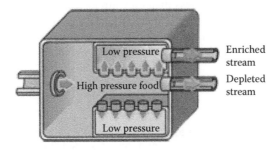

FIGURE 22.2 Gaseous diffusion enrichment process. (Adapted from U.S. Nuclear Regulatory Commission, http://www.nrc.gov/materials/fuel-cycle-fac/ur-enrichment.html#2.)

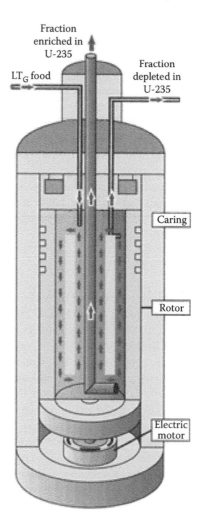

Fraction
enriched in
U-235

Fraction
depleted in
U-235

LT$_G$ food

Caring

Rotor

Electric
motor

FIGURE 22.3 Gaseous centrifuge enrichment process. (From U.S. Nuclear Regulatory Commission, http://www.nrc.gov/materials/fuel-cycle-fac/ur-enrichment.html#2.)

advantage of very small differences between U^{235} and U^{238} in their absorption peaks for photons. Because of this difference, finely tuned lasers are able to selectively excite ^{235}U ions, which are then electrostatically deflected to a collector, while the neutral ^{238}U passes through. Some proliferation concerns have been expressed about laser isotope enrichment technologies, because the space and energy requirements are substantially less than they are for current enrichment technologies, and therefore, it might be easier for a rogue government or other organization to hide a laser enrichment facility. The greater efficiency makes them attractive economically for the commercial sector, and several variant laser isotope enrichment processes have been and are being tested for their commercial viability.

NUCLEAR REACTOR WASTE

In most countries, it has proved difficult to establish a path for the permanent disposal of the high-level radioactive waste from nuclear reactor fuel. The long half-lives of some of the radioactive elements create a requirement for very long-term sequestration of the waste. This requirement has led to worries about the integrity of disposal sites over the hundreds of thousands of years the facility must operate, and the risk to any future surrounding population should the sites ever be compromised.

When used nuclear fuel is first removed from a reactor, it is initially stored on the reactor site in what is known as a spent fuel pool. The fuel in the pool is arranged in racks incorporating neutron absorbers to assure that the fuel stays subcritical (i.e., it cannot undergo a chain reaction), and is submerged in a circulating water supply. There, it is allowed to decay to reduce the heat and radio-activity. When it is sufficiently cooled that the circulating water coolant is no longer necessary, it may be moved to dry storage facilities, either on- or off-site. Dry storage structures are air-cooled, and are passive facilities in that there is no forced circulation.

While the used fuel can be stored for decades in such dry storage facilities, the fuel remains radioactive for thousands of years. Thus, most countries are seeking to build long-term disposal facilities, known as repositories, for the permanent disposal of the fuel. For reprocessed fuel, the demands on such a repository are not as great. Nevertheless, some long-term radioactivity remains in the waste from the reprocessing process, so a long-term repository is still needed.

The concept of permanent, underground disposal is that the waste repository could seques-ter nuclear wastes for many thousands of years. (In the United States, the U.S. Environmental Protection Agency set a requirement of 1 million years for a proposed repository at Yucca Mountain in Nevada.) Such long-term sequestration requires a combination of stable geologic formations and "engineered barrier systems." Geologic formations that have been considered for nuclear waste dis-posal include clay, salt, and rock. Engineered barriers include the waste form itself, as well as waste canisters, buffer materials, backfill, and seals.

No country is yet operating a permanent repository for civilian reactor wastes. It has proved dif-ficult in most places to obtain public support for such long-term repositories because of the require-ment that the facility remain intact for hundreds of thousands of years to assure isolation of the used fuel until the radioactivity has fully decayed to background levels. That requires confidence in the understanding of complex and long-term geologic and hydrologic behavior. Even where the scien-tific evidence has been strong, public opposition has stalled several initiatives.

The most successful efforts to date, in Finland and Sweden, have selected sites based on a lengthy process incorporating public input and the ultimate agreement of the local communities, called a consent-based process, and this same process is now being recommended in other countries. Finland is in the middle of construction of a permanent waste repository, the Onkalo spent nuclear fuel repository, at the site of its Olkiluoto nuclear power plant, and Sweden has approved a site for a permanent waste repository at Forsmark.

The United States has an operating repository called the Waste Isolation Pilot Plant that is lim-ited by law to taking only transuranic (TRU) wastes from the U.S. military program. This waste contains very long-lived plutonium isotopes, but does not have the highly radioactive fission prod-ucts that reactors generate. However, this facility was shut down for an extended period in 2014 because of leakage from one cask delivered to the facility. At the time of this writing, it is still shut down. The efforts of the United States to construct a civilian waste repository at Yucca Mountain in Nevada were stalled due to strong political opposition. The United States is currently consider-ing options for storage of the nuclear waste. These options could include Yucca Mountain, but also include other options. In particular, they include the idea of one or more centralized, temporary facilities that could hold the waste for decades or longer, pending the development of other options.

Some people have proposed that an alternative means of high-level radioactive waste disposal might be possible using "boreholes," holes drilled several miles deep into the earth in which cyl-inders of waste could be placed. To date, the feasibility of using boreholes for nuclear waste has not been examined in great detail, but a few observations are possible. Since these would be much deeper than repositories, there would likely be many more sites that would be geologically suitable, and it might be possible to spread the waste disposal over many locations. Ideally, a lot of the waste could be emplaced near where it is used. The hypothesis is that this approach might reduce the pub-lic opposition that has arisen whenever centralized waste sites are discussed.

For countries that, like the United States, are not currently reprocessing their spent fuel, a disadvantage is that boreholes would preclude the chance of recovering and reprocessing the fuel later. (In the case of repositories, it is often assumed that they can be designed for later access, if desired, although the details of how that would be accomplished need further development.)

Still other concepts have been proposed, such as launching waste into deep space, or disposing of waste in the seabed, but none of these concepts has advanced very far.

Reprocessing

Reprocessing was originally developed to extract plutonium for weapons programs. Therefore, early efforts explored a number of alternative reprocessing technologies, some of which were used in the past, but have now become obsolete. This discussion will focus only on existing and advanced technologies. For commercial use, reprocessing makes use of the plutonium that is created as a by-product of the primary fission reaction, thereby extracting more energy from the fuel. The plutonium recovered by reprocessing is usually mixed with uranium to create a mixed-oxide fuel. Reprocessing does not remove the need for a long-term repository, but can reduce the demands on the repository.

At present, nuclear fuel is reprocessed in only a few countries. These include France, the United Kingdom, Russia, Japan, and India. The United States explored the development of reprocessing technology in the past, but currently does not reprocess reactor fuel. The reasons for the U.S. decision to cease reprocessing activities are complex. They include the fact that reprocessing was not considered economical when it was first considered, as well as the concern that the separation of uranium and plutonium during reprocessing could be a proliferation risk.

All commercial reprocessing plants today use the plutonium uranium extraction (PUREX) process. In this process, the used fuel is chopped up and dissolved in nitric acid. First, the uranium and plutonium in the aqueous nitric acid stream is separated from the fission products and minor actinides by a solvent extraction process, using tributyl phosphate in a solvent such as kerosene. Following that, the uranium is separated from the plutonium by a reduction process. Several variations of the PUREX process have been proposed.

A more advanced technique that is being developed is called "pyroprocessing." This is an electrolytic/electrometallurgical technique that involves several stages, including electrorefining and solvent–solvent extraction. To date, it has been used experimentally, but is not being used for reprocessing any commercial fuel. A number of different variations have been, or are being, explored in several countries, including the United States, South Korea, and Russia.

CURRENT NUCLEAR REACTOR TECHNOLOGIES

Nuclear reactors are very complex systems, comprising the basic components to produce power as well as numerous backup systems to assure safe operation in the event of a failure of one of the systems. In addition, there are a number of different types of reactors, and even within each type, there are variations in design from different manufacturers and over time. This section will cover some of the key characteristics that distinguish the designs currently in operation around the world (El-Wakil, 1982; Murray and Holbert, 2014).

Classifying Nuclear Reactors

Nuclear reactors are generally classified by one or more of their key characteristics. The broadest and most common classification is by the type of coolant used. Thus, one can speak of light-water reactors (LWRs), heavy-water reactors (HWRs), gas-cooled reactors (GCRs), liquid-metal-cooled reactors, or molten salt reactors (MSRs).

These categories can be subdivided further. For example, there are currently two types of LWRs—boiling water reactors (BWRs) and pressurized water reactors (PWRs). Sometimes, reactors are characterized by the specific coolant used; for example, sodium-cooled reactors and lead-cooled reactors are both liquid metal reactors.

Reactors can also be characterized by their neutron "spectrum"—that is, thermal reactors or fast reactors. Thus, liquid metal reactors are sometimes called liquid metal fast reactors or liquid metal fast breeder reactors. Occasionally, reactors are identified by the moderator used. This is particularly true for graphite-moderated reactors.

There are a large number of possible materials and configurations that can be used in reactor designs. Indeed, research and test facilities have been built using many different options. However, to date, the majority of operating power reactors—276 of the 435 in operation in 2015—are, and have been, thermal light-water reactors (often abbreviated LWRs). There has also been a significant number of HWRs. These will be described below. Only a few GCRs and graphite-moderated reactors and one liquid metal reactor have been built for power production in the past and continue to operate today. However, these and other concepts have attracted considerable interest for future generations of reactors and will be discussed later.

REACTOR DESIGN: GENERAL

The philosophy of nuclear reactor design is to provide redundant and diverse features and systems to assure safety, even in the event of the failure of one or more systems.

Therefore, all current reactors have multiple layers of protection against the release of radioactivity. The layers of protection include the physical structure of the fuel, the reactor, and the reactor building. In addition, backup systems are incorporated in the design to assure the removal of heat in the event that the primary cooling system is lost.

The major physical barriers in a reactor include the fuel itself, the cladding of the fuel rods, the reactor vessel, and the containment. The uranium oxide ceramic fuel serves as the first barrier to the release of the radioactive fission products. The zirconium fuel rod cladding provides a second barrier. The core and other essential components of a reactor, such as the control rods and the moderator, are housed in a strong steel container called the reactor pressure vessel (RPV), and the RPV in turn is housed in a containment building, a thick-walled steel and concrete structure. This structure is designed both to contain the radioactive material from the core and to prevent penetration of the reactor from external sources.

In addition to these physical layers of protection, reactors have an emergency core cooling system, a separate system designed to inject water into a reactor core to remove the heat in the event that the normal cooling system should fail to function. In addition, every reactor has emergency diesel generators to provide power in the event that the normal source of electrical power is lost. Still other systems, features, and procedures contribute to the monitoring and control of the reactor operations at all times.

BOILING WATER REACTORS

BWRs are perhaps the easiest to understand. In BWRs, very pure water serves as both the moderator and the coolant. The water is pumped past the fuel rods to cool the fuel, and in the process, the water is heated and turned to steam (Figure 22.4). The resulting steam is passed through steam separator and dryer plates above the core in order to remove the water before the steam is transported through pipes to the turbine generator, where it turns the turbine to produce electricity. About 80 of the reactors currently operating around the world are BWRs.

As described earlier, as for fossil-fired plants, the steam exiting the turbine is transported to condensers, where it is returned to a liquid state before being recirculated to the reactor core.

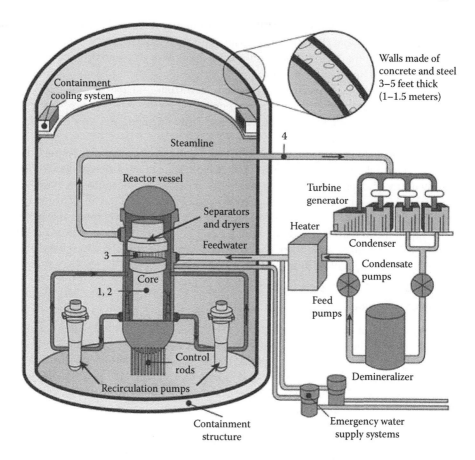

FIGURE 22.4 BWR schematic. (Adapted from U.S. Nuclear Regulatory Commission, http://www.nrc.gov/reactors/bwrs.html.)

PRESSURIZED WATER REACTORS

By far, the largest number of reactors in operation today are PWRs—about 276 of the 435 in operation in 2015. There are also several hundred additional PWRs used for naval propulsion. The operation of the PWR is a little more complex than that of the BWR and requires a couple of additional steps (Figure 22.5).

There are two loops of water required in this design. The water in the first loop serves as both the moderator and coolant. It circulates through the reactor core, but unlike in the BWR, the system is highly pressurized so that the water does not boil. Control is achieved through the use of steam from a separate vessel, called the pressurizer, which is heated electrically. A constant pressure of about 150 times atmospheric pressure (150 bar) is maintained in the reactor. As in a home pressure cooker, the pressure allows the water to reach a temperature higher than the normal boiling point of water, in this case, about 350°C.

The water in this loop, called the primary loop, is circulated through a steam generator, where it transfers the heat to a secondary loop. The use of separate loops minimizes the possibility of radioactive contamination being transferred out of the core to the turbine.

The higher temperature of the pressurized water makes the energy transfer in the heat exchanger more efficient. The water in the secondary loop does boil and it is the water in the secondary loop that therefore drives the turbine generator.

Like BWRs, PWRs use a condenser to cool the steam from the turbine. In this case, of course, the condenser cools the water in the secondary loop. In the case of PWRs, there is no mixing of the

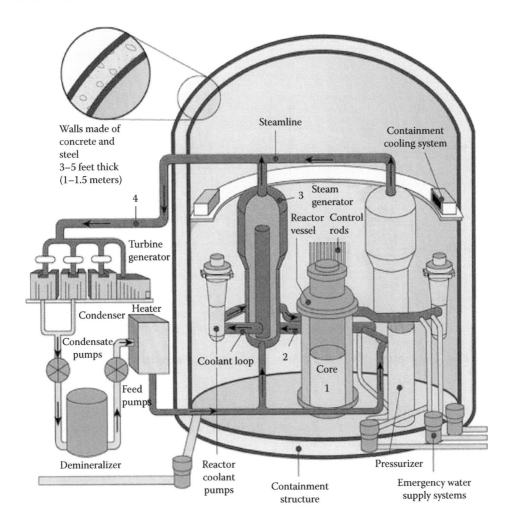

FIGURE 22.5 PWR schematic. (Adapted from U.S. Nuclear Regulatory Commission, http://www.nrc.gov/reactors/pwrs.html.)

water between any of the loops. Also like BWRs, cooling towers are used when there is insufficient cooling water.

Heavy-Water Reactors

HWRs are the third most frequently used reactors today. About 48 are currently in operation. Most of them are pressurized heavy-water reactors (PHWRs) (Figure 22.6). PHWRs are a good example of how alternative technological solutions can be used to address a particular technical constraint. As previously noted, thermal neutrons mainly fission ^{235}U atoms. LWRs therefore need enriched uranium to permit a sustained fission reaction. The reason this is necessary is that, although water is a very good moderator (because the mass of the hydrogen in water is very close to that of a neutron, the momentum transfer is very efficient), water also absorbs neutrons. Therefore, a self-sustaining reaction is not possible with the amount of ^{235}U in natural uranium.

An alternative to using enriched uranium and a water moderator is to use natural uranium and heavy water (D_2O). Deuterium does not absorb neutrons as much as water does (because the hydrogen atoms already have an extra neutron), so there are more neutrons available to be absorbed

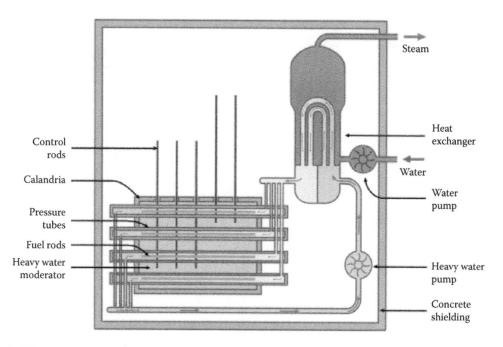

FIGURE 22.6   Pressurized HWR. (Adapted from http://commons.wikimedia.org/wiki/File:CANDU_reactor_schematic.png.)

by the smaller fraction of U^{235}. Because of the greater difference in the masses, the deuterium in heavy water is not quite as good a moderator as the hydrogen atoms in water, but that shortcoming is offset by the lower absorption of neutrons.

Heavy water is very expensive, but HWRs were developed in the early days of nuclear power and provided a way of avoiding the need to build and operate uranium enrichment facilities, which were difficult to operate, and even more expensive. There are a number of other trade-offs between LWR and HWR fuel cycles. The fuel for LWRs is enriched only to about 5%, well below levels useful for weapons production. However, the same facilities used to enrich uranium to 5% can potentially be used to enrich uranium to the much higher levels needed for weapons, and thus presents a proliferation risk. HWRs do not require enrichment facilities, and so avoid this possible proliferation path. On the other hand, HWRs produce more plutonium than LWRs and the plutonium is readily extractable from the uranium, so this fuel cycle presents a different proliferation risk.

While not unique to HWRs, one feature of the current generation of HWRs is that their design, which incorporates multiple pressurized tubes instead of a single large pressure vessel, facilitates on-line refueling. On-line refueling reduces the need for reactor shutdowns, although of course, some shutdowns are still required for periodic maintenance.

ADVANCED NUCLEAR REACTOR TECHNOLOGIES AND CONCEPTS

Advanced reactor development is moving in two directions (Marcus and Levin, 2002; Blake, 2014). One focus is on the potential for advanced versions of light-water reactor designs. This has the advantage that there is a substantial knowledge base for these designs, both among the operators and the regulators. The other focus is on exploring other advanced concepts, mostly involving non-LWRs. In both cases, alternatives to the current large reactors, as well as alternatives to current aboveground, land-based siting are being examined as well.

SMALL MODULAR REACTORS: GENERAL CONSIDERATIONS

Small modular reactors (SMRs) are generally considered to be reactor modules of about 350 MWe or less. This size compares to current designs that have trended to over 1500 MWe in recent years. It is envisioned that a given site could operate with one or more such modules. When more than one module is operated on a given site, most concepts assume that only one control room and one set of operators would be required, thus achieving some operating economies.

It must be remembered that the first commercial reactors were small compared to the current designs. Reactors were built in larger and larger sizes over time to meet demand and to achieve economies of scale. Thus, it initially seems counter-intuitive to return to smaller designs.

However, current thinking attributes a number of advantages to smaller designs. Multiple identical modules can be built, and the modules can be fabricated in a centralized factory, thus leading to economies of mass production. (Mass production, of course, was not employed in the early days.) The smaller size also allows for use in a broader range of applications, including in remote areas, such as in small, remote towns in Alaska or other places, or at remote mining sites. It also permits incremental installation to meet demand as it grows, thus avoiding the need for a large investment made years in advance of actual need.

On the technical side, the small size facilitates the use of more passive safety features, both increasing the level of safety and avoiding the need for mechanical systems that are expensive to build and maintain.

While all these advantages appear possible, at present, SMR designs are in the early stages of development, and extensive analytical studies and physical tests will be required to prove whether these advantages are achievable, as well as to determine whether there are any unanticipated weaknesses to these design concepts.

SMR versions of LWRs are being designed, as well as SMR versions of non-LWR concepts.

INTEGRAL PWRS

One approach to SMRs seeks to improve the safety and economics of PWRs by developing smaller PWRs and putting most of the major components, including the steam generator, the reactor coolant pumps, and the drive mechanisms for the control rods all inside the RPV. Such a design is called an integral PWR (iPWR) (Figure 22.7). Several organizations have developed concepts for iPWRs. These differ in many details, but all have in common that they have fewer penetrations of the RPV. Most also have lower power density than conventional reactors, and have passive cooling systems. These features result in several safety and economic advantages, including an increased operating margin (i.e., the difference between the operating level and the level at which the fuel would fail), lower core damage frequency, and lower radiation doses to workers.

OTHER TYPES OF REACTORS: GENERAL

As noted, to date, a few non-water reactors have operated commercially. However, for various reasons, the field of commercial power production has so far been dominated by light- and heavy-water reactors.

This mix could change in the future, as interest turns to some of the potential benefits of advanced reactor designs using gas, liquid metals, and molten salts as coolants. At present, there are a number of initiatives underway, both government and privately funded, involving a variety of alternative reactor designs, as well as alternative sizes, siting options, and other variations. It is unclear which concepts will emerge from the multiple avenues being pursued. Success will depend on several factors, including the competitive advantages and disadvantages of each concept, the funding made available for development, and the interest of the operator community.

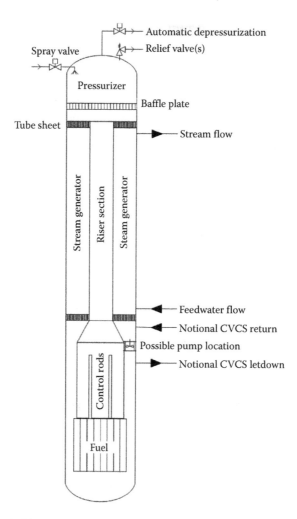

FIGURE 22.7   Integral PWR. (Courtesy of R. Belles, Oak Ridge National Laboratory.)

Specific designs that are presently being pursued may change over time. Therefore, rather than describing particular concepts, this discussion will focus on generic versions of potential concepts under consideration. The bulk of this discussion is drawn from an international initiative known as the Generation IV International Forum (GIF, 2014), that identified six major design concepts out of over 100 variants. The six concepts are:

- Very-high-temperature reactor (VHTR)
- Gas-cooled fast reactor (GFR)
- Sodium-cooled fast reactor (SFR)
- Lead-cooled fast reactor (LFR)
- Supercritical water-cooled reactor (SCWR)
- Molten salt reactor (MSR)

It should be noted that, although the GIF effort is primarily focusing on non-water reactor designs, one of the six major concepts, the SCWR, is an advanced light-water reactor. The VHTR and GFR are both gas-cooled reactors. The SFR and LFR are both liquid-metal-cooled reactors. One of the six concepts, the MSR, is focused on the use of molten salt. Initially, three of the concepts were considered thermal reactor concepts (VHTR, SCWR, and MSR) and the other three were designed to be fast-spectrum reactors (GFR, SFR, and LFR). However, SCWRs can be configured either as thermal or fast-spectrum reactors and MSRs can be configured either as thermal or epithermal reactors, and there is even a variation being explored that could operate as a fast-spectrum reactor.

All six of the GIF reactor concepts are in the preliminary stages of development. In fact, as noted, there are various versions of some concepts, using different moderators, coolants, or other features. Some would likely be SMRs, while others would be large reactors. All are intended to achieve higher levels of safety than current reactors. The various concepts are also intended to help achieve other objectives, such as improved proliferation resistance, improved economics, higher-temperature outputs to facilitate their use for process heat, and more flexibility and efficiency in the use of fuel (to extend the available resources). The fast reactors could also be used to breed new fuel. In such cases, some of the design concepts would likely include a co-located facility for reprocessing the fuel.

VERY-HIGH-TEMPERATURE REACTOR

The VHTR concept (Figure 22.8) is a gas-cooled, thermal reactor design optimized to produce a high outlet temperature, typically around 1000°C. Such temperatures make them useful as a source of process heat for various industrial applications. One application that has received some attention is that of thermochemical hydrogen production. In addition, of course, the reactors can be used to produce electricity. GCRs are considered passively safe in that the core can be cooled by natural circulation.

VHTR concepts use graphite as the moderator and helium as the coolant. Some of the earliest reactors used graphite blocks as a moderator, and graphite has been, or can be, used with a variety of coolants, including water, carbon dioxide, and molten salt. Many of the newer graphite reactor concepts envision the graphite moderator either in the form of prismatic blocks or in the form of "pebbles."

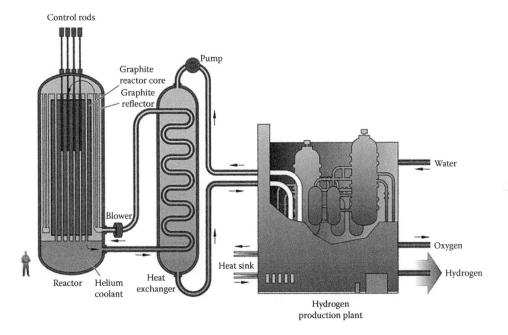

FIGURE 22.8 Very-high-temperature reactor. (Courtesy of the U.S. Department of Energy, https://www.gen-4.org/gif/upload/docs/application/pdf/2013-09/genivroadmap2002.pdf.)

In a pebble-bed reactor (PBR), the "pebbles" are spherical fuel elements consisting of a graphite matrix containing thousands of small fuel particles called TRISO particles. TRISO fuel, which stands for tri-structural-isotropic fuel, consists of a fissile fuel surrounded by a ceramic coating, such as silicon carbide. This arrangement contains the fission products. One novel feature of the PBR is that the pebbles can cycle continuously through the reactor. This allows the pebbles to be removed and replaced as the fissile fuel is used up, so refueling is continuous. However, it may be harder to recycle the silicon carbide coated fuel. TRISO fuel can also be used in prismatic graphite blocks.

For electricity production, the high-temperature helium coolant can be used to drive a gas turbine directly (Brayton cycle). For hydrogen production, the heat from the helium coolant can be used via an intermediate heat exchanger, with electricity cogeneration.

GAS-COOLED FAST REACTOR

The GFR (Figure 22.9) is similar in a number of respects to the VHTR. Most of the concepts would operate at a high temperature, although perhaps not quite as high as that of the VHTR. The technology would therefore also be suitable for electricity generation, thermochemical hydrogen production, or other process heat. For electricity production, the helium could directly drive the turbine.

The primary difference between the GFR and the VHTR is that, for the GFR, the neutrons would not be moderated; that is, the reactor would have a fast neutron spectrum. The advantage of this is that the reactor would use the ^{238}U and be able to breed fuel. The expectation is that the fuel would be reprocessed on site and recycled into the reactor in a "closed cycle." This would create more fuel for the reactor and would minimize the production of long-lived radioactive wastes. Another benefit is that the reactor could make further use of the fuel after it has been used in light-water reactors and removed from those reactors.

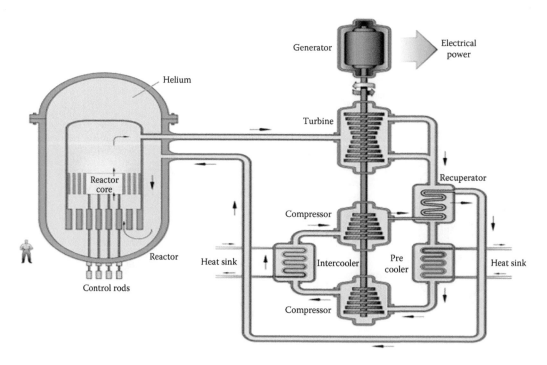

FIGURE 22.9 Gas-cooled fast reactor. (Courtesy of the U.S. Department of Energy, https://www.gen-4.org/gif/upload/docs/application/pdf/2013-09/genivroadmap2002.pdf.)

SODIUM-COOLED FAST REACTOR

The SFR (Figure 22.10) is another fast reactor concept. This one is of the class known as liquid metal reactors, which use metal in a liquid form as the coolant. In the case of the SFR, the reactor coolant is liquid sodium. The use of sodium allows the reactor to operate at low pressure and with a high power density. As is the case with the GCR, the fast reactor spectrum permits breeding and facilitates the reuse of the fuel on-site, or the use of fuel from light-water reactors. Because of the chemical reactivity of sodium, special precautions are necessary to prevent fires.

There are several different designs being explored, including integral, pool-type and loop-type designs, and different methods of reprocessing, depending on the type of fuel used. (Pool-type reactors are currently in use mainly as research reactors. In most of the pool-type research reactors today, the liquid is water and serves as both moderator and coolant. For a sodium reactor, the liquid is sodium, and would serve only as the coolant. As the name implies, the fuel rods sit in a pool of the coolant, rather than having the coolant flow past them, and cooling is accomplished by natural convection.)

LEAD-COOLED FAST REACTOR

The LFR (Figure 22.11) is another liquid metal reactor. In this case, the coolant is lead. There are several possible variants of lead-cooled reactors, including the use of either molten lead or lead–bismuth alloys as a coolant, and the use of different fuels, including uranium and thorium. Like the SFR, the LFR is a fast neutron spectrum reactor and can be used to breed fuel.

Different sizes are also possible, as well as different coolant outlet temperatures. Depending on the coolant outlet temperature, the reactor could be used just for electricity production, or for process heat and hydrogen production, as well as for electricity production. One LFR concept envisions a long-life, factory-fabricated core that is often called a battery because it would operate for multiple years and then be removed intact for disposal, thus simplifying on-site operations.

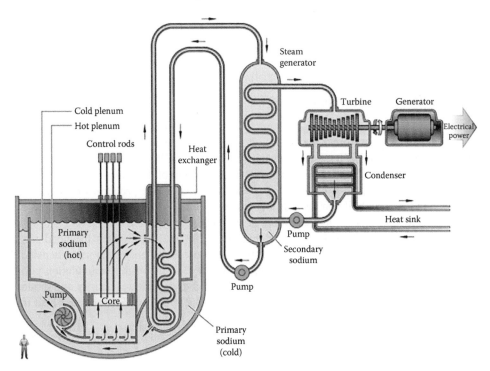

FIGURE 22.10 Sodium-cooled fast reactor. (Courtesy of the U.S. Department of Energy, https://www.gen-4. org/gif/upload/docs/application/pdf/2013-09/genivroadmap2002.pdf.)

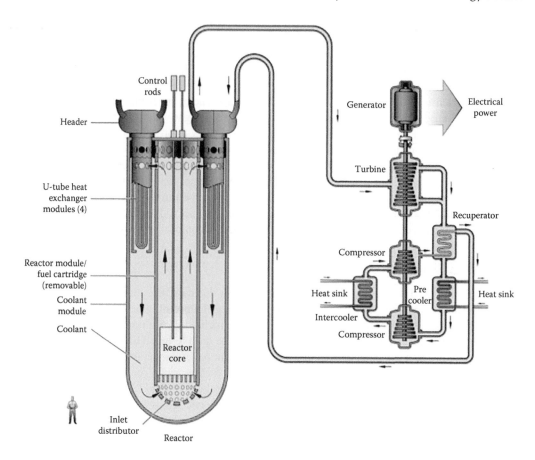

FIGURE 22.11   Lead-cooled fast reactor. (Courtesy of the U.S. Department of Energy, https://www.gen-4. org/gif/upload/docs/application/pdf/2013-09/genivroadmap2002.pdf.)

Supercritical Water-Cooled Reactor

SCWRs (Figure 22.12) build on LWR technology. They are similar to LWRs, but they operate at higher pressure and temperature. Like BWRs, they have a single coolant loop. However, the water does not boil; rather, as in PWRs, it remains in a single state. SCWRs would operate at a temperature and pressure above the critical point of water (the end point of the pressure–temperature curve where the liquid and gas phases coexist). As a result, it would have properties between those of water and steam. SCWRs build on the considerable experience with LWR technologies, as well as experience with supercritical steam generators in nonnuclear applications.

The use of supercritical water as the moderator and coolant eliminates the need for some components used in PWRs and BWRs. Since the steam and water are of the same density, there is no need for the pressurizers or steam generators of a PWR, or for the recirculation pumps or steam separators and dryers in BWRs. The supercritical water has a lower density and moderating effect than liquid water, which allows for a faster neutron spectrum. It also has better heat transfer properties.

Molten Salt Reactor

For MSRs (Figure 22.13), the primary coolant is a molten salt mixture and the moderator is a graphite core. The fuel may be a solid or it may be dissolved in the coolant and circulate with the

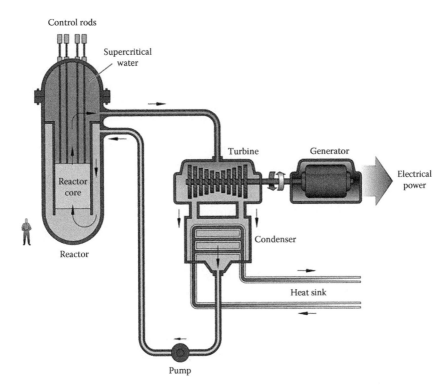

FIGURE 22.12   Supercritical water reactor. (Courtesy of the U.S. Department of Energy, https://www.gen-4. org/gif/upload/docs/application/pdf/2013-09/genivroadmap2002.pdf.)

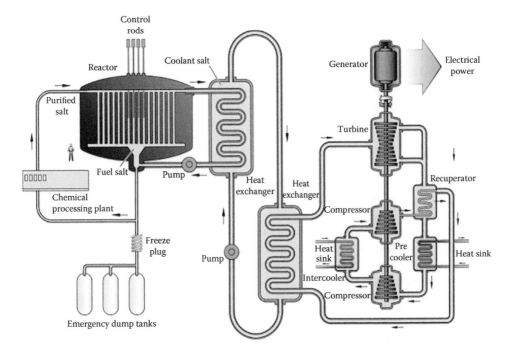

FIGURE 22.13   Molten salt reactor. (Courtesy of the U.S. Department of Energy, https://www.gen-4.org/gif/ upload/docs/application/pdf/2013-09/genivroadmap2002.pdf.)

coolant. MSRs operate at higher temperatures than LWRs and therefore have higher thermodynamic efficiency, and the molten salts are more efficient than helium at removing heat from the core, thereby reducing the size of the core and the need for pumping.

Fluoride salts have been used in experimental MSRs in the past, and most design concepts envision the use of fluoride salts such as lithium fluoride and lithium-beryllium fluoride. In concepts where the fuel is dissolved in the coolant, uranium fluorides, such as UF_4, would be used.

Thorium can also be used in an MSR, either dissolved with the uranium in a single-loop arrangement known as a homogeneous design, or in a second loop, which would allow the reactor to operate as a breeder reactor generating ^{233}U. The latter arrangement is known as a heterogeneous design. In either case, since the fission products are dissolved in the salt, they can be removed continuously in an on-line reprocessing system. The on-line system also allows new fissile material to be added during operation.

The heat generated in the molten salt is transferred to a secondary nonradioactive molten salt loop through an intermediate heat exchanger. That loop in turn transfers heat to a water loop through another heat exchanger to generate the steam to run the turbine.

Although most MSRs are thermal or epithermal reactors, schemes have been proposed that would allow a fast spectrum by eliminating the graphite core and substituting heavier, less moderating salts, such as chlorides (KCl, $RbCl$, and $ZrCl_4$), for the lighter fluorides (LiF and BeF_2).

SITING CONSIDERATIONS

To date, all commercial power reactors have been sited on land and above ground. Most, although not all, have also been sited adjacent to large bodies of water, such as oceans or lakes, for access to sufficient cooling water. Such siting often runs into public opposition, as it tends to favor sites that are also scenic or recreational, and/or are desirable for residences. Furthermore, for such reactors, the cooling water is usually returned to the lake or ocean at a higher temperature than the surrounding water. This has led to concerns about the impact of the heated cooling water on the local marine life. In addition, marine life is often caught in the reactor intakes. Screens protect the reactor from infiltration by marine species, but there are concerns that some marine animals are killed in the intake canals. Others point to positive impacts of the discharged water—marine animals are often attracted to the warmer temperatures around the discharge areas and thrive there. Nevertheless, the concerns have led to increasing difficulties in siting reactors.

A number of reactors are sited in inland areas without access to large bodies of cooling water. These typically use large cooling towers to provide sufficient and reliable cooling for reactors. The cooling towers use water, but instead of returning the water to a large, natural body of water, they cool the water by heat exchange with the air in the cooling tower and recycle it in a closed system. These cooling towers, which are similar to cooling towers used for other types of power plants and for industrial activities generating large amounts of heat, have drawn criticism as well. In the first place, they are visually large and intrusive. Furthermore, there is concern about the impacts of the heat exchange on the local environment. In fact, nuclear opponents often use images of cooling towers as a symbol of a nuclear reactor, even though such a one-to-one correlation is not accurate.

All these factors have led to consideration of different options for siting reactors. Several ideas are being considered, including locating reactors underground or on the water (Murray and Holbert, 2014). For water-based siting options, both floating and underwater concepts, and both docked and offshore concepts are being discussed.

Underground reactors have the potential advantage of being lower in public visibility, and therefore, less likely to offend the aesthetics of an area. In an age of heightened security concerns, they also appear to be easier to protect, since there is not as large a perimeter as there is in the case of surface-based reactors. On the minus side, the cost of excavating a site to put a reactor underground is greater than the cost of a surface site and could be considerable. However, a smaller reactor would require a smaller excavation than a large reactor, so the interest in SMRs could fit well with the interest in underground siting. There are other potential issues that have not yet been explored, such as the potential for flooding, the cost of maintenance, and emergency accessibility.

Of course, the underground siting removes a portion of the visibility of the plant, but does not remove the need for cooling, or the facilities and impacts of the cooling process. However, for small reactors, the facilities and impacts of the cooling process would not be as large.

There are several concepts that involve floating or submerged reactors currently being explored (Murray and Holbert, 2014). However, the concept is not new. There was a test of a floating reactor in the Panama Canal Zone between 1968 and 1975 (Marcus, 2010). One advantage of floating reactors is that they have a ready supply of water. Another advantage is that they do not have the land requirements of a fixed site, and therefore, would be suitable for very small countries. One concept would be similar to the concept in the Panama Canal Zone in that a reactor would be mounted on a barge that could be towed to a site and docked there to provide power. This would allow the reactor to be connected to a grid on shore very easily and to be towed away for refueling or other maintenance activities. Conceptually, the reactor could even be towed away in the event of an accident in the plant in order to limit the potential danger to the surrounding population, although doing so might raise other concerns among other countries.

Other concepts would tether the reactor far offshore, either on the surface or under water. Such concepts could be useful in tsunami-prone areas, as tsunami waves build toward the shore, and areas several miles from the coastline would be minimally impacted. Once again, there are a number of issues that need to be considered, such as protection of a remote facility, and the need for rapid accessibility in the case of an emergency.

ISSUES RELATING TO NUCLEAR POWER

There are a number of issues related to the use of nuclear power (Holt, 2014). Some of these have already been covered, to some extent, in the discussions of the technology. This section will summarize the key public policy considerations for issues already noted, and will address several other issues. The public policy issues most often raised for nuclear power are: waste disposal, proliferation concerns, the potential for accidents and their consequences, and economics and government support.

WASTE DISPOSAL

Opponents of nuclear power assert that nuclear power is not viable because there is no viable plan for disposal of the spent fuel, which remains radioactive for thousands of years. Proponents of nuclear power note that the volume of nuclear waste is very small, and that disposal is technically feasible. The problem, they claim, is a political one.

Some of the options and issues associated with waste disposal were described above in general terms. To be more specific, several countries, including the United States, have had major policy battles over the decision on where to site a nuclear waste repository. No jurisdiction likes the idea of taking the waste from somewhere else, especially when the waste will remain radioactive for thousands of years. In the United States, the first site selected as a repository for civilian wastes, Yucca mountain in Nevada, has been mired in a political battle for years. Other countries have had similar battles.

However, as noted, Finland and Sweden have used a consent-based process to make decisions with much less public controversy. Such a process may work in other countries as well. Also, there are other options, such as boreholes, that could be explored, and measures such as reprocessing that could reduce the requirements for a repository.

PROLIFERATION

As mentioned several times in this chapter, nuclear power grew out of weapons development programs. Even though nuclear power development was separated from weapons development long ago, it remains true that some of the nuclear power technologies can be used to create materials that could be diverted for use in weapons. Some people argue against nuclear power on those grounds, or at least against the use of nuclear technology options that might make it easier to create weapons-usable materials.

The problem with foregoing the use of certain technologies is that doing so does not prevent others from using—or misusing—a technology. President Carter cancelled a breeder reactor program in the United States in hopes that other countries would follow the U.S. lead (Boudreau, 1981). They did not. Several countries now have active reprocessing programs, and other countries continue to be interested in reprocessing.

Currently, most governments interested in limiting the spread of nuclear weapons are focusing on the use of international inspections to try to prevent weapons-related activities, and diplomacy to try to persuade other countries to restrict their nuclear activities to peaceful applications. The thinking behind this approach is that the technology is now known, so it is no longer possible to sequester the technology. It follows that the best option is to be certain that new weapons development programs cannot be spun off covertly from civilian activities.

The international nonproliferation activities are too extensive and complex to cover in detail here. Nonproliferation starts with international commitments, such as the Treaty on the Non-Proliferation of Nuclear Weapons (NPT), to which most nations in the world are signatory. It is supported by a variety of programs designed to assure signatory nations access to the benefits of nuclear power without having to have a full fuel cycle themselves. The IAEA in Vienna, Austria, plays a major role in such activities. The IAEA also supports nonproliferation goals by organizing multinational teams to conduct inspections of nuclear facilities in member countries. The inspections check materials inventories and facilities, and are designed to identify whether nuclear materials are being diverted from civilian applications. In particularly difficult cases, groups of countries have imposed harsh economic sanctions against other countries to try to pressure them to abandon potentially weapons-related activities. The most recent major example of this approach was the effort to reign in Iran's nuclear activities.

ACCIDENTS

Probably the greatest public concern, especially for those living anywhere near a nuclear power plant, is the potential consequences of a serious accident. There have been two cases so far in which an accident at a nuclear power plant led to evacuations and the long-term loss of habitable land. These were the accident at the Chernobyl Nuclear Plant in Ukraine (formerly part of the Soviet Union) in 1986 and the accident at the Fukushima Daiichi Nuclear Power Plant in Japan in 2011. While neither accident resulted in the loss of life of a large number of members of the public (the Chernobyl accident did result in the deaths of some of the firefighters who responded to the incident, and the Fukushima accident led to the deaths of a few members of the public as a result of the evacuation process), the accidents did disrupt the lives of thousands of individuals. In the case of Chernobyl, there was also an increase in the incidence of thyroid cancer. Because this risk was known, individuals suffering from thyroid cancer have received early treatment, which usually prevents the disease from being fatal. Furthermore, both accidents spread smaller amounts of radiation over a wide radius, raising concerns in other countries.

From a regulatory and operational perspective, these concerns have been addressed in part by enhanced regulations requiring modifications in plant facilities, by procedures to assure that similar accidents do not occur at other nuclear power plants, and by other actions operators have taken. However, these accidents and their consequences continue to raise concerns in areas surrounding other nuclear power plants.

Despite the concerns, public opinion is clearly mixed. In the United States, there is strong opposition to local nuclear power plants in some regions of the country, while there is equally strong approval of local plants in other regions of the country. Local support comes from the high-paying jobs at the power plant, the secondary economic benefits the employees have on the local economy, including the generation of other employment in the local service industry, and the taxes paid by the facility to the regional jurisdiction.

Likewise, as previously noted and as discussed further below, while some countries are phasing out their nuclear programs, others are expanding theirs, or are even starting brand-new nuclear power programs. For the countries embarking on new or expanded nuclear power development, the need for clean, reliable sources of electricity is a big attraction. There is also a growing recognition of the role nuclear power is playing, and can play, in reducing GHG emissions.

ECONOMICS AND GOVERNMENT SUPPORT

The economic case for nuclear power has proved to be a difficult one for all parties. Opponents of nuclear power raise concerns about the high cost of building nuclear power plants, and the frequent cost overruns and delays in construction. They also object to any semblance of government financial support for nuclear power, whether through loan guarantees by the government, cost recovery of construction work in progress, price guarantees for the power produced, or provisions designed to pool or restrict liability.

Proponents of nuclear power raise concerns about the electric industry's move away from traditional rate-based regulation toward increased competition in a deregulated marketplace. These regulations create spot markets in which nuclear power plants can be underbid by other sources. Natural gas is currently cheap and abundant, and can follow loads better than nuclear power, but in the long term, the price of natural gas is not guaranteed to stay low. Wind and solar power receive special consideration because they are renewable energy sources, but they do not operate all the time.

Therefore, the current electricity market gives insufficient credit to nuclear plants (or other base load facilities) for their reliability, round-the-clock availability, and long-term stability of their prices. Proponents also note that all energy sources have had, and continue to have, different government benefits in terms of taxes, loan guarantees, and other mechanisms, and consider that an appropriate part of the government role in assuring the public adequate and reliable supplies of energy.

All these financial issues are very complex and cannot be probed fully in this chapter. It is sufficient to note that several nuclear power plants have been forced to close prematurely in recent years because of these adverse financial markets. These closures have resulted in the loss of sunk costs in infrastructure and are having unintended negative consequences on efforts to reduce carbon emissions.

OVERVIEW OF CURRENT NATIONAL POLICIES AND ACTIVITIES

At present, the policies and activities of different countries with respect to nuclear power are widely divergent. Several countries with existing power programs are planning to shut down all their plants in the relatively near future, while other countries have active construction programs underway or are gearing up to expand their programs dramatically, and a number of new countries are considering developing nuclear power programs (International Status and Prospects for Nuclear Power 2014, 2014).

It is impossible to cover all countries, and furthermore, some countries have changed their positions on nuclear power several times over the past decades, so a detailed analysis could well be rendered obsolete following an election or some other event. Therefore, only a few selective policies will be highlighted, mainly as examples.

UNITED STATES

The United States was one of the earliest and most active countries to develop nuclear power, and it presently has the largest number of operating reactors of any country in the world (99 in 2015).

Nevertheless, for the last several decades, the development of new nuclear power plants has been limited. There are several reasons for this. They range from economic slowdowns that reduced projected demands and put many projects on hold, to delays resulting from the reviews, analysis, and new requirements developed following each nuclear accident and major incident.

In recent years, there have been promising signs of a renewal of interest in nuclear power, and there are presently several new nuclear power plants under construction. On the other hand, as noted above, the economic deregulation of power production has resulted in the closure of several plants.

Much recent attention has focused on new designs that proponents hope will prove a better economic fit to the market, and will provide improved safety, which it is hoped, will improve the public acceptance. There are a number of new design options being discussed, but most of the attention is focused on the SMRs discussed earlier. It is unclear at present which of these designs will prove the most attractive. The result will depend on obtaining sufficient funding to develop the concepts. That development also requires proving the claims about new technologies, achieving regulatory approval, and gaining the commitment of utilities. The costs to develop new designs and bring them to market are likely to be considerable. The U.S. Department of Energy is providing matching funding for several efforts, and there are a number of bilateral, multilateral, and international initiatives related to advanced reactor development.

GERMANY

The accident at the Fukushima Daiichi reactors in Japan in 2011 triggered several countries to decide to phase out nuclear power. Germany is perhaps the most well-known example. There had been significant opposition to nuclear power in Germany even before the Fukushima accident. The first strong moves against nuclear power in Germany resulted from the Chernobyl accident in the U.S.S.R. (now in Ukraine) in 1986. In August 1986, Germany passed a resolution to abandon nuclear power within 10 years. That did not happen, but the issue was addressed several subsequent times over the years.

Following the March 2011 accident at Fukushima, the German government ordered the immediate shutdown of eight plants had begun operation in 1980 or earlier, and later, announced plans to shut all nine remaining reactors down by 2022. The capacity is to be replaced by renewable energy plus the construction of new coal and gas-fired plants. The immediate consequence of the shutdowns has been an increase in carbon emissions due to the substitution of coal and gas plants.

FRANCE

After the oil embargo of 1973, France, which has virtually no indigenous fuel resources, committed heavily to nuclear power. Currently, it is among the countries with the highest proportion of its electricity coming from nuclear power (about 75%). It made no immediate changes following the Fukushima accident. However, the present government has announced that it wants to reduce the reliance on nuclear power. Despite that, France continues to market its reactors abroad, and like the United States, is engaged in efforts associated with advanced reactor design concepts.

UNITED KINGDOM

The United Kingdom, which was one of the earliest countries to develop and use nuclear power, is presently pursuing plans to build a new reactor at an existing site based on the latest available current technology. The country is also engaging in efforts to develop advanced nuclear designs. In particular, they have signed an agreement with NuScale, a U.S. firm developing one of the integral, pressurized light-water SMR concepts.

OTHER EUROPEAN NATIONS

The policies of other European countries range from that of Austria, which has had a long-standing opposition to nuclear power that continues to this day; to Poland, which has previously opposed nuclear power but is now considering starting a nuclear power program for the first time. Perhaps the most extreme reaction to Fukushima occurred in Italy, which conducted a national referendum shortly after the accident that resulted in the closure of all its nuclear power plants.

Several other European countries with nuclear power programs have announced plans to phase out nuclear power. These include Sweden, Denmark, Belgium, and Switzerland. The details and certainty of their plans differ. In some cases, they have a moratorium on new plants; in other cases, they also expect to shut down existing reactors at some point in the future. Some of these countries have had a long history of opposition to nuclear power and have made previous attempts to terminate the use of nuclear power, but political changes and other issues have either stalled or reversed previous decisions. Because of the differences in the approaches used in different countries and the history of shifts in policy, it is difficult to provide specific projections for all countries.

RUSSIA

Despite the nuclear accident, possibly the worst in the world (Chernobyl, in 1986), Russia has emerged as a leader in the world's nuclear market. Their nuclear program dates back to the earliest days of nuclear power (under the banner of the Soviet Union). In fact, it was a Russian reactor that first put power onto a local grid—at Obninsk, in 1954 (Marcus, 2010). Like the United States, their early efforts were initially aimed at developing nuclear weapons, but they later developed an active nuclear power program. This included the Chernobyl design, but reactors of this design have mostly been shut down and are no longer being built.

Russia has long been at the forefront of developing new technologies as well. Recently, they appear to be making great strides in marketing their reactors abroad, and they have penned agreements to work with a number of countries. Their initiatives include schemes to help finance nuclear power plants in other countries, in some cases, by a scheme where they build, own, and operate the plant in the host country. They are also offering to take back used fuel, removing the need for countries with small nuclear programs to have to deal with the nuclear waste issue themselves.

CHINA AND INDIA

China and India both have large populations and insufficient electricity supplies. Furthermore, their existing electrical plants rely heavily on coal, and air pollution sometimes reaches dangerous levels in highly populated areas. Therefore, both countries are poised to significantly increase their use of nuclear power.

China presently has the world's most ambitious and aggressive program to expand their use of nuclear power. They now have 26 reactors in operation, 23 under construction, and more being planned. Their construction program is by far the largest in any country in the world. They have also begun marketing their reactor designs to other countries. In addition, they have been active in both domestic and international advanced reactor research and development initiatives.

To date, India's expansion plans have been inhibited by the fact that it is not a signatory to the NPT. In the past, this has prevented vendors from countries like the United States from selling reactors or fuel to India. As a result, India developed an indigenous capability based on thorium fuel, as it had domestic sources of thorium, but not of uranium. This embargo has nonetheless limited India's expansion. Recently, it appears there has been a political breakthrough that should permit India to purchase nuclear supplies from U.S. vendors, but the details of this program are still emerging. The U.S. policy, of course, has not prohibited India from establishing ties with Russia and other nuclear suppliers.

JAPAN

Japan is where the Fukushima accident occurred, so the accident has understandably had a profound effect on public opinion there. Before the accident, nuclear power provided approximately one-third of Japan's electricity. Shortly after the accident, all operating nuclear power plants in the country were shut down while a new regulatory organization was started and new measures were developed to check the safety of each plant. A number of plants have now gone through safety assessment, and the first plant was restarted in 2015. However, public opposition remains strong and local government approval is needed for the plant restarts, so uncertainties remain about when, and even whether, all the plants approved for restart by the national regulator will actually be restarted.

SOUTH KOREA

South Korea has been experiencing considerable domestic opposition to its nuclear power program. Some of this opposition arises from the proximity to Japan and concerns raised about potential impacts in Korea of the Fukushima accident. However, revelations about the falsification of safety-related documents associated with components for existing reactors by Korean utility officials have probably exacerbated public concerns. Nevertheless, South Korea remains committed to its domestic nuclear power program, and in the last couple of years, has emerged as a nuclear supplier. In particular, it won the competition to supply reactors for the United Arab Emirates.

AFRICA, ASIA, AND SOUTH AMERICA

Prior to the Fukushima accident, the United Arab Emirates had made a decision to develop a nuclear power program. They continued this program after the accident and are currently in the process of constructing their first nuclear power plants, purchased from South Korea. Prior to this project, most countries that now use nuclear power began their commercial nuclear power programs by first building and operating a research reactor. This allowed them to build up expertise and institutions (particularly educational and regulatory) before embarking on a commercial program. The UAE is the first country to begin a commercial program without having a research reactor, so their initiative has required the parallel development of a regulatory framework and other institutional infrastructure.

Around the world, other countries are in various stages of developing nuclear power programs. In Asia, Bangladesh has two plants in the early stages of construction. In South America, Argentina and Brazil, which already operate nuclear power plants, are in the process of building more. In addition, a number of other countries in Asia, Africa, and South America are considering staring new nuclear programs. The programs in these countries are in various stages, and like other such initiatives, not all of them can be expected to come to fruition. Changing politics and the extensive commitments needed (financial, human resource, regulatory, etc.) may slow or halt some of the programs. Nevertheless, since there are reported to be several new dozen countries now considering nuclear power, it is likely that some of these efforts will result in more countries operating nuclear power plants in the next couple of decades and beyond.

CONCLUSIONS

Nuclear power development has had a complex history. It progressed extremely rapidly in the early years, to a significant extent because it was an offshoot of military initiatives to develop weapons and nuclear propulsion. The military origins both created pressure for rapid progress and provided the funding to facilitate that progress. For a variety of reasons, nuclear power has drawn a lot of opposition in some countries and some segments of the public. Nuclear power may not be unique in this regard, as the second half of the twentieth century in which the early growth of nuclear power occurred was also a period of considerable social unrest, and spawned the growth of movements in opposition to other industrial facilities, as well as to the implementation of scientific advances in other areas (such as genetically modified organisms in foods).

Nevertheless, nuclear power was embraced by a number of countries for different reasons. In France, it effectively gave them energy independence. (Although countries like France may have to import uranium, the small quantities and long refueling periods make them more immune to political or weather-related disruptions than is the case for fossil fuels.) Where nuclear power plants have been built and operated, they have proven to provide a reliable supply of electricity, suitable for meeting base load requirements.

In recent years, the low emissions from nuclear power plants have been recognized as another positive feature. There are no particulate or GHG emissions at the nuclear power plants themselves, and only a relatively small amount associated with the fuel cycle (mining, milling, enrichment, etc.). Nuclear power production even compares well to renewable energy sources in this regard; when the mining and processing of materials needed to capture the renewable energy are fully taken into consideration, nuclear power has emissions in a similar range to those from renewable energy sources (Weisser, 2007; Lifecycle Assessment Literature Review of Nuclear, Wind, and Natural Gas Generation, 2014).

Cost has been another issue for nuclear power. Nuclear power plants are generally costly to build compared to other electric power plant technologies. This has made financing them difficult in some cases. However, their fuel costs are very low, making their operating costs competitive with other technologies, and furthermore, making the operating costs more stable than is possible for some other technologies.

The dichotomy of national and individual views on nuclear power continues to this day. The accident at the Fukushima Daiichi plants in Japan crystallized opposition to nuclear power in some countries and among some groups of people and resulted in decisions to abandon nuclear power in several countries. Nevertheless, at the same time, nuclear power development continues in other countries, and there are a number of countries where it is under active consideration for the future. These efforts are continuing because the potential benefits of nuclear power are perceived to outweigh the small probability of an accident. Furthermore, every country operating nuclear power plants, as well as every country considering nuclear power plants is very mindful of the accidents that have occurred, and measures are being taken to improve existing plants and to modify designs for future plants in order to reduce the risks of such an accident in the future.

Finally, technological developments currently underway hold the promise of better nuclear power plant designs in the future. These designs would permit the continued use of nuclear power to provide reliable base load power with very low emissions, and would offer many potential improvements, including: passive designs that are more robust in the case of accidents; smaller plants that can meet the needs of small or remote sites, or that can be built incrementally to better match demand growth; recycling of fuel that can greatly extend the effective nuclear fuel resource and reduce waste disposal requirements; higher-temperature outputs that can allow nuclear power to be used for a variety of industrial needs; improved economics; improved proliferation resistance; reduced need for cooling water; and expanded siting options.

REFERENCES

1992. Controlled Nuclear Chain Reaction: The First 50 Years. American Nuclear Society, La Grange Park, Illinois.

2015. Nuclear News world list of nuclear power plants. *Nuclear News*, 58(3): 39–72, http://www.ans.org/pubs/magazines/nn/y_2015/m_3.

2010. Public Attitudes to Nuclear Power. OECD/Nuclear Energy Agency, https://www.oecd-nea.org/ndd/reports/2010/nea6859-public-attitudes.pdf/.

2014. International Status and Prospects for Nuclear Power. IAEA, https://www.iaea.org/About/Policy/GC/GC58/GC58InfDocuments/English/gc58inf-6_en.pdf.

2014. Lifecycle Assessment Literature Review of Nuclear, Wind, and Natural Gas Generation. Canadian Nuclear Association, H345621-236-02, Rev. E, October 9, https://cna.ca/wp-content/uploads/2014/05/Hatch-CNA-Report-RevE.pdf.

Blake, M. 2014. Advanced reactors, an endless landscape. *Nuclear News*, 57(13): 44–45, in Special Section on Advanced Reactors, 43–87, December 2014, http://www.ans.org/pubs/magazines/nn/y_2014/m_12.

Boudreau, J. 1981. The American breeder reactor program gets a second chance. Los Alamos Science, 2(2): 118–119, summer/fall, https://www.fas.org/sgp/othergov/doe/lanl/pubs/00416678.pdf.

El-Wakil, M. M. 1982. Nuclear Energy Conversion. American Nuclear Society, La Grange Park, Illinois.

Flyvbjerg, B., N. Bruzelius, and W. Rothengatter. 2003. *Megaprojects and Risk: An Anatomy of Ambition*. Cambridge: Cambridge University Press.

GIF. 2014. Technology Roadmap Update for Generation IV Nuclear Energy Systems. Issued by OECD Nuclear Energy Agency for the Generation IV International Forum, January, https://www.gen-4.org/gif/upload/docs/application/pdf/2014–03/gif-tru2014.pdf.

Holt, M. 2014. Nuclear Energy Policy. Congressional Research Service, October 15, https://www.fas.org/sgp/crs/misc/RL33558.pdf.

IAEA. 2014. Uranium 2014: Resources, Production and Demand. Joint Report by the OECD/NEA and the IAEA, http://www.oecd-nea.org/ndd/pubs/2014/7209-uranium-2014.pdf.

International Atomic Energy Agency (IAEA) Power Reactor Information System (PRIS), https://www.iaea.org/PRIS/WorldStatistics/OperationalReactorsByCountry.aspx (retrieved September 15, 2015).

Marcus, G. H. 2010. Nuclear Firsts: Milestones on the Road to Nuclear Power Development. American Nuclear Society, La Grange Park, Illinois.

Marcus, G. H. and A. E. Levin. 2002. New designs for the nuclear renaissance. *Physics Today*, 55(4): 54–60.

Murray, R. and K. Holbert. 2014. *Nuclear Energy, Seventh Edition: An Introduction to the Concepts, Systems, and Applications of Nuclear Processes*. Butterworth-Heinemann, Oxford, United Kingdom.

Röhrlich, E. 2013. Eisenhower's atoms for peace. *IAEA Bulletin*, 54: 3–4, December 4, https://www.iaea.org/sites/default/files/bull54_4_dec2013.pdf.

Tsoulfanidis, N. 2014. The Nuclear Fuel Cycle. American Nuclear Society, La Grange Park, Illinois.

Weisser, D. 2007. A guide to life-cycle greenhouse gas (GHG) emissions from electric supply technologies. *Energy*, 32(9): 1543–1559.

World Nuclear Association. 2015. World Nuclear Power Reactor and Uranium Requirements, London, United Kingdom. http://www.world-nuclear.org/info/Facts-and-Figures/World-Nuclear-Power-Reactors-and-Uranium-Requirements/ (retrieved September 15, 2015).

23 Carbon Capture and Storage

Jim Underschultz, Kevin Dodds, Karsten Michael,
Sandeep Sharma, Terry Wall, and Steve Whittaker

CONTENTS

ABSTRACT

Carbon dioxide capture and storage (CCS) involves separating CO_2 from an emission source gas, transporting it to a storage location, and securing it in long-term isolation from the atmosphere. The three main capture technologies are (1) postcombustion capture, (2) integrated gasification combined cycle, and (3) oxyfuel. All three (first-generation) technologies are associated with higher generation costs with energy penalties of approximately 25%. Current research is focused on cost reduction across the full range of technology.

Theoretically, any captured CO_2 can be stored in the geological subsurface, in oceans, or it can be used for mineral carbonation or for industrial uses. Realistically, capture from large point-source emitters such as high-CO_2 natural gas production or power stations (to a lesser extent liquefied natural gas [LNG] plants, mineral processing plants, or cement plants) coupled with geological storage is the most likely large-scale commercial implementation of CCS where projects have four phases: (1) site selection and development, (2) operations, (3) site closure, and (4) postclosure stewardship. Site selection includes assessment that focuses on characterizing injectivity, storage capacity, and containment security. This then dictates the purpose and design of monitoring and the particular choice of the technology dependent on the role it plays in managing the project risk, with emphasis on the reporting to the various stakeholders.

While there are now a number of carbon capture and storage projects at a variety of scales operational in the world, including many mature CO_2 enhanced oil recovery operations in North

America, more broad deployment of CCS technology is largely dependent on a business case to proceed in the future. The business case is dependent on factors such as the price of carbon, regulatory constraints on the emissions of carbon, and technology advances that reduce the energy penalty and cost of CCS.

CARBON CAPTURE AND STORAGE DEFINED

Carbon dioxide capture and storage (CCS) involves separating CO_2 from an emission source gas, transporting it to a storage location, and securing it in long-term isolation from the atmosphere. This technology was identified as integral to climate change policy with the publication of the Intergovernmental Panel on Climate Change (IPCC) Special Report on CCS (2005). In the most recent assessment, the 5th IPCC Assessment Report (2014) concluded that many of the global mitigation models could not meet the 2°C target without CCS and that mitigation costs would increase significantly if CCS technology was not included as part of a total mitigation strategy.

Theoretically, any captured CO_2 can be stored in the geological subsurface, in oceans, or it can be used for mineral carbonation or for industrial uses (IPCC, 2005). Realistically, capture from large point-source emitters such as high-CO_2 natural gas fields or power stations (to a lesser extent LNG plants, mineral processing plants, or cement plants) coupled with geological storage is the most likely large-scale commercial implementation of CCS. The term "CCUS" is often used in the literature and refers to *carbon capture utilization and storage*, where "utilization" normally means enhanced oil recovery (EOR) operations. These components are what we discuss in this chapter.

CAPTURE TECHNOLOGIES FOR POWER PLANTS

The CO_2 (molar concentration) levels of flue gas from fuel combustion in power plants are typically only approximately 15%–20%. Various "capture technologies" use different strategies to concentrate the CO_2 to levels of approximately 75%–80% or more making subsequent CO_2 compression to the supercritical state viable for geological storage. Recent R&D on CO_2 capture from power stations has focused on three technologies that have all progressed through pilot-scale and semicommercial scale (Stanger and Wall, 2007), which are

1. CO_2 capture from conventional pulverized fuel (PF) technology with scrubbing of the flue gas for CO_2 removal termed postcombustion capture (PCC), which uses the same technology used to remove CO_2 from natural gas in the oil and gas industry. A flow sheet is given in Figure 23.1.
2. Integrated gasification combined cycle (IGCC) with a shift reactor to convert CO to CO_2, followed by CO_2 capture, which is called precombustion capture (Figure 23.2).
3. Oxyfuel utilizes oxygen rather than air for combustion, where the oxygen is diluted with an external recycled flue gas to reduce its combustion temperature and add volume to carry the combustion energy through the heat transfer operations (Figure 23.3).

PCC and oxyfuel can be retrofitted to an existing plant, whereas IGCC requires a new build. PCC and IGCC can utilize partial capture from flue gas. IGCC may use O_2 rather than air as the oxidant to establish higher proportions of CO_2, but only oxyfuel does not require CO_2 capture prior to compression. All plants use similar combustion temperatures of 1300–1700°C, but IGCC with CO_2 capture uniquely also requires high pressures of approximately 7–8 MPa.

All three (first-generation) technologies are associated with higher generation costs with energy penalties of approximately 25%. Important contributors to efficiency losses are PCC—solvent regeneration and CO_2 compression, IGCC—oxygen production and CO_2 compression, and

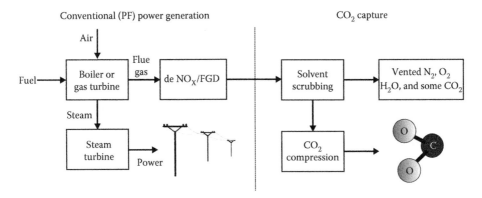

FIGURE 23.1 Illustrative PCC technology, with new unit operations required for a retrofit of an existing PF power station, with gas cleaning, solvent scrubbing, and compression.

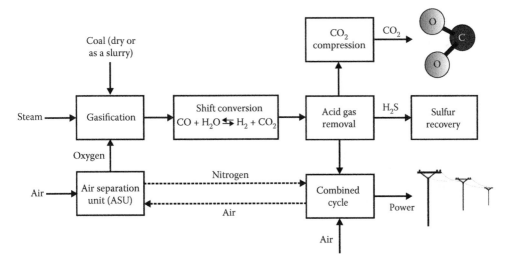

FIGURE 23.2 Illustrative IGCC technology adapted for capture, to include a shift reactor, and CO_2 compression.

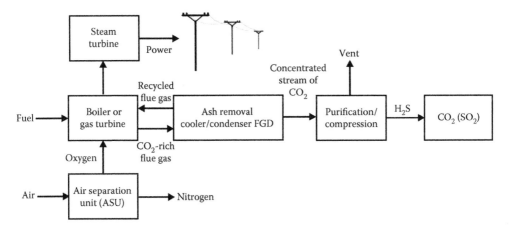

FIGURE 23.3 Illustrative oxyfuel technology, to include an air separation unit (ASU) for oxygen supply, recycled flue gas (RFG) and CO_2 purification and compression.

oxyfuel—oxygen production and CO_2 compression. Second-generation technologies target lowering the energy penalty but are currently only at pilot scale.

CAPTURE TECHNOLOGIES FOR OTHER INDUSTRIAL PLANT

Industrial separation of CO_2 is common practice in chemical industries. Of note is that the separation of CO_2 from natural gas is of a similar scale to that of a single power station unit and this is required in order to have saleable natural gas. However, the status of CO_2 capture for climate abatement in other industries lags behind that of power generation and natural gas processing, with some plants at pilot scale and most at laboratory or concept scale. Early applications are likely to be PCC and oxyfuel technologies applied to blast furnaces used in ironmaking and the rotary kilns of the cement industry.

RECENT CAPTURE TECHNOLOGY ADVANCEMENT

In 2015, the International Energy Agency (IEA) made an assessment of where significant progress has been made in the last 10 years. While the separation of CO_2 with amine-based solvents was already considered a mature technology in 2005 (IPCC, 2005), great advancements had been made in terms of scaling up the technology (Liang et al., 2015). A commercial system is now operating at the 110 MWe Boundary Dam coal power plant in Canada, which captures approximately 1 million tons CO_2 per year. A variety of solvents has been tested at laboratory and pilot scales for application in CO_2 capture systems. Corrosion prevention strategies using specialized corrosion inhibitors, solvent degradation mechanisms, solvent reclamation methods, and analysis of solvent emissions are active areas of R&D. As PCC can be applied partially on a slip stream of flue gas, PCC has the largest number of pilot-scale development units operating. Today, the main technology challenge remains the reduction of the energy required for regeneration of the solvent, but substantial development has been achieved with mixed solvents.

CO_2 capture by oxyfuel combustion has also experienced significant advances in the last decade (Stanger et al., 2015), underpinned by developing knowledge in the field of combustion science and power technology. Different burner and boiler designs, process configuration involving different gas recycles, mitigation of SO_x, NO_x, and Hg species, agglomeration, fouling and corrosion, management of air ingress, and other operational challenges have been the subject of ongoing research. Oxyfuel combustion technologies could be particularly well suited for some large-scale industrial applications (cement, steel, or oil refining) and several oxyfuel combustion large-scale projects in coal power plants (PF and circulating fluidized bed designs) are under consideration for investment decisions (such as the White Rose Project in the United Kingdom, for a supercritical power plant at 426 MWe), although the lack of economic incentives for CCS remains a detractor. Heat integration of the boiler and cryogenic purification unit (CPU), together with gas cleaning in the CPU rather than the boiler has been identified as ways to reduce cost and energy penalty.

IGCC CO_2 capture process can use commercially mature equipment for producing a syngas by reforming and partial oxidation or solid fuel gasification, or for separating the H_2/CO_2 resulting from the water gas shift reaction. The first fully integrated coal-based precombustion large-scale system is the Kemper County-Mississippi Power Plant in the United States, rated at 582 MWe, which will be ready for operation in 2016 (Jansen et al., 2015). The cost of IGCC has recently proven to be an issue for broad commercial deployment.

GEOLOGICAL CARBON STORAGE AND BASIN RESOURCE MANAGEMENT

Geological storage of carbon dioxide is the process whereby CO_2 captured and separated from a source is transported and injected into the geological subsurface for long-term removal from the atmosphere. It is analogous to natural gas being trapped in the deep subsurface in hydrocarbon pools for geological time periods. The main geological constraints for finding the right place to store CO_2 include a porous and permeable reservoir rock to allow injection at the desired rates (injectivity) and storage of the desired volume (storage capacity), overlaid by an impermeable seal rock to retain the injected CO_2 in the geological subsurface (containment security). Two major categories of the storage target are (1) deleted oil and gas fields and (2) saline aquifers. But geological carbon storage must be conducted where it will not adversely affect other sedimentary basin resources. Sedimentary basins can host different natural resources that may occur in isolated pockets, across widely dispersed regions, in multiple locations, within a single layer of strata, or at various depths. The primary resources to consider are groundwater, conventionally and unconventionally trapped oil, gas, coal, and geothermal energy. Other resources that may also need consideration include gas hydrates, mineral and oil sands, potash, uranium, diamonds, and other sediment hosted mineral deposits. Understanding the nature of how these resources are distributed and hydraulically connected in the subsurface is fundamental to managing basin resource development together with carbon dioxide storage. Surface infrastructure and land use (e.g., national park, towns, and cities) also impact subsurface resource development options and must be considered in any basin resource management strategy.

So how can injection of carbon dioxide attain the desired injectivity, storage capacity, and containment security without adverse impacts on another basin resource? Two general processes need to be considered: (1) unintended migration of carbon dioxide and (2) increase of formation pressure (Figure 23.4). Carbon dioxide may migrate laterally or vertically outside the planned storage complex. If migration did occur, it is possible that some carbon dioxide may comingle with a hydrocarbon reservoir or enter a coal seam. Once a reservoir is used for geologic storage, it may also limit the use of that strata for future resource development such as for production of as yet undiscovered hydrocarbons or geothermal energy. Injecting carbon dioxide increases reservoir pressure that could push saline water into groundwater sources. Alternatively, increased pressure may provide support for oil or gas fields that have had their pressure reduced by production, and may even limit the decline of groundwater levels in stressed aquifer systems.

The area of potential impact (Bandilla et al., 2012) or area of review (AOR) is an important regulatory concept that defines the monitored area of carbon dioxide storage. The AOR includes both the projected extent of the carbon dioxide plume (the physical presence of carbon dioxide in the subsurface) and the area in the reservoir subjected to pressure increase beyond the plume itself (Figure 23.4). The focus of a resource management strategy should be within the defined AOR, and a tiered approach to monitoring and site characterization was proposed by Birkholzer et al. (2014).

INJECTIVITY

Injectivity relates to the ability of a reservoir to accept fluids pumped into it and is a function of the thickness (or length of injection interval) and permeability of the storage reservoir. These factors primarily determine the rate at which CO_2 can be pumped down an injection well and into the reservoir before reaching a bottom-hole pressure that fractures the rock or causes a preexisting fault to slip and become transmissive. The permeability is determined by the size of the pore throats in the storage reservoir, which is related to the ability of the rock to allow CO_2 to flow through it. Engineering methods can address limitations of reservoir injectivity through such means as horizontal drilling or by drilling multiple bores spaced sufficiently far apart to avoid pressure interference. Injectivity can potentially be increased by producing formation water out of the storage reservoir to maintain a lower formation pressure. Each of these engineered solutions has a cost/benefit trade-off that needs to be considered on a case-by-case basis.

STORAGE CAPACITY

For ease of transport and greater storage capacity, CO_2 is best injected as a dense supercritical fluid. The critical point where CO_2 has the density typical of a liquid and the viscosity of a gas is at $31.1°C$

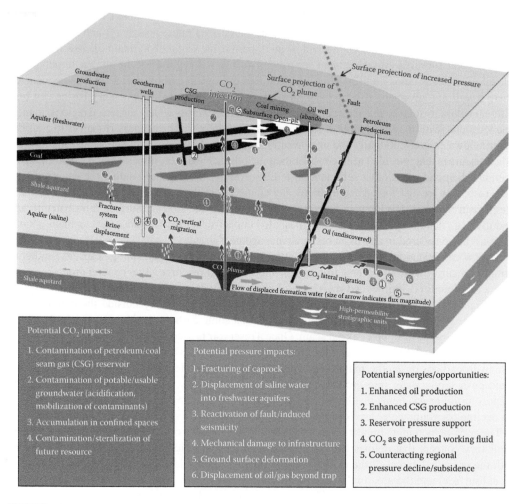

FIGURE 23.4 Schematic diagram summarizing potential subsurface impacts as a result of CO_2 injection. (After Michael, K. et al. 2016. *Environmental Science: Processes & Impacts*. 18: 164–175.)

and 7.38 MPa (Figure 23.5). Based on average geothermal and hydrostatic pressure conditions, this equates to an approximate minimum subsurface depth of 600–800 m.

Below this depth (under normal sedimentary basin conditions), supercritical CO_2 is 30%–40% less dense than a typical saline formation water under the same conditions. This means that the lighter CO_2 will naturally rise by buoyancy through the reservoir until trapped by various physical or geochemical trapping mechanisms. The pore volume available for storage (G_{CO_2}) can be calculated using

$$G_{CO_2} = A \times h \times \phi \times \rho \times E$$

where
    A is the area of storage site
    h is the storage formation thickness
    ϕ is the rock porosity (% of voids per bulk rock volume)
    ρ is the density of CO_2
    E is the storage efficiency

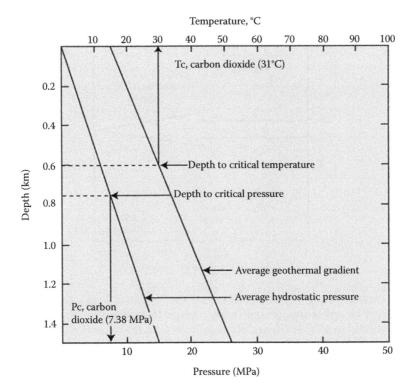

FIGURE 23.5   Carbon dioxide phase pressure–depth diagram. (After Holliday, D. W. et al. 1991. A preliminary feasibility study for the underground disposal of carbon dioxide in the UK. British Geological Survey Technical Report 1991, Nottingham.)

In this case, storage efficiency is defined as the fraction of pore volume actually saturated by CO_2 and it accounts for net-to-total area ("net" refers to the portion of the total area actually containing CO_2), net-to-gross thickness ("net" refers to the portion of the total thickness actually containing CO_2), effective-to-total porosity, and displacement terms for storage efficiency. Displacement terms include the fraction of pore volume that is not contacted by CO_2 as a consequence of the density difference between CO_2 and formation water (volumetric displacement efficiency) and the fraction of pore space unavailable due to immobile *in situ* fluids (microscopic displacement efficiency). These factors define the volume of CO_2 that can be stored within a given storage complex.

CONTAINMENT SECURITY

Containment of the CO_2 can occur in various ways and over various time scales as described in Figure 23.6. The first and perhaps simplest process is conventional trapping by buoyancy in a closed structure below an impervious seal in a manner similar to naturally occurring oil, gas, and carbon dioxide accumulations. This process is termed "structural trapping."

As the supercritical CO_2 moves through the reservoir rock, capillary forces occur between the CO_2 as a nonwetting phase and the saline formation water as a wetting phase. Once the main CO_2 plume has passed through the rock, small droplets of the CO_2 will remain trapped in the center of each of the pores. This droplet of CO_2 is immobilized by the wetting phase, closing across the pore throats by a process known as imbibition (Holtz, 2003). This suggests that during the time period of CO_2 migration, a significant proportion of the CO_2 may be trapped by the second trapping mechanism termed "residual trapping."

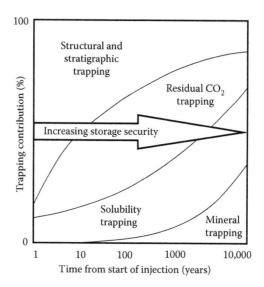

FIGURE 23.6 Diagram of trapping mechanisms with time in carbon storage reservoirs. (After IPCC. 2005. Special report on carbon dioxide capture and storage. In *Prepared by Working Group III of the Intergovernmental Panel on Climate Change*, edited by B. Metz, O. Davidson, H. Coninck, M. Loos, and L. Meyer. Cambridge, UK and New York: Cambridge University Press, 442 p.)

CO_2 is weakly soluble in water, so when it is injected into a saline aquifer, it will gradually dissolve into the formation water resulting in a CO_2-enriched water. As such a solution is slightly denser than CO_2-free water, buoyancy forces will induce the CO_2-saturated water to migrate downwards. This process may occur when the CO_2 is contained within a structural trap or when it is migrating through the pores of the reservoir rock as it rises by buoyancy from the injection point. The process occurs over slightly longer time periods and is referred to as "solubility trapping."

Finally, CO_2 that has been contained for a long period of time (i.e., hundreds to thousands of years) may, depending on the mineralogy of the reservoir rock, react to precipitate new minerals (mainly as carbonate cements) that will essentially trap the carbon permanently. This process is called "mineral trapping." Understanding the way in which these different mechanisms may operate at a specific injection site requires the construction of detailed reservoir models and the reiteration of a number of reservoir simulation runs (refer to http://www.netl.doe.gov/technologies/carbon_seq/corerd/storage.html).

Loss of Containment: Leakage Pathways

Loss of CO_2 containment is possible through one of four leakage mechanisms (see Figure 23.4): (1) leakage through faults may be a result of either insufficient capillary sealing capacity or mechanical fault reactivation caused by the increase in pore pressure as the CO_2 is injected. (2) Leakage through unidentified permeable zones within the seal may occur. (3) Leakage via well bores where there may be a failure in well bore integrity (i.e., via the cement, casing, or the bonding with formation either due to poor well construction or reactive geochemical changes from prolonged exposure of the casing and cement to CO_2). (4) Exceeding the storage capacity of the reservoir where the volume of injected CO_2 exceeded the capacity of the storage container. Here, leakage would occur through the spill points of the trap with migration into overlying formations or into adjacent regions of the same formation. The loss of containment risk can be mitigated and managed by a comprehensive risk assessment and a robust monitoring system. Monitoring within the storage reservoir and immediately above allows early detection of nonconformance such that injection can be stopped and pressure dissipated well in advance of CO_2 having the opportunity to reach the atmosphere or other basin resources.

RISK ASSESSMENT

A good understanding is needed of the likely risk events associated with possible leakage pathways, their probability, and consequences established through a quantitative risk assessment process and a robust engineering design methodology. A comprehensive monitoring and verification (M&V) plan can then be designed as an outcome of this risk assessment analysis. The plan should define trigger points that identify conformance to, or departure from, agreed performance criteria. Contingency actions with a remediation plan can also be identified as part of the overall storage plan.

Risk assessment and management are critical elements of developing a CO_2 storage site and assuring its viability throughout the project lifetime. It informs the selection of an appropriate site, establishes the benchmarks for project viability in the initial phases, provides a strategy to mitigate possible adverse events, and enables the design of appropriate surveillance with considered intervention options should an event occur. The risk assessment and management need to establish the AOR and be an effective communication tool that captures the complexities and uncertainties of the site to the satisfaction of the regulator and public (refer to http://www.netl.doe.gov/technologies/carbon_seq/corerd/simulation.html).

MONITORING AND VERIFICATION

The accepted industry practice is to describe a CCS project as four project phases (Figure 23.7): site selection and development, operations, site closure, and postclosure stewardship. These dictate the purpose of the monitoring and the particular choice of the technology dependent on the role it plays in managing the project risk, with emphasis on the reporting to the various stakeholders. At the heart of the monitoring process is understanding the risk of leakage and the iterative process of reducing risk at each project stage. Information gained from the monitoring references key identifiable events such as deviations from expectation. This is then used to update the risk analysis and provide performance indicators.

Site selection consists of activities that screen several sites with selection of the best option. This also includes the initial reservoir characterization and monitoring of baseline conditions. The operating phase consists of project planning, iterative risk assessment, monitoring to calibrate dynamic modeling, and assessment of expected performance. It also allows repeated monitoring for model validation. The operational phase concludes at cessation of injection, for

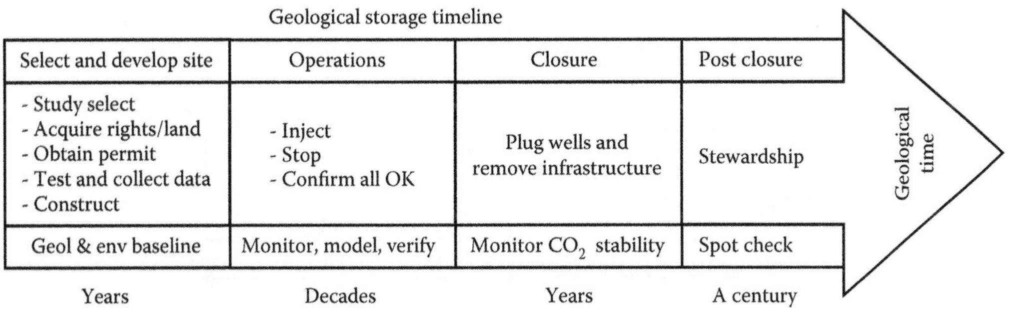

Geological storage timeline

Select and develop site	Operations	Closure	Post closure	
- Study select - Acquire rights/land - Obtain permit - Test and collect data - Construct	- Inject - Stop - Confirm all OK	Plug wells and remove infrastructure	Stewardship	Geological time
Geol & env baseline	Monitor, model, verify	Monitor CO_2 stability	Spot check	
Years	Decades	Years	A century	

FIGURE 23.7 Four phases of CCS projects. (After Cooper, C. et al. 2009. A technical basis for carbon dioxide storage. CO_2 Capture Project, http://www.co2captureproject.org/viewresult.php?downid=123.)

example, after approximately 30 years. The closure phase provides the means to establish storage stability over a suitable period, and complete the abandonment of wells. Closure is completed on meeting key performance criteria to hand the site over to national or state control. Throughout each operation and closure mode, an iterative series of monitoring, evaluation, reporting, and validation exercises are conducted to establish conformance and performance. Postclosure is where liability has been mutually agreed and the responsibility of the site is relinquished to the nation or state.

There is considerable uncertainty early in a project, which gradually reduces as more information is added and the dynamic performance is verified. Equally, the project risk of leakage rises in the operation phase as the injection increases, formation pressure increases, and the CO_2 plume becomes more exposed to possible leakage paths. This risk starts to reduce at cessation of injection, although there may still be remnant risk as the plume migrates within its containment structure until it reaches stability. These transient levels of uncertainty and risk and how they relate to the project stages and associated monitoring relevance are summarized in Figure 23.8.

The defined AOR includes the storage container (where the CO_2 resides) and also the surrounding region and overburden domains where there may be a change of formation pressure as a result of injection. The AOR may involve monitoring from surface and borehole using various fluid sampling and geophysical sensing technologies. Monitoring the storage container provides the means to establish its integrity and that of the seal (see Figure 23.8). The overburden domain comprises the layers above the seal and extends to the near surface. The role of monitoring in this domain is verifying nonseepage of the injected CO_2 above the seal. This establishes a body of information that assures no leakage has occurred. The near-surface and atmospheric domain requires a diverse range of geochemical and gas sensing technologies deployed in grids, down shallow boreholes, and through soil gas accumulators and atmospheric sensing towers. The role of monitoring here

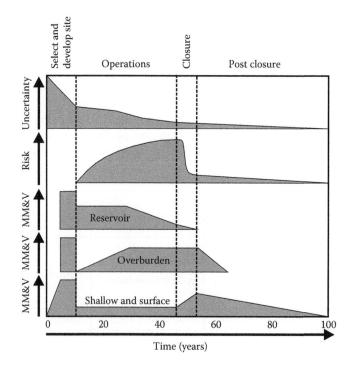

FIGURE 23.8 CCS project phases with associated uncertainty, risk, and monitoring relevance. (After Dodds, K. 2010. Demonstrating iterative risk assessments using In-Salah experience. *IEA Risk Network Meeting Denver 2010*, Denver, Colorado.)

is to provide a baseline characterization of seasonal and diurnal variation of existing gas, including CO_2 derived from biological processes. This baseline provides the norm against which sources of measurable anomalous CO_2 can be identified and from which a minimum detectable level of emission from the storage complex can be set, usually with the regulator. Again, this assures the public and stakeholders that any observed CO_2 did not leak from the container. Consequently, the required technology is quite different to that for subsurface surveillance. In a well-designed program, the CO_2 will not leak into these domains and thus the monitoring will likely comprise only an "assurance" role. As such, it is important to establish the limit of detection of these technologies in detecting small quantities, so that confidence is gained in the absence of a signal and the assurance it brings. The various terminology commonly referred to in monitoring domains are summarized in Figure 23.9.

Regulatory Reporting

The role of M&V in establishing safe storage also provides the means to verify storage of the CO_2 to meet the regulatory conditions and for recovery of carbon credits assigned to emissions abatement. The M&V plan should be consistent with and meet the requirements of the appropriate regulatory authority under which the storage program is conducted. The plan should provide for the generation of clear, comprehensive, timely, and accurate information that is used to effectively and responsibly manage environmental, health, safety, and economic risks and to ensure that set performance standards are being met. It should also determine the quantity, composition, and

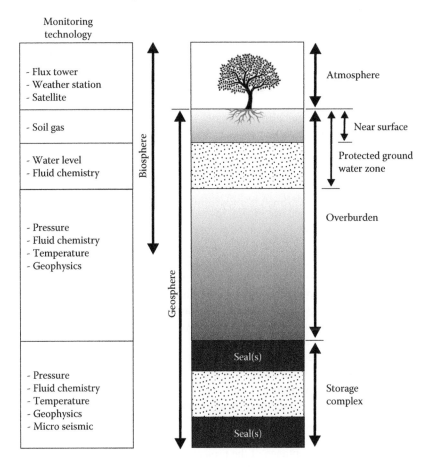

FIGURE 23.9 Categorization of monitoring domains. (After Alberta Government. 2012. CCS Summary Report of the Regulatory Framework Assessment, http://www.energy.alberta.ca/CCS/pdfs/CCSrfaNoAppD.pdf.)

location of CO_2 captured, transported, injected, and stored to an appropriate level of accuracy and the net abatement of emissions. This should include identification and accounting of fugitive emissions.

Public Acceptance

The M&V program in conjunction with regulatory compliance is the basis for a communication plan for effective engagement with the community and stakeholders through a defined outreach program that will inform them of the nature, progress, and outcomes of the project. This is fundamental to gaining the "social license to operate."

The objectives of the "stakeholder engagement plan" are to inform the local community and other stakeholders about the study, its viability, sustainability, and about CCS technology in general. It will aim to ensure that stakeholders understand the concept and stages of the project and how it will affect them. Experience from research and pilot projects has shown that a detailed stakeholder engagement and communications plan targeting the stakeholders should be developed as early as possible and should be an integral component of project management (NETL, 2009).

The suggested consultation structure should be based on the use of parallel consultative processes maximizing direct communication backed by managed information to the wider community. A key principle is that any news is communicated to the community directly through the project and before they have a chance to hear of the same through other sources (Sharma et al., 2006). Trust is to be built by having consistent messages, providing relevant information through multiple means (without creating an overload), listening to the feedback provided, being accessible, and dealing with any issues quickly and with transparency.

The plan should address key issues related to potential concerns and perceptions regarding CCS as a viable technology. Community workshops should be conducted and the input used to shape future communication activities. Typical key community issues addressed by the stakeholder engagement include

- CCS and its importance—globally and locally
- Site selection for the demonstration project. Why this location?
- Landowner concerns about potential impacts on potable underground water supplies
- Community concerns about potential impacts on property values
- Technical questions about the way sequestration works
- Concerns about the implications of technical failure
- Concerns about the continued use of coal instead of increased renewable energy
- Perceived risks to people living above the storage sites and along the selected pipeline route

The consultation program should support a proactive strategy of which key elements are

- Identification of project spokespeople and impartation of media training
- The formation of a project-specific consultative group from the community
- Media information programs based on consultative group deliberations
- Employee and contractor information programs
- Briefings for a wide range of community groups
- Risk assessment and issues definition
- Development of a project website
- Direct mail in the form of a bulletin or newsletter to people on a stakeholder database
- A feedback system

The matter of community acceptance of the project should be given great importance as several CCS projects around the world have failed because of community backlash. The stakeholder management plan should be treated as a live document and regularly updated to include the suggestions and learnings from the various meetings and feedback sessions. Effective stakeholder management can create community goodwill and facilitate project success.

CURRENT CCS PROJECTS

The Global CCS Institute provides regular updates of the status of CCS technology (GCCSI, 2014). Large-scale integrated CCS projects are defined as those which involve the capture, transport, and storage of CO_2 at a scale of

- At least 800,000 t of CO_2 annually for a coal-based power plant
- At least 400,000 t of CO_2 annually for other emission-intensive industrial facilities (including natural gas-based power generation)

About 60 projects under this definition were listed in 2014 of which 12 projects were in operation globally, nine under construction and another 39 in various stages of development planning. The 21 projects in operation or under construction represent a 50% increase since 2011, indicating growth in the application of CCS technology at a large scale. Although there is a growth in active projects, there is a decline of planned projects since 2011. Currently, North America (United States and Canada) is leading in the implementation of CCS/CCUS technology and China is quickly increasing in importance. Australia has a significant carbon storage project at Chevron's Barrow Island gas processing facility set to start injection of 3.4–4 million tonnes per year in late 2015 or early 2016. England is active though prospective project start dates are toward the end of this decade and Europe has lost its previous leadership position. Of note is the Canadian PCC project as part of the refurbished brown-coal power plant at Boundary Dam which is now operating. Also noted is the dominance of enhanced oil recovery (EOR) projects in North America.

With the lack of a global price on carbon, there is yet to be a general business case for CCS, with a few niche exceptions. Many projects remain in the category of research or pilot scale and of those that extend to "commercial scale," many are highly government subsidized in order to make them financially feasible. One exception to this is EOR where there is an economic incentive of increased oil production, yet strictly EOR operations inevitably involve cycling of CO_2 and often are not subject to the same rigor of establishing what percentage of CO_2 injected remains "stored" in the subsurface. As such, it is important to examine EOR commercial operations separately.

ENHANCED OIL RECOVERY

Injecting carbon dioxide into oil fields to enhance oil recovery has been used commercially for more than 40 years. In the EOR process, compressed carbon dioxide gas is injected (usually as supercritical) into the oil reservoir where it acts as a solvent to increase the amount of oil that can be produced from the field. Typically, CO_2 EOR is a tertiary method applied to reservoirs that have declining oil production and that have progressed through primary and secondary production stages (Godec et al., 2013).

CO_2 EOR can be applied to a range of reservoirs, including sandstone, limestone, and dolostone; in structural or stratigraphic traps; in small isolated buildups or giant fields; and onshore or offshore (although no commercial offshore CO_2 EOR has yet been performed). Limitations to deployment are largely influenced by depth (temperature) of reservoir, oil composition, previous oil recovery practices, and internal reservoir features that may hinder effective distribution of the injected CO_2. Access and proximity to a relatively pure and consistent stream of low-cost CO_2, however, is among the more critical factors limiting wider deployment of CO_2 EOR. Not all oil fields, however, are suitable for CO_2 injection as oil composition, depth, temperature, and other reservoir characteristics significantly influence the effectiveness of this method (Melzer, 2012).

CO_2 EOR is often referred to as a miscible flood reflecting the mixing of CO_2 and oil to become a single phase (which differentiates this process from a water flood). The effect of this miscibility is to cause the oil to swell slightly and become less viscous so that it flows within and through the reservoir pores more easily. Most CO_2 EOR projects operate in fully miscible conditions; however, partially immiscible CO_2 floods may also be effective at increasing oil production. Either geologically sourced CO_2 or anthropogenic CO_2 can be used in CO_2 EOR although CO_2 purity of greater than 95% is a rule-of-thumb requirement. Other components in the CO_2 stream may reduce (or enhance) miscibility, so most operators prefer to work with relatively pure CO_2.

As oil nears the surface in the production well, the pressure decreases and the once miscible CO_2 begins to unmix with the oil. At the surface, the CO_2 can be collected, dehydrated, compressed, and reinjected into the reservoir. This recycling reduces the need to purchase additional CO_2 and effectively establishes a closed-loop use of CO_2. Significantly, it also avoids emitting this CO_2 to the atmosphere. A large portion of the CO_2 injected into an oil reservoir, generally around 30%–40%, will not return to the surface as it gets residually trapped. This "loss" of CO_2 to the oil production cycle is a form of geologic storage as the CO_2 will be contained within the reservoir indefinitely and is sometimes referred to as "incidental" storage. Moreover, as the recycled CO_2 is reinjected in numerous cycles, more of it will be progressively retained by incidental storage so that through this process essentially all the purchased CO_2 will eventually reside within the geologic reservoir. Melzer (2012) indicates that almost all of the purchased CO_2 for a CO_2 EOR project will eventually be securely trapped in the subsurface and that losses are minor and related to surface activities such as maintenance procedures.

During the course of a CO_2 EOR operation, the daily amount of CO_2 purchased will remain somewhat constant to offset that lost through incidental storage and support expansion of the flood. As the cumulative injection of purchased CO_2 grows, however, the amount of recycled CO_2 correspondingly increases, so that the total CO_2 injected daily (recycled plus newly purchased) increases during the initial to mid-stages of the flood. At some point, the amount of recycled CO_2 may surpass that of the fresh CO_2 and eventually the flood will begin to taper off purchase of new CO_2 and rely increasingly on recycled CO_2 (Figure 23.10).

Storage estimates associated with CO_2 EOR in North America are over 20 billion metric tonnes CO_2 (Carpenter, 2012). Globally, Carpenter (2012) has suggested storage potential of several hundreds of billion tonnes indicating that there is significant storage potential associated with CO_2 EOR and that it can be important for future CCS development.

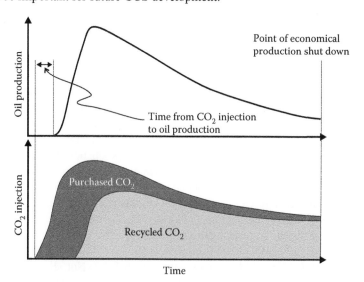

FIGURE 23.10 Depiction of the relation between oil production, purchased CO_2, and recycled CO_2 during the project life of a CO_2 EOR operation. (After Jakobsen, V. et al. 2005. CO_2 for EOR on the Norwegian shelf—A case study. Bellona Foundation Report, Oslo, Norway, 108 p.)

FUTURE

While there are now a number of carbon capture and storage projects at a variety of scales operational in the world, the future deployment of CCS technology is largely dependent on a business case to proceed. The business case is dependent on factors such as the price of carbon, regulatory constraints on the emissions of carbon, and technology advances that reduce the energy penalty and cost of CCS.

REFERENCES

Alberta Government. 2012. CCS Summary Report of the Regulatory Framework Assessment, http://www. energy.alberta.ca/CCS/pdfs/CCSrfaNoAppD.pdf.

Bandilla, K. W., S. R. Kraemer, and J. T. Birkholzer. 2012. Using semi-analytic solutions to approximate the area of potential impact for carbon dioxide injection. International Journal of Greenhouse Gas Control, 8: 196–204, doi:10.1016/j.ijggc.2012.02.009.

Birkholzer, J., A. Cihan, and K. Bandilla. 2014. A tiered area-of-review framework for geological carbon sequestration. *Greenhouse Gases Science and Technology*, 4: 1–16, doi: 10.1002/ghg.

Carpenter, S. 2012. CCUS—Opportunities to utilize anthropogenic CO_2 for enhanced oil recovery and CO_2 storage. AAPG Search and Discovery Article, 80269, http://www.searchanddiscovery.com/documents/2012/80269carpenter/ndx_carpenter.pdf.

Cooper, C., W. Crow, K. Dodds, S. Imbus, P. Ringrose, A. Simone, and I. Wright. 2009. A technical basis for carbon dioxide storage. CO_2 Capture Project, http://www.co2captureproject.org/viewresult.php?downid=123.

Dodds, K. 2010. Demonstrating iterative risk assessments using In-Salah experience. *IEA Risk Network Meeting Denver 2010*, Denver, Colorado.

GCCSI. 2014. The global status of CCS—The February 2014 update from the GCCSI, http://www.globalccsinstitute.com; http://www.globalccsinstitute.com/projects/browse.

Godec, M. L., V. A. Kuuskraa, and P. Dipietro. 2013. Opportunities for using anthropogenic CO_2 for enhanced oil recovery and CO_2 storage. *Energy Fuels*, 27(8): 4183–4189, doi: 10.1021/ef302040u.

Holliday, D. W., G. M. Williams, S. Holloway, D. Savage, and M. P. Bannon. 1991. A preliminary feasibility study for the underground disposal of carbon dioxide in the UK. British Geological Survey Technical Report 1991, Nottingham.

Holtz, M. H. 2003. Optimization of CO_2 sequestration as a residual phase in brine-saturated formations (abs.). In *Briefing Book: Second Annual Conference on Carbon Sequestration: Developing & Validating the Technology Base to Reduce Carbon Intensity*, Alexandria, Virginia, 5–9 May, unpaginated. GCCC Digital Publication Series #03–03.

IPCC. 2005. Special report on carbon dioxide capture and storage. In *Prepared by Working Group III of the Intergovernmental Panel on Climate Change*, edited by B. Metz, O. Davidson, H. Coninck, M. Loos, and L. Meyer. Cambridge, UK and New York: Cambridge University Press, 442 p.

IPCC. 2014. Climate change 2014: Mitigation of climate change. In *Contribution of Working Group III to the Fifth Assessment Report of the Intergovernmental Panel on Climate Change*, edited by O. Edenhofer, R. Pichs-Madruga, Y. Sokona, E. Farahani, S. Kadner, K. Seyboth, A. Adler et al. Cambridge, UK and New York: Cambridge University Press.

Jakobsen, V., F. Hauge, M. Holm, and B. Kristiansen. 2005. CO_2 for EOR on the Norwegian shelf—A case study. Bellona Foundation Report, Oslo, Norway, 108 p.

Jansen, D., M. Gazzani, G. Manzolini, E. van Dijk, and M. Carbo. 2015. Precombustion CO_2 capture. *International Journal of Greenhouse Gas Control*, 40: 167–187.

Liang, Z., W. Rongwong, H. Liu, K. Fu, H. Gao, F. Cao, R. Zhang et al. 2015. Recent progress and new developments in post-combustion carbon-capture technology with amine based solvents. *International Journal of Greenhouse Gas Control*, 40: 26–54, doi: http://dx.doi.org/10.1016/j.ijggc.2015.06.017.

Melzer, S. L. 2012. Carbon dioxide enhanced oil recovery (CO_2 EOR): Factors involved in adding carbon capture, utilization and storage (CCUS) to enhanced oil recovery. Report Prepared for National Enhanced Oil Recovery Initiative (NEORI), Center for Climate and Energy Solutions, Arlington, 17 p.

Michael, K., S. Varma, E. Bekele, B. Ciftci, J. Hodgkinson, L. Langhi, B. Harris, C. Trefry, and K. Wouters. 2013. Basin Resource Management and Carbon Storage—Part I. ANLEC R&D Report 3-05115-0057, Canberra, 117 p.

Michael, K., S. Whittaker, S. Varma, E. Bekele, L. Langhi, J. Hodgkinson, and B. Harris. 2016. Framework for the assessment of interaction between CO_2 geological storage and other sedimentary basin resources. *Environmental Science: Processes & Impacts*. 18: 164–175.

NETL. 2009. Public Outreach and Education for Carbon Storage Projects DOE/NETL-2009/1391. National Energy Technology Laboratory, December, www.netl.doe.gov.

Sharma, S., P. J. Cook, S. Robinson, and C. Anderson. 2006. Regulatory challenges and managing public perception in planning a geological storage pilot project in Australia. In *Proceedings of the 8th International Conference on Greenhouse Gas Control Technologies*. Trondheim, Norway, 19–22 June.

Stanger, R., T. Ting, C. Spero, and T. Wall. 2015. Oxyfuel derived CO_2 compression experiments with NO_x, SO_x and mercury removal—Experiments involving compression of slip-streams from the Callide Oxyfuel Project (COP). International Journal of Greenhouse Gas Control, 41: 50–59, doi: 10.1016/j.ijggc.

Stanger, R. and T. F. Wall. 2007. Combustion processes for carbon capture. Invited plenary lecture and review. In *31st International Symposium Combustion*, Vol. 31. University of Heidelberg, Proceedings of the Combustion Institute, Elsevier, pp. 31–47.

24 Photovoltaics
Current and Emerging Technologies and Materials for Solar Power Conversion

Venkata Manthina, Alexander Agrios, and Shahzada Ahmad

CONTENTS

ABSTRACT

Enough solar energy reaches the earth's surface to cover humanity's energy needs thousands of times over, yet it accounts for <1% of global energy consumption. This chapter presents an overview of technologies that have been developed for the direct conversion of sunlight to electrical power along with discoveries of materials and achievements. This includes silicon technologies, which are commercially dominant today; III–V technologies that offer the best performance at high cost; thin-film semiconductor materials such as copper indium gallium (di)selenide and cadmium telluride, which offer less power per area than silicon but more power per dollar; and emerging technologies such as copper zinc tin sulfide, organic, dye sensitized, and perovskite solar cells. Advantages and disadvantages of each approach are discussed including cost, power conversion efficiency, scalability, scarcity of critical elements, and environmental concerns such as toxicity. The outlook for future directions are discussed.

INTRODUCTION

The quest for renewable energy began in the 1970s with an appreciation of the ultimate scarcity of fossil energy sources, and has been given renewed impetus by concerns of climate change. The solar energy resource is vast and the only one of all the nonfossil energy sources with the potential to power human energy needs indefinitely (Lewis, 2005). Substantial sunlight is available throughout inhabited regions of the globe and is especially abundant in the most densely habited zones close to the equator. There are many ways that sunlight can be converted into energy usable by humans, including biomass (where photosynthesis stores sunlight as chemical energy), solar thermal (where solar heating is used to drive a heat engine that generates electricity), and solar fuels (where sunlight drives chemical reactions that produce fuels). This chapter, however, is an overview of photovoltaic (PV) technology, where sunlight is directly converted to electrical energy. In particular, the focus is on the materials challenge of obtaining high solar power conversion efficiency (PCE) in a material that is inexpensive, durable, of limited toxicity, and reasonably abundant. The emphasis is on alternatives to the commercially dominant silicon-based technology. Some aspects of policy, cost, and abundance will also briefly be discussed.

PV panels have existed for decades but played a negligible role in overall human energy usage until recently. A very strong acceleration of manufacture and installation of PV systems, driven by a combination of lowered cost and increased social awareness and policy incentives, has increased total installed capacity to more than 177 GW by the end of 2014 (International Energy Agency, 2014a). This has finally exceeded 1% of the total human energy usage, which stands at 17.2 TW (BP, 2015). Clearly, PV will have to make up a far more substantial portion of energy generation to alleviate climate change and other environmental problems. Great progress is being made, with PV growing at 100 MW/day (International Energy Agency, 2014b) and forecast to account for a third of new power generation capacity worldwide between now and 2030 (Turner, 2013). The obstacles to PV are many, including the energy storage that is necessary due to the diurnal nature of sunlight in addition to less predictable fluctuations due to weather. Here, however, we restrict ourselves to the PV materials themselves.

PRINCIPLES GOVERNING EFFICIENCY

The PCE of a solar cell is defined as the electrical power output by a cell divided by the incident power of sunlight. For comparison, they are measured under simulated sunlight with prescribed conditions: a total power of 1000 W/m^2 (equal to 100 mW/cm^2), with a defined spectrum called air mass 1.5 global (AM1.5G), which is the spectrum of sunlight when the sun is at a zenith angle of 48.2°.

Most of the technologies described here are based on semiconductors, which have as a critical parameter the band gap (E_g), the gap in energy between the highest-energy occupied band (the valence band) and the lowest-energy unoccupied band (the conduction band). Incoming photons with energy greater than E_g can excite an electron from the valence band to the conduction band and are thereby absorbed. Photons with lower energy cannot be absorbed. Photon energy in excess of E_g is lost due to nonradiative relaxation of "hot carriers." The value of E_g therefore sets the maximum conversion efficiency possible for a particular semiconductor, as the material will be able to recover energy E_g from photons having an energy of E_g or higher, and no energy from photons with lower energy.

For example, Figure 24.1 shows the AM1.5G spectrum and the fraction of this light (shown in green) that can be absorbed by a bulk Si device. In bulk Si solar cells with E_g of 1.1 eV, the losses due to spectral mismatch is close to 50%. The up conversion (UC) and down conversion (DC) spectrum availability of the Si is shown in Figure 24.1, where DC is the generation of more than one lower-energy photon being generated per incident high-energy photon, while UC is absorbing more low-energy photons to emit less quantity of higher energy.

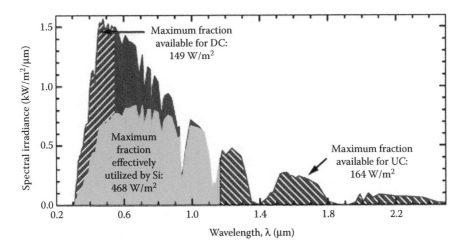

FIGURE 24.1 AM1.5G spectrum showing the fraction that is currently absorbed by a thick silicon device and the additional regions of the spectrum that can contribute toward UC and DC. (Adapted from Richards, B. S. 2006. *Sol. Energy Mater. Sol. Cells*, 90(15): 2329–2337.)

The classic Shockley—Queisser limit (Shockley and Queisser, 1961) establishes the highest possible efficiency for a single-junction solar cell, taking into account the above considerations together with the fact that a solar cell will also reemit (akin to blackbody radiation) some of the light it absorbs. More accurate updates of their calculation have shown the maximum efficiency (33.3%) to lie at $E_g = 1.16$ eV, with a secondary maximum (33.1%) at $E_g = 1.36$ eV. But there is a general zone of maximum efficiencies near 33% lying between the band gap values of 1.1 and 1.4 eV (Tiedje et al., 1984).

It is possible to exceed the Shockley—Queisser limit by (1) combining materials with different band gaps, (2) concentrating the sunlight, or (3) capturing the energy of "hot carriers" before they can relax. The only one of these that has been practically implemented in PV is (1), which has been used in double- and triple-junction cells based on combinations of III–V materials such as GaAs. Regardless of the scheme used, the ultimate maximum efficiency for conversion of sunlight to electrical energy, based on an entropy balance as well as reradiation of absorbed light by the solar cell, is about 80% (Würfel, 2002). The timeline of efficiency records for different technologies, compiled by the U.S. National Renewable Energy Laboratory, is shown in Figure 24.2.

CRYSTALLINE SILICON

The PV industry to date has been dominated by crystalline silicon (c-Si) technology. Based on 2013 data, this mature technology accounts for 90% of the total PV market (Fraunhofer ISE, 2014b). Its advantages include the very high abundance of the element Si, good stability, and the wealth of prior research on Si-based devices for computing applications. Its band gap of about 1.1 eV is at the edge of the ideal zone according to Shockley—Queisser theory. Its main disadvantage is its relatively high cost, which results from two facts. First, a good Si PV panel requires extremely pure (99.99999%) silicon that is nearly perfectly crystallized. Achieving this is energy-intensive and requires sophisticated fabrication facilities, imposing high capital costs on the production of the material for Si PV. Second, Si is an indirect-band gap material, and is therefore a relatively weak absorber of light. Efficient harvesting of sunlight therefore requires substantial amounts of the precious refined Si material. Typical Si layer thicknesses are on the order of hundreds of micrometers, which is much thicker than competing technologies as described below.

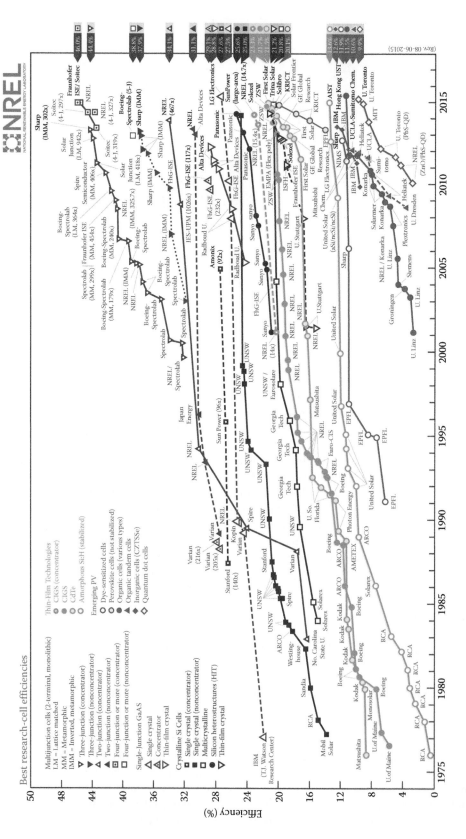

FIGURE 24.2 Progress of research scale efficiency of variety of solar technologies. (Courtesy of National Renewable Energy Laboratory, Golden, Colorado.)

c-Si PV exists in two main forms: monocrystalline (mono-Si) and multi- or polycrystalline (mc-Si). Efficiencies for mono-Si modules surpassed 22%, while that of the lower-cost mc-Si modules is above 18% (Green et al., 2015). The latter has captured a majority of the c-Si market due a better cost-per-watt ratio, but mono-Si is gaining market share as the cost gap closes and due to the value of a higher-efficiency product requiring less total area for the same delivered power (IHS Technology, 2015).

The generally high cost of Si PV has motivated the development of lower-cost materials, which underlie the competing technologies described below. However, in addition to its advantages mentioned above, Si PV today benefits from its own success, as steady technological advancement has combined with the enormous economies of scale to result in sustained, impressive reductions in cost per watt of installed power. Module prices have plummeted from 3.2 USD/W_p in mid-2009 to only 0.7 USD/W_p in late 2014 (GlobalData, 2014; IRENA, 2015). As a result, competing technologies have been increasingly squeezed by Si PV.

THIN-FILM SOLAR CELLS

Thin-film technologies are based on materials that have direct band gaps and therefore absorb light more strongly than silicon. The result is that a smaller amount of the active material is needed. In addition, they tend to use simpler manufacturing processes. Generally, the thin-film technologies have lower efficiencies than c-Si, but attempt to make up for this with substantially reduced costs. The first of these is based on amorphous silicon (a-Si). Copper indium gallium (di)selenide (CIGS) and cadmium telluride (CdTe) are the leading technologies of this type, but both include somewhat rare elements. Alternatives such as copper zinc tin sulfide (CZTS) and pyrite, which use highly abundant materials, have attracted significant research interest.

Amorphous Si

Silicon in a noncrystalline form is characterized by a random network of interconnected Si atoms without long-rage crystal structure. As formed, a significant fraction of Si atoms have three rather than four bonds with neighbors, leading to "dangling bonds" that ruin device performance. These dangling bonds can be passivated by hydrogenation, leading to hydrogenated amorphous silicon, a-Si:H, with much better performance. This material has several favorable properties. Compared to c-Si, it has a fundamentally different electronic structure that renders it much more strongly light-absorbing, so that much less material is needed for effective light harvesting. It can also be made with much simpler processing. In particular, it can be deposited at low temperatures, enabling the use of flexible plastic substrates. These advantages combine to make a-Si:H much lower in cost than c-Si. On the other hand, hydrogenation of a-Si results in photoinduced degradation of the material by the Staebler—Wronski effect (Staebler and Wronski, 1980). In addition, the PCE of a-Si:H is significantly lower than c-Si. This is partly alleviated by the fact that a-Si:H can be made transparent to the photons that it does not absorb, unlike the thick opaque c-Si layers. This allows the construction of tandem (multijunction) a-Si:H devices, which is the technology chiefly used in large modules, where different layers have different band gaps due to alloying with carbon or germanium, or by using a mixed phase intermediate between amorphous and polycrystalline (Green, 2001). Amorphous silicon technology has found widespread application in powering small devices, such as calculators. For larger-scale modules, a-Si:H was gaining traction as a significant fraction of the thin-film PV market, but has been in retreat for the last few years as its PCE has been left behind by improvements in competing thin-film materials. In 2013, a-Si:H accounted for 2% of the overall PV market, and 22% of the thin-film segment (Fraunhofer ISE, 2014b). However, one technology that has been on the rise recently is a hybrid of mc-Si and a-Si, frequently referred to as a HIT (heterojunction with intrinsic thin-layer) cell (Figure 24.3).

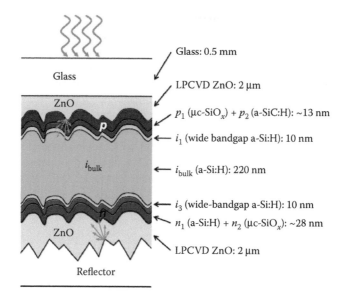

FIGURE 24.3 Solar cell structure used for all p–i–n solar cells. (Adapted from Stuckelberger, M. et al. 2013. *J. Appl. Phys.*, 114(15): 154509.)

COPPER INDIUM GALLIUM (DI)SELENIDE

Copper indium gallium (di) selenide (CIGS) technology is based on an active layer consisting of those four elements made by alloying $CuInSe_2$ and $CuGaSe_2$. The band gaps of each of these pure materials are 1.0 and 1.7 eV, respectively. The two can be mixed with an arbitrary ratio of In to Ga to form $CuIn_{(1-x)}Ga_xSe_2$, with x set to any value between 0 and 1, giving rise to a band gap anywhere between 1.0 and 1.7 eV. The mixed material thus allows tuning the band gap to the observed optimal value. Some of the selenium can also be substituted by sulfur. The technology is sometimes called CIS, reflecting its origins as a $CuInSe_2$ material that was later modified with Ga.

CIGS is a strongly absorbing material, allowing the active material to have a thickness between 2 and 3 µm (Repins et al., 2008; Jackson et al., 2011). Completing a CIGS solar cell requires a molybdenum conductor on one side, while the opposite side of the CIGS layer contacts a thin buffer layer (typically CdS), a doped zinc oxide transparent conducting layer, and a metal grid contact with optional antireflection coatings. The first thin-film solar cells of this type were gallium-free (CIS) devices having efficiencies between 4% and 5% (Kazmerski et al., 1976). Recent CIGS efficiency records have reached 21.7% for laboratory-scale (small) devices (Jackson et al., 2015), and 16.5% for a module of more than 1 m² (TSMC Solar, 2015).

CIGS technology is unusually flexible in the diversity of manufacturing processes than can be used to produce it. In one method, the metals Cu, In, and Ga are deposited by room-temperature techniques such as sputtering or electrodeposition, followed by a high-temperature "selenization" in which selenium reacts with the metals to form CIGS. However, since the Se enters the film by diffusing from one side of it, this leads to a selenium gradient in the film. For better control of the concentration profiles throughout the film, all elements can be applied together by "co-evaporation" under vacuum. Although it is more costly, to date coevaporation has produced the highest cell efficiencies. Alternatively, the cost of manufacturing at large scale can be reduced by using solution deposition or ink jet printing techniques. All-solution processed CIGS devices have exceeded 15% PCE (Todorov et al., 2013). A number of start-up companies were generated based on CIGS technology in this millennium, but most have failed, leaving the Japanese company Solar Frontier as today's dominant commercial producer of CIGS solar cells. Its market share in 2013 was the same as that of a-Si:H in 2013, at 2% of total and 22% of thin-film technologies (Fraunhofer ISE, 2014b).

CdTe Solar Cells

CdTe has a band gap of 1.5 eV, close to the optimal range for PV materials of 1.1–1.4 eV. The standard techniques used for synthesis of high crystalline quality CdTe are chemical spraying, screen printing, chemical vapor deposition (CVD), sputtering, and electrodeposition (ED). An important step in the fabrication of CdTe solar cells is annealing the CdTe/CdS p–n structure in the presence of $CdCl_2$. This step is referred to as the "activation step," since it induces valuable structural and electrical properties in the CdTe and CdS layers, converting a cell with <2% efficiency before activation to >10% efficiency after activation. $CdCl_2$ treatment introduces electrical doping, recrystallization of smaller grains, and passivation of the grain boundaries and interface states (Major et al., 2014).

Uniquely among the major PV technologies, the fortunes of CdTe have been directly tied to those of a single company, namely, First Solar, based in Tempe, Arizona. In the competition among thin-film technologies, CdTe for most of the last 15 years lagged behind CIGS in terms of lab-scale record efficiencies, with the best efficiency for CdTe cells in the range of 15%–16% while CIGS achieved 18%–19%. (Both were well above a-Si:H, at 12%–13%.) However, First Solar was able to develop a low-cost process for large-scale manufacture of CdTe panels, resulting in the lowest-cost technology in terms of dollar per peak watt ($/W_p$) of any PV manufacturer. CdTe has since become an important part of the overall PV market. First Solar has focused on large-scale installations; many of the world's largest PV "solar farms" consist of CdTe panels manufactured by First Solar. Conversely, the technology has been generally absent from the rooftop market. CdTe has led the thin-film PV market with greater practical developments in the field with building solar farms.

Recent years have seen significant changes. Continued improvements in c-Si, in terms of both cost and efficiency, have allowed that technology to undermine First Solar's claim to the lowest $/$W_p$ figure. At the same time, improvements in CdTe technology have rapidly improved the record cell efficiencies, beginning a climb in 2012 that has surpassed 21% in 2015, roughly equaling the best CIGS cells, which have also seen recent improvement. As First Solar attempts to implement these advances in its manufacturing, its best module efficiencies have surpassed 18%.

Both of the elements that make up CdTe have been cause for concern. Cadmium is a well-known heavy metal environmental toxin. Its presence in solar modules has led to fears of toxic exposure in the event of breakage, and raises end-of-life disposal issues. In response, First Solar has instituted a recycling program to reclaim their own panels at end of life and recover and reuse most of the cadmium. First Solar and other CdTe advocates have also argued that (1) the solid CdTe, sandwiched between glass plates in a solar cell, is extremely stable and in fact an excellent way to sequester cadmium, which is after all produced as a by-product of zinc mining and would otherwise require disposal and (2) given the energy a CdTe panel produces in its lifetime, it contains less cadmium that would be emitted into the atmosphere if an equivalent amount of energy were obtained by burning coal instead (Fthenakis, 2004). However, the activation step uses a substantial amount of dissolved Cd^{2+} which requires costly disposal after use (Fthenakis, 2004). Research has shown that it may be possible to replace $CdCl_2$ with $MgCl_2$, which is nontoxic and far cheaper, with similar efficiency and stability (Major et al., 2014). Meanwhile, telluride is the ionic form of tellurium, a metal that is rather rare on earth, having an abundance similar to that of platinum. In the near to medium term, this is not expected to be problematic. But if one looks farther ahead and imagines scaling CdTe PV to the extent of supplying a large fraction of human energy needs, the availability of Te may become limiting (Figure 24.4).

Copper Zinc Tin Sulfide

Concerns regarding the toxicity of cadmium in CdTe and the use of rare elements in both CdTe and CIGS has motivated research toward thin-film PV based entirely on sustainable materials. In CIGS, both indium and selenium are of low earth abundance. The indium can be replaced with zinc, if the gallium is also replaced with tin. Replacing selenium with sulfur then gives rise to copper zinc tin sulfide (CZTS), a material comprising exclusively nontoxic, abundant, and cheap elements. Owing

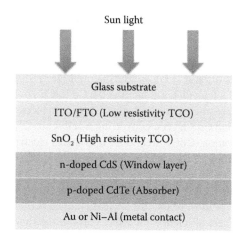

FIGURE 24.4   Cross section of CdTe thin-film solar cell.

to the toxicity of cadmium and limited reserves of indium and tellurium used in CdTe and CIGS solar cells, the search for alternative materials for thin-film solar cells started. CZTS was developed with nontoxic and earth-abundant materials. CZTS is an excellent absorber because of its ideal band gap of 1.45 eV and high absorption coefficient (1×10^4 cm^{-1}) in the visible wavelength range (Ito and Nakazawa, 1988). CZTS solar cells were fabricated using CZTS p-type absorber and CdS n-type layer. Aluminum-doped zinc oxide (AZO) and tin-doped indium oxides (ITO) are used as transparent conducting oxide (TCO) layers (Yang et al., 2008; Ding et al., 2013). Silver/chromium is used as a finger electrode.

When compared to CIGS, these cells replace the indium and gallium by less expensive and earth-abundant zinc and tin, respectively. CZTS band gap can be tuned between 1 and 1.5 eV by incorporating Se ($Cu_2ZnSn(S_{1-x}Se_x)_4$). CST(S,Se) solar cells achieved 9% efficiency by changing the synthesis technique which yielded nanoparticles with superior compositional uniformity and improved selenization (Miskin et al., 2015). The Mitzi group at IBM achieved 12.6% efficiency with better bulk CZTSSe quality and improved optical architecture (Wang et al., 2014).

Similar to CIGS, 12.6% CST(S, Se) fabrication used hydrazine-based metal inks as precursors which favor solution processing, but not large-scale production due to the toxicity and flammability of hydrazine. Alternative methods such as colloidal nanocrystal dispersions and other ink-based techniques have to be developed for roll-to-roll large-scale production. The present challenge with the CST(S,Se) films is cracking during the drying due to capillary stress. Strategies such as polymer binders (Todorov et al., 2009; Jasieniak et al., 2011), deposition of films thinner than the critical cracking thickness (Chiu et al., 1993), intermediate heating, and/or chemical treatments can be used to prevent cracking (Jasieniak et al., 2011).

In 2010, IBM, DelSolar, and Solar Frontier partnered to develop the CZTS technology and large-scale manufacturing.

PYRITE SOLAR CELLS

The iron sulfide known as pyrite (FeS_2) has been a promising material for PV applications since the first pyrite PV device was made in 1984 (Ennaoui and Tributsch, 1984). It is a semiconductor with a band gap of 0.95 eV (Ennaoui et al., 1985), which is lower than optimal but not far from the ideal zone. It absorbs strongly across the visible solar spectrum (Ennaoui et al., 1985). It is nontoxic. Most appealingly, it is beyond abundant: it is routinely disposed of in large amounts as mining waste. Such is its abundance that Puthussery et al. (2011) estimated that if a pyrite PV device could be made with 10% efficiency and a reasonable 5-μm-thick active layer, the entire United States could be powered using only 10% of the mining waste from six states alone. Efforts to achieve such device efficiency have been motivated by an appreciation of pyrite as potentially the ultimate sustainable PV material (Wadia et al., 2009).

Unfortunately, the performance of pyrite devices has been poor. Pyrite PV devices were first attempted in the 1980s but largely abandoned after several years of little progress. While large currents (>40 mA/cm^2) can be extracted from pyrite (Ennaoui et al., 1990), the voltage it produces is low—around 200 mV (Ennaoui and Tributsch, 1986). Pyrite materials for PV have been revisited in recent years amid the drive for sustainable PV and hope that new tools will allow improved performance. So far, this goal has not been achieved. Despite advances in methods for depositing the pyrite layer, including hot injection (Zhu et al., 2015) and a printable pyrite ink (Puthussery et al., 2011), the efficiency remains below 3% (Cabán-Acevedo et al., 2014). Instead, the new research has produced a better understanding of the material's shortcomings. FeS$_2$ inherently forms with sulfur vacancies on the surface, leading to a material that behaves more like a metal than a semiconductor (Steinhagen et al., 2012). Studies in vacuum show that the sulfur vacancies lead to ionization of deep donor states in the bulk, with complex effects on the space-charge layer that result in a barrier height of about 230 mV, limiting the device voltage (Cabán-Acevedo et al., 2014). These findings give some indication as to how the material might be improved, especially by passivating the sulfur vacancies. Whether this can be practically achieved remains an open question (Figure 24.5).

ORGANIC PHOTOVOLTAICS

Metal-free molecules can serve as the active layer in a solar cell, giving rise to organic photovoltaics (OPV). Metal electrodes are often used as contacts to one or both sides, but in OPV, by definition, the light absorption and initial charge (or exciton) separation occur in a purely organic phase (or phases). OPV has its roots in discoveries of molecular conductivity, and in particular the findings that certain polymers and some small molecules can have conducting and/or semiconducting properties. The technology is generally held to have originated with C. W. Tang (1986), who built a planar bilayer device with an electron-accepting material in contact with an electron-donating material. That combination of materials has driven essentially all OPV devices. However, a planar device must be very thin to allow efficient separation of charges, since they typically form as excitons which can diffuse little more than 100 nm before recombining. But a film this thin is not able to absorb much incident light. Instead, the electron donor and acceptor materials can be blended together, so that the two phases are intermixed, with domains on the order of 10 nm within a film that can be several microns thick. This allows for good light absorbance combined with efficient charge separation. This structure, called a bulk heterojunction (BHJ), has become the dominant

FIGURE 24.5 FeS$_2$ called fool's gold for solar PVs. (Adapted from Tenenbaum, D. 2014. Scientists get to the heart of fool's gold as a solar material. http://news.wisc.edu/scientists-get-to-the-heart-of-fools-gold-as-a-solar-material/#sthash.SHDESy62.dpuf (Last modified November 2014).)

form of organic solar cell. It requires excellent control of the morphology of the intermingled phases to ensure that they form interpenetrating networks, since each charge must have a route to its collector to not be stranded and lost to recombination.

The major breakthrough in organic solar cells occurred in 1986 when an electron acceptor and electron donor were brought together by Tang (1986). The discovery of photoinduced electron transfer conjugated polymer to fullerene molecule led to the development of BHJ polymer solar cells (Sariciftci et al., 1992). Poly (3-hexylthiophene) (P3HT) application and studies in organic solar cells resulted in power conversion efficiencies (PECs) of around 5% (Dang et al., 2011). With the development of new low band gap polymers such as PBDTTT, an efficiency of 6.77% was reached in 2009 (Chen et al., 2009). A new alternating copolymer of dithienosilole and thienopyrrole-4,6-dione (PDTSTPD) possesses both a low optical band gap (1.73 eV) and a deep highest occupied molecular orbital energy level (5.57 eV) and resulted in 7.3% efficiency (Chu et al., 2011). Konarka Technologies reported an efficiency of 9% for single-junction organic solar cell. An efficiency of 11.5% was achieved by Yang Yang et al. with a triple-junction tandem design using three distinct organic donor materials having band gap energies ranging from 1.4 to 1.9 eV (Chen et al., 2014). The record OPV efficiency is 12%, set by Heliatek (Dresden, Germany) in 2013 (Heliatek, 2013).

DYE SENSITIZED SOLAR CELLS

The dye sensitized solar cell (DSSC) is a nanotechnology-based PV device that represents a radical departure from previous technologies for solar energy conversion. Early work by Tributsch and others (Gerischer et al., 1968; Tributsch, 1972) established that certain dye molecules (by definition, molecules that absorb visible light) adsorbed onto a titanium dioxide (TiO_2) surface can, following photoexcitation, inject the photoexcited electron into the TiO_2, whence the electron can be extracted as current. However, the current was minimal due to the poor light harvesting efficiency of a molecular layer of dye. Further work toward a working device was pursued by Grätzel (Desilvestro et al., 1985; Kalyanasundaram et al., 1987; Vlachopoulos et al., 1988; O'Regan et al., 1990), including using a rough TiO_2 surface to increase the interfacial contact area between the TiO_2 and dye, but efficiencies remained very low. The culmination was a breakthrough (O'Regan and Grätzel, 1991) in which the flat TiO_2 was replaced by a mesoporous film, about 10 µm thick, of nanoparticles (ca. 20 nm in diameter) of TiO_2. This increases the interfacial area by three orders of magnitude compared to a flat film and allows a corresponding increase in the amount of dye that can be present and in contact with TiO_2.

The photoanode is sealed against a counterelectrode with a gap between them, which is filled with a liquid electrolyte solution containing a redox couple. Upon photoinjection, a dye molecule becomes oxidized. Electrons withdrawn from the photoanode, after moving through the external circuit (the load), are returned to the counterelectrode, where they reduce the oxidized form of the electrolyte redox couple, which diffuses across the gap and reduces an oxidized dye molecule, completing the circuit. The counterelectrode is typically lightly coated with Pt nanoparticles (about 5 µg/cm²) to catalyze electron transfer from the conducting glass surface to the dissolved redox couple.

A key advantage of the technology is that charges are immediately separated into different phases upon photoexcitation. The semiconductor (TiO_2) only carries electrons, so electron—hole recombination is not a problem; this allows the semiconductor to be of low crystal quality and hence inexpensive. It also allows each component to do one task: electron transport (the semiconductor), light absorption (the dye), and hole transport (the redox electrolyte), allowing each component to be developed and optimized. This contrasts with solid-state semiconductor systems in which one material must perform all three duties.

Upon its invention in 1991, the DSSC was already claimed to have a solar PCE better than 7% under simulated sunlight, and (erroneously) 12% in actual outdoor sunlight (O'Regan and Grätzel, 1991). Hopes were high for 20% efficiency with additional research. Unfortunately, after quickly improving to 11% (Green et al., 1997), the record efficiency stagnated there for more than a decade, although this masked underlying progress in stability and scientific understanding. Recent developments, however, have allowed new advances in device efficiency (Hardin et al., 2012).

Much of the research on DSSCs has concerned the development of new dyes, accounting for thousands of different molecules and publications. The dye N3 emerged as an early standard (Nazeeruddin et al., 1993). It has a ruthenium atom coordinated to two dicarboxyl-substituted bipyrines and two thiocyanate ligands. That dye and many variants on its essential ruthenium-complex structure dominated the DSSC for more than a decade. All-organic (metal-free) dyes achieved comparable efficiencies beginning in the mid-2000s, avoiding concerns regarding the cost and relative rarity of ruthenium. Since then, porphyrin dyes, usually containing an atom of zinc (an abundant and cheap metal) have arisen as the highest-efficiency dyes (Yella et al., 2011; Mathew et al., 2014).

Equally important has been work on the redox couple in the electrolyte. Since the invention of the DSSC, the standard redox couple was for many years iodide/tri-iodide. This couple has many good properties: it is cheap, small (good for mass transport), fast at reducing a dye molecule, and uniquely slow at stripping electrons from the TiO_2 nanoparticles. However, its complex electrochemistry results in the need for an overpotential of ca. 0.5 V or more to reduce an oxidized dye molecule, resulting in energy loss (Boschloo and Hagfeldt, 2009). Alternatives with more facile electrochemistry and smaller overpotentials tend to rapidly remove electrons from TiO_2, internally shortcircuiting the cell (Nusbaumer et al., 2001, 2003; Hamann et al., 2008; Klahr and Hamann, 2009). The breakthrough was to combine these alternatives with bulky dyes that protect the TiO_2 surface from the redox couple (Feldt et al., 2010). Out of this finding, cobalt(II/III)—bipyridine complexes have emerged as the leading redox couple and have enabled a significant jump in record DSSC efficiency to 13% (Yella et al., 2011; Hardin et al., 2012; Mathew et al., 2014).

Much research has explored modifications or alternatives to the TiO_2 nanoparticle film as the electron-transporting layer. These have mainly been addressed at two problems of TiO_2. First, the reaction of electrons in the TiO_2 with the redox electrolyte can result in a significant loss of current. A variety of insulating coatings, such as silica (Feldt et al., 2010), aluminum oxide (Palomares et al., 2002), and magnesium oxide (Green et al., 2005), around TiO_2 nanoparticles have been employed to slow this reaction. Unfortunately, they also tend to impede electron injection from the dye into the TiO_2, and these coatings generally have not been incorporated into the highest-efficiency cells. Second, transport of electrons throughout the TiO_2 film is slow due to the many hops between nanoparticles that are involved in the "random walk" to the conducting substrate. Solutions to this have improved more radical redesigns of the photoanode. The most intensively investigated structure has been zinc oxide nanorods, which benefit from the high electron mobility of ZnO and, more importantly, the direct pathway for a photoinjected electron to the conducting glass substrate through a single-crystal ZnO rod (Law et al., 2005). However, they suffer from low surface area and unfavorable properties of the ZnO/dye interface compared to that of TiO_2/dye (Keis et al., 2002). These problems have been ameliorated coatings over the nanorods by TiO_2 (Manthina et al., 2012; Xu et al., 2012) or high-surface area ZnO (Park et al., 2010), but they have not surpassed the basic TiO_2 nanoparticle film for PCE. Ultimately, today's best DSSCs use essentially the same mesoporous TiO_2 film as the original 1991 invention of the DSSC, save for two modifications. One is treatment with titanium tetrachloride (Nazeeruddin et al., 1993), which modifies the surface of the TiO_2 nanoparticles in a way that boosts current and voltage (Sommeling et al., 2006; O'Regan et al., 2007). The other is a "scattering layer" of large (hundreds of nanometers in diameter) particles of TiO_2 coated on the side of the nano-TiO_2 film opposite the light source, such that photons that pass through the nanoparticulate film are scattered back through it for a second pass, for improved light harvesting efficiency (Hore et al., 2006).

The liquid layer component of the DSSC has been a concern due to the practical difficulties of sealing a liquid layer for 20 years. Efforts to address this have included gels, nonvolatile ionic liquids, and even clays. In 1998, an entirely solid-state version of the DSSC was invented, in which the liquid electrolyte was replaced with an organic layer of the molecule spiro-OMeTAD, which is capable of relaying holes from the dye to an evaporated metal contact that serves as the counterelectrode

(Bach et al., 1998). Its efficiencies were well below liquid-containing DSSCs for many years, limited by fast recombination between the spiro-OMeTAD and TiO_2 and by the difficulty of infiltrating the organic molecule throughout the pores of the photoanode. The record solid-state DSSC in this configuration reached 6% efficiency (Cai et al., 2011). However, the efficiency jumped to 8.5% when spiro-OMeTAD was replaced with an inorganic perovskite material, F-doped $CsSnI_3$ (Chung et al., 2012). The use of different perovskites as the light-absorbing material has led to even greater gains, which will be treated in a separate section.

Commercialization of the DSSC has initially targeted niche applications such as solar chargers for cellular phones or wireless keyboards. The DSSC is promising for building-integrated photovoltaics (BIPV), where the PV devices are typically not aimed at an optimal angle toward the sun. One advantage here is that DSSCs absorb light isotropically, unlike silicon panels, whose crystal orientation makes them inefficient at absorbing light that does not arrive perpendicular to the surface. Another is that DSSCs can be made transparent and use a variety of dyes for different colors, enabling tinted or gray "solar windows" or architectural glass. Today, however, much commercializing effort is being redirected toward the related perovskite solar cell (PSC) technology.

PEROVSKITE SOLAR CELLS

The PSC emerged out of research on the DSSC. Inorganic pigments had been of interest for years as alternatives to dye molecules in DSSCs due to the pigments' broad and strong light absorbance, and potential advantages in cost and stability over organic molecules. Many quantum dots were studied for this purpose, but their efficiencies were low. Instead, organometal halide materials have emerged to fill this role, nearly two decades after their development chiefly by David Mitzi (Mitzi et al., 1994, 1995; Mitzi, 2004). These have the form BCX_3, where B is an organic cation such as methylammonium (MA) or formamidinium (FA), C is a metal (either Sn(II) or Pb(II)), and X is a halogen (Cl, I, or Br). The perovskite $MAPbI_3$ has a band gap of about 1.55 eV, slightly higher than the ideal range. It can be tuned by judicious substitution of formamidinium for methylammonium, or of bromide or chloride for iodide, giving a range from 1.48 to 2.23 eV (Eperon et al., 2014). The low end of that range is suitable for a single-junction solar cell, while higher values are useful as the wide band gap part of a tandem cell, for example, in conjunction with c-Si (Kojima et al., 2009; Im et al., 2011).

Initial attempts to use $MAPbI_3$ to replace the dye in a DSSC yielded respectable efficiencies (up to 6.5%), but only for several minutes, since the perovskite dissolves in polar solvents, which are needed for the liquid electrolyte (Kojima et al., 2009; Im et al., 2011). In nearly simultaneous findings by Henry Snaith and a collaboration between Michael Grätzel and Nam-Gyu Park (Kim et al. 2012a; Lee et al., 2012), using $MAPbI_3$ perovskites in a solid-state DSSC resulted in a great breakthrough with efficiencies quickly surpassing 10%, far beyond previous solid-state DSSCs and approaching the records for liquid DSSCs. Surprisingly, instead of simply absorbing light and injecting charge into TiO_2, the perovskite itself was found to transport electrons faster than the TiO_2 can, and in fact the TiO_2 can be replaced with an inert scaffold, resulting in even better performance of nearly 11% (Kim et al., 2012b; Lee et al., 2012).

Astounding progress has followed those breakthrough discoveries. The perovskite layer in those studies was formed by spin-coating a solution of MAI and PbI_2 (or $PbCl_2$) dissolved in N,N-dimethylformamide (DMF), and then annealing at 100°C for <1 hour. Instead, coating the two salts in two sequential steps boosted the efficiency to 15% (Burschka et al., 2013). Co-evaporating the salts to form a planar perovskite layer, free of any nanoparticles or mesoporous structure, also achieved 15% (Liu et al., 2013). This established two tracks for PSC research: (1) including a mesoporous structure, in keeping with the technology's DSSC roots or (2) moving to a flat perovskite layer with electron- and hole-extracting contacts on either side, more akin to inorganic semiconductor devices such as Si or CdTe. The Seok group incorporated some bromide into the perovskite to optimize its band gap, used a poly(triarylamine) hole-transporting material (HTM) instead of spiro-OMeTAD, and developed a technique for dripping toluene on the spin-coated perovskite film during spinning to improve crystallization, and achieved 16.5% (Jeon et al., 2014), including a mesoporous layer. The crystal growth can also be improved by preheating the substrate and solution before spin-coating, giving an efficiency near 18% (Nie et al., 2015). The planar-type PSC was re-engineered with careful attention to device physics, yielding an efficiency over 19% (Zhou et al., 2014). Recently, a mesoporous device reclaimed the lead, as the Seok group grew a formamidinium lead iodide ($FAPbI_3$) perovskite through an intermolecular exchange process, surpassing the 20% mark of PCE (Yang et al., 2015). The performance of the perovskite materials is all the more remarkable given that they are formed from solution at low temperatures (~100°C), in a very facile and low-cost process.

The fantastic gains in PSC efficiency come with important caveats. All of the efficiencies cited above are for very small devices (the record holder is about 0.1 cm^2), and larger cells suffer large drops in performance. Part of the problem is the nonscalability of spin coating. Efforts to produce large-area cells by other means have achieved 10.4% for a 10.1-cm^2 module, and 4.3% for a 100-cm^2 module (Razza et al., 2015). Also, PSC devices are not stable, tending to react with moisture in air and morph into nonphotoactive phases (Niu et al., 2015). However, a device design from the Grätzel group replacing the organic hole conductor with carbon (Mei et al., 2014) can withstand prolonged heat and light exposure while maintaining 13% efficiency (Li et al., 2015). Further work on device design and encapsulation will be needed to withstand humidity. There are also some concerns regarding the toxicity of the lead. Many electronics, including many Si solar cells, contain lead in the solder, although in that zerovalent form it is far less soluble (and therefore less dangerous) than the Pb(II) in a PSC (Green, 2015). There have been attempts to replace lead with tin in the perovskite materials, but so far the efficiencies are much lower (Noel et al., 2014).

PSC technology is highly promising given the high efficiencies that have been obtained. It remains very new, however, and more fundamental research will be needed before the devices can find real-life application. At a minimum, commercialization will require finding scalable methods for producing uniformly good perovskite crystals over large areas at low cost, and ensuring stability in hot and humid environments.

CONCENTRATED PV

Concentrated PV (CPV) refers to the focusing of light onto a PV device of smaller area than the focusing lens or mirrors. (This is distinct from using concentrated sunlight to heat a working fluid for powering steam turbines, which is a non-PV technology.) The concept has been investigated since the surge in renewable energy research that followed the 1973 oil crisis (Spratt and Schwarz, 1975). Concentrating sunlight in a PV system has two benefits. First, if the light can be focused using low-cost materials (e.g., cheap mirrors or plastic Fresnel lenses), a large area of sunlight can be collected with a small area of expensive PV materials. Second, theoretical maximum solar efficiencies increase under concentrated sunlight. Indeed, in extreme implementation in a laboratory, the absolute world record solar PCE of 46% was obtained by illuminating a quadruple-junction cell with a light intensity equivalent to 508 suns (Fraunhofer ISE, 2014a). Such a cell would be prohibitively expensive over large areas but may be feasible if only small areas are used to convert a large area of sunlight via concentration.

However, there are downsides. Keeping the sunlight focused on the solar cells as the sun moves requires a solar tracking system, substantially adding to balance of system (BoS) costs. Also, while theoretical maximum efficiencies increase with solar concentration, the added heat can significantly increase the temperature of the solar cell, which may decrease its actual efficiency. CPV therefore frequently includes mechanisms for cooling, including at the least heat sinks. Finally, only light reaching the focusing apparatus directly from the sun is focused onto the solar cell. The diffuse part of sunlight, that is, the blue light scattered by the sky, is lost. In areas of high insolation, using high-efficiency multijunction solar cells, CPV can be competitive with conventional PV. The amount of CPV currently installed is 330 MW_p (Philipps et al., 2015), small compared to the aforementioned

177 GW$_p$ of total PV. Nevertheless, advances in CPV technology and long-running demonstrations of stability suggest future growth in the CPV share of the overall PV market (IHS Technology, 2015).

TRANSPARENT CONDUCTING OXIDES

Solar cells generally separate charges to be collected at opposite sides of the device, requiring current collection on both sides. However, sunlight must enter from (at least) one side, requiring that side to be both conducting and transparent. In many PV technologies, this is achieved by coating glass (which is electrically insulating) with a thin layer of a TCO. Transparent conducting oxides (TCOs) are normally wide band gap semiconductors with high concentration of free electrons in the conduction band. These electrons are present due to defects in the material or a high degree of doping of the material. The defect states lie near the conduction band edge of the semiconductor.

The first TCO thin film was deposited over 108 years ago by Badeker (Bädeker, 1907; Philipps, 2015), who deposited CdO with a vapor deposition system. The most commonly used TCOs in solar cells today are based on tin oxide or zinc oxide. They include indium tin oxide (ITO), aluminum-doped zinc oxide (AZO), boron-doped zinc oxide (ZnO:B or BZO) and fluorinated tin oxide (FTO). Critical criteria for TCO materials are electrical conductivity, optical transparency, stability, and cost. There is a trade-off between the first two criteria as more free carriers conduct current better but tend to absorb some visible light. Many TCO materials have good stability over time, but they differ in the processing temperatures they can withstand before conductivity is degraded. ITO offers the best combination of transparency and conductivity and is therefore the most widely used TCO, despite its relatively high cost due to the indium content (Table 24.1).

TABLE 24.1
TCOs Employed in Photovoltaics Needs and Their Goals

Cell Type	TCO in Current Use	TCO Needs	Materials Goals
CIGS	Intrinsic-ZnO/Al:ZnO	Interfacial stability to CdS, low-temperature deposition, resistance to diffusion and shorting, need to make/improve the junction	Single-layer TCO to replace two layers and CdS layers
CdTe	(SnO_2) Zn_2SnO_4/ Cd_2SnO_4	Stable interface to CdTe/CdS at temperature, barrier diffusion	Doping of $ZnSnO_x$ materials, single-layer TCO
Polymer solar cells	ZnO, TiO$_2$, SnO$_2$	Nanostructure with right thickness, interface with organic, right doping level for carrier transport	Self-organized nanostructures, core-shell nanostructures, new nonconventional TCO (e.g., MoO$_3$)
DSSC	ZnO, TiO$_2$, SnO$_2$	Nanostructures with high electron mobility, Core-shell nanostructures	Doped ZnO, TiO$_2$, and SnO$_2$ Band engineered core-shell materials
Amorphous Si	SnO, ITO, and ZnO many cells employ two TCOs	Temperature stability, chemical stability, and appropriate texture for both TCO layers	Higher conductivity, texture, and ohmic contact for both TCO layers, two-dimensional morphology
CZTS	Aluminum-doped ZnO (AZO), ITO, ZnO, FTO employ two TCOs	Temperature stability, chemical stability, and appropriate texture for both TCO layers	Higher conductivity, texture, and ohmic contact for both TCO layers, nanostructures Indium-free materials
Perovskite Solar	ZnO, TiO$_2$, SnO$_2$	Nanostructures with high electron mobility, Core-shell nanostructures	Doped ZnO, TiO$_2$, and SnO$_2$ Band engineered core-shell materials
Pyrite Solar	ZnO, TiO$_2$, I:SnO$_2$	Nanostructures with high electron mobility Core-shell nanostructures	Doped ZnO, TiO$_2$, and SnO$_2$ Band engineered core-shell materials

Source: Fortunato, E. et al. 2007. *MRS Bull.*, 32(03): 242–247.

RAW MATERIAL ABUNDANCE AND PRICE

Raw material costs are important to PV deployment, especially in long-term scenarios expecting PV to scale up to the point of displacing most other energy sources. A study in 2009 estimated maximum efficiencies for different materials, and related this to the materials' cost to arrive at limiting cost per watt figures for 23 compounds. The results are shown in Figure 24.6. The range of costs are between 0.327¢/W for Ag_2S to <0.000002¢/W for FeS_2. A reliable and low-cost supply of raw materials ensures that solar will continue to become an increasingly attractive source of electricity.

The two elements of concern for PV applications are tellurium and indium. Te and In are by-products of Cu and Zn mining, respectively. The In recovered ranges from <1 to 100 parts per million from Zn deposits. More than 90% of the Te is recovered from Cu refining. Increased demand for photovoltaics (PVs), electronics, thermionics, and display applications has increased the prices of Te and In. If demand continues to rise, the direct mining of these materials may be necessary. Even if Te and In are available in required quantities in the earth's crust, their low concentration will make extraction expensive. Other than solar use, which consumes 40% of Te production, other consumption of Te includes 30% for thermoelectric production, 15% for metallurgy, 5% for rubber applications, and 10% for other applications (U.S. Geological Survey, 2015). Indium is more abundant than Te, Au, and Pt, but most of available In is used for the production of ITO used in liquid crystal displays (LCDs), smart phones, organic light-emitting diodes (LEDs), display panels, and tablets (Green, 2009). The demand for In is increasing at 25%/year, but the production has increased only at 0.02%/year from 2013 to 2014 (U.S. Geological Survey, 2015).

Gallium used in CIGS has an annual production of 430 metric tons in 2014, a 26% increase from 2013. Gallium is extracted from the metal ores such as bauxite and zinc processing in ppm concentrations. Metal production has to be increased significantly to produce more gallium for solar power

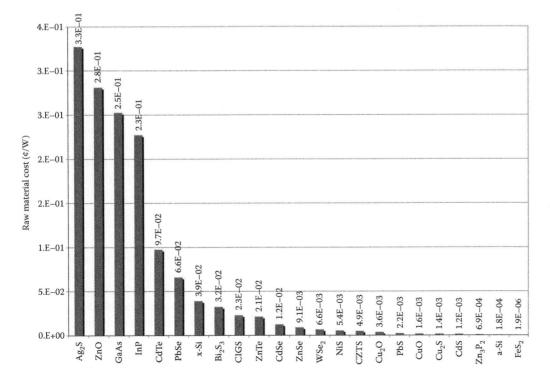

FIGURE 24.6 Minimum ¢/W for 23 inorganic PV materials. (Adapted from Wadia, C. et al. 2009. *Environ. Sci. Technol.*, 43(6): 2072–2077.)

generation. Gallium use in integrated circuits and optoelectronic devices (LEDs, laser diodes, and photodetectors) account to 80% of the consumption and is more required for these applications.

Recent reports suggest that if thin-film PV account for 25% of the electricity production by 2030, the production of thin-film PV metals need to grow at 15%–30%/year. The silicon-based PV cells can meet the silicon production range for large-scale deployment (Kavlak et al., 2014).

POLICY

The PV policy varies from one region to the other based on the national context and regional economic viability. The PV policy can be on the demand side and the supply side. The demand side includes incentives for PV energy such as subsidies for electricity production and installation and tax credits. The supply side includes support for research and development, materials import, and large-scale production (Finon, 2008). The government policy on the toxicity of individual materials in PV manufacturing and carbon policy may affect the large-scale deployment of CIGS and CdTe solar cells (MIT, 2015). In the past decade, module production has shifted from the United States to Japan, from Japan to Europe, from Europe to China. In 2012, China produced 67% of the world's total PV modules. Europe's PV production decreased from 12% in 2012 to 9% in 2013. Japan's production stayed at 5% for 2012 and 2013. While the U.S. production share is 2.6%, leading in thin-film PV production accounting to 39% in 2013, an increase from 36% in 2012 (International Energy Agency, 2013; REN21, 2014). In India the manufacturing was idle due to lack of low-cost financing and underdeveloped supply chains. In China, local governments provided the firms with incentives such as low-interest loans to purchase equipment, land transfer price refunds, electricity price refunds, and multiple-year corporate tax reductions. In the United States, federal policies such as investment tax credit (ITC) provides financial incentives for solar power at a rate of 30%, advanced energy manufacturing tax credit (MTC), Loan Guarantee Program directed funds to manufacturing facilities that employ "new or significantly improved" technologies, Treasury Cash Grant Program for solar projects, and Sunshot Initiative were implemented for supporting PV manufacturing industry (Platzer, 2015).

In 2012. the U.S. Department of Commerce and International Trade Commission decided to impose antidumping duties of 26.71%–78.42% and antisubsidy duties of 27.64%–49.79% on solar panels made in China (Cardwell, 2014). In 2013, China imposed antidumping tariffs against polysilicon imports from the United States and South Korea, as a direct response to the U.S. antidumping tariff on Chinese solar products.

Minimum sustainable price (MSP) is the term generally used to predict factory location and to take decisions. MSP estimates the long-term market-clearing price for the product assuming competitive equilibrium (the minimum price of modules that will provide an adequate rate of return for a company). The study reported by National Renewable Energy Laboratory (NREL), Golden, Colorado showed that a 2-GW Chinese factory enjoys a 23% MSP advantage over a 500-MW U.S. factory (Goodrich et al., 2013).

CONCLUSIONS

The solar PV industry has been in a state of flux, with a number of competing technologies and companies rising and falling. Through it all, crystalline silicon technologies have been dominant, with thin-film alternatives playing a significant role. Continued refinement of all technologies, from fundamental science to engineering process improvements, continues to raise efficiencies while lowering costs. Spurred and assisted by government incentives, these improvements have led to an explosive growth in PV as a whole that shows no signs of slowing. Still, PV currently accounts for barely 1% of global energy consumption. Further research on materials challenges and manufacturing techniques, with considerations of materials sustainability in terms of abundance, cost, and toxicity, will be needed to allow PV to significantly displace fossil energy sources in a reasonable amount of time.

REFERENCES

Bach, U., D. Lupo, P. Comte, J. E. Moser, F. Weissörtel, J. Salbeck, H. Spreitzer, and M. Grätzel. 1998. Solid-state dye-sensitized mesoporous TiO_2 solar cells with high photon-to-electron conversion efficiencies. *Nature*, 395(6702): 583–585.

Bädeker, K. 1907. Über die elektrische Leitfähigkeit und die thermoelektrische Kraft einiger Schwermetallverbindungen. *Annalen der Physik*, 327(4):749–766. doi: 10.1002/andp.19073270409.

Boschloo, G. and A. Hagfeldt. 2009. Characteristics of the iodide/triiodide redox mediator in dye-sensitized solar cells. *Accounts of Chemical Research*, 42(11): 1819–1826.

BP. 2015. BP Statistical Review of World Energy.

Burschka, J., N. Pellet, S.-J. Moon, R. Humphry-Baker, P. Gao, M. K. Nazeeruddin, and M. Gratzel. 2013. Sequential deposition as a route to high-performance perovskite-sensitized solar cells. *Nature*, 499: 316–319.

Cabán-Acevedo, M., N. S. Kaiser, C. R. English, D. Liang, B. J. Thompson, H.-E. Chen, K. J. Czech, J. C. Wright, R. J. Hamers, and S. Jin. 2014. Ionization of high-density deep donor defect states explains the low photovoltage of iron pyrite single crystals. *Journal of the American Chemical Society*, 136(49): 17163–17179. doi: 10.1021/ja509142w.

Cai, N., S.-J. Moon, L. Cevey-Ha, T. Moehl, R. Humphry-Baker, P. Wang, S. M. Zakeeruddin, and M. Grätzel. 2011. An organic D-π-A dye for record efficiency solid-state sensitized heterojunction solar cells. *Nano Letters*, 11(4): 1452–1456. doi: 10.1021/nl104034e.

Cardwell, D. 2014. U.S. imposes steep tariffs on chinese solar panels. *The New York Times*, December 16. http://www.nytimes.com/2014/12/17/business/energy-environment/-us-imposes-steep-tariffs-on-chinese-solar-panels.html

Chen, C.-C., W.-H. Chang, K. Yoshimura, K. Ohya, J. You, J. Gao, Z. Hong, and Y. Yang. 2014. An efficient triple-junction polymer solar cell having a power conversion efficiency exceeding 11%. *Advanced Materials*, 26(32): 5670–5677. doi: 10.1002/adma.201402072.

Chen, H.-Y., J. Hou, S. Zhang, Y. Liang, G. Yang, Y. Yang, L. Yu, Y. Wu, and G. Li. 2009. Polymer solar cells with enhanced open-circuit voltage and efficiency. *Nature Photonics*, 3(11): 649–653.

Chiu, R. C., T. J. Garino, and M. J. Cima. 1993. Drying of granular ceramic films: I. Effect of processing variables on cracking behavior. *Journal of the American Ceramic Society*, 76(9): 2257–2264. doi: 10.1111/j.1151-2916.1993.tb07762.x.

Chu, T.-Y., J. Lu, S. Beaupré, Y. Zhang, J.-R. Pouliot, S. Wakim, J. Zhou et al. 2011. Bulk heterojunction solar cells using thieno[3,4-c]pyrrole-4,6-dione and dithieno[3,2-b:2′,3′-d]silole copolymer with a power conversion efficiency of 7.3%. *Journal of the American Chemical Society*, 133(12): 4250–4253. doi: 10.1021/ja200314m.

Chung, I., B. Lee, J. He, R. P. H. Chang, and M. G. Kanatzidis. 2012. All-solid-state dye-sensitized solar cells with high efficiency. *Nature*, 485(7399): 486–489.

Dang, M. T., L. Hirsch, and G. Wantz. 2011. P3HT:PCBM, best seller in polymer photovoltaic research. *Advanced Materials*, 23(31): 3597–3602. doi: 10.1002/adma.201100792.

Desilvestro, J., M. Graetzel, L. Kavan, J. Moser, and J. Augustynski. 1985. Highly efficient sensitization of titanium dioxide. *Journal of the American Chemical Society*, 107(10): 2988–2990. doi: 10.1021/ja00296a035.

Ding, G., M. F. A. Hassan, H. M. H. Le, and Z. W. Sun. 2013. TCO materials for solar applications. Google Patents.

Ennaoui, A., S. Fiechter, H. Goslowsky, and H. Tributsch. 1985. Photoactive synthetic polycrystalline pyrite (FeS_2). *Journal of the Electrochemical Society*, 132(7): 1579–1582. doi: 10.1149/1.2114168.

Ennaoui, A., S. Flechter, G. Smestad, and H. Tributsch. 1990. Preparation of iron disulfide and its use for solar energy conversion. In *Energy and the Environment: Into the 1990s—Proceedings of the 1st World Renewable Energy Congress*, edited by A. A. M. Sayigh. Oxford, UK: Pergamon Press, pp. 458–464.

Ennaoui, A. and H. Tributsch. 1986. Energetic characterization of the photoactive FeS_2 (pyrite) interface. *Solar Energy Materials*, 14(6): 461–474. doi: 10.1016/0165-1633(86)90030-4.

Ennaoui, A. and H. Tributsch. 1984. Iron sulphide solar cells. *Solar Cells*, 13(2): 197–200. doi: 10.1016/0379-6787(84)90009-7.

Eperon, G. E., S. D. Stranks, C. Menelaou, M. B. Johnston, L. M. Herz, and H. J. Snaith. 2014. Formamidinium lead trihalide: A broadly tunable perovskite for efficient planar heterojunction solar cells. *Energy & Environmental Science*, 7(3): 982–988. doi: 10.1039/C3EE43822H.

Feldt, S. M., U. B. Cappel, E. M. J. Johansson, G. Boschloo, and A. Hagfeldt. 2010. Characterization of surface passivation by poly(methylsiloxane) for dye-sensitized solar cells employing the ferrocene redox couple. *Journal of Physical Chemistry C*, 114(23): 10551–10558. doi: 10.1021/jp100957p.

Feldt, S. M., E. A. Gibson, E. Gabrielsson, L. Sun, G. Boschloo, and A. Hagfeldt. 2010. Design of organic dyes and cobalt polypyridine redox mediators for high-efficiency dye-sensitized solar cells. *Journal of the American Chemical Society*, 132(46): 16714–16724. doi: 10.1021/ja1088869.

Finon, D. 2008. L'inadéquation du mode de subvention du photovoltaïque à sa maturité technologique. Working Paper, CIRED & Gis LARSEN.

Fortunato, E., D. Ginley, H. Hosono, and D. C. Paine. 2007. Transparent conducting oxides for photovoltaics. *MRS Bulletin*, 32(03): 242–247. doi:10.1557/mrs2007.29.

Fraunhofer ISE. 2014a. New world record for solar cell efficiency at 46%. Press release, December 1.

Fraunhofer ISE. 2014b. Photovoltaics report, October 24.

Fthenakis, V. M. 2004. Life cycle impact analysis of cadmium in CdTe PV production. *Renewable and Sustainable Energy Reviews*, 8(4): 303–334. doi: 10.1016/j.rser.2003.12.001.

Gerischer, H., M. E. Michel-Beyerle, F. Rebentrost, and H. Tributsch. 1968. Sensitization of charge injection into semiconductors with large band gap. *Electrochimica Acta*, 13(6): 1509–1515. doi: 10.1016/0013-4686(68)80076-3.

GlobalData. 2014. *Power Generation Technologies Capacities, Generation and Markets Database.* GlobalData, London.

Goodrich, A. C., D. M. Powell, T. L. James, M. Woodhouse, and T. Buonassisi. 2013. Assessing the drivers of regional trends in solar photovoltaic manufacturing. *Energy & Environmental Science*, 6(10): 2811–2821. doi: 10.1039/c3ee40701b.

Green, A. N. M., E. Palomares, S. A. Haque, J. M. Kroon, and J. R. Durrant. 2005. Charge transport versus recombination in dye-sensitized solar cells employing nanocrystalline TiO_2 and SnO_2 films. *Journal of Physical Chemistry B*, 109(25): 12525–12533. doi: 10.1021/jp050145y.

Green, M. A. 2001. Photovoltaic physics and devices. In *Solar Energy: The State of the Art*, edited by J. M. Gordon. New York: Routledge, pp. 291–356.

Green, M. A. 2009. Estimates of Te and In prices from direct mining of known ores. *Progress in Photovoltaics: Research and Applications*, 17(5): 347–359. doi: 10.1002/pip.899.

Green, M. A. 2015. Mainstream photovoltaics: Where can perovskites make impact? *Hybrid and Organic Photovoltaics Conference (HOPV15)*, Rome, May 10–13.

Green, M. A., K. Emery, K. Bücher, D. L. King, and S. Igari. 1997. Solar cell efficiency tables (version 10). *Progress in Photovoltaics: Research and Applications*, 5(4): 265–268. doi: 10.1002/(SICI)1099-159X(199707/08)5:4<265::AID-PIP177>3.0.CO;2-4.

Green, M. A., K. Emery, Y. Hishikawa, W. Warta, and E. D. Dunlop. 2015. Solar cell efficiency tables (Version 45). *Progress in Photovoltaics: Research and Applications*, 23(1): 1–9. doi: 10.1002/pip.2573.

Hamann, T. W., O. K. Farha, and J. T. Hupp. 2008. Outer-sphere redox couples as shuttles in dye-sensitized solar cells. Performance enhancement based on photoelectrode modification via atomic layer deposition. *Journal of Physical Chemistry C*, 112: 19756–19764.

Hardin, B. E., H. J. Snaith, and M. D. McGehee. 2012. The renaissance of dye-sensitized solar cells. *Nature Photonics*, 6(3): 162–169.

Heliatek. 2013. Heliatek consolidates its technology leadership by establishing a new world record for organic solar technology with a cell efficiency of 12%. Press release, January 16.

Hore, S., C. Vetter, R. Kern, H. Smit, and A. Hinsch. 2006. Influence of scattering layers on efficiency of dye-sensitized solar cells. *Solar Energy Materials & Solar Cells*, 90: 1176–1188.

IHS Technology. 2015. *Top Solar Power Industry Trends for 2015.* Wellingborough, United Kingdom: IHS Technology.

Im, J.-H., C.-R. Lee, J.-W. Lee, S.-W. Park, and N.-G. Park. 2011. 6.5% efficient perovskite quantum-dot-sensitized solar cell. *Nanoscale*, 3(10): 4088–4093.

International Energy Agency. 2013. *World Energy Outlook 2013.* London: International Energy Agency.

International Energy Agency. 2014a. *Snapshot of Global PV Markets.* Paris, France: International Energy Agency.

International Energy Agency. 2014b. *Technology Road Map Solar Photovoltaic Energy.* Paris: International Energy Agency.

IRENA. 2015. Renewable Power Generation Costs in 2014. Abu Dhabi, United Arab Emirates: International Renewable Energy Agency.

Ito, K. and T. Nakazawa. 1988. Electrical and optical properties of stannite-type quaternary semiconductor thin films. *Japanese Journal of Applied Physics*, 27(11R): 2094.

Jackson, P., D. Hariskos, E. Lotter, S. Paetel, R. Wuerz, R. Menner, W. Wischmann, and M. Powalla. 2011. New world record efficiency for $Cu(In,Ga)Se_2$ thin-film solar cells beyond 20%. *Progress in Photovoltaics: Research and Applications*, 19(7): 894–897. doi: 10.1002/pip.1078.

Jackson, P., D. Hariskos, R. Wuerz, O. Kiowski, A. Bauer, T. M. Friedlmeier, and M. Powalla. 2015. Properties of Cu(In,Ga)Se$_2$ solar cells with new record efficiencies up to 21.7%. *Physica Status Solidi (RRL)— Rapid Research Letters*, 9(1): 28–31. doi: 10.1002/pssr.201409520.

Jasieniak, J., B. I. MacDonald, S. E. Watkins, and P. Mulvaney. 2011. Solution-processed sintered nanocrystal solar cells via layer-by-layer assembly. *Nano Letters*, 11(7): 2856–2864. doi: 10.1021/nl201282v.

Jeon, N. J., J. H. Noh, Y. C. Kim, W. S. Yang, S. Ryu, and S. I. Seok. 2014. Solvent engineering for high-performance inorganic–organic hybrid perovskite solar cells. *Nature Materials*, 13(9): 897–903. doi: 10.1038/nmat4014.

Kalyanasundaram, K., N. Vlachopoulos, V. Krishnan, A. Monnier, and M. Graetzel. 1987. Sensitization of titanium dioxide in the visible light region using zinc porphyrins. *Journal of Physical Chemistry*, 91(9): 2342–2347. doi: 10.1021/j100293a027.

Kavlak, G., J. McNerney, R. L. Jaffe, and J. E. Trancik. 2014. Growth in metals production for rapid photovoltaics deployment. *Photovoltaic Specialist Conference (PVSC), 2014 IEEE 40th*. June 8–13.

Kazmerski, L. L., F. R. White, and G. K. Morgan. 1976. Thin-film CuInSe$_2$/CdS heterojunction solar cells. *Applied Physics Letters*, 29(4): 268–270. doi: 10.1063/1.89041.

Keis, K., C. Bauer, G. Boschloo, A. Hagfeldt, K. Westermark, H. Rensmo, and H. Siegbahn. 2002. Nanostructured ZnO electrodes for dye-sensitized solar cell applications. *Journal of Photochemistry and Photobiology A: Chemistry*, 148: 57–64.

Kim, H.-S., C.-R. Lee, J.-H. Im, K.-B. Lee, T. Moehl, A. Marchioro, S.-J. Moon et al. 2012a. Lead iodide perovskite sensitized all-solid-state submicron thin film mesoscopic solar cell with efficiency exceeding 9%. *Scientific Reports*, 2. doi: http://www.nature.com/srep/2012/120821/srep00591/abs/srep00591.html#supplementary-information.

Kim, H.-S., C.-R. Lee, J.-H. Im, K.-B. Lee, T. Moehl, A. Marchioro, S.-J. Moon et al. 2012b. Lead iodide perovskite sensitized all-solid-state submicron thin film mesoscopic solar cell with efficiency exceeding 9%. *Scientific Reports* 2. doi: 10.1038/srep00591.

Klahr, B. M. and T. W. Hamann. 2009. Performance enhancement and limitations of cobalt bipyridyl redox shuttles in dye-sensitized solar cells. *Journal of Physical Chemistry C*, 113(31): 14040–14045. doi: 10.1021/jp903431s.

Kojima, A., K. Teshima, Y. Shirai, and T. Miyasaka. 2009. Organometal halide perovskites as visible-light sensitizers for photovoltaic cells. *Journal of the American Chemical Society*, 131(17): 6050–6051. doi: 10.1021/ja809598r.

Law, M., L. E. Greene, J. C. Johnson, R. Saykally, and P. Yang. 2005. Nanowire dye-sensitized solar cells. *Nature Materials*, 4: 455–459.

Lee, M. M., J. Teuscher, T. Miyasaka, T. N. Murakami, and H. J. Snaith. 2012. Efficient hybrid solar cells based on meso-superstructured organometal halide perovskites. *Science*, 338(6107): 643–647. doi: 10.1126/science.1228604.

Lewis, N. 2005. *Basic Research Needs for Solar Energy Utilization*. Bethesda, Maryland: Basic Energy Sciences Workshop on Solar Energy Utilization.

Li, X., M. Tschumi, H. Han, S. S. Babkair, R. A. Alzubaydi, A. A. Ansari, S. S. Habib, M. K. Nazeeruddin, S. M. Zakeeruddin, and M. Grätzel. 2015. Outdoor performance and stability under elevated temperatures and long-term light soaking of triple-layer mesoporous perovskite photovoltaics. *Energy Technology*, 3(6): 551–555. doi: 10.1002/ente.201500045.

Liu, M., M. B. Johnston, and H. J. Snaith. 2013. Efficient planar heterojunction perovskite solar cells by vapour deposition. *Nature*, advance online publication. doi: 10.1038/nature12509.

Major, J. D., R. E. Treharne, L. J. Phillips, and K. Durose. 2014. A low-cost non-toxic post-growth activation step for CdTe solar cells. *Nature*, 511(7509): 334–337. doi: 10.1038/nature13435.

Manthina, V., J. P. C. Baena, G. Liu, and A. G. Agrios. 2012. ZnO–TiO$_2$ nanocomposite films for high light harvesting efficiency and fast electron transport in dye-sensitized solar cells. *Journal of Physical Chemistry C*, 116(45): 23864–23870. doi: 10.1021/jp304622d.

Mathew, S., A. Yella, P. Gao, R. Humphry-Baker, F. E. CurchodBasile, N. Ashari-Astani, I. Tavernelli, U. Rothlisberger, K. NazeeruddinMd, and M. Grätzel. 2014. Dye-sensitized solar cells with 13% efficiency achieved through the molecular engineering of porphyrin sensitizers. *Nature Chemistry*, 6(3): 242–247. doi: 10.1038/nchem.1861. http://www.nature.com/nchem/journal/v6/n3/abs/nchem.1861.html#supplementary-information.

Mei, A., X. Li, L. Liu, Z. Ku, T. Liu, Y. Rong, M. Xu et al. 2014. A hole-conductor-free, fully printable mesoscopic perovskite solar cell with high stability. *Science*, 345(6194): 295–298. doi: 10.1126/science.1254763.

Miskin, C. K., W.-C. Yang, C. J. Hages, N. J. Carter, C. S. Joglekar, E. A. Stach, and R. Agrawal. 2015. 9.0% efficient Cu$_2$ZnSn(S,Se)$_4$ solar cells from selenized nanoparticle inks. *Progress in Photovoltaics: Research and Applications*, 23(5): 654–659. doi: 10.1002/pip.2472.

MIT. 2015. *The Future of Solar Energy.* Boston: Massachusetts Institute of Technology.

Mitzi, D. B., C. A. Feild, W. T. A. Harrison, and A. M. Guloy. 1994. Conducting tin halides with a layered organic-based perovskite structure. *Nature*, 369(6480): 467–469.

Mitzi, D. B., S. Wang, C. A. Feild, C. A. Chess, and A. M. Guloy. 1995. Conducting layered organic-inorganic halides containing <110>-oriented perovskite sheets. *Science*, 267(5203): 1473–1476. doi: 10.2307/2886541.

Mitzi, D. B. 2004. Solution-processed inorganic semiconductors. *Journal of Materials Chemistry*, 14(15): 2355–2365.

Nazeeruddin, M. K., A. Kay, I. Rodicio, R. Humphry-Baker, E. Müller, P. Liska, N. Vlachopoulos, and M. Grätzel. 1993. Conversion of light to electricity by *cis*-X$_2$bis(2,2'-bipyridyl-4,4'-dicarboxylate)ruthenium(II) charge-transfer sensitizers (X = Cl$^-$, Br$^-$, I$^-$, CN$^-$, and SCN$^-$) on nanocrystalline titanium dioxide electrodes. *Journal of the American Chemical Society*, 115(14): 6382–6390. doi: 10.1021/ja00067a063.

Nie, W., H. Tsai, R. Asadpour, J.-C. Blancon, A. J. Neukirch, G. Gupta, J. J. Crochet et al. 2015. High-efficiency solution-processed perovskite solar cells with millimeter-scale grains. *Science*, 347(6221): 522–525. doi: 10.1126/science.aaa0472.

Niu, G., X. Guo, and L. Wang. 2015. Review of recent progress in chemical stability of perovskite solar cells. *Journal of Materials Chemistry A*, 3(17): 8970–8980. doi: 10.1039/C4TA04994B.

Noel, N. K., S. D. Stranks, A. Abate, C. Wehrenfennig, S. Guarnera, A.-A. Haghighirad, A. Sadhanala et al. 2014. Lead-free organic-inorganic tin halide perovskites for photovoltaic applications. *Energy & Environmental Science*, 7(9): 3061–3068. doi: 10.1039/c4ee01076k.

Nusbaumer, H., S. M. Zakeeruddin, J. E. Moser, and M. Grätzel. 2003. An alternative efficient redox couple for the dye-sensitized solar cell system. *Chemistry—A European Journal*, 9(16): 3756–3763.

Nusbaumer, H., J.-E. Moser, S. M. Zakeeruddin, M. K. Nazeeruddin, and M. Grätzel. 2001. Co^{II}(dbbip)$_2$$^{2+}$ complex rivals tri-iodide/iodide redox mediator in dye-sensitized photovoltaic cells. *Journal of Physical Chemistry B*, 105: 10461–10464.

O'Regan, B. C., J. R. Durrant, P. M. Sommeling, and N. J. Bakker. 2007. Influence of the TiCl$_4$ treatment on nanocrystalline TiO$_2$ films in dye-sensitized solar cells. 2. Charge density, band edge shifts, and quantification of recombination losses at short circuit. *Journal of Physical Chemistry C*, 111: 14001–14010.

O'Regan, B. C., J.-E. Moser, M. Anderson, and M. Grätzel. 1990. Vectorial electron injection into transparent semiconductor membranes and electric field effects on the dynamics of light-induced charge separation. *Journal of Physical Chemistry*, 94: 8720–8726.

O'Regan, B. and M. Grätzel. 1991. A low-cost, high-efficiency solar cell based on dye-sensitized colloidal TiO$_2$ films. *Nature*, 353(6346): 737–740.

Palomares, E., J. N. Clifford, S. A. Haque, T. Lutz, and J. R. Durrant. 2002. Slow charge recombination in dye-sensitised solar cells (DSSC) using Al$_2$O$_3$ coated nanoporous TiO$_2$ films. *Chemical Communications*, 0(14): 1464–1465.

Park, J. Y., S.-W. Choi, K. Asokan, and S. S. Kim. 2010. Growth of ZnO nanobrushes using a two-step aqueous solution method. *Journal of the American Ceramic Society*, 93(10): 3190–3194. doi: 10.1111/j.1551-2916.2010.03899.x.

Philipps, S. P., A. W. Bett, K. Horowitz, and S. Kurtz. 2015. *Current Status of Concentrator Photovoltaic (CPV) Technology*. Fraunhofer ISE, National Renewable Energy Laboratory, Freiburg, Germany, Golden, Colorado, USA.

Platzer, M. D. 2015. U.S. solar photovoltaic manufacturing: Industry trends, global competition, federal support. *Congressional Research Service*, Washington, DC.

Puthussery, J., S. Seefeld, N. Berry, M. Gibbs, and M. Law. 2011. Colloidal iron pyrite (FeS$_2$) nanocrystal inks for thin-film photovoltaics. *Journal of the American Chemical Society*, 133(4): 716–719. doi: 10.1021/ja1096368.

Razza, S., F. D. Giacomo, F. Matteocci, L. Cinà, A. L. Palma, S. Casaluci, P. Cameron et al. 2015. Perovskite solar cells and large area modules (100 cm^2) based on an air flow-assisted PbI$_2$ blade coating deposition process. *Journal of Power Sources*, 277: 286–291. doi: 10.1016/j.jpowsour.2014.12.008.

REN21. 2014. *Renewables 2014 Global Status Report*. Paris: Renewable Energy Policy Network for 21st Century.

Repins, I., M. A. Contreras, B. Egaas, C. DeHart, J. Scharf, C. L. Perkins, B. To, and R. Noufi. 2008. 19·9%-efficient ZnO/CdS/CuInGaSe^2 solar cell with 81·2% fill factor. *Progress in Photovoltaics: Research and Applications*, 16(3): 235–239. doi: 10.1002/pip.822.

Richards, B. S. 2006. Enhancing the performance of silicon solar cells via the application of passive luminescence conversion layers. *Solar Energy Materials & Solar Cells*, 90(15): 2329–2337. doi: http://dx.doi.org/10.1016/j.solmat.2006.03.035.

Sariciftci, N. S., L. Smilowitz, A. J. Heeger, and F. Wudl. 1992. Photoinduced electron transfer from a conducting polymer to buckminsterfullerene. *Science*, 258(5087): 1474–1476. doi: 10.1126/science.258.5087.1474.

Shockley, W. and H. J. Queisser. 1961. Detailed balance limit of efficiency of *p-n* junction solar cells. *Journal of Applied Physics*, 32(3): 510–519.

Simon, P. P., A. W. Bett, K. Horowitz, and S. Kurtz. 2015. Current status of concentrator photovoltaic (Cpv) technology. Fraunhofer Institute for Solar Energy Systems ISE, Freiburg, Germany, National Renewable Energy Laboratory NREL, Golden, Colorado.

Sommeling, P. M., B. C. O'Regan, R. R. Haswell, H. J. P. Smit, N. J. Bakker, J. J. T. Smits, J. M. Kroon, and J. A. M. van Roosmalen. 2006. Influence of a $TiCl_4$ post-treatment on nanocrystalline TiO_2 films in dye-sensitized solar cells. *Journal of Physical Chemistry B*, 110: 19191–19197.

Spratt, J. P. and Schwarz, R. F. 1975. Concentrated photovoltaic power generation systems. In *Annual Intersociety Energy Conversion and Engineering Conference 10th*, Newark, Delaware, August 18–22, pp. 404–412.

Staebler, D. L. and C. R. Wronski. 1980. Optically induced conductivity changes in discharge-produced hydrogenated amorphous silicon. *Journal of Applied Physics*, 51(6): 3262–3268. doi: 10.1063/1.328084.

Steinhagen, C., T. B. Harvey, C. J. Stolle, J. Harris, and B. A. Korgel. 2012. Pyrite nanocrystal solar cells: Promising, or fool's old? *Journal of Physical Chemistry Letters*, 3(17): 2352–2356. doi: 10.1021/jz301023c.

Stuckelberger, M., M. Despeisse, G. Bugnon, J. -W. Schüttauf, F. -J. Haug, and C. Ballif. 2013. Comparison of amorphous silicon absorber materials: Light-induced degradation and solar cell efficiency. *Journal of Applied Physics*, 114(15): 154509. doi: 10.1063/1.4824813.

Tang, C. W. 1986. Two-layer organic photovoltaic cell. *Applied Physics Letters*, 48(2): 183–185. doi: 10.1063/1.96937.

Tenenbaum, D. 2014. Scientists get to the heart of fool's gold as a solar material. http://news.wisc.edu/scientists-get-to-the-heart-of-fools-gold-as-a-solar-material/#sthash.SHDESy62.dpuf (Last modified November 2014).

Tiedje, T., E. Yablonovitch, G. D. Cody, and B. G. Brooks. 1984. Limiting efficiency of silicon solar cells. *Electron Devices, IEEE Transactions on*, 31(5): 711–716. doi: 10.1109/t-ed.1984.21594.

Todorov, T., M. Kita, J. Carda, and P. Escribano. 2009. Cu_2ZnSnS_4 films deposited by a soft-chemistry method. *Thin Solid Films*, 517(7): 2541–2544. doi: 10.1016/j.tsf.2008.11.035.

Todorov, T. K., O. Gunawan, T. Gokmen, and D. B. Mitzi. 2013. Solution-processed $Cu(In,Ga)(S,Se)_2$ absorber yielding a 15.2% efficient solar cell. *Progress in Photovoltaics: Research and Applications*, 21(1): 82–87. doi: 10.1002/pip.1253.

Tributsch, H. 1972. Reaction of excited chlorophyll molecules at electrodes and in photosynthesis. *Photochemistry and Photobiology*, 16(4): 261–269. doi: 10.1111/j.1751-1097.1972.tb06297.x.

TSMC Solar. 2015. TSMC solar commercial-size modules (1.09 m^2) set CIGS 16.5% efficiency record. Press release, April 29.

Turner, G. 2013. Global renewable energy market outlook 2013.

U.S. Geological Survey. 2015. *Mineral Commodity Summaries 2015*. Reston, Virginia: U.S. Geological Survey.

Vlachopoulos, N., P. Liska, J. Augustynski, and M. Graetzel. 1988. Very efficient visible light energy harvesting and conversion by spectral sensitization of high surface area polycrystalline titanium dioxide films. *Journal of the American Chemical Society*, 110(4): 1216–1220. doi: 10.1021/ja00212a033.

Wadia, C., A. P. Alivisatos, and D. M. Kammen. 2009. Materials availability expands the opportunity for large-scale photovoltaics deployment. *Environmental Science & Technology* 43(6): 2072–2077. doi: 10.1021/es8019534.

Wang, W., M. T. Winkler, O. Gunawan, T. Gokmen, T. K. Todorov, Y. Zhu, and D. B. Mitzi. 2014. Device characteristics of CZTSSe thin-film solar cells with 12.6% efficiency. *Advanced Energy Materials*, 4(7): n/a-n/a. doi: 10.1002/aenm.201301465.

Würfel, P. 2002. Thermodynamic limitations to solar energy conversion. *Physica E*, 14(1–2): 18–26.

Xu, C., J. Wu, U. V. Desai, and D. Gao. 2012. High-efficiency solid-state dye-sensitized solar cells based on TiO_2-coated ZnO nanowire arrays. *Nano Letters*, 12(5): 2420–2424. doi: 10.1021/nl3004144.

Yang, C.-h., S.-c. Lee, T.-c. Lin, and S.-c. Chen. 2008. Electrical and optical properties of indium tin oxide films prepared on plastic substrates by radio frequency magnetron sputtering. *Thin Solid Films*, 516(8): 1984–1991. doi: 10.1016/j.tsf.2007.05.093.

Yang, W. S., J. H. Noh, N. J. Jeon, Y. C. Kim, S. Ryu, J. Seo, and S. I. Seok. 2015. High-performance photo-voltaic perovskite layers fabricated through intramolecular exchange. *Science*, 348(6240): 1234–1237. doi: 10.1126/science.aaa9272.

Yella, A., H.-W. Lee, H. N. Tsao, C. Yi, A. K. Chandiran, M. K. Nazeeruddin, E. W.-G. Diau, C.-Y. Yeh, S. M. Zakeeruddin, and M. Grätzel. 2011. Porphyrin-sensitized solar cells with cobalt (II/III)–based redox electrolyte exceed 12 percent efficiency. *Science*, 334(6056): 629–634. doi: 10.1126/science.1209688.

Zhou, H., Q. Chen, G. Li, S. Luo, T.-b. Song, H.-S. Duan, Z. Hong, J. You, Y. Liu, and Y. Yang. 2014. Interface engineering of highly efficient perovskite solar cells. *Science*, 345(6196): 542–546. doi: 10.1126/science.1254050.

Zhu, L., B. J. Richardson, and Q. Yu. 2015. Anisotropic growth of iron pyrite FeS_2 nanocrystals via oriented attachment. *Chemistry of Materials*, 27(9): 3516–3525. doi: 10.1021/acs.chemmater.5b00945.

25 Improving Process Efficiency by Waste Heat Recuperation

An Application of the Limaçon Technology

Ibrahim A. Sultan, Truong H. Phung, and Ali Alhelal

CONTENTS

ABSTRACT

Energy resources are under constant pressure to grow to match the rate of energy consumption which has been steadily increasing since the last 50 years. The expected depletion of fossil resources and the global climate change are the major factors that have added pressures to the task of global energy planning. These pressures, however, have helped fuel research and industry drives to find alternative and renewable energy resources and modify the current practices in the energy industry.

Recuperation of waste heat is a rather new concept emerging in the field of process engineering. Waste heat offers a low-grade thermal resource which has previously been overlooked or considered of no value. To obtain work or electrical power from such a resource, an organic Rankine cycle or a trilateral flash cycle can be used in a plant which consists of heat exchangers, pumps and gas expanders. The performance of these plants depends mainly on the efficiency of the gas expander used in them for energy generation. A number of gas expanders are available, such as scroll expanders, screw expanders, reciprocating expanders and rotary expanders. The limaçon expander is a new addition to the existing designs. The limaçon design, however, offers some thermodynamical and mechanical aspects beyond what other designs can offer. This chapter will introduce the limaçon technology and shed some light on its potential to operate thermal power plants fuelled by low-grade heat resources. The application of this endeavour should readily extend to various scenarios such as in residential-scale power generation, to improve the process efficiency in existing power plants, food processing plants and mineral processing plants to name a few.

INTRODUCTION

The growth in world population has escalated global manufacturing activities to unprecedented levels. These activities have placed pressures on the mineral and energy resources of our planet. Consequently, these resources are being rapidly depleted in rates that put at risk the opportunity of future generations to realise their dreams and ambitions. Energy is predominantly generated by burning fossil fuels; a process that causes depletion of resources and pollution of the environment. Currently, there is an increase in the production of energy dictated by the desire of some nations to push manufacturing to higher levels and improve the quality of life for their citizens. As such, it is not practical to propose energy moderation policies which incorporate measures intended to reduce energy generation. Instead, proposals to moderate energy production and use should feature two aspects: more effective reliance on renewable resources and more efficient industrial processes. Both aspects will result in being able to use less fossil fuel without impacting the amount of energy generated. One way of improving process efficiency is by capturing the large amounts of heat which escape from most industrial processes and use of this heat to generate electricity. This chapter intends to investigate the available technologies that can be used for that purpose.

To produce electricity from waste heat, a power plant that uses the concept of organic Rankine cycle (ORC) or trilateral flash cycle (TFC) can be utilised. Owing to their ability to extract energy from low-grade heat sources, these specific cycles are more suitable for heat-recuperation applications than the conventional thermodynamic cycles, which are featured in gas and steam power plants. More details on the structure of the ORC and the TFC plants will be given later in the chapter; for now, it suffices to point out that these plants generally consist of four main components: an evaporator, an expander, a condenser and a pump. Owing to their small size, gas expanders are utilised in these plants to generate mechanical power instead of the conventional turbines used in traditional installations. These expanders are often of the positive displacement type which can handle small mass flow rates albeit at the expense of various thermodynamic efficiencies. As the characteristics of the gas expander control the performance of the whole thermal plant, considerable amount of research and development work has been dedicated to optimise and improve the workings of these expanders. The authors of this chapter have been working to develop and optimise a gas expander technology with promising potentials. This is the limaçon technology which will be detailed later in the chapter.

In the next section, a brief discussion is given on the use of heat in mineral and material processing as an example of how heat is utilised in industrial plants.

HEAT IN METAL PRODUCTION

Heat is a vital element in the processing and preparation of engineering components which are manufactured out of metal, plastic or rubber. Typically, about 2%–15% of the overall production cost is attributed to heating. Of note is aluminum whose smelting process features the production of gases at temperature in the range of 1200–1300°C (U. S. Department of Energy, 2008).

Heat usage in the material industry can be classified into three basic types: fuel-based heating, electric-based heating and steam-based heating. In the first type, heat can be produced from the burning of solid, liquid and gaseous fuel and transferred to the material directly or indirectly depending on the process design. In the electric-based heating, materials are treated by using an electric current either directly by inducing an eddy current into the material or indirectly via conduction, convection and radiation. Steam-based heating utilises vapour, which can make a considerable amount of heat accessible at a constant temperature. This is desirable in heating applications (Lawrence Berkeley National Laboratory, 2007). Unfortunately, in all the above-mentioned heating processes, heat loss is inevitable, a fact which adversely impacts efficiency and cost. Heat escapes in material processing with combustion gases and through the walls of the equipment.

According to the U.S. Department of Energy (2008), industrial waste heat is estimated to vary from 20% to 50% of the energy consumption during the process. This ratio increases to 60% in the case of aluminum smelting, which requires 211 GJ energy per tonne (Brooks, 2012). It is worth mentioning here that it is not possible to eliminate waste heat entirely but it is possible to capture some of that heat and utilise it either for heating or for power generation. This chapter intends to offer an insight into the thermodynamical and practical considerations which determine the limiting parameters in the process of heat harvesting and utilisation.

WASTE HEAT RECUPERATION

In industry, more than 50% of the generated heat is considered as low-grade waste heat and generally discarded. Additionally, the efficiency of a thermal power generation system becomes particularly low when the exhaust temperature of the working fluid drops below 370°C (Hung et al., 1997; Wang et al., 2010). Recently, however, the trend of using energy effectively and efficiently has seen power generation that utilises low-grade or waste-heat sources become more and more popular.

When utilising a low-grade heat source, the theoretical Carnot cycle is considered impractical while the traditional Rankine cycle performs poorly due to its low thermal efficiency and large volume flow rate. Hence, the ORC and the TFC, which utilise organic gases as working fluids, are able to operate at relatively low speed, and attain comparatively good efficiency over a wide range of temperatures from 120°C to 370°C (Hung et al., 1997; Wang et al., 2010).

THERMODYNAMIC SYSTEMS

A thermodynamic system plays a very important role in the process of extracting work or electrical power from available heat sources. As described in Figure 25.1, such a system consists of four main components: a pump, an evaporator, an expander or a turbine and a condenser. It is necessary to have those four components interconnected as a closed cycle in which a working fluid is utilised to transfer heat from one component to another as well as produce work in the expander. To start the cycle, a pump is used to feed the working fluid to the evaporator in which heat from the heat source is utilised to vaporise the fluid. The pressurised high-temperature vapour is then directed towards the expander to transfer its potential energy to kinetic energy and produce power. The expanded fluid exits the expander and enters the condenser to expel its remaining heat to the cooling reservoir and returns to the liquid phase before entering the pump again to complete the cycle. The aforementioned cycle will continue to operate and generate power as long as the temperature difference between the evaporator and the condenser is still significant (Maizza and Maizza, 2000; Wei et al., 2006; Hung et al., 2009; Bombarda et al., 2010).

Within the four main components of a thermodynamic system, which have been previously mentioned, the expander is considered to be the most important part of such a system. Even though every component is vital to the workings of the thermodynamic system, the expander is the only part that actually extracts power from the heat source via the expanding fluid process and converts that potential energy to the rotational motion of the output shaft. It is also worth noting that the overall efficiency of the thermodynamic cycle depends considerably on the performance of the expanders (Fukuta and Yanagisawa, 2009; Chen et al., 2010; Lemort et al., 2010; Sultan, 2012). There are various expander designs such as scroll (Xiaojun et al., 2004; Peterson et al., 2008), screw (Smith

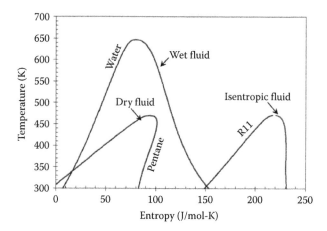

FIGURE 25.1   Wet, dry and isentropic fluid. (Adapted from Chen, H., D. Yogi Goswami, and E. K. Stefanakos. 2010. *Renewable and Sustainable Energy Reviews*, 3059–3067.)

et al., 2005), reciprocating piston (Gimstedt and Karlsson, 1979; Zhang et al., 2006), revolving vane (Fukuta and Yanagisawa, 2009; Subiantoro and Ooi, 2010) and limaçon (Sultan, 2005b; Sultan and Schaller, 2011) to name only a few.

Organic Rankine Cycles

Basic Rankine Cycles

A typical Rankine cycle has four main modules, a pump, a boiler, an expander (aka turbine) and a condenser, and utilises water as a working fluid. The working fluid within a Rankine cycle is used to transfer thermal energy from one component to another. With medium- and large-scale Rankine cycle plant, one of the three groups of working fluids is used: wet fluids, dry fluids and isentropic fluids. As shown in Figure 25.2, the fluid classification is based on the slope of the saturation vapour curve on the temperature, T, versus entropy, s (T–s) diagram. Small-scale Rankine cycle plant, on the other hand, makes use of organic fluids such as Honeywell synthetic refrigerants R123, R410a and R143a, which are largely categorised as isentropic fluid (Gao et al., 2012).

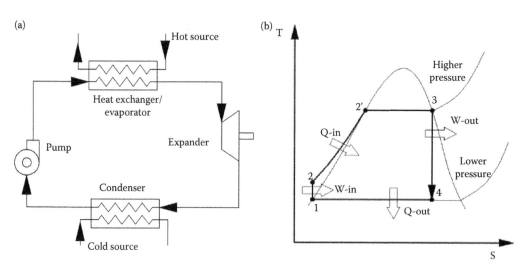

FIGURE 25.2   (a, b) A basic Rankine cycle.

Chen et al. proposed to utilise dry or isentropic fluids for ORCs to avoid the impact between the liquid droplets and the turbine blades during the expansion process. However, Chen et al. (2010) argued that if the liquid is too dry, usable heat might still remain in the expanded fluid, which is not only is a waste but also increases the cooling loads to the condenser.

A basic single-fluid Rankine cycle with four stages of change is depicted in Figure 25.2, where

- 1–2: Work is added by pressurising liquid in a feed pump (W-in).
- 2–3: Heat is added to the working fluid in a boiler or a heat exchanger (Q-in).
- 3–4: Work is extracted by expanding the fluid isentropically in a turbine or an expander(s) (W-out).
- 4–1: Excess heat is rejected in a condenser (Q-out).

Superheated, Transcritical and Supercritical Rankine Cycles

Illustrated in Figure 25.3 are the typical superheated and supercritical Rankine cycles shown on *T–s* diagrams.

In a superheated Rankine cycle, the saturated vapour is further heated towards the dry vapour region shown as point 3 in Figure 25.3a. The superheating process results in a shift of the expansion stage of wet fluid Rankine cycles towards the superheated region, which in turn will increase the output power at the expander's shaft (1′–2′ > 1–2) as shown in Figure 25.4a. On the other hand, the expansion stage of dry fluids has already taken place in the superheated region; therefore, superheating is not necessary and will bring no benefit to the performance of the dry fluid systems as shown in Figure 25.4b. The performance of cycles with is entropic fluid is not affected by superheating due to the processes 1′–2′ ≡ 1–2 shown in Figure 25.4c (Hung, 2001; Mago et al., 2008; Tchanche et al., 2014).

If the highest temperature of a Rankine cycle falls below the critical temperature of the fluid, that cycle is described as 'subcritical'. As the highest temperature of a cycle approaches the critical temperature, the cycle efficiency increases. This motivated the further increase in highest temperature to exceed the critical value thus producing the transcritical cycle. When both the highest and lowest temperatures of a cycle exceed the critical temperature (Tchanche et al., 2014), the cycle is referred to as 'supercritical' as shown in Figure 25.5. The efficiency of the transcritical cycle is known to be up to 40%; however, shifting from subcritical cycle to transcritical cycle results in only about 8% efficiency increase. Such upturn is not significant compared with the requirements of specific materials to manufacture the transcritical cycle components and high operational safety precautions

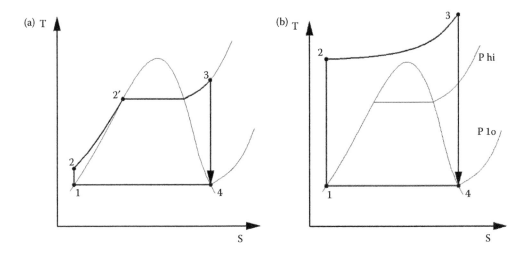

FIGURE 25.3 (a) Superheated RC and (b) transcritical RC.

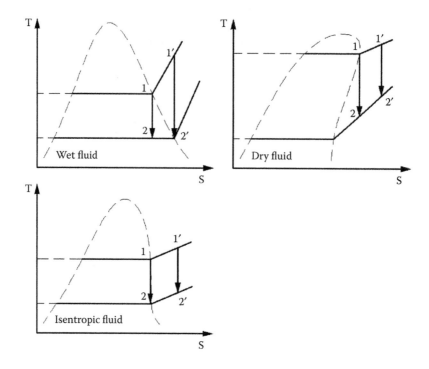

FIGURE 25.4 (a) Wet, (b) dry and (c) isentopic fluids.

owing to very high pressures. Nevertheless, the supercritical cycle has not been fully investigated with any working fluid (Shengjun et al., 2011; Kim et al., 2012; Tchanche et al., 2014).

The next four sections will demonstrate the application of different types of cycles in the heat recovery processes.

Rankine Cycles with Reheating

Rankine cycles with reheating have been widely used in water steam power plants. These can use a series of high-pressure (HP), intermediate-pressure (IP) and low-pressure (LP) expanders or

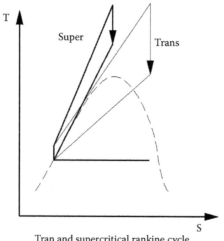

Tran and supercritical rankine cycle

FIGURE 25.5 Trans and supercritical RC.

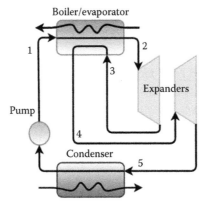

Rankine cycle with re-heater

FIGURE 25.6 Rankine cycle with re-heater.

turbines. As the partially expanded working fluid exits the HP turbine(s), it returns to the boiler, to be reheated and directed to the intermediate expansion process in the IP turbine(s). The same process is repeated for the following stage turbines as depicted in Figure 25.6. The working fluids can be reheated 2, 3 or more times depending on the number of turbine levels of a specific system. Other than water, CO_2 can also be used as a working fluid. Organic working fluids, however, have yet to be tested with reheating Rankine cycles (Acar, 1995; Tuo, 2012; Tchanche et al., 2014).

Rankine Cycles with Heat Regeneration

Also known as regenerative ORC, the Rankine cycle with heat regeneration has been developed from the basic cycle. As shown in Figure 25.7, a heat regenerator is used to extract heat from the expanded fluid and preheat the fresh charge of the working fluid before it enters the boiler. The purpose of integrating a heat regenerator to a cycle is to increase the system efficiency, especially when dry fluids are utilised. The effectiveness of this method has been asserted by the work of Mago et al. (2008) particularly in relation to ORCs. An important constraint of the regeneration cycle is that the

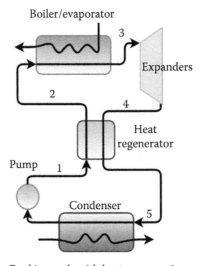

Rankine cycle with heat regeneration

FIGURE 25.7 Rankine cycle with heat regeneration.

temperature of the outlet of the expander is considerably high. This constraint has been discussed by Vanslambrouck et al. (2011) (Mago et al., 2008; Gang et al., 2010; Vanslambrouck et al., 2011; Roy and Misra, 2012).

Rankine Cycle with Integrated Feed-Liquid Heater

The open-feed-liquid and the closed-feed-liquid cycles belong to the same category as the conventional Rankine cycle. However, these two cycles utilise partially expanded fluid (i.e. extracted from the expander) or fully expanded fluid to preheat the feed liquid before it enters the evaporator (Figure 25.8). If the extracted fluid is mixed with the feed liquid, we have an open-feed-liquid cycle; otherwise, it is called a closed-feed-liquid cycle. The integrated feed-liquid process helps increasing the fluid temperature in the expander and hence increases the system efficiency (DiPippo, 2004; Tchanche et al., 2014).

Binary Fluids Rankine Cycles

A binary fluid Rankine cycle consists of two separate cycles that use different working fluids. Those are called the high-temperature cycle and the low-temperature cycle. The pre-expanded low-temperature liquid absorbs heat from the expanded high-temperature liquid through a shared heat exchanger, and if necessary can receive additional heat from the boiler before entering its expander.

Another type of binary fluid cycle utilises the fluid mixture instead of single-component fluids (i.e. a mixture of water and ammonia). Depicted in Figure 25.9 are the Maloney and Robertson, and the Kalina cycles. These are two of the mixture fluid cycles which have been developed and tested

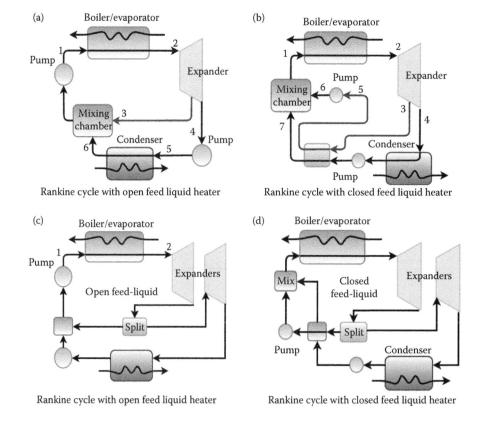

FIGURE 25.8 (a) Rankine cycle with open feed liquid heater, (b) Rankine cycle with closed feed liquid heater, (c) Rankine cycle with open feed liquid heater – two expanders and (d) Rankine cycle with closed feed liquid heater – two expanders.

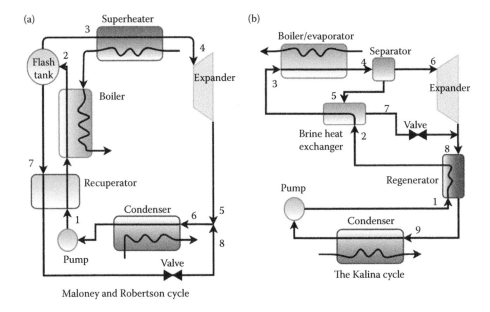

FIGURE 25.9 (a) Maloney and Robertson cycles and (b) Kalina cycle.

over the years. The Maloney and Robertson cycle do not offer an advantage over the Rankine cycle; hence, the investigation on such a cycle has been terminated. The Kalina cycle, on the other hand, offers an improvement of thermal efficiency in the order of 3% above what the standard Rankine cycle can offer. However, the Kalina cycle is not cost effective in comparison with the standard ORC because of the high evaporator pressure and large evaporator surface requirements (Tchanche et al., 2014). With the ambition to achieve higher thermal efficiency, 'Uehara cycle', a modification of the Kalina cycle, has emerged, which yields higher efficiency than both the Kalina and Rankine cycles. However, the structure of the Uehara cycle has to be very complex to convey such efficiency; hence it has been abandoned (DiPippo, 2004; Tchanche et al., 2014).

TRILATERAL FLASH CYCLE

A trilateral flash cycle (TFC), a modification of the ideal Carnot cycle, is formed with three processes: continuous heat addition to the working fluid in an evaporator, isentropic expansion to produce work in an expander and isothermal heat rejection to a cool medium. Dipippo (2007) argued that an ideal TFC is not possible to achieve, and instead, a practical four-stage version of TFC is often considered. Figure 25.10 depicts the *T–s* diagrams of an ideal TFC and a practical TFC.

Steffen et al., Fischer and Zamfirescu et al. along with other researchers highlight TFC as one of the three processes that can be used to convert low-temperature and waste heat resources to energy: the ORC, the Kalina cycle and the TFC. With its ability to recover most of the heat from the heat source, the TFC can easily outperform the ORC and the Kalina cycle. Additionally, with purposefully selected inlet temperatures at various points of the plant, it has been proven that the exergy efficiency for power production of the TFC, which is the ratio of the net power output to the incoming exergy flow of the heat carrier, is between 14% and 20% higher than that of the ORC. Besides, the total efficiency of the TFC, which is the ratio of all the outgoing exergy flows to all incoming exergy flows, is from 1% to 9% higher than that of ORCs in most cases. Moreover, Zamfirescu and Dincer have demonstrated that with certain set-ups and at given temperatures the calculated exergy efficiency of the TFC could be twice as much as that of the ORC. However, TFC is a fairly new technology which still requires further research and development (Zamfirescu and Dincer, 2008; , Fischer, 2011; Steffen et al., 2013).

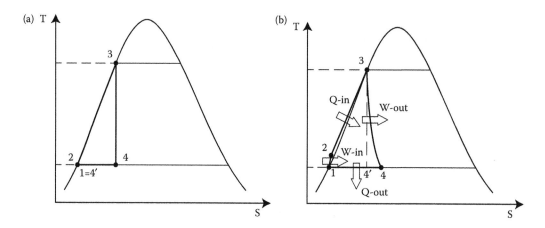

FIGURE 25.10   (a) Ideal TFC and (b) practical TFC.

LIMAÇON TECHNOLOGY AND ITS APPLICATIONS

LIMAÇON TECHNOLOGY

Limaçon technology and limaçon machines, which can be utilised as pumps, compressors or expanders, belong to the rotary positive displacement family with the earliest design proposed by Wheildon (1896). A limaçon machine consists of two main components: a machine housing with inlet and outlet (discharge) ports to which the manifolds are attached and a two-lobe rotor. The profiles of the housing and rotor should ideally be manufactured to the limaçon curve; however, according to Sultan (2008), they can also be manufactured to the circular curve. The rotation of the rotor inside the housing follows the limaçon motion in which the rotor chord, p1p2 of length 2L, slides and rotates about the pole, o; the profile of the limaçon housing is formed by the trajectory of the apex, p1 or p2, as depicted in Figure 25.11. During rotations, the centre point, m, of the rotor chord always falls on to the circumference of the base circle; when the rotor chord is at the lowest horizontal position, the centre point, m, coincides with the pole, o. As shown in Figure 25.11 the stationary frame XY is attached to the pole, o, while the XrYr frame is attached to the rotor centre point, m and moves with it. The angular displacement of the rotor chord, $\theta \in [0, 2\pi]$, is measured from the Xr-axis to the X-axis in a right-hand sense. The instantaneous sliding distance, s, between the chord centre point, m and the pole, o, can be expressed in terms of the chord angle, θ and the base circle radius, r, as follows:

$$s = 2r\sin(\theta) \tag{25.1}$$

The radial distance, R_h, which is measured along the chord to define the position of the apex p1 with respect to the pole, o. R_h may be calculated as follows:

$$R_h = 2r\sin(\theta) + L \tag{25.2}$$

where L is half the rotor chord length.

The radial distance can also be expressed in the Cartesian frame XY as follows:

$$\begin{cases} x_h = r\sin(2\theta) + L\cos(\theta) \\ y_h = r - r\cos(2\theta) + L\sin(\theta) \end{cases} \tag{25.3}$$

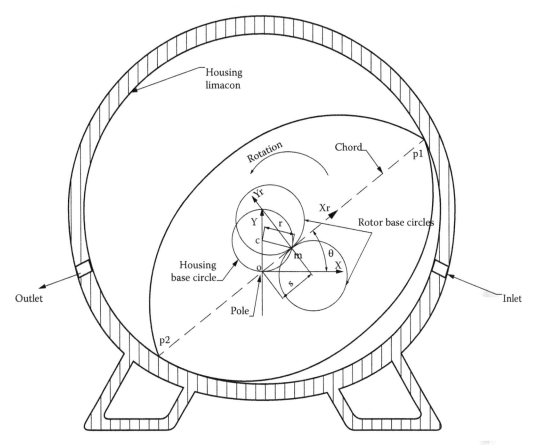

FIGURE 25.11 A typical limaçon-to-limaçon machine. (Adapted from Sultan, I. A. 2005a. *Journal of Mechanical Design*, 128(4): 787–793.)

or

$$\begin{cases} x_h = L[b\sin(2\theta) + \cos(\theta)] \\ y_h = L[b[1 - \cos(2\theta)] + \sin(\theta)] \end{cases} \tag{25.4}$$

where $b_r = (r/L) \leq 0.25$ is the limaçon aspect ratio. The value of b is restricted to less than 0.25 to ensure that the limaçon curve is free from dimples and loops.

In the following sections, the limaçon machine and its thermodynamic performance will be closely investigated.

Limaçon Expander Embodiments

Currently, there are three embodiments of the limaçon machines that are named after the shape of the housing and rotor profile. Such embodiments are limaçon-to-limaçon machines, circolimaçon machines and limaçon-to-circular machines. Although being constructed considerably differently, these three machines share some common characteristics mentioned in section 'Limaçon Technology' and developed based on Equations 25.1 through 25.4 (Sultan, 2005b, 2008, 2012; Sultan and Schaller, 2011).

Geometric Model of the Limaçon-to-Limaçon Machines

Machine Profile

The term 'limaçon-to-limaçon' refers to the embodiment where the rotor and the housing are both machined to the limaçon curve. Recently, this embodiment has been studied by researchers such as Sultan (2005a,b) with a vision of utilising this technology for low-grade and/or waste-heat recovery. To prevent housing-rotor interference while the machine is in motion, a clearance value, C, is introduced so that the rotor chord is actually shorter than p1p2 by a small amount of $2C$. In a manner similar to Equation 25.2, the radial distance of the rotor, R_r, which measured from the moving pole of the rotor to a point on its profile, is described as follows:

$$R_r = 2r_r \sin(\phi) + (L - C) \tag{25.5}$$

or

$$R_r = L[2b_r \sin(\phi) + (1 - C_L)] \tag{25.6}$$

where $b = (r_r/L) \le 0.25$ and $C_L = C/L$ are two unitless constants, and $\phi \in [\pi, 2\pi]$. The rotor radial distance, R_r, can also be expressed in the Cartesian coordinates as

$$\begin{cases} x_{rl} = L[b_r \sin(2\phi) + (1 - C_L)\cos(\phi)] \\ y_{rl} = L[b_r(1 - \cos(2\phi)) + (1 - C_L)\sin(\phi)] \end{cases} \tag{25.7}$$

Equations 25.5 through 25.7 are necessary to develop the profile for the lower half of the rotor. The upper half of the rotor is a mirror image of the lower half reflected about the chord line. As such, this upper half can be described as follows:

$$\begin{cases} x_{ru} = L[b_r \sin(2\phi) + (1 - C_L)\cos(\phi)] \\ y_{ru} = L[b_r(\cos(2\phi - 1)) - (1 - C_L)\sin(\phi)] \end{cases} \tag{25.8}$$

The volumetric relationships of the limaçon-to-limaçon machine will be described in the following section.

Volumetric Relationship

To calculate the area bounded by the limaçon curve we employ Figure 25.12 which depicts an infinitesimal section of this area subtended by two rays originating from the pole, o and separated by an angle $\delta\theta$. These two rays represent the chord of the limaçon at positions θ and $\theta + \delta\theta$. The shaded region in Figure 25.12 can be expressed by the following equation:

$$\delta A = \frac{1}{2} R_h^2 \delta\theta \tag{25.9}$$

where R_h is given by Equation 25.2. The areas which fall above and below the rotor centreline, A_t and A_b, can be calculated by integrating Equation 25.9 in the range from θ to $\theta + \pi$ and from $\theta + \pi$ to $\theta + 2\pi$, respectively. Therefore, those areas can be expressed as follows:

$$A_t = \pi\left(r^2 + \frac{1}{2}L^2\right) + 4rL\cos(\theta) \tag{25.10}$$

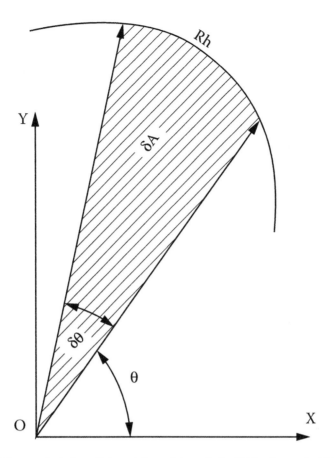

FIGURE 25.12 Volume calculation. (Adapted from Sultan, I. A. 2005a. *Journal of Mechanical Design*, 128(4): 787–793.)

and

$$A_b = \pi\left(r^2 + \frac{1}{2}L^2\right) - 4rL\cos(\theta) \tag{25.11}$$

With a similar method, the cross-sectional area of the rotor, A_r, can be determined by the following equation:

$$A_r = \pi\left(r_r^2 + \frac{1}{2}L^2\right) - \pi C\left(L - \frac{1}{2}C\right) - 4r_r(L - C) \tag{25.12}$$

As such, the net volume above the rotor profile, V_t, which can be used for fluid processing is expressed as follows:

$$V_t = H\left[\pi(r^2 - r_r^2) + \pi C\left(L - \frac{1}{2}C\right) + 4r_r(L - C) + 4rL\cos(\theta)\right] \tag{25.13}$$

where H is the rotor depth measured perpendicular to the plane of Figure 25.11. The corresponding net volume below the rotor profile, V_b, can be calculated as

$$V_b = H\left[\pi(r^2 - r_r^2) + \pi C\left(L - \frac{1}{2}C\right) + 4r_r(L - C) - 4rL\cos(\theta)\right] \tag{25.14}$$

At the rotor dead centre, where $\theta = 0$, V_b becomes the minimum clearance volume that separates the rotor profile from the housing; likewise, at this $\theta = 0$, V_t is the maximum chamber volume. In the case of the gas expander, the volumetric ratio, R_{vol}, which is defined as the minimum clearance volume, V_b, to the maximum clearance volume, V_t, has to be kept as small as possible in order to maximise the volumetric efficiency of the machine. The volumetric ratio, R_{vol}, can be expressed as follows:

$$R_{vol} = \frac{\pi(r^2 - r_r^2) + \pi C(L - (1/2)C) - 4r_r C - 4L(r - r_r)}{\pi(r^2 - r_r^2) + \pi C(L - (1/2)C) - 4r_r C + 4L(r + r_r)} \tag{25.15}$$

R_{vol} can also be expressed in terms of limaçon housing aspect ratio $b = (r/L) \leq 0.25$ and $C_L = C/L$ as follows:

$$R_{vol} = \frac{\pi\left(b^2 - (r_r^2/L^2)\right) + \pi C_L(1 - (1/2)C_L) - 4(r_r/L)C_L - 4L(b - (r_r/L))}{\pi\left(b^2 - (r_r^2/L^2)\right) + \pi C_L(1 - (1/2)C_L) - 4(r_r/L)C_L + 4L(b + (r_r/L))} \tag{25.16}$$

Equation 25.16 shows that in order for a limaçon machine to have a small volumetric ratio, R_{vol}, the value of C has to be very small and r_r has to be as close to r as possible. However, such setting may result in housing-rotor interference during operation. The next section is intended to investigate the interference aspects of the limaçon-to-limaçon machine from the perspective of curve tangents.

Interference

In the limaçon-to-limaçon machine, rotor-housing interference may take place during the rotational motion of the rotor if proper design precautions have not been taken. Sultan (2005a) suggested that the base circle of the rotor limaçon, r_r, can be slightly larger than the base circle of the housing limaçon, r. As such, the following design formula has been suggested:

$$r_r = \begin{cases} \dfrac{a(1 - C/L)}{\sqrt{1 - 4b^2}}r & \text{for } a \geq 1 \\[2mm] r & \text{for } a < 1 \end{cases} \tag{25.17}$$

The design factor, a, in the above formula could be set by the designer to create favourable running conditions for the machine. The higher the value of a, the larger the radius of the rotor limaçon base circles and the narrower would the resulting lenticular profile be.

Alternatively, Sultan (2007) suggested the following surrogate model to calculate a suitable value for the clearance C:

$$C = \Delta_{all}F_1 + L(F_2 + F_3\Delta_{all}) \tag{25.18}$$

where functions F_1, F_2 and F_3 are given as follows:

$$F_1 = K_0 + K_1 b + K_2 b^2 + K_3 b^4 + K_4 b e^{-K_5/b}$$
$$F_2 = K_6 + K_7 b + K_8 b^2 + K_9 b^4 + K_{10} b e^{-K_{11}/b} \qquad (25.19)$$
$$\text{and}$$
$$F_3 = K_{12} + K_{13} b + K_{14} b^2 + K_{15} b^4$$

The coefficients K_0 to K_{15} were calculated using a suitable regression technique. The values of these coefficients can be found in the paper by Sultan (2007).

Port Area Calculations

The limaçon machine has two ports: one for fluid admission and other for discharge. The concepts presented in this section will apply to both ports without discrimination and as such, no specific reference will be made to either port in the section.

Figure 25.13 depicts a port on a limaçon machine with two position vectors, $\mathbf{R}_l$ and $\mathbf{R}_t$, used to define the radial positions of the port leading and trailing edges, respectively. The leading edge is the one which the rotor's leading apex, p_l, meets first during one cycle of operation. The port edge vectors can be expressed as follows:

$$\mathbf{R}_l = L(2b\sin\theta_l + 1)\hat{\mathbf{R}}_l \quad \text{and} \quad \mathbf{R}_t = L(2b\sin\theta_t + 1)\hat{\mathbf{R}}_t \qquad (25.20)$$

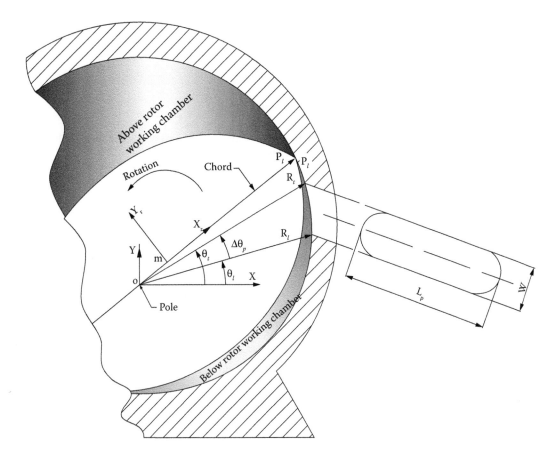

FIGURE 25.13 Port geometric particulars. (Adapted from Sultan, I. A. and C. G. Schaller. 2011. *Journal of Engineering for Gas Turbines and Power.*)

where the angular locations of the leading and trailing edges are, respectively, defined by the angles, θ_l and θ_t. Usually, θ_l is assigned first and then a port angular width, $\Delta\theta_p$, is determined. The angular position of the trailing edge is then calculated from $\theta_t = \theta_l + \Delta\theta_p$. If θ_t is found to be greater than 2π, then 2π has to subtracted from the calculated value of θ_t. The unit vectors, $\hat{\mathbf{R}}_l$ and $\hat{\mathbf{R}}_t$, used in Equation 25.20 are given as follows:

$$\hat{\mathbf{R}}_l = [\cos\theta_l \quad \sin\theta_l \quad 0]^{\mathrm{T}} \quad \text{and} \quad \hat{\mathbf{R}}_t = [\cos\theta_t \quad \sin\theta_t \quad 0]^{\mathrm{T}} \tag{25.21}$$

The port length, L_p, is normally given within the constraints imposed by the rotor depth. The width, W, is calculated from the two edge vectors as follows:

$$W = |\mathbf{R}_t - \mathbf{R}_l| \tag{25.22}$$

Taking the port ends to be semicircular, the full area, A_f, of the port may be expressed as follows:

$$A_f = L_p W - W^2\left(1 - \frac{\pi}{4}\right) \tag{25.23}$$

The rotor–port interaction is studied here in relation to the location of the leading apex of the rotor with respect to the port edges. For this purpose, we define the position of the rotor's leading apex using the vector $\mathbf{P}_l$ as follows:

$$\mathbf{P}_l = L(2b\sin\theta + 1)\hat{\mathbf{X}}_r \tag{25.24}$$

where the unit vector $\hat{\mathbf{X}}_r$ is given by

$$\hat{\mathbf{X}}_r = [\cos\theta \quad \sin\theta \quad 0]^{\mathrm{T}} \tag{25.25}$$

The relative locations of the rotor leading apex with respect to the port edges can now be defined using the signs of two scalar quantities, s_l and s_t, which are given as follows:

$$s_l = (\hat{\mathbf{R}}_l \times \hat{\mathbf{X}}_r)\cdot(\hat{\mathbf{X}}_r \times \hat{\mathbf{Y}}_r) \quad \text{and} \quad s_t = (\hat{\mathbf{R}}_t \times \hat{\mathbf{X}}_r)\cdot(\hat{\mathbf{X}}_r \times \hat{\mathbf{Y}}_r) \tag{25.26}$$

where the unit vector $\hat{\mathbf{Y}}_r$ is given by

$$\hat{\mathbf{Y}}_r = [-\sin\theta \quad \cos\theta \quad 0]^{\mathrm{T}} \tag{25.27}$$

Sultan and Schaller (2011) suggest the following algorithm to calculate the instantaneous value of the port area, A_p:

$$\text{for} \quad s_l \geq 0 \begin{cases} \text{if } s_t \geq 0 & A_p = A_f \\ \text{if } s_t < 0 & A_p = (W_l/W)A_f \end{cases} \tag{25.28}$$

and

$$\text{for} \quad s_l < 0 \begin{cases} \text{if } s_t \leq 0 & A_p = 0 \\ \text{if } s_t > 0 & A_p = (W_t/W)A_f \end{cases} \tag{25.29}$$

where $W_l = |\mathbf{P}_l - \mathbf{R}_l|$, $W_t = |\mathbf{P}_t - \mathbf{R}_t|$ and the position vector $\mathbf{P}_t$ is given by

$$\mathbf{P}_t = L(2b\sin\theta - 1)\hat{\mathbf{X}}_r \qquad (25.30)$$

The velocity of flow blowing through the downstream side of a port is presented in the next section.

Thermodynamic Model of the Limaçon-to-Limaçon Machines

Velocity of Flow through Ports

The velocity, U_p, of flow blowing through the downstream side of a port may be expressed as follows:

$$U_p = \begin{cases} \left[\left(\left(\dfrac{\rho_{pd}D_{pd}}{\mu_{pd}} \right)^{0.1} \sqrt{\dfrac{2(h_{pu}-h_{pd})}{0.184N_p}} \right)^{10/9} \right] & \text{if } \sqrt{2(h_{pu}-h_{pd})} < U_{ps} \\[4mm] \left[\left(\dfrac{\rho_{pd}D_{pd}}{\mu_{pd}} \right)^{0.1} \dfrac{U_{ps}^{1.1}}{\sqrt{0.184N_p}} \right] & \text{otherwise} \end{cases} \qquad (25.31)$$

where the letter p which appears in the subscripts may be replaced by i for the inlet port and o for the outlet port. Moreover, u and d signify upstream and downstream, respectively. The variable h_{pu} is the specific stagnation enthalpy on the upstream side of a port and h_{pd} is the specific enthalpy on its downstream side. The downstream density, viscosity and hydraulic diameter are given, respectively, as ρ_{pd}, μ_{pd} and D_{pd}. In the above equation, h_{pd} and ρ_{pd} are calculated by considering an isentropic expansion process from the upstream to the downstream; and U_{ps} is the speed of sound on the downstream side of the port. The coefficient N_p is selected based on the geometry of the port. Equation 25.31 is used for both the single-phase and two-phase flow situations. For a two-phase flow, however, μ_{pd} may be calculated in accordance with the McAdams formula:

$$\frac{1}{\mu_{pd}} = \frac{x}{\mu_{vap}} + \frac{1-x}{\mu_{liq}} \qquad (25.32)$$

where x is the dryness fraction and the subscripts *vap* and *liq* refer to vapour and liquid, respectively. Also, for a two-phase flow, the speed of sound may be calculated based on the assumption that the fluid is in equilibrium, over pressure and temperature, as detailed in the excellent paper by Lund and Flätten (2010).

As suggested by Equation 25.31 the use of loss coefficient, while being a reasonable approximation, makes it possible to calculate flow velocities in a non-iterative manner. Besides, the use of enthalpy difference in the velocity equation instead of pressure difference eliminates the need to calculate the expansion factor. This is useful for situations in which a two-phase flow may take place during the expansion process. In the following section, the flow velocity on the downstream side of a port, U_p, will be used in the continuity and energy equations to obtain a differential thermodynamic model for the system. In that context, U_p will be referred to as U_i and U_o to associate the velocity with either the inlet or outlet ports, respectively.

Thermodynamic Model

The flow rate, dm_c/dt, at which the mass inside a chamber varies in relation to density and available volume, V_c, can be obtained from the following equation:

$$\frac{dm_c}{dt} = \omega \left[V_c \left(\frac{d\rho_c}{d\theta} \right) + 4\rho_c bL^2 H_r \sin\theta \right] \qquad (25.33)$$

where the subscript c signifies the chamber fluid; m_c and ρ_c are the mass and density of this fluid, respectively, at any crank angle θ. In the model presented here, dm_c/dt denotes the derivative with respect to time and ω is a constant rotor velocity given in rad/s. The change occurring to the chamber mass is created by the flow through the inlet and outlet ports. As such, the continuity equation can be written for the chamber which falls below the rotor in the following form:

$$\frac{d\rho_c^B}{d\theta} = \left(\frac{1}{\omega V_c^B}\right)\left\{A_i^B\rho_{id}^B U_i^B - A_o^B\rho_{od}^B U_o^B - \dot{m}_s^B \pm \dot{m}_{ap} - 4\rho_c^B b\omega L^2 H_r \sin\theta\right\} \tag{25.34}$$

where the superscript, B, denotes 'below rotor'. The inlet and outlet port areas, A_i and A_o, respectively, are calculated as described above. The density and velocity on the downstream side of the inlet port are given, respectively, as ρ_{id} and U_i. The corresponding quantities on the downstream side of the outlet port are given as ρ_{od} and U_o, respectively. Both U_i and U_o are calculated as described by Equation 25.31. In the above equation, $\dot{m}_s$ and $\dot{m}_{ap}$, respectively, signify leakage past the side and apex seals. The $\pm$ sign is used in front of $\dot{m}_{ap}$ to highlight the fact that the direction of leakage, from one chamber to another, is determined by the instantaneous value of pressure in each chamber. The corresponding continuity equation for the working chamber which falls above the rotor is written as follows:

$$\frac{d\rho_c^A}{d\theta} = \left(\frac{1}{\omega V_c^A}\right)\left\{A_i^A\rho_{id}^A U_i^A - A_o^A\rho_{od}^A U_o^A - \dot{m}_s^A \mp \dot{m}_{ap} + 4\rho_c^A b\omega L^2 H_r \sin\theta\right\} \tag{25.35}$$

Taking the energy transfer to and from a working chamber fluid as an adiabatic process will result in the following equation:

$$\frac{dH_i}{dt} - \frac{dH_o}{dt} = \frac{d(m_c e_c)}{dt} + \frac{P_c\, dV_c}{dt} \tag{25.36}$$

where H_i and H_o are the total enthalpies moving in and out of a working chamber, respectively; and e_c is the specific internal energy available in the working chamber. The term $d(m_c e_c)/dt$ can be substituted by $(d(m_c e_c - P_c\, dV_c))/dt$ and $m_c\, dh_c/dt$ which in turn is given by $(m_c T_c\, ds_c/dt) + (V_c\, dP_c/dt)$, where T_c and s_c are, respectively, the temperature and entropy inside the working chamber and h_c is the specific enthalpy of the chamber fluid. For the working chamber below the rotor, it should now be possible to manipulate Equation 25.36 into the following form:

$$\frac{ds_c^B}{d\theta} = \frac{1}{\omega\rho_c^B T_c^B V_c^B}\left(A_i^B U_i^B \rho_{id}^B\left(h_{iu} - h_c^B\right) - A_o^B U_o^B \rho_{od}^B\left(h_{ou}^B - h_c^B\right)\right.$$
$$\left. - \dot{m}_s^B\left(h_{su}^B - h_c^B\right) \pm \dot{m}_{ap}\left(h_{apu} - h_c^B\right)\right) \tag{25.37}$$

where h_{iu} and h_{ou} refer to the specific enthalpies on the upstream side of the inlet and outlet ports, respectively. Moreover, h_{su} and h_{tu} signify the specific enthalpies on the upstream sides of the side and apex seals, respectively. The corresponding equation for the working chamber which falls above the rotor is given as follows:

$$\frac{ds_c^A}{d\theta} = \frac{1}{\omega\rho_c^A T_c^A V_c^A}\left(A_i^A U_i^A \rho_{id}^A\left(h_{iu} - h_c^A\right) - A_o^A U_o^A \rho_{od}^A\left(h_{ou}^A - h_c^A\right)\right.$$
$$\left. - \dot{m}_s^A\left(h_{su}^A - h_c^A\right) \mp \dot{m}_{ap}\left(h_{apu} - h_c^A\right)\right) \tag{25.38}$$

Now the four simultaneous differential equations (25.34), (25.35), (25.37) and (25.38) can be solved numerically to find instantaneous values for s_c^A, s_c^B, ρ_c^A and ρ_c^B at the corresponding values for the crank angle, θ. At every step, the corresponding working chamber pressure can be obtained using the function, Pressure(s_c, ρ_c), offered by REFPROP 9.0 (Lemmon et al., 2010). This function produces accurate results for both single- and two-phase flows. The working chamber temperature and specific enthalpy can, respectively, be obtained by Temperature(P_c, s_c) and Enthalpy(P_c, s_c). While all the given conditions in the inlet manifold are taken to be constant, only the pressure, P_o, is assumed to be maintained constant in the outlet manifold. The entropy in this manifold, s_o, can be obtained by mixing the flows blowing out of the two working chambers. As such, the temperature is calculated by Temperature(P_o, s_o) and the density is obtained from Density(P_o, s_o). The same approach has been applied to the conditions inside the rotor side cavity which houses the driving mechanism.

At every iteration, the instantaneous value of shaft torque, τ_{sh}, which results from the fluid pressure has been given by Sultan (2005b) as follows:

$$\tau_{sh} = 4bL^2 H_r \left(P_c^B - P_c^A \right) \sin\theta \tag{25.39}$$

The iterative approach proceeds by calculating the thermodynamic model at small intervals in the range $\theta_{li} \le \theta \le \theta_{li} + \pi$, where θ_{li} defines the angular position of the inlet port leading edge. The values, $P_c^B(\theta_{li})$ and $\rho_c^B(\theta_{li})$, assumed for the pressure and density below the rotor, respectively, at the start of the cycle are compared with the corresponding values, $P_c^A(\theta_{li} + \pi)$ and $\rho_c^A(\theta_{li} + \pi)$, calculated above the rotor. Moreover, the values, $P_c^A(\theta_{li})$ and $\rho_c^A(\theta_{li})$, assumed for the pressure and density above the rotor, respectively, at the start of the cycle are compared with the corresponding values, $P_c^B(\theta_{li} + \pi)$ and $\rho_c^B(\theta_{li} + \pi)$, below the rotor using the following dimensionless error expression:

$$\sigma = \left(\left(\frac{P_c^B(\theta_{li}) - P_c^A(\theta_{li} + \pi)}{\overline{P}_{c1}} \right)^2 + \left(\frac{\rho_c^B(\theta_{li}) - \rho_c^A(\theta_{li} + \pi)}{\overline{\rho}_{c1}} \right)^2 \right.$$
$$\left. + \left(\frac{P_c^A(\theta_{li}) - P_c^B(\theta_{li} + \pi)}{\overline{P}_{c2}} \right)^2 + \left(\frac{\rho_c^A(\theta_{li}) - \rho_c^B(\theta_{li} + \pi)}{\overline{\rho}_{c2}} \right)^2 \right)^{1/2} \tag{25.40}$$

where $\overline{P}_{c1}$, $\overline{P}_{c2}$, $\overline{\rho}_{c1}$ and $\overline{\rho}_{c2}$ are, respectively, the pressure and density values used for error calculations. These values as obtained as follows:

$$\overline{P}_{c1} = \frac{P_c^B(\theta_{li}) + P_c^A(\theta_{li} + \pi)}{2} \tag{25.41}$$

$$\overline{\rho}_{c1} = \frac{\rho_c^B(\theta_{li}) + \rho_c^A(\theta_{li} + \pi)}{2} \tag{25.42}$$

$$\overline{P}_{c2} = \frac{P_c^A(\theta_{li}) + P_c^B(\theta_{li} + \pi)}{2} \tag{25.43}$$

$$\overline{\rho}_{c2} = \frac{\rho_c^A(\theta_{li}) + \rho_c^B(\theta_{li} + \pi)}{2} \tag{25.44}$$

If the outcome of Equation 25.40 is larger than a small pre-defined value, $P_c^B(\theta_{li})$, $\rho_c^B(\theta_{li})$, $P_c^A(\theta_{li})$ and $\rho_c^A(\theta_{li})$ are set, respectively, equal to $P_c^A(\theta_{li} + \pi)$, $\rho_c^A(\theta_{li} + \pi)$, $P_c^B(\theta_{li} + \pi)$ and $\rho_c^B(\theta_{li} + \pi)$

to repeat the procedure again over the π range of θ. This is iterated until the calculated error, σ, falls within an acceptable range to reflect the cyclical nature of the thermodynamic process. Once convergence has been achieved, the total energy, E_{cyc}, per cycle is calculated using the numerical integration as follows:

$$E_{cyc} = 2\delta\theta\left(\sum_{n=0}^{N}(\tau_{sh})_n - \frac{(\tau_{sh})_N + (\tau_{sh})_0}{2}\right) \quad (25.45)$$

where $\delta\theta$ is the size of the angular interval, n is a counter for successive points on the curve being integrated and N is the total number of intervals on the curve. In a similar manner, the total mass flow through the machine in one cycle, M_{cyc}, is calculated by the following expression:

$$M_{cyc} = 2\frac{\delta\theta}{\omega}\left\{\sum_{n=0}^{N}\left(A_i^B\rho_{id}^B U_i^B\right)_n + \left(A_i^A\rho_{id}^A U_i^A\right)_n\right.$$
$$\left. - \frac{\left(A_i^B\rho_{id}^B U_i^B\right)_N + \left(A_i^B\rho_{id}^B U_i^B\right)_0}{2} - \frac{\left(A_i^A\rho_{id}^A U_i^A\right)_N + \left(A_i^A\rho_{id}^A U_i^A\right)_0}{2}\right\} \quad (25.46)$$

It is worth noting here that the expressions in Equations 25.45 and 25.46 feature a multiplication by 2 in order to account for the fact that a limaçon machine does produce two cycles for every full shaft rotation.

Case Study 1: Ported Expander

For this study, an expander is to be designed to utilise a supply of R-245fa available, as a working fluid, at an absolute pressure of 10 bar and a temperature of 150°C. The expander, which is not equipped with an inlet control valve, will be running at a speed of 1000 rpm and will expand the working fluid to 1 bar maintained in the outlet manifold. The geometric particulars of the expander are given as follows:

$$L = 46.45\,\text{mm}, b = 0.172, L_i = 13.35\,\text{mm}, L_o = 21.48\,\text{mm}$$
$$\theta_{li} = -24.96°, \Delta\theta_i = 19.1°, \theta_{lo} = 140° \text{ and } \Delta\theta_o = 35.78°$$

The performance metrics produced are as follows:

$$\eta_i = 32.53\%, \ \psi = 0.92 \quad \text{and} \quad P_{ind} = 4.03\,\text{kW} \text{ at a mass flow rate of } 13.95\,\text{kg/min}$$

Figures 25.14 and 25.15 represent the PV diagram and the pressure–angle diagram, respectively, for the expander in this study. The considerable pressure fall which starts, as shown in Figure 25.15, at $\theta = 140°$ signifies exposure to the discharge port.

Case Study 2: Expander with a Control Valve

It is now required to improve the performance of the expander designed in case study 1. For this purpose, a cam-operated inlet control valve, which has been pre-manufactured to open and close periodically every 90° of the crank rotation, is being considered for the expander. Such a valve increases the isentropic efficiency by expanding the fluid in the working chamber instead of being allowed to escape at high pressure through the discharge port. The geometric particulars of the expander are given as follows:

$$L = 66.14\,\text{mm}, b = 0.064, L_i = 32.9\,\text{mm}, L_o = 38.53\,\text{mm}, \theta_{li} = -21.67°$$
$$\Delta\theta_i = 7.4°, \theta_{lo} = 146.47° \text{ and } \Delta\theta_o = 13.27°$$

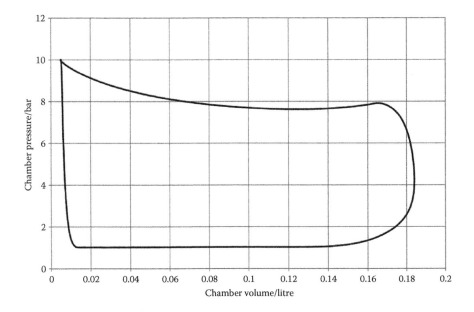

FIGURE 25.14   PV diagram (case study 1). (Adapted from Sultan, I. A. 2012. *Applied Thermal Engineering*, 188–197.)

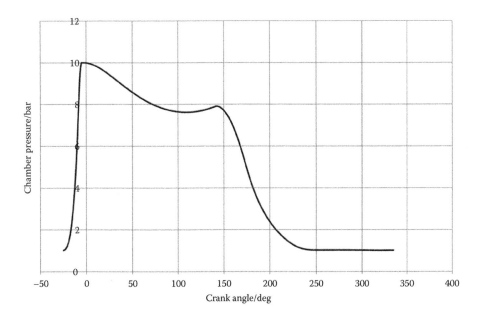

FIGURE 25.15   Variation of chamber pressure with crank angle (case study 1). (Adapted from Sultan, I. A. 2012. *Applied Thermal Engineering*, 188–197.)

The resulting performance metrics are as follows:

$$\eta_i = 58.67\%, \quad \psi = 0.96 \quad \text{and} \quad P_{ind} = 4.10\,\text{kW} \quad \text{at a mass flow rate of } 7.87\,\text{kg/min}$$

Figures 25.16 and 25.17 represent the PV diagram and the pressure–angle diagram, respectively, for the new expander.

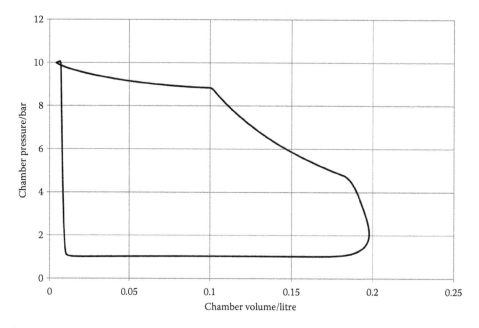

FIGURE 25.16   PV diagram (case study 2). (Adapted from Sultan, I. A. 2012. *Applied Thermal Engineering*, 188–197.)

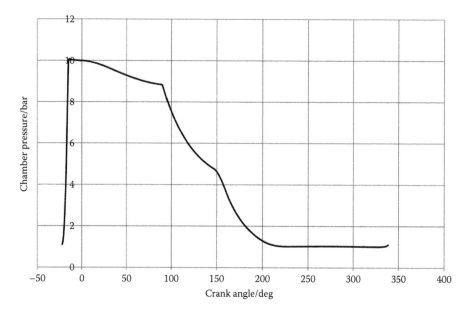

FIGURE 25.17   Variation of chamber pressure with crank angle (case study 2). (Adapted from Sultan, I. A. 2012. *Applied Thermal Engineering*, 188–197.)

CONCLUSION

Improving the thermal efficiency of industrial processes will lead to a reduction in fossil fuel consumption, which will in turn result in reduction in emission, improved environmental conditions and sustainable use of resources. One method to improve process efficiency is by capturing waste heat to utilise it for power generation. The present chapter considered different methods to achieve that outcome and shed some light on the structures of thermodynamic plants which can be utilised for the purpose. It has been pointed out that the gas expander is the most critical component of a power generation plant. Gas expanders are usually of the positive displacement type, which is meant to handle low flow rates at high pressures. The performance of the whole plant is limited by what the expander can offer.

The limaçon technology offers considerable potential for gas expanders as has been described in the chapter with the aid of geometric and thermodynamic models. The results suggest that this technology readily lends itself to future optimisation endeavours. For future work in this area, various rotor-housing configurations of the limaçon expander will be explored and described mathematically with the purpose of conducting optimisation studies and creating design methods for practitioners.

REFERENCES

Acar, H. I. 1995. Second law analysis of the reheat–regenerative Rankine cycle. *Elsevier Science*, 38(7): 647–657.

Bombarda, P., C. M. Invernizzi, and C. Pietra. 2010. Heat recovery from diesel engines: A thermodynamic comparison between Kalina and ORC cycles. *Applied Thermal Egineering*, 30(2–3): 212–219.

Brooks, G. 2012. *theconversation.com*. 25 May. http://theconversation.com/the-trouble-withaluminium-7245.

Chen, H., D. Y. Goswami, and E. K. Stefanakos. 2010. A review of thermodynamic cycles and working fluids for the conversion of low-grade heat. *Renewable and Sustainable Energy Reviews*, 14(9): 3059–3067.

DiPippo, R. 2004. Second Law assessment of binary plants generating power from low-temperature geothermal fluids. *Geothermics*, 33(5): 565–586.

DiPippo, R. 2007. Ideal thermal efficiency for geothermal binary plants. *Geothermics*, 36(3): 276–285.

Fischer, J. 2011. Comparison of trilateral cycles and organic Rankine cycles. *Energy*, 36(10): 6208–6219.

Fukuta, M. and T. Yanagisawa. 2009. Performance of vane-type CO_2 expander and characteristic of transcritical expansion process. *HVAC&R Research*, 15(4): 711–727.

Gang, P., L. Jing, and J. Jie. 2010. Analysis of low temperature solar thermal electric generation using regenerative organic Rankine cycle. *Applied Thermal Engineering*, 30(8–9): 998–1004.

Gao, H., C. Liu, C. He, X. Xu, S. Wu, and Y. Li. 2012. Performance analysis and working fluid selection of a supercritical organic Rankine cycle for low grade water heat recovery. *Energies*, 5(9): 3233–3247.

Gimstedt, L. J. and K. Karlsson. 1979. An arrangement for gas expanders. US Patent WO1979000444 A1. 26 July.

Hung, T.-C. 2001. Waste heat recovery of organic Rankine cycle using dry fluids. *Energy Conversion and Management*, 42(5): 539–553.

Hung, T. C., T. Y. Shai, and S. K. Wang. 1997. A review of organic Rankine cycle (ORCs) for the recovery of low grade waste heat. *Energy*, 22(7): 661–667.

Hung, T. C., S. K. Wang, C. H. Kuo, B. S. Pei, and K. F. Tsai. 2009. A study of organic working fluids on system efficiency of an ORC using low-grade energy sources. *Energy*, 35(3): 1403–1411.

Kim, Y. M., C. G. Kim, and D. Favrat. 2012. Transcritical or supercritical CO_2 cycles using both low- and high temperature heat sources. *Energy*, 43(1): 402–415.

Lawrence Berkeley National Laboratory. 2007. *Improving Process Heating System Performance*. Colorado: U.S. Department of Energy.

Lemmon, E. W., M. L. Huber, and M. O. McLinden. 2010. *NIST Reference Fluid Thermodynamic and Transport Properties—REFPROP Version 9.0*. National Institute of Standards and Technology, Gaithersburg, Maryland: U.S. Department of Commerce – Technology Administration – National Institute of Standards and Technology.

Lemort, V., J. Lebrun, and S. Quoilin. 2010. Experimental study and modeling of an organic Rankine cycle using scroll expander. *Applied Energy*, 87(4): 1260–1268.

Lund, H. and T. Flätten. 2010. Equilibrium conditions and sound velocities in two-phase flows. *SIAM Annual Meeting (AN10)*. Pittsburgh, Pennsylvania, pp. 1–15.

Mago, P. J., L. M. Chamra, K. Srinivasn, and C. Sosamyaji. 2008. An examination of regenerative organic Rankine cycles using dry fluids. *Applied Thermal Engineering*, 28(8–9): 998–1007.

Maizza, V. and A. Maizza. 2000. Unconventional working fluids in organic Rankine-cycles for waste energy recovery systems. *Applied Thermal Engineering*, 21(3): 381–390.

Peterson, R. B., H. Wang, and T. Herron. 2008. Performance of a small-scale regenerative Rankine power cycle employing a scroll expander. *Journal of Power and Energy*, 222(3): 217–282.

Roy, J.P. and A. Misra. 2012. Parametric optimization and performance analysis of a regenerative organic Rankine cycle using R-123 for waste heat recovery. *Energy*, 39(1): 227–235.

Shengjun, Z., W. Huaixin, and G. Tao. 2011. Performance comparison and parametric optimisation of subcritical organic Rankine cycle (ORC) and transcritical power cycle systems for low temperature geothermal power generation. *Applied Energy*, 88(8): 2740–2754.

Smith, I. K., N. Stosic, and A. Kovacevic. 2005. Screw expanders increase output and decrease the cost of geothermal binary power plant systems. *Transactions of Geothermal Resource Council*, 29: 25–28.

Steffen, M., M. Löffler, and K. Schaber. 2013. Efficiency of a new triangle cycle with flash evaporation in a piston engine. *Energy*, 57: 295–307.

Subiantoro, A. and K. T. Ooi. 2010. Design analysis of the novel revolving vane expander in a transcritical carbon dioxide refrigeration system. *Science Direct*, 33(4): 675–685.

Sultan, I. A. 2005a. Profiling rotors for limacon-to-limacon compression–expansion machines. *Journal of Mechanical Design*, 128(4): 787–793.

Sultan, I. A. 2005b. The limaçon of Pascal: Mechanical generation and utilization for fluid processing. *The Journal of Mechanical Engineering Science*, 219(8): 813–822.

Sultan, I. A. 2007. A surrogate model for interference prevention in the limaçon-to-limaçon machines. *Engineering Computation*, 22(4): 437–449.

Sultan, I. A. 2008. A geometric design model for the Circolimacon positive displacement machines. *Journal of Mechanical Design*, 130(5): 1–8.

Sultan, I. A. 2012. Optimum design of limaçon gas expanders based on thermodynamic performance. *Applied Thermal Engineering*, 39: 188–197.

Sultan, I. A. and C. G. Schaller. 2011. Optimum positioning of ports in the Limaçon gas expanders. *Journal of Engineering for Gas Turbines and Power* (The American Society of Mechanical Engineers), 133(10): 1–11.

Tchanche, B. F., M. Pétrissans, and G. Papadakis. 2014. Heat resources and organic Rankine cycle machines. *Renewable and Sustainable Energy Reviews*, 39: 1185–1199.

Tuo, H. 2012. Thermal-economic analysis of a transcritical Rankine power cycle with reheat enhancement for a low-grade heat source. *International Journal of Energy Research*, 37(8): 857–867.

U.S. Department of Energy. 2008. *Waste Heat Recovery: Technology and Opportunities in U.S. Industry*. U.S. Department of Energy. http://www1.eere.energy.gov/manufacturing/intensiveprocesses/pdfs/waste_heat_recovery.pdf.

Vanslambrouck, B., I. Vankeirsbilck, S. Gusev, and M. De Paepe. 2011. Turn waste heat into electricity by using an organic Rankine. *2nd European Conference on Polygeneration*. Tarragona, Spain, pp. 1–14.

Wang, H., R. Peterson, K. Harada, E. Miller, R. Ingram-Goble, L. Fisher, J. Yih, and C. Ward. 2010. Performance of a combined organic Rankine cycle and vapor compression cycle for heat activated cooling. *Elsevier Energy*, 36(1): 447–458.

Wei, D., X. Lu, Z. Lu, and J. Gu. 2006. Performance analysis and optimization of organic Rankine cycle (ORC) for waste heat recovery. *Energy Conversion and Management*, 48(4): 1113–1119.

Wheildon, W. M. 1896. Rotary engine. United States Patent US553086 A. 14 January.

Xiaojun, G., L. Liansheng, Z. Yuanyang, and S. Pengcheng. 2004. Research on a scroll expander used for recovering work in a fuel cell. *Journal of Thermodynamics*, 7(1): 1–8.

Zamfirescu, C. and I. Dincer. 2008. Thermodynamic analysis of a novel ammonia–water trilateral Rankine cycle. *Thermochimica Acta*, 477(1–2): 7–15.

Zhang, B., X. Peng, Z. He, Z. Xing, and P. Shu. 2007. Development of a double acting free piston expander for power recovery in transcritical CO_2 cycle. *Applied Thermal Engineering*, 27(8–9): 1629–1636.

26 Urban Waste (Municipal Solid Waste—MSW) to Energy

Moshfiqur Rahman, Deepak Pudasainee, and Rajender Gupta

CONTENTS

ABSTRACT

The level of municipal solid waste (MSW) generation has been increasing with increasing population and urbanization. Consequently, environmental concerns related to it have increased. MSW has been utilized as a valuable renewable energy sources due to advancements in technology and awareness. In this chapter, the global trend of MSW generation, its characterization, handling and management practices, and thermal and nonthermal conversions of MSW to energy have been presented. Further, the advantages of an integrated waste management approach "reduce, reuse, and recycle" and its proper practice has been discussed. The chapter also discusses environmental issues during landfilling and thermal conversion.

Environmental issues during MSW handling, processing, transportation, disposal, and treatment can also be reduced by adapting proper technology and mitigation measures. Waste-to-energy (WtE) technologies are primarily aimed at recovering energy from waste and increasing the benefits, however, these technologies also involve environmental concerns to a certain extent. Control devices play a crucial role in controlling pollutants generated in WtE facilities and reduce the associated hazards to human beings and ecosystems. In addition, several best practice measures and control strategies have been presented that can be adopted to mitigate the release of pollutants during its thermal conversion to energy.

INTRODUCTION

The term "urban waste" is referred to as municipal solid waste (MSW), which is commonly known as trash or garbage in the United States and Canada. In European countries, as in the United Kingdom, "urban waste" is referred to as rubbish or refuse. Urban waste consists of packaging boxes (cardboard), grass clippings (yard waste), broken or unusable furniture, clothing, glass bottles, food waste, and newspapers. Small appliances, metals, aluminum, paint, electronics, and batteries are also under the same category but in most developed countries these materials are separated and recycled. There are other types of waste such as industrial waste, construction and demolition waste, and commercial waste. Hospital waste is treated in a separate category. However, in most developing countries, different forms of waste are often mixed up with urban waste.

The world's cities generated around 1.3 billion tons of MSW/year in 2010; this is predicted to rise to 2.2 billion tonnes by 2025. MSW generation in 2010 was 1.2 kg per capita per day, nearly half of which came from Organization for Economic Cooperation and Development (OECD) countries (World Bank, 2012). China is the number one waste producer (430,000 tons/day in 2009) as a whole and is facing major challenges in terms of waste management (World Bank, 2014). India produces 188,500 tons/day of MSW (Waste Management World, 2015). Per capita waste production per year for China and India is in the order of 120 and 60 kg, respectively. In comparison, Canada produces 777 kg per capita, making it the highest MSW producer per capita per annum (Stantec, 2011).

The current world population is 7.3 billion, and it is expected to be 8.9 billion by 2050. With the rise in population, the demand for resources increases as does waste generation. To meet the materials demand of society, cost-effective innovative waste management approaches are needed. In this context, the concept of "reuse and recycle" must be promoted, as for example, the recovered materials from waste can be used for remanufacturing processes. The MSW remaining after reuse and recycle can be thermally converted to power or chemicals which is a very promising option to mitigate greenhouse gas (GHG) emissions. Generating power or producing chemicals from MSW may be more expensive compared to production from conventional raw materials. However, reducing the landfill area and reducing GHG from methane emissions make thermal treatment an attractive and feasible option. According to a report from the U.S. Environmental Protection Agency (EPA) (2011), 10% of MSW was converted to power by incineration with about 15% of efficiency (U.S. EPA, 2011). One of the world's major MSW gasification plants producing biofuels was officially started on June 4, 2014 in Edmonton, Alberta, Canada. This plant is built through collaboration of the city of Edmonton and Enerkem Alberta Biofuel (EAB) to address waste disposal challenges and turn MSW into biofuels like ethanol and methanol. This facility will be able to convert 100,000 tons of MSW into 38 million liters of methanol and/or ethanol annually and will also help to reduce GHG emissions in Alberta.

GLOBAL MSW GENERATION

Waste generation, worldwide, is increasing continuously. The rate of waste generation depends on a number of factors, including economic development, degree of industrialization, human habits, and region. Countries with more economic development and urbanization also have a higher production of MSW. In general, global waste generation is directly proportional to economic growth and urbanization. This also depends on human lifestyle, income level, population, economic developmental status, industrialization resources, consumption, etc. (World Bank, 2012). Urban communities produce almost twice as much MSW as rural communities.

The generation of waste varies from country to country and region to region as a function of prosperity. In 2005, OECD countries produced almost half of the world's waste (44%), while African and South Asian countries produced the least. OECD countries generate 572 million tonnes of solid waste per year with an average of 2.2 kg per capita per day.

Figure 26.1 illustrates global waste generation for four income levels, classified according to the World Bank's estimate (2005) of the gross national income (GNI) per capita. High-income countries produced the most solid waste per capita, while low-income countries produced the least solid waste per capita. The total waste generated by lower-middle-income countries is higher than that of the waste generated by upper-middle-income countries. This result was probably influenced by China, which has a large population that falls into the lower-middle-income group. The other possible explanation is that upper-middle-income populations might be more cautious and possess good habits of waste management and utilization.

Table 26.1 shows estimates of solid waste generation by urban cities for the year 2005 and a projection for 2025, based on the current trends of population growth and the countries' income levels. It is evident that waste generation is directly proportional to the population of urban areas (cities) and relates to income level (World Bank, 2012).

In Indian cities, around 90% of MSW is disposed of unscientifically in open dumps and landfills (Sharholy et al., 2008). In other developing countries, such as Bangladesh, Nepal, and Bhutan, open dumping in landfills is common and there is no formal recycling of waste materials. However,

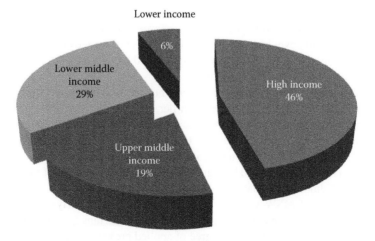

FIGURE 26.1 Solid waste generation by different income levels. (Adapted from World Bank. 2012. Hoornweg D., and P. Bhada-Tata. What a waste: A global review of solid waste management, March 2012, No. 15.)

TABLE 26.1

Solid Waste Generation in 2005 and Projection for 2025 by Income Level

	World Bank Data (2005)			Projection for 2025			
		Urban Waste Generation		Projected Population		Projected Urban Waste Generation	
Region	Total Urban Population (Millions)	Per Capita (kg/capita/day)	Total (ktons/day)	Total Population (Millions)	Urban Population (Millions)	Per Capita (kg/capita/day)	Total (ktons/day)
Lower income	343	0.60	205	1637	676	0.86	585
Lower middle income	1293	0.78	1012	4010	2,080	1.30	2618
Upper middle income	572	1.16	666	888	619	1.60	987
High income	774	2.13	1650	1112	912	2.10	1880
Total	2982	1.19	3532	7647	4287	1.40	6070

Source: World Bank. 2012. Hoornweg D., and P. Bhada-Tata. What a waste: A global review of solid waste management. March 2012, No. 15.

Note: Countries are classified into four income levels according to World Bank estimates of 2005 GNI per capita. High: $10,726 or above; upper middle: $3,466–10,725; lower middle: $876–3,465; and lower: $875 or less.

because of a lack of resources in general, there is large-scale reuse of material in an informal way. Urban waste can be considered a renewable energy source as it contains about 50%–60% organic matter. Urban waste materials can be turned into valuable resources and energy through proper waste management practices (Tanigaki et al., 2015).

Characterization of MSW

Physical Properties

Some of the important physical properties of MSW include density, moisture content, permeability, and particle size distribution.

The density varies with the geographic location, season, and length of storage time. MSW in a compacted landfill without cover has a specific density within the range of 440–740 kg/m^3, whereas loose MSW without any processing has a density of 90–150 kg/m^3. Table 26.2 shows the density of the major MSW components (www.pwut.ac.ir).

The characteristics of the MSW components are not uniform; they vary a lot, since different materials are mixed with MSW. As mentioned earlier, the rate of MSW generation depends on a number of factors, such as population growth, socioeconomic development and economic prosperity, and industrialization, all of which affect waste composition and characteristics.

The inorganic matter in MSW has significant quantities of heavy/trace metals such as Cr, Ni, Zn, Hg, As, and Se (Paradelo et al., 2011). Figure 26.2 presents an example of major components of MSW, sourced from Edmonton, Canada.

The moisture content of MSW varies greatly and depends completely on the source of waste materials. Field moisture absorption capacity is critically important for measuring the leachable components in landfills. It varies with the degree of applied pressure and the rate of decomposition of waste over a period of time, but typical values for uncompacted waste generated from residential and commercial sources are in the range of 50%–60%.

TABLE 26.2

Specific Weight of the Major MSW Components

Components	Density (kg/m³)	
	Range	Typical
Food wastes	130–480	290
Paper	40–130	89
Yard wastes	65–225	100
Plastics	40–130	64
Glass	160–480	194
Aluminum	65–240	160
Cans	50–160	89

Source: http://www.pwut.ac.ir/FA/Adjutancies/Adj2/Office4/
Risk/khadamat/pdf/133.pdf

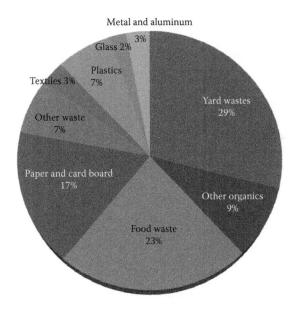

FIGURE 26.2 Major components of MSW after recycling in Edmonton, Alberta, Canada. (Adapted from www.edmonton.ca/waste. The Edmonton sustainability papers. May 2010. Discussion paper 10—Sustainable waste management. Accessed May 2015.)

The permeability of condensed waste directs the movement of liquids and gases generated in the landfill. It depends on the pore size distribution, surface area, and porosity of the deposited MSW materials.

The particle size distribution of MSW materials is very important for separating and recovering valuable materials, especially during mechanical separation with a magnetic separator and trammel screens.

Chemical Properties of Individual Components

Table 26.3 presents the typical proximate analysis data of residential MSW and its segregated materials. It is very important to evaluate the chemical properties of MSW in order to determine its

TABLE 26.3

Typical Proximate Analysis (wt%) of Residential MSW and Its Components

Waste	Moisture	Volatiles	Ash	Fixed Carbon
Residential MSW	21.0	52.0	20.0	7.0
Mixed food	70	21.4	5.0	3.6
Mixed paper	10.2	75.9	5.4	8.4
Mixed plastics	0.2	95.8	2.0	2.0
Yard wastes	60.0	42.3	0.4	7.3
Glass	2.0	0	96–99	0

Source: http://www.pwut.ac.ir/FA/Adjutancies/Adj2/Office4/Risk/khadamat/pdf/133.pdf

proper utilization and energy recovery, which includes proximate analysis, ultimate analysis, ash fusion temperature, and energy content or heating value (HV).

Proximate analysis provides information about the MSW materials, including the percentage of volatile ash and fixed carbon contents and the percentage of moisture. Proximate analysis results vary significantly and depend on the type of components present in the mixed material.

MSW contains a significant amount of ash. While recovering energy from MSW, ash may cause deposits and slag formations in the reactor. Ash fusion temperature affects ash behavior and combustion and gasification performances; if the temperature is low it can lead to slagging and fouling in industrial boilers or reactors.

An ultimate analysis is the percentage of elemental composition (carbon, hydrogen, oxygen, nitrogen, and sulfur) and ash present in the fuel. Elemental composition varies greatly, depending on the developmental status of the society and type of material. Figure 26.3 presents the chemical composition of typical MSW. These elemental compositions are used to characterize the chemical composition of the organic matter in MSW. They are also used to define the proper mix of waste materials to achieve suitable C/N ratios for biological conversion processes as well as to determine the H/C ratios in the fuel.

Energy content or HV is the amount of heat released by a complete combustion of fuel and it is measured as a unit of energy per unit mass or volume of substance (kcal/kg or kJ/kg). The heat of combustion of fuels is expressed as higher and lower heating values (HHV and LHV). The HHV

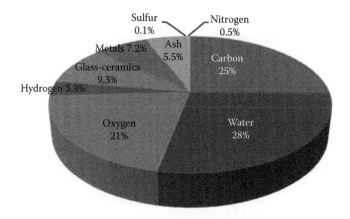

FIGURE 26.3   Chemical composition of typical MSW generated in Iran. (Adapted from http://www.pwut. ac.ir/FA/Adjutancies/Adj2/Office4/Risk/khadamat/pdf/133.pdf.)

includes the heat of condensation of water formed during combustion. This can also be calculated using the respective fuel's ultimate and proximate analysis data (Parikh et al., 2005).

LHV assumes that the latent heat of water vaporization in the fuel and the reaction products are not recovered. It is useful to compare fuels where the combustion products' condensation is impossible, or heat at a temperature below 150°C (302°F) cannot be used. When the LHV is determined by cooling at 150°C, the heat of reaction is only partially recovered. The LHV is also defined as the net calorific value and can be determined by subtracting the heat of vaporization of the water content (generated during fuel combustion) from the HHV (Meriçboyu et al., 1998). Usually, the same types of fuels are compared with respect to their HHV, whereas different types of fuels are usually compared according to their LHV. Since the hydrogen contents of different types of fuels are different from each other, it is necessary to determine the hydrogen content of the fuel when calculating the LHV. For this reason, it is better to express the LHV of MSW. The LHV of different components in MSW has been reported as paper: 7.07 and 9.79; wood: 10.55 and 9.74; food: 0.93 and 1.82; and plastics: 23.31 and 29.24 in the United States and South Korea, respectively (U.S. DOE, 2007; Johnke, 2000, referred from Ryn and Shin, 2013).

CURRENT STATUS OF MSW HANDLING

The management of solid waste is extremely important and one of the most challenging issues in larger cities. Sustainable waste management can protect human health, property, and the environment, and preserve valuable natural resources. A sustainable waste management system protects future generations from environmental degradation and associated financial burdens. Waste management is a critical issue for growing municipalities worldwide. Improper waste management can cause:

- Environmental and health-related problems
- GHGs emission and climate change, which can result when certain forms of waste decompose
- A burden for future generations, as it can be costly to manage waste disposal sites

A solid waste management system includes the collection, storage, transportation, processing, and final disposal or treatment of waste. Not all municipalities are well equipped to handle these issues effectively, especially in developing countries, where shortcomings include limited knowledge of technology, limited manpower and, most important, inadequate funding. An integrated solid waste management approach covers the whole process of generation, reuse, recycling, composting, WtE and landfilling of residual waste, with a view toward reducing waste generation and safely disposing of waste to mitigate the impact on human health and the environment (U.S. EPA, 2002; Seo, 2014). Therefore, managing waste is essential for maintaining a sustainable world. Waste management is not only how we dispose garbage into the landfill, and how much or what we recycle, it is also how we can reduce waste production. The "3R" approach—"reduce, reuse, and recycle" of waste is gaining momentum (Henry, 2006). The success of any long-term sustainable waste management program also depends on public awareness, technical feasibility, and available funds (Reddy, 2011).

There have been several initiatives worldwide from government and private sectors. For example, Edmonton, Alberta, is one of the first major cities in Canada to implement a sustainable integrated waste management system, the City of Edmonton Waste Management Centre (EWMC). The Centre, which promotes reducing waste at the source, operates several collection services; diverts waste from the city landfill through reuse, recycling, and composting; and recovers products and energy from residual waste materials. The city built a waste-to-biofuels plant in collaboration with EAB, which will increase Edmonton's waste diversion rate from 60% to 90% (Enerkem, 2014). It will also reduce landfill usage and associated costs significantly, and help to mitigate GHG emissions and improve air and groundwater quality, promoting integrated solid waste management. The hierarchy

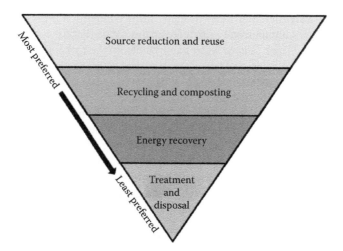

FIGURE 26.4 EPA waste hierarchy pyramid. (Adapted from U.S. EPA. 2011. Inventory of U.S. greenhouse gas emissions and sinks: 1990–2009; EPA 430-R-11-005, April 15, 2011. N.W. Washington, DC: USEPA.)

of a sustainable waste management system, which is widely used as a reference to sustainable solid waste management and disposal, suggests (in pyramid from bottom to top priority): Landfill and open burn < Landfill without CH_4 capture < Modern landfills recovering and flaring CH_4 < Modern landfills recovering and using CH_4 < Waste to energy < Aerobic composting < Anaerobic composting < Recycling < Waste reduction (Subramani et al., 2014).

Since there is no universal waste management approach capable of managing all aspects of waste streams, the U.S. EPA developed a hierarchy ranking the most environmentally friendly strategies for MSW management. It emphasizes the reducing, reusing, and recycling the majority of waste. Reducing the use of materials and reusing them are the most environmentally friendly methods and are also recognized by the hierarchy of the waste management approach. Source reduction begins with reducing the amount of waste generated and reusing materials to prevent them from entering into the waste stream (U.S. EPA, 2011). Figure 26.4 demonstrates the key components of the EPA's Sustainable Materials Management Program, which is trying to reduce the use of certain materials and their environmental impacts over their entire life cycles, starting with source reduction in extraction of natural resources, recycling, composting, energy recovery, treatment, and final disposal.

SOURCE REDUCTION, REUSE, AND RECYCLING

Source reduction is one of the key methods for reducing solid waste generation. This term is also known as "waste prevention," and includes reducing waste generation from houses by composting household waste; reusing industrial by-products for valuable purposes; changing product design and packaging commodities to assist; and aiming to reduce, reuse, and recycle these items before they become MSW. Reusing and recycling electronic goods also plays a significant role in waste reduction and prevention. Reducing and reusing are the most active approaches to prevent waste generation and are the most effective ways to save financial and natural resources. Source reduction can also lower costs and improve productivity by targeting wasteful processes and products. The advantages of overall source reduction are that it: (a) saves natural resources; (b) conserves energy; (c) reduces pollution; (d) reduces the toxicity of waste; and (e) saves money for consumers and businesses.

The practices of reuse have been working well in developed countries. In some developing countries, drinks are now packaged in reusable glass bottles and cans instead of nonreusable plastic or paper. For source reduction and materials reuse, the U.S. EPA has introduced "Pay-As-You-Throw" and "Waste Wise" programs, which are most suitable for developed countries.

The MSW recycling process involves a series of activities that includes sorting and processing recyclable products and remanufacturing the recycled raw materials, which can later be used to produce new products. Making pulp and paper from waste papers and recycling plastic waste rather than using polyvinylchloride (PVC) reduces the use of virgin raw materials. It also causes less environmental pollution and imposes fewer burdens on natural resources and biodiversity. Recycling also significantly reduces waste in landfills. For example, in the United States, MSW landfill decreased from 89% to about 54% from 1980 to 2012, because recycling increased from 14.5% to 34.5% (U.S. EPA, 2014). Apart from this, recycling MSW rather than disposing off it in a landfill has environmental and economic benefits. Some of these are as follows (U.S. EPA, 2011, 2014):

- Recycling 1 ton of paper can save 168 gallons of gasoline
- Recycling 1 ton of aluminum cans is equivalent to saving 1665 gallons of oil
- Recycling 1 Mt of MSW reduces slightly less than 2 Mt (1.93) of GHG emissions, which is equivalent to removing 176,000 cars from roadways
- It also creates new jobs, develops greener technologies, conserves natural resources for our future generations, and reduces the need for new landfills and combustors
- Prevents the emission of greenhouse gases and water pollutants, saves energy, and supplies valuable raw materials to industry

Energy Recovery

Energy recovery from MSW is an attractive approach that is receiving significant attention as it will reduce the use of landfills. Moreover, it is reducing environmental pollution and producing greener energy. MSW will always result in a nonrecyclable waste stream, which can neither be recycled nor composted and may need to be combusted in refuse derived fuel (RDF) plants. Energy recovery reduces landfilling and generates useable energy, in terms of heat, electricity, or fuel through a variety of processes, including combustion, gasification, pyrolysis, and anaerobic digestion. This process is also often called WtE. Landfill gas (LFG) can be utilized as an energy source; this is discussed in detail in section "Environmental Issues."

Treatment and Disposal of MSW

WtE processing involves thermal treatment and can release particulates and toxic metals including Hg, in contrast to the other energy recovery options from natural resources. Major advantages of the thermal treatment method are that it reduces the volume of disposed waste and destroys harmful microorganisms, helping to keep society healthy. MSW treatment can be divided into two sections: (a) nonthermal treatment and (b) thermal treatment. Nonthermal treatment processes are classified as dumping and landfilling and aerobic and anaerobic composting processes (described in section "Nonthermal Conversion"). Thermal treatment methods are pyrolysis, gasification, and incineration technologies (described in section "Thermal Conversion"). It must be noted that the purpose of thermal WtE is not to produce power for the sake of producing power (electricity or fuels), rather it is a by-product of reducing the volume of MSW as well as environmental issues. WtE may produce more harmful emissions that may be expensive in terms of treatment.

NONTHERMAL CONVERSION

Dumping and Landfilling

Dumping and landfilling are the most common practices worldwide for disposing of MSW. MSW disposal and landfilling has been carried out without considering the environmental risks in most developing countries. Modern landfills are not just dump sites; they are well engineered, well monitored, and secure facilities designed to protect the environment from contaminants. Disposing of solid waste this way reduces the hazards associated with public health or safety, such as gas emissions, disease transmission, or insect breeding and groundwater contamination. Most developed countries practice modern dumping and landfilling processes, while developing countries are often lacking in these areas. Landfilling processes have progressed from open dumping, controlled dumping, and controlled landfilling to sanitary landfilling. Modern landfills are now built in suitable geographical areas, far from residential areas, faults, wetlands, flood plains, and other restricted zones. Landfill gases (LFGs) generated are collected from landfill sites and used to produce energy. The risk of dumping and landfilling processes and managing the associated risks are described in detail in section "Environmental Issues."

AEROBIC COMPOSTING

MSW organics can be treated with aerobic composting to prepare compost. Aerobic composting is a biological process by which organic waste decomposes by microorganisms (bacteria, fungi, and actinomycetes) in the presence of oxygen. This biological decomposition process requires moist conditions and temperatures ranging from 55°C to 65°C. Most human pathogens, plant pathogens, and weed seeds in the feedstock are killed during the composting process. Biological decomposition occurs with the help of microorganisms and through the oxidation of carbonaceous components present in organic waste. Energy released during oxidation increases the temperature during composting. Typically, a 50% mass reduction (80% by volume) may occur during the process. For example, 100 tons of feedstock can produce 50 tons of composted materials and 50 tons of moisture and carbon dioxide during a successful composting process (Michael, 2006). Once the composting process is complete, the compost can be used as organic fertilizer (Khan, 1994; Ahsan, 1999; Mufeed et al., 2008), as it is rich in plant nutrients including nitrogen, phosphorous, potassium, and other essential micronutrients.

The quality of the compost materials depends on the quality of the MSW. When MSW is composted aerobically, the inorganic materials (mainly heavy metals) and organic substances will contaminate the compost materials. These inorganic materials can cause serious environmental issues and harm to public health. This is the main reason that MSW compost cannot be used as a fertilizer in agricultural operations (KEEI [Korea Energy Economics Institute], 2010). Therefore, mixed waste composting is not an option for all waste management, although it has been widely practiced.

ANAEROBIC COMPOSTING (BIOGAS PRODUCTION)

Anaerobic digestion is a naturally occurring biological process for breaking down organic materials in the absence of oxygen. Anaerobic processes are typically slower than aerobic processes and generate more odors. This process is also referred to as the bio-methanation process, which is basically a method of converting separated organic waste through a biodegradable process and recovering energy in the form of biogas, and compost in the form of liquid slurry. Biogas consists mainly of methane (55%–70%) and carbon dioxide (30%–45%) (Michael, 2006; Jørgensen, 2009) and can be utilized as fuel or be converted to electricity (on-site) by using a generator. The amount of biogas production depends on several key factors including the process design, composition of the feedstock, its volatile contents, and the carbon/nitrogen (C:N) ratio (Dioha et al., 2013). The liquid slurry can be used as organic fertilizer. It has been estimated that using the controlled anaerobic digestion process, 1 ton of MSW produces 2–4 times the amount of methane in 3 weeks in comparison with waste deposited in a landfill, which will take 6–7 years to generate the same amount (Khan, 1994; Ahsan, 1999; Mufeed et al., 2008).

The electricity required for a small community can be generated by this process of anaerobic digestion waste. This form of landfilling also reduces GHG emissions because it uses methane as an energy source, otherwise it would be emitted into the atmosphere.

THERMAL CONVERSION

Thermal treatment technologies are used to convert MSW, utilizing thermochemical processes such as pyrolysis, combustion or incineration, and gasification. During the thermal treatment of MSW, energy is released. Pyrolysis and gasification technologies are well established. Incineration technology is also widely used for MSW treatment and has been applied for energy recovery from MSW.

Pyrolysis

Pyrolysis is the thermal decomposition of organic substances into micro-molecular components in an inert atmosphere or in the absence of oxygen. Pyrolysis is carried out at a relatively low-temperature range (typically from 300°C to 800°C). This temperature range is good enough for chemical reactions to occur. The main components of pyrolysis products are carbon monoxide, hydrogen, methane, and a mixture of hydrocarbons including condensable tars or liquid products such as waxes and bio-oils. These liquid and gaseous products can be burned more cleanly than MSW and generate fewer secondary pollutants, which can be captured more easily. The major product of fast pyrolysis (60%–75%) is in liquid form (bio-oils) and is easy to transport and can also be used to synthesize petrochemicals (Knox, 2005; Mohan et al., 2006). In general, through a conventional slow-heating pyrolysis process (slow pyrolysis), a solid material is formed (charcoal), whereas rapidly heating a carbonaceous feedstock, also termed as "fast pyrolysis," produces liquid fuel (Garcia-Perez et al., 2008; Authier et al., 2009; Simone et al., 2009; Kwon et al., 2010).

The residual solid products consisting of combustible materials usually known as char, consist of a substantial amount of carbon (~40%), representing a significant amount of the energy from the waste and some inorganic impurities including some pollutants that are not volatilized at lower temperatures. This char can subsequently undergo high-temperature processing such as gasification or combustion to recover remaining energy efficiently.

Pyrolysis liquid products are rich in hydrocarbons and may be toxic or corrosive in nature. These products can be refined to produce high-quality value-added products such as motor fuels, chemicals, and adhesives. These liquid products can be directly used as boiler fuel and in some stationary engines (Mohan et al., 2006). The overall pyrolysis process transforms less chemical energy into thermal energy compared with combustion and gasification, and occurs at a lower temperature than combustion and gasification. Figure 26.5 demonstrates the overall schematic diagram of the MSW pyrolysis process.

There are several advantages to the pyrolysis process. Among them, it is

- Safe and environmentally friendly compared to incineration, gasification, and landfilling.
- Suitable for a mixed (heterogeneous) waste stream and produces value-added products.
- Possible to recover metals and glasses either before or after the overall process.
- Preserves land for agricultural or other uses that would be taken up by landfills.
- It has the potential to reduce the carbon footprint.

Combustion and Incineration

The first MSW incinerator was built in the eighteenth century, not with the goal of energy recovery but with a view toward sanitation and a cleaner environment. The incinerator was designed to put an end to the practice of disposing waste in the agricultural fields, as such waste consisting of glass, ceramics, and other constituents that were harmful to agricultural practices. When large cities such as Hamburg and London could not cope with the volume of MSW in their territories for hygienic reasons, their municipal officials decided to construct simple incinerators

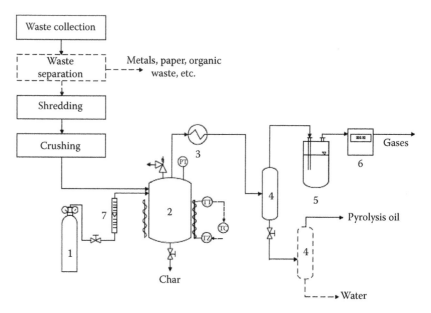

(1. Gas bottle, 2. Reactor, 3. Cooling, 4. Separation, 5. Water trap, 6. Gas flow meter, 7. Rotameter)

FIGURE 26.5 Schematic diagram of MSW pyrolysis process. (Reprinted from *Bioresource Technology*, 133, Ates, F., N. Miskolczi, and N. Borsodi, Comparison of real waste (MSW and MPW) pyrolysis in batch reactor over different catalysts. Part I: Product yields, gas and pyrolysis oil properties, 443–454, Copyright 2013, with permission from Elsevier.)

(de Fodor, 1911; Brunner and Rechberger, 2015). The incinerators contributed to a reduction in rates of cholera and typhoid fever (Brunner and Rechberger, 2015). It also helped to decrease the volume of MSW. Although modern WtE facilities are transforming the organic components of MSW into energy in the form of valuable heat and/or electricity, 100 years ago some incinerators produced electricity and steam from MSW as well (de Fodor, 1911). Nowadays, incineration is a very common thermal treatment method utilized in WtE facilities. An MSW incineration plant is an expensive solid waste management option and requires highly skilled personnel and careful maintenance. However, the cost of an incineration facility and control devices depends on the air pollution regulations existing in a given country; such regulations are usually less strict in developing countries (World Bank, 1999).

Incineration is a process of complete conversion of waste materials into ash, usually at high temperatures. The incineration temperature depends on the type of equipment—grate, fluidized bed, or pulverized waste combustion or the regulation; for example, in the United Kingdom, the minimum combustion temperature for MSW is 850°C (DEFRA [Department of Environment Food & Rural Affairs, UK], 2013). Incomplete incineration can produce toxic gases, dioxins, and other harmful substances, which may lead to serious environmental issues. Incineration removes water from waste materials and reduces the mass and/or volume and produces noncombustible ashes that can be disposed of safely in a landfill by 80%–90%. Some modern incinerators are designed to operate even at higher temperatures with the purpose of producing molten materials and extracting secondary resources including iron, aluminum, copper, zinc, and other metals. In such cases, the volume of combustible waste materials can be reduced to 5% or even less (Peavey et al., 1985; Ahsan, 1999; Jha et al., 2003). Thus, incineration with energy recovery is a good approach to preserve natural resources by utilizing MSW as source of both energy and recycling materials. However, environmental emissions must be monitored. Developed countries have the emission limit for many pollutants such as particulates, SO_x, NO_x, CO, dioxin/furans, and acid gases. Also, these pollutants must be monitored continuously, by law. To maintain a clean environment and to reach optimum performance, the following conditions must be met:

- Incineration temperatures below 1000°C reduce the release of Cl, reduce NO_x formation, and increase thermal efficiency.
- Solid residues must remain in the incinerator for a minimum of 30 minutes at a high-temperature, and flue gases must remain in the postcombustion chamber for at least 2 seconds.
- Excess oxygen must be maintained in the incinerator for complete combustion.

Figure 26.6 shows energy production and balances for two typical cases, A (electricity production) and B (energy output), from MSW (OWAV, 2013; Brunner and Rechberger, 2015). In case A, the energy lost was ~20% through off-gas emission, discharge of ashes and radiation, convection, and heat conduction. The major energy lost (56%) was via steam condensation and a minor loss in the generator. The facility itself consumes 2.6% of the produced electricity, but the overall efficiency for electricity production from MSW was 21%. In case B, the steam is not fully expanded in the turbine, resulting in lower electricity production but a higher steam temperature after the turbine (150–200°C instead of 30–40°C on case A). The total thermal efficiency of this plant can be as high as 74%.

According to a report prepared by the International Solid Waste Association (ISWA, 2012), based on WtE, there are 558 MSW treatment plants in 19 countries. Most of these, 472, are in Europe, and 86 are in the United States. Each of these 558 plants has a capacity of more than 15 tons/day or 10,000 tons/year (Lombardi et al., 2015). In Sweden, the landfill taxation rule was introduced on January 1, 2000, and played a vital role in diverting MSW from the landfill in favor of recycling and incineration. Successive increases in taxation in the following years prompted a continuous increase in material recycling of MSW. Sweden is recycling 50% and incinerating 49% of the MSW it generates, using 32 incineration plants and producing heat and electricity. The amount of electricity produced daily in one plant is sufficient to power 250,000 houses. In addition, Sweden is importing

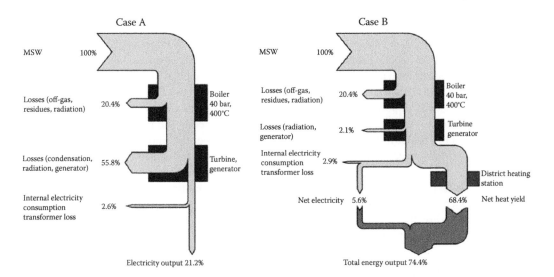

FIGURE 26.6 Energy production and balances for two typical cases A (electricity production) and B (energy output) from MSW. (Reprinted from OWAV (Osterreichischer Wasser- und Abfall wirtschaftsverb and). 2013. Energetische Wirkungsgrade von Abfall verbrennungsanlagen, OWAV Regelblatt 519. 1020 Vienna, Heinestrase 2: Austrian Standards plus Publishing, OWAV; *Waste Management*, 37, Brunner, P. H. and H. Rechberger, Waste to energy—Key element for sustainable waste management, 3–12, Copyright 2015, with permission from Elsevier.)

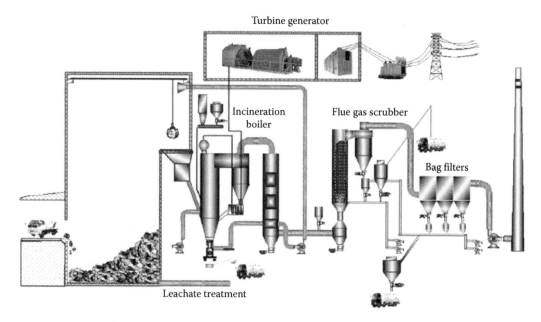

FIGURE 26.7   Schematic diagram of MSW incinerator to generate energy. (Reprinted from *Renewable and Sustainable Energy Reviews*, 36, Zheng, L., J. Song, C. Li, Y. Gao, P. Geng, B. Qu, and L. Lin, Preferential policies promote municipal solid waste (MSW) to energy in China: Current status and prospects, 135–148, Copyright 2014, with permission from Elsevier.)

8 million tonnes of MSW every year from neighboring countries including Britain, Italy, Ireland, and Norway, to fulfill its energy demand (Milios and Reichel, 2013). Figure 26.7 shows the schematic diagram of a modern MSW incinerator for power generation.

MSW generated in most cities in India is rich in organic materials (40%–60%), high moisture (40%–60%), and high inert contents (30%–50%), which possesses low calorific value (800–1100 kcal/kg) (Chakrabarty et al., 1995; Jalan and Srivastava, 1995; Sudhire et al., 1996; Bhide and Shekdar, 1998; Joardar, 2000; Kansal, 2002; Sharholy et al., 2008). This could be one of the main reasons that the incineration process is not well practiced in Indian cities. The first large-scale MSW incineration plant was constructed by Miljotecknik Volunteer (Denmark) in Timarpur, New Delhi in 1987 with a capacity of 300 tons/day. The plant operated for a few months but was shut down due to poor performance. Another incineration plant was constructed at the Bhabha Atomic Research Centre (BARC) in Trombay, to burn business waste, mostly paper (Sharholy et al., 2008).

Apart several advantages described earlier, there are some disadvantages as well, such as

- Incineration plants involve heavy investments and high operating costs.
- There is a possibility of environmental pollution from the flue gas cleaning residues, unless the residues are disposed of in controlled and well-operated landfills to prevent ground and surface-water pollution.

GASIFICATION TO ENERGY

Gasification is an established technology. For more than 75 years, industries around the world have used it to produce chemicals and refine fuels and fertilizer. For more than 35 years, it has been used to generate electric power. The technology is playing an important role in meeting worldwide energy demands. Energy recovery from MSW gasification has several benefits which include substantial reduction in the total quantity of waste, significant reduction in environmental pollution (it limits the formation of dioxins and large quantities of SO_x and NO_x), and a reduction in landfill waste disposal.

Gasification is a process that converts organic or fossil fuel-based carbonaceous materials into carbon monoxide, hydrogen, and methane in a controlled reactor. The resulting products also depend on the composition of feedstock. The following are possible reactions inside a gasifier (Swanson, 2010):

$$C + H_2O \rightarrow CO + H_2 \text{ (Water gas)}$$
$$C + CO_2 \rightarrow 2CO \text{ (Boudouard)}$$
$$CO + H_2O \rightarrow CO_2 + H_2 \text{ (Water – Gas – Shift)}$$
$$CO + 3H_2 \rightarrow CH_4 + H_2O \text{ (Methanation)}$$

Gasification involves the reaction of carbonaceous feedstock at high temperatures ranging from 800°C to 1650°C, without complete combustion and with a controlled amount of oxygen, steam, or carbon dioxide. This is an exothermic reaction process but some heat may be required to initialize and maintain the gasification process. It is a very efficient technology and the energy recovery from gasification can reach up to 60% (Knox, 2005). Gaseous fuels are more homogeneous than MSW and burn more cleanly to produce energy. This technology is very popular in the European Union (EU) countries. Germany has installed the maximum number of gasification plants (20), followed by Finland, with 14. Austria and the United States have installed 10 plants, Denmark has installed nine, and Sweden has seven (Knox, 2005).

There is a high temperature (1500–10,000°C) (Anyaegbunam et al., 2014) gasification process to deal with hazardous MSW. This process is called plasma gasification, and it converts organic waste materials into syngas, electricity, and slag using a plasma torch (Moustakasa et al., 2003). Plasma gasification usually uses an inert gas such as argon (Ar), and the electrodes may include copper (Cu), tungsten (W), hafnium (Hf), or zirconium (Zr). A high-voltage electric current passes through the two electrodes as an electric arc, and the high-pressurized inert gas is ionized while passing through the plasma. The conversion rate of plasma gasification is very high (≥99%) (U.S. DOE, 2015) and produces a high caloric value synthesis gas. The slag produced at a higher temperature is chemically inert and safe to handle and can be used to prepare value-added products (Hajime, 1996) or can be sold as a commodity (Mountouris et al., 2008). Plasma gasification occurs in the absence of oxygen at a high temperature. This process is environmentally friendly since a high temperature in the reactor prevents the formation of toxic compounds such as dioxins and furans.

The main advantages of plasma gasification technologies for waste treatment are the following:

- Hazardous MSW materials are destroyed efficiently (Anyaegbunam et al., 2014)
- Hazardous waste materials do not reach landfills
- There are no toxic emissions from MSW plasma gasification (Bert et al., 2007)
- Organic waste materials are converted into combustible syngas for electric power generation and thermal energy production (Edbertho, 2004)

Typically, the gas generated from gasification has a net calorific value in the range of 4–10 MJ/N m^3. The noncombustible material from gasification is a solid residue (ash), which contains a relatively low level of noncombusted carbon and can be utilized for making construction materials. The clean syngas (carbon monoxide and hydrogen), which is also called "synthesis gas," can be transferred to a boiler, internal combustion engine, or a gas turbine to produce power, or it can be further transformed into chemicals, fertilizers, and transportation fuels utilizing suitable technologies. WtE plants based on gasification technology utilize MSW as their fuel and conserve conventional natural sources of energy such as coal, oil, or natural gas. Figure 26.8 shows a number of possible ways to use syngas generated from MSW.

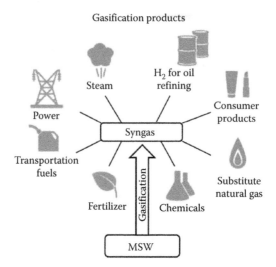

FIGURE 26.8 MSW Gasification products from syngas. (Redrawn after Kerester, A. 2015. Gasification can help meet the world's growing demand for cleaner energy and products. *Cornerstone, An Official Journal of the World Coal Industry.* Spring. http://cornerstonemag.net/gasification-can-help-meet-the-worlds-growing-demand-for-cleaner-energy-and-products/.)

Figure 26.9 shows a schematic diagram of MSW gasification to produce combustible gas that can be burned either in a gas turbine or in a boiler system to generate energy (Breeze, 2014).

MSW gasification involves a large amount of mechanical processing and close supervision of the process, which greatly impacts the operational costs and accounts for the overall plant capital costs. Using a simpler shredding and sorting process at the waste management facility can make the gasification process more cost-effective, as can raising community awareness about recycling

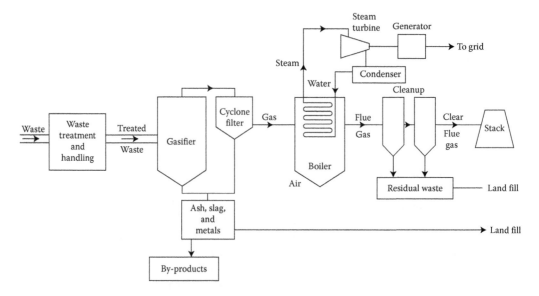

FIGURE 26.9 Schematic diagram of MSW gasification to produce energy. (Reprinted from *Power Generation Technologies*, 2nd ed., Breeze, P., Power from waste, 335–352, Copyright 2014, with permission from Elsevier.)

and disposal practices. There are several examples of MSW gasification projects. In July 2014, Edmonton's EAB started a gasification plant to produce biofuels and chemicals from nonrecyclable urban waste. EAB is diverting around 90% of the City of Edmonton's waste for its biofuel production. Implementing such thermal treatment technologies can reduce the quantity of waste diverted to landfills. A large-scale commercial MSW gasification plant in Ebara TwinRec Kawaguchi, Japan has a capacity of 420 tons/day. This plant has been operational since 2002 (Hester and Harrison, 2013).

ENVIRONMENTAL ISSUES

Environmental concerns arise in every step of urban waste handling, processing, transportation, and disposal. Whichever waste treatment methods are applied, environmental concerns exist; all that differs is the magnitude. Landfill, one of the conventional methods of solid waste treatment methods, generates many environmental impacts. Even though WtE technologies are primarily aimed at recovering energy from waste resources and increasing benefits, environmental concerns exist to a lesser extent. The energy generated in WtE units contributes to primary energy savings and consequently to reducing GHG emissions and other pollutants to the extent comparable to the energy produced from the biomass (Pavlas et al., 2010). The following section describes environmental issues related to WtE facilities and their mitigation measures.

LANDFILL

Landfill is the most commonly used solid waste disposal method, more widely practiced in developing countries. Landfill is the cheapest among waste disposal methods; however, it involves many environmental problems which depend on the type of landfill and environmental management practice currently in use. Environmental impacts are noticed in the landfill's construction, operation (~20 years) and postmanagement phases (5–30 years, even to centuries). More aspects of these impacts are described in the following paragraphs.

Local Pollution Problems
- The location of the landfill often has an impact on property values and the well-being of the surroundings, due to unpleasant odors, dust, noise, and pollution.
- Landfill waste can attract rodents, insects, and flies, which transmit parasites and communicable diseases to humans.
- Informal waste recycling often involves waste pickers who collect plastic, glass, metals, and other materials that can be merchandized for a profit. They are helping to reduce waste volume in landfills, but negative social-economic issues such as disease and poverty are often associated with these activities.
- Many of the environmental impacts of landfilling (local pollution and resource degradation) are faced by marginalized groups, such as people with a lower socioeconomic standing in the community.

Landfill Gas and GHG Emissions
The organic portion of waste dumped in the landfill undergoes decomposition naturally and forms gaseous products called LFG, composed of CO_2, methane, hydrogen, nitrogen, etc. GHG emissions, mainly methane from landfill disposal of urban waste are major environmental concerns. Since urban waste contains a large portion of organic waste, disposing of it in a landfill releases methane into the atmosphere. Methane is 21 times more powerful than CO_2 in global warming effect. Figure 26.10 shows the general trend of CH_4 emissions from landfills.

It has been reported that 0.350 N m^3/kg of biogas is produced from urban solid waste (Desideri et al., 2003). The global methane emission from MSW was estimated to be 760.6 million tonnes

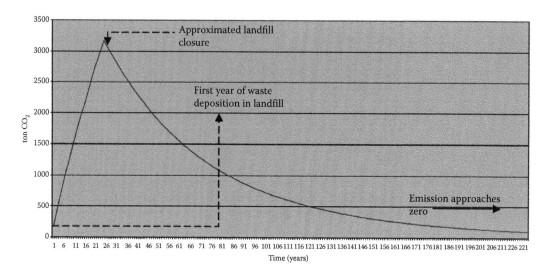

FIGURE 26.10 CH_4 emission trend from landfills in their operating post closure years. (Reprinted from *Bioresource Technology*, 100, Lou, X. F. and J. Nair, The impact of landfilling and composting on greenhouse gas emissions—A review, 3792–3798, Copyright 2009, with permission from Elsevier.)

CO_2 eq in 2010 (U.S. EPA, 2006). Globally, the waste sector contributes approximately 5% of the global greenhouse emissions (IPCC, 2006), mainly as CH_4 and CO_2. The waste management sector in the United States represents about 4% of total U.S. anthropogenic GHG emissions (i.e., 260 Tg of CO_2 equivalents) (U.S. EPA, 2001).

Soil and Water Pollution from Leachate

Solid waste disposed in a landfill undergoes chemical and physical changes and releases leachate. Leachate can be defined as liquid draining out from the landfill site, which contains significantly elevated concentrations of toxic and harmful materials. The quantity of leachate generation and its toxicity depends on several factors such as waste properties, composition, climatic conditions, precipitation, and temperature. Soil and groundwater pollution from leachate depends primarily on geological and groundwater conditions at the landfill site. Leachate contains high chemical oxygen demand, biochemical oxygen demand, total organic carbon, suspended solids, dissolved solids, organic pollutants, and heavy metals (Lou and Nair, 2009). About 200 hazardous compounds have been identified in heterogeneous landfill leachate (Jensen et al., 1999). Typical leachate composition from MSW is presented in Table 26.4.

Leachate released from a landfill site contaminates aquifers (Rosa et al., 1996), and pollutes groundwater and soil. The environmental impact and human health hazards depend significantly on leachate properties. Contamination of groundwater from toxic leachate is a problem in landfill sites that are not engineered properly and do not follow environmental management guidelines. The leaking of the leachate in groundwater sources is affected by how far the source is from the site and the geology of the landfill site. For example, if sand, gravel, and limestone are lying beneath the landfill, the leachate may easily pass into the water sources. The leachate production rate increases with rainfall. Disposing of solid waste and water in the landfill is practiced in most Jordanian landfills, and increases leachate production (Aljaradin and Persson, 2012).

THERMAL TREATMENT OF WASTE

Conventionally, most solid waste used to end up in landfills and incinerators. The waste management policy for MSW in many developed countries shifted from landfill to incineration, recently.

TABLE 26.4

Typical Leachate Composition from Municipal Solid Wastes

Parameter	Typical Range (mg/L, Unless Otherwise Noted)	Upper Limit (mg/L, Unless Otherwise Noted)
Total alkalinity (as $CaCO_3$)	730–15,050	20,850
Calcium	240–2330	4080
Chloride	47–2400	11,375
Magnesium	4–780	1400
Sodium	85–3800	7700
Sulfate	20–730	1826
Specific conductance	2000–8000 µmhos/cm	9000 µmhos/cm
Total dissolved solids (TDS)	1000–20,000	55,000
Chemical oxygen demand (COD)	100–51,000	99,000
Biochemical oxygen demand (BOD)	1000–30,300	195,000
Iron	0.1–1700	5500
Total nitrogen	2.6–945	1416
Potassium	28–1700	3770
Chromium	0.5–1.0	5.6
Manganese	Below detection limit—400	1400
Copper	0.1–9.0	0 9.9
Lead	Below det ection limit—1.0	14.2
Nickel	0.1–1.0	7.5

Source: Mukherjee, S. et al. 2015. *Crit. Rev. Environ. Sci. Technol.*, 45: 472–590. Reprinted with copyright from Taylor & Francis, 2015.

Waste incinerators contribute to protecting human health and the environment by destroying hazardous microorganisms and pests, but at the same time they can create air pollution. It took several decades to overcome this issue and to improve the furnace and flue gas cleaning technologies. The first MSW incinerator equipped with a filter was installed in the 1950s, although it was not entirely sufficient. Incineration technology has improved significantly, but MSW incineration plants still pose environmental challenges (Brunner and Zobrist, 1983; Giusti, 2009).

With the increase in waste incineration, the amount of pollutants emitted into the atmosphere has increased, as have solid residue and wastewater release from incineration facilities. For instance, sewage sludge from the treatment of municipal wastewater has become a problem. Conventional methods of treatment such as disposing of sludge in agricultural fields and landfills have been restricted in most developed countries, because of their adverse effect on the environment. Several sewage sludge combustion and cocombustion technologies with energy recovery are in use (Werther and Ogada, 1999). Converting sewage sludge into useful energy by utilizing pyrolysis and gasification, combustion, and cocombustion process are getting more popular (Groß et al., 2008).

The handling, collection, storage, and transportation of heavy metal-contaminated waste are similar to those for hazardous wastes. Two types of heavy metals waste are collected from the point of generation: (a) toxic metal-containing products; and (b) toxic metal-contaminated materials. The Basel Convention (1989) identified and provided suggestions for treatment of toxic metals, including a mercury collection procedure for metal-contaminated waste from spills and manufacturing sites. However, these are not strictly practiced in reality; accordingly, this toxic waste ends up in waste and is released into the environment once the waste is treated in the incinerator or WtE facilities.

Waste has been used as an alternative fuel in cement kilns as well. Using waste in the cement industry as an alternative fuel and for raw materials is acceptable in the EU (Cembureau, 1999).

A large amount of combustible waste has also been burned in incinerators and cement kilns. This practice needs to be regulated, as it might increase the release of pollutants into the environment.

WtE thermal technologies have been recently reviewed by Astrup et al. (2015). Mass-burn incineration based on moving grate systems was the most widely used (Figure 26.11) among the available technologies. Other WtE technologies such as pyrolysis, gasification, cocombustion, and cocombustion of waste in cement kilns are less utilized. Incineration is practiced widely. Compared to all the technologies mentioned earlier, incineration has the most environmental issues in terms of emission into the environment, release of waste water, and solid waste residue (ash) disposal. On the other hand, gasification and cocombustion have less environmental impacts than incineration. The impact of WtE facilities on the environment depends on the technologies, pollution control devices, and environmental practices in use.

Particulate, NO_x, SO_x Emission

The uncontrolled burning of waste, which has been widely practiced in developing countries, emits large volumes of particulates, NO_x, and SO_x. Particulates are of microscopic solid particles or liquid droplets. The formation and emission of fine inhalable or breathable particles enriched with toxic metals is a major concern of waste incineration.

NO_x emission from waste incineration is mainly composed of nitric oxide (NO), with a lesser amount of nitrogen dioxide (NO_2). NO_x is emitted in two ways: (1) oxidation of N in waste (fuel NO_x) and (2) NO_x produced from N_2 and O_2 during high-temperature incineration (thermal NO_x). Incineration of sulfur contained in waste results in SO_x formation. In the presence of an excess amount of oxygen at a higher temperature (800°C), SO_2 can be further oxidized to SO_3. The CO_2 and SO_x are emitted due to the oxidation of carbon, nitrogen, and sulfur present in the waste. Given that the waste composition varies, CO_2, NO_x, and SO_x emission concentrations in flue gas can vary significantly.

GHG Emissions

The incineration, cocombustion, and gasification processes of MSW release GHGs, mainly CO_2, CH_4, and N_2O. Among these processes, MSW incineration plays a dual role, as it emits GHGs and, in contrast, it saves GHGs, mainly when recovering heat and electricity. The incineration of 1 ton of municipal waste in MSW incinerators produces/releases about 0.7–1.2 tons of CO_2 (Johnke,

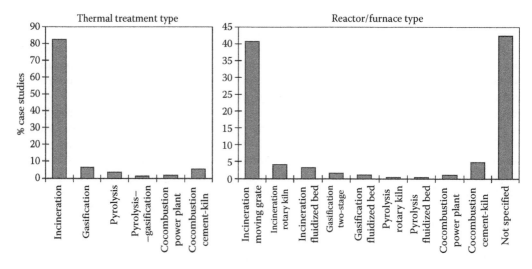

FIGURE 26.11 Overview of WtE thermal treatment technologies. (Reprinted from *Waste Management*, 37, Astrup, T. F., D. Tonini, R. Turconi, and A. Boldrin, Life cycle assessment of thermal waste-to-energy technologies: Review and recommendations, 104–115, Copyright 2015, with permission from Elsevier.)

2000). GHG emissions from the thermal treatment of MSW technologies are lower in comparison to the GHG emissions from landfills. The estimated current GHG emissions from waste incineration are small, around 40 $MtCO_2$-eq/year, or less than one-tenth of landfill CH_4 emissions (IPCC [Intergovernmental Panel on Climate Change], 2015). The plastic and food waste in the MSW are reported to be the critical factors influencing GHG emissions from MSW incinerators (Yang et al., 2012). Yang argues that decreasing the food waste content in MSW by half will significantly reduce the GHG emissions from MSW incinerators.

Studies on GHG emissions from MSW incinerators in European countries have shown that MSW incinerators work as GHG sinks. Damgaard et al. (2010) have shown that energy recovery from the production of both heat and electricity can save much more GHG emissions than the production of electricity alone. Recovering the excess heat after the generation of electricity would be a good way to convert MSW incinerators from GHG sources to sinks.

Toxic Metals and Dioxin/Furans

Waste is made up from a variety of materials containing toxic metals such as mercury in consumer products (found in thermometers, fluorescent light bulbs, and batteries) and products where mercury is added (such as in dental amalgam and manometers). In many countries there is regulation requiring these materials to be placed in a separate stream, but in many cases this is poorly practiced and the materials do end up in waste. During incineration, metals in waste are oxidized and transformed into metal oxides. If the vapor pressure of the metals in waste is low, they can vaporize and are removed within the flue gas cleaning system and can be emitted into the atmosphere. The emission concentrations of heavy metals in flue gas are related to their concentration in waste. Metals distribution and speciation during combustion are affected by several factors such as temperature, chlorine content, feeding conditions, and oxygen availability. Metals distribution during incineration is a complex process; some species like Ni and Cr interact with other ash constituents, which affect metals enrichment behavior (Yan et al., 2001). Fly ash contaminated with harmful pollutants generated from coal combustion often creates a solid waste disposal problem. These pollutants can be vitrified at high temperatures so as to avoid leaching into the environment.

Residual Solid Waste

The thermal treatment of waste generates large volumes of solid waste in the form of bottom ash, fly ash, and waste from air pollution control devices (APCDs), for example, gypsum. Collection, transport, and disposal of these forms of waste need to be done with great care. Many countries have guidelines for dumping such waste in landfills. Solid waste must meet certain criteria before it can be dumped in landfills. For example, in South Korea, bottom ash can be dumped into a general landfill site after a pH test (<13), a loss-of-ignition test (<5 wt.%), and a leaching test of heavy metals (Pb and Cu [3 mg/L], Cr^{6+}[1.5 mg/L], Cd [0.3 mg/L]) and other heavy metals and organic compounds such as polychlorinated dibenzodioxin (PCDD), polychlorinated dibenzofurans (PCDFs), and polychlorinated benzene (PCB) in bottom ash are not regulated to be dumped in general landfill sites (referred from Song et al., 2008).

Wastewater

Wastewater from thermal treatment facilities can be released from cooling water, pollution control devices, ash transport and storage sites, and run-off from waste storage areas. The properties and toxicity of wastewater depend on where the water was released. Tests for odor, temperature, pH, total suspended solids, chemical oxygen demand, and biological oxygen demand are carried out as a first approach in measuring pollution and to prepare for the necessary treatment methods. Wastewater from boiler cooling, if released directly in natural water bodies, increases the temperature. Wastewater released from surface run-off might be contaminated with dissolved and suspended particles, whereas wastewater released from pollution control devices may contain toxic pollutants including heavy metals.

MITIGATION MEASURES

A large amount of organic waste going to landfills and/or thermal treatment plants can be reduced by practicing "4R" methodology, that is, the "reduce, reuse, recycle, and rethink" approach: reduce the waste generation, reuse the materials, recycle, and rethink before wasting anything. Buying products with less packaging materials, using recyclable products, donating items that are of no use to the community, etc., may help to decrease waste generation. The following are some of the good practices that may help to reduce waste generation and mitigate associated negative impacts:

- Composting the organic portion of solid waste instead of dumping it into a landfill reduces the volume of waste generated and decreases the transportation and disposal costs, as well as their associated impacts.
- The economic cost of urban waste management is high and paid mainly by municipal governments. The proper handling and transportation can economize the process. Environmental strategies such as "volume-based payment" systems, as practiced in some developed countries, can reduce the cost of management and reduce waste quantities.
- Reuse and recycling can reduce economic costs by avoiding the extraction of raw materials. Both practices often cut transportation costs.
- Source reduction, reuse, and recycling prevent and/or reduce the generation of waste, save natural resources, reduce pollution, and economize the waste management process. These practices also prevent GHG emissions, reduce water pollutants, save energy, and conserve natural resources.
- The MSW also contains nonrecyclable waste, which can neither be recycled nor composted; they can also be combusted or gasified to recover energy.

Mitigating the Negative Effects of Landfilling

The landfill site must be selected in such a way that it will minimize pollution and has the least impact on the human population and the natural environment during the construction and operation phases.

1. Lining system: Lining the landfill is important to avoid the release of leachate to the surrounding area. In particular, for landfills with codisposal of wastewater and/or hazardous waste, a liner is necessary.
2. Leachate collection system: Leachate should be collected properly to ensure that it is not released in the environment. Management of leachate and to ensure that it meets the limits of contaminants prior to disposing is a challenging job for regulatory bodies and municipalities. Leachate should be managed, regulated during its formation/release, collected and transported properly, and properly treated prior to disposal.
3. Covering: The waste should be covered properly to avoid being exposed to the environment and to prevent access by birds and animals.
4. A drainage collection system should be properly built to collect waste water from the landfill site.
5. Measures for gas emission control: LFG such as methane can be collected and used as a renewable energy source for heating and even for generating electricity. This has been well practiced: for example, 21 landfills in Denmark captured about 5800 N m^3 of biogas/hour, equivalent to 27.4 MW of thermal energy (Willumsen, 2003; Themelis and Ulloa, 2007). It has been reported that 0.350 N m^3/kg of biogas is produced from urban solid waste (Desideri et al., 2003). Proper utilization of this can help to mitigate GHG emissions and meet energy needs.

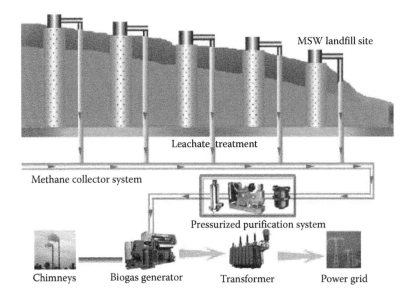

FIGURE 26.12 Electricity generation process from LFG. (Reprinted from *Renewable and Sustainable Energy Reviews*, 36, Zheng, L., J. Song, C. Li, Y. Gao, P. Geng, B. Qu, and L. Lin, Preferential policies promote municipal solid waste (MSW) to energy in China: Current status and prospects, 135–148, Copyright 2014, with permission from Elsevier.)

Figure 26.12 demonstrates the electricity generation from LFG. The LFG is collected and purified through a pressurized purification system and subsequently placed into a biogas generator for electricity generation via combustion (Liu, 2010; Zheng et al., 2014). The most common mitigation measures for LFG are to capture and burn it using flares or to recover energy for heating and cooking purposes; the latter having significant environmental and socioeconomic implications (El-Fadel and Sbayti, 2000). LFG collection in landfills are increasing worldwide; there were approximately 955 such units in 2001 (Themelis and Ulloa, 2006).

One way to reduce the amount of methane released from a landfill is to reduce the organic portion of waste in urban waste. GHG emissions from waste decomposition are much higher from landfills than from composting since anaerobic decomposition facilitates the production of CH_4. In this respect, composting is a simpler and more effective means of reducing GHG emissions.

6. Landfill stabilization:
   a. Gas removal, ventilation: by adding a gravel layer or using a vertical wall or forced ventilation by gas pumping, blower installation, etc.
   b. Maintain an aerobic condition
   c. Make use of landfill compost
7. Postlandfill management:
   a. Periodic inspection and testing of ground water
   b. Effluent management
   c. Management of landfill cover
   d. LFG management

Once a landfill is full, it can be filled in or covered and then utilized to build a recreation facility such as a park, playground, golf course, or ski resort.

Control of Particulate, NO$_x$, and SO$_x$

APCDs have a crucial role in controlling pollutants generated in WtE facilities and reducing the hazards in the environment. The control devices for pollutants differ. Figure 26.13 shows the APCDs available in currently operated in WtE facilities (Astrup et al., 2015) for dust, acid gases, NO$_x$, and PCDD/F removal.

Particulates can be controlled using a cyclone, bag filter, and electrostatic precipitator (ESP). Cyclones are simpler and most widely used for particulate control. Cyclones work well for coarse particles, mainly for the removal of particles of 50 μm or larger. ESP has a high collection efficiency (>99%) and operates with a low-pressure drop. The overall process in ESP can be classified into (a) ionization of particles in flue gas; (b) charge, migration, and collection of particles; (c) removal of particles from plate; and (d) collection in hopper.

The bag filter captures particles mainly by direct interception, inertial impaction, and diffusion (Miller, 2011). Particles collected on the filter surface are periodically removed by blowing air in the opposite direction through the filter, and then particles are collected in a hopper. Bag filters are 99.9% more efficient at removing particulates, which means they are good for removing fine particles.

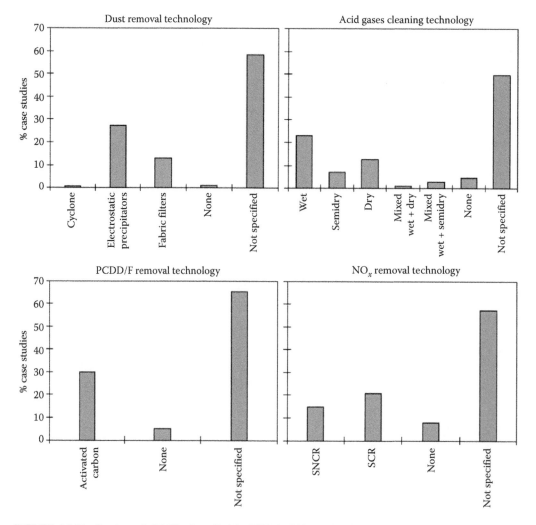

FIGURE 26.13 Review of APCDs installed in WtE facilities. (Reprinted from *Waste Management*, 37, Astrup, T. F., D. Tonini, R. Turconi, and A. Boldrin, Life cycle assessment of thermal waste-to-energy technologies: Review and recommendations, 104–115, Copyright 2015, with permission from Elsevier.)

Particulates are also controlled by wet scrubbers. Examples of such are bubbling scrubbers, spray towers, wet ESPs, and venture scrubbers. Wet scrubbers are efficient at removing coarse particles (their efficiency is up to 99%); however, their record at removing fine particles is poor.

NO_x can be controlled using a low NO_x burner, controlling fuel–air mixing rates, or by installing a selective catalytic reactor (SCR) and selective noncatalytic reactor (SNCR). SCRs have been widely used in coal-fired power plants and incinerators because of their higher efficiency (80%–90%). In the SCR process, ammonia (NH_3) or urea reduces NO_x to N_2 and H_2O. Metal oxide catalysts such as titanium dioxide-supported vanadium pentoxide (TiO_2/V_2O_5) are used in the SCR process. SNCR can reduce NO_x emissions by 30%–50%. To achieve similar NO_x reductions, the SNCR process requires 3–4 times more reagent (ammonia or urea) than the SCR process (IEA, 2014).

SO_x can be controlled by dry scrubbers and wet scrubbers, in a process known as flue gas desulfurization (FGD). The FGD system can be classified into wet or dry, depending on the water used to spray the reagent into the flue gas. Wet FGD uses an alkaline reagent such as calcium-, sodium- and ammonium-based sorbents to capture SO_2. In a lime/limestone wet scrubber, $CaCO_3$ reacts with SO_2 and H_2O and forms calcium sulfite ($CaSO_3 \cdot 1/2H_2O$). Wet FGD can achieve 90%–98% of SO_2 control (DePriest and Gaikwad, 2003).

Green House Gases Capture

CO_2 emitted in flue gas from the thermal treatment of waste can be captured in three ways: precombustion capture, oxy-fuel combustion, and postcombustion capture of CO_2 (IPCC, 2005). In precombustion capture, carbon is removed from the fuel prior to combustion. In postcombustion CO_2 capture, CO_2 is removed from the flue gas. The flue gas composition released from air combustion consists of N_2, NO_x, CO_2, and SO_2. The concentration of CO_2 in waste combustion is about 13%–15%, which is difficult to capture. In oxy-fuel combustion, waste is combusted in an oxygen- and CO_2-enriched environment. Flue gas from oxy-fuel combustion mainly consists of CO_2 (with lesser amounts of N_2, SO_2, and O_2), which can be captured easily. These technologies are still in the developmental stage and have not been applied in WtE facilities.

CO_2 capture technologies using sorbents, solvents, and membranes are emerging. The amine-based chemical absorption process is one of the most widely used methods to capture CO_2 in flue gas from refineries (Sreenivasulu et al., 2015).

Toxic Metals and Dioxin/Furans

A major portion of the toxic heavy metals (except Hg, Se) emitted during the thermal treatment of waste is removed as particulates. Since most of the treatment plants are installed with particulate control devices, particulate removal efficiency in general is reasonably good. The removal of toxic heavy metals in flue gas depends on how the metals are distributed in the particle and gas phase, which is affected by several possible factors such as operating temperature, chlorine, and oxygen availability (Pudasainee et al., 2012). The best available technologies to control emissions of heavy metals are high-efficiency particulate control devices. In addition, particle collection wet scrubbers also have the benefit of controlling acid gases, bases, and other elements that are soluble in water and/or scrubbing solutions.

Modern waste incinerators are equipped with advanced APCDs, as shown in Figure 26.13. The toxic heavy metals are captured well in existing APCDs as a cobeneficial control. In a study of three hazardous waste incinerators, Kim et al. (2009) found that APCDs were >99.9% successful at efficiently removing heavy metals (except Hg). The concentration of particulate matter and heavy metals was controlled 90%–95% in the existing APCDs (e.g., Stoker (SNCR) + waste spray dry absorber + activated carbon injection + bag filter) (Jang et al., 2007).

The emission concentrations of dioxin and furan can be controlled by keeping the temperature higher than 850°C in the boiler, and reducing the temperature to less than 200°C before the beginning of APCDs, and keeping a residence time of more than 2 s in the incinerator (Jang et al., 2007).

Residual Solid Waste

With the increase of thermal treatment technologies for solid waste, the amount of fly ash generated from incinerators and gasifiers is increasing. The solid wastes, mainly fly ash, bottom ash, and waste from pollution control devices, are highly contaminated with toxic elements and should be properly treated. Fly ash from MSW, in general, is disposed of in landfills. Depending on the nature and properties of ash generated, if the ash is not contaminated with toxic pollutants, it can be used in the construction of roads and embankments, etc. Many countries have regulatory guidelines for dumping ash from thermal treatment plant landfills.

Wastewater Management

The amount of wastewater generation from thermal WtE technologies can be reduced and/or managed as follows:

- Use dry or semidry pollution control devices
- Conserve water by maximizing the recirculation of polluted wastewater in wet pollution control devices
- Properly collect and treat run-off from waste storage sites
- Control odor by chemical sterilization
- Ensure that wastewater is treated and is safe (i.e., meets regulatory standards in terms of physical and chemical makeup) before it enters natural bodies of water

SUMMARY

Urban waste contains materials discarded for no further use; however, these waste materials can transform to wealth in terms of energy and resources, if we can utilize them properly. The level of global waste generation is increasing. Most developing countries follow open dumping or disposal of MSW in landfills, mostly in an uncontrolled manner. Such disposal practices lead to problems related to the environment, which possess a risk to human and animal health. On the other hand, most developed countries practice proper waste management, even though they are facing challenges maintaining environmental friendliness while producing energy from MSW.

Integrated waste management practice reduces waste generation, promotes recycling, and plays a significant role in reducing GHGs emission and environmental pollution. Thermal treatment technologies are useful for reducing the volume of waste and destroying hazardous pathogens. Although most EU countries use modern incinerator technology for burning waste to produce clean energy with less pollution, a recent report (Howard, 2009) revealed that modern incinerators in EU countries are the major source of ultrafine particulates in the atmosphere.

The MSW incineration technology is the most widely used among the thermal treatment technologies, which is well adapted in most developed countries. Adapting integrated waste management systems with proper recycling, reuse, and remanufacturing would create more employment and business opportunities, even with fewer investments (Seldman, 2008).

Even though thermal technologies are technically feasible, it is better to reuse, recycle, reprocess, or remanufacture to maintain a healthy environment. At the same time, it is necessary to develop cost-effective new approaches for integrated WtE technologies based on MSW.

Environmental concerns arise in every step of MSW handling, processing, transportation, disposal, and treatment. Landfill is the cheapest among waste disposal methods; however, it leads to many environmental problems. Environmental impacts depend on the type of landfill and environmental management practice currently in use. Environmental impacts occur in the construction, operation, and postmanagement phases of a landfill.

WtE technologies are primarily aimed at recovering energy from waste and increasing the benefits; however, and to a lesser extent, these technologies also involve environmental concerns. With the increase in waste incineration, the amount of solid residue, wastewater, and pollutants emitted into the atmosphere from incineration facilities has increased, causing a severe impact on ecosystems and human health. Incineration, cocombustion, pyrolysis, and gasification of waste release particles, NO_x, SO_x, GHGs, toxic metals, and dioxin/furans into the atmosphere.

A large amount of organic waste going to a landfill and/or thermal treatment plants can be reduced by practicing the "4Rs" approach, known as reduce, reuse, recycle, and rethink: reduce waste generation, reuse materials, recycle, and rethink before wasting anything. There are several ways to mitigate the negative impacts of landfilling, and recover energy. APCDs play a crucial role in controlling pollutants generated in WtE facilities and reduce the associated hazards in the environment. Currently, most APCDs are highly efficient at capturing pollutants released (except heavy metals emission such as Hg) from WtE facilities. Also, GHG emission control technologies have not been commercially demonstrated. In addition to the application of technologies, there are several best practices that help to mitigate the formation and release of pollutants in the environment.

ACKNOWLEDGMENTS

The authors acknowledge the financial support from Helmholtz Alberta Initiative (HAI) and the University of Alberta, Edmonton, Canada.

REFERENCES

Ahsan, N. 1999. Solid waste management plan for Indian megacities. *Indian Journal of Environmental Protection*, 19(2): 90–95.

Aljaradin, M. and K. M. Persson. 2012. Environmental impact of municipal solid waste landfills in semi-arid climates—Case study—Jordan. *Open Waste Management Journal*, 5: 28–39.

Anyaegbunam F. N. C. 2014. Hazardous waste vitrification by plasma gasification process, *IOSR Journal of Environmental Science, Toxicology and Food Technology (IOSR-JESTFT)*, 8(3): 15–19.

Astrup, T. F., D. Tonini, R. Turconi, and A. Boldrin. 2015. Life cycle assessment of thermal waste-to-energy technologies: Review and recommendations. *Waste Management*, 37: 104–115.

Ates, F., N. Miskolczi, and N. Borsodi. 2013. Comparison of real waste (MSW and MPW) pyrolysis in batch reactor over different catalysts. Part I: Product yields, gas and pyrolysis oil properties. *Bioresource Technology*, 133: 443–454.

Authier, O., M. Ferrer, G. Mauviel, A.-E. Khalfi, and J. Lede. 2009. Wood fast pyrolysis: Comparison of Lagrangian and Eulerian modeling approaches with experimental measurements. *Industrial & Engineering Chemistry Research*, 2009, 48(10): 4796–4809.

Basel Convention. 1989. http://www.basel.int/portals/4/basel%20convention/docs/text/baselconventiontext-e.pdf (assessed October 2015).

Bert, L., H. Elslander, I. Vanderreydt, and K. Peys. 2007. Assessment of plasma gasification of high caloric waste streams. *Waste Management*, 27(11): 1562–1569.

Bhide, A. D. and A. V. Shekdar. 1998. Solid waste management in Indian urban centers. *International Solid Waste Association Times (ISWA)*, 1(1): 26–28.

Breeze, P. 2014. Power from waste. In *Power Generation Technologies,* Chapter 16. 2nd ed. Elsevier Ltd., Amsterdam, pp. 335–352.

Brunner, P. H. and H. Rechberger. 2015. Waste to energy—Key element for sustainable waste management. *Waste Management*, 37: 3–12.

Brunner, P. H. and J. Zobrist. 1983. Die Mullverbrennung als Quelle vonMetallen in der Umwelt. Mull and Abfall, 9: 221–227.

Cembureau. 1999. Environmental benefits of using alternative fuels in cement production. The European Cement Association, Brussels.

Chakrabarty, P., V. K. Srivastava, and S. N. Chakrabarti. 1995. Solid waste disposal and the environment—A review. *Indian Journal of Environmental Protection*, 15(1): 39–43.

Damgaard, A., C. Riber, T. Fruergaard, T. Hulgaard, and T. H. Christensen. 2010. Life cycle assessment of the historical development of air pollution control and energy recovery in waste incineration. *Waste Management*, 30: 1244–1250.

de Fodor, E. 1911. *Elektrizitat aus Kehricht*. Budapest: MABEG (reprinted).

DEFRA (Department of Environment Food & Rural Affairs, UK). 2013. *Incineration of Municipal Solid Waste*. DEFRA: UK, 56pp.

De Priest, W. and R. Gaikwad. 2003. Economics of lime and limestone for control of sulfur dioxide. In *Proceedings of Combined Power Plant Air Pollutant Control Mega Symposium*, Washington, DC, p. 1.

Desideri, U., F. Di Maria, D. Leonardi, and S. Proietti. 2003. Sanitary landfill energetic potential analysis: A real case study. *Energy Conversion and Management*, 44: 1969–1981.

Dioha, I. J., C. H. Ikeme, T. Nafi'u, N. I. Soba, and M. B. S. Yusuf. 2013. Effect of carbon to nitrogen ratio on biogas production, *International Research Journal of Natural Sciences*, 1(3): 1–10.

Edbertho, L.-Q. 2004. Plasma processing of municipal solid waste. *Brazilian Journal of Physics*, 34(4B): 1587–1593.

El-Fadel, M. and H. Sbayti. 2000. Economics of mitigating greenhouse gas emissions from solid waste in Lebanon. *Waste Management & Research*, 18: 329–340.

Enerkem 2014. *Enerkem Alberta Biofuels*. http://enerkem.com/facilities/enerkem-alberta-biofuels/

Garcia-Perez, M., X. S. Wang, J. Shen, M. J. Rhodes, F. Tian, W.-J. Lee, H. Wu, and C.-Z. Li, 2008. Fast pyrolysis of oil mallee woody biomass: Effect of temperature on the yield and quality of pyrolysis products. *Industrial & Engineering Chemistry Research*, 47(6): 1846–1854.

Giusti, L. 2009. A review of waste management practices and their impact on human health. *Waste Management*, 29: 2227–2239.

Groß, B., C. Eder, P. Grziwa, J. Horst, and K. Kimmerle. 2008. Energy recovery from sewage sludge by means of fluidized bed gasification. *Waste Management*, 28: 1819–1826.

Hajime, J. 1996. Plasma melting and useful application of molten slag. *Waste Management*, 16(5): 417–422.

Henry, R. K., Z. Yongsheng, and D. Jun. 2006. Municipal solid waste management challenges in developing countries—Kenyan case study. *Waste Management*, 26: 92–100.

Hester, R. E. and R. M. Harrison. 2013. *Waste as a Resource*. Royal Society of Chemistry, London, UK. ISBN: 978-1-84973-668-8.

Howard, C. V. 2009. Statement of evidence, particulate emissions and health. Proposed Ringaskiddy Waste-to-Energy Facility, June 2009.

IEA. 2014. http://www.iea-coal.org.uk/site/ieacoal/databases/ccts/selective-catalytic-reduction-scr-for-nox-control (assessed May 2015).

International Solid Waste Association (ISWA). 2012. Report. http://www.iswa.org/fileadmin/galleries/Publications/ISWA_Reports/ISWA_Report_2012.pdf

IPCC. 2005. Working Group III of the Intergovernmental Panel on the Climate Change (IPCC). In: *Carbon Dioxide Capture and Storage, Intergovernmental Panel on the Climate Change (IPCC)*, Metz, B., O. Davidson, H. Coninck, M. Loos, and L. Meyer (Eds.). Cambridge, UK: Cambridge University Press.

IPCC. 2006. Hoornweg D. and P. Bhada-Tata, 2006. IPCC Guidelines for National Greenhouse Gas Inventories programme. Prepared by the Produced by the World Bank's Urban Development and Local Government Unit of the Sustainable Development Network, the Urban Development Series, March 2012, No. 15.

IPCC (Intergovernmental Panel on Climate Change). 2015. https://www.ipcc.ch/publications_and_data/ar4/wg3/en/ch10-ens10-3-4.html. Accessed May 2015.

Jalan, R. K. and V. K. Srivastava. 1995. Incineration, land pollution control alternative—Design considerations and its relevance for India. *Indian Journal of Environmental Protection*, 15(12): 909–913.

Jang, H. N., Y. C. Seo, J. H. Lee, J. H. Kim, D. Pudasainee, J. M. Park, K. J. Song, Y. H. Mun, and K. S. Park. 2007. Implementation of US MACT regulation to Korean municipal waste incinerators. *Journal of Korea Society of Waste Management*, 12(1): 31–35.

Jensen, D. L., A. Ledin, and T. H. Christensen. 1999. Speciation of heavy metals in landfill-leachate polluted groundwater. *Water Research*, 33: 2642–2650.

Jha, M. K., O. A. K. Sondhi, and M. Pansare. 2003. Solid waste management—A case study. *Indian Journal of Environmental Protection*, 23(10): 1153–1160.

Joardar, S. D. 2000. Urban residential solid waste management in India. *Public Works Management and Policy*, 4(4): 319–330.

Johnke, B. 2000. IPCC. In: *Good practice guidance and uncertainty management in national greenhouse gas inventories, emission from waste Incineration*, pp. 455–468.

Jørgensen, P. J. 2009. PlanEnergi, *Biogas-Green Energy*, 2nd edition, Editor: A. B. Nielsen, Digisource Danmark A/S, ISBN 978-87-992243-2-1.

Kansal, A. 2002. Solid waste management strategies for India. *Indian Journal of Environmental Protection*, 22(4): 444–448.

KEEI (Korea Energy Economics Institute). 2010. *Yearbook of Energy Statistics*, Seoul, S. Korea.

Kerester, A. 2015. Gasification can help meet the world's growing demand for cleaner energy and products. *Cornerstone, an Official Journal of the World Coal Industry*. Spring. http://cornerstonemag.net/gasification-can-help-meet-the-worlds-growing-demand-for-cleaner-energy-and-products/

Khan, R. R. 1994. Environmental management of municipal solid wastes. *Indian Journal of Environmental Protection*, 14(1): 26–30.

Kim, J. H., Y. C. Seo, D. Pudasainee, S. H. Lee, S. J. Cho, H. N. Jang, and J. M. Park. 2009. Efforts to develop regulations in Korea similar to the US maximum achievable control technology (MACT) regulations for hazardous waste incinerators. *Journal of Material Cycles and Waste Management*, 11: 183–190.

Knox, A. 2005. An Overview of Incineration and EFW Technology as Applied to the Management of Municipal Solid Waste (MSW). http://www.durhamenvironmentwatch.org/Incinerator%20Files%20II/OverviewOfIncinerationAndEFWKnox.pdf.

Kwon, E., K. J. Westby, and M. J. Castaldi. 2010. Transforming Municipal solid waste (MSW) into fuel via gasification/pyrolysis process, *Processing of the 18th Annual Waste-to-Energy Conference NAWTEC 18*, Orlando, Florida, May 11–13, 2010.

Liu, B. 2010. Research on garbage power. *Applied Energy Technology*, 7b.2: 25–28.

Lombardi, L., E. Carnevale, and A. Corti. 2015. A review of technologies and performances of thermal treatment systems for energy recovery from waste. *Waste Management*, 37: 26–44.

Lou, X. F. and J. Nair. 2009. The impact of landfilling and composting on greenhouse gas emissions—A review. *Bioresource Technology*, 100: 3792–3798.

Meriçboyu, A. E., U. G. Beker, and S. Küçükbayrak. 1998. Important futures determining the use of coal. In: *Coal*, edited by O. Kural. Kurtis Press, Istanbul-Turkey, pp. 149–165.

Michael, C. 2006. Municipal solid waste (MSW) options: Integrating organics management and residual treatment/disposal. Technical report, Natural Resources Canada, April 2006.

Milios, L. and A. Reichel. 2013. Municipal waste management in Sweden. European Environment Agency, February 2013.

Miller, B. G. 2011. *Clean Coal Engineering Technology*. New York: Elsevier.

Mohan, D., C. U. Pittman, and P. H. Steele. 2006. Pyrolysis of wood/biomass for bio-oil: A critical review. *Energy & Fuels*, 20: 848–889.

Mountouris, A., E. Voutsas, and D. Tassios. 2008. Plasma gasification of sewage sludge: Process development and energy optimization. *Energy Conversion and Management*, 49(8): 2264–2271.

Moustakasa, K., D. Fattab, S. Malamisa, and K. Haralambousa. 2003. Demonstration plasma gasification/vitrification system for effective hazardous waste treatment. *Journal of Hazardous Materials*, 123(1–3): 120–126.

Mufeed, S., A. Kafeel, M. Gauhar, and R. C. Trivedi. 2008. Municipal solid waste management in Indian cities—A review, *Waste Management*, 28: 459–467.

Mukherjee, S., S. Mukhopadhyay, M. A. Hashim, and B. Sengupta. 2015. Contemporary environmental issues of landfill leachate: Assessment and remedies. *Critical Reviews in Environmental Science and Technology*, 45: 472–590.

OWAV (Osterreichischer Wasser- und Abfall wirtschaftsverb and). 2013. Energetische Wirkungsgrade von Abfall verbrennungsanlagen, OWAV Regelblatt 519. 1020 Vienna, Heinestrase 2: Austrian Standards plus Publishing, OWAV.

Paradelo, R., A. Villada, R. Devesa-Rey, A. B. Moldes, M. Domínguez, J. Patiño, and M. T. Barral. 2011. Distribution and availability of trace elements in municipal solid waste composts. *Journal of Environmental Monitoring*, 13: 201–211.

Parikh, J., S. A. Channiwala, and G. K. Ghosal. 2005. A correlation for calculating HHV from proximate analysis of solid fuels. *Fuel*, 84: 487–494.

Pavlas, M., M. Tou, L. Bébar, and P. Stehlík. 2010. Waste to energy—An evaluation of the environmental impact. *Applied Thermal Engineering*, 30: 2326–2332.

Peavey, H. S., R. R. Donald, and G. Gorge. 1985. *Environmental Engineering*. Singapore: McGraw-Hill Book Co.

Pudasainee, D., J. H. Kim, Y. S. Yoon, and Y. C. Seo. 2012. Oxidation, reemission and mass distribution of mercury in bituminous coal fired power plants with SCR, CS-ESP and wet FGD. *Fuel*, 93: 312–318.

Reddy, P. J., 2011. *Municipal Solid Waste Management: Processing—Energy Recovery—Global Examples*. CRC Press, USA. ISBN 9780415690362 - CAT# K13873.

Rosa, E. D., D. Rubel, M. Tudino, A. Viale, and R. J. Lombardo. 1996. The leachate composition of an old waste dump connected to groundwater: Influence of the reclamation work. *Journal of Environmental Monitoring and Assessment*, 40: 239–252.

Ryn, C. and D. Shin. 2013. Combined heat and power from municipal solid waste: Current status and issues in South Korea. *Energies*, 6: 45–57.

Seldman, N. 2008. Recycling first—Directing federal stimulus money to real green projects. *E Magazine*.

Seo, Y. 2014. Current MSW Management and Waste-to-Energy Status in the Republic of Korea. MSc Thesis. Department of Earth and Environmental Engineering, Columbia University, USA.

Sharholy, M., K. Ahmad, G. Mahmood, and R. C. Trivedi. 2008. Municipal solid waste management in Indian cities—A review. *Waste Management*, 28: 459–467.

Simone, M., E. Biagini, C. Galletti, and L. Tognotti. 2009. Evaluation of global biomass devolatilization kinetics in a drop tube reactor with CFD aided experiments. *Fuel*, 88(10): 1818–1827.

Song, G.-J., S. H. Kim, Y. C. Seo, and S. C. Kim. 2008. Dechlorination and destruction of PCDDs/PCDFs in fly ashes from municipal solid waste incinerators by low temperature thermal treatment. *Chemosphere*, 71: 248–257.

Sreenivasulu, B., D. V. Gayatri, T. Sreedhar, and K. V. Raghavan. A journey into the process and engineering aspects of carbon capture technologies. *Renewable and Sustainable Energy Reviews*, 41(2015): 1324–1350.

Stantec. 2011. A technical review of municipal solid waste thermal treatment practices. Final report, Waste to Energy, Project No.: 1231–10166, March 2011, British Columbia, Canada.

Subramani, T., R. Umarani, and S. K. B. Devi. 2014. Sustainable decentralized model for solid waste management in urban India. *Internationl Journal of Engineering Research and Applications*, 4(6): 264–269.

Sudhire, V., V. R. Muraleedharan and G. Srinivasan. 1996. Integrated solid waste management in urban India: A critical operational research framework. *Journal of Socio-Economic Planning Science*, 30(3): 163–181.

Swanson, R. M. 2010. Techno-economic analysis of biofuels production based on gasification. National Renewable Energy Laboratory Technology Report. Golden, Colorado: National Renewable Energy Laboratory.

Tanigaki, N., Y. Ishida, and M. Osada, 2015. A case-study of landfill minimization and material recovery via waste co-gasification in a new waste management scheme. *Waste Management*, 37, 137–146.

Themelis, N. J. and P. A. Ulloa. 2006. Methane Generation in Landfills. http:// www.aseanenvironment.info/ Abstract/41014160.pdf (accessed March 2008).

Themelis, N. J. and P. A. Ulloa. 2007. Methane generation in landfills. *Renewable Energy*, 32, 1243–1257.

U.S. DOE. 2007. Energy Information Administration. In: *Methodology for Allocating Municipal Solid Waste to Biogenic and Non-Biogenic Energy*. Washington, DC: U.S. Department of Energy.

U.S. DOE. 2015. Plasma gasification. United States Department of Energy. Retrieved May 10, 2015. http:// www.netl.doe.gov/research/coal/energy-systems/gasification/gasifipedia/westinghouse

U.S. EPA (US Environmental Protection Agency). 2001. *Inventory of U.S. Greenhouse Gas Emissions and Sinks: 1990–1999*. EPA-236-R-01-001. Washington, DC: Office of Atmospheric Programs, U.S. Environmental Protection Agency, April 2001.

U.S. EPA. 2002. What is integrated solid waste management? Report # EPA530-F-02-026a, May 2002. http:// www.epa.gov/climatechange/wycd/waste/downloads/overview.pdf

U.S. EPA. 2006. *US Environmental Protection Agency. Global anthropogenic non-CO$_2$ greenhouse gas emissions: 1990–2020*. Washington, DC: USEPA.

U.S. EPA. 2014. Municipal solid waste generation, recycling, and disposal in the United States: Facts and figures for 2012. EPA-530-F-14-001. http://www.epa.gov/solidwaste/nonhaz/municipal/pubs/2012_msw_fs.pdf

U.S. EPA. 2011. Inventory of U.S. greenhouse gas emissions and sinks: 1990–2009; EPA 430-R-11-005, April 15, 2011. N.W. Washington, DC: USEPA.

Waste Management World. 2015. A billion reasons for waste to energy in India. http://www.waste-management-world.com/articles/print/volume-14/issue-6/wmw-special/a-billion-reasons-for-waste-to-energy-in-india.html. Accessed April 28, 2015.

Werther, J. and T. Ogada. 1999. Sewage sludge combustion. *Progress in Energy and Combustion Science*, 25: 55–116.

Willumsen, H. 2003. Energy recovery from landfill gas in Denmark and worldwide; www.lei.lt/Opet/pdf/ Willumsen.pdf.

World Bank. 2005. Countries are classified into four income levels according to World Bank estimates of 2005. http://siteresources.worldbank.org/INTURBANDEVELOPMENT/Resources/336387-1334852610766/ What_a_Waste2012_Final.pdf.

World Bank. 2012. Hoornweg D. and P. Bhada-Tata. What a waste: A global review of solid waste management, March 2012, No. 15.

World Bank. 2014. GEF grant to enhance the environmental performance of municipal solid waste incinerators in Chinese cities. http://www.worldbank.org/en/news/press-release/2014/11/14/gef-grant-to-enhance-the-environmental-performance-of-municipal-solid-waste-incinerators-in-chinese-cities. Accessed April 28, 2015.

World Bank. 1999. Municipal solid waste incineration. Technical guidance report. http://www.worldbank.org/urban/solid_wm/erm/CWG%20folder/Waste%20Incineration.pdf. Accessed May 19, 2015.

www.edmonton.ca/waste. The Edmonton sustainability papers. May 2010. Discussion paper 10—Sustainable waste management. Accessed May 2015. http://www.edmonton.ca/city_government/documents/Discussion_Paper_10_Sustainable_Waste_Management.pdf

Yan, R., D. Gauthier, and G. Flamant. 2001. Partitioning of trace elements in the flue gas from coal combustion. *Combustion and Flame*, 125(1–2): 942–954.

Yang, N., H. Zhang, M. Chen, L. M. Shao, and P. J. He. 2012. Greenhouse gas emissions from MSW incineration in China: Impacts of waste characteristics and energy recovery. *Waste Management*, 32: 2552–2560.

Zheng, L., J. Song, C. Li, Y. Gao, P. Geng, B. Qu, and L. Lin. 2014. Preferential policies promote municipal solid waste (MSW) to energy in China: Current status and prospects. *Renewable and Sustainable Energy Reviews*, 36: 135–148.

27 Electrochemical Energy Systems and Efficient Utilization of Abundant Natural Gas

Manoj K. Mahapatra, Boxun Hu, and Prabhakar Singh

CONTENTS

ABSTRACT

Demand for affordable clean energy is ever growing for energy security and sustainable environment as it relates to the societal and economic well-being and prosperity of the current and future generations. Electrochemical systems such as fuel cells are at the forefront of the emerging energy technologies to fulfill the demand. Compared to other fossil fuels, utilization of abundant natural gas in the fuel cells further decreases the carbon footprint and emission of other pollutants such as nitrogen oxide (NO_x), sulfur oxide (SO_x), and heavy metal particulates. Economical and reliable processing of natural gas for efficient use in fuel cell systems is one of the crucial steps for the commercial success of the fuel cell systems. In this contribution, we have introduced the advantages of the use of natural gas in fuel cell systems followed by the discussion on different fuel processing methods and challenges.

INTRODUCTION

The International Energy Agency (IEA) has projected that the world's energy demand will increase from about 12 billion t.o.e. (ton oil equivalents) in 2009 to 17–18 billion t.o.e. by 2035. Carbon dioxide emissions are expected to increase from 29 gigatonnes (Gt)/year to 36–43 Gt/year (International Energy Agency in World Energy Outlook, 2011; Chu and Majumdar, 2012). Conventional energy

TABLE 27.1
Global Climate Change Estimate around 2050

Scenario	Temperature Increase (°C)	Sea Level Rise (cm)	CO_2 Concentration (ppm)
Low	0.9	12	467
Medium low	1.5	18	443
Medium high	2.1	25	554
High	2.4	67	528

Source: Reprinted from *Renewable and Sustainable Energy Reviews*, 15(9), Stambouli, A. B., Fuel cells: The expectations for an environmental-friendly and sustainable source of energy, 4507–4520, Copyright 2011, with permission from Elsevier.

production and consumption of fossil fuels have significant environmental impact due to emission of greenhouse gases, organic volatile matter, and particulate matters, leading to health hazards and global climate change. An estimate for the global climate change around 2050 is given in Table 27.1 (Stambouli, 2011). Air pollution costs more than 1,700,000 deaths each year worldwide.

Maintaining CO_2 concentration in atmosphere at 450 ppm will require an 80% reduction in global emissions by 2050 relative to a 2005 baseline. Utilization of natural gas as fuel emit less CO_2 and other toxic pollutants than other conventional fossil fuel such as coal and oil as shown in Table 27.2. Other advantages include less water consumption than coal and petroleum and its abundance in nature (Stambouli, 2011; Center for Climate and Energy Solutions, 2013).

Along with the utilization of low-emission natural gas in stationary and transportation energy sectors, increase in the efficiency of the conventional power plants, vehicles, and turbines, modernizing electrical grids, and adoption of combustion-free advanced energy systems such as fuel cells, photovoltaics, and wind turbines, and renewable energy resources such as sunlight and wind can fulfill the demand for energy security and sustainable environment. Energy systems using renewable energy resources (sunlight and wind) have two major disadvantages: low efficiency and geographical location dependency (Stambouli, 2011, Chu and Majumdar, 2012). Fuel cells, in principle, converts available chemical energy present in fuels (hydrogen, hydrocarbons) to useful electrical energy by electrochemical interaction with oxidant while emission of pollutants is low (Table 27.3) (Wang et al., 2010). Absence of direct chemical combustion in a fuel cell produces clean electricity. Fuel cells are considered as the best choice for clean energy generation due to high efficiency (40%–50%), fuel flexibility, and modularity for diverse applications (Haile, 2003; EG&G Technical Services, Inc., 2004). Utilization of natural gas in a fuel cell has the potential to fulfill the demand of clean energy. The efficiency of fuel cells can be increased to 85%–90% with the use of combined heat and power (CHP) (Ellamla et al., 2015). The development and expansion of CHP systems can reduce the CO_2 emissions by ~4% (170megatonnes[Mt]/year) in

TABLE 27.2
Average Fossil Fuel Power Plant Emission Rates (lb/MW h)

Fuel Type	CO_2	SO_x	NO_x
Coal	2249	13	6
Natural gas	1135	0.1	1.7
Oil	1672	12	4

Source: Center for Climate and Energy Solutions. 2013. Leveraging natural gas to reduce greenhouse gas emissions. http://www.c2es.org/publications/leveraging-natural-gas-reduce-greenhouse-gas-emissions

TABLE 27.3
Different Energy Systems Emission Rates (lb/MW h)

Energy Systems	Greenhouse Gas (CO_2)	NO_x	Particulates	CO
Average U.S. grid baseload	1655	1.88	2.16	0.42
Natural gas microturbine	903	0.75	0.06	0.32
Natural gas internal combustion engine (NG ICE)	963	3.76	0.05	6.2
Fuel cell power plant (46% efficiency)	752	0.17	0.01	0–0.07

Source: Wang, M.Q., A. Elgowainy, and J. Han. 2010. Life-cycle analysis of criteria pollutant emissions from stationary fuel cell systems. *US DOE Annual Merit Review Proceedings.*

2015 and ~10% (950 Mt/year) by 2030. Other advantages include (a) uninterrupted electricity (with continuation of fuel supply) with minimal attention during operation, (b) reliability, (c) low levelized cost, (d) noise-free operation for indoor applications, (e) centralized as well as distributed power generation (grid free) even suited for remote applications, (f) adaptability to grid applications for stable power transmission, and (g) potential reuse of the produced water (EG&G Technical Services, 2004; Wang et al., 2010; Stambouli, 2011; Tibaquirá et al., 2011; Chick et al., 2015; Ellamla et al., 2015).

A number of fuel cells, depending on the electrolyte materials and operating temperature, exist in literature and market place despite very similar working principles. Among them, polymer electrolyte fuel cell (PEFC), molten carbonate fuel cell (MCFC), and solid oxide fuel cells (SOFCs) are attractive for commercialization due to the versatility of the technology and applications (EG&G Technical Services, Inc., 2004). The shipments of PEFC, MCFC, and SOFCs have, respectively, been grown by 87.2%, 600%, and 300% in 2010 and 2011 period. The growth is anticipated to continue (The Fuel Cell Today, 2012).

WORKING PRINCIPLE OF FUEL CELLS

Fuel cell is an electrochemical device that operates on Nernst's principle

$$V_{Nernst} = \frac{RT}{4F} \ln\left(\frac{P_{O_2} \text{ oxidant}}{P_{O_2} \text{ fuel}}\right) \tag{27.1}$$

A single fuel cell consists of anode, cathode, and electrolyte. The electrode materials consist of electro-catalysts for fuel oxidation at the anode and oxygen reduction at the cathode. The electrolyte, which is essentially an ionic conductor, facilitates electrochemical interaction between fuel and oxidant while prevents direct chemical combustion. Ions generated at the electrodes due to electrochemical reactions migrates through the ionic conductor electrolyte, simultaneously, electronic charges flow through an external circuit and produces electrical power (Haile, 2003). The basic reaction of a fuel cell using hydrogen as the fuel and oxygen as the oxidant is exothermic and water is the by-product (Larminie and Dicks, 2003):

$$H_2 + \frac{1}{2}O_2 \rightarrow H_2O, \quad \Delta H_{298} = -241.83 \text{ kJ / mol if } H_2O \text{ is in steam form}$$
$$= -285.84 \text{ kJ / mol if } H_2O \text{ is in liquid form} \tag{27.2}$$

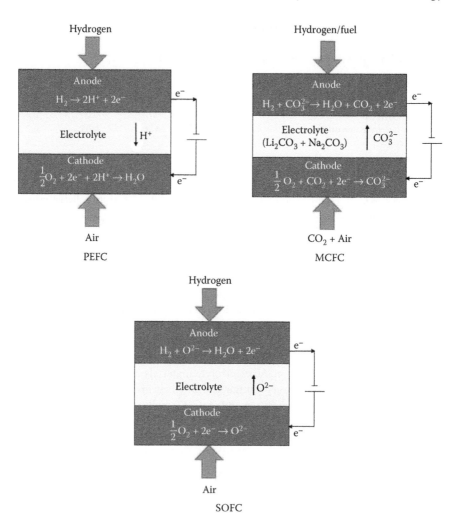

FIGURE 27.1 Schematic showing working principle of PEFC, MCFC, and SOFC.

The ideal cell voltage can vary between 0.90 and 1.23 V depending on the fuel cell type (EG&G Technical Services, Inc., 2004).

In PEFC, hydrogen is oxidized in the anode and the protons (H^+) conducts through electrolyte and combines with oxygen at the cathode to produce water. In MCFC, carbonate ion forms at the cathode, conducts through molten carbonate electrolyte, and oxidizes fuel at the anode. Carbon dioxide produced in the anode can be streamed to the cathode. In SOFC, oxygen ion forms at the cathode, conducts through solid electrolyte, and oxidizes fuel at the anode. Proton conducting electrolyte is also being developed for SOFC. The working principle and the electrochemical reactions of the fuel cells are shown in Figure 27.1 (EG&G Technical Services, Inc., 2004). The characteristics of the different kinds of fuel cells are given in Table 27.4 (EG&G Technical Services, Inc., 2004; Stambouli, 2011; Mahapatra and Singh, 2013; Kalmula and Kondapuram, 2015).

FUEL CELL POWER SYSTEM

A typical fuel cell power system, schematically shown in Figure 27.2, consists of a fuel processor, fuel cell power module, power conditioning equipment, and process gas heat exchanger for thermal management. Fuel processor is required to convert commercially available fuels into hydrogen rich

TABLE 27.4

Characteristics of PEFC, MCFC, and SOFC

	Polymer Electrolyte Fuel Cell (PEFC)	Molten Carbonate Fuel Cell (MCFC)	Solid Oxide Fuel Cell (SOFC)
Operating temperature (°C)	40–80	600–700	600–1000
Charge carrier	H^+	CO_3^{2-}	O^{2-}
Electrolyte	Hydrated polymeric ion exchange membrane	Liquid molten carbonate in $LiAlO_2$	Ion conducting ceramics (yttria-stabilized zirconia, gadolinia-doped ceria, lanthamgallate)
Electrodes	Carbon	Nickel and nickel oxide	Perovskite, cermet (perovskite/fluorite and metal cermet)
Fuel	Hydrogen or methanol	Hydrogen, hydrocarbons	Hydrogen, hydrocarbons
Oxidant	O_2/air	CO_2/O_2/air	O_2/air
Ideal cell voltage (Nernst potential) in air-fed system and pure hydrogen fuel	~1.17 V at 80°C	~1.03 V	~0.99 V at 800°C
Heat quality	–	High	High
Advantages	• Solid electrolyte reduces corrosion and electrolyte management problems • Low operating temperature • Quick start up	• Fuel flexibility • Low-cost catalyst due to high operating temperature	• All solid-state components • Fuel flexibility • Low-cost catalysts due to high operating temperature • No electrolyte flooding • Highest efficiency
Disadvantages	• Poisoning by trace contaminants in fuel • High-cost platinum catalyst • Difficulties in thermal and water management	• Corrosive electrolyte • High operating temperature • Long-term reliability of materials due to high temperature	• High operating temperature • High manufacturing cost • Long-term reliability of materials due to high temperature
Impurity tolerance level	<20 ppm H_2S; <50 ppm H_2S + COS (carbonyl sulfide); <10 ppm CO	<10 ppm H_2S	<1 ppm H_2S and <3000 ppm H_2S at 1000°C operating temperature for all ceramic fuel cell
Applications	Portable and transport applications	Distributed and centralized stationary applications	Distributed and centralized stationary applications

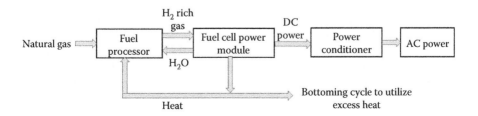

FIGURE 27.2   Schematic of a fuel cell power system.

gas. A fuel cell power module consists of several cells connected in series through interconnect (electrical conductor) to obtain useful power. Power conditioning unit is required to convert fuel cell generated DC power to usable AC power. Heat exchangers are integral parts in a fuel cell system for thermal management. Heat exchangers are required for preheating fuel, steam, reformer, and air, waste-heat recovery from the exhaust gas (Larminie and Dicks, 2003; EG&G Technical Services, Inc., 2004).

Fuel Processor

Fuel processor is required to convert commercially available hydrocarbon fuels into syngas and/or hydrogen used as fuel in a fuel cell. No fuel processor but a hydrogen storage is required if hydrogen is the fuel feedstock. Hydrogen is preferred over hydrocarbons because (EG&G Technical Services, Inc., 2004)

1. The kinetics of hydrogen oxidation on the anode is significantly faster than that of CO, CH_4, and other higher hydrocarbon oxidation.
2. Mass-transfer coefficient of hydrogen to the electrochemically active triple phase boundary (TPB) and on the interface of the anode and electrolyte through the porous anode is at least 10 times higher than that of CO, CH_4, and other higher hydrocarbons.

Natural gas primarily consists of methane, very small amount of higher hydrocarbons such as ethane, propane, and butane and negligible sulfur or halogen impurities. It can either be processed to produce hydrogen fuel or can be used directly as a fuel for high-temperature fuel cells. However, sulfur-containing odorants such as mercaptans, disulfides, or commercial odorants are added in pipeline natural gas (PNG) for leak detection (Economides and Oligney, 2000; Larminie and Dicks, 2003; EG&G Technical Services, Inc., 2004; Demirbas, 2010; Center for Climate and Energy Solutions, 2013). Presence of sulfur above tolerance level (Table 27.4) deactivates the catalysts and/ or electro-catalysts used for fuel processing and fuel oxidation at the fuel cell anode as well. Fuel processing primarily involves three steps: (1) fuel cleaning to remove impurities such as sulfur and halogens, (2) conversion of fuels into hydrogen rich fuel gas, and (3) water–gas shift (WGS) reforming for conversion of CO to hydrogen rich fuel stream.

Fuel Cleaning

There are two main processes for desulfurization of fuel: hydrodesulfurization (HDS) and sulfur adsorption (Singh and George, 1997; Farrauto et al., 2003; Larminie and Dicks, 2003; EG&G Technical Services, Inc., 2004; Van herle et al., 2004; Aasberg-Petersen et al., 2011; Wang et al., 2013), HDS takes place at 300–450°C and the sulfur-containing compounds are converted into hydrogen sulfide (H_2S) as per the below chemical reaction:

$$(C_2H_5)_2S + 2H_2 = 2C_2H_6 + H_2S \qquad (27.3)$$

Nickel-molybdenum oxide or cobalt-molybdenum oxide is commonly used as catalysts for desulfurization. The HDS process can decrease the sulfur content possibly up to ~10 ppm; suited for PEFC system. Activated carbon, zeolites, and ceria adsorbents are probably suited to lowering sulfur level below 10 ppm. To decrease the sulfur content to ~0.02 ppm in the fuel, the gas is passed through zinc oxide absorbent at ~400°C where the H_2S gas is converted into stable zinc sulfide:

$$H_2S + ZnO = ZnS + H_2O \qquad (27.4)$$

Alkali and alkaline earth carbonate adsorbents are used to remove halogen impurities such as chlorine (Dolan et al., 2012). Sodium aluminate is also being investigated as the adsorbents for halogen removal (Kobayashi et al., 2015).

FUEL CONVERSION

There are three methods for processing and conversion of gaseous hydrocarbons to hydrogen rich fuel for a fuel cell system. The available processes are partial/preferential oxidation (POX), steam reforming (SR), and auto-thermal reforming (ATR) (Larminie and Dicks, 2003; EG&G Technical Services, Inc., 2004; Cheekatamarla and Finnerty, 2006; Aasberg-Petersen et al., 2011).

Partial Oxidation

Combustion of natural gas in the presence of substoichiometric air or oxygen is known as partial oxidation (POX). The POX temperature is generally 1200–1400°C in the absence of a catalyst. The POX onto a catalyst bed, popularly known as catalytic partial oxidation (CPOX), can decrease the oxidation/reaction temperature up to 800–900°C. Hydrogen-rich gas is produced according to the following exothermic reaction:

$$CH_4 + \frac{1}{2}O_2 = CO + 2H_2, \quad \Delta H_{298} = -35.7\,\text{kJ/mol} \tag{27.5}$$

Platinum, palladium, and rhodium catalysts are used for CPOX. Heat generation due to exothermic reaction increases the temperature of the reactor. Steam is usually added to cool down the reactor temperature and promote SR and WGS reactions. CPOX process is preferred for auxiliary power unit (APU) and mobile applications due to simple design, compact size, and lightweight as well as fast response (Larminie and Dicks, 2003; EG&G Technical Services, Inc., 2004; Aasberg-Petersen et al., 2011).

Steam Reforming

In SR, the oxidizer is steam instead of air or oxygen. Hydrogen concentration and conversion efficiency are the highest for SR. This technique may save ~30% fuel than other processing technique (Hubert and Metkemeijer, 2006). SR is preferable for stationary applications where start up time is not a major concern (Larminie and Dicks, 2003; EG&G Technical Services, Inc., 2004; Matsumura and Nakamori, 2004; Cheekatamarla and Finnerty, 2006; Aasberg-Petersen et al., 2011; LeValley et al., 2014). SR of CH_4 or other hydrocarbons (C_nH_m) is endothermic and requires a heat source. SR reaction generally occurs at high temperature (>800°C) and can be expressed as

$$CH_4 + H_2O = CO + 3H_2, \quad \Delta H_{298} = 206.2\,\text{kJ/mol} \tag{27.6}$$

$$C_nH_m + nH_2O = nCO + \left(\frac{m}{2} + n\right)H_2 \tag{27.7}$$

Nickel is commonly used as the SR catalyst. Along with SR, exothermic WGS reaction occurs at/above 500°C according to

$$CO + H_2O \rightarrow CO_2 + H_2, \quad \Delta H_{298} = -41\,\text{kJ/mol} \tag{27.8}$$

WGS reaction forms further hydrogen. CO_2 in the reaction product may further react with the unreacted methane to produce CO and H_2 according to

$$CO_2 + CH_4 \rightarrow 2CO + 2H_2 \quad \Delta H_{298} = 247 \text{ kJ/mol} \tag{27.9}$$

$$CO_2 + 2CH_4 + H_2O \rightarrow 2CO + 2H_2 \tag{27.10}$$

Despite CO can being used as a fuel, it is preferred to lower its quantity because

1. The electrochemical oxidation rate of CO is ~2 times lower than that of H_2 (Matsuzaki and Yasuda, 2000).
2. The presence of CO above tolerance level (Table 27.4) deactivates the catalysts by adsorption of platinum metals in PEFC anode and deposition of carbon by Boudouard reaction ($2CO \rightarrow C + CO_2$) in high-temperature fuel cell (MCFC, SOFC) anodes.

Auto-Thermal Reforming

ATR is a combination of POX and SR carried out in an adiabatic reactor as illustrated in Figure 27.3 (Aasberg-Petersen et al., 2011). Reaction kinetics, product hydrogen concentration, conversion efficiency, and response time can be tailored. A conventional ATR consists of a burner, a combustion reactor, and a catalyst bed in a refractory lined pressure vessel where CH_4 is converted into CO and H_2. To further increase the H_2 content, an adiabatic reactor is used for WGS reaction in two stages: high-temperature WGS (HTWGS) at 350–450°C and low-temperature WGS (LTWGS) reaction at 190–235°C. Use of adiabatic reactor increases its temperature due to exothermic reaction. Consequently, CO and H_2O concentration decreases with decrease in temperature but CO_2 and H_2 increases. HTWGS increases the reaction rate while LTWGS increases H_2 purity by decreasing the CO and H_2O content. Iron-8%–14% chromium-based catalyst is typically used for HTWGS reaction. Chromium acts as a stabilizer in the catalyst. However, addition of high amount

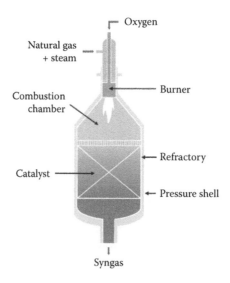

FIGURE 27.3 Schematic of a conventional auto-thermal reformer. (Reprinted from *Journal of Natural Gas Science and Engineering*, 3, Aasberg-Petersen, K. et al., Natural gas to synthesis gas—Catalysts and catalytic processes, 423–459, Copyright 2011, with permission from Elsevier.)

of chromium deactivates the catalyst due to segregation of Cr_2O_3 as a separate phase. Addition of Al_2O_3 and CeO_2 in the Fe–Cr catalyst enhances the catalytic activity and stability (Escritori et al., 2009). On the other hand, the $ZnO/ZnAl_2O_4$-based catalysts are also being investigated as chromium-free HTWGS catalyst. Copper-based catalysts supported on Al_2O_3 or ZnO/Al_2O_3 are used for LTWGS (Larminie and Dicks, 2003; Aasberg-Petersen et al., 2011).

During fuel processing, carbon can be deposited in several areas of metal reactor due to pyrolysis of hydrocarbon fuels (Larminie and Dicks, 2003; Wang et al., 2013; Aasberg-Petersen et al., 2011). For example, natural gas (CH_4) decomposes above 650°C in the absence of air or steam according to the following equation:

$$CH_4 \rightarrow C + 2H_2, \quad \Delta H = 75\,kJ/mol \qquad (27.11)$$

The carbon deposition by pyrolysis and Boudouard reactions causes "metal dusting" of the metal reactors made of nickel-containing stainless steel. Higher hydrocarbons decompose more easily than CH_4 increasing the risk of "metal dusting." Addition of steam above a critical steam/carbon ratio (2.0–3.0) eliminates the risk of carbon formation by not only reducing CO but also direct gasification of carbon:

$$C + H_2O \rightarrow CO + H_2 \qquad (27.12)$$

"Prereforming" of fuel is generally performed to avoid carbon deposition due to higher hydrocarbons. Prereforming is simply a SR process at 250–500°C in an adiabatic reactor. This process is beneficial to remove higher hydrocarbons from the natural gas since they convert preferentially into hydrogen. The exit gas stream from the prereformer reactor mainly consists of CH_4 and H_2O together with small quantity of hydrogen and carbon dioxide (Larminie and Dicks, 2003).

Carbon monoxide (CO) is a potential fuel for MCFC and SOFC but detrimental for PEFC. Removal of CO to 10–50 ppm is essential after the shift reactors for PEFC since it deactivates the platinum catalysts. CO removal can be done by three ways (Larminie and Dicks, 2003; Farrauto et al., 2003):

- In the selective oxidation, a small amount of air (typically around 2%) is added to the fuel stream in the reactor and then passes over a precious metal catalyst. This catalyst preferably absorbs CO rather than H_2 and reacts with the oxygen in air.
- The methanation of the CO is an approach that reduces the danger of producing explosive gas mixtures. The reaction is the opposite of the steam reformation

$$CO + 3H_2 \rightarrow CH_4 + H_2O, \quad \Delta H_{298} = -206\,kJ/mol \qquad (27.13)$$

- Palladium/platinum membranes can be used to separate and purify the hydrogen.
- A reformer product gas is passed into a reactor containing hydrogen gas absorbent material. After a set time, the reactor is isolated and the feed gas is diverted into a parallel reactor. At this stage, the first reactor is depressurized, allowing pure hydrogen gas to desorb from the materials. The process is known as pressure swing adsorption (PSA).

The fuel processing steps are shown schematically in Figure 27.4.

External and Internal Reforming

Natural gas can be processed into hydrogen-enriched gas either in a stand-alone reforming unit—"external reforming" or being integrated with fuel cell stacks (fuel cell power module)—"internal reforming." System compactness, higher overall fuel cell efficiency, high methane conversion,

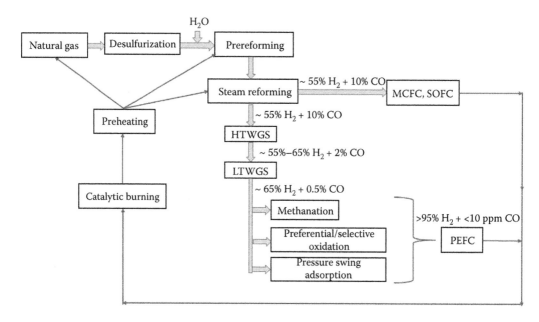

FIGURE 27.4 Schematic showing natural gas processing steps. (Larminie, J. and A. Dicks: *Fuel Cell Systems Explained*. 2003. 2nd ed. England. Copyright Wiley-VCH Verlag GmbH & Co. KGaA. Reproduced with permission. Farrauto, R. et al. 2003. *Annu. Rev. Mater. Res.*, 33, 1–27.)

economical energy management, and lower cost are the benefits of the internal reforming (Dicks, 1998; Larminie and Dicks, 2003; EG&G Technical Services, Inc., 2004; Qi and Karan, 2007; Campanari and Chiesa, 2013). Internal reforming is suited only for MCFC and SOFC because the heat generated due to hydrogen oxidation at high temperature can be directly used for fuel processing and CO can be used as a fuel (Matsuzaki and Yasuda, 2000). There are two types of internal reforming: indirect internal reforming (IIR) and direct internal reforming (DIR). In IIR, the fuel conversion reactors are in thermal contact with the fuel cell stacks and the reforming reactions and electrochemical reactions are separated. In IIR, the heat transfer from fuel cell compartment can be used only for reforming reaction but recycling of spent fuel or external heat source is required for rising the steam temperature. In DIR, the fuel conversion (reforming reactions) occurs directly in the anode compartment of the fuel cell stack by using heat produced by exothermic fuel cell reaction. The advantages of the DIR compared to IIR include better heat transfer, ease of chemical integration, and direct use of product steam from hydrogen oxidation at the anode for SR without use of external heat source.

For MCFC, IIR is preferred to avoid the deactivation of catalysts due to alkali attack. However, alkali resistant catalysts such as nickel supported on MgO, $LiAlO_2$, γ-Al_2O_3, core-shell catalysts (nickel catalysts by encapsulating Ni/SiO_2 and Ni/Al_2O_3 within zeolite shells), and precious metal catalysts (Ru/ZrO_2) are being developed for DIR (Larminie and Dicks, 2003; Wang et al., 2012; Antolini, 2011).

For SOFC, DIR is preferred and nickel in the anode is the catalyst for SR of CH_4 (Wang et al., 2013; Dicks, 1998; Qi et al., 2007). There are two major challenges for DIR in the SOFC anode:

1. Fast reforming rate of methane at SOFC operating temperature (~800°C) causes temperature drop at the fuel inlet. Thermal stress due to temperature gradient between fuel inlet and outlet can lead to mechanical failure of the SOFC. Robust design and thermal management address the problem. For example, recycling of anode tail gas, fuel supply in multiple stages, and decreasing SR rate decrease the temperature gradient. Temperature drop at the fuel inlet is also a problem for MCFC (Dicks, 1998).

2. Decomposition of CH_4 at the operating temperature and subsequent carbon deposition deactivate the catalytic activity of nickel (Wang et al., 2013). Three types of carbon, pyrolytic, encapsulating, and whisker, have been observed on nickel surface. Whisker carbon formation can cause "metal dusting" of nickel. Several approaches can be followed to avoid carbon deposition as discussed below (Wang et al., 2013; Hanna et al., 2014).

   a. Introduction of sufficient steam or other oxygen-containing gases into the fuel can inhibit the carbon deposition as shown in Figure 27.5. Carbon deposition increases with the operating temperature if CH_4 to H_2O molar ratio is less than 1:0.5. Maintaining CH_4 to H_2O molar ratio at 1:1.5 completely inhibits the carbon deposition (Sasaki and Teraoka, 2003a, b; Wang et al., 2013). However, high steam content decreases the Nernst potential, oxidizes nickel particles, and assists sintering of nickel. Volume change of nickel due to oxidation and cycling stress due to nickel oxidation and further reduction in H_2 (redox cycle) can cause cracks at the nickel–YSZ (yttria-stabilized zirconia) interface, disintegrate nickel particles, and detach the nickel particles from the YSZ. Both oxidation and sintering of nickel decreases the TPB for electrochemical oxidation of hydrogen. Sintering of nickel particles also favors carbon deposition and subsequent catalytic deactivation.

   b. Addition of hydrogen in the fuel can suppress carbon deposition by increasing the concentration of adsorbed hydrogen at nickel surface. However, an excessive amount of hydrogen slightly decreases the catalytic activity due to competition between active sites on nickel particles.

   c. Addition of alloying elements such as Sn and Cu interact with 3d electron of Ni can suppress carbon deposition by inhibition of nickel carbide formation.

   d. Controlling the O^{2-} flux ion from electrolyte to anode through polarization converts the deposited carbon on the nickel, if any, into CO and/or CO_2.

   e. Surface modification and decoration of nickel surface by homogeneous distribution of alloying elements such as Au, Cu, Ru, and Rh can reduce carbon deposition.

   f. Deposition of a catalyst layer (Ru-CeO_2) with high activity for reforming/partial oxidation of CH_4 can resist carbon deposition.

   g. Nickel catalyst, fuel electrode, and catalyst support impregnated with metal oxide forming chemicals consisting Mg, Ca and Al, Sr, and Al, Zr, Y, Ce, and their mixtures can resist carbon formation (Singh et al., 1990a, b, 1996).

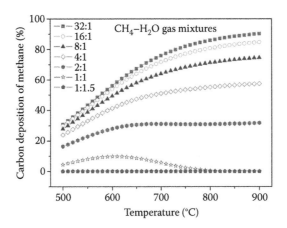

FIGURE 27.5 Thermodynamically predicted coke formation toward methane as a function of temperature. (Reprinted with permission from Wang, W. et al. Progress in solid oxide fuel cells with nickel-based anodes operating on methane and related fuels. *Chemical Reviews*, 113(10), 8104–8151. Copyright 2013 American Chemical Society.)

CHALLENGES

There are several challenges for the adoption of fuel cell as well as natural gas for tapping the commercial energy market. The initial investment cost of a fuel cell system is high even for a combined heat and power fuel cell (CHP-FC) system. For example, the price of a 1-kW PEFC or SOFC in Japan was US$21,000–27,000, which is 30–50 times higher than U.S Department of Energy (USDOE) targets. However, mass production will significantly reduce the cost as evident from ~20% cost reduction just by doubling in production of PEFC. Larger system also decreases the cost due to better thermal and water management (Ellamla et al., 2015). The investment for infrastructure required for utilization of natural gas in transportation sector either in conventional vehicles or in fuel cell vehicles is very expensive. For example, the infrastructure for compressed natural gas vehicles in United States alone will cost more than $100 billion (Chu and Majumdar, 2012).

Expensive platinum and platinum group materials (palladium, ruthenium, and rhodium) are used for fuel processing and the electro-catalyst for the PEFC anode as well. Reduction in use of precious metal catalysts is believed to decrease the cost. However, alternate metal catalysts such as Cu–Zn, Cu–Zn–Al, and Fe–Cr are pyrophoric in nature and require stringent reactor design for safe use. Decrease in the platinum loading as the electro-catalyst by controlling the microstructure and innovative processing is an attractive approach (Farrauto et al., 2003). Rare earth elements such as yttrium, lanthanum, and cerium are the constituents of the SOFC ceramic components. The global reserves of these elements are limited and USDOE has listed yttrium as the critical material. Recycling of these materials and alternate materials need to be explored (Thijssen, 2011).

Achieving a lifetime for 5–10 years for larger unit (kW–MW), high efficiency, and negligible carbon footprint are the trade-off for the very expensive initial investment. However, materials degradation during the fuel cell operation and their incompatibility with each other not only degrades system performance but also leads to stack failure. The system undergoes several thermal cycles due to system shut down and start up for replacing failed stacks. The start up time for SOFC varies from 2.5 to 20 h, which is unsuited for mobile applications (Ellamla et al., 2015). Fuel cleaning, preheating, and prereforming add up extra cost for fuel cell operation. Sulfur and carbon resistant mixed ionic-electronic conductors (perovskite and fluorite-perovskite composite materials) are being investigated to replace conventional nickel catalysts but low electrical conductivity and thermo-mechanical strength are their limitations (Mahapatra and Singh, 2013; Wang et al., 2013).

Continuous supply of fuel is essential for uninterrupted operation of fuel cell systems and may be a problem for auxiliary power units (APUs) and stand-alone stationary units in remote areas with lack of continuous fuel supply.

Lack of sufficient knowledge exchange between the commercial developers and R&D professionals in the academic institutions and national laboratories is another hindrance for mitigating the existing technical issues. As a result, the recent progress in the fuel cell technology commercialization is only incremental. Commercial developers contain the wealth of knowledge and vast experience at large-scale system level. However, very few developers can afford for in-depth study for system failure and mitigation while the community in the research universities and the national laboratories can focus on the fundamental study. Fortunately, several agencies across the world such as USDOE, European Union, New Energy and Industrial Technology Development Organization (NEDO) in Japan are fostering public–private partnership in terms of sharing knowledge and technology and business collaboration by regular coordination and financial support.

CONCLUDING REMARKS

Natural gas utilization in a fuel cell system has tremendous scope for sustainable energy security and clean environment but several issues, cost in particular, need to be overcome to compete with the conventional technologies. Government policies and support at domestic and international levels such as stringent carbon abatement tax, subsidies, and easy access to required resources influence the commercial success of the technology along with the R&D progress. Multiorganizational extensive collaboration across the industrial developers, national laboratories, research universities, and the state and federal agencies is the need of the hour for gaining the share in commercial energy market. Infrastructure development, business model to withstand the cost variation of natural gas and other resources such as rare earth elements and expensive noble metals, further improvement in technology through innovative design, breakthrough in materials discovery, thermal management, life cycle analysis, mitigation of system degradation, and competitive cost will determine the future of fuel cell systems expansion using natural gas in diverse applications.

REFERENCES

Aasberg-Petersen, K., I. Dybkjær, C. V. Ovesen, N. C. Schjødt, J. Sehested, and S. G. Thomsen. 2011. Natural gas to synthesis gas—Catalysts and catalytic processes. *Journal of Natural Gas Science and Engineering*, 3: 423–459.

Antolini, E. 2011. The stability of molten carbonate fuel cell electrodes: A review of recent improvements. *Applied Energy*, 88: 4274–4293.

Campanari, S., G. Manzolini, and P. Chiesa. 2013. Using MCFC for high efficiency CO_2 capture from natural gas combined cycles: Comparison of internal and external reforming. *Applied Energy*, 112: 772–783.

Center for Climate and Energy Solutions. 2013. Leveraging natural gas to reduce greenhouse gas emissions. http://www.c2es.org/publications/leveraging-natural-gas-reduce-greenhouse-gas-emissions

Cheekatamarla, P. K. and C. M. Finnerty. 2006. Reforming catalysts for hydrogen generation in fuel cell applications. *Journal of Power Sources*, 160: 490–499.

Chick, L., M. Weimar, G. Whyatt, and M. Powell. 2015. The case for natural gas fueled solid oxide fuel cell power systems for distributed generation. *Fuel Cells*, 15(1): 49–60.

Chu, S. and A. Majumdar. 2012. Opportunities and challenges for a sustainable energy future. *Nature*, 488: 294–303.

Demirbas, A. 2010. *Methane Gas Hydrate*. London: Springer, pp. 57–76.

Dicks, A. L. 1998. Advances in catalysts for internal reforming in high temperature fuel cells. *Journal of Power Sources*, 71(1–2): 111–122.

Dolan, M. D., A. Y. Ilyushechkin, K. G. McLennan, and S. D. Sharma. 2012. Halide removal from coal-derived syngas: Review and thermodynamic considerations. *Asia-Pacific Journal of Chemical Engineering*, 7: 171–187.

Economides, M. J. and R. E. Oligney. 2000. Natural gas: The revolution is coming. In *the Society of Petroleum Engineers SPE Annual Technical Conference and Exhibition*, Dallas, Texas, 2000. Document ID: SPE-62884-MS.

EG&G Technical Services, Inc. 2004. *Fuel Cell Handbook*, 7th ed. EG&G Technical Services, Inc. Morgantown, West Virginia, USA.

Ellamla, H. R., I. Staffell, P. Bujlo, B. G. Pollet, and S. Pasupathi. 2015. Current status of fuel cell based combined heat and power systems for residential sector. *Journal of Power Sources*, 293: 312–328.

Escritori, J. C., S. C. Dantas, R. R. Soares, and C. E. Hori. 2009. Methane autothermal reforming on nickel–ceria–zirconia based catalysts. *Catalysis Communications*, 10(7): 1090–1094.

Farrauto, R., S. Hwang, L. Shore, W. Ruettinger, J. Lampert, T. Giroux, Y. Liu, and O. Ilinich. 2003. New material needs for hydrocarbon fuel processing: Generating hydrogen for the PEM fuel cell. *Annual Review of Materials Research*, 33: 1–27.

Haile, S. M. 2003. Fuel cell materials and components. *Acta Materialia*, 51(19): 5981–6000.

Hanna, J., W. Y. Lee, Y. Shi, and A. F. Ghoniem. 2014. Fundamentals of electro- and thermochemistry in the anode of solid-oxide fuel cells with hydrocarbon and syngas fuels. *Progress in Energy and Combustion Science*, 40: 74–111.

Hubert, C., P. Achard, and R. Metkemeijer. 2006. Study of a small heat and power PEM fuel cell system generator. *Journal of Power Sources*, 156: 64–70.

International Energy Agency in World Energy Outlook. 2011, pp. 546–547. http://www.worldenergyoutlook.org

Kalmula, B. and V. R. Kondapuram. 2015. Fuel processor—Fuel cell integration: Systemic issues and challenges. *Renewable and Sustainable Energy Reviews*, 45: 409–418.

Kobayashi, M., H. Akiho, and Y. Nakao. 2015. Performance evaluation of porous sodium aluminate sorbent for halide removal process in oxy-fuel IGCC power generation plant *Energy* 92(3): 320–327.

Larminie, J. and A. Dicks. 2003. *Fuel Cell Systems Explained*. 2nd ed. England: John Wiley & Sons Ltd.

LeValley, T. L., A. R. Richard, and M. Fan. 2014. The progress in water gas shift and steam reforming hydrogen production technologies—A review. *International Journal of Hydrogen Energy*, 39(30): 16983–7000.

Mahapatra, M. K. and P. Singh. 2013. Fuel cells: Energy conversion technology. In *Future Energy*, 2nd ed. edited by Trevor M. L. Elsevier, pp. 511–547.

Matsumura, Y. and T. Nakamori. 2004. Steam reforming of methane over nickel catalysts at low reaction temperature. *Applied Catalysis A: General*, 258(1): 107–114.

Matsuzaki, Y. and I. Yasuda. 2000. Electrochemical oxidation of H_2 and CO in a H_2-H_2O-CO-CO_2 system at the interface of a Ni-YSZ cermet electrode and YSZ electrolyte. *Journal of the Electrochemical Society*, 147(5): 1630–1635.

Qi, A., B. Peppley, and K. Karan. 2007. Integrated fuel processors for fuel cell application: A review. *Fuel Processing Technology*, 88(1): 3–22.

Singh, P. and R. A. George. 1997. System for operating solid oxide fuel cell generator on diesel fuel. US patent 5686196, November 11, 1997.

Singh, P., R. J. Ruka, and R. A. George. 1990a. Electrochemical generator apparatus containing modified fuel electrodes for use with hydrocarbon fuels. US Patent 4894297, January 16, 1990.

Singh, P., R. J. Ruka, and R. A. George. 1990b. Electrochemical generator apparatus containing modified high temperature insulation and coated surfaces for use with hydrocarbon fuels. US Patent 4898792, February 6, 1990.

Singh, P., L. A. Shockling, R. A. George, and R. A. Basel. Hydrocarbon reforming catalyst material and configuration of the same. US Patent 5527631, June 18, 1996.

Sasaki, K. and Y. Teraoka. 2003a. Equilibria in fuel cell gases II. The C-H-O ternary diagrams. *Journal of the Electrochemical Society*, 150(7): A885–A888.

Sasaki, K. and Y. Teraoka, Y. 2003b. Equilibria in fuel cell gases I. Equilibrium compositions and reforming conditions. *Journal of the Electrochemical Society*, 150(7): A878–A884.

Stambouli, A. B. 2011. Fuel cells: The expectations for an environmental-friendly and sustainable source of energy. *Renewable and Sustainable Energy Reviews*, 15(9): 4507–4520.

The Fuel Cell Today. 2012. *The Fuel Cell Industry Review*. The Fuel Cell Today, London.

Thijssen, J. 2011. LLC, solid oxide fuel cells and critical materials: A review of implications. Report number: R102 06 04D1, 2011.

Tibaquirá, J. E., K. D. Hristovski, P. Westerhoff, and J. D. Posner. 2011. Recovery and quality of water produced by commercial fuel cells. *International Journal of Hydrogen Energy*, 36(6): 4022–4028.

Van herle, J., A. Schuler, L. Dammann, M. Bosco, T. Truong, E. D. Boni, F. Hajbolouri, F. Vogel, and G. G. Scherer. 2004. Fuels for fuel cells: Requirements and fuel processing. *Chimia*, 58(12): 887–895.

Wang, M. Q., A. Elgowainy, and J. Han. 2010. Life-cycle analysis of criteria pollutant emissions from stationary fuel cell systems. In *US DOE Annual Merit Review Proceedings*. Washington, DC, June 7–11.

Wang, W., C. Su, Y. Wu, R. Ran, and Z. Shao. 2013. Progress in solid oxide fuel cells with nickel-based anodes operating on methane and related fuels. *Chemical Reviews*, 113(10): 8104–8151.

Wang, P., L. Zhou, G. Li, H. Lin, Z. Shao, X. Zhang, and B. Yi. 2012. Direct internal reforming molten carbonate fuel cell with core-shell catalyst. *International Journal of Hydrogen Energy*, 37(3): 2588–2595.

28 Unconventional Gas

Jim Underschultz

CONTENTS

ABSTRACT

Despite the commonly used term 'unconventional gas', the gas itself is not unconventional but rather methane and other hydrocarbon compounds, the same as are found in conventional reservoirs. The gas is however trapped in an unconventional manner such as by capillarity or adsorption where the gas occurs as a continuous phase over large areas independent of trap geometry. There are a number of categories of unconventional gas we consider, including coal bed methane, shale gas, tight gas, basin-centred gas and gas hydrates. Initially, the commercial development of unconventional gas was motivated by a desire to improve the safety of coal mining through degassing, coupled with the OPEC (Organization of the Petroleum Exporting Countries) oil crisis of the mid-1970s raising concerns of energy security. This drove the U.S. government to provide tax incentives and invest in the adaptation of horizontal drilling technology and micro-seismic monitoring that ultimately unlocked the shale gas potential in North America.

The very nature of unconventional gas resources being widely distributed but with low technical recovery rates makes resource and reserve estimation difficult to define and subject to high uncertainty. Recovery is fundamentally technology dependent and thus as advances occur the resulting available commercially viable reserves could vary widely. The International Energy Agency (IEA) estimates total global technically recoverable gas resources (conventional and unconventional) to be >30 million PJ (>30,000 tcf) of which roughly half is unconventional gas. Despite the rapid development of unconventional gas resources in the United States, closely followed in Canada, and later in

Australia; and despite the impact on reducing the cost of energy, reducing greenhouse gas (GHG) emissions and bringing the United States towards energy self-sufficiency, the shift to unconventional gas has not been without its challenges. There have been concerns about the long-term impact of gas development on the environment (particularly groundwater and surface water resources), there has been speculation that methane emissions may result in higher life cycle emissions for power generation, and the historical surface footprint of unconventional gas operations has led to local community concern. Research has focused on addressing these challenges.

The IEA predict that between 2010 and 2035 world energy demand will increase between 16% and 47% depending on the policy scenario, and various parts of the energy mix will grow or shrink to various degrees to match this demand. In 2010, the global primary energy consumption consisted of oil (32%), coal (27%), gas (21%) and all other energy sources combined (20%). For electricity generation the mix is coal (40%), gas (22%), renewables (19%, mainly hydro), nuclear (12%) and other sources combined (7%). In all three IEA forward scenarios the demand for gas is expected to grow. This is because of transport flexibility (pipeline by land or liquefied natural gas by sea), a geographically widely dispersed resource base, the cost competitiveness of gas as a fuel, GHG emissions advantage over coal and the flexibility of gas fired power to rapidly match the variable output of distributed renewable energy sources (i.e. achieves grid stability). Public pressure to solve air quality issues in China may also drive down the utilisation of coal fired power aiding the uptake of gas.

UNCONVENTIONAL GAS RESOURCES DEFINED

It is important to first define what we mean by 'unconventional gas'. Firstly, the gas itself is not unconventional, it is simply methane and other higher hydrocarbon compounds. In the literature there are two main criteria for definition that can be found. The first is based on the trapping mechanism. Conventional reservoirs are where gas is trapped by the buoyancy of the hydrocarbon phase beneath a low permeability seal and with a downdip formation water leg. In such cases a discrete hydrocarbon pool or field can be defined. In contrast, unconventional gas does not rely on buoyancy and trap geometry as the primary trapping mechanism. Rather, low permeability strata hold gas in place via capillarity or adsorption. In these cases the gas occurs as a continuous phase over large areas independent of trap geometry (Schmoker, 2002). A second definition often quoted in the literature is based on the method of extraction required to obtain commercial production. Here, conventional gas either flows to surface on its own accord or can be pumped to surface via production wellbores. Conversely, unconventional gas requires some sort of reservoir stimulation (mechanical, chemical, thermal or biological) and/or specialised drilling techniques such as horizontal wells (Burwen and Flegal, 2013).

The second definition is less satisfying since one can find a continuum between resources that may or may not require reservoir stimulation to obtain commercial flows. For example, some reservoirs with gas trapped by buoyancy within a structural closure may have parts of the reservoir that are in a 'sweet spot' that flows to surface unaided, while other parts of the same reservoir may only achieve commercial flows with reservoir stimulation. Similarly, roughly two-thirds of coal bed methane (CBM) wells produce without the need of stimulation. In this chapter we adopt the first definition, where unconventional gas is defined by the trapping mechanism not being reliant on a combination of buoyancy and trap geometry to contain the gas. In this case there are a number of categories of unconventional gas we can consider including: CBM, shale gas, tight gas, basin-centred gas and gas hydrates.

COAL BED METHANE

Coal bed methane (CBM) is natural gas that is adsorbed into the matrix of coal and held in place by weak chemical bonds (van der Waals) that are related to the organic content of the coal (Brunauer et al., 1940). The contained natural gas may be thermogenic (formed by heat and pressure) or biogenic (formed by microbial action) in origin, or a mix of both, and it may have been generated *in situ* or migrated into the coal from elsewhere. The gas is predominantly dry methane. The cleat system (natural fractures) of the coal is normally saturated with formation water. To produce the gas via a wellbore, formation water is produced from the cleat system first and this reduces the formation pressure within the cleats. The resulting pressure gradient induced between the coal matrix and the cleats creates a sufficient hydraulic driving force to overcome the van der Waals forces holding the gas adsorbed into the coal matrix, and the coal matrix begins to degas. A typical production curve is shown in Figure 28.1 and shows that initial water production tails off as gas production ramps up. Depending on the details of the coal reservoir, wells may be constructed as either vertical or horizontal either with or without stimulation.

It should be noted that the terms 'coal bed methane' (CBM) and 'coal seam gas' (CSG) are used interchangeably in the literature although CBM is normally used in the North American context whilst CSG is used in the Australian context.

SHALE GAS

Shale gas occurs in what is often considered to be the source rock for conventional hydrocarbon accumulations. Organic-rich mudstones and fine grained sediments that have been buried to the point at which the *in situ* pressure and temperature have passed through the hydrocarbon generation window where the original preserved organic matter is first converted to immobile kerogen and then mobile hydrocarbons (Leythaeuser et al., 1979). In the hydrocarbon generation process some of the oil and gas is expelled into adjacent carrier beds where it can migrate and become trapped in conventional reservoirs by buoyancy; however, there is a large portion of the hydrocarbon retained in the source rock and held in place by adsorption and capillarity associated with extremely low permeability. There are various factors that impact our ability to commercially extract gas from these shale reservoirs (source rocks). These include factors that determine gas content such as total organic content and others that influence the effectiveness of well stimulation such as porosity, mineralogy, rheology, pore pressure and *in situ* stress (Close et al., 2012). Shale gas is normally produced via horizontal wells that have been stimulated.

TIGHT GAS

Tight gas resources are defined as quartz-rich siliciclastic reservoirs with extremely low permeability (<0.1 mD as defined by the U.S. federal government legislation Section 29 tax credit, Burwen

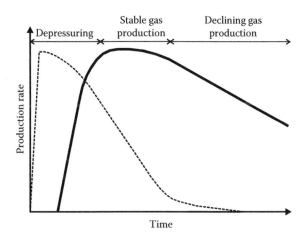

FIGURE 28.1 A generic CBM production profile with initial water production followed by gas production.

and Flegal, 2013). Tight gas reservoirs have low organic content and are not source rocks. Gas has migrated into the pore space over geological time and become trapped by capillarity. The low permeability may be primary or secondary due to post-depositional diagenetic (mainly chemical) processes (Rushing et al., 2008). Tight gas may be produced via vertical or horizontal wells that have been stimulated.

BASIN-CENTRED GAS

Basin-centred gas was a term developed in the mid-1980s and initially associated with the Elmworth field in Western Canada in the Cadomin Formation (Masters, 1984). The characteristics of basin-centred gas are: very large areas of continuous gas saturation, low permeability, anomalous formation pressure (often underpressured) and lacking a downdip water leg (Law, 2002). As such, there is a certain amount of overlap with the definition of tight gas. Basin-centred gas resources were associated with a number of Rocky Mountain foreland basins in North America and the term has since been applied to other tight gas deposits around the world. For example, the Geological Survey of South Australian identifies basin-centred gas potential in the Cooper basin (http://www.petroleum. statedevelopment.sa.gov.au/prospectivity/basin_and_province_information/unconventional_gas/ tight_gas).

GAS HYDRATES

Gas hydrates are a natural form of 'trapped' methane typically occurring in shallow marine sediments or beneath permafrost. Gas hydrate is an ice-like crystalline structure where a solid lattice of water molecules forms a cage that can hold various-sized gas molecules, typically methane but also CO_2, N_2, H_2S, C_2H_6 and many others (Boswell and Collett, 2011). It remains uncertain as to (1) detailed mechanism of formation, (2) how best to characterise gas hydrate resources and reserves and (3) what the most appropriate extraction techniques that are environmentally acceptable might be. Significant research programmes conducted in Canada, Japan, the United States and elsewhere are reported on by the United Nations Environment Programme (Beaudoin et al., 2014). Currently, commercial production is only known from the Messoyakha field in western Siberia (BREE, 2014). Here the gas hydrates unusually occur overlying permafrost and are contributing to the gas produced from the underlying conventional gas field (Pearce, 2009).

HISTORY

The history of unconventional gas development is one that is rooted in two key motivational drivers that happened to closely coincide in time. The first was a desire to improve the safety record of underground coal mining. A study commissioned by the NSW Department of Primary Industries (MacNeill, 2008) searched public records to build a database of historical coal mining fatalities and were able to identify a cumulative total of 12,146 deaths globally to 2008, the majority of which occurred in underground mines. Nearly 10,000 of the deaths were from explosions due to gas ignition. The coal mining industry, government regulators and scientists were interested in developing technology that could improve the safety record of underground coal mining. One way to increase safety was to circulate air through the underground mine to dilute methane concentrations. It was found that general air circulation of about 100 times the volume of methane emissions was required to adequately improve explosive risks in underground mines. In 1971, it was estimated that in the United States alone there was in the order of 300,000 GJ/day (~300 MMcfd) of methane being vented from underground coal mines (Elder and Deul, 1974). It was recognised that this vented methane had intrinsic value and the U.S. Bureau of Mines began experiments to degas coal with vertical bore holes where the methane could be captured and used. One experiment in the Mary Lee coal bed at Jefferson County Alabama (1971–1973) also stimulated the coal by inducing fractures (now known colloquially as 'fracking', or 'fraccing'. This was found to greatly enhance the effective degassing of the coal (Elder and Deul, 1974).

A second key motivation for unconventional gas development materialised in 1973 with the OPEC (Organization of the Petroleum Exporting Countries) oil crisis where shortages of gasoline resulted in people queuing at petrol stations (Figure 28.2), the price of oil more than doubled (from <\$20US to >\$40US/barrel) and the general public in the developed world became worried about their energy security. By 1976, the United States faced declining domestic gas production and in the wake of the OPEC oil crisis the U.S. Congress funded the 'Unconventional Gas Research Program' with the intent to help rebuild the U.S. domestic natural gas resource base (Burwen and Flegal, 2013). The programme ran from 1976 to 1992 with \$92 million U.S. funding to encourage research and industry collaboration. The programme included (1) an 'Eastern Gas Shales Project' targeting an estimate of Devonian shale gas resources and research into an economically effective extraction methodology, (2) a 'Western Gas Sands Project' targeting low permeability tight gas and (3) a 'Methane Recovery from Coalbeds' project. That same year the Federal Energy Regulatory Commission approved a charge on interstate gas sales to fund R&D through the Gas Research Institute (Burwen and Flegal, 2013).

The U.S. Federal Government passed the Windfall Profits Tax Act in 1980 which was effectively a production tax credit (Section 29) for gas from unconventional reservoirs calculated on a formula based on oil price (Drapkin and Verleger, 1981). The Section 29 tax credit favoured CBM development since by and large the wells did not require stimulation (LaFehr, 1993) and the resource could be exploited through what has become known as 'factory drilling' where shallow (<1000 m) vertical wellbores on a close spaced grid (on the order of ~700 m spacing) are designed to reduce drilling costs. By 1991 the value of the Windfall Profits credit was approximately half the price of gas at the time. In 1992 the Section 29 credit ended and covered production had ended by 2002.

Out of the federal government-funded research–industry collaboration, there were investigations in improved horizontal drilling with application to Devonian shale formations in the eastern United States. In 1986 the Department of Energy completed with industry the first successful truly horizontal shale well in the Appalachian Basin where they also tested wellbore imaging, new well completion technology and hydraulic stimulation (Yost, 1988) (Note: The earlier horizontal drilling was successful in non-shale formations). In parallel with the shale gas drilling development the eastern United States, federal government-funded research on real-time micro-seismic monitoring of fracture development during well stimulation was advancing. Sandia laboratories field tested this technology

FIGURE 28.2 A photo of a gasoline station queue in New York in 1973 during the OPEC oil crisis. (From http://hannahscribbles.com/tag/oil-crisis. With permission.)

aided by 'measured while drilling' techniques on research wells at the 'M-Site' in Colorado. Whilst the initial test was conducted in the mid-1980s it was not until the mid-1990s that Sandia was able to validate the successful application of this technique (Warpinski et al., 1997). From the mid-1990s to the mid-2000s both horizontal drilling with fracture stimulation and micro-seismic monitoring of stimulation jobs moved through to full commercialisation and industrial application.

These key events related to unconventional gas development are displayed in Figure 28.3 on a timeline and overlaid with U.S. CBM production, U.S. shale gas production and the Henry Hub gas spot price. A combination of rising gas prices and the Section 29 tax credit of the Windfall Profits tax drove a steady development of CBM resources in the United States with a corresponding increase in production. The economic effect of the Section 29 tax credit ending on production in 2002 did not influence CBM development in the United States as the gas price was increasing sufficiently to offset the end of the tax credit. However, by the mid- to late 2000s the commerciality of shale gas on the back of improved horizontal drilling and stimulation techniques began to displace CBM. This was influenced by many shale gas resources also having a liquid-rich component that added significant value to the production, whereas CBM was simply dry gas production. This effect was solidified by the dramatic event of the collapse of Lehman Brothers and the ensuing global financial crisis that helped drive down the price of gas in the domestic U.S. market despite hurricanes Gustav and Ike temporarily interrupting supply of 7 bcf/day from the Gulf of Mexico (Towler, 2014). From that point onwards, the economics of shale gas with its liquids component and the low U.S. domestic gas price effectively halted any new CBM development in the United States but saw the dramatic expansion of shale gas. With the rise in gas consumption, the United States considered importing LNG (liquefied natural gas); however, domestic shale gas production grew so rapidly that LNG imports never came to fruition.

An unanticipated trend, partially influenced by the drop in domestic gas price in the United States associated with the rapid shift of the energy sector to CSG and shale gas, is the post 2007 reversal in U.S. GHG emissions. Figure 28.4, based on data from the U.S. Environmental Protection Agency (EPA), shows the long-term trend in U.S. GHG emissions that is dominated by CO_2. The continued

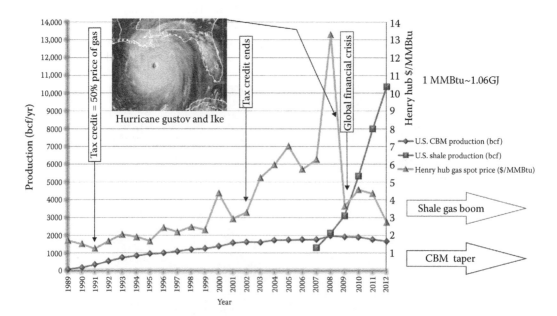

FIGURE 28.3 A plot of U.S. CBM and shale gas production (bcf/year) together with the historical U.S. Henry Hub domestic gas price ($US/MMBtu). Key events are marked. Note that 1 MMBtu ~ 1.06 GJ and 1 bcf ~ PJ. (Adapted from U.S. Energy Information Administration.)

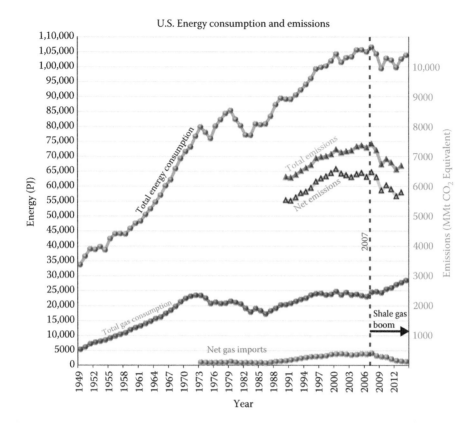

FIGURE 28.4 U.S. energy consumption and emissions. (Adapted from U.S. Energy Information Administration and the U.S. Environmental Protection Agency.)

decline from 2007, despite flat total energy consumption, is thought to be due to a combination of the shift from coal to gas for power generation, consecutive mild winters and a decrease in electricity consumption in some sectors of the economy. At the same time as decreasing emissions, the United States is rapidly moving towards oil and gas self-sufficiency (see gas net imports in Figure 28.4).

CANADA

Since about 2000, Canada's natural gas annual consumption is ~3000 PJ (source: http://www.eia. gov/beta/international/analysis.cfm?iso = CAN) with production roughly twice that, and where exports have historically gone into the North American gas market. As such, the development of unconventional gas in Canada was heavily influenced by the same technology and economic conditions as prevailed in the United States. It is the regulatory environment and mineral rights ownership in Canada that differ. With a few minor exceptions, mineral rights are owned by the government (crown) rather than private ownership as in the United States. The federal government regulates offshore production as well as most aspects of international import and export, whilst onshore production is regulated by provincial governments. With Alberta and British Columbia being long established oil and gas producers both the regulatory and physical surface infrastructure were in place that allowed more rapid development of unconventional gas than in other parts of Canada where unconventional gas potential exists (Popp, 2014). Since basin-centred gas was discovered first in Canada, tight gas production has dominated the mix of unconventional gas production from the mid-1980s (Masters, 1984). CBM was produced in small volumes since 2000 but production began to expand in the mid-2000s when the Henry hub gas price went over $5/GJ in the United States (see Figure 28.3). This lag from the U.S. CBM production reflects the impact of the Section 29 tax credit in the United States. Shale gas production began in 2009 with a 2–3-year lag behind that in the United States (source: http://www.iea.org/ugforum/ugd/).

AUSTRALIA

CSG has been produced in Australia's east coast domestic gas market since 1996; however, the domestic market is relatively small with gas consumption being 1000–1200 PJ/year since the mid-2000s (source: http://www.eia.gov/beta/international/analysis.cfm?iso = AUS). The rise in international LNG prices above $10/GJ as well as the QLD government's Electricity Act which mandated 15% of domestic electricity generation to be by gas, both stimulated an increased exploration and appraisal programme in the mid-2000s, culminating in the proposal of four large CSG–LNG projects (six trains) in 2008/2009. Japan's earthquake and tsunami in March 2011, as well as the tragedy that struck the Fukushima nuclear power plant resulted in a rise in demand for LNG. It is on the basis of this demand (largely in SE Asia), the availability of long-term contracts with Japan, China and Korea, and Australia's geographic location that resulted in three mega projects (of the four initially proposed) being approved to export LNG from CSG production in Queensland (BREE, 2014). This is the first time globally that unconventional gas has been developed for LNG export. The first shipment was made in January 2015 with full production from the three projects expected to reach 1,500 PJ/year by 2025. Australia has two commercial shale gas production wells drilled by Santos located in the Cooper Basin, but more substantial shale gas resources are still in the process of resource valuation.

REST OF THE WORLD

The state of unconventional gas development in the rest of the world is currently of much smaller volumes and where it occurs, the gas produced is used in the domestic market. The World Energy Outlook (IEA, 2011) indicates that CBM is now produced in more than a dozen countries with the bulk being produced in the United States, Canada, Australia, India and China. There is limited world production data for shale and tight gas and tight gas production is often included in conventional gas production values. The unprecedented shift in the U.S. energy landscape with the advent of significant shale gas production and its associated liquids components has led other jurisdictions around the world to evaluate their unconventional gas potential. This is currently in process at various stages of advancement across the globe.

GLOBAL PRODUCTION/RESERVES

The very nature of unconventional gas resources being widely distributed but with low technical recovery rates makes resource and reserve estimation difficult to define and subject to high uncertainty. Recovery is fundamentally technology dependent and thus as advances occur the resulting available commercially viable reserves could vary widely. For this reason, when comparing published resource and reserves estimates it is important to compare estimates made at approximately the same time and based on the technology at that time. Even still, published resource and reserves estimates vary widely, reflecting the degree of inherent uncertainty. The Society of Petroleum Engineers provides guidance on definitions for resources and reserves including subcategories (SPE, 2007, 2011) and they have actively guided developments of the petroleum resources management systems tuned to unconventional gas developments over the last 10 years. They suggest (SPE, 2007, 2011):

> TOTAL PETROLEUM INITIALLY-IN-PLACE is that quantity of petroleum that is estimated to exist originally in naturally occurring accumulations. It includes that quantity of petroleum that is estimated, as of a given date, to be contained in known accumulations prior to production plus those estimated quantities in accumulations yet to be discovered (equivalent to 'total resources').

DISCOVERED PETROLEUM INITIALLY-IN-PLACE is that quantity of petroleum that is estimated, as of a given date, to be contained in known accumulations prior to production.

RESERVES are those quantities of petroleum anticipated to be commercially recoverable by application of development projects to known accumulations from a given date forward under defined conditions. Reserves must further satisfy four criteria: they must be discovered, recoverable, commercial and remaining (as of the evaluation date) based on the development project(s) applied. Reserves are further categorized in accordance with the level of certainty associated with the estimates and may be sub-classified based on project maturity and/or characterized by development and production status.

As of the beginning of 2011 the total global gas reserves (conventional and unconventional) were estimated to be more than 7.3 million PJ or >7,300 tcf (BREE, 2014). The Russian Federation, Iran and Qatar together account for more than half. The International Energy Agency (IEA) estimates total global technically recoverable gas resources (conventional and unconventional) to be >30 million PJ (>30,000 tcf) of which roughly half is unconventional gas (IEA, 2011).

The United States ranks fourth globally for technically recoverable shale gas resources behind China, Algeria and Argentina. Canada ranks fifth (Popp, 2014). An extensive review of published unconventional gas reserves and resources estimates for various countries globally was synthesised into a single table by McGlade et al. (2013). Despite these vast global resources, the United States, Canada, Australia, India and China (note that China counts coal mine degassing as CBM production) account for the bulk of global production (IEA, 2011). This is because there are challenges still facing unconventional gas production, both technical and societal, which we will explore next.

CHALLENGES

Despite the rapid development of unconventional gas resources in the United States, closely followed in Canada and later in Australia; and despite the impact on reducing the cost of energy, reducing GHG emissions and bringing the United States towards energy self-sufficiency, the shift to unconventional gas has not been without its challenges. There have been concerns about the long-term impact of gas development on the environment (particularly groundwater and surface water resources), there has been speculation that methane emissions from the unconventional gas industry may result in high life cycle emissions for power generation, and the historical surface footprint of unconventional gas operations has led to local community concern. Outside of North America, in addition to the challenges already mentioned, the cost of capital and operations for unconventional gas development is often much higher due to remoteness of resource and lack of infrastructure as well as local markets. The nature of each challenge is slightly different depending on the type of unconventional gas.

ENVIRONMENT AND WATER RESOURCES

Since North America is really the only location where there is a statistically significant historical record of unconventional gas development and subsequent operations, this is the best source of information to assess the risk of development on the environment. There have been a number of studies by various organisations looking at exactly this issue, including EPA (2004, 2015), National Research Council (2010). An extensive review of the unconventional gas industry in the United States was recently conducted by the U.S. EPA (EPA, 2015) that draws on the historical evidence of development in the United States and previous reviews. It concludes that contamination of drinking water resources incidents have mainly been due to

- Surface spills of frack fluid and produced water
- Deliberate or accidental discharge of treated flowback water
- Gas migration to aquifers via production wells (loss of well integrity)
- Stimulating reservoirs that are also used for domestic water supply

but they also set the context of these conclusions by stating 'The number of identified cases where drinking water resources were impacted are small relative to the number of hydraulically fractured wells'. The incidents of drinking water contamination sources are categorised by number of incidents and by volume of water contaminated in Figure 28.5.

Methane Emissions

For context, methane as a substance is non-toxic but can pose an explosive risk if present with oxygen and an ignition source in concentrations between the lower and upper explosive limits (5%–15% by volume at room temperature and atmospheric pressure). Methane is also a greenhouse gas. There have been many studies and publications regarding global methane emission estimates from various sources, both natural and anthropogenic. Methane is thought to make up 16%–20% of the total anthropogenic greenhouse gas emissions (Karakurt et al., 2012; Yusuf et al., 2012). While estimates vary between studies and over time (Karakurt et al., 2012; Yusuf et al., 2012; Kelly et al., 2015); the 'ballpark' relative contribution for various source categories remains valid. From the published literature, the range of global anthropogenic methane emissions to the atmosphere by industrial source is 19%–21% from waste (primarily landfills and municipal waste water such as sewage), 28%–29% from energy production and utilisation and 50%–53% from agriculture. Natural sources of methane include wetlands, termites, wildfires, grasslands, coal outcrops and sub-crops and water bodies (Yusuf et al., 2012).

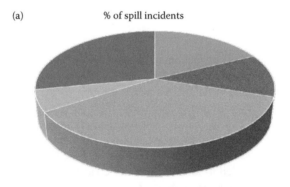

(a)   % of spill incidents

■ Equipment ■ Hose or line ■ Storage ■ Well or wellhead ■ Unknown

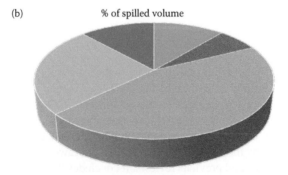

(b)   % of spilled volume

■ Equipment ■ Hose or line ■ Storage ■ Well or wellhead ■ Unknown

FIGURE 28.5 U.S. EPA investigated ~35,000 spills incidents of which 457 (~1%) were demonstrably linked to hydraulic fracking operations. (a) Percentage of incidents by source. (b) Percentage of volume spilled by source. (Adapted from EPA. 2015. *Assessment of the Potential Impacts of Hydraulic Fracturing for Oil and Gas on Drinking Water Resources*. Washington DC: Office of Research and Development, 599 pp.)

The terminology 'fugitive methane emissions' associated with onshore gas development has been used in various contexts throughout the scientific literature and popular media. It remains a topic of general public concern (evident in the media by the number of articles published) and while the physical processes are reasonably well understood (e.g. see the U.S. EPA STAR programme: http://www.epa.gov/gasstar/index.html), sufficient density of measurement has yet to be obtained in some jurisdictions to characterise the process thoroughly. As a result, one can find a wide range of speculation, model estimates and uncertainty recorded in the peer reviewed scientific literature as to how severe fugitive methane emissions may be from oil and gas (O&G) development. Some work is published on fugitive methane emissions from coal mining; however, most O&G industry methane emissions estimates are linked to shale gas development and its associated reservoir stimulation (e.g. Rigby et al., 2008; Cathles, 2012; Karakurt et al., 2012; Caulton et al., 2014; Kort et al., 2014; Schwietzke et al., 2014; Kelly et al., 2015). As a result, it is important to carefully consider the context of published data and reports. Two major categories of potential fugitive emissions from onshore gas development are (1) surface plant and well heads and (2) subsurface methane released from the reservoir but not captured by the production wells. Most public domain data on gas industry fugitive emissions is based on examples from North America; however, some work has been published in Australia that is specific to coal seam gas development (Day et al., 2012, 2014; Kelly et al., 2015) that measured methane emissions from coal seam gas infrastructure to be slightly less than that reported for the United States.

Historically, the first hydrocarbon discoveries have been made by observing hydrocarbon occurrences either as surface seeps or in water wells (Cadman et al., 1998; Finch and Tippett, 2014). Dyck and Dunn (1986) describe conducting a survey of nearly 1000 water wells in the late 1970s across 18,000 km^2 area in Alberta, analysing for CH_4, He and other constituents to determine a relationship between shallow groundwater chemistry and the occurrence of deeper petroleum accumulations. They then used this as an exploration tool to find as yet undiscovered pools.

In more recent times, particularly with the rise in unconventional gas development, there has been speculation that hydrocarbon presence in groundwater may be at least in part anthropogenic, induced by gas resource development. However, with hydrocarbon in groundwater known to also be naturally occurring, it has led to debate on the relative share in provenance (natural thermogenic, natural biogenic or anthropogenic). As was described earlier for methane in the atmosphere, hydrocarbon content in groundwater and seeping to the ground surface is a topic that has recently been widely published in the peer-reviewed literature and in the media. Much of the attention to hydrocarbons in groundwater is associated with concern about the environmental impacts of fracking and shale gas development (Llewellyn et al., 2015) which is covered below.

Water Requirements

Depending on the unconventional gas resource in question there are various water use challenges. For unconventional gas that requires stimulation, there is a need to utilise water for making up the fracturing fluids that are used to stimulate the well. Shale gas development commonly utilises long horizontal wells that are fractured in multiple stages starting at the toe of the well and working back to the heel where the well deviates from vertical to horizontal. After examining data from nearly 40,000 wells in the United States (Jackson et al., 2015) determined that the average shale gas well used ~9 million litres (9 ML) of water, although there is a range of values depending on the nature of the specific geology. Historically this water was obtained from local surface or groundwater sources. The U.S. EPA (EPA, 2015) reports that water requirement for shale gas development nationally use on average 1% of the total local demand for groundwater resources, but in some areas such as Texas the requirement can be as high as 38% and this can locally lead to overutilisation (Nicot and Scanlon, 2012). If gas fired power replaces coal fired power there are some life cycle water savings to be made in the power generation stage. Recent technology developments has allowed for flowback water recovered to surface after a fracture stimulation is complete, to be recycled with water treatment technology. This reduces the requirement of groundwater or surface water resources. Also, new fracture fluid technology can utilise lower quality brackish water to make up the fracturing fluid (Reyes et al., 2014). These options are now available in regions where surface and ground water supplies are scarce or where environmental values require water savings.

CSG has the opposite challenge where the production of gas first requires depressuring the coal seams by producing water. Although highly variable, the water volume produced can initially be between 7 and 300 ML of water per year and a well may produce water for more than a year before the methane flows from the coal reservoir (Klohn Crippen Berger [KCB], 2012). Produced water is often of low quality (100s to 1000s of ppm TDS) and may require treatment before it can be beneficially used (Davies et al., 2015). Although the ratios vary from basin to basin, in North America ~45% is disposed of in evaporation/infiltration ponds, ~15% goes to surface discharge, ~25% is reinjected into deeper saline aquifers and ~15% is treated and used on the surface. In Australia evaporation ponds are no longer allowed by the regulations in Queensland that require the beneficial use of produced water where possible (Department of Environment and Heritage Protection [DEHP], 2012). The process of purifying the produced water creates a second smaller waste stream of more concentrated salinity. This is normally reinjected into deep saline aquifers but some locations do not have this option and the brine needs to be processed at surface. In Australia the industry current preferred solution is to crystallise a solid salt product that is placed in regulated landfills (Davies et al., 2015).

Beneficial use of the produced water can include agricultural irrigation, livestock watering, environmental flows in surface waters, industrial application for dust suppression or cooling or managed aquifer recharge (Department of Environment and Heritage Protection [DEHP], 2012).

Reservoir Stimulation

An area that has attracted a great deal of public attention since the rapid development of shale gas in the United States over the last 10 years is reservoir stimulation, although hydraulic fracturing of underperforming conventional reservoirs has been occurring since ca. 1950 with the Society of Petroleum Engineers estimating that some 2.5 million fracture stimulations have been performed worldwide with over 1 million of these occurring in the United States (King, 2012). The process of reservoir stimulation includes injecting a combination of water with proppant and chemicals under pressure to induce hydraulic fractures in low permeability reservoir rock. The well trajectory relative to the orientation of both *in situ* stress and natural fractures can make an important contribution to successful stimulation (Johnson et al. 2010). The proppant (often sand but could also be synthetic material) is meant to hold open the fracture as the injection pressure is subsequently decreased. An ideal fracture stimulation will allow the bulk of the injected fluid to be recovered and sufficient proppant placed in the fracture to maintain high permeability but not impede subsequent gas flow. The chemicals used in the fracture fluid are variable and optional depending on the local regulations and the specific requirements of the reservoir. Some key factors that influence the effectiveness of the fracture stimulation are: (1) viscosity of the fracture fluid needs to be adequate to entrain the proppant and deliver it to the induced fracture is sufficient quantity, (2) stability of the fracture fluid properties between the well head and formation under variable pressure and temperature conditions, (3) effective fracture fluid recovery after proppant has been emplaced is dependent on the wettability and surface tension of the fracture fluid in the reservoir, (4) solubility of minerals such as calcite in the reservoir can further improve the effective permeability after a fracture stimulation, (5) the introduction of oxygen and microbial bacteria in the fracture fluid can cause biofouling of the reservoir and wellbore, (6) reactive chemistry between the metal casing, formation water and fracture fluid can cause corrosion and (7) hardness components of the fracture fluid and formation fluid can result in precipitation of carbonate minerals as scale on the formation and infrastructure. There are chemicals that can be included in the fracture fluid to address each of these operational issues and improve performance (Batley and Kookana, 2012). There are a number of publications and public domain documents provided by various government agencies and energy companies that describe the various compositional components of fracture fluids, and the formulation of fracture fluid is a rapidly changing technology that is striving to match community expectations of achieving the lowest environmental risk (Batley and Kookana, 2012; EPA, 2012, 2015; Vengosh et al., 2014; Davies et al., 2015).

The produced water after a fracture stimulation (flowback water) can include compounds and elements naturally residing in the reservoir that have been liberated by the stimulation process. These can include volatile organic compounds such as benzene, toluene, ethylbenzene and xylene (BTEX) and naturally occurring radioactive materials (NORM). Where these occur, they can be treated and removed from the flowback water before it is reused. Shallow aquifers are protected from exposure to the produced fluids by well design based on multiple engineered barriers (King, 2012).

Scanning the popular and news media over the last 10 years, there is a certain anxiety in the public regarding the risk of fracture fluid and its contained chemicals escaping from the gas reservoir and into adjacent aquifers. The evidence of this occurring is scant (King, 2012; EPA, 2015) and this discrepancy with the level of public concern is partly a function of definition. King (2012) pointed out that reservoir engineers refer to a very specific operation 'hydraulic fracturing' to have an extremely low occurrence of contaminating aquifers, whereas the public might consider *any inci- dent* with a gas well that has been hydraulically fractured to be related to 'fracking'. As concluded by the U.S. EPA (EPA, 2015) and discussed earlier, there are a number of groundwater contamination incidents from other aspects of oil and gas development that are not related to hydraulic fracturing specifically. Many of these types of incidents and the controls required to reduce and mitigate the risk apply to both conventional and unconventional gas resource development.

SOCIAL PERFORMANCE

Community concern over the impact of unconventional gas development on the environment, particularly surface and groundwater resources have a strong influence on government policy and resulting regulations. There have been four countries that placed either a ban or a moratorium on hydraulic fracturing (France in 2011, Bulgaria in 2012, Germany in 2012 and Scotland in 2015) and there are many more states or municipalities that have also done so. Achieving improved social awareness for unconventional gas, its advantages and its actual risks will require better communication, improved understanding of societal values, a regulatory regime that provides the public with confidence and the industry with stable governance and demonstrated environmental and economic performance (Everingham et al., 2014; Raufflet et al., 2014; Uhlmann et al., 2014).

COST

Outside of North America, the costs of unconventional gas production are higher due to the lack of existing infrastructure (pipelines, roads, drill rigs, service rigs, etc.), a lack of local trained professionals and a lack of domestic markets (unconventional seaborne gas export requires the additional investment in LNG and port facilities). Major cost categories include: (1) pipeline infrastructure, (2) oil field service supply chain and (3) LNG processing. Continued technology breakthroughs and long term contracts for LNG will help monetise the large global reserves of unconventional gas in the future, although with LNG prices linked to the oil price and increasing competition for supply (Canada, United States, Middle East) and with Russia potentially supplying China with gas by pipeline, there remains a downward pressure on price and the need to find cost savings to be competitive.

FUTURE

The IEA predicts future global demand for energy, and does this based on a number of different scenarios: (1) based only on current policies, (2) based on new policy development and (3) based on policies that limit atmospheric concentration of CO_2 equivalent greenhouse gases to 450 ppm. They predict that between 2010 and 2035 world energy demand will increase between 16% and 47%

depending on the policy scenario and that various parts of the energy mix will grow or shrink to various degrees to match this demand. In 2010 the global primary energy consumption consisted of oil (32%), coal (27%), gas (21%) and all other energy sources combined (20%) (BREE, 2014). For electricity generation the mix is coal (40%), gas (22%), renewables (19%, mainly hydro), nuclear (12%) and other sources combined (7%). In all three IEA scenarios the demand for gas is expected to grow. This is because of transport flexibility (pipeline by land or LNG by sea), a geographically widely dispersed resource base, its cost competitiveness as a fuel, GHG emissions advantage over coal, and flexibility of gas fired power to rapidly change power output to match the variable output of distributed renewably energy sources that allows for grid stability (IEA, 2011). Domestic public pressure to solve air quality issues in China may also drive down the utilisation of coal fired power aiding the uptake of gas.

REFERENCES

Batley, G. E. and R. S. Kookana. 2012. Environmental issues associated with coal seam gas recovery: Managing the fracking boom. *Environmental Chemistry*, 9(5): 425–428. doi: 10.1071/En12136.

Beaudoin, Y. C., R. Boswell, S. R. Dallimore and W. Waite. 2014. *Frozen Heat: A UNEP Global Outlook on Methane Gas Hydrates*. Norway: United Nations Environment Programme, pp. 32.

Boswell, R. and Collett, T. S. 2011. Current perspectives on gas hydrate resources. *Energy & Environmental Science*, 4(4): 1206–1215. doi: 10.1039/c0ee00203h.

BREE. 2014. *Australian Energy Resource Assessment*. Canberra: Bureau of Resources and Energy Economics, pp. 364.

Brunauer, S., L. S. Deming, W. E. Deming and E. Teller. 1940. On a theory of the van der waals adsorption of gases. *Journal of the American Chemical Society*, 62(7): 1723–1732. doi: 10.1021/ja01864a025.

Burwen, J. and J. Flegal. 2013. *Case Studies on the Government's Role in Energy Technology Innovation*. American Energy Innovation Council, Washington, 16pp.

Cadman, S. J., L. Pain and V. Vuckovic. 1998. Australian petroleum accumulations report. Bowen and Surat Basins, Clarence-Moreton Basin, Sydney Basin, Gunnedah Basin and other minor onshore basins, Qld, NSW and NT. Retrieved from: http://www.ga.gov.au/metadata-gateway/metadata/record/37058/

Cathles, L. M. 2012. Assessing the greenhouse impact of natural gas. *Geochemistry, Geophysics, Geosystems*, 13(6), n/a–n/a. doi: 10.1029/2012GC004032.

Caulton, D. R., P. B. Shepson, R. L. Santoro, J. P. Sparks, R. W. Howarth, A. R. Ingraffea, M. O. L. Cambaliza et al. 2014. Toward a better understanding and quantification of methane emissions from shale gas development. *Proceedings of the National Academy of Sciences*, 111(17): 6237–6242. doi: 10.1073/pnas.1316546111.

Close, D., M. Perez, B. Goodway and G. Purdeu. 2012. Integrated workflows for shale gas and case study results for the Horn River Basin, British Columbia, Canada. *The Leading Edge: Society of Exploration Geophysicists*, 11: 556–569.

Davies, P. J., D. B. Gore and S. J. Khan. 2015. Managing produced water from coal seam gas projects: Implications for an emerging industry in Australia. *Environmental Science and Pollution Research International*, 22(14): 10981–11000. doi: 10.1007/s11356-015-4254-8.

Day, S., L. Connell, D. Etheridge, T. Norgate and N. Sherwood. 2012. *Fugitive Greenhouse Gas Emissions from Coal Seam Gas Production in Australia*. Australia: CSIRO, pp. 36.

Day, S., Dell'Amico, Fry, R. and H. Javanmard Tousi. 2014. *Field Measurements of Fugitive Emissions from Equipment and Well Casings in Australian Coal Seam Gas Production Facilities*. Australia: CSIRO, pp. 48.

Department of Environment and Heritage Protection (DEHP). 2012. *Coal Seam Gas Water Management Policy 2012*. Brisbane, Australia: Queensland Government. Retrieved from https://www.ehp.qld.gov.au/management/non-mining/documents/csg-water-management-policy.pdf

Drapkin, D. B. and P. K. Verleger. 1981. The windfall profit tax origins development implications. *Symposium on the Crude Oil Windfall Profit*, Boston.

Dyck, W. and C. E. Dunn. 1986. Helium and methane anomalies in domestic well waters in southwestern Saskatchewan, Canada, and their relationship to other dissolved constituents, oil and gas fields, and tectonic patterns. *Journal of Geophysical Research*, 91(B12): 12343. doi: 10.1029/JB091iB12p12343.

Elder, C. H. and M. Deul. 1974. *Degasification of the Mary Lee Coalbed Near Oak Grove, Jefferson County, Ala., by Vertical Borehole in Advance of Mining*. Pittsburgh PA: United States Department of the Interior: Bureau of Mines, pp. 25.

EPA. 2004. *Final Study Design for Evaluating of Impacts to Underground Sources of Drinking Water by Hydraulic Fracturing of Coalbed Methane Reservoirs*. Washington DC: U.S. Environmental Protection Agency

EPA. 2012. *Study of the Potential Impacts of Hydraulic Fracturing on Drinking Water Resources Progress Report*. Washington DC: U.S. Environmental Protection Agency.

EPA. 2015. *Assessment of the Potential Impacts of Hydraulic Fracturing for Oil and Gas on Drinking Water Resources*. Washington DC: Office of Research and Development, 599pp.

Everingham, J., N. Collins, W. Rifkin, D. Rodriguez, T. Baumgartl, J. Cavaye and S. Vink. 2014. How farmers, graziers, miners, and gas-industry personnel see their potential for coexistence in rural queensland. *SPE Economics and Management*, July, 9.

Finch, D. and C. Tippett. 2014. Remembering a rich history at Turner valley. *AAPG Explorer*. 35 (5).

IEA. 2011. *World Energy Outlook 2011*. International Energy Agency, Paris, 666pp.

Jackson, R. B., E. R. Lowry, A. Pickle, M. Kang, D. DiGiulio and K. Zhao. 2015. The depths of hydraulic fracturing and accompanying water use across the United States. *Environmental Science & Technology*, 49(15): 8969–8976. doi: 10.1021/acs.est.5b01228.

Johnson, R. L., M. P. Scott, R. G. Jeffrey, Z. Chen, L. Bennett, C. B. Vandenborn and S. Tcherkashnev. 2010. Evaluating hydraulic fracture effectiveness in a coal seam gas reservoir from surface tiltmeter and microseismic monitoring. *SPE Annual Technical Conference and Exhibition*, SPE-133063-MS, Florence, Italy, 19–22 September.

Karakurt, I., G. Aydin and K. Aydiner. 2012. Sources and mitigation of methane emissions by sectors: A critical review. *Renewable Energy*, 39(1): 40–48. doi: http://dx.doi.org/10.1016/j.renene.2011.09.006.

Kelly, B. F. J., C. P. Iverach, D. Lowry, R. E. Fisher, J. L. France and (2), E. G. Nisbet. 2015. Fugitive methane emissions from natural, urban, agricultural, and energy-production landscapes of eastern Australia. *EGU General Assembly*, Vienna, Austria, 12–17 April 2015.

King, G. E. 2012. Hydraulic fracturing 101: what every representative, environmentalist, regulator, reporter, investor, university researcher, neighbor and engineer should know about estimating frac risk and improving frac performance in unconventional gas and oil wells. *SPE Hydraulic Fracturing Technology Conference*, The Woodlands, Texas. https://www.onepetro.org/download/conference-paper/SPE-152596-MS?id = conference-paper%2FSPE-152596-MS

Klohn Crippen Berger (KCB). 2012. Forecasting coal seam gas water production in Queensland's Surat and southern Bowen basins. *Technical Report*. Brisbane, Australia: Klohn Crippen Berger .

Kort, E. A., C. Frankenberg, K. R. Costigan, R. Lindenmaier, M. K. Dubey and D. Wunch. 2014. *Geophysical Research Letters*, 41: 5. doi: 10.1002/2014GL061503. Kort_et_al-2014-Geophysical_Research_Letters.pdf

LaFehr, E. D. 1993. The effects of Section 29 tax credit on energy and the environment: A cost-benefit analysis. *The Journal of Energy and Development*, 17(1): 1–22.

Law, B. E. 2002. Basin-centered gas systems. *American Association of Petroleum Geologists Bulletin*, 86(11): 1891–1919.

Leythaeuser, D., H. W. Hagemann, A. Hollerbach and R. G. Schaefer. 1979. PD 1(4) Hydrocarbon Generation in Source Beds as a Function of Type and Maturation of Their Organic Matter: A Mass Balance Approach. 10th World Petroleum Congress. 9–14 September, Bucharest, Romania, pp. 31–41.

Llewellyn, G. T., F. Dorman, J. L. Westland, D. Yoxtheimer P. Grieve, T. Sowers, E. Humston-Fulmer, and S. L. Brantley. 2015. Evaluating a groundwater supply contamination incident attributed to Marcellus Shale gas development. *Proceedings of the National Academy if Sciences U S A*, 112(20): 6325–6330. doi: 10.1073/pnas.1420279112.

MacNeill, P. 2008. *International Mining Fatality Database*. Sydney: University of Wollongong, 33pp.

Masters, J. A. 1984. Deep basin gas trap, Western Canada. *American Association of Petroleum Geologists Bulletin*, 62(2): 29.

McGlade, C., J. Speirs and S. Sorrell. 2013. Unconventional gas – A review of regional and global resource estimates. *Energy*, 55: 571–584. doi: 10.1016/j.energy.2013.01.048.

National Research Council. 2010. *Management and Effects of Coalbed Methane Produced Water in the Western United States*. Washington, DC: The National Academies Press.

Nicot, J. P. and B. R. Scanlon. 2012. Water use for Shale-gas production in Texas, *U.S. Environmental Science & Technology*, 46(6): 3580–3586. doi: 10.1021/es204602t.

Pearce, F. 2009. Ice on fire. *New Scientist*, 27: 4.

Popp, S. M. 2014. Unconventional gas regulation in Canada. *Oil and Gas Energy Law*, 12(3): 28.

Raufflet, E., L. Barin Cruz and L. Bres. 2014. An assessment of corporate social responsibility practices in the mining and oil and gas industries. *Journal of Cleaner Production*, 84: 256–270. doi: 10.1016/j.jclepro.2014.01.077.

Reyes, F. R., M. Engle, M. Jacobs, L. Jin and J. G. Konter. 2014. Suitability of Brackish water for use in hydraulic fracturing: Hydrogeochemistry of Dockum Group Groundwater, Midland Basin, Texas. *AAPG Annual Convention and Exhibition*, Houston.

Rigby, M., R. G. Prinn, P. J. Fraser, P. G. Simmonds, R. L. Langenfelds, J. Huang, D. M. Cunnold et al. 2008. Renewed growth of atmospheric methane. *Geophysical Research Letters*, 35(22), n/a-n/a. doi: 10.1029/2008GL036037.

Rushing, J. A., K. E. Newsham and T. A. Blasingame. 2008. *Rock Typing: Keys to Understanding Productivity in Tight Gas Sands.*

SPE 114164 – Paper presented at the 2008 SPE Unconventional Reservoirs Conference, Keystone, Colorado, 10–12 February, 31pp.

Schmoker, J. W. 2002. Resource-assessment perspectives for unconventional gas systems. *American Association of Petroleum Geologists Bulletin*, 86(11): 1993–1999.

Schwietzke, S., W. M. Griffin, H. S. Matthews and L. M. Bruhwiler. 2014. Natural gas fugitive emissions rates constrained by global atmospheric methane and ethane. *Environmental Science & Technology*, 48(14): 7714–7722. doi: 10.1021/es501204c.

SPE. 2007. *Petroleum Resources Management System*. Society of Petroleum Engineers, 49pp.

SPE. 2011. *Guidelines for Application of the Petroleum Resources Management System*. Society of Petroleum Engineers, Richardson, Texas, pp. 222.

Towler, B. F. 2014. *The Future of Energy*. Academic Press, Laramie, Wyoming.

Uhlmann, V., W. Rifkin, J.-A. Everingham, B. Head and K. May. 2014. Prioritising indicators of cumulative socio-economic impacts to characterise rapid development of onshore gas resources. *The Extractive Industries and Society*, 1(2): 189–199. doi: 10.1016/j.exis.2014.06.001.

Vengosh, A., R. B. Jackson N. Warner, T. H. Darrah and A. Kondash. 2014. A critical review of the risks to water resources from unconventional shale gas development and hydraulic fracturing in the United States. *Environmental Science & Technology*, 48(15): 8334–8348. doi: 10.1021/es405118y.

Warpinski, N. R., J. E. Uhl, B. P. Engler, J. C. Lorenz and C. J. Young. 1997. *Development of Stimulation Diagnostic Technology: Annual Report*. Sandia National Laboratories, Albuquerque, New Mexico, pp. 65.

Yost, A. B. 1988. Eastern gas shales research: Recovery efficiency test. *Eastern Gas ShalesRresearch*, Morgantown Energy Technology Centre, Morgantown, West Virginia, pp. 35–115.

Yusuf, R. O., Z. Z. Noor A. H. Abba, M. A. A. Hassan and M. F. M. Din. 2012. Methane emission by sectors: A comprehensive review of emission sources and mitigation methods. *Renewable and Sustainable Energy Reviews*, 16(7): 5059–5070. doi: http://dx.doi.org/10.1016/j.rser.2012.04.008.

Section V

Socio-Economic, Regulatory

29 A Research and Education Framework to Support the Development of a Sustainable and Socially Responsible Mining Industry in Africa

Jennifer Broadhurst, Susan T. L. Harrison, Jochen Petersen, Jean-Paul Franzidis, and Dee Bradshaw

CONTENTS

ABSTRACT

A sustained programme of research and human capacity development in the context of the extraction and beneficiation of mineral resources in Africa is considered critical in ensuring that the minerals sector continues to thrive and contribute to socio-economic development in a manner consistent with sustainability principles. Of key importance in this regard is the need to promote inter- and trans-disciplinary research activities and to generate highly skilled people who have a broad understanding and awareness of the complexities involved. To this end, the Minerals to Metals Signature Research Theme was established at the University of Cape Town in 2007, with a view to addressing the key sustainability challenges in the minerals sector through the establishment of integrated and holistic research and education programmes. This innovative and dynamic initiative draws together the skills of world-renowned academic and research staff in existing groupings both within and beyond the University of Cape Town, to advance the provision of metal and mineral-related products and services to society in a manner which maximises development opportunities for its stakeholders, and which minimises adverse environmental impacts due to mining, processing and related activities.

INTRODUCTION

Minerals and metals are essential for modern life and their reliable and responsible supply is critical to maintain and develop a sustainable world. However, the global mining industry currently faces multi-faceted, internal and external challenges. These include technical challenges such as declining ore grades, more complex ores and remote ore bodies; financial and economic challenges, with cost escalations and delays in the development of new projects and rising operating costs with lower productivity; and social and environmental challenges, which include heightened competition for water, energy and land, growing scrutiny of the industry's social and environmental performance, and an increasingly complex policy and regulatory environment (Hajkowicz et al., 2012; Venter, 2013; Development Partner Framework, 2014; World Economic Forum, 2014). Addressing these challenges requires both greater process integration and enhanced systems thinking in a more risk-sensitive operating environment (Cutifani, 2013; Edraki et al., 2014). Deloitte (2013) reported a widening skills and talent gap in the mining industry that now extends up to the executive level. The next generation of leaders in the mining industry will require a broad range of skills that can best be learned in a dynamic, inter-disciplinary, industry-focused education environment (Veiga and Tucker, 2014).

Africa is blessed with abundant mineral wealth and has significant reserves of the world's major mining commodities (metals, industrial minerals and fuels). According to Lane and Ndlovu (2012), the African continent is known to hold about 30% of the world's mineral reserves, including 40% of its gold, 60% of its cobalt and 90% of its platinum-group mineral reserves. Development of these mineral resources has the potential to serve as a vehicle for sustainable economic growth and development on the continent. However, Lane and Ndlovu (2012) also note the significant and multi-dimensional risks that need to be navigated to realise this potential: it is critical to develop strategies and policies that involve all stakeholders, underpinned by a well-developed understanding of the challenges, opportunities and trade-offs involved. To achieve this, we must develop the appropriate technology through research and the appropriate skills through education and capacity building.

The World Economic Forum (2014) has outlined a roadmap to a sustainable extractive industry. It emphasises the need for appropriate research and development of technologies to operate in a clean and safe environment in frontiers previously considered inaccessible. It also recognises that the skills and capabilities needed to operate in a sustainable world are different from those required currently, and need to be cultivated.

The drive towards education for a sustainable world was articulated during the 2002 World Summit on Sustainable Development (WSSD) which emphasised the educational objectives of the Millennium Development Goals. The summit also proposed the Decade of Education for Sustainable Development for the period 2005–2014, with UNESCO as the leading agency.

Sustainability science is an emerging field that is 'defined by the problem it addresses rather than the disciplines it employs' (Clark, 2010, p. 88). It is acknowledged as an appropriate platform to develop this capability. Uwasu et al. (2009) present an example of such a response and also note that during the 2002 WSSD, the world-leading educational and scientific organisations signed the Ubuntu Declaration on Education, Science and Technology for Sustainable Development (UNU-IAS, 2005). The main goals of the Ubuntu Declaration are the following

- Strengthening of collaboration between educators and researchers
- Better integration of science and technology into educational programmes for sustainable development at all levels
- Problem-based approach for education and scientific research
- Innovation in knowledge transfer to bridge the gaps and inequalities in knowledge

This is aligned with continued adherence to the Milos declaration, articulated at the first Sustainable Development Indicators in the Minerals Industry (SDIMI) conference in 2003, and presented as a living document by Shields et al. (2015) at SDIMI in 2015. This states that 'the engineers, scientists, technical experts, and academics who work in, consult for, educate, study, or are in some other manner associated with the minerals industry, share a mutual responsibility with all individuals to ensure that our actions meet the needs of today without compromising the ability of future generations to satisfy their own needs'. (http://sdimi2015.com/milos-declaration/.)

As part of the response to this opportunity, the University of Cape Town (UCT) in 2007 established the 'Minerals to Metals' (MtM) Signature Theme as an integrated research and education framework with a strong emphasis on the inter-disciplinarity. It forms an active collaboration across four research groupings in chemical engineering but extends beyond these boundaries, engaging other fields and disciplines in the sciences, commerce and humanities at UCT and at other universities both in South Africa and internationally. The initiative was strengthened with the award of the SARChI (South African Research Chairs Initiative) Chair in Minerals Beneficiation to Professor Jean-Paul Franzidis in 2008. The SARChI programme is a national intervention for knowledge and human resource development led by the Department of Science and Technology and managed by the National Research Foundation (NRF) to enhance research and innovation capacity in South Africa (http://hicd.nrf.ac.za/). In 2014 the research-based master of philosophy programme specialising in sustainable mineral resource development was inaugurated, hosted by MtM, to provide an inter-disciplinary postgraduate qualification that highlights the critical factors of sustainable development in the context of mining and minerals processing in Africa, and promotes experimentation with innovative and systemic approaches to meet such challenges (Broadhurst et al., 2015). These aims, and the philosophy underpinning them, are directly aligned with those of the emerging field of sustainability science.

MtM has a problem-orientated-approach in which the technical design of mineral beneficiation processes is integrated with consideration of its interaction with the environmental, economic and socio-political spheres that surround them. In this context, technology choices can be evaluated not only in terms of conventional economic returns, but also on measures of their impact on natural and human environments, which allows stakeholders to make more holistically informed decisions. Inversely, this approach can make meaningful contribution to the development of new process technologies that feature more favourably on the balance of scores. Thus it aims to develop research solutions focused on enhanced value addition and resource productivity through the conversion of MtM in a manner congruent with providing a sustainable future for the African people and their environment.

RESEARCH FRAMEWORK

The continually evolving research programme which falls under MtM is underpinned by a number of industry-based projects and case studies which together explore the sustainability challenges facing the minerals sector from both a systemic and fundamental perspective. These serve to establish linkages between traditionally separate but cognate research areas. Systemic research projects are concerned with the performance of minerals processing and beneficiation systems as a whole, considering their interface with the environment and society as well as their technical or economic merit: examples of these include acid mine drainage mitigation; deriving added value from mine waste; developing tools for holistic evaluation and optimisation of mineral extraction and beneficiation systems and their impact on their environment; improving energy efficiency; minimising carbon footprint, water usage and land degradation; and improving mine safety and mining company–community stakeholder engagement. The fundamental research portfolio is currently science and technology based, aimed at developing an understanding of the underlying physical and chemical principles that govern processes within the minerals extraction and beneficiation chain and their interactions with or impact on the remainder of the system. Fundamental research projects incorporate techniques and methodology from the fields of electrochemistry, mineralogy, rheology and turbulent multi-phase processing, amongst others. Thus research is aimed simultaneously at *reducing* the environmental and social *impacts* of mineral beneficiation operations, developing safe and sustainable operational practices and *increasing* the amount of mineral or *metal extracted* from each ore body beneficiated.

Research Vignette

Systemic Research Themes

One of the key focus areas of MtM over the past 8 years, covering the themes of both *acid rock (ARD) mitigation* and *value from waste*, has entailed the development of an integrated approach to the management of sulphide mine tailings that not only removes risks of ARD pollution, but also provides opportunities for value recovery and re-allocation of unavoidable wastes as feedstock for other uses (Harrison et al., 2010; Hesketh et al., 2010b; Harrison et al., 2013; Broadhurst and Harrison, 2015). This approach is based on the concept of pre-disposal desulphurisation of mine waste to produce a large volume sulphide-lean fraction which can be land disposed without posing an ARD risk, and a smaller volume sulphide-rich fraction which can be disposed with containment or, preferably, used as backfill or processed further into useful by-products such as sulphuric acid (Figure 29.1).

Studies have demonstrated that desulphurisation flotation of a porphyry-type copper sulphide tailings can be conducted in stages through manipulation of the reagent regime, producing an additional metal-rich stream that could be recycled to the conventional metal recovery plant (Hesketh et al., 2010b).

The concept of using froth flotation to both recover value and remove ARD risks from mine waste was subsequently extended to fine coal wastes in a two-stage process that recovers coal by means of oily collectors in stage 1, followed by desulphurisation flotation to remove pyritic sulphur in stage 2 (Figure 29.2).

Case studies on a variety of coal wastes from South African and Brazilian collieries have resulted in an upgraded coal product and a sulphide-lean tailings stream, which is non-acid generating, thus effectively eliminating long-term ARD risks (Harrison et al., 2010, 2013; Amaral Filho et al., 2011; Kazadi Mbamba et al., 2012, 2013). A modification to the two-stage coal desulphurisation process, in which the sulphide flotation step is replaced with reflux classification, has also been investigated in collaboration with researchers from the North West University in South Africa. Preliminary testwork has shown that the combined application of froth flotation and reflux classification has the potential to generate high yields of low ash coal and high grades of sulphur in the sulphide-rich

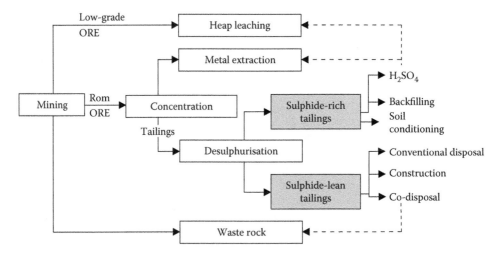

FIGURE 29.1 Conceptual approach for the integrated management of sulphide mine tailings. (Modified from Hesketh, A. H. et al. 2010a. *Hydrometallurgy* 104(3): 459–464.)

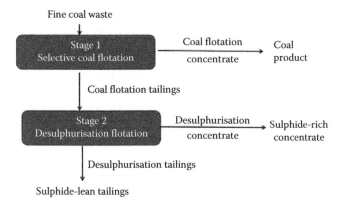

FIGURE 29.2   Two-stage flotation approach for the management of fine coal waste.

stream. Laboratory-scale feasibility testwork on the proposed desulphurisation flotation approaches has been supplemented by studies on the economic viability and broader environmental implications of the proposed desulphurisation flotation approaches. Selected case studies have demonstrated the potential economic viability of the desulphurisation flotation treatment process when combined with value recovery (Harrison et al., 2013; Jera, 2013). Holistic environmental studies using life cycle approaches and assessment tools have highlighted the environmental benefits of the desulphurisation flotation tailings management approach in terms of reduced aqueous emissions and waste burden as well as enhanced resource recovery. However, the flotation process also consumes additional electricity, which results in higher climate change, fossil fuel depletion and acidification impacts in cases where the production of electricity is coal based (Broadhurst et al., 2014). Further research is focusing on the potential downstream utilisation of the separated tailings streams, either with or without additional processing (Harrison et al., 2013). This approach is consistent with the principles of a circular economy which looks at reducing the quantities of material disposed of as waste, as well as maximising the efficient use of processed materials.

Another related research area that has received considerable attention within MtM is the development of tools and protocols for the accurate and reliable characterisation and prediction of ARD generation associated with solid mine wastes (Broadhurst et al., 2013). In an attempt to address the shortcomings of standard static chemical and kinetic ARD prediction tests, a microbial shake flask test was developed by Hesketh et al. (2010a). Studies on a range of sulphide-bearing mine wastes from different sources and origins have demonstrated the use of the biokinetic test both validate and enhance the results of simple static tests, providing valuable information on the effect of microbial activity and the relative kinetics of the acid forming and neutralising reactions under stagnant and dynamic flow conditions in a relatively short period of time (Hesketh et al. 2010b; Amaral Filho et al., 2011; Kazadi Mbamba et al., 2012; Broadhurst et al., 2013; Chimbganda et al., 2013; Dyantyi et al., 2013; Harrison et al., 2013; Becker et al., 2015; Opitz et al., 2015). The studies have also highlighted the important role of mineralogy, both in characterising the acid generating potential of a sample directly and in assisting in the interpretation of the geochemical reactions and governing species controlling the acid generating behaviour (Dyantyi et al., 2013; Becker et al., 2015). These studies have been used in combination with abiotic and biotic column studies to assess further the potential for the onset of ARD generation and its mitigation (Opitz 2013; Kotsiopoulos and Harrison, 2015). Another project for predicting the potential long-term ARD risks associated with sulphidic waste disposal (Simunika et al., 2013) showed that the PHREEQC-based mass transport model can provide reasonably accurate estimates of the time-related concentration profiles of key elements under continuous flow-through conditions, as well as identify the controlling geochemical processes and parameters. The assessment of potential health and environmental risks due to the release of metals from mine wastes, either during land disposal (Broadhurst and Petrie, 2010; Opitz et al., 2015) or in downstream applications is also currently the subject of both under- and post-graduate research projects within MtM.

Also falling under the theme of *value from waste* is the investigation into the potential for using mine waste for CO_2 sequestration through mineral carbonation. Initial studies by Vogeli et al. (2011) showed that, in theory, tailings from the primary processing of PGM (platinum group metal) ores in South Africa have the potential to sequester ~14 Mt of CO_2 per annum through mineral carbonation. Subsequent laboratory-scale testwork has, however, indicated that, in practice, the mineral carbonation of these tailings will require relatively aggressive and high energy process conditions (such as high temperatures, pressures and/or extensive pre-conditioning), due to the inert nature of orthopyroxene, which accounts for 88% of the available Mg (Meyer et al., 2014). Further studies are currently in progress to investigate the feasibility of potential mineral carbonation process options on the basis of their carbon neutrality. Other work in the area of *value from waste* has entailed the development of a novel flow sheet that employs flotation to recover monazite from a zircon reject stream at Namakwa Sands heavy minerals plant in the Northern Cape.

In the general area of *systemic performance analysis and optimisation* of minerals beneficiation, systems, the Aspen® plus simulation package was used as a simulation tool to facilitate the optimisation of the water balance of the Skorpion Zinc refinery, with the ultimate aim of reducing water consumption (Bhikha et al., 2011). Aspen is also being used to simulate potential processing systems for the mineral carbonation of PGM tailings, in the study on carbon neutrality. Another systemic tool that has been used fairly extensively in the evaluation of the environmental performance of minerals beneficiation and industrial systems is that of life cycle assessment (see e.g. Ras and Blottnitz, 2012; Broadhurst et al., 2014). Models for the simultaneous evaluation of technical, environmental and economic performance have been developed and applied in the comparison of both copper (Lusinga et al., 2011) and zinc processing options, whilst a study by Chiloane et al. (2013) provided a basis for developing a framework for analysing the co-location of utility-scale solar power plants within metallurgical operations. Guma et al. (2009) explored the strengths and limitations of eco-efficiency indicators as performance metrics in guiding environmentally sustainable decision-making during minerals process design. The results of this study confirmed that the meaningful application of eco-efficiency indicators is strongly dependent on the decision context in which the process data is generated.

In the area of *mine safety*, a framework was developed for investigating the causes of accidents in the South African platinum mining industry, using a combination of the Mark III version of James Reason's Swiss Cheese model with the Nertney Wheel and safety management principles. This study is being extended to copper and coal mining by master of philosophy students in the new sustainable minerals resource development programme.

Also in the area of systemic research, a project commenced in 2013 to develop a *minerals beneficiation strategy* for the KwaZulu-Natal province in collaboration with The Green House, a niche consultancy specialising in sustainability projects. This project is supported by four master's projects, drawn from the disciplines of anthropology, economics and chemical engineering, encompassing the development of a framework for the evaluation and selection of minerals beneficiation options using multi-criteria decision analysis; macro-economic modelling to evaluate the knock-on economic effects of mineral beneficiation in the province; the exploration of models for promoting and expanding multi-stakeholder engagement in minerals beneficiation; and the impact of mineral beneficiation on communities in the areas surrounding the Richards Bay area.

Fundamental Research Themes

One of the key fundamental research themes has entailed the development and application of fine particle activation for *positron emission particle tracking* (*PEPT*) for the *in situ* characterisation of particulate flow within mineral processing units, with particular emphasis on tumbling mills. A number of projects aimed at developing techniques related to studies on interpreting the data from

PEPT experiments, and improving the sampling rates in these experiments, have been supported as a part of an ongoing collaboration between the chemical engineering and physics departments at UCT, using the facilities at iThemba laboratories (Buffler et al., 2010; Govender et al., 2011; Volkwyn et al., 2011; Cole et al., 2012). A current project entails the development of a new technique to extract and purify 64-Cu from a waste stream, for labelling tracers for use in PEPT and PET (particle emission tomography) experiments, so as to allow for longer experiments to be carried out. Yet another project is developing a technique to remove contaminants from the 68-Ge/68-Ga generator product, so that it is possible to extend the life of this expensive device (van Heerden, 2015).

In the area of *particle dynamics and fluid flow*, a computational fluid dynamics (CFD) model of a pilot-scale autoclave was developed based on the fundamental principles of hydrodynamics, gas dispersion and mass transfer and validated with experimental data (Appa et al., 2013, 2014). In a separate study by Bakker et al. (2009, 2010), a CFD model was used to simulate the behaviour of non-Newton slurry in a mechanical flotation cell, whilst a current study developed a granular flow model for the passage of particles across moving screen surfaces, using a DEM (discrete element method)-based model.

Another cross-cutting fundamental theme that is related to particle and fluid flow is that of *rheology*. A study of the rheology of select phyllosilicate minerals that commonly exist as problematic gangue elements during the beneficiation of low-grade ores led to an enhanced understanding of the effects of mineral structure and surface chemistry on the behaviour of three phyllosilicate minerals: chrysotile, mica and vermiculite (Ndlovu et al., 2011a, b; Becker et al., 2013). Other studies under this theme include the technical feasibility of transporting coal as coal–water mixtures, and the development of a numerical model for the study of particulate suspension rheology, which is a major problem on many mineral processing plants.

In the general area of *hydrometallurgy*, some focus has been placed on heap leaching as a technology for the extraction of value form low-grade ores as well as secondary materials (Petersen, 2015). The similarity of heap bioleaching with the processes leading to AMD generation is also noted, especially in terms of suitable modelling tools (Simunika et al., 2013). A doctoral project investigated the chemical and bioleaching behaviour of large particles in a heap leach scenario, using a combination of mineralogical, chemical and physical characterisation tools, and column leach experiments (Ghorbani et al., 2011a–c, 2102, 2013a–c). Mineralogical characterisation tools have also been used extensively in the investigation of the effect of high pressure grinding rolls (HPGR) on the flotation performance of PGM ores (Solomon et al., 2011) and the leaching of a low-grade Witwatersrand gold ore in a flow-through scenario to characterise its performance in heap leaching (Nwaila et al., 2013). A recent MSc project (Dlamini, 2015), investigated the techno-economic feasibility of heap leaching of a zinc sulfide ore in comparison to conventional processes and found the technology would be competitive, if operated optimally (some of this data is presented in Petersen, 2015). A substantial review of heap leach technology (Ghorbani et al., 2015) describing the *status quo* and highlighting the remaining challenges and future opportunities of this technology has been prepared under MtM auspices to guide future research efforts in this direction.

Also in the general area of hydrometallurgy, an electrochemical and leaching study of the dissolution of chalcopyrite in ammonia–ammonium sulphate solutions, using pure mineral electrodes, has indicated that copper (II), and not dissolved oxygen, is the effective oxidant for chalcopyrite under the conditions of the study (Moyo et al., 2015). Moving further down the minerals beneficiation chain, a study is currently in progress to characterise and model the decay of ion-exchange resins after protracted use (Nesbitt et al., 2014).

EDUCATION FRAMEWORK

The learning outcomes from the research projects feed into the framework for education in sustainable and socially responsible development of mineral resources in an integrated way to achieve maximum benefit for all stakeholders. Much of the research described above forms part of PhD and masters studies, including the MSc(Eng), carried out in an inter-disciplinary environment. Further to this, the research-based master of philosophy (MPhil) programme specialising in sustainable mineral resource development is of key importance in this framework. It is designed to produce meaningful targeted research outcomes, while generating highly skilled people with a good integrated understanding of the critical issues involved in developing mineral resources sustainably, and the sensitivity to project and harness this understanding in the context of different stakeholders. The MPhil programme aims to produce what has been termed 'T-shaped' individuals (Brown, 2015), who have depth (vertical dimension, from the analytical expertise developed in their specific postgraduate research topic) as well as breadth (horizontal dimension, achieved through working alongside and interacting with other individuals from a wide range of disciplines and backgrounds). Having such individuals available to the industry will ensure that industry-stakeholders have access to leading-edge technologies, inherently safe processes and high-level skills to enhance profitability and benefits to local communities whilst simultaneously minimising socio-environmental impacts.

The 2-year MPhil programme was developed as part of the Education for Sustainable Development in Africa (ESDA) project of the United Nations University Institute for Sustainability and Peace (UNU-ISP), in collaboration with seven other African universities and the University of Tokyo. The programme targets students from all over the world, but from Southern Africa in particular, and is offered and delivered collaboratively by UCT and the University of Zambia (UNZA). The target number of students accepted into the programme between the two universities each year is 25.

Selection is based on an applicant's academic record, and the duration, level and relevance of any work experience. The basic entry requirement is a 4-year university Bachelors or Honours degree, but a higher national diploma or equivalent qualification is also considered as a basis for entry, subject to appropriate experience. The programme is open to graduate professionals from across a spectrum of disciplines, including geologists, engineers, planners, strategists, lawyers, regulators, health professionals, safety specialists, environmental officers, economists and social scientists, who seek a broader understanding of what is involved in sustainable mineral resource development in order to promote its application in the most meaningful way. One of the criteria applied when selecting students for the programme is to assemble cohorts with a diversity of backgrounds, to provide opportunities for students to learn from one another's disciplines and experiences as well as from the course material. The first two cohorts (2014 and 2015 intakes) have included students with first qualifications in chemical, mining, civil and mineral process engineering; forestry; geography; geology; social anthropology; social science; psychology; law; business science and economics.

The course work programme for the MPhil comprises four core course modules that students from the two universities (UCT and UNZA) attend as one class. In line with the inter-disciplinary and inter-institutional nature of this programme, the courses are delivered across four different faculties at three different academic institutions:

- Sustainable Development, convened by the Sustainability Institute at the University of Stellenbosch in South Africa
- Strategic Social Engagement Practice, convened by the Graduate School of Business at the UCT
- Environmental Stewardship in Mining and Minerals Beneficiation, convened by the Zambian School of Mines at UNZA
- Research Communication and Methodology, convened by the Department of Chemical Engineering at UCT

In addition to coursework, students are required to undertake a non-credit bearing field-based internship which entails structured engagement with a real problem in the developmental setting of a host organisation. This comes from the realisation that the complexity of coupled social-ecological systems can only partly be learned in the classroom. The specific aims and content of the internship are tailored to the needs of individual students. In the case of those students employed in or around the industry, opportunity is provided to experience their working environments from a different perspective, by spending time in another part of their organisation or in another sector completely. Students are required to keep a logbook and complete an internship report for the host organisation and for submission to their respective university.

Finally, each student is required to complete a dissertation, typically incorporating a research project of a theoretical or practical nature, in one or other of the areas covered by the core courses. The topic is decided in consultation with the academic course convenor, and depends on the student's background expertise and professional interests. Cross-disciplinary research is promoted through the joint supervision of student dissertations across faculties and universities.

At the time of writing, in September 2015, after nearly 2 years of operation, the programme has a total cohort of 33 students, 18 of whom are registered at UCT and 15 at UNZA. The students hail from four southern African countries (South Africa, Zimbabwe, Malawi and Zambia), as well as Australia and Japan. They range from 21 to 51 years in age, and 13 of the 33 are female. Of the UCT students, four are full-time, while the remaining 14 represent government, academia, the mining industry, consulting firms and the economic and business sectors.

The first cohort of 15 students completed the four core courses in October 2014, and have commenced their research projects through their respective universities. Research topics include

- Community involvement in rehabilitation of degraded mine land
- Performance analysis and decision-making frameworks for mineral value chains
- Life cycle-based indicators for eco-efficient processing of PGMs
- Entrepreneurship in communities around mining and minerals beneficiation operations
- Measuring the sustainability signature of mining assets by integrating models
- A systemic approach to mining accident causality analysis
- Legal frameworks for encouraging sustainable communities post-mining
- Broader implication of deforestation by mining companies
- Downstream uses of mine wastes: opportunities, challenges and implications
- Reconciling different stakeholders: A Zambian case study
- Challenges and opportunities for revenue collection from mining companies
- The effectiveness of EIA protocols and legislation in relation to mining

The workshop sessions to gauge the development and application of integrative knowledge at the end of the third block of courses showed a noticeable improvement in the students' understanding of the terms ('jargon') typically associated with each other's disciplines, with the student cohort having, to a large extent, developed a 'common language' which transcended their individual fields of expertise. Effective communication and interaction between students was further facilitated by an enhanced respect and tolerance for each other's perspectives and viewpoints. The students also emphasised the extent to which the programme had enhanced their self-awareness, both in terms of their personal attributes and beliefs, as well as their potential capacity to contribute to the society in which they function. A high level of self-awareness was, furthermore, considered to be of key importance in addressing the complex challenges of sustainable development, particularly in terms of effective stakeholder engagement.

Concepts and principles considered by the students to be of key relevance to the sustainable development of mineral resources include

- Value-based leadership and governance
- Shared value as a business model
- Constructive and inclusive stakeholder engagement
- Respect for nature (deep ecology)
- Effective planning and monitoring
- Systemic perspectives and integrated approaches to the complex challenges facing the minerals industry

At the time of writing (September 2015), the second cohort of 18 students was completing the final block of coursework, and beginning work on proposals for their dissertations. The first cohort is in the process of completing their research dissertations. Their feedback on the coursework component of the programme has been extremely positive. The students have clearly integrated the learnings from the different course modules, and developed a new appreciation for the complexity of sustainable minerals resource development and the need for value-based leadership and governance. This integrated knowledge and understanding underpins the research projects in progress.

The academic staff involved with this programme have also benefitted through engagement with academics from other disciplines, as well as external stakeholder organisations (minerals industry, government bodies and NGOs). In time, it is envisaged that this programme will generate a number of new research areas involving an as-yet unexplored synthesis of disciplines and collaborative partnerships, both within and external to UCT. The benefits gained from this experience are, furthermore, likely to extend beyond the mining sector through the development of trans-disciplinary research capacities for solving complex sustainability problems in general, as well as practical guidelines in how such research should be practiced.

CONCLUSIONS AND THE WAY FORWARD

There is significant potential for Africa's mineral wealth to increase its contribution to the sustainable development of the continent. This potential can be harnessed by integrating sustainable development principles into the development of technology and future leadership capability that can be used by all stakeholders, and will ensure the supply of minerals and metals to the modern world.

The research undertaken by the MtM Signature Theme at UCT focuses on providing solutions that can be directly applied to contribute to the mining industry's sustainability. Since the establishment of the MtM initiative as a UCT Signature Theme in 2007, minerals beneficiation research in the Department of Chemical Engineering and at UCT has developed from a narrow focus on the deep understanding of separate extraction processes to incorporate a novel holistic approach to the entire minerals beneficiation chain, and its interactions with the broader society and environment. This has been achieved by forming exciting new trans-disciplinary staff collaborations and innovative inter-disciplinary research, based on a strong fundamental foundation in the areas of natural, engineering, social and economic sciences.

The MPhil programme specialising in sustainable minerals resource development has, furthermore, demonstrated the desired attributes of education in sustainable development (ESD) put forward by UNESCO in 2002 (UNESCO, 2007). These include inter-disciplinarity; holistic, values-driven thinking; critical thinking; a problem-solving approach; utilisation of multi-methods for teaching; and, most importantly, being participation-oriented and locally relevant. The programme has gathered considerable interest from the community of stakeholders and is set to deliver a new generation of professionals for whom sustainability thinking in a mining context is no longer an add-on but integral to all activities. It is important to note that collaboration with the other stakeholders is crucial to its relevance and future impact, as is alignment with other courses offered around the world.

Through this inter-disciplinary research and education framework, MtM has created a unique platform for addressing the sustainability challenges facing the minerals industry sector using deep technical and integrated trans-disciplinary approaches, while actively participating in its transformation to a more socially responsible entity. It is expected that future decision makers will come from this platform and be equipped to realise the promise of Africa.

ACKNOWLEDGEMENTS

The launch of this progressive and innovative programme has not been without its challenges, particularly administratively, and would not have been possible without the extensive support and assistance of Professor Danie Visser and Dr. Marilet Sienaert from the UCT executive, and Professor

Francis Petersen, Mr Billy Daubenton and Ms Gita Valodia of the Faculty of Engineering and the Built Environment at UCT. Our appreciation and thanks go also to Professor Masafumi Nagao and Dr. Emmanuel Mutisya from the United Nations University, and Professor Stephen Simukanga and Dr. Jewette Masinja from UNZA, who have participated in the establishment of the new MPhil degree. The participation and support of other UCT research supervisors and MtM executive members is also recognised and appreciated, namely Professors Harro von Blottnitz, Alison Lewis, David Deglon, Aubrey Mainza, Andy Buffler, Anthony Black, Tony Leiman and Drs. Megan Becker, Niyi Isafiade, Indresan Govender, Divine Fuh and Brett Cohen.

REFERENCES

Amaral Filho, J., C. Kazadi Mbamba, J.-P. Franzidis, J. Broadhurst, S. T. L. Harrison and I. A. H. Schneider. 2011. In: *Froth Flotation for Desulphurization and Coal recovery: A Comparative Study of South African and Brazilian Colliery Wastes, Flotation '11*, Cape Town, 14–17 November 2011.

Appa, H., D. A. Deglon and C. J. Meyer. 2013. Numerical modelling of hydrodynamics and gas dispersion in an autoclave. *Hydrometallurgy*, 131–132: 67–75.

Appa, H., D. A. Deglon and C. J. Meyer. 2014. Numerical modelling of mass transfer in an autoclave. *Hydrometallurgy*, 147–148: 234–240.

Bakker, C. W., C. J. Meyer and D. A. Deglon. 2009. Numerical modelling of non-Newtonian slurry in a mechanical flotation cell. *Progress in Computational Fluid Dynamics*, 9(6/7): 308–315.

Bakker, C. W., C. J. Meyer and D. A. Deglon. 2010. The development of a cavern model for mechanical flotation cells. *Minerals Engineering*, 23(11–13): 968–972.

Becker, M., N. Dyantyi, J. L. Broadhurst, S. T. L. Harrison and J.-P. Franzidis. 2015. A mineralogical approach to evaluating laboratory scale acid rock drainage characterisation tests. *Minerals Engineering*, 80: 33–36.

Becker, M. E., G. A. Yorath, B. Ndlovu, M. C. Harris, D. A. Deglon and J.-P. Franzidis. 2013. A rheological investigation of the behaviour of two Southern African platinum ores. *Minerals Engineering*, 49: 92–97.

Bhikha, H., A. E. Lewis and D. A. Deglon. 2011. Reducing water consumption at Skorpion Zinc. *Journal of the South African Institute of Mining and Metallurgy*, 111: 437–442.

Broadhurst, J., J. -P. Franzidis, S. T. L. Harrison and H. von Blottnitz. 2015. Educating managers and leaders for responsible mining in Africa: A post-graduate degree programme. In: *Conference Proceedings, Sustainable Development in the Minerals Industry*, Vancouver, Canada, 12–15 July 2015.

Broadhurst, J. L., C. G. Bryan, M. Becker, J. -P. Franzidis and S. T. L. Harrison. 2013. Characterising the acid generating potential of mine wastes by means of laboratory-scale static and biokinetic tests. In: *Reliable Mine Water Technology*, edited by A. Brown, L. Figueroa and C. Wolkersdorfer. International Mine Water Association (IMWA), Denver, Colorado, Vol. I, pp. 275–281. ISBN 978-0-615-79385-6.

Broadhurst, J. L. and S. T. L. Harrison. 2015. A desulfurization flotation approach for the integrated management of sulfide wastes and acid rock drainage risks. In: *10th International Conference on Acid Rock Drainage & IMWA Annual Conference,* Santiago, Chile, 20–25 April 2015. Chapter 2: *Applied Mineralogy and Geoenvironmental Units*, edited by A. Brown, C. Bucknam, M. Carballo et al. Gecamin.

Broadhurst, J. L., M. C. Kunene and H. von Blottnitz. 2014. Life cycle assessment of the desulfurisation flotation process to prevent acid rock drainage: A base metal case study. *Minerals Engineering*, 76: 126–134.

Broadhurst, J. L. and J. G. Petrie. 2010. Ranking and scoring potential environmental risks from solid mineral wastes. *Minerals Engineering*, 23: 182–191.

Brown, T. 2015. T-shaped stars: The backbone of IDEO's collaborative culture. http://chiefexecutive.net/ideo-ceo-tim-brown-t-shaped-stars-the-backbone-of-ideoae%E2%84%A2s-collaborative-culture/. Accessed 14 September 2015.

Buffler, A., I. Govender, J. J. Cilliers, D. J. Parker, J. -P. Franzidis, A. Mainza, R. T. Newman, M. Powell, A. Van der Westhuizen. 2010. PEPT Cape Town: A new positron emission particle tracking facility at iThemba LABS. In: *Proceedings of International Topical Meeting on Nuclear Research Applications and Utilization of Accelerators*, Vienna, 4–8 May 2009. Vienna: IAEA. STI/PUB/1433, ISBN 978-92-0-150410-4, ISSN 1991-2374.

Chiloane, L., J. Petersen, H. Von Blottnitz and J.-P. Franzidis. 2013. Towards a framework for analysing co-location of utility scale solar power plants within metallurgical operations. In *Proceedings of the 3rd International Seminar on Environmental Issues in Mining* (*Enviromine 2013*), Santiago, Chile, 4–6 December 2013, edited by S. Winchester, F. Valenzuela and D. Mulligan. . Santiago, Chile: Gecamin. ISBN 9789569393044.

Chimbganda, T., M. Becker, J. L. Broadhurst, S. T. L. Harrison and J. -P. Franzidis. 2013. A comparison of pyrrhotite rejection and passivation in two nickel ores. *Minerals Engineering*, 46–47: 38–44.

Clark, B. 2010 Sustainable development and sustainable science. In: *Report from Toward a Science of Sustainability Conference*, Airlie Center, Warrenton, Virginia, November–December 2009, edited by S.A. Levin and W.C. Clark, Center for International Development at Harvard University, pp. 82–104.

Cole, K. E., A. Buffler, N. P. van der Meulen, J. J. Cilliers, J. -P. Franzidis, I. Govender, C. Liu and M. R. van Heerden. 2012. Positron emission particle tracking measurements with 50 micron tracers. *Chemical Engineering Science*, 75: 235–242.

Cutifani, M. 2013. A critical imperative – Innovation and a sustainable future. Speech to the *World Mining Congress*, Montreal, Canada. http://www.angloamerican.co.za/~/media/Files/A/Anglo-American-South-Africa/Attachments/media/presentation/mark-cutifani-speech-at-WMC.pdf

Deloitte. 2013. Tracking the trends. https://www2.deloitte.com/content/dam/Deloitte/global/Documents/Energy-and-Resources/dttl-er-Tracking-the-trends-2014_EN_final.pdf. Accessed 6 December 2014.

Development Partner Framework. 2014. Mining company of the future. http://www.kinglobal.org/catalyst.php. Accessed 6 December 2014.

Dlamini, Z. 2015. Techno-economical comparison of three process routes for the treatment of Gamsberg zinc ore. MSc thesis, University of Cape Town, Cape Town, South Africa.

Dyantyi, N., M. Becker, J. L. Broadhurst, S. T. L. Harrison and J.-P. Franzidis. 2013. Use of mineralogy to interpret laboratory-scale acid rock drainage prediction tests – A gold case study. In *World Gold*. Brisbane: AusIMM, pp. 519–525.

Edraki, M., T. Baumgartl, E. Manlapig, D. J. Bradshaw, D. M. Franks and C. J. Moran. 2014. Designing mine tailings for better environmental, social and economic outcomes: A review of alternative approaches. *Journal of Cleaner Production*, 84: 411–420.

Ghorbani, Y., M. E. Becker, A. N. Mainza, J. -P. Franzidis and J. Petersen. 2011a. Large particle effects in chemical/biochemical heap leach processes – A review. *Minerals Engineering*, 24: 1172–1184.

Ghorbani, Y., M. E. Becker, J. Petersen, A. N. Mainza and J.-P. Franzidis. 2013a. Investigation of the effect of mineralogy as rate-limiting factors in large particle leaching. *Minerals Engineering*, 52: 38–51.

Ghorbani, Y., M. E. Becker, J. Petersen, S. Morar, A. N. Mainza and J.-P. Franzidis. 2011b. Use of X-ray computed tomography to investigate crack distribution and mineral dissemination in sphalerite ore particles. *Minerals Engineering*, 24: 1249–1257

Ghorbani, Y., J. -P. Franzidis and J. Petersen. 2015. Heap leaching technology – Current state, innovations and future directions: A review. *Minerals Processing and Extractive Metallurgy Review* (in press). Accepted 19 October 2015.

Ghorbani, Y., A. N. Mainza, J. Petersen, M. E. Becker, J.-P. Franzidis and J. T. Kalala. 2013b. Investigation of particles with high crack density produced by HPGR and its effect on the redistribution of the particle size fraction in heaps. *Minerals Engineering*, 43–44: 44–51.

Ghorbani, Y., J., Petersen, M. E. Becker, A. N. Mainza and J. -.P. Franzidis. 2013c. Investigation and modelling of the progression of zinc leaching from large sphalerite ore particles. *Hydrometallurgy*, 131–132: 8–23.

Ghorbani, Y., J. Petersen, S. T. L. Harrison, O. Tupikina M. E. Becker, A. Mainza and J.-P. Franzidis. 2012. An experimental study of the long-term bioleaching of large sphalerite ore particles in a circulating fluid fixed-bed reactor. *Hydrometallurgy*, 129–130: 161–171.

Govender, I., N. Mangesana, A. N. Mainza and J.-P. Franzidis. 2011. Measurement of shear rates in a laboratory tumbling mill. *Minerals Engineering*, 24: 225–229.

Guma, M., H. Von Blottnitz and J. L. Broadhurst. 2009. A systems approach for the application of eco-efficiency indicators for process design in the minerals industry. In: *Sustainable Development Indicators in the Minerals Industry Conference (SDIMI 2009)*, Gold Coast, Australia, 6–8 July 2009. AusIMM, pp. 301–311.

Hajkowicz, S., H. Cook and A. Littleboy. 2012. Global megatrends that will change the way we live. CSIRO FUTURES. http://www.csiro.au/Portals/Partner/Futures/Our-Future-World.aspx. Accessed 6 December 2014.

Harrison, S. T. L., J. L. Broadhurst, R. P. van Hille, O. O. Oyekola, C. Bryan, A. Hesketh and A. Opitz. 2010. A systematic approach to sulphidic waste rock and tailings management to minimise ARD formation. WRC K5/1831/3, June 2010.

Harrison, S. T. L, J. -P. Franzidis, R. P. van Hille, T. Mokone, J. L. Broadhurst, C. Kazadi Mbamba and A. Opitz et al. 2013. Evaluating approaches to and benefits of minimising the formation of acid rock drainage through management of the disposal of sulphidic waste rock and tailings. WRC 2015/1/2013, February 2013.

Hesketh, A. H., J. L. Broadhurst, C. G. Bryan, R. P. van Hille and S. T. L. Harrison. 2010a. Biokinetic test for the characterisation of AMD generation potential of sulfide mineral wastes. *Hydrometallurgy*, 104(3): 459–464.

Hesketh, A., J. Broadhurst and S. T. L. Harrison. 2010b. Mitigating the generation of acid mine drainage from copper sulfide tailings impoundments in perpetuity: A case study for an integrated management strategy. *Minerals Engineering*, 23(3): 225–229.

Jera, M. K. 2013. An economic analysis of coal desulphurisation by froth flotation to prevent acid rock drainage (ARD) and an economic review of capping covers and ARD treatment processes. MSc thesis, University of Cape Town, Cape Town, South Africa.

Kazadi Mbamba, C., J. -P. Franzidis, S. T. L. Harrison and J. L. Broadhurst. 2013. Flotation of coal and sulphur from South African ultrafine colliery wastes. *Journal of the South African Institute of Mining and Metallurgy*, 133: 399–405.

Kazadi Mbamba, K., S. T. L. Harrison, J. -P. Franzidis and J. Broadhurst. 2012. Mitigating acid rock drainage risk while recovering low-sulphur coal from ultrafine colliery wastes using froth flotation. *Minerals Engineering*, 29: 13–21.

Kotsiopoulos, A. and S. T. L. Harrison. 2015. Enhancing ARD mitigation by application of benign tailings to reduce the permeability of waste rock dumps. *Journal of Advanced Materials Research*, 1130: 560–563.

Lane, A. and J. Ndlovu. 2012. The promise of Africa. In: *Conference Proceedings, XXVI International Mineral Processing Congress – XXVI IMPC 2012*, Delhi, 23–27 September 2012, pp. 3784–3791.

Lusinga, D., J. Petersen and J. Broadhurst. 2011. A multi criteria analysis and comparison of primary copper processing options. In *2nd International Seminar on Environmental Issues in the Mining Industry*, Santiago, Chile, 23–25 November 2011, edited by M. Sánchez, D. Mulligan and J. Wiertz. Gecamin, pp. 14–15.

Meyer, N. A., J. U. Vögeli, M. Becker, J. L. Broadhurst, D. L. Reid and J. -P. Franzidis. 2014. Mineral carbonation of PGM mine tailings for CO_2 storage in South Africa: A case study. *Minerals Engineering*, 59: 45–51.

Moyo, T., J. Petersen, J.-P. Franzidis and M. Nicol. 2015. An electrochemical study of the dissolution of chalcopyrite in ammonia–ammonium sulphate solutions. *Canadian Metallurgical Quarterly*, 54: 268–277.

Ndlovu, B., M. E. Becker, E. Forbes, D. A. Deglon and J.-P. Franzidis. 2011a. The influence of phyllosilicate mineralogy on the rheology of mineral slurries. *Minerals Engineering*, 24: 1314–1322.

Ndlovu, B., E. Forbes, M. E. Becker, D. A. Deglon, J.-P. Franzidis and J. S. Laskowski. 2011b. The effects of chrysotile mineralogical properties on the rheology of chrysotile suspensions. *Minerals Engineering*, 24: 1004–1009.

Nesbitt, A., J. Petersen and J. -P. Franzidis. 2014. Decoupling intra-particle diffusion from lumped parameters to determine in-service decay of an acid ion exchange resin. In: *Proceedings of Hydrometallurgy* 2014 22–25 June 2014. Victoria, BC, Canada: Canadian Institute of Mining, Metallurgy & Petroleum, Vol. II, pp 213–222. ISBN: 978-1-926872-22-3.

Nwaila, G., M. Becker, Y. Ghorbani, J. Petersen, D. Reid, C. Bam, F.C. de Beer and J-P. Franzidis. 2013. A geometallurgical study of the Witwatersrand gold ore at Carletonville, South Africa. In: *Proceedings of the Second AUSIMM International Geometallurgy Conference*. Brisbane: AusIMM, pp. 77–85.

Opitz A. K. B. 2013. An investigation into accelerated leaching for the purpose of ARD mitigation. MSc(Eng) dissertation, University of Cape Town, Cape Town, South Africa.

Opitz, A. K. B., J. L. Broadhurst and S. T. L. Harrison. 2015. Assessing environmental risks associated with ultrafine coal wastes using laboratory-scale tests. *Journal of Advanced Materials Research*, 1130: 635–639. doi:10.4028/www.scientific.net/AMR.1130.635.

Petersen, J. 2015. Heap Leaching as a Key Technology for Recovery of Values from Low-grade Ores – A Brief Overview. Hydrometallurgy (in press); published online: 26 September 2015. doi: 10.1016/j.hydromet.2015.09.001.

Ras, C. and H. Von Blottnitz. 2012. A comparative life cycle assessment of process water treatment technologies at the Secunda industrial complex, South Africa. *Water SA*, 38(4): 549–554.

Shields D., Z. Agioutantis, M. Karmis and P. N. Martens. 2015. The Milos Declaration: creating a living document. In: *Conference Proceedings, Sustainable Development in the Minerals Industry*, Vancouver, Canada. 12–15 July 2015.

Simunika, N., J. L. Broadhurst, J. Petersen, S. T. L. Harrison and J.-P. Franzidis. 2013. Predicting the time-related generation of acid rock drainage from mine waste: A copper case study. In *Proceedings of the 3rd International Seminar on Environmental Issues in Mining (Enviromine 2013)*, Santiago, Chile, 4–6 December 2013, edited by S. Winchester, F. Valenzuela and D. Mulligan. Santiago, Chile: Gecamin. ISBN 9789569393044.

Solomon, N., M. E. Becker, A. N. Mainza, J. Petersen and J. -P. Franzidis. 2011. Understanding the influence of HPGR on PGM flotation behavior using mineralogy. *Minerals Engineering*, 24: 1370–1377.

UNESCO, 2007. The UN Decade of Education for Sustainable Development (DESD 2005–2014). The first two years. http://unesdoc.unesco.org/. Accessed 20 October 2015.

UNU-IAS. 2005. Mobilizing for education for sustainable development: Towards a global learning space based on regional centres of expertise. United Nations University, Institute of Advanced Studies, Yokohama, Japan.

Uwasu, M., H. Yabar, K. Hara, Y. Shimoda and T. Saijo. 2009. Educational initiative of Osaka University in sustainability science: Mobilizing science and technology towards sustainability. *Sustainability Science*, 4: 45–53.

Van Heerden, M. 2015. Optimization of the radio-labelling of ion exchange resin tracers for positron emission particle tracking. MSc thesis, University of Cape Town, Cape Town, South Africa, 2015.

Veiga, M. and C. Tucker. 2014. Sustainability and public engagement in mining: The role of Engineers. In: *Conference Proceedings, XXVII International Mineral Processing Congress* – IMPC 2014, Santiago Chile, 19–23 October 2014, pp. 3–13.

Venter, E. 2013. FFF Exxaro's Next Journey. Ernst Venter Executive Head, Growth, Technology, Projects, Services and Ferrous. 13 February 2013. http://www.fossilfuel.co.za/presentations/2014/FFF-March-2014-Ernst-Venter-v9.pdf. Accessed 6 December 2014.

Vogeli, J., D. L. Reid, M. E. Becker, J. L. Broadhurst and J. -P. Franzidis. 2011. Investigation of the potential for mineral carbonation of PGM tailings in South Africa. *Minerals Engineering*, 24: 1348–1356.

Volkwyn, T. S., A. Buffler, I. Govender, J.-P. Franzidis, A. Morrison, A. Odo, N. P. van der Meulen and C. Vermeulen. 2011. Studies of the effect of tracer activity on time-averaged positron emission particle tracking measurements on tumbling mills at PEPT Cape Town. *Minerals Engineering*, 24: 261–266.

World Economic Forum. 2014. Scoping paper: Mining and metals in a sustainable world. *World Economic Forum Mining & Metals Industry Partnership in Collaboration with Accenture*, February 2014. http://www3.weforum.org/docs/WEF_MM_Mining MetalSustainableWorld_ScopingPaper_2014.pdf. Accessed 6 December 2014.

30 Electronic Waste Processing through Copper Recycling
An Economic Analysis

Maryam Ghodrat, M. Akbar Rhamdhani, Geoffrey Brooks, Syed Masood, and Glen Corder

CONTENTS

ABSTRACT

E-waste is a complex waste category containing both hazardous and valuable substances. It demands for a cost-efficient treatment system which simultaneously liberates and refines target fractions in an environmentally sound way.

This study compiles data to allow cost evaluations for metal recovery from e-waste treatment through copper smelting process and compares that with the secondary copper smelting without adding electronic waste to the feed. To formulate a preliminary assessment of likely project economics and viability, a theoretical techno-economic study was attempted. The data given focuses on the process plant and main consumption data. Site-specific cost factors have been eliminated as far as possible. Costs are mainly presented on the basis of cost per annual tonne of copper production.

The outcome of the project confirms that e-waste recycling process through secondary copper smelting includes considerable potential value and therefore should be taken into consideration. The cost–benefit analysis demonstrates how the e-waste recycling market can grow profitability over the first 3 years of the plant operation.

The results of this study provide a reliable estimate of future e-waste infrastructure requirement to support product stewardship schemes that meet municipal expectations.

INTRODUCTION

The wastes of electrical and electronic equipment (hereinafter referred to as e-waste) have ambiguous characteristics. They can be viewed as resources due to high concentration of valuable metals. Meanwhile, they can be dangerous to human health and ecosystems if certain hazardous substances exist in them are not treated appropriately. The issue of e-waste treatment attracted much attention in developed countries because its generation constituted one of the fastest growing fractions of waste in the last decade (Bertram, 2002). Recycling of electronic waste is an important subject not only from the point of waste treatment but also from the recovery aspect of valuable metals. The U.S. Environmental Protection Agency (EPA) has identified a number of benefits, such as saving in energy and reduction in pollutions when recycled materials are used instead in place of virgin materials (Cui, 2003; ISRI, 2003).

From the point of material composition, electronic waste can be defined as a mixture of various metals, particularly copper, aluminum, and steel, attached to, covered with, or mixed with various types of plastics and ceramics (Hoffmann, 1992). Precious metals (PMs) such as gold, silver, and palladium, have a wide application in the manufacture of electronic products and appliances, serving as contact materials due to their high chemical stability and their good conducting properties. The value distribution of PMs in printed circuit boards (PCBs) and calculators is more than 80%. After precious metals, copper is the next highest value metal to be extracted from e-waste (Teller, 2006). It is worth noting that sustainable resource management demands the isolation of hazardous metals from e-waste and also maximizes the recovery of PMs. The extraction of PMs (Au, Ag, and Pd) and base metals (BMs) such as Cu, Pb, and Zn from e-waste is a major economic driver due to their associated value. Typically, PCBs in e-waste contained 20 wt% copper, 1000 ppm Ag, and 250 ppm Au (Teller, 2006).

Australia is a significant producer of e-waste in the world. According to the recent study by Institute for the Advanced Study of Sustainability (UNU-IAS), in 2014, 0.47 million tonnes of e-waste was generated, which are composed of discarded televisions, computers and their products, and mobile phones. Only 10% of the total waste generated in Australia is recycled; the rest is exported to other countries or disposed in landfills. In theory, there is a significant value in waste that is disposed and could be exploited by developing or improving current e-waste recycling technologies (Khaliq, 2014).

There is potential for direct recycled metal manufacturing as a part of integrated smelting and refining facility. This could reduce the burden on the refining process by preparing alloys with a broad range of compositions. In this study presented in this chapter the metal recycling out of electronic scrap through secondary copper smelting process has been targeted and economic viability of the process has been assessed in detail. Several economic and performance evaluations of recycling systems can be found within the economics literature (Gunter, 2007). These evaluations typically take the form of cost–benefit analyses and comprehend three major benefits: (a) increased materials recovery, (b) reduced landfilling or incineration, and (c) reduced solid waste collection; and two major costs: (a) recyclable collection and (b) reprocessing (Ackerman, 1997). Some analyses incorporate environmental externalities (Craighill, 1996) or indirect socioeconomic implications (Callan, 1997; Hong, 1993). In the current literature, there are several reviews that summarize the benefits versus costs debate (Callan 1997; Fullerton 2002; Pearce 2004; Kinnaman 2006; Gunter 2007). Notably, these studies have focused almost exclusively on municipal waste, predominantly packaging, and paper. Studies on the economics of e-waste recycling systems do exist, particularly on personal computers and electronic and electric households (Caudill 2003; Lu 2003; Macauley 2003; Kang 2006; Product Stewardship Institute, 2007) but these studies have mainly focused on the costs associated with specific recycling activities without evaluating system-wide costs.

One of the major existing industrial routes of processing and recovery of value metals from e-waste is through large-scale metallurgical processes (Khaliq, 2014). E-waste is processed through combination of pyrometallurgical, hydrometallurgical, and electrometallurgical processes embedded in base metals: primary or secondary productions (copper, zinc, lead). The overall objective of this study is therefore to investigate and quantify the relative economic performance of recycling operations for end-of use PCBs through secondary copper smelting process. Detailed economic analysis of PCBs recycling helps understand the business models and drivers as well as the operational strategies and challenges that underlie and surround this processing route.

PROCESS OVERVIEW

SECONDARY COPPER PRODUCTION

Secondary copper can be produced by combination of metallurgical processes. The process stages and routes depend on the copper content of the secondary raw material to be processed, its size distribution, and other constituents (Traulsen, 1981). Inputs for secondary copper production can be divided into mainly metallic, oxidic, and sulfuric raw materials, fluxes, and additives together with necessary reductants (e.g., coal, coke, fuel oil, and natural gas). The most common materials for secondary copper production are metallic raw materials (different scraps from copper and copper alloys) and oxidic materials (copper-rich slags, ashes, drosses, dust, and sludges). E-waste can be added as a source of copper, carbon, and oxides. Depending on the metal content and the type of the feed, the smelting step to produce copper metal from secondary materials may comprise several stages like reduction, oxidation/slagging and volatilization, fire refining, and electrolytic refining. The fire refining is mainly employed when the concentration of oxygen and sulfur is high in the copper. Electrolytic refining process is used to produce high purity copper similar to the primary copper production. Depending on their purity and composition, copper containing materials are added at different stages of the process. Precious and other valuable metals in e-waste are segregated to different phases along the process chains, for example, gaseous (volatilization), liquid slag, and liquid copper. Precious metals such as gold and silver mainly report to copper, and are separated into anode slime during electrolytic refining, which can be further processed for their recovery.

FLOW-SHEET OPTIONS

Main Process

There are a number of flow-sheeting options available for treating secondary copper materials. Typically, lower-grade feeds are smelted under reducing conditions to generate a "black copper" intermediate product which may then be converted under oxidizing conditions to a clean raw copper product either in the same furnace or in a separate vessel(s) such as the Pierce-Smith (PS) converter in the primary copper processing route. High-grade ("clean") feeds may be introduced during the converting stage in the furnace, with the copper containing slag and fume/dust from this stage recycled to the subsequent reduction cycle. Ultimately, flow-sheet selection is based on the grade of secondaries being treated, the type and concentration of impurities in the feed, the nature and capacity of existing processing/refining infrastructure, and the preferred product stream/s.

In a separate study by the authors, a number of possible process routes for e-waste processing have been developed. A selected flow sheet from the study was selected and further analyzed in this study as shown in Figure 30.1. The method consists of an e-waste processing embedded in a secondary copper pyrometallurgical process.

A number of furnaces such as the blast, top blown rotary furnace (TBRC), sealed submerged arc electric furnace, Ausmelt/ISA Smelt/KRS (Kayser recycling system) furnace, reverberator, and rotary furnace can be used for low and medium grade material in reduction and oxidation stage (Rentz, 1999). The additional fire refining step involves the addition of air and then a reducing agent (e.g., hydrocarbons) to reduce sulfur and oxygen in the copper (Traulsen, 1981). This step is only used when the sulfur and oxygen contents are still high in the charge. Fire refining is carried out

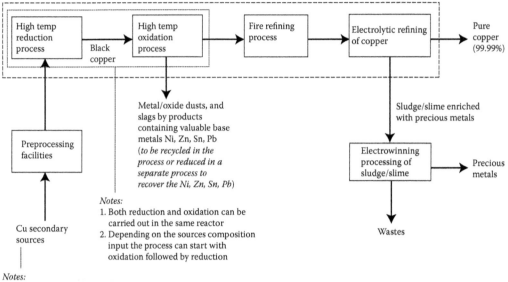

FIGURE 30.1   Selected flow sheet for e-waste processing embedded in secondary copper recycling plant.

in rotary refining furnaces or in hearth furnaces (fixed or tiltable type) or even in the same furnace used for previous step of reduction and oxidation. Further removing of impurities and the extraction of elements, such as noble metals, nickel, selenium, and tellurium, takes place at the electrolytic refining. The electrolytic refining of the copper anodes is carried out through similar process in a conventional electrolytic plant. The minor valuable metals can be recovered from the sludge from the electrolytic refining of copper through either hydrometallurgy or electrowinning.

Preprocess

The preprocess flow diagram of the e-waste treatment technology selected in this study is presented in Figure 30.2. This flow sheet is based on a Japanese e-waste mechanical preprocessing plant. The major equipment includes breaking and crushing equipment such as crushers and hammer mills. Crushed material is separated according to density, granulate size, and magnetic properties, while the milled material is separated by magnetic and eddy current separation systems. Based on the flow sheet below, the purchased equipment costs of the preprocessing phase of the e-waste recycling plant are estimated in "section Capital Cost of Major Equipment."

METHODOLOGY

For the purpose of the estimation of the economics of the processing and recovery of valuable metals from e-waste, thermodynamic modeling, and material and energy balance of the process were carried out focusing on the metallurgical plant (excluding the mechanical preprocessing). The flow sheet and material balance of the e-waste processing through the secondary copper smelting plant were developed using the HSC Chemistry Sim 8.0 package. The distribution of elements in the various streams (molten metal, off gas, and slag) was calculated through a combination of equilibrium modeling carried out using the FactSage 6.4 thermochemical package and information in the open literature. The results of the material balance and thermodynamic modeling show that 6.56 tonnes

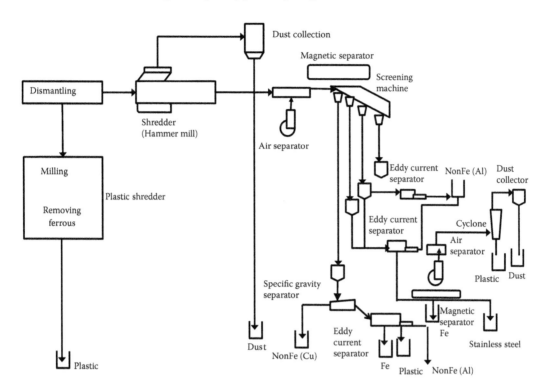

FIGURE 30.2 Flow sheet of a Japanese e-waste mechanical preprocessing plant reproduced from E-waste Management Manual, Volume II, UNEP DTIE/ITC.

of copper, 1.81 kg of gold, and 3.57 kg of silver can be extracted from 12 tonnes feed of both copper scrap and e-waste (PCBs). In this case, the recovery of Cu is 83.3 wt%. However, the extraction of precious metals is highly economical for the e-waste recycling. The predicted recovery of Ag and Au is 88.5% and 96.5%, respectively. For the purpose of this study, the amount of e-waste (PCBs) input into the process was 30% (in weight) of the total solid input. Figure 30.3 shows the general material balance flow sheet of the process within the metallurgical plant showing the flow of the different streams. This flow sheet was developed assuming total 100,000 tonnes/year input into the process.

ECONOMIC ANALYSES

COST CALCULATION

Based on the literature review, it can be determined that preliminary cost estimates are usually based on the purchased equipment cost with all additional cost components being estimated through different "factors", that is, certain percentages of the purchased equipment cost. The accuracy of the estimate will vary depending on the level of detail known about the design of the plant. In early stages of any plant development projects, only preliminary estimates can be made (Towler, 2013). Another important consideration is the quality of the cost data (i.e., prices, scaling up or down for the specific technology), which is one of the key factors in determining the final cost estimate. From the preliminary flow sheets, the major pieces of equipment are selected and the purchased equipment cost is estimated. Usually prices of equipment are given by manufacturers and vendors as free on board (FOB) delivery charges and installation expenses should be accounted for separately. The best source for this information is of course the specific manufacturers and sellers of the equipment, however, for preliminary estimates, such quotations can be too difficult or time-intensive to obtain. Once the cost of the delivered equipment is known, correlations can be applied

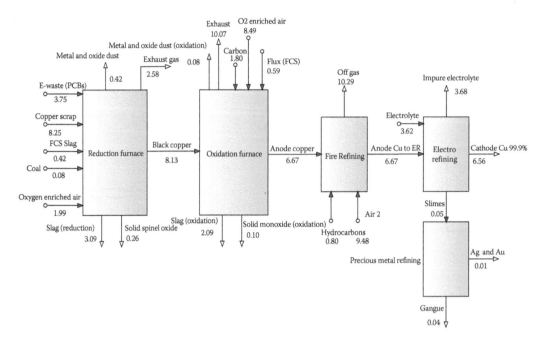

FIGURE 30.3 Material balance flow sheet of e-waste processing through secondary copper recycling with 30% e-waste in the feed (the amount values shown are material flows in tonne/hour).

to determine the fixed-capital investment, which could be used as a base for estimating other costs. There are also some costs that cannot be estimated based on the fixed-capital investment since there is no correlation between them. Examples are the costs for raw materials and utilities. Those expenses are estimated from the material and energy balances (e.g., raw materials demand, utilities).

CAPITAL COST OF MAJOR EQUIPMENT

To perform the economic analysis, it is necessary to estimate the capital cost of major equipment. The most accurate method is to have a price list from vendors. The next best alternative is to use cost data based on previously purchased equipment of the same type. Another technique, sufficiently accurate for study and preliminary cost estimates, is using correlations and graphs available for various types of common equipment in some references. In this study, the cost calculations are based on the third option using data from Turton (2008) and Sinnott and Towler (2013).

The base condition equipment price is calculated using Equation 30.1:

$$Log_{10}C_p^0 = K_1 + K_2 log_{10}(A) + K_3[Log_{10}(A)]^2 \tag{30.1}$$

where C_p^0 is the purchased cost for base conditions and A is the size parameter of the equipment. K1, K2, and K3 are given by Turton (2008) for different equipment. The cost data must then be adjusted for any differences in unit capacity and also for any elapsed time since the cost data were generated. The most common relationship to account for attribute differences is given by Equation 30.2.

$$C_a = C_b(A_a / A_b)^n \tag{30.2}$$

where A is the equipment capacity, C is the purchased cost, and n is the cost exponent. The subscripts a and b refer to the equipment with the required and the base attribute, respectively. The estimated cost is for base equipment made from carbon steel and for atmospheric pressure. To calculate the purchased equipment cost with different materials and different operating pressure, we can use material factor (FM) and pressure factor (FP), respectively.

A simple technique to estimate the capital cost of a chemical plant is using the Lang factor method, where the total cost is determined by multiplying the total purchased cost for all the major equipment by a constant (Lang, 1947). The Lang factor takes on one of either three values for fluid processing plants 4.74, solid–fluid processing plants 3.63, and for solid processing plants 3.10.

Another technique that is introduced in Turton (2008) is the calculation of module cost for the main equipment. For some equipment like heat exchangers, vessels, and pumps, bare module cost is calculated from Equation 30.3, and for other equipment like packing and compressors use Equation 30.4.

$$C_{BM} = C_P^0 (B_1 + B_2 F_M F_p) \tag{30.3}$$

$$C_{BM} = C_P^0 \times F_{BM} \tag{30.4}$$

$$C_{TM} = \sum C_{TM,i} = 1.18 \sum C_{BM,i} \tag{30.5}$$

$$C_{GM} = C_{TM} + 0.5 \sum C_{BM}^0, \tag{30.6}$$

where C_{BM} is the bare module price, B_1 and B_2 are constants, F_M is the material factor, F_P is the pressure factor, F_{BM} is the bare module factor, C_{TM} is the cost of making small-to moderate expansion, and C_{GM} is the cost of completely new facilities in which we start the construction in a gross field.

The capital cost estimates used in this study were based on a combination of the price list from vendors, literature data, particularly from Turton (2008) and Towler (2013), and The Chemical Engineering Plant Cost Index (CEPCI). A list of equipment used for the main process and preprocess stages for the e-waste recycling embedded in secondary copper smelting plant are given in Tables 30.1 and 30.2.

Table 30.1 summarizes a list of the purchased and installed equipment presented in the proposed flow sheet of e-waste processing (Figure 30.3). Figure 30.4 illustrates how the total purchased costs for the equipment of the main process can be determined as a percentage of one of the main equipment cost (here the reduction furnace).

Table 30.2 presents a list of purchased preprocessing equipment of the plant. As discussed earlier, the capital costs for the preprocessing equipment can be obtained based on data from various sources including industry equipment suppliers, published literature, and local or international vendor quote.

The economics of this process were determined based on the assumptions derived from literature—refer to Table 30.3. Reported values from the literature were scaled using the capacity power-law expression, cost2/cost1 = (capacity2/capacity1)^m, and adjusted to 2015 AU dollars using the Chemical Engineering Plant Cost Index. A discounted cash flow rate of return (DCFROR) analysis was used to determine the profitability of the plant. The selling price of the recycled precious metal over the plant's lifetime at which the net present value (NPV) is positive was also calculated.

The amounts of raw materials required for the recycling of the metal and recovery of precious metals out of e-waste through secondary copper smelting were estimated based on the mass balance calculation, which is a 100,000 tonnes/year capacity. A list of used raw materials and their selling

TABLE 30.1

Estimation of the Equipment Investment for E-Waste Processing through Copper Recycling (Main Process)

Unit	Type	Feed (kg/h)	Operating Conditions	Power (kWh)	Dimensions/Material	Number
Furnace (reduction)	Hearth	8250	1300°C, $pO_2 = 10^{-8}$ atm.	5000	D: 1.5 m	1
Conveyor	Belt	–	–	–	Length: 10.5 m Wide: 0.5 m	5
Furnace (oxidation)	Cylindrical	12,500	1300°C, $pO_2 = 10^{-8}$ atm.	3000	D: 3.5 m H = 12 m	1
Bag-house	BH-Large		2750 Nm³/h		Carbon steel	3
Scrubber	Venturi Medium	–	6780 Nm³/h	–	Carbon steel	3
Fire refining Anode furnace	Rotary	9000	1200°C, $pO_2 = 10^{-8}$ atm.	700	Drum diameter = 3.5 m Length: 6 m	1
Electrolytic refining (Cu)	Kettle, jacketed and agitated	4500	65°C, H_2SO_4 solutions	350	3 m³ 34 cells	1
Precious metals refining	Electrowinning cells-rubber lined carbon steel (CS)	61	65°C, acidic solutions	–	1 m³ 10 cell	1

TABLE 30.2

Estimation of the Equipment Investment of Preprocessing Equipment for E-Waste Processing through Copper Recycling Based on Flow Sheet Given in Figure 30.2

Unit	Specification	Dimensions (m)/Capacity
Shredder (Hammer mill)	Input: 6 × 4 cm Output: 3–4 mm	Diameter: 1.2 m
Dust collector	Type: Venturi	0.94 m³/s
Conveyor	Open-long	Length: 4.25 m Wide: 0.3 m
Air separator	Air sweep-dry w/o motor	D = 1.2 m
Magnetic separator	Screw classifier – 2 screws	Separator diameter = 1.2 m
Screening machine	Vibratory, one deck, large	Deck area = 7.4 m²
Hopper	Hopper with bottom bolted	Bin volume = 34 m³
Eddy current separator	Cyclones, wet, rubber line C1	Diameter: 1.2 m
Cyclone	Turbo screen	D = 0.9 m
Close conveyor	Open-long	Length: 6 m Wide: 0.3 m
Stainless-steel bin	Cone bottom, small	Volume = 28 m³
Specific gravity separator	Cyclone, wet, ceramic lined	Diameter: 0.6 m

price is presented in Table 30.4. The updated feedstock costs were obtained by exploring the most recent literature (e.g., Labys et al., 1971; Kentaka and Managi, 2011) and London Metal Exchange (LME) accessed on June 2014. In addition, for certain chemicals or minerals, large seasonal price fluctuations may exist and the average price has been considered over a period of 1 year. The copper scrap fed into the reduction furnace was assumed to be a low-grade copper scrap.

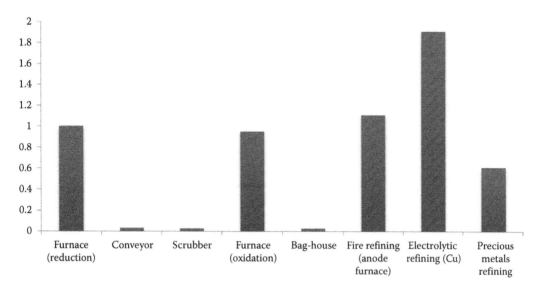

FIGURE 30.4 Main process equipment cost breaks down as a percentage of reduction furnace price.

TABLE 30.3

Economic Parameters and Indirect Cost Basis Used in Analysis

Assumption Description	Assumed Value
Internal rate of return	10%
Plant financing debt/equity	50%–50%
Plant life	
Income tax rate	30%
Interest rate for debt financing	20 years
Term for debt financing	10%
Working capital cost	5.0% of fixed capital investment (excluding land)
Depreciation schedule	7 years MACRS[a] schedule
Construction period	3 years (20% Y1, 45% Y2, 35% Y3)
Plant salvage value	No value
Start-up time	1 year [28]
Revenue and cost during start-up	Revenue = 50% of normal
	Fixed cost = 100% of normal
On stream factor	90% (7884 operating hours per year)
Inflation rate	2.14%[28]
On stream percentage	90% (7884 h/year)
Land	6.5% of total purchased equipment cost (TPEC)[28]
Royalties	6.5% of TPEC[28]
Working capital	5% of fixed capital investment (excluding land)
Engineering and supervision	32% of TPEC[29]
Construction expenses	34% of TPEC[29]
Contractor fee and legal expenses	23% of TPEC[29]
Contingencies	20.4% of TPEC[28]

[a] Modified accelerated cost recovery system.

TABLE 30.4

Estimated Raw Material Cost of E-Waste Processing through Copper Recycling

Items	Consumption (tonne)	Unit Price (AUD/tonne)	Source	Price per Hour of Consumption (AUD/h)
Cu scrap (low grade)	8.25	3500	http://www.copperrecycling.com.au/scrap-copper-prices/	28,875
PCBs High-grade boards (green boards with gold corners on IC chips)	3.75	4000	http://scrapforum.com.au/forum/the-scrap-rooms/the-business-of-scrapping/67-scrap-metal-price-guide	15,000
FCS slag	0.42	100		42
Coal	0.08	65		5.2
Total				43,922

ECONOMIC ANALYSIS RESULTS

The total capital investment cost of the plant is presented in Table 30.5 and the total operating cost calculation is illustrated in Table 30.6.

Fixed operating costs are calculated using correlations from Seider (2009) and include line items such as labor-related operations, maintenance, operating overhead, property tax, and insurance.

Figure 30.5 shows capital investment results for the e-waste recycling plant production. The major capital cost of the plant is the cost of equipment followed by construction expenses. The variable operating costs used for this analysis are detailed in Table 30.6. The prices listed in the table were converted into 2015 Australian dollars using a report from the Australian Bureau of Statistics A.U. Bureau of Labor.

The operating costs break down used for this analysis is shown in Figure 30.6. The energy cost is the highest operating cost of the plant marginally followed by the operating labor cost. The market price of utilized energy for the plant in 2015 is based on 2014 Annual Report of Australian Institute of Energy information average price for 2013–2014. The variable operating costs used

TABLE 30.5

Total Capital Investment Calculation

Quality	Calculation	Million (AUD)
Total purchased equipment cost (TPEC)		10.5
Installation factor	2	
Total direct cost (TDC)	TPEC × installation factor	21
Engineering and supervision	32% of TPEC	3.4
Construction expenses	34% of TPEC	3.6
Contractor fee and legal expenses	23% of TPEC	2.4
Contingencies	20.4% of TPEC	2.15
Total indirect cost (TIC)		11.55
Total depreciation capital (TDep)	TDep = TDC + TIC	32.55
Royalties	2% of TDep	0.651
Land (pure real state)	2% of TDep	0.651
Fixed capital investment (FCI)	FCI = TDep + royalties + land	33.852
Working capital (WC)	5% of FCI (excluding land)	1.04
Total capital investment (TCI)	TCI = FCI + WC	34.893

TABLE 30.6
Total Operating Cost Calculation in This Study

Description			Million AUD/year
Operating labor	C_{OL}		5.85
Utilities (energy)	C_u	Based on energy flow sheet	6.14
Maintenance and repair	C_m	0.069 FCI (fixed capital investment)	2.335
Local taxes and insurance	$C_{T\&I}$	0.032 FCI	1.083
Plant overhead	C_{po}	$0.708C_{OL} + 0.009FCI$	4.446
General operating expenses	C_{GOE}	$0.31 \times C_{OL}$	1.813
Total			21.670

for this analysis are detailed in Table 30.4. The prices listed in the table were converted into 2015 Australian dollars using average stockpile prices available in Australia.

The copper scrap feedstock cost of 3500 AUD/tonne is selected based on the low-grade copper scrap price excluding feedstock logistics.

The annual benefit gained from sale of recycled silver, gold, and copper is given in Table 30.7. As can be seen from this table, it is assumed that the working load of the main process (copper production) is 12 h and for precious metal recycling is 6 h per day.

As shown in the mass balance flow sheet of the valuable metals extraction from e-waste presented in Figure 30.3, 6.56 tonnes of copper, 1.81 kg of gold, and 3.75 kg of silver can be extracted from 12 tonnes feed of both copper scrap and e-waste (PCBs) as illustrated in Figure 30.7. The recovery of Cu is 83.3 wt%, however, the extraction of precious metals is highly economical for the e-waste recycling. The predicted recovery of Ag and Au is 88.5% and 96.5%, respectively.

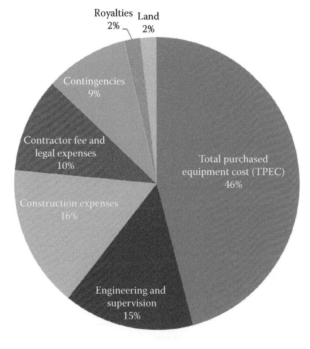

FIGURE 30.5 The capital investment breakdown calculated in the current study.

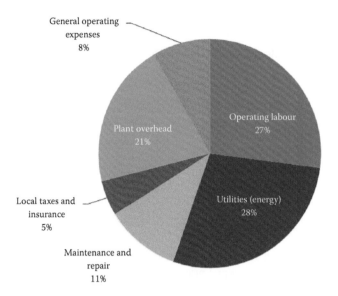

FIGURE 30.6 The operating cost breakdown calculated in this study.

The net benefit of 116 million Australian dollar obtained after 20 year operation of the plant is favorable when compared to the fixed and variable cost required for the plant start-up from the grass root. A DCFROR analysis was used in this study to evaluate the selling price of the recycled metallic materials from waste printed circuit board (PCB) over the plant's lifetime. It can be found from DCFROR analysis that the internal rate of return of the plant over 20 years of operation is 22% and the net present value over the same period of the plant operation is a $116 million AUD as shown in the Figure 30.8. Overall the plant is in the status of positive cash inflow in the calculated time. The payback period was calculated to be 4 years which is a short period and indicates the investment is highly desirable. In order to measure the performance and compare the magnitude and timing of gains from investment directly to the magnitude and timing of investment costs, the ROI (return on investment) was calculated to be 0.03 for the studied plant, using the following equation:

$$(\text{ROI}) = \frac{(\text{Gains from Investment} - \text{Cost of Investment})}{\text{Cost of Investment}} = 0.03 \qquad (30.7)$$

TABLE 30.7
Revenue for the Recovered Copper, Gold, and Silver

Description

Raw PCB fed to the furnace	3750 kg		
Recycled metals	Gold	Silver	Copper
Gold/silver weight	1.88 kg	4.03 kg	
Gold/silver recovery	96.50%	88.50%	
Recovered gold/silver/copper	1.81 kg	3.57 kg	6560 kg
Gold/silver/copper unit price	42,000 $/kg	720 $/kg	7.4 $/kg
Gold/silver/copper value per cycle	75,994 $/cycle	2569 $/cycle	49 $/cycle
Running cycle per day	6 hour	6 hour	12 hour
Working day per annum	365	365	365
Annual benefit	$166,426,313	$5,625,480	$212,623
Annual production	4 tonne	8 tonne	29 tonne

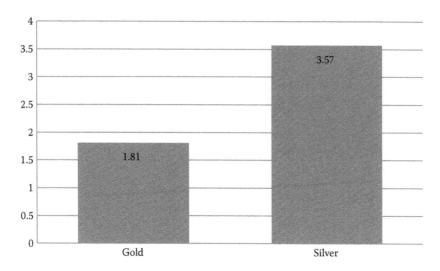

FIGURE 30.7   Precious metal recovery (kg/hour).

As it can be seen from the above formula, the investment has a positive ROI resulting in the net profit. Another indicative parameter used to evaluate the plant efficiency for converting sale to cash was operating cash flow margin which is cash from operating activities as a percentage of sales in a given period. The operating cash flow margin is generally calculated using the following formula:

$$\text{Operating cash flow margin} = \frac{(\text{Cash flow from operating activities})}{\text{Net sale in 20 years}} = 1.2 \qquad (30.8)$$

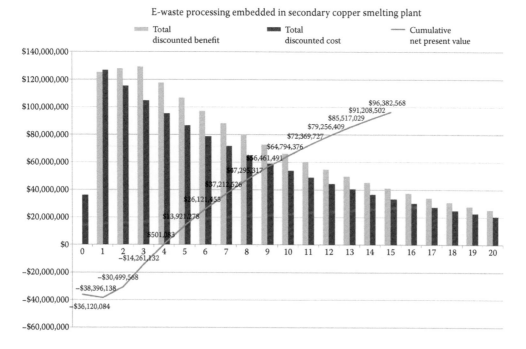

FIGURE 30.8   Cash flow diagram for e-waste processing embedded in the secondary copper smelting.

A relatively high operating cash flow margin gained here indicates that the plant is efficient at converting sales to cash, and also is an indication of high earnings quality of the proposed plant. The time period required for the amount invested in the plant to be repaid by the net cash outflow generated by the plant (payback period) is 4 years as illustrated in Figure 30.8.

It is worthwhile to note that there is a wide range of prices for raw material fed to the furnaces; for example, for the waste PCBs and copper scrap. This is mainly a result of assumptions and key parameters related to feedstock costs and by product prices. For the current study, this fact highlights the importance of a rigorous sensitivity analysis to investigate the impact of key parameters on thermochemical and techno-economic analysis of electronic waste processing through secondary copper smelting.

EFFECT OF THE SCALE OF PRODUCTION

Figure 30.9 shows the impact of changes in feed capacity on total cost or benefit ratio and actual internal rate of return (IRR) of e-waste metal recycling plant. Three feed capacities for the plant capacity have been considered starting from 50,000 tonnes/annum to 200,000 tonnes/annum. The plant with feed material capacity of 100,000 tonnes per annum considered as the base case in this study. The total cost to the cost of the base case along with the same ratio for benefit has been calculated and shown at the left side of vertical axis (Y axis) in Figure 30.9. The right side of vertical axis

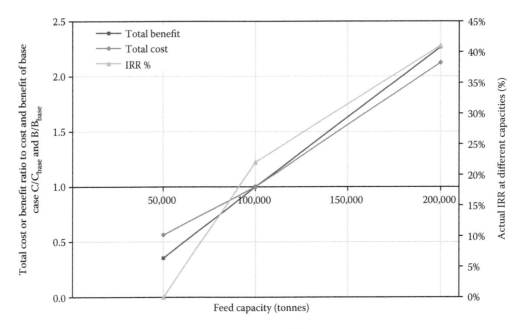

FIGURE 30.9   Effect of feed capacity on cost/benefit: Benefit/Benefit$_{base case}$, Cost/Cost$_{base case}$, and IRR%.

TABLE 30.8
Changing Production Scale on Cost/Cost$_{base case}$, Benefit/Benefit$_{base case}$, and IRR%

Feed Capacity (tonnes/a)	Cost/Cost$_{base case}$	Benefit/Benefit$_{base case}$	IRR% Over 20 years
200,000	2.1	2.3	41%
100,000 (base case)	1.0	1.0	22%
50,000	0.6	0.4	0%

in Figure 30.9 shows the actual rate of return of the project for each feed capacity. Table 30.8 lists the effect of increasing/decreasing of feed capacity on the cost and benefit ratio and IRR.

CONCLUSION

An estimated economic analysis for a proposed 100,000 tonnes per annum e-waste recycling facility has been carried out based on capital and operating cost estimates. The ROI value is relatively high and the payback period for the project is 4 years. The designed plant is most sensitive to energy and raw material costs followed by the interest rate. Financial viability is also sensitive to price fluctuations, that is, prices with respect to both input raw material as well as the output recovered metals. A comparative analysis of economic indicators shows that the project feasibility improves with the improved capacity utilization. Capacity utilization depends upon the availability of the raw material. In order to make the project more achievable an efficient e-waste collection and transportation system and a set of incentives like lower interest rates or duty exemption would be highly desirable. The main outcome of this study is the confirmation that the e-waste recycling process embedded in the secondary copper smelting project has considerable potential value and demonstrates how the e-waste recycling market can grow profitability over the first few years of the plant's operation.

REFERENCES

Ackerman, F. 1997. *Why Do We Recycle?*; Island Press: Washington, DC.

Australian Bureau of Statistics, http://www.abs.gov.au/websitedbs/c311215.nsf/web/labour

Bertram, M. 2002. The contemporary European copper cycle: Waste management subsystem. *Ecological Economics*, 42(1–2): 43–57.

Callan, S. J. 1997. The impact of state and local policies on the recycling effort. *The Eastern Economic Journal*, 23: 411–423.

Caudill, R. 2003. Assessing base level of service for electronics collection and recycling programs: Seattle-Tacoma case study. *IEEE International Symposium on Electronics and the Environment*. Boston, MA: Institute of Electrical and Electronics Engineers, pp. 309–314.

Craighill, A. L. 1996. Lifecycle assessment and economic evaluation of recycling: A case study. *Resources, Conservation and Recycling*, 17: 75–96.

Cui, J. 2003. Mechanical recycling of waste electric and electronic equipment: A review. *Journal of Hazardous Material*, 99: 243–263.

Flnkenrath, M. 2011. Cost and Performance of Carbon Dioxide Capture from Power Generation; Working Paper No. 2011/5; Organization for Economic Co-operation and Development (OECD).

Fullerton, D. 2002. *The Economics of Household Garbage and Recycling Behavior*. Northampton, MA: Edward Elgar.

Gunter, M. 2007. Do economists reach a conclusion on household and municipal recycling?. *Econ Journal Watch (EJW)*, 4: 83–111.

Hoffmann, J. E. 1992. Recovering precious metals from electronic scrap. *Journal of the Minerals Metals & Materials Society (JOM)*, 44: 43–48.

Hong, S. 1993. An economic analysis of household recycling of solid wastes: The case of Portland, Oregon. *Journal of Environmental Economics and Management*, 25: 136–146.

ISRI. 2003-11-06. Scrap recycling: Where tomorrow begins, Institute of scrap recycling industries Inc. (ISRI) Report, http://www.isri.org/isri-downloads/scrap2.pdf.

Kang, H. Y. 2006. Economic analysis of electronic waste recycling: Modeling the cost and revenue of a materials recovery facility in California. *Environmental Science and Technology*, 40: 1672–1680.

Kentaka, A. and S. Managi. 2011. Price linkages in the copper futures, primary, and scrap markets. *Resources Conservation and Recycling*, 56(1): 43–47.

Khaliq, A. 2014. Metal extraction processes for electronic waste and existing industrial routes: A review and Australian perspective. *Journal of Resource*, 3: 152–179.

Kinnaman, T. C. 2006. Policy watch—examining the justification for residential recycling. *Journal of Economic Perspectives*, 20: 219–232.

Labys, W. C., H. J. B. Rees, and C. M. Elliott. 1971. Copper price behaviour and the London metal exchange. *Applied Economics*, 3(2): 99–113.

Lang, H. J. 1947. Engineering approach to preliminary cost estimates. *Chemical Engineering*. 54(9): 130–137. Brisbane, QLD, Australia.

Lu, Q. 2003. An investigation of bulk recycling planning for an electronics recycling facility receiving industrial returns versus residential returns. *IEEE Transactions on Electronics Packaging and Manufacturing Journal*, 26: 320–327.

Macauley, M. 2003. Dealing with electronic waste: Modeling the costs and environmental benefits of computer monitor disposal. *Journal of Environmental Management*, 68: 13–22.

National Centre for Electronics Recycling. 2006. *A Study of the State by-State e-Waste Patchwork*. National Centre for Electronics Recycling, West Virginia, USA.

Pearce, D. 2004. Does European Union environmental policy pass a costs benefit test? *World Economic*, 5: 115–137.

Peters, M. S. 2003. *Plant Design and Economics for Chemical Engineers*. 5th ed. New York: McGraw-Hill Science/ Engineering/Math.

Product Stewardship Institute. 2007. The collection and recycling of used computers using a reverse distribution system: A pilot project with Staples, http://www.productstewardship.us/associations/6596/files/Staples_final_rep.doc

Rentz, O. K. M. 1999. Report on Best Available Techniques (BAT) in Copper Production. DeutschFranzösisches. Institut für Umweltforschung (DFIU) and University of Karlsruhe (TH).

Seider, W. D. 2009. *Product & Process Design Principles: Synthesis, Analysis and Evaluation*. 3rd ed. New York: John Wiley & Sons.

Teller, M. 2006. *Recycling of Electronic Waste Material*. Sustainable Metals Management Springer, 563–576, Dordrecht, The Netherlands.

The Global E-Waste Monitor. 2014. United Nations University, UNU-IAS. Institute for the advance study of sustainability, http://i.unu.edu/media/unu.edu/news/52624/UNU-1stGlobal-E-Waste-Monitor-2014-small.pdf

Towler, G. 2007. *Chemical Engineering Design: Principles, Practice and Economics of Plant and Process Design*. Butterworth-Heinemann, Waltham, MA, USA.

Towler, R. S. 2013. Chemical engineering design: Principles, practice, and economics of plant and process design.books.google.com

Traulsen, H. R. 1981. *Copper Smelting—An Overview*. J.C. Taylor: Copper Smelting. An Update, AIME, Warrendale, PA, USA.

Turton, R. B. 2008. *Analysis, Synthesis and Design of Chemical Processe*. books.google.com

United States Environmental Protection Agency. 2004. *Final Report on the Mid-Atlantic States' Electronics Recycling Pilot*. Washington, DC: U.S. EPA.

Westcott. 2012. *E-waste; Queensland Parliamentary*. Brisbane, QLD, Australia.

31 The Importance of Risk Management in the Extractive Sector for a Sustainable Economy

Samanthala Hettihewa

CONTENTS

ABSTRACT

This study uses a qualitative descriptive analysis to demonstrate the importance of risk management in extractive enterprises, where overall sustainability is often profoundly dependent on cross-linked environmental, social, corporate and financial issues. Risk and sustainability have been well studied over the past few decades. However, there appear to be gaps in resolving the *a priori* and/or *a posteriori* drivers and outcomes of risk in the extractive industry (EI). Much of the research focus in EI sustainability has been on environmental factors in contrast; research in corporate sustainability has tended to focus on economic risk. In keeping with modern precepts of corporate social responsibility, this study considers four pillars of overall sustainability: (i) environmental, (ii) social, (iii) financial and (iv) economic.

INTRODUCTION

OBJECTIVE

Using a qualitative descriptive analysis (Sandelowski, 2000), this study seeks to demonstrate how managing key risks in extractive firms can help achieve the objective of creating, maximising and sustaining stakeholder wealth and in turn the sustainability of the global economy. This study includes environmental, operational, strategic, compliance, project, corporate, market, price, financial and political risk. These risk aspects in EI are interwoven into a potentially convoluted system of risk with possible adverse outcomes. It is essential to identify and manage these risks in order to attain the objectives of firms and to achieve the overall sustainability that depend on four pillars: (i) environmental, (ii) social, (iii) financial* and (iv) economic sustainability. While sustainability is defined and explained with many different emphases, in the end, it comes to one objective that is to optimise the trade-off between benefits and the costs/harm associated with those benefits.

IMPORTANCE OF RISK ANALYSIS IN EXTRACTIVE SECTOR

Compared to other industries, the unique characteristics – non-renewability, vital instability, proclivity for large-scale environmental disasters, massive investment from exploration to the final stage and investors' and researchers' continuous efforts to find proper substitutes through energy and materials innovations – make extractive industry (EI) vulnerable to many risks directly as well as indirectly.

High demand for natural resources in the past few decades, mainly driven by the countries that demonstrated a rapid economic development, shows that EI-related activities are rising. This situation stimulated many companies to expand their supply as well as new junior extractive companies to enter the industry, escalating interest in the EI, resulting in a global supply expansion (Iddon et al., 2015). As such, attempts to identify and minimise the risks relate to the industry are essential. Perrini and Tencatti (2006) stated that when a firm implements sustainability-oriented policies, it creates value to the economy. This study argues that to achieve EI company objectives and to attain overall sustainability, it is essential to identify and manage risks (see the section 'Objective' for different types of risks). A large part of managing socio-political risk, for an extractive firm, involves maintaining a healthy relationship with society.

Importance of Size and Level of Development

As the firms in EI are in different sizes and stages of excavation (exploration, mine development, mining, processing, smelting and refining, etc.), one could argue against the generic treatment of risk management warranting a brief explanation on the validity of the generic measures. While a proper understanding of industry/firm/situation-specific risks is paramount to risk management, an overall analysis is important. The term 'EI' includes all levels of firms, regardless of their development level. The literature on EI can be broadly categorised based on size and their level of development (Moon and Evans, 2006). The attributes of large and small firms form a continuum and thus any division is, by necessity, arbitrary. However, such a division is needed to meet this study's objectives. Iddon et al. (2015) suggest that the vast majority of large-scale firms (LSFs) in EI are engaged in excavation (mining, processing and/or marketing processed or semi-processed ore domestically and/or internationally). These firms are often funded by thousands of investors via shares listed in stock exchanges. In contrast, most junior mining companies (JMCs) are focussed on exploration of commercial deposits of natural resources (Iddon et al., 2014).

---

* Though traditionally, financial sustainability is considered a sub-set of economic sustainability, the vast development in the finance field made it appropriate to consider it as a separate segment from economic sustainability while bearing in mind the dependency with each other.

These significant differences cause these two EI groups to diverge in their nature, particularly in their view and management of risk. In terms of sheer firm numbers, JMCs represent the vast majority of the EI (≈98%). Most of the LSFs in EI have diversified across a wide range of mining activities, commodity types and geographical regions. A small sub-group of active miners, at the lower end of the LSF, generally focus on a single mine or a small collection of mines (i.e. one or a few mineral/energy deposits), in a narrow geographical area. Their small scale and limited access to resources make JMCs less able than LSF to access cost-effective risk management tools and that lack can expose them to savage negative outcomes (see Iddon et al., 2014). This suggests that awareness of, and access to, risk management tools can be essential for a JMC's continuity.

GAP IN LITERATURE AND RESEARCH QUESTIONS

While studies on risk management in the EI have been undertaken by researchers focussing on a specific risk component (e.g. environmental and operational risks are well examined), a comprehensive analysis with many different types of risks and their interactions to sustainability are still needed along with comprehensive research on interlinked risk and risk management. There is a clear gap in the literature on how each of the four pillars of sustainability influences sustainable development in the EI. This study seeks to lessen this gap and enhance policies towards sustainable global economy.

Research Question

The research question in this chapter is: How different types of risks jointly and severally (i.e. together and separately) affect the EI and how that knowledge helps good risk management to mitigate those effects and, thereby, enhance sustainability in the EI?

This chapter is organised as follows: After the research methodology and conceptual framework are explored, an overview of key terms – extractive sector, sustainability, stakeholder concept, corporate social responsibility (CSR) and economic and financial sustainability – is provided, which is followed by a discussion on the EI, including risks, risk management and sustainability issue, and the need for enhanced policies to achieve sustainability in the EI. The chapter concludes with conclusions, limitations and suggestions for future research.

METHODOLOGY AND CONCEPTUAL FRAMEWORK

This study uses a qualitative descriptive analysis. 'In quantitative research, there is a sharper line drawn between exploration (finding out what is there) and description (describing what has been found) than in qualitative descriptive studies. In summary, qualitative descriptive studies offer a comprehensive summary of an event in the everyday terms of those events. Researchers conducting such studies seek descriptive validity, or an accurate accounting of events that most people (including researchers and participants) observing the same event would agree is accurate, and interpretive validity, or an accurate accounting of the meanings participants attributed to those events that those participants would agree is accurate (Maxwell, 1992)'. (Sandelowski, 2000, p. 336).

Risks and sustainability are considered from corporate and environmental sustainability perspectives that combine into socio-political sustainability. The previous literature is reviewed, gaps are identified and the essential concepts for proper risk management are examined. Figure 31.1 helps understand the structure of the study with a conceptual framework. While a complete comprehensive risk management evaluation requires empirical assessments based on explicit situations, the analysis in this study looks to the broad generic notions that form a framework for risk management evaluation. The notions in this study are derived using qualitative descriptive analysis.

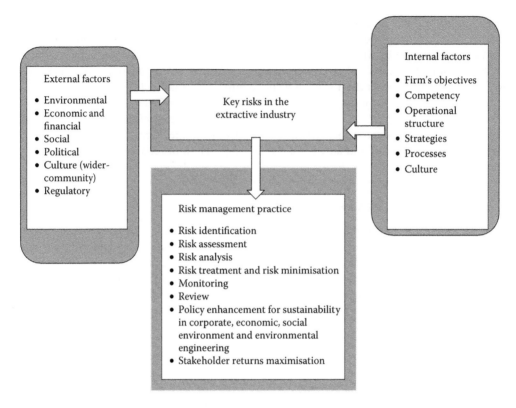

FIGURE 31.1 Conceptual framework of risk analysis in the extractive industry.

KEY TERMS AND LITERATURE REVIEW

EXTRACTIVE SECTOR

The population of EI comprises of activities that include exploration and excavation of many different types of natural minerals as well as oil and gas. Natural resources include massive amounts of diversified mineral types (for different types of commodities, see SNL Metals & Mining 2012).[*] The Joint Ore Reserves Committee's (JORC Code, 2012, pp. 11 and 16) definition of mineral resources is as follows:

> A 'Mineral Resource' is a concentration or occurrence of solid material of economic interest in or on the Earth's crust in such form, grade (or quality), and quantity that there are reasonable prospects for eventuate economic extraction. The location, quantity, grade (or quality), continuity and other geological characteristics of a Mineral resource are known, estimated or interpreted from specific geological evidence and knowledge, including sampling. Mineral Resources are sub-divided, in order of increasing geological confidence, into Inferred, Indicated and Measured categories.... [It defines the ore reserves as] the economically mineable part of a measured and/or Indicated Mineral Resources. It includes diluting materials and allowances for losses, which may occur when the material is mined or extracted and is defined by studies at Pre-feasibility levels as appropriate that include application of Modifying Factors.

About 100 countries contribute to the global economy by engaging in mineral production-related activities and 56 of these contribute via exports. Iron ore, gold and copper lead mining projects in

---

[*] SNL Metals & Mining is the leading information resource for the global resources industry. See http://www. ga.gov.au/ scientific-topics/minerals/mineral-resources (accessed 8/7/2015).

terms of value, cumulatively contributing US$606 billion in 2011, which is 68% of the total value of all non-fuel mineral production in 2011 (MCA, 2014).

Importance of EI

Besides its direct contribution in fiscal revenue, EI contributes to national economic development through employment creation, education and training, leading to skill and knowledge transfer, supplying inputs for other sectors, research and development, and flow-on effect to other sectors (banking, logistics, etc.). Also, EI has a major contribution to the advancement of CSR programs of the broader society. McKinsey (2013, p. 7) asserts that '…to meet soaring demand for resources and replace rapidly depleting supply, the world should invest a total of up to US$17 trillion in oil and gas and in minerals by 2030, double the historical rate. …estimates that out of the 843 million people living in extreme poverty in resource-driven economies in 2010, around 540 million people could be lifted out of poverty by 2030, if reserves were to be used effectively, as in the most successful resource-driven economies' (for case studies, see MCA (2014)). The World Bank Group (2012, pp. 6–10), demonstrating the significant linkages to local economies through EI-related activities, states that

> In support of private sector investment, [the International Finance Corporation] IFC provided US$490.6 million of financing and MIGA provided US$119.5 million of risk coverage. In addition, IBRD/IDA provided grants funded by partners of roughly US$21 million.…During the reporting period of FY2012, IFC's oil, gas and mining client companies contributed approximately US$6.2 billion to government revenues, created or sustained about 102,000 direct jobs and supported local communities with US$100 million of dedicated community related spending. Total spending by these companies on goods and services from local and national suppliers approached US$5.4 billion, demonstrating both significant linkages to local business and making a major contribution to local economies…Overall, IFC holds an EI portfolio of US$2.5 billion, roughly 75% in oil and gas and 25% in mining in terms of US$ million committed…investments.…Total spending by these [EI] companies on goods and services from local and national suppliers approached US$5.4 billion, demonstrating both significant linkages to local business and making a major contribution to local economies'. MCA emphasises on the private sector stake in working with governments and global and regional organisations in improving border processes, making the vast investments in physical infrastructure needed to link the provinces/states more effectively to improve the network of value chains and attempts were made to focus on *resource blessing* instead of *resource curse* in policy issues by showing examples from Australia and other.

Also, after examining EI activities and macro-economic contributions in low- and middle-income countries, the International Council on Mining and Metals (ICMM, 2012) claim that it is relatively higher compared to other sectors with increased FDI, exports and fiscal revenue.

Oil and Gas

Investigations in natural resources often separated the principal commodity-type mining/exploration from oil and gas due to clear differences in these two industries. Rudenno (2009) and Australian Securities Exchange (ASX, 2011) explain the differences throughout the stages of development. Such differences include very high risks and high costs associated with offshore activities, high levels of labour involvement (initially) and relatively rapid adaptive production process with low operating cost after reserve discovery. As these differences do not deter but rather encourage the objective of this study – risk management for sustainability – both industries (minerals and oil and gas) are considered in this study.

Importance of Oil and Gas

The extraction of modern global liquefied natural gas (LNG) started more than 50 years ago in 1964 in Algeria (Ernst & Young, 2013). LNG is fast growing and has become a key source of energy, with consumption forecasted to be 137.7 trillion cubic feet for 2017 (International Energy Agency, 2010). Globally, the growing desire for clean energy fuelled the demand for LNG (Langdon, 1994; Jacob, 2011; Kumar et al., 2011). Most of this increased capacity was expected to be in the Middle East and

Australia (International Energy Agency, 2010). While EI's importance is undisputed, it is critical to determine whether that importance is sustainable in terms of benefits to stakeholders.

SUSTAINABILITY

Sustainability and its related notions have been gaining in importance over the past few decades and it has encompassed ideas from a broad array of disparate disciplines. Sustainability notion has matured and gained further substance with typological improvements (pollution prevention, cleaner production, green chemistry, corporate social responsibility and governance, etc.). The term 'sustainable development' (SD) appeared often in research, reports and related literature (Labuschagne et al., 2005). However, in EI, the main focus is on the environmental aspect, though a broader perspective is essential if overall sustainability is to be achieved (MCA, 2014). In achieving overall SD, the United Nations and national governing bodies have provided much of the impetus. Labuschagne et al. (2005, p. 374) state that

> Far less work has been done on a company level to develop and implement sustainability performance assessment practices… . An analysis of more than 50 business-related initiatives revealed that frameworks do exist to assess business sustainability but 'in most of the (six) integrated (sustainability measurement) initiatives the primary focus is the environmental dimension… . Furthermore, the focus of the available integrated sustainability frameworks is mostly on a product level within an organization. …with the Global Reporting Initiative (GRI) the only recognized international initiative that focuses on reporting the sustainability of the entire organization…'.

The SD notion is still evolving with branching semantic similarities and differences that are often discipline driven and sustained. This study attempts to merge the different extant focusses into a generic concept that encompasses environmental engineering, economics and corporate business ideals and social concerns. Though there is no uniquely accepted definition of SD, the definition given in the Report of the World Commission (1987, p. 16) – 'Humanity has the ability to make development sustainable to ensure that it meets the needs of the present without compromising the ability of future generations to meet their own needs' – shows an appropriate meaning of sustainability.

Glavic and Luckman (2007, p. 1884) assert that '…sustainable systems introduce inter-connections between environmental protection, economic performance and social welfare, guided by political will, and ethical and ecological imperatives'. This study proposes a comprehensive framework with four sustainability pillars (environmental, finance, economic and social) on which overall sustainability is linked and supported.

Sustainable Development

Stressing the opacity in SD definitions, Glavic and Luckman (2007, p. 1878) surveyed the literature and investigated 51 terms in sustainability and found that SD encompasses environmental/ecological, economic and societal elements (principles); environmental tactics (approaches) and environmental sub-strategies (systems). Under the environmental principles, renewable resources, minimisation of resource usage, source reduction (dematerialisation), recycling/reuse/repair, regeneration/recovery/remanufacturing, purification and degradation are explained in an attempt to describe the process to SD. In their attempt to explain the economic principles, they focused on environmental accounting to get the attention of corporate stakeholders to reduce environmental costs by improving environmentally responsible corporate practices as part of enhancing the sustainable profitability of their organisation/s. These principles build a more complex process, demonstrating different approaches, which include pollution control, cleaner production, eco-design, green chemistry, life cycle assessment and waste minimisation importance of approaches in achieving environmental sustainability. These approaches then encourage environmental sub-systems that incorporate appropriate plans and policies with a perspective on environmental engineering and technology.

In a proper risk assessment, the linkages of IE activities, and the process in which EI connects to the economic and finance side of sustainability should be vital factors.

EI Key Linkages of Risk and Risk Management for Sustainability

ICMM (2012), showing the contribution of mineral and mining activities to the global economy and future trends of the industry over the next decade, says that the contribution of EI rests directly and indirectly on the overall value of production, the payments to factors of production and return to shareholders (reward them for investing in risky, long-term mining assets). As risk emerges, coming not only from one isolated section of EI but also from multiple sections, the integrated risk components and their linkages and cross-linkages should be examined to achieve overall sustainability.

Co-dependent elements of sustainability: An overall view of sustainability draws upon coherent processes from prior co-dependent stages, where sustainable production and sustainable consumption are essential elements. The UN Division for Sustainable Development (1992, Chapter 35.1) at the Rio Declaration on Environment and Development that was embraced by 178 governments asserts that

> …the sciences should be to provide information to better enable formulation and selection of environ-ment and development policies in the decision-making process. In order to fulfil this requirement, it will be essential to enhance scientific understanding, improve long-term scientific assessments, strengthen scientific capacities in all countries and ensure that the sciences are responsive to emerging need.

In explaining the co-dependent stages, Hailu et al. (2014) describe four linkages in the process – fiscal, production, infrastructure and consumption (see Figure 31.2).

However, there are limitations to these links, which are more severe in the developing markets whose market power is relatively less (price takers). Limits to fiscal linkage include (a) difficulty in identifying the earnings of rent by producers (as it is driven by the degree of market power), (b) tem-porary under-supply, (c) absolute shortage of the resource, (d) efficacy of production over the market and (e) capturing revenue by governments (the ability to tax the EI depends on the negotiating power of governments and the degree of information asymmetry in the market). Developing countries are vulnerable as many of them have large contracts with extractive companies who have the ability/power to lock-in low taxes and royalties, or wide-ranging exclusions.

Progressive tax gains popularity in policy issues with an ability to mitigate cyclicality throughout a commodity price boom. However, tying fiscal revenue to commodity prices can send government budgetary planning into turmoil during a period of declining prices. In policy fronts of fiscal reve-nue, this risk can be managed by setting up stabilisation funds, improving prudential administration

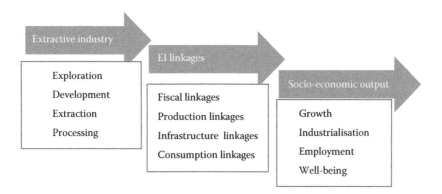

FIGURE 31.2   Links between extractive industry and socio economic output. (Adapted from Hailu, D., U. Gankhuyag and C. Kipgen. 2014. How does the extractive industry promote growth, industrialization and employment generation? Dialogue on the extractive sector and sustainable development – Enhancing public-private-community cooperation in the context of the post-2015. December 3–5, Paper presented to the UN Program and Government of Brazil.)

and promoting structural transformation, including infrastructure and tax and subsidy incentives into the industrial policy and in relation to the current issues of the economy. Production linkages are important in risk management and sustainability process as the EI operates as 'enclave' and contributes to the economy through limited backward linkages (providing other essential goods and services to the sector) and forward linkages (getting input to value add to the sector). Infrastructure contribution to the sustainability is paramount, as nearly 40% of capital expenditure in mining projects goes towards infrastructure development, of which 80% is spent on the development of rail and ports (McKinsey Global Institute, 2013). Although oil and gas projects require fewer infra-structures than mining projects, they still require ports and network of pipelines. McKinsey Global Institute (2013) estimates that by 2030, US$1.3 trillion will be invested in infrastructure develop-ment in resource-rich countries. EI generates relatively higher income via workers' income, (e.g. in Mongolia, wage levels in mining sector were 73% higher than the national average), directly leading to consumption linkage while capital reinvestment also links to growth (Hailu et al., 2014, p. 22). Limitations to growth via these links in the process include the outflow of income and profits, tax avoidance, capital flight, illegal financial flows, etc. to other economies, specifically

> Zimbabwe reports that the country loses over USD 50 million worth of gold every month due to smug-gling activities – resulting in lost government revenues of 42 million annually.

However, there are other large amounts of indirect contributions from EI, for the development of other non-EI sectors such as enhancing property rights and regulatory requirements and *horizontal* skill development. For instance:

> ...knowledge gained in developing control lines for Angola's oil sector is creating additional demand for metal working capabilities and is expected to help boost other manufacturing sectors as well as the construction and infrastructure sectors. In Nigeria, IT skills initially developed to serve the needs of the oil sector are being applied to other sectors. Technologies developed for South Africa's mining sec-tor are also helping develop capabilities in biofuels and have gone so far as to have global implications (Hailu et al. 2014, p. 26).... .

Emphasising on the concept sustainability, Deloitte and Touche, IISD (1992, p. 1) state that

> adopting business strategies and activities that meet the needs of the enterprise and its stakeholders today while protecting, sustaining and enhancing the human and natural resources that will be needed in the future. ...There are some forms of development that are both environmentally and socially sus-tainable. They lead not to a trade-off but to an improved environment, together with development that does not draw down our environmental capital. Sustainable development is good business in itself. It creates opportunities for suppliers of 'green consumers', developers of environmentally safer materials and processes, firms that invest in eco-efficiency, and those that engage themselves in social well-being. These enterprises will generally have a competitive advantage. They will earn their local community's goodwill and see their efforts reflected in the bottom line.

STAKEHOLDER

As ever more studies focus on EI and stakeholder engagement, it is important to consider the stake-holder notion as it relates to risk management in the EI. Orts and Strudler (2002, p. 218) suggest that stakeholders, under an economic risk-based approach, must encompass shareholders, creditors, employees and suppliers and customers. All of these participants in a business have some kind of economic stake directly at risk. Help achieving stakeholder satisfaction has become an integral part of corporate governance and CSR has become an integral part in achieving sustainability. As such, this study further explains the concept of CSR with *stakeholder engagement* below.

CSR and Stakeholder Engagement for Sustainability

A host of often competing definitional ambiguities have been spawned throughout the often exhaustive debate on CSR. Further, while many studies on CSR focus on economic, financial, and competitive advantages, they often miss the effect of a firm's environmental impact on its sustainability. The co-dependence of sustainability on CSR was often studied in isolation to their interrelated links, but has received research interest in the recent past (Linnenluecke and Griffith, 2013). Well-known papers on CSR go back to the 1950s with a debate on whether the firms should purely focus on the production of economic goods and profits or should it be firms following an array of broader social goals (Bowen, 1953) with a concentration on stakeholder concept over the concept of shareholders leading to stakeholder theory (Freeman, 1984). Also, the previous concentration that narrowly defined *firm environment*, which was limited to a firm's economic environment, is now discounted to include a much broader environment reconciling the firm's economic growth, financial performance and environmental protection to demonstrate the direct and indirect outcomes. Stakeholder engagement is the practice that organisations undertake to get stakeholders involved positively in organisational undertakings (Greenwood, 2007).

Orts and Strudler (2002, p. 216) consider traditional stakeholder theory to be unserviceable in interpreting environmental issues because its focus is on the interests of human participants in business enterprise and ignores the interests of environmental issues. World Business Council for Sustainable Development, WBCSD (2000) which was formed with 30 countries to develop co-operation between business, government and all organisations to achieve SD considered many aspects of CSR (see Prahalad and Ramaswamy, 2004; Sarmah et al., 2015 for details). CSR must be aligned with wider social objectives to improve the reputation. Greenwood (2007) emphasising on this said that Royal Dutch Shell moving from low reputation in 1995 to high in 2000 is reputational change example for achieving stakeholder objectives via social engagement from the *trust me* approach to *join me* approach. Stakeholder objectives were analysed by many in the business field, looking at the relationship between social and financial performances, as they focus their studies on the concept CSR from a single aspect (Wood, 1991; Carrol, 1999). In highlighting the importance of integrated broader aspect of stakeholder value, Donaldson (2002, p. 108) stated that

> No well-known writer on stakeholder theory has questioned the importance of share-holder value, but many have written that theory and practice should at times balance the importance of the value of money with that of other values.

It is vital to understand the level of engagement of stakeholders as an element in achieving SD and that process must incorporate and balance the stakeholder objectives. Koll et al. (2005) grouped stakeholders into the (i) primary group, with more influence on firm decisions and performances, which include employees, investors, customers and suppliers and (ii) secondary interest groups that have less influence on company activities. Although it is common to assume that primary stakeholders have to be incorporated in the process of achieving sustainability, and engagement of employees was considered as a prerequisite for CSR in the past, Greenwood (2007, p. 317) notes that employee consultation is not, per se, a requirement to fulfil this need and asserts that

> … just because an organisation does not engage with employees does not mean that the organisation is not responsible toward them. Such assumptions do not account for the propensity of the organisation to act in self-interest, particularly where there is a large power imbalance in the favour of the organisation.

Baumgartner and Ebner (2010), addressing a common problem, say communicating sustainability orientation to internal stakeholders is problematic than that to external stakeholders due to the hassles involved in identification and understanding of stakeholders' needs, interests and the value meaningful to them. Labuschangne et al. (2005, p. 373), proposing a new charter to evaluate the sustainability in the manufacturing sector, suggests that CSR requires integration of objectives – social equity, economic efficiency and environmental – into a company's operational practices for overall SD. Still the controversy is going on as some found a positive relationship between social and financial performances of firms (Cochran and Wood, 1984) while others Alexander and Buchholz (1978) found that there is no clear relationship. Salzmann et al. (2005) stated that sustainability is not a generic argument but the case that has been identified with the specific circumstances and the position of the firm within its industry. Arlow and Gannon (1982) studied CSR dependency upon factors, including organisation size, issue relevancy and industry characteristics. Husted (2000) focussed on social issues that increase social performance. Wood (1991) states that performance has to be viewed in the context of drivers, processes and outcomes, rather than just an outcome.

RISK

Risk is a key but diversely defined notion in many fields and a common definition describes it as the probability of moving from a desired expected value (Holton, 2004). Risk perception varies from person to person. A given action and/or inaction, anticipated/not anticipated foreseen/unforeseen, can increase/decrease the value, highlighting the importance of the concept uncertainty. As risk and uncertainty go hand in hand, it is more a subjective judgement and finding measures to eliminate risk are not possible. However, the important task is to find tools to minimise risks. Thus, the next section is devoted to discussing different types of risks in EI and possible ways to manage them.

Environmental risk is a wide concept that includes natural, business, social and political risks (see Ernst & Young, 2013 for details) making it difficult for researchers to put it within exact boundaries. Though the emphasis is often on natural environmental risk associated with EI-related activities, it is extremely essential to examine the links to other risk components to achieve overall sustainability. When an EI seeks to achieve its corporate objectives (directly/indirectly and/or foreseen/ unforeseen), it exposes itself to environmental, economic, financial, social, political and other risks that arise with operationalising those objectives. The high risks inherent in EI make that industry more susceptible to the compounding and potentiation complications associated with interlinked risks. This also highlights the necessity to comply with existing rules and regulations of the government. Though the external and internal risk elements are not entirely detached from each other, they are reviewed here for simplicity of understanding (see Figure 31.1). Internal risk elements are often considered to be direct elements in risk management and vary with a firm's objectives, competency, operational structure, level of development in EI activities, strategies, processes and culture. Though project risks are often considered internal, unless the risk levels are correctly identified with company objectives, it can lead to many different interrelated negative outcomes. A clear firm structure is needed to identify whether the primary stakeholders are provided with all the information needed to review internal risks attached in the production, management and process stages – for example, when a natural disaster strikes, resolving contiguous risks (such as operational and legal etc.) in a vigorous manner requires a clear understanding of company structure and strategies with clear lines of responsibility. In their empirical analysis of one of the largest EI companies in the world (NLNG Ltd. in Nigeria), Andeobu et al. (2015) found that the firm was able to manage and reduce risk and achieve stakeholder objectives by (1) carefully balancing the ownership structure, (2) profit satisficing balanced with national aspirations, (3) health and safety and environmental practices that conform to international standards, (4) use of appropriate insurance coverage and (5) using an employee structure that is exceptionally reliant on temporary employees. The competency (capability and experience, etc.) of the EI company can vary vastly with its size, development stage, ownership structure, market power, location, etc. However, it is vital for firms to understand their competency levels before a key risk event occurs and pre-plan quick and correct steps to mitigate the aftermath. Large extractive firms, as per their company charter, are often well equipped with risk management tools, rules and regulations; the JMC often find it cheaper to risk bankruptcy and dissolution. Financial risk and market and price risk are apparent elements, as the EI attributes include massive capital investments and it requires risk management tools such as financial derivatives and other instruments that minimise price, market, interest rate, and foreign exchange risks. Further, establishing a stabilisation fund to soak up the fluctuations from unexpected and/or unforeseen events may also be desirable. As mentioned previously, the interdependency of various risks make risk management difficult and complex. However, an understanding of co-dependency provides forewarning and insight to how to resolve such risks. In any EI risk assessment, resolving reputational risk can be paramount. Company objectives must be matched with the social engagement and the approach of *join me* instead of *trust me* (Greenwood, 2007) has been a valuable *a priori* risk management tool. Political risk has been a major impairment in EI activities in politically unstable but resource-rich developing countries. When the social structure and the culture of the wider community seem to be drastically different to the primary objective of the company, the political risk has to be observed to build up the social engagement and to enhance the economic benefit to the community. It is advisable to have objectives to be consistent with the regulatory and legal requirements while engaging the primary and secondary stakeholders to share the wider theme of the company to achieve overall sustainability.

As a first step towards minimising political risk, it is essential for EI firms and organisations to work together with governments and international organisations and to be cognizant of the international commitments and agreements of the nations involved. Highlighting the importance of those agreements (free trade agreements, e.g. ASEAN–Australia–New Zealand–Free Trade Agreement [AANZFTA]) and concurrence of rules and regulations to have international consistency to strengthen the regional and global sustainability, MCA (2014) asserts that 'In an optimistic reading of events,[...] and other proposed regional agreements are not about creating trading blocs but are based on a common understanding that the region needs to be integrated more closely in order to work better'.

Owing to heavy financial investment in EI, excavating firms and JMCs (includes prospecting) have widespread ownership. Any negative environmental impact of EI damages the company objective of improving owner's wealth due to share price decline showing a link between environmental risk and financial risk (FR). Showing the limitations of identifying FR, Holton (2004) notes that while risk is definable operationally in specific instances, an effective operational definition for risk in general has not yet been found. Thus, financial risk is intuitively understood rather than definitively understood. Specific problems include incorporating the notions of exposure and uncertainty, neither of which can be defined operationally. FR is often associated with the financial structure and transactions of the particular market incorporating the concept business environment risk which includes economic risk, compliance/regulatory risk and political risk. Though FR is often considered being an external risk element, it is *a priori* and/or *a posteriori* to the operational and other internal and external risks. Many environmental incidents have perturbed EI over the past few decades. It is essential to identify, assess, analyse and manage risks with respect to the objectives of the firm, its stakeholders and its host countries.

Link between Risk Management Tools, CSR and SD

CSR is no longer identified only with a firm's economic and financial objectives; rather it is undisputedly acknowledged that proper CSR is an essential step for achieving overall SD, compelling corporations to focus on fulfilling primary and secondary stakeholders' objectives of environmental, financial socio-economic and political sustainability. In doing so, the CSR must take actions to educate its stakeholders on the company objectives, structure, strategies, operation, processes, culture and competency.

CONCLUSION, POLICY ISSUES, LIMITATION, FUTURE RESEARCH

The unique characteristics of EI make it vulnerable to many risks directly as well as indirectly. As such, it is essential to identify and minimise the risks to achieve a firm's objectives and to attain sustainability. As the interdependency of various risks makes risk management difficult and complex, understanding of such co-dependencies facilitates the risk management task. Balancing overall sustainability can be achieved only by working together with stakeholders. This requires proper plans and policy implementation. Making the regulations simple to understand and facilitate the engaged companies to implement the existing policies and engage them in policy changes are essential to achieve sustainable EI. Company charter and policy should be well articulated among

the stakeholders to avoid misunderstandings and misguided steps in risky events. Though eliminating risk is not possible when it comes to political risk, it is essential for the EI companies and
organisations to work together with governments and international organisations to achieve overall
sustainability which depends on four pillars: environmental, finance, economic and social SD. In
attaining global sustainability, firms must be aware of and appreciate international laws, regulations
and agreements.

The main limitation of this study is, as the risk elements in EI are widely spread and interconnected, that it requires individual risk appraisals based on the specific situations for a targeted
analysis calling for quantitative empirical analyses on specific company events. This study incorporates the risks that can be identified in the EI in general. However, there are company-specific
as well as sub-sector-specific risks (oil and gas vs. other minerals) that can only be identified with
specific empirical studies opening the path for future studies.

REFERENCES

Alexander, G. H. and R. A. Buchholz. 1978. Corporate social responsibility and stock market performance. *Academy of Management Journal*, 21: 479–486.

Andeobu, L., S. Hettihewa and C. S. Wright. 2015. Risk management in the extractive industry: An empirical investigation of the Nigerian oil and gas industry. *Journal of Applied Business and Economics*, 17(1): 86–102.

Arlow, P. and M. J. Gannon. 1982. Social responsiveness, corporate structure and economic performance. *Academy of Management Executive*, 7(2): 235–241.

ASX. 2011. Metals and Mining Sector Profile. Retrieved from http://www.asx.com.au/documents/research/metals_and_mining_sector_factsheet.pdf. Accessed 23 July 2014.

Baumgartner, R. J. and D. Ebner. 2010. Corporate sustainability strategies: Sustainability profiles and maturity levels. *Sustainable Development*, 18(2): 76–89.

Bowen, H. R. 1953. *Social Responsibilities of the Businessman*. New York: Harper.

Carroll, A. B. 1999. Corporate social responsibility: Evolution of a definitional construct. *Business and Society*, 38(3): 268–295.

Cochran, P. L. and R. A. Wood. 1984. Corporate social responsibility and financial performance. *Academy of Management Journal*, 27: 42–56.

Deloitte and Touche, IISD. 1992. Business strategy for sustainable development: Leadership and accountability for the 90s. https://www.iisd.org/business/pdf/business_strategy.pdf. Accessed May 25, 2015.

Donaldson, T. 2002. The stakeholder revolution and the Clarkson principles. *Business Ethics Quarterly*, 12(2): 107–111.

Ernst & Young. 2013. *Global LNG: Will New Demand and New Supply Mean New Pricing?* Australia: EYGM Limited, pp. 1–20.

Freeman, R. E. 1984. *Strategic Management: A Stakeholder Approach*. Boston: Pitman.

Glavic, P. and R. Lukman. 2007. Review of sustainability terms and their definitions. *Journal of Cleaner Production*, 15: 1875–1885.

Greenwood, M. 2007. Stakeholder engagement: Beyond the myth of corporate responsibility. *Journal of Business Ethics*, 74: 315–327.

Hailu, D., U. Gankhuyag and C. Kipgen. 2014. How does the extractive industry promote growth, industrialization and employment generation? Dialogue on the extractive sector and sustainable development –
Enhancing public-private-community cooperation in the context of the post-2015. December 3–5, Paper
presented to the UN Program and Government of Brazil.

Holton G. A. 2004. Defining risk. *Financial Analysts Journal*, 60(6): 19–25.

Husted, B. W. 2000. A contingency theory of corporate social performance. *Business and Society*, 39(1): 24–48.

ICMM. 2012. The role of mining in national economies – Mining's contribution to sustainable development.
http://www.icmm.com. Accessed May 22, 2015.

Iddon, C., S. Hettihewa and C. S. Wright. 2015. Value relevance of accounting and other variables in the
junior-mining sector. *Australasian Accounting Business and Finance Journal*, 9(1): 25–42.

Iddon, C., C. S. Wright and S. Hettihewa. 2014. Externalising intolerable risk and uncertainty: The mining
sector as a strategic-cost leader. *Cost Management*, (Nov/Dec): 42–48.

International Energy Agency. 2010. *World Energy Outlook*. France: IEA Publications.

Jacob, D. 2011. *The Global Market for LNG*. Canberra: Reserve Bank of Australia.

JORC Code. 2012. The Australasian Institute of Mining and Metallurgy, Australian Institute of Geoscientists and Minerals Council of Australia.

Koll, O., A. G. Woodside and H. Mühlbacher. 2005. Balanced versus focused responsiveness to core constituencies and organizational effectiveness. *European Journal of Marketing*, 39(9/10): 1166–1183.

Kumar, S., H. Kwon, K. Choi, W. Lim, J. Cho and K. Tak. 2011. LNG: An eco-friendly cryogenic fuel for sustainable development. *Applied Energy*, 88: 4264–4273.

Labuschagne, C., A. C. Brent and R. P. G. van Erck. 2005. Assessing the sustainability performances of industries. *Journal of Cleaner Production*, 13: 373–385.

Langdon, T. 1994. LNG prospects in the Asia-Pacific region. *Resource Policy*, 20(4): 257–264.

Linnenluecke, M. K. and A. Griffith. 2013. Firms and sustainability: Mapping the intellectual origins and structure of the corporate sustainability field. *Global Environmental Change*, 23: 382–391.

Maxwell, J. A. 1992. Understanding and validity in qualitative research. *Harvard Educational Review*, 62: 279–299.

MCA. 2014. Minerals Council of Australia joint parliamentary inquiry into the role of the private sector in promoting economic growth and reducing poverty in the Indo-Pacific region, May.

McKinsey Global Institute. 2013. Reverse the curse: Maximizing the potential of resource-driven economies, December. Report:1–164.

Moon C. and A. Evans 2006. Ore, mineral economics, and mineral exploration. In *Introduction to Mineral Exploration*. 2nd ed., edited by C. Moon, M. Whateley and A. Evans. Carlton, Victoria: Blackwell Publishing, pp. 3–18.

Orts, E. W. and A. Strudler. 2002. The ethical and environmental limits of stakeholder theory. *Business Ethics Quarterly*, 12(2): 215–233.

Perrini, F. and A. Tencatti. 2006. Sustainability and stakeholder management: The need for new corporate performance evaluation and reporting systems. *Business Strategy and the Environment*, 15: 296–308.

Prahalad, C. K. and V. Ramaswamy. 2004. Co-creating unique value with customers. *Strategy and Leadership*, 32(3): 4–9.

Rudenno, V. 2009. *The Mining Valuation Handbook*. Milton, Queensland: Wrightbooks.

Report of the World Commission. 1987. Report of the World Commission on Environment and Development: Our common future. http://www.un-documents.net/our-common-future.pdf. Accessed July 12, 2015.

Salzmann, O., A. Ionescu-Somers and U. Steger. 2005. The Business C case for corporate sustainability: Literature review and research options. *European Management Journal*, 23(1): 27–36.

Sandelowski, M. 2000. Focus on research methods whatever happened to qualitative description? *Research in Nursing and Health*, 23: 334–340.

Sarmah, B., J. U. Islamb and Z. Rahman. 2015. Sustainability, social responsibility and value co-creation. *Social and Behavioral Sciences*, 189: 314–319.

The World Bank Group. 2012. http://documents.worldbank.org/curated/en/2012/01/17743538/world-bank-group-extractive-industries-2012-annual-review. Accessed July 20, 2015.

UN Division for Sustainable Development. 1992. https://sustainabledevelopment.un.org/content/documents/Agenda21.pdf. Accessed July 15, 2015.

WBCSD. 2000. http://www.wbcsd.org. Accessed June 16, 2015.

Wood, D. 1991. Corporate social performance revisited. *Academy of Management Review*, 16(4): 691–719.

32 Integrating Sustainability for Long-Term Business Value

Liza Maimone and Kerryn Schrank

CONTENTS

ABSTRACT

The importance of sustainability goes far beyond environmental issues, as the need to behave responsibly becomes a key aspect of strategy and operations, maintaining brand and reputation and seeking good growth. This is true irrespective of the sector in which a company operates.

Traditionally, 'heavy' industries such as oil and gas, mining, transport and manufacturing have attracted most scrutiny. But no company, no matter what it does, can now afford to take its eye off its environmental, economic, tax and social impact.

Financial markets are increasingly assigning a greater value to effective risk management of non-financial issues, and non-governmental organisations, activists, social networks, the media and customers are increasingly holding companies accountable. A company's reputation and social license to operate is at stake.

Leading mining and energy companies that operate sustainably recognise that social, environmental, economic and ethical factors affect their core business strategy. They know that operating sustainably creates and protects long-term stakeholder value, and they identify and respond to the changing needs and demands of society. They are integrating sustainability into their overall strategies by

- Building trust with their key stakeholders through actively engaging them and actively managing their risks
- Setting and reporting against performance targets
- Maximising the positive effects of their operations
- Embracing a growing body of international best practice on non-financial matters

This chapter outlines the drivers for sustainability, the journey for progressing a sustainability agenda and a leading practice model for integrating sustainability consideration into day-to-day operations and reporting.

SUSTAINABILITY AGENDA

Many companies have long viewed sustainability as a matter of corporate philanthropy, with no relevance to their company's core strategies. The cost of sustainability activities (e.g. waste and recycling initiatives, community partnerships, product and social marketing) was seen as detracting from profitability and was often accounted for in a public relations or marketing budget.

But those days are gone, or at least fading from view. Most progressive companies understand that the traditional trade-off between sustainability and profitability is an outdated perspective. Mining and energy companies in particular have come to realise that their ability to prosper depends upon their responses to the challenges of a carbon-constrained world and other issues critical to sustainability.

Business leaders who operate sustainably recognise that social, environmental, economic and ethical factors affect their core business strategy. They know that operating sustainably creates and protects long-term stakeholder value, and they identify and respond to the changing needs and demands of society.

These leaders evaluate the range of sustainability issues and respond by mitigating risks and leveraging opportunities. They also understand that to operate in a sustainable manner, they must work collaboratively with all stakeholders, including suppliers, shareholders, employees and regulators.

Figure 32.1 represents the potential linkages between sustainability activity and stakeholder value.

SUSTAINABILITY JOURNEY

Sustainability for mining and energy companies encompasses a broad range of issues such as water quality and quantity, pollution, climate change, talent attraction and retention, health, safety, community engagement and human rights. While the global challenges relating to sustainability are evident, defining how companies can meet the challenges can be daunting. Not surprisingly, companies often wonder where to begin when it comes to progressing sustainability.

Companies around the world are at different points in addressing sustainability issues. PricewaterhouseCoopers (PwC) developed a framework (Figure 32.2) that identifies three phases in

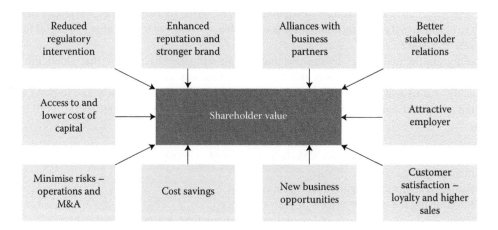

FIGURE 32.1 Sustainability drivers of stakeholder value. (Adapted from ABN AMRO Morgans.)

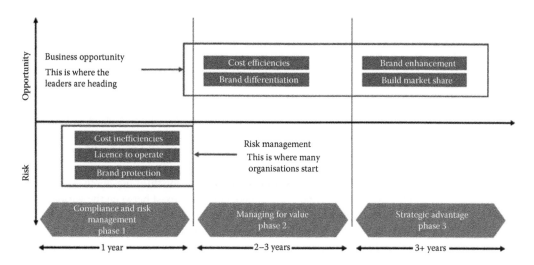

FIGURE 32.2 Sustainability maturity framework. (Courtesy of PwC.)

response to sustainability, from simple regulatory compliance to making sustainability a competitive advantage.

PHASE 1: RECOGNITION AND UNDERSTANDING

Companies in Phase 1 are beginning to recognise and understand the implications of not addressing sustainability issues effectively. They realise that issues such as climate change and safety can present material risks and potentially impact their profit and brand if not effectively managed. At the same time, regulations might compel them to develop a sustainability strategy.

PHASE 2: MANAGEMENT

Companies in Phase 2 are managing sustainability risks and issues by implementing programmes that are able to cut costs (e.g. by using less energy, reducing waste and recycling more). These companies start to reorganise their operations in order to exploit sustainability-related opportunities and can thereby promote their brand as more sustainable.

PHASE 3: PERFORMANCE

As companies in Phase 2 start seeing the benefits in cost-efficiencies and brand differentiation, companies in Phase 3 develop a deeper understanding of the long-term value of sustainability and how it can become a strategic advantage. They are able to measure their sustainability performance internally and externally and have a true competitive advantage often signalled by increasing market share.

There is no quick fix that will make a company sustainable; it is a journey that takes time. The time to move through the phases depends on a company's strategy and the level of investment they are willing to make.

Most Australian mining and energy companies are in Phases 1 and 2 with only a few in Phase 3. Other industry sectors – such as finance, retail and telecommunications – have not been as quick as their global peers to seize sustainability opportunities. In general, many Australian companies either do not have a stance on sustainability or are only focussing on specific issues such as water and energy consumption, packaging rather than product innovation, ethical sourcing or supply chain issues.

Global experience indicates that sustainability will soon be a mainstream concern and a core part of business rather than an 'end game'.

INTEGRATED APPROACH

One of the biggest challenges that companies face is integrating their sustainability approach with the broader objectives of the business. Sustainability programmes run the risk of being short lived and viewed as 'nice to have' unless they demonstrate tangible business value and are hardwired into the day-to-day operations of the company. Essentially, it is about making the shift from 'doing' sustainability to 'being' sustainable.

The model set out in Figure 32.3 provides a logical structure for thinking about a company's information needs and the critical links and interdependencies that exist between the various information sets – external, strategic, business and performance. The suggested model is grounded in years of PwC research and work with investors and companies from broad industry sectors.

None of the information that flows from each category of the model is new – the real insight and value comes when each category is considered in the context of another, rather than in isolation. It succeeds when governance interfaces with remuneration and risk; when strategy is designed to exploit a changing market environment; and when strategic priorities align with key performance indicators (KPIs).

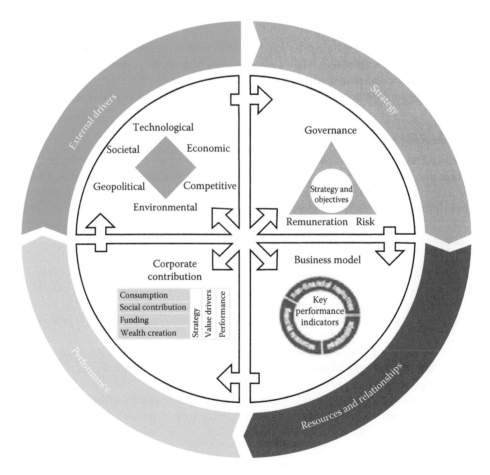

FIGURE 32.3 Integrated approach. (Courtesy of PwC.)

It is this cohesive and integrated way of thinking that is critical to the model and to being a sustainable company. Few companies excel in this space, at this stage. But we believe that it is now a key determinant of corporate success.

The four core categories are common to all industries and companies, but their relative importance and the information that sits beneath each category will need to flex depending on how these dynamics impact industries and influence business strategies and operating models.

Each category is discussed in more detail below.

EXTERNAL DRIVERS

Some shifts in strategy are the result of internal activities, but many are inspired by shifts in the external environment (Figure 32.4).

> 76% of Energy and Power & Utility CEOs put resource scarcity and climate change in their top three megatrends to transform their business with Mining CEOs not far behind at 62%.
>
> *PwC's 17th Annual Global CEO Survey of 1,344 CEOs in 68 Countries*

The more a company can identify and respond to external drivers and then explain this process – including its connectivity to the market environment and the ecosystem on which it relies – the better able it will be to secure a competitive advantage.

The ability to identify the main 'emerging' risks to which companies are increasingly exposed is a key driver. There are a number of external drivers that present sustainability-related risks specific to mining and energy companies:

- Scarcity of fossil fuels and raw materials.
- Regulatory, political and economic factors. Escalating costs arise from the need to explore remote regions and develop smaller-scale reserves under adverse and hazardous conditions.
- Changing societal expectations and growing demand for clean, renewable energy.
- Climate change/physical risks to facilities and infrastructure.
- Long-standing challenge of protecting the health and safety of employees working in extreme environments.
- Public concerns and opposition to large-scale infrastructure projects.

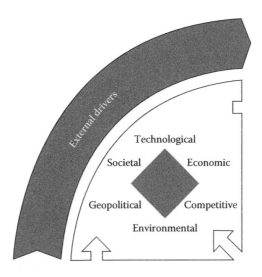

FIGURE 32.4 External drivers.

Operating sustainably is broader than identifying and containing risks. For leading companies, risk identification can yield rich opportunities if they are identified, assessed and managed for competitive advantage.

Companies must focus on enhancing their market intelligence, leveraging it in their decisions and making this visible in their stakeholder engagement and reporting processes. Keeping abreast of market developments and ensuring that a longer-term perspective is brought to this process will require most companies to rethink their current processes. Scenario planning and stress testing will become more commonplace to help organisations identify their sustainability risks and issues. For example, the growing number of consumers seeking healthy and sustainable lifestyles constitutes a potentially vast market for new products and services. Consumers' environmental, ethical and social concerns mean that their definition of quality is increasingly extending to the product lifecycle.

Investors Tell Oil and Gas Sector to 'Stress Test' for 450 ppm

The Investor Group on Climate Change (IGCC), along with counterpart organisations from Europe, the US and Asia, released a statement which expects oil and gas companies to embed 'stress testing' of various scenarios into their capital spending plans, including successful attempts to limit CO_2 levels to 450 ppm.

The statement also says investors expect oil and gas companies to have clearly-defined processes in place to ensure boards and management understand climate risk 'and the strategic implications of a transition to low-carbon energy systems'.

Thorough scenario testing, risk analysis and transparency by fossil fuel companies will help investors avoid misplaced faith in both endless fossil fuels growth and universal divestment.

IGCC (2014)

Wherever environmental, social or ethical issues can be addressed, companies have an opportunity to innovate, differentiate, create value and attract more customers. These issues can also create opportunities to attract and motivate employees.

Mining and energy companies have a number of opportunities on both the supply and demand sides of energy production and consumption. Many companies have long-standing, though modest, initiatives in the development of energy alternatives from wind and solar, to fuel cells and tidal. Biofuel additives to transportation fuels are being mandated in the EU and many Latin American countries, with the hope of reducing reliance on gasoline and diesel, adapting the vehicle and fuelling station infrastructure, and seeding new industries in biofuel production and processing. All of these areas will require technologies to advance and capital is indeed flowing towards 'clean tech'. In addition, the payoff for significant advances in the area of carbon capture and storage could be substantial.

On the demand side, energy companies can invest in the development of innovative programmes to encourage and enable businesses, governments and individuals to reduce their energy consumption. Under one such new business model, companies guarantee energy supplies at a fixed price to customers, undertake a range of efficiency measures for their operations and then split the resulting savings with them.

Although many mining and energy companies now understand that pursuing sustainability is a long-term investment, they constantly face the challenge of balancing short-term shareholder concerns with long-term business objectives. The global reach and complexity of supply chains, raw materials required for production, the nature of products and the special characteristics of their workforce, all determine the risks that play the most significant roles in any given company, and the opportunities that exist.

STRATEGY

Until recently, most companies' sustainability efforts operated outside of their core business strategy (Figure 32.5). But as companies become increasingly aware of the strategic implications of

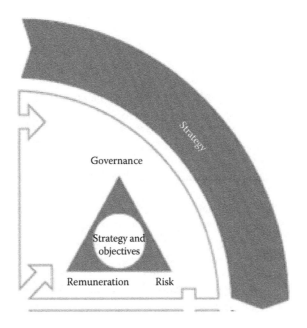

FIGURE 32.5   Strategy.

climate change, population growth and resource use, this situation is changing. In fact, in the future, we expect the term 'sustainability' to disappear altogether as sustainability issues are integrated into core business thinking.

> If you want to continue to succeed as an energy company in the coming decades, you need to understand and meet people's expectations for environmental and social performance, as well as delivering solid technical and financial performance. That means putting solid business principles, including sustainable development, at the heart of how you do your business.
>
> *Jeroen Van der Veer, Former CEO of Shell*

Sustainability issues pose a unique set of challenges and afford a distinct set of opportunities to mining and energy companies. However, there are components to every strategy that are critical to all businesses, some of which must be reconsidered in the context of sustainability, for example, risk and governance.

A clear understanding of the culture, values and governance practices of any company is important when determining how a sustainability strategy should be embedded. Ideally, it should be developed with consideration of how it will be integrated and cascaded through the company, with reference to risks and opportunities, people and relationships and performance measurement. Failure to consider the capacity of these areas when developing a sustainability strategy will limit its chances of success. For example, a well-considered strategy that is aligned to risks and opportunities may not yield results if there is no one in the company to champion the strategy, or if there are no mechanisms in place to measure the success or outcomes of the strategy.

Good strategies are built on a foundation of clear goals and objectives. A goal is a statement that clearly describes desired long-term improvement/change in the company through the integration of sustainability into the business. Once goals have been defined, then objectives can be developed. Objectives describe the specific changes within the business, which are required to achieve the goals, and ideally should relate to risk factors. Objectives are the stepping stones to achieving set goals.

Resources and Relationships

The growing complexity of business coupled with companies' desire to increase operational flexibility has given rise to many new business models over the past decade (Figure 32.6). In mining and energy sectors, we have seen fly-in–fly-out workforces and a trend towards outsourcing, off-shoring and more recently a reversal to on-shoring, elements of the business.

Unlike the functional areas of a company – such as marketing, logistics, information technology and human resources – the responsibility of sustainability seldom fits discreetly within a single group. Sustainability issues are often all encompassing and impact a number of different areas.

For example, sustainability activities could be facilitated by health, safety and environment teams with a focus on regulatory compliance. It could be leveraged by the marketing team to differentiate a company's product from the competition, the logistics team could use sustainability to improve the efficiency of the supply chain and the human resources team could use the company's commitment to sustainability as a point of differentiation to attract and retain employees.

As such, an effective sustainability strategy needs to be integrated across all functions and operations. KPIs need to be established for particular sustainability issues and cascaded across the business and a comprehensive range of communication and engagement initiatives are needed to help employees understand how sustainability is relevant to them.

As stakeholders adopt a more active and intrusive approach to monitoring business activity, companies will have to be able to explain elements of their business and operating models. This will extend to explaining the key dependencies on which companies rely and giving insight into the health and development of the key business relationships deemed to be critical to long-term success.

Many companies are in the early stages of the maturity curve when it comes to integrating sustainability within the business. This is similar to where many mining and energy companies were in relation to safety culture in the 1980s and 1990s. Safety was once seen as the responsibility of safety professionals within a company but nowadays, it is largely part of corporate cultures.

Just as safety has done in the past 20 years, the concept of sustainability is expected to become a shared responsibility that delivers benefits across all areas of a company. However, companies that are currently in the early stages of their sustainability journey will find that significant investment

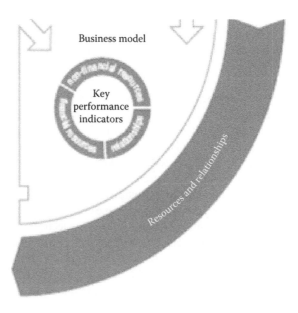

FIGURE 32.6 People and relationships.

into education and integration of sustainability will be required to create a culture of shared responsibility and strategic advantage.

PERFORMANCE MEASUREMENT

Performance measurement is the final link in the integrated approach to sustainability (Figure 32.7). Assessing performance allows management to evaluate the implementation of the sustainability strategy and determine targeted actions for the future.

Accountable personnel and reliable data are both critical factors, which will influence the effectiveness of an organisation's sustainability strategy. Systems and resources must be capable of delivering the strategic objectives of the organisation's sustainability vision. Specifically, this requires personnel with accountability to performance and data that are complete, accurate and timely.

BHP Links Sustainability to Remuneration

The Sustainability Committee assists the Remuneration Committee in determining appropriate HSEC metrics to be included in GMC scorecards and in assessing performance against those measures. HSEC performance is measured as a balanced scorecard within the overall annual bonus assessment. In FY2014, six elements were considered: no work-related fatalities or significant environmental or community incidents; reduction in total recordable injury frequency (TRIF) and occupational illness; Health; Environment; Community; and HSEC risk management.

Excerpt from BHP Billiton Sustainability Report 2014

Performance can be measured on an individual level or in the context of the company as a whole. In leading companies, KPIs of individuals in the organisation are linked to the overall goals and objectives of the business and are used to assess an individual's annual performance and often linked to remuneration. Individual sustainability-related KPIs are an effective way to assist a company to deliver on its overall sustainability strategy.

Effective performance measurement allows for scrutiny of a sustainability strategy. Based on this process, management can assess the effectiveness of the strategy and the relevance of the goals and objectives leading into the next performance period.

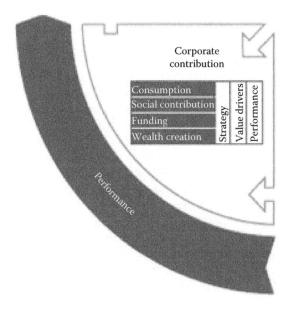

FIGURE 32.7   Performance measurement.

REPORTING

Over the recent years, the demand for greater transparency in corporate environmental, social and ethical performance reporting has risen significantly. So much so, there is now a large and growing movement worldwide towards mandatory public company disclosure of sustainability information, often referred to by the investor community as non-financial or environmental, social and governance (ESG) data.

The Australian Stock Exchange (ASX) released a new edition of the *ASX Corporate Governance Council Principles & Recommendations* (Third Edition) in March 2014, applying to all ASX-listed entities' full financial year reporting periods commencing on or after 1 July 2014. Of particular interest with respect to sustainability reporting is the new Principle 7, Recommendation 7.4, which recognises that there is an increasing demand from investors for greater transparency on environmental and sustainability risks to ensure they are able to adequately make investment decisions.

Recommendation 7.4 states that

> A listed entity should disclose whether it has any material exposure to economic, environmental and social sustainability risks and, if it does, how it manages or intends to manage those risks.
>
> 74% CEOs told us that measuring and reporting their total impact (financial and non-financial impacts) contributes to their long term success.

PwC's 17th Annual Global CEO Survey of 1,344 CEOs in 68 Countries

The Australian Council of Super Investors (ACSI) has been surveying the extent to which ASX200 companies report on ESG risks over the recent years. As of 31 March 2014, 170 companies provided some level of public sustainability risk reporting; 123 of these companies have been included in ACSI's research since 2009.

Despite this progress on reporting, sustainability issues are often discussed in isolation from information on corporate strategy. Many companies state their commitment to sustainability policies and strategies but do not provide an indication of their performance against their policy or provide an indication of the materiality of factors included in such a policy. Companies must recognise that information reported in any public report is only meaningful to investors and other report users if it is linked to corporate strategy and outcomes, and companies often fail to make this critical link.

'Integrated' reporting, where both financial and non-financial (i.e. sustainability) data are reported together in mainstream annual reports, is being proposed as the best way of reporting. It recognises that financial information alone does not provide a complete view of the company's overall performance or respond to the varying interests and concerns of multiple-stakeholder groups.

The International Integrated Reporting Framework (International IR Framework), which was released in December 2013, was developed by the International Integrated Reporting Council collaboration with leading companies, investors, regulators, professional advisers and NGOs in 25 jurisdictions worldwide. The aim of Integrated Reporting is to improve the standard and depth of company reporting by addressing *all* dimensions of the value creation process within organisations – including strategy, governance, business model and forward prospects, in addition to the traditional focus on historical financial reporting alone.

Integrated reporting does not mean adding the sustainability report as a section of the annual report. The process demands the integration of information, which can flex with the changing dynamics and can deliver truly valuable strategic and operational insights to management and stakeholders. Figure 32.8 depicts the integration journey.

A company cannot move to integrated reporting in a single step. Companies must first manage their business in an integrated manner, in order to monitor and assess together performance of both financial and non-financial KPIs.

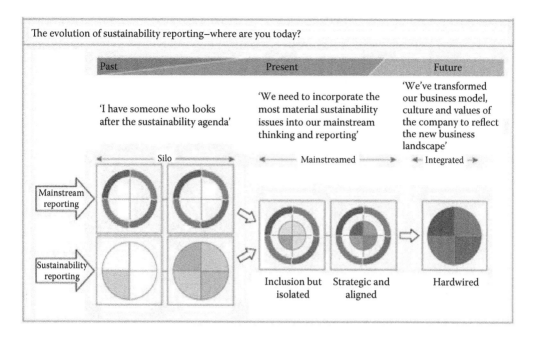

The evolution of sustainability reporting—where are you today?

FIGURE 32.8 Integration journey. (Courtesy of PwC.)

While the concept of integrated reporting is gaining more attention, the benefits have been known for a number of years. In 2003, the Australian Government commissioned Corporate Sustainability – an Investor Perspective (The Mays Report). It found that those companies that integrated sustainability activities into their core business had a lower risk profile while enhancing their brand and reputation.

The Global Reporting Initiative (GRI) surveyed a group of 756 reports – published between 2010 and 2012 – that were self-declared to be 'integrated'. The GRI report *The Sustainability Content of Integrated Reports – A Survey of Pioneers* found that globally, the financial sector self-declared more integrated reports than any other sector, followed by the energy utilities, energy and mining sectors.

Australian organisations published 44 self-declared integrated reports over the 3 years that were the subject of the GRI analysis, comprising nearly 6% of the total 756 reports included in the full sample. Commensurate with global trends, the financial services sector in Australia publishes by far the largest volume (9%–11%) of self-declared integrated reports, followed by energy utilities (5%). The non-profit/services sector is the third highest reporting sector in Australia, while globally it holds the sixth place.

This trend continues today with more and more companies integrating ESG and financial data in performance reporting and wider corporate communications.

CONCLUSION

The integration of sustainability into the core business requires a fundamental rethink of a company's business model, culture and values. It requires companies to identify and address the most material sustainability issues into day-to-day operations and business decisions.

A short-term focus on business survival will remain important for many, but the shift to a longer-term strategic perspective will be essential in the battle to build trust and market value. Companies that intend surviving and prospering long term will need to more clearly demonstrate that they are currently *thinking* long term.

Such a perspective allows shareholders, and other key stakeholders, to understand the quality and sustainability of performance by providing insights into external influences, strategic priorities and the dynamics of the chosen business model, as well as the key drivers of success.

REFERENCE

IIGCC. 2014. IIGCC Supporting Investor Expectations: Oil and Gas Company Strategy, Supporting investor engagement on carbon asset risk.

33 Challenging the Challenging Regulatory Environment in Papua New Guinea

Mohan Singh

CONTENTS

ABSTRACT

Papua New Guinea (PNG) has some of the biggest and most advanced opencast mines in the world. The country also awaits one of the world's largest underground mine with block-caving method in the future. PNG already stands on the threshold of having the world's first hard rock deep sea mine. The nation also holds enormous potential in the area of geothermal energy, which would bring another dimension to the regulation of its robust and world-class mining industry.

The mining sector is an important sector for the economic and social well-being of the people of PNG. This is because PNG's economy has been largely dependent on the mining sector for revenue generation since independence and continues to do so. Nonetheless, no progress can be considered worthwhile if it ultimately does not result in a safe and healthy society—a vision that also very well aligns with PNG's long-term national goals. Occupational health and safety (OHS) for the mining industry in PNG is governed by an outdated legislation, and the regulators lack manpower resources and capacity building.

The long absence of a full-time chief inspector till late 2008 rendered the Mines Inspectorate inefficient, directionless, and disintegrated, leading to high rates of accidents and incidents. On the international platform, PNG virtually had no recognition as a mine safety regulator!

This chapter illustrates how the Mines Inspectorate, with such a dismal and disappointing backdrop, tunneled its way through a mountain of odd challenges during all these years and turned the tables on the other side on all fronts.

Resultantly, not only did the safety performance improve drastically, but the whole industry also knitted into one big family that now shares many common platforms to deliver various safety and health programs as a cohesive and committed team. Going forward, PNG's profile in the international arena is also lifted to an entirely new level.

INTRODUCTION

Minerals play a very vital and essential role in changing the fate of civilizations in every nation and thus remain one of the critical drivers of a nation's economy, in particular, and the whole world, in general. We live in a material world where the economy is fueled by consumerism and sustained primarily by consumers' lust, and their capacity to spend. However, the world also witnesses changes to its economic order from time to time due to the roller coaster behavior of the commodity prices. Owing to a decline in prices of most of the minerals in recent years, the mining industry, in general, had been passing through some rough weather, forcing the mining fraternity to find more pragmatic strategies and solutions to keep its head above the water.

Amidst such upheavals, a mining regulator from around the Equator had also been tunneling its way through the rugged terrains and been successful in conquering many a peak, which the reader would be taken a brief tour of in this chapter.

The Independent State of Papua New Guinea (PNG) is proud to have some of the biggest and most advanced opencast mines in the world in terms of engineering, technology, mechanization, processes, and systems. It already stands on the threshold of having the world's first hard rock deep sea mine to extract massive sea floor sulfide deposits from a depth of about 1.6 km. PNG also awaits to host one of the largest underground mine with sublevel/block-caving method in the future.

In the backdrop of such a state-of-the-art technical environment, the subject of OHS for the mining industry in PNG is governed by an outdated legislation, viz., the Mining (Safety) Act 1977 and Mining (Safety) Regulation 1935 through the Mines Inspectorate Branch of the Regulatory Operations Division of the Mineral Resources Authority (MRA), which was established under the Mineral Resources Authority Act 2005 and formally inaugurated to commence functioning on June 1, 2007.

As per the provisions of the Mining (Safety) Act 1977, the statutory appointments of the chief inspector and all inspectors are made by the mining minister through gazettal notification and the Act obligates the chief inspector to report any matter under the Act directly to the minister. However, the Mines Inspectorate banks wholly on MRA for its administrative support and resource requirements.

The organization structure of the MRA vis-à-vis the Office of the Chief Inspector is depicted in Figure 33.1.

The author had a novice team of only three inspectors in the MRA in late 2008, which reduced to only one in 2010, then rose to two in 2012, and three in 2013, but remained so till the beginning of 2014. The Mines Inspectorate inherited nothing in terms of systems, procedures, and records for steering such a critical regulatory institution!

The old and obsolete legislation of 1935 and 1977 looked like a dwarf before the modern-day and highly sophisticated mining industry in PNG. Owing to the long absence of a full-time chief inspector, the Mines Inspectorate was ineffective and directionless and the mining industry had

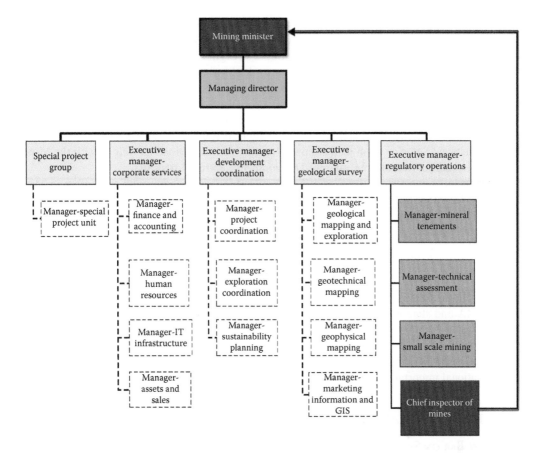

FIGURE 33.1 Organization structure of the MRA vis-à-vis the Office of the Chief Inspector.

virtually disintegrated and drifted to an alien territory with high rates of accidents and incidents. On the international platform, PNG virtually had no recognition as a mine safety regulator. Listed below are the significant milestones achieved over the last 6 years in the area.

ORGANIZATIONAL LEVEL

STRATEGY, SYSTEMS, AND PROCEDURES

With a view to define the future road map, the Mines Inspectorate developed the "Strategic Plan 2009–2014," which captured the various essential elements for action and execution. Over all these years, many areas were streamlined, including standardization and upgradation of monthly and quarterly OHS formats, inspection format, inquiry format, yearly book format, inspector's monthly performance format, office registry, archiving, safety alert to the industry, chief inspector's annual report format, as well as regular intradepartmental review meetings and protocol procedures, etc.

RESOURCING AND CAPACITY BUILDING

There were only three inspectors and seven sanctioned staff positions in the Mines Inspectorate in December 2008. Acute manpower shortage was a major constraint that the inspectorate had to deal with apart from very poor remuneration packages to attract and retain qualified and engineering professionals. Consequently, the staff strength was dwindling and seriously impacting our regulatory obligations and satisfactory service to the industry. Upon various submissions and persuasions, the MRA Board considered 18 additional positions and an attractive pay package for the inspectors. The board increased the inspectorate strength to 34. This increase was considered primarily in the light of proposed amendments to the Mining (Safety) Act and Regulation that would exponentially increase the inspectorate's regulatory jurisdiction thus requiring commensurate resources.

Retention of inspectors was a challenge, and the new recruits needed professional training. Inspectors were deputed to different mines from time to time for their exposure to the current industry practices. Some of them also attended training courses abroad as well as with the Mines Inspectorates and International Mining for Development Centre in Australia.

The financial support from the World Bank enabled the engagement with the Safety in Mine Testing and Research Station, a training and research arm of the Government of Queensland, to organize a formal training for our inspectors, stage one of which commenced on May 25, 2015 for a duration of 4 weeks.

Lateral Support

The Mines Inspectorate availed the opportunity to significantly and meaningfully contribute in various MRA and state-level meetings, including the Mining Advisory Council as well as in the extensive exercise of reviewing the entire mining legislation and policies and would soon be leading the review of the obsolete Mining (Safety) Act 1977 and Regulation 1935. The Mines Inspectorate regularly offers its valuable professional advice on various regulatory matters of national interest and international importance to the Office of the Mining Minister.

INDUSTRY (NATIONAL) LEVEL

Lifting the Bar on Statutory Management of Mines

Previously, all mines in the country were being managed by mine managers who held junior positions in the management hierarchy at the project level, subordinating the role of a registered mine manager, potentially compromising with the statutory requirements under the Mining (Safety) Act. The mines are now operating under the charge of general managers who are required to hold the statutory role of the registered mine manager under the Act.

Redefining the Process of Competency Examinations

The competency examinations with only one theory paper were transformed into a written examination with few practical questions to assess the managerial and decision-making capabilities, followed by an oral examination to gauge the communication and presentation skills of the candidates. This pragmatic shift in the examination and assessment process attracted many senior candidates every year as depicted in Figure 33.2.

Encouraging Self-Regulation and Collaboration

To redefine the chapters on self-regulation and collaboration, three industry bodies were constituted, viz., National Apex Mining Safety Council, National Mining Safety Forum, and National Mining Emergency Response Forum. These bodies are represented by the senior mining executives and meet regularly on a quarterly basis under the presidency of the chief inspector to steer the mining industry for increasing mutual cooperation, participation, and promoting safety and health to tens of thousands of persons working in PNG mines. These fora are vibrant and extremely fruitful to deliver on many new frontiers.

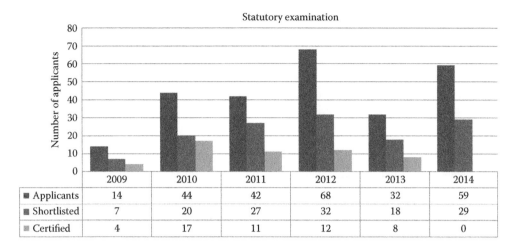

Statutory examination

	2009	2010	2011	2012	2013	2014
■ Applicants	14	44	42	68	32	59
■ Shortlisted	7	20	27	32	18	29
▨ Certified	4	17	11	12	8	0

FIGURE 33.2   Chart showing the yearly number of applicants and certificates issued.

NATIONAL MINING EMERGENCY RESPONSE CHALLENGE

As an outcome of the above endeavors, the 1st National Mining Emergency Response Challenge was held in Lae in 2011 followed by the 2nd and 3rd ER Challenges held in Port Moresby in 2012 and 2013, respectively. The 4th ER Challenge was held in Kavieng, New Ireland Province last year with plans underway to host the *5th ER Challenge* in Madang in August this year. These challenges have enabled various mine ER teams to participate in mock exercises to keep them abreast to handle real-life situations during any emergency at their mines. The petroleum sector is showing keen interest to participate in the forthcoming *ER Challenge* in Madang. It should be noted that ladies team also participate in these challenges. A national team across to Australia for participating in similar challenges in those jurisdictions is being considered. Mutual Aid Agreement signed by senior executives of all mines on this occasion is another initiative that provides a framework to collectively respond in the unlikely event of an emergency at any mining site in the country.

NATIONAL MINING SAFETY WEEK

In March 2012, the nation witnessed its 1st National Mining Safety Week, which not only embraced the mine workforce into its strong fold but also the surrounding communities as well as the school children, thus rewriting chapters on promoting safety and health across a wide spectrum. The 2nd, 3rd, and 4th National Mining Safety Weeks were celebrated in March 2013, 2014, and 2015, respectively.

Through the week-long celebrations, various events were organized at each mine to involve and remind the mine employees and their families about the importance of safety and health in their daily lives. Mines inspectors also visited the mines during these celebrations to oversee and encourage the participants and also to host special sessions on yoga for promoting health and fitness.

HEALTH AND FITNESS

Acknowledging the importance of health and fitness not only in our professional life but also in our personal lives, the Mines Inspectorate introduced a highly effective and proven technique of yoga for the mining workforce for their overall well-being. This practice was taken to the surrounding communities and to the school children in the mining project areas to harness its miraculous benefits for a wider populace.

In a huge global endorsement for yoga last year, 175 out of 193 members of the UN agreed to declare June 21 as the International Yoga Day while the mining industry in PNG formally embraced yoga on September 22, 2014 at Kavieng and adopted the day as the PNG Mining Industry Yoga Day.

Regular yoga classes are undertaken at the Mining Haus, POM Technical College, St. Mary Catholic Church, and Bomana Prison, and the public in thousands now turn up for yoga at the Ela Beach every Sunday during the "Walk for Life" program initiated by the governor of National Capital District.

1st PNG Mining and Petroleum OHS Conference

With constant thrust and encouragement from the Mines Inspectorate, the mining and petroleum industry collectively took another leap and hosted the *1st PNG Mining and Petroleum OHS Conference* in July 2013 in Port Moresby. The mining minister inaugurated the conference, which was attended by about 200 delegates. The Mines Inspectorate is persuading the mining industry to host the *2nd PNG Mining and Petroleum OHS Conference* soon.

Chief Inspector's Annual Reports

In July 2013, the mining minister formally released the *Chief Inspector's Report 2012*, the first of its kind, at the inaugural function of the *PNG Mining and Petroleum OHS Conference* held in Port Moresby. This compilation, *inter alia*, put the spotlight on the activities of the Mines Inspectorate as well as the OHS performance of the mining industry over the last 5 years.

In September 2014, the mining minister formally released the *Chief Inspector's Annual Report 2013* during the *National Mining Emergency Response Challenge* in Kavieng, and presented these reports to the National Executive Council (NEC) on November 3, 2014 and to the Parliament vide NEC Decision No. 318/2014. The *Chief Inspector's Annual Report 2014* will be published shortly.

Industry OHS Performance Reviews

The Mines Inspectorate holds regular performance reviews with the industry to continuously oversee its performance and impart necessary direction. These exercises, working as a team, have generated a new environment of mutual trust and partnership with the industry that is writing new chapters of governance in PNG.

Impact on Safety Performance

Figures 33.3 through 33.6 depict safety performance over the years.

As is evident, the higher trends in fatalities and injuries till 2007–2008 were not only arrested but also nosedived in the subsequent years while employment in the industry kept rising.

A significant drop took place in the "fatality rate" from 0.43 in 2008 to just 0.02 in 2013 (a reduction of more than 95%!), "total injury frequency rate" from 40.55 in 2007 to 12.63 in 2013 (a reduction of around 69%!), "serious injury frequency rate" from 2.22 in 2007 to 0.90 in 2013 (a reduction of more than 83%), and "severity rate" from 0.32 in 2007 to 0.04 in 2013 (a reduction more than 87%), while the industry reported 56,372,577 man-hours (average employment of 22,110 persons) in 2013 as against 24,760,216 man-hours (average employment of 10,424 persons) in 2007.

Though there was some spike in the fatality rate toward the end of the year, the "total injury frequency rate" further dropped from 12.63 in 2013 to 9.01 in 2014, "serious injury frequency rate" from 0.90 in 2013 to 0.50 in 2014, and "severity rate" from 0.04 in 2013 to 0.03 in 2014. The longest fatality-free period of 689 days from January 2, 2013 to November 22, 2014 was also achieved.

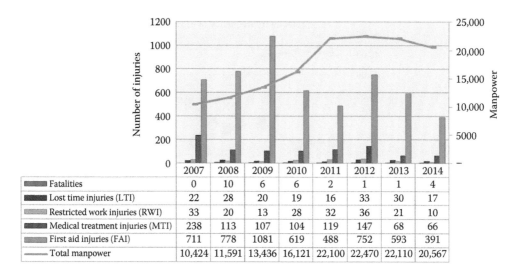

	2007	2008	2009	2010	2011	2012	2013	2014
Fatalities	0	10	6	6	2	1	1	4
Lost time injuries (LTI)	22	28	20	19	16	33	30	17
Restricted work injuries (RWI)	33	20	13	28	32	36	21	10
Medical treatment injuries (MTI)	238	113	107	104	119	147	68	66
First aid injuries (FAI)	711	778	1081	619	488	752	593	391
Total manpower	10,424	11,591	13,436	16,121	22,100	22,470	22,110	20,567

FIGURE 33.3 Graphs showing safety performance over the years.

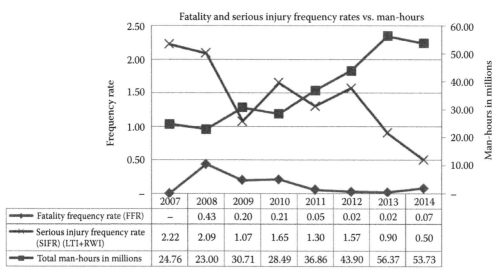

Fatality and serious injury frequency rates vs. man-hours

	2007	2008	2009	2010	2011	2012	2013	2014
Fatality frequency rate (FFR)	–	0.43	0.20	0.21	0.05	0.02	0.02	0.07
Serious injury frequency rate (SIFR) (LTI+RWI)	2.22	2.09	1.07	1.65	1.30	1.57	0.90	0.50
Total man-hours in millions	24.76	23.00	30.71	28.49	36.86	43.90	56.37	53.73

LTI: Lost time injury   RWI: Restricted work injury

MTI: Medical treatment injury   FAI: First aid injury

Frequency rate = No. of injuries x 1,000,000/Man-hours

Severity rate = Man-days lost/No. of LTIs

FIGURE 33.4 Graphs showing safety performance over the years.

INTERNATIONAL LEVEL

CONFERENCE OF CHIEF INSPECTORS OF MINES

For the first time in the known history of PNG, the *Conference of the Chief Inspector of Mines* was hosted in the country in September 2012. The conference was attended by chief inspectors from all states and territories of Australia as well as New Zealand. During the conference, few mines were visited by the chief inspectors. This international event lifted the profile of the country and the mining industry to a new level.

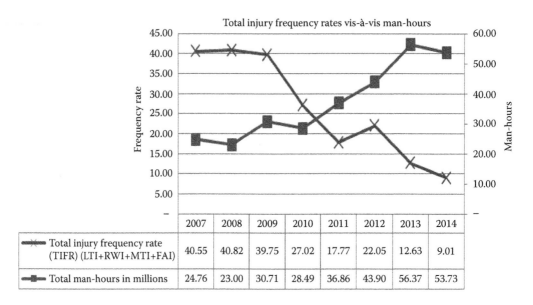

FIGURE 33.5   Graphs showing safety performance over the years.

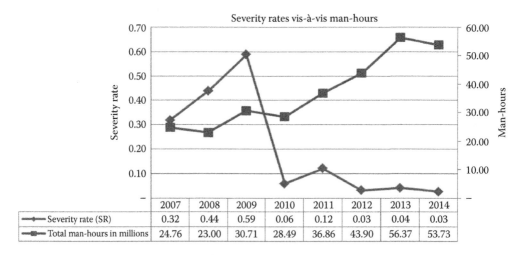

FIGURE 33.6   Graphs showing safety performance over the years.

In September 2013, the Mines Inspectorate as representative of PNG chaired the CCIM in Western Australia. It is worth noting that PNG used to be merely an "observer" on this forum for over half a century till 2011 and it had been a great turnaround in just 2 years.

PNG is on the threshold of becoming the world's first nation to mine massive sulfides from the deep sea floor. While this development would put PNG in the forefront of this new technology, it would also bring new challenges to effectively regulate these operations. The Mines Inspectorate had to secure membership for PNG on the International Ocean Mining Committee of the International Society of Offshore and Polar Engineers, which is represented by the United States, Germany, the Russian Federation, India, Korea, Japan, China, Bulgaria, Czech Republic, Slovakia, and Cuba.

PNG would now be able to interact and collaborate with other nations for better understanding of the dynamics of this new frontier and keep abreast with the latest developments. The Mines Inspectorate represents PNG at the Ocean Mining Symposia held every 2 years across the world.

INTERNATIONAL PROMOTION AND COLLABORATION

The Mines Inspectorate had also been projecting and promoting PNG on various international platforms from time to time to make it an attractive destination for investment and business. The Mines Inspectorate also engaged in the process of formalizing an memorandum of understanding (MOU) with India in the mining sector that would give PNG a distinct advantage in the fields of skilled manpower, technical education, professional training, mining regulation, trade, and business.

In January 2014, the Mines Inspectorate hosted two consultant mining regulators from Canada who had expressed their interest in visiting mines in PNG, thus making the country a preferred destination for the regulatory bodies of the developed world as well, and open new frontiers for international collaboration in the area of health and safety regulation.

SOME SPEAKING ACCOUNTS

The picture projected so far is only a tip of the iceberg as a significant segment still remains submerged in the souls, minds, and hearts of our countless stakeholders. Fully dedicated to bring a perceptible change in the regulatory environment in PNG, the Mines Inspectorate has scaled many a height in the last 5 years under the dynamic and visionary leadership of the chief inspector of mines.

THE NEAR FUTURE

While continuing with the current crusade, the Mines Inspectorate is positioning itself to undertake some more key initiatives over the next couple of years in the areas of

1. *Capacity building* in which national inspectors shall continue to receive specialized training, visit mines/other inspectorates, attend international conferences and workshops, etc.
2. *Development of safety regulation system* to enable capturing and processing of OHS data and information through an IT framework with a view to enhance performance, efficiency, and effectiveness
3. *Promoting health and fitness* in the mining workforce, school children, and surrounding communities through the application of an effective technique of yoga and meditation, which has the ability to holistically address one's physiological, psychological, and spiritual needs and can thus completely transform a character

Needless to emphasize, the Mines Inspectorate would need adequate budgetary and administrative support to paint the picture it has envisioned.

VISION AND THE VISIONARY

The Mines Inspectorate very much aspires to write many more new chapters to unlock the PNG potential and realize its mission of building a formidable regulatory institution that is adequately and appropriately resourced to effectively serve and lead the mining industry with a view to be reckoned as a force that inspires the rest of the world to emulate. Propelled by conviction, courage, and faith, the mission of change and transformation shall continue to make lasting impacts.

BIBLIOGRAPHY

Chief Inspector of Mines Report. 2012. Mineral Resources Authority, Papua New Guinea.
Chief Inspector of Mines Annual Report. 2013. Mineral Resources Authority, Papua New Guinea.
Chief Inspector of Mines Annual Report. 2014. Mineral Resources Authority, Papua New Guinea.

Section VI

Sustainable Materials, Fleets

34 Sustainability Allusion, Societal Marketing, and the Environmental Footprint of Hybrid Autos

Christopher S. Wright, Alex Kouznetsov, and Sarah Kim

CONTENTS

ABSTRACT

Those seeking to *tread lightly on this good earth* are encouraged, via *societal marketing*, to achieve that worthy goal by replacing their carbon-belching petrol autos with hybrid autos. The idea is that hybrid autos gain efficiencies-of-scope by combining the best aspects of petrol and electric motors. However, while the hybrid auto design may have been intended by auto manufacturers to be an efficient and elegant interim response to rising environmental concerns and the need for alternative-energy transportation, the hybrid reality may be much less propitious. Specifically, a strong and rising body of literature suggests that hybrid autos tend to involve so many trade-offs (in technology, safety, materials and efficiency) that they (for the moment) fall short of a viable, efficient and sustainable solution. Specifically, literature, product reviews and other analysis clearly show that (when a pairwise comparison of hybrid autos is made with equivalent petrol models) hybridisation of autos appears to deliver little to no significant fuel efficiencies. Further, given the current state of the relevant technologies, much of the vaunted efficiencies of hybrid autos should (*ceteris paribus*) be overwhelmed by significant increases in vehicle curb weight arising from an extra engine/generator, large battery packs, four regenerative braking mechanisms to recapture braking energy and the added circuit breakers/fuses and other advanced electronics needed for the electric- and/or charging-phase of hybrid autos. While these added items are becoming ever more efficient, it is unlikely that they are sufficiently efficient to deliver their services at the reported relatively small or negative weight differentials. One way to achieve these efficiencies is via a lightening of the auto (via reduced fuel tanks, smaller tyres, lighter materials, lighter seats, dashes, insulation/coatings, etc.) that are neither unique to hybrid autos nor are precluded from being done to petrol autos. However, these changes are not widely adopted in non-hybrid autos because consumers tend to object to the way such changes can also inflate an auto's cost and/or increase wear and tear and/or reduce the comfort of the ride, the quality of handling of the car and occupant safety in an accident.

It is equally important to note that all autos are becoming more efficient in raw material use, energy and in being recycled. Further, the unutilised scope in extant technology to enhance the *cradle-to-grave* green of autos suggests that exotic evolving technology (like hybridisation) is less about cost-effectiveness than some other inducement.

If malice and/or cupidity are discounted as the driving forces behind the overly enthusiastic societal marketing of hybrid autos, what remains is a common vital limitation of *green societal marketing*. Specifically, it often has a myopic appeal to limited aspects of consumer use that ignores the less appealing aspects and conflicts of a product that may arise during its life cycle (e.g. from creation through use and on to senescence/disposal).

This chapter uses an iterative process to illustrate how some sustainability claims can be illusory in nature. Sustainability in consumer products should involve a cradle-to-grave process with the cradle being resource sourcing of the consumer product, which then flows into product production, distribution and consumer use, and (ideally) be completed via closed-system recycling loops. Unfortunately, the product termination is often via leaky or illusory recycling loops that lead to disposal of the product, its parts and/or its more toxic elements in landfill tips, boneyards and/or via abandonment. The concept of life-cycle sustainability has proven the green credentials of such ephemeral (and presumed wasteful) products as paper cups and disposable diapers. This chapter will show the importance of life-cycle sustainability in evaluating the greenness of durable goods. In this case, hybrid autos from three manufacturers (i.e. BMW, Hyundai and Toyota) are pairwise contrasted with their full-petrol (non-hybrid) analogues. While hybrid autos appear to be an ideal example of societal marketing (e.g. they are marketed at a premium *as being an efficient and viable part of the sustainability solution*), hybrid autos appear to be *long on image* and *short on performance*. It is hoped that this chapter will incite and expand discussion on the net contribution of hybrid autos as part of environment sustainability.

INTRODUCTION

The hybrid automobile (auto) is relatively old technology. Its 1900 invention by Ferdinand Porsche predates, but is roughly contemporaneous with Karl Benz's (1885) release of the first practical internal combustion (petrol) auto (Cho, 2004; Daimler, 2015). The latter auto-motivation process overshadowed that of its rivals primarily because petrol (an inescapable by-product of separating crude oil into kerosene and other highly valued fractions) is a highly energy-to-weight efficient fuel, has no other significant use, and can produce more noxious and dangerous waste product if used for non-fuel related purposes. The internal combustion engine has, during the 115 years to 2015, driven from success to greater success (Pulkrabek, 2003) and other modes of auto motivation have tended to either wither away (e.g. steam power and hydrogen) or to be relegated to boutique-niche markets (e.g. electric/battery and LNG/LPG).[*]

Cost and general inefficiency of batteries contributed greatly to electric/battery and/or hybrid/battery autos failing to gain significant share in the auto market batteries (i.e. see Westbrook, 2001; Anderson and Anderson, 2005). In recent decades, ongoing debate over the relative influence of various human activities on global warming has encouraged three immensely powerful perceptions and revived interest in battery-dependent autos:

1. The myth that peak oil (output) was reached near the beginning of the twenty-first century
2. The perception that the price of petrol is rising exponentially
3. The efficiency gains and cost reductions in batteries arising from the ongoing shift from lead-acid batteries to nickel hydride to lithium to lithium–sulphur batteries

---

[*] LNG – liquid natural gas and LPG – liquid petroleum gas.

This study scrutinises *jointly and severally* the above syndrome of myths, perceptions and premises that spur faith in the superiority of battery-based autos. Also explored in this study is whether or not:

1. Hybridisation of autos can and does reduce the net harm to the environment
2. Any net reduction in environmental harm is cost effective?

REVIEW OF THE BASIC PREMISES BEHIND THE HYBRID AUTO REVIVAL

PREMISE 1: PEAK OIL HAS BEEN REACHED

While proto-peak oil adherents have warned about the collapse of oil production since as early as 1885, the current obsession with peak oil draws inspiration from the mid-twentieth century work of Hubbert (1949, 1950a,b, 1956). While Hubbert's forecast (i.e. that world oil output had peaked and would decline thereafter) tracked reality fairly well until 2000, Figure 34.1a shows that the U.S. oil

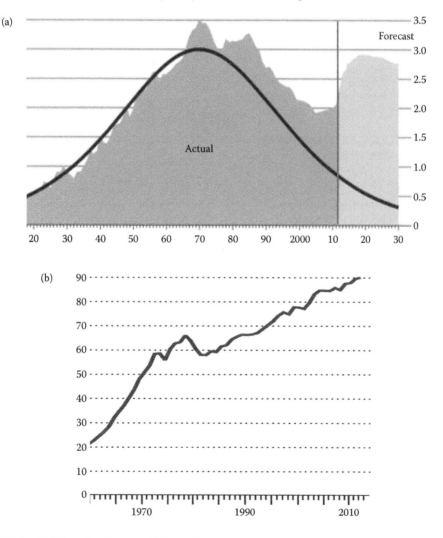

FIGURE 34.1 (a) US crude oil output (billions of barrels per day). (Adapted from *The Economist*, 5 March, 2013.) (b) World oil output (millions of barrels/day 1950–2013). (The output includes crude oil, shale oil, oil shale, oil sands and natural-gas liquids.) (Adapted from Gold, R., *Wall Street Journal*, 29 September, 2014.)

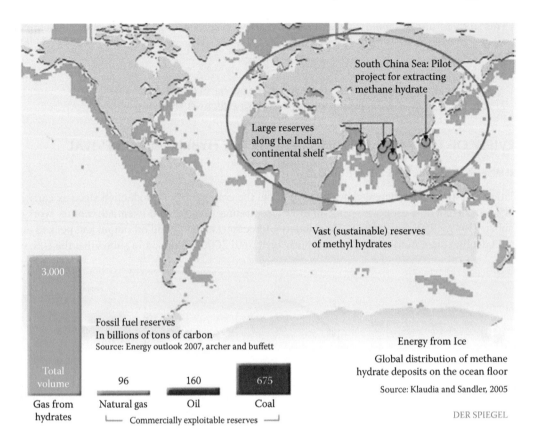

FIGURE 34.2 Known methyl (methane) hydrate deposits: estimated as 300% of currently exploitable fossil fuel reserves. (Adapted from Traufetter, G. 2007. Warning signs on the ocean floor: China and India exploit icy energy reserves. *Spiegel Online International*, available at http://www.spiegel.de/international/world/ warning-signs-on-the-ocean-floor-china-and-india-exploit-icy-energy-reserves-a-523178.html (accessed 22 June 2015); World Ocean Review. 2010. Methane hydrates. Living with the Oceans: A Report on the State of the World's Oceans, available at http://worldoceanreview.com/wp-content/downloads/worl/WOR1_english. pdf (accessed 22 February 2014).)

output rises exponentially over and above Hubbert's forecast. Figure 34.1b shows world oil production was in decline in the 1970s, but reversed into growth in the mid-1980s. Together Figure 34.1a and b clearly show that, while peak oil may one day occur that day has not yet arrived (see Simon, 1981 and 1996 for an explanation). Further, given the vast reserves of methyl hydrates (Figure 34.2), the sustainability of carbon-based fossil fuels is not likely to be a serious issue for generations.* However, global warming or even running out of oxygen are still potentially serious issues – perhaps more so, if the vast reserves of methyl hydrates are heavily exploited.

PREMISE 2: THE PRICE OF PETROL IS RISING EXPONENTIALLY

If peak oil is not yet an issue, why do prices appear to be accelerating (USDOE, 2015)? Key contributors to the perceived rises are inflation and taxes. Specifically, Figure 34.3 shows that when adjusted for the effects of inflation and taxes, the often touted exponential rise in petrol prices

---

* Methyl hydrate is a frozen form of methanol that forms under great pressure and/or cold. When warmed, methyl hydrate (methanol hydrate) releases methane gas and water.

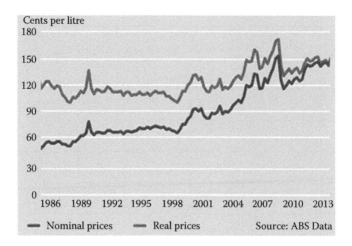

FIGURE 34.3   Australian retail petrol prices, tax excluded (1983–2013). (Reprinted with permission from *Downstream Petroleum 2013*. Australia: Australian Institute of Petroleum, available at http://www.aip.com.au/industry/dp2013/index.htm (accessed 26 April 2015). Copyright 2013, American Institute of Physics.)

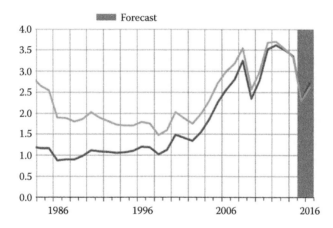

FIGURE 34.4   US retail petrol prices ($/US gal), tax included (1976–2015). (Adapted from EIA. 2015. *Short-Term Energy Outlook*. Washington, DC: US Energy Information Administration, available at http://www.eia.gov/forecasts/steo/realprices/ (accessed 22 February 2015).)

flattens out to range from a mere −0.42% to 0.53% per year in Australia (depending on whether the run is from 1983 to 2016 or is from 1983 to 2013). Figure 34.4 shows that in the United States, real pump price for petrol during the four decades to 2015 fell by an annual rate of 0.22%.

Figures 34.3 and 34.4 show that the real price of petrol has been essentially static for a very long time – it is taxes and inflation that cause the exponential rise in the after-tax normative price of petrol. The tax issue is important, in that if an alternative fuel becomes popular, it is likely to also be loaded up with road and other taxes and, as a result, will lose any tax-related relative economic advantage. In many countries, when auto owners switched from petrol to LNG or LPG, the governments imposed road tax on those fuels when used in an auto and those who switched then found that the after-tax LNG and LPG were inefficient (having a low energy-to-weight ratio), uneconomic and more dangerous than petrol.[*]

---

[*] High pressure LNG or LPG tanks can burst in accidents or due to maintenance failures and cause impact injuries and frostbite and any released gas is not only highly flammable, but also can disperse faster and into more places than liquid petrol (which must first evaporate to become flammable).

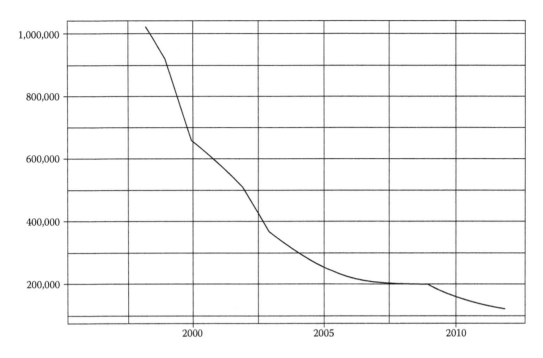

FIGURE 34.5 Cost of Li-ion batteries (¥/kAh). (Adapted from SR. 2012. Automotive lithium-ion battery costs aren't going to fall 'dramatically'. Selective Rationality, available at http://selectiverationality.com/auto-lithium-ion-battery-costs-arent-going-to-fall-dramatically/ (accessed 23 February 2015).)

Premise 3: Batteries Have Experienced Major Efficiency Gains and Cost Reductions

As noted previously, the domination of the auto market by internal combustion engines was largely due to the high cost and general inefficiency of batteries. As Figure 34.5 shows, the cost of lithium-ion (Li-ion) batteries has fallen and average of 13.6% (compounded) per year from the early 1990s to 2012. While this ongoing drop in costs is spectacular, it is important to note that the cost of generating 1 kW of energy is nearly 50% higher for a Li-ion battery than it is for petrol (Table 34.1).

TABLE 34.1
Cost of Generating 1 kW of Energy in the United States

Fuel Type	Equipment to Generate 1 kW	Life Span	Cost of Fuel per kWh	Total Cost per kWh
Li-ion	$1,000/kW (based on 10 kW	2,500 h (replacement	$0.10	$0.50 (replacement and
For vehicular use	battery at $10,000)	cost $0.40/kW)		$0.10/kWh)
Gasoline engine	$30/kW (based on IC engine	4,000 h (replacement	$0.33	$0.34
For vehicular use	at $3,000/100 kW)	cost $0.01/kW)		
Fuel cell	$3,000–7,500	2,000 h	$0.35	
• Portable use		4,000 h	>	$1.85–4.10
• Mobile use		40,000 h	>	$1.10–2.25
• Stationary use			>	$0.45–0.55
Electricity	All inclusive	All inclusive	$0.10	$0.10
Electric grid				

Source: Adapted from Buchmann, I. 2015b. Learning the basics about batteries. *Battery University*, Cadex Electronics Inc. Available at http://batteryuniversity.com/learn/article/cost_of_power. Accessed February 13, 2015.

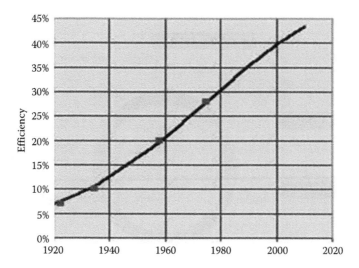

FIGURE 34.6 Available efficiency of the internal combustion engine (1920–2020). (Adapted from Ferreira, O. C. 2015. Efficiency of internal combustion engines, available at http://www.ecen.com/content/eee7/motoref. htm (accessed 23 February 2015).)

Also, the most common battery used in hybrid autos is the nickel-metal hydride (NiMH) battery at $2000–$3000 per battery pack (Buchmann, 2015a). While NiMH batteries are in the late-maturity or early-decline stage of their product life cycle, they are not being phased out because, at $10,000 each, Li-ion battery packs are still far too costly. For now, the more efficient cobalt Li-ion batteries are still too dangerous for general commercial use (Buchmann, 2015a). It should be noted that battery packs for hybrid autos are almost always designed for extended life rather than maximum efficiency. Further, as illustrated by Figure 34.6, significant improvements to internal combustion engines are also in play. One such improvement is the ceramic engine on which the Figure 34.6 extrapolations (past 1980) are based. Ceramic engines and/or ceramic-coated engines (per LF, 2015) bring a number of benefits: an expected efficiency of up to twice the current 27%–30%; a power-output boost to 150%; reduced wear and tear; a further bonus of not needing a cooling system, and a weight savings of over 90% (from 600 kg to under 60 kg). A key difficulty with the ceramic internal combustion engines, however, is the hardness of their parts tends to make them brittle – the brittleness issue has been reduced but not fully resolved. In summary, hybrid/battery tech is competing against a moving target and likely to remain *a dollar short and a day late* for the near-to-intermediate future. While hybrid/battery technology and costs are improving, improvements made on the internal combustion technology also should not be overlooked (e.g. Schäfer and Jacoby, 2006; Youngmin et al., 2009; Berggren and Magnusson, 2012). One area that requires much further study is the cradle to grave cost of the battery packs used in both the hybrid and non-hybrid autos.

ALL AUTOMOBILES ARE BECOMING GREENER

As noted in Figures 34.5 and 34.6, batteries and petrol engines are becoming ever more efficient and the potential efficiency is far above what is currently applied. This later observation leads to the notion that the scope for automobiles (hybrid, non-hybrid and electric) to be greener is currently very large (as shown in Figure 34.7) and is growing.

Figure 34.8 shows that while steel is still the most important material in autos, its share of the curb weight has been declining by a compounded rate of around 1% per annum for the past four decades and, if the projections of accelerating aluminum use are valid, that decline is likely to accelerate over the next two decades. Given that engine blocks and cooling systems currently comprise 15%–20% of the steel use in autos, the projected steel use of 40% and 16% for, respectively, 2020

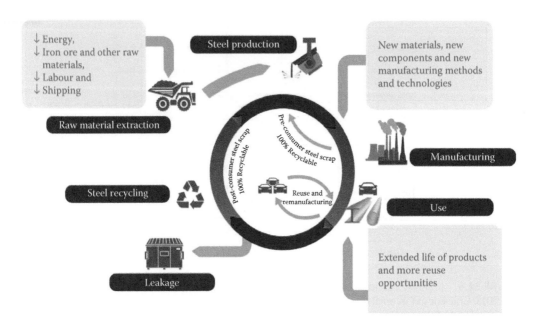

FIGURE 34.7 Increasing environmental efficiency of steel and other auto inputs. (Adapted from Worldsteel. 2010. *Life Cycle Assessment in the Steel Industry: A Position Paper Issued by the World Steel Association,* p. 2, available at http://www.worldsteel.org/dms/internetDocumentList/downloads/publications/LCA-position-paper/document/LCA%20position%20paper.pdf (accessed 7 July 2015).)

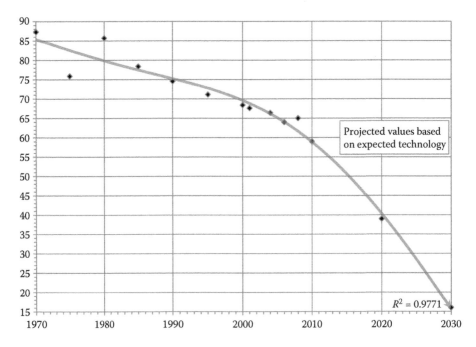

FIGURE 34.8 Contribution of steel to the curb weight of North American automobiles. (Data from USGS. 2006. Steel stocks in use in automobiles in the United States. United States Geological Services, United States Department of the Interior, available at http://pubs.usgs.gov/fs/2005/3144/fs2005_3144.pdf (accessed 22 June 2015); Schnatterly, J. 2012. *Trends in Steel Content of N. American Auto,* available at http://www.autosteel. org/~/media/Files/Autosteel/Great%20Designs%20in%20Steel/GDIS%202012/Trends%20in%20Steel%20 Content%20of%20North%20American%20Auto.pdf (accessed 2 July 2015).)

and 2030 should be achievable if ceramic engines become viable. In counter-point, steel is becoming stronger and thinner and is much cheaper and stronger than aluminium. However, composites become ever cheaper and they may prove to be a materials wildcard in the composition of autos. Thus, from a cradle-to-grave perspective, as automobiles become lighter, they should *ceteris paribus* become greener. Further, there is a lot of scope for further greening of the automobile within existing technologies (see While Book, 2004). Finding out why that scope remains unexplored while hybridisation is embraced is well worth examining in future research.

GREEN SOCIETAL MARKETING OF ENVIRONMENTAL BENEFITS

There have been extensive discussions concerning green matters for more than 40 years and marketing is increasingly focussed on buyers' awareness of societal and/or environmental issues (e.g. Prothero, 1990; Kang and James, 2007; Matear and Dacin, 2010; Davis, 2013).

> Firms have long recognized that it is also important to evaluate products and practices from a societal perspective. 'Should a product be sold?' The first idea reflects a managerial orientation and what must be done to sell a product; the second idea reflects a societal orientation and the impact of selling a product. In relation to the second idea, the societal marketing concept (SMC) was introduced in 1972 (Kang and James, 2007, p. 301).

The societal marketing concept (SMC) was introduced as a notion that an organisation should, after establishing the a fuller picture of the needs and wants of their consumers, deliver satisfaction to those customers more effectively and efficiently than their rivals, by also maintaining or improving the well-being of society (Philip, 1972). In the decades since its introduction, SMC has become a central concept in contemporary marketing.

Crane and Desmond (2002, p. 553) suggest that SMC is an extension of classical marketing theory:

> …who's (largely unstated) moral basis rests on psychological egoism towards a more morally robust basis of ethical egoism. …[Furthermore] while the SMC may hold out some possibilities for developing a more moral approach in some instances, its fundamental assumptions, coupled with its very dependence on a set of rational-instrumentalist justifications, will not prevent, and may even lead to, morally dubious and even dangerous outcomes.

In this regard, *greenwashing*, which is the selective '…disclosure of positive information about a company's environmental or social performance, while withholding negative information on these dimensions' (Lyon and Maxwell, 2011, p. 5), may be used as a way of convincing consumers of the firm's concern for environmental sustainability. In other words, firms may subvert the consumer disposition towards using a greener-and-environmentally safer product to meet a more important goal of satisfying their shareholders. In this case, the latter could become the dominant goal and the former can be turned into a tool of achieving this goal. This appears to be pertinent in automobile manufacturing companies as they increasingly seek to be seen to be serving society's increasing awareness of the environmental risks of driving.

Gärling and Thøgersen (2001), in their review of the *marketing of electric vehicles*, argue that people with heightened concern for the environment may form a rich market for manufacturers as such consumers may be less sensitive to product deficiencies. This effect implicitly suggests that a *green emphasis* may serve as an enticing selling point to this growing segment of the population. Dwyer (2008) gives several examples of car manufacturers overemphasising the environmental attractiveness of their vehicles, one of which was a newspaper advert from Lexus (part of Toyota) claiming that the firm's petrol-electric hybrid sports utility vehicle, the RX-400 h, made 'environmental sense' when in fact its CO_2 emissions, at 192 g/km, were not exactly low. Furthermore, Toyota and Honda, as Dwyer (2008) also argued, have gained public relations (PR) benefits from making petrol-electric hybrid cars such as the Prius and Civic.

Equally, it should be noted that petrol-auto manufacturers may also make positive environmental assertions, often without much substantiation of those claims. However, consumer environmental sensitivity may give even more incentive for hybrid auto manufacturers to overstress the positive environmental aspects of their hybrid autos. Specifically, environmental sustainability and fuel efficiency are primary selling points of hybrid autos, particularly in justifying higher the price and often lower or even marginal performance of those vehicles (Oliver and Rosen, 2010).

A number of blogs and popular magazines (e.g. Deaton, 2008; George and Hall-Geisler, 2009; Edmunds, 2014; Almeida, 2015; Autos, 2015; Cars Direct, 2015; Elton, 2015; Jones, 2015), also, suggest that environmental claims for hybrid auto tend to be more of a *greenwash* than a focus on *genuine performance*. While the academic rigor of such commentary *often leaves much to be desired*, it is current, topical and often useful as reflection of evolving attitudes/trends in the marketplace.

This chapter conducted a pairwise comparison of hybrid autos and their full-petrol (non-hybrid) analogues from three global auto manufacturers, namely BMW, Hyundai and Toyota. These manufacturers were selected for study based on their representativeness in terms of region of origin, types and popularity of autos manufactured and perceived quality of their autos. The analysis of the data collected focusses on external size, vehicle weight, engine type and size, fuel efficiency, tyres and rims and interior quality and capacity.

Several (unpublished) informal comparisons done by management accounting students in 2010, suggested that the hybridisation of auto motors not only failed to deliver net savings on petrol use, but also caused more harm than their petrol analogue autos. Specifically, the informal studies attributed most of the reduced petrol use to the hybrids being artificially lightened to more than offset the increased weight from two motors, four mechanisms to recapture braking energy and a large battery. The artificial lightening of the hybrid autos was suggested by smaller fuel tanks, smaller tyres and an under-powered petrol motor; along with a mean-meagre interior and light-weight body and interior (via the use of alloys and composite materials, etc.) and/or over-inflating the tyres to reduce the rolling resistance of hybrid autos (Cobb, 2012).

The current analysis, in Table 34.2, shows that hybrid autos have improved greatly and/or their manufacturers and retailers have become much better at obfuscating evidence of their trade-offs.

TABLE 34.2

Relative Performance of Hybrid Autos (per Tables 34A.1 through 34A.5, Appendix 34A)

Vehicle size	The reviewed hybrid and non-hybrid pairs are identical in terms of length, width, height and wheelbase for Toyota and BMW, but for the Hyundai pair, the hybrid was smaller in each of the dimensions by 0.4%–1.6%
	While the curb weight for each of the three manufacturers was slightly higher for the hybrid, the weights (all a little over an extra 100 kg) were significantly lower than the 200–300 kg expected – given the hybrid reality of an extra motor, four braking-energy-recapture mechanisms and a large heavy battery pack
Engine	The reviewed hybrid and non-hybrid pairs have near identical petrol engines
Fuel efficiency	The claims for city driving appear extravagant with Toyota and BMW each claiming to nearly halve the fuel usage in the city. The highway driving is moderate at best and a small increase in fuel usage for BMW
	The combined and extra urban are so dependent on assumptions as to be meaningless
Fuel tank	The BMW and Hyundai hybrids have fuel tanks which are 7.1% smaller than their petrol analogues and the BMW is a close 5.0% smaller
	While the smaller tank weight difference is small (about 3.5 kg in fuel), the small amounts add up
Tyres	The reviewed hybrid and non-hybrid pairs have identical tyres
Interior volume	The interior volumes: Toyota = were not given; BMW were identical and Hyundai were significantly smaller for the hybrid
Boot capacity	Are 18%–26% smaller in the hybrid autos

While the increased fuel efficiency for hybrid autos appears to be very large for use within the city, it is still very unclear how much of it is due to the hybridisation of the engines and how much is a trade-off from lightening the hybrid autos in ways which also can be done to petrol autos, that may compromise the integrity and/or handling of hybrid autos, and may compromise the safety and comfort of the occupants of hybrid autos. In summary, hybrid autos could well be a societal marketing allusion.

WHAT DRIVES THE SUSTAINABILITY ALLUSION

Who benefits from the creation and marketing of an allusion of sustainability? If malice and/or cupidity are discounted as the driving forces behind the overly enthusiastic societal marketing of hybrid autos, what remains is a vital limitation of *green societal marketing*. Specifically, it often has a myopic appeal to limited aspects of consumer use that ignores the less appealing aspects and conflicts of a product that may arise during its life cycle (e.g. from creation through use and on to senescence and disposal). This is particularly true for the very environmentally unfriendly metals used in the rechargeable battery packs of hybrid autos.

Further, when a decision-making process is dominated by fear and worry, as evident in the four premises challenged earlier, those processes are often easy to manipulate and the outcomes can often be far less than ideal. Those marketing on the basis of green claims and credentials in the form of greenwash should have a duty to clearly display those credentials and not hide behind statistical exaggerations, guile, obfuscation and other mendacities (Key and Popkin, 1998).

CONCLUSIONS

Hybridisation of autos is being promoted as a cost-effective solution to *peak oil* shortages and greenhouse gas reduction. This research used extant literature to challenge and dismiss the claims/beliefs underpinning the marketing of hybridised autos. What conclusions can be drawn from this study?

1. World oil output continues to rise. Thus, the many claims that peak oil (output) has occurred or will soon occur are currently without substance. The rising world oil output, when combined with ever more carbon-based energy substitutes (e.g. methyl hydrate) and increasing energy efficiency mean that the ever-deepening scarcity suggested by peak oil predictions is unlikely to be an issue for the foreseeable future.
2. The price of petrol has in real terms (after taxes are blacked out) remained essentially constant over the past three decades and (as noted in item a, above) is likely to remain (in absolute, tax-free terms) constant for the foreseeable future.
3. Energy delivery by rechargeable batteries may one day be sufficiently cost effective to represent a competitive threat to the internal combustion engine, but that day is not today, nor does it look likely anytime within the immediate or mid-term future.
4. All automobiles are becoming more efficient and, therefore, more green in relative terms. The observation that automobiles are becoming larger and heavier reflects a personal choice among trade-offs made by consumers rather than an economic or a technological exigency.
5. On balance, whether hybridising autos is a cost effective way to reduce greenhouse gas emissions is, at best, debateable with the debate leaning towards hybrid autos appearing to be more of an allusion to, than the reality of, sustainability.

The extravagant claims made for the emissions reduction capability of hybrid autos are not as easily refuted as they were only a few years ago when the hybrid autos were significantly lighter than their petrol counterparts. Specifically, hybrid autos are created by adding elements which tend to be very heavy (e.g. a second motor, a battery pack, four regenerative braking mechanisms to recapture braking energy and added circuit breakers/fuses and other advanced electronics) and should in total add 300–500 kg to the curb weight. As this research shows, that expected extra weight has been offset to a great degree. Such lightening of autos is not exclusive to hybrid autos but it can also be done to petrol autos, may involve potential compromises to the integrity and handling of hybrid autos and may compromise the safety and comfort of occupants of hybrid autos. Yet, these claims enable marketers to successfully appeal to environmentally concerned consumers.

A key issue for future research is isolating the effect of hybridisation: how much of the vaunted fuel efficiency of hybrid autos is due to hybridisation of the engines and how much arises from measures unrelated to the hybridisation? It is also vital that the unspecified measures used in lightening hybrid autos (e.g. light alloys and composite materials) be examined for undesirable side effects (e.g. reduced structural integrity; premature aging and wear and tear and flammability, toxic gas formation when burnt). Another consideration for future research is the cradle-to-grave life-cycle sustainability of hybrid autos and their petrol analogues in terms of the costs, emissions and other environmental harm.

APPENDIX 34A: TABLES

Data from the technical brochures of the auto manufacturers.

TABLE 34A.1
Dimension and Weight

Attributes		Toyota		BMW		Hyundai	
		Hybrid/Petrol	**Difference**	**Hybrid/Petrol**	**Difference**	**Hybrid/Petrol**	**Difference**
Dimension	Length (mm)	4815/4815	0	4624/4624	0	4820.92/4853.94	−33.02
	Width (mm)	1825/1825	0	1811/1811	0	1833.88/1864.36	−30.48
	Height (mm)	1470/1470	0	1429/1429	0	1465.58/1475.74	−10.16
	Wheelbase (mm)	2775/2775	0	2810/2810	0	2794/2804.16	−10.16
Weight	Curb weight (kg)	1555/1465	90	1655/1520	135	1591–1648/1475	116–173

TABLE 34A.2
Engine

Models	Toyota		BMW		Hyundai	
	Hybrid Camry H	Non-Hybrid Camry Altise Sedan	Hybrid Active Hybrid 3	Non-Hybrid 3 Sedan (335i)	Hybrid SONATA HYBRID	Non-Hybrid SONATA SE
Engine	• Engine type petrol • Engine code 2AR-FXE • Configuration in-line • Light alloy head • Variable valve timing • Standard and type VVT-i • Valve gear type DOHC • Compression ratio 12.5:1 • Secondary engine power • Maximum power kW 118	• Engine type petrol • Engine code 2AR-FE • Configuration in-line • Light alloy head • Variable valve timing standard and dual VVT-i • Valve gear type DOHC • Compression ratio 10.4:1	Max speed (km/h) 250 BMW TwinPower Turbo six-cylinder in-line petrol engine, one twin-scroll turbocharger with valvetronic, double-VANOS and high precision injection	Max speed (km/h) 250 BMW TwinPower Turbo six-cylinder in-line petrol engine combining one twin-scroll turbocharger with valvetronic, double-VANOS and high precision injection	• Compression ratio: 13.0:1 • Displacement (L): 2.4	• Compression ratio: 11.3:1 • Displacement (L): 2.4
Cylinders	Four cylinders, in-line, chain driven DOHC, variable valve timing, two balance shafts, alloy block and cross-flow alloy head	Four cylinders, in-line, chain driven DOHC, variable valve timing, two balance shafts, alloy block and cross-flow alloy head	6	In-line/6	• 2.4 L Atkinson cycle DOHC 4-cylinder engine: standard	Inline 4-cylinder
Valves	16	16	24	24	16	16
Capacity (cc)	2494	2494	2979	2979		
Maximum power (kW/hp/rpm)	118 kW at 5700 rpm	133 kW at 6000 rpm	250/306/5800–6000	225/306/5800–6000	Horsepower at RPM: 159 at 5500	Horsepower at RPM: 185 at 6000

(Continued)

TABLE 34A.2 (*Continued*)
Engine

Models	Toyota		BMW		Hyundai	
	Hybrid Camry H	Non-Hybrid Camry Altise Sedan	Hybrid Active Hybrid 3	Non-Hybrid 3 Sedan (335i)	Hybrid SONATA HYBRID	Non-Hybrid SONATA SE
Max torque (N m/rpm)	213 Nm at 4500 rpm	231 Nm/4100 rpm	450/1200–5000	400/1200–5000	Torque (lb.-ft.) at RPM: • 154 at 4500	Torque (lb.-ft.) RPM: 178 at 4000
Electric motor/ battery	• Series parallel THS II • Max voltage: 650 V • Max power 105 kW • Max torque 270 Nm • Three driving modes: 'normal', 'ECO' and 'EV' • NiMH battery		• 40 kW electric engine • Li-ion battery		• Interior-permanent magnet synchronous motor standard • Horsepower at RPM 47 at 1630–3000 • Torque at RPM 151 at 0–1630 • Lithium polymer battery 270 V	
Hybrid system combined	Combined power maximum output 151 kW				Combined hybrid system horsepower at RPM 199 at 5500	

Note: DOHC = Double overhead camshaft.

TABLE 34A.3
Fuel Efficiency

Models	Toyota			BMW			Hyundai		
	Hybrid Camry H	Non-Hybrid Camry Altise Sedan	% Difference	Hybrid Active Hybrid 3	Non-Hybrid Match 3 Sedan (335i)	% Difference	Hybrid SONATA HYBRID (US)	Non-Hybrid Match SONATA SE (US)	% Difference
City	5.7 L/100 km	10.9 L/100 km	−47.7	5.3 L/100 km	10.2 L/100 km	−48.0	7.8	11.3	−30.9
Highway				6.4 L/100 km	5.5 L/100 km	+0.2	7.1	7.6	−9.3
Combined	5.2 L/100 km	7.8 L/100 km	−33.3	5.9 L/100 km	7.2 L/100 km	−18.1	7.4	9.7	−23.7
Extra urban	4.9 L/100 km	6 L/100 km	−18.3	–	–		–	–	
Emission control level	• Combined CO_2 g/km 121	• Combined CO_2 g/km 183	−33.9	CO_2 emission (g/km) 139	CO_2 emission (g/km) 169	−17.8	CO_2 emission (g/km) 147	CO_2 emission (g/km) 160	−33.7
Fuel system	• Multi-point injection • Full mgmt ctrl ignition	• Multi-point injection • Full mgmt ctrl ignition					Multi-point fuel injection: standard	Gasoline direct injection: standard	
Fuel tank capacity	65	70	−7.1	57	60	−5.0	65	70	−7.1

TABLE 34A.4
Tyres

	Toyota Camry H and Altise Sedan		BMW Hybrid 3 and Sedan 335i		Hyundai	
	Hybrid	**Non-hybrid**	**Hybrid**	**Non-hybrid**	**Hybrid**	**Non-hybrid**
Tyres and rims	Tyres – front • Tyre width (mm) 215 • Tyre profile 60 • Tyre rating V Tyres – rear • Tyre width (mm) 215 • Tyre profile 60 • Tyre rating V Wheels – front • Rim type alloy • Rim diameter (inch) 16 • Rim width (inch) 6.5 Wheels – rear • Rim type alloy • Rim diameter (inch) 16 • Rim width (inch) 6.5 Spare wheel – internal • Spare wheel type full size • Rim type steel • Spare wheel tyre size 215/60 R16	Tyres – front • Tyre width (mm) 215 • Tyre profile 60 • Tyre rating V Tyres – rear • Tyre width (mm) 215 • Tyre profile 60 • Tyre rating V Wheels – front • Rim type alloy • Rim diameter (inch) 16 • Rim width (inch) 6.5 Wheels – rear • Rim type alloy • Rim diameter (inch) 16 • Rim width (inch) 6.5 Spare wheel – internal • Spare wheel type full size • Rim type steel • Spare wheel tyre size 215/60 R16 95V	18" BMW light alloy; • Front: 8 J × 19", 225/40, R19 • Rear: 8.5 J ×19" with 255/35 R19 run-flat safety tyres	18" BMW light alloy • Front: 8 J × 19", 225/40, R19 • Rear: 8.5 J × 19" with 255/35 R19 run-flat safety tyres	16" alloy wheels with 205/65R16 tyres	16" alloy wheels with 205/65R16 tyres

TABLE 34A.5
Interior Volume

Interior Volume (mm or L)	Toyota Hybrid/ Petrol	Toyota % Difference	BMW Hybrid/ Petrol	BMW % Difference	Hyundai Hybrid/ Petrol	Hyundai % Difference
Shoulder room front	–	–	1400/1400	0.0	1471/1471	0.0
Shoulder room rear	–	–	1400/1400	0.0	1440/1435	5.0
Elbow room front	–	–	1451/1451	0.0	–	–
Elbow room rear	–	–	1458/1458	0.0	–	–
Head room front	–	–	1023/1023	0.0	1016/1026	−1.0
Head room rear	–	–	957/957	0.0	960/965	−0.5
Leg room front	–	–	–	–	1156/1156	0.0
Leg room rear	–	–	–	–	879/904	−2.8
Hip room front	–	–	–	–	1402/1402	0.0
Hip room rear	–	–	–	–	1394/1425	−2.2
Boot capacity (L)	421/515	−18.6	390/480	−18.8	342/462	−26.0

REFERENCES

AIP. 2013. *Downstream Petroleum 2013*. Australia: Australian Institute of Petroleum, available at http://www.aip.com.au/industry/dp2013/index.htm (accessed 26 April 2015).

Almeida, A. 2015. Do hybrid cars get better gas mileage? *Cars Direct*, available at http://www.carsdirect.com/green-cars/do-hybrid-cars-get-better-gas-mileage (accessed 30 January 2015).

Anderson, J. and C. D. Anderson. 2005. *Electric and Hybrid Cars: A History*. London, UK: McFarland & Co.

Autos. 2015. Hybrid car cons: Drawbacks to know before buying, available at http://www.autos.com/car-buying/hybrid-car-cons-drawbacks-to-know-before-buying (accessed 30 January 2015).

Berggren, C. and T. Magnusson. 2012. Reducing automotive emissions – The potentials of combustion engine technologies and the power of policy. *Energy Policy*, 41: 636–643.

Buchmann, I. 2015a. Cadex Electronics Inc., available at http://batteryuniversity.com/learn/article/are_hybrid_cars_here_to_stay (accessed 19 June 2015).

Buchmann, I. 2015b. Learning the basics about batteries. *Battery University*, Cadex Electronics Inc., available at http://batteryuniversity.com/learn/article/cost_of_power (accessed 13 February 2015).

Cars Direct. 2015. The pros and cons of hybrid cars. *Cars Direct*, available at http://www.carsdirect.com/green-cars/major-pros-and-cons-of-buying-used-hybrid-cars (accessed 30 January 2015).

Cho, D. 2004. Prescient Porsche. *Technology Review*, 107(5): 84.

Cobb, V. 2012. Low rolling resistance tire primer. *Hybrid Cars*, available at http://www.hybridcars.com/low-rolling-resistance-tire-primer-50256 (accessed 30 January 2015).

Crane, A. and J. Desmond. 2002. Societal marketing and morality. *European Journal of Marketing*, 36(5/6): 548–569.

Daimler, 2015. *The Birth of the Automobile: Benz Patent Motor Car, The First Automobile (1885–1886)*. Available at http://www.daimler.com/dccom/0-5-1322446-1-1323352-1-0-0-1322455-0-0-135-0-0-0-0-0-0-0-0.html. Accessed February 13, 2015.

Davis, I. 2013. How (not) to market socially responsible products: A critical research evaluation. *Journal of Marketing Communications*, 19(2): 136–150.

Deaton, J. P. 2008. Are some hybrids just greenwashed? *HowStuffWorks*, available at http://auto.howstuffworks.com/hybrid-greenwash.htm (accessed 30 January 2015).

Dwyer, J. 2008. Trail of hot air [industrial pollution]. *Engineering and Technology*, 3(20): 62–65.

Edmunds. 2014. The real costs of owning a hybrid: Do fuel savings offset a higher price? *Edmunds*, available at http://www.edmunds.com/fuel-economy/the-real-costs-of-owning-a-hybrid.html (accessed 30 January 2015).

EIA. 2015. *Short-Term Energy Outlook*. Washington, DC: US Energy Information Administration, available at http://www.eia.gov/forecasts/steo/realprices/ (accessed 22 February 2015).

Elton, B. 2015. The truth about hybrids, available at http://www.motorists.org/ma/1100027274.html (accessed 30 January 2015).

Ferreira, O. C. 2015. Efficiency of internal combustion engines, available at http://www.ecen.com/content/eee7/motoref.htm (accessed 23 February 2015).

Gärling, A. and J. Thøgersen. 2001. Marketing of electric vehicles. *Business Strategy and The Environment*, 10(1): 53–65.

George, P. E. and K. Hall-Geisler. 2009. 5 reasons not to buy a hybrid 31. *HowStuffWorks*, available at http://auto.howstuffworks.com/5-reasons-to-not-buy-hybrid.htm (accessed 30 January 2015).

Gold, R. 2014. Why peak-oil predictions haven't come true: More experts now believe technology will continue to unlock new sources. *Wall Street Journal*, 29 September, available at http://www.wsj.com/articles/why-peak-oil-predictions-haven-t-come-true-1411937788 (accessed 22 February 2015).

Hubbert, M. K. 1949. Energy from fossil fuels. *Science*, New Series, 109(2823), 103–109. http://www.jstor.org/stable/1676618?seq=1#page_scan_tab_contents (Accessed 13 June 2016).

Hubbert, M. K. 1950a. Discussion: Estimates of undiscovered petroleum reserves by A.I. Levorsen. *Proceedings of the United Nations Scientific Conference on the Conservation and Utilization of Resources*, Vol. 1. pp. 103–104.

Hubbert, M. K. 1950b. Energy from fossil fuels. *Centennial*, Washington, DC: American Association for the Advancement of Science, pp. 171–177.

Hubbert, M. K. 1956. *Nuclear Energy and the Fossil Fuels*. Publication No. 95, Houston, Texas: Shell Development Company, 40 pp., http://www.oilcrisis.org/CC950B74-C5D3-458A-939A-14A644EDEE0C/FinalDownload/DownloadId-92719B724DF90B78C1F540F3D2C649B7/CC950B74-C5D3-458A-939A-14A644EDEE0C/hubbert/1956/1956.pdf (retrieved 5 March 2015).

Jones, T. 2015. Problems with hybrid cars. *Street Directory*, available at http://www.streetdirectory.com/travel_guide/49132/cars/problems_with_hybrid_cars.htm (accessed 5 September 2014).

Kang, G. and J. James. 2007. Revisiting the concept of a societal orientation: Conceptualization and delineation. *Journal of Business Ethics*, 73(3): 301–318.

Key, S. and S. J. Popkin. 1998. Integrating ethics into the strategic management process: Doing well by doing good. *Management Decision*, 36(5): 331–338.

LF. 2015. Uncooled engines. *Global Warming Mitigation and Adaptation*. The Litus Foundation, available at http://litusfoundation.org/uncooled-engines/ (accessed 23 February 2015).

Lyon, T. and J. Maxwell. 2011. Greenwash: Corporate environmental disclosure under threat of audit. *Journal of Economics and Management Strategy*, 20: 3–41.

Matear, M. and P. A. Dacin. 2010. Marketing and societal welfare: A multiple stakeholder approach. *Journal of Business Research*, 63(11): 1173–1178.

Oliver, J. D. and D. E. Rosen. 2010. Applying the environmental propensity framework: A segmented approach to hybrid electric vehicle marketing strategies. *Journal of Marketing Theory and Practice*, 18(4): 377–393.

Philip, K. 1972. What consumerism means for marketers. *Harvard Business Review*, 50: 48–57.

Prothero, A. 1990. Green consumerism and the societal marketing concept: Marketing strategies for the 1990's. *Journal of Marketing Management*, 6(2): 87–103.

Pulkrabek, W. 2003. *Engineering Fundamentals of the Internal Combustion Engine*. Upper Saddle River, New Jersey: Prentice-Hall, available at http://www.rmcet.com/lib/E-Books/Mech-auto/Engineering%20Fundamentals%20of%20IC%20Engines%20(WW%20Pulkrabek).pdf (accessed 22 February 2015).

Schäfer, A. and H. D. Jacoby. 2006. Vehicle technology under CO_2 constraint: A general equilibrium analysis. *Energy Policy*, 34(9): 975–985.

Schnatterly, J. 2012. *Trends in Steel Content of N. American Auto*, available at http://www.autosteel.org/~/media/Files/Autosteel/Great%20Designs%20in%20Steel/GDIS%202012/Trends%20in%20Steel%20Content%20of%20North%20American%20Auto.pdf (accessed 2 July 2015).

Simon, J. L. 1981. *The Ultimate Resource*. Princeton, New Jersey: Princeton University Press.

Simon, J. L. 1996. *The Ultimate Resource II*. Princeton, New Jersey: Princeton University Press, available at http://www.juliansimon.com/writings/Ultimate_Resource/.

SR. 2012. Automotive lithium-ion battery costs aren't going to fall 'dramatically'. Selective Rationality, available at http://selectiverationality.com/auto-lithium-ion-battery-costs-arent-going-to-fall-dramatically/ (accessed 23 February 2015).

The Economist. 2013. Focus: Peak oil. *The Economist*, 5 March.

Traufetter, G. 2007. Warning signs on the ocean floor: China and India exploit icy energy reserves. *Spiegel Online International*, available at http://www.spiegel.de/international/world/warning-signs-on-the-ocean-floor-china-and-india-exploit-icy-energy-reserves-a-523178.html (accessed 22 June 2015).

USDOE. 2015. Alternative Fuels Data Center. US Department of Energy, available at http://www.afdc.energy. gov/fuels/prices.html (accessed 22 February 2015).

USGS. 2006. Steel stocks in use in automobiles in the United States. United States Geological Services, United States Department of the Interior, available at http://pubs.usgs.gov/fs/2005/3144/fs2005_3144. pdf (accessed 22 June 2015).

Westbrook, M. H. 2001. *The Electric Car: Development and Future of Battery, Hybrid and Fuel-Cell Cars.* Warrendale, Pennsylvania: Society of Automotive Engineers Inc.

While Book. 2004. European Autos: Emissions and the Future of Powertrains. EBSCO. *Black Book–The Long View: 2004 Edition–European Perspectives*, 83–90.

World Ocean Review. 2010. Methane hydrates. Living with the Oceans: A Report on the State of the World's Oceans, available at http://worldoceanreview.com/wp-content/downloads/worl/WOR1_english.pdf (accessed 22 February 2014).

Worldsteel. 2010. Life Cycle Assessment in the Steel Industry: A Position Paper Issued by the World Steel Association, available at http://www.worldsteel.org/dms/internetDocumentList/downloads/publications/LCA-position-paper/document/LCA%20position%20paper.pdf (accessed 7 July 2015).

Youngmin, W., L. Youngjae, and L. Yonggyun. 2009. The performance characteristics of a hydrogen-fuelled free piston internal combustion engine and linear generator system. *International Journal of Low Carbon Technologies*, 4(1): 36–41.

35 Developing Novel Polymer Composites with Typical Applications in Transport Systems Exposed to High Wear and Corrosion Such as Mining Conveyors

Juan Gimenez and Sri Bandyopadhyay

CONTENTS

ABSTRACT

Increasing the wear endurance of transport systems reduces the environmental footprint of mining operations; however, their wear mechanisms differ considerably from the commonly studied closed tribosystems. Initially, basic concepts of friction and wear mechanisms in polymers are introduced followed by techniques of friction characterization. Recent results of the authors' work on fatigue wear of thermoplastic semicrystalline materials under nontransfer film formation conditions, and zinc oxide nanowires interfaces are included. Two novel methods to produce wear-resistant surfaces in polymers are introduced. The influence of interface strength in this type of materials is examined, together with common methods to produce high-strength interfaces including nanocomposites. High-performance dry bearings composites for extreme environmental and corrosive conditions often seen in the mineral industry are presented. Recent studies on wear resistance processes and mechanisms present in natural (bio) systems applicable to transport systems are reviewed.

INTRODUCTION

Owing to the nature of mining processing operations, handling vast amounts of materials under highly demanding conditions, in increasingly efficient ways, is necessary to improve the sustainability of operations. This increased efficiency is sought by extending the life cycle of conveying components using tribological analysis, with resulting improvements in costs, energy consumption, noise generation, and decreased downtime.

Components with longer life cycles reduce material and manufacturing costs, and their associated energy consumption when assessed per unit conveyed. Reductions in energy consumption will be obtained either by retardation of surface damage which maintains mating surfaces under low friction states longer, or by using low friction surfaces. Decreased noise emission levels will result from lower sliding friction and from components maintaining their original dimensions longer. Lastly, maintenance and downtime will decrease by the use of rugged and durable materials.

The use of polymers in transport system applications presents several benefits, which span from low cost and small footprint to resistance of extreme mechanical, tribological, and corrosive conditions.

Since the field of polymer tribology is broad and covers several types of materials and many different phenomena, applications, and length scales, we will concentrate on semicrystalline thermoplastic polymer composites applicable to bulk handling and unit handling conveyor systems.

Four different approaches to increase wear resistance will be reviewed: polymer surfaces, bulk morphology, composites interfaces, and the wisest: nature. Current results from the authors on a novel method to increase wear resistance by increasing interface strength in glass fibers–polymer matrix composites are presented.

While most of the literature available on polymer tribology covers closed tribosystems where mating surfaces are continually in contact, and therefore the formation of transfer films is decisive to system performance; conveying surfaces in mining and other industries are open systems with minimal possibility of transfer film formation, as surfaces in contact are continually changing (Friedrich, 1986b). Hence all subsequent approaches to reduce wear rate will be surveyed focused on tribosystems with conditions similar to those of mining transport systems.

TRIBOLOGY OF POLYMERS

Increasing the wear resistance of polymer composites without changing their operating environment requires adequate understanding of all processes taking place and ultimately causing material loss.

Friction and wear are the result of several independent and sometimes interrelated mechanisms acting on the surface of the material, and depend on the tribological system where the material is applied (Friedrich and Reinicke, 1998).

While friction and wear are not directly related, a reduction in friction for the same material and operating environment will result in less energy required to be dissipated, leading to less damage in the material.

A two-component model of friction, adhesion and ploughing, is generally accepted to quantify friction (Sinha and Briscoe, 2009b; Zeng, 2013).

The adhesion component includes the energy required to separate the adhesive junctions formed at the real contact area, and the ploughing component groups the energy required for hard asperities to produce material deformations (elastic, viscoelastic, and plastic) and creation of new surfaces. The friction force is given by (Zeng, 2013)

$$F_{\parallel} = H'A_0 + sA_{real} \qquad (35.1)$$

Where, the ploughing component defined by H′ the mean stress required to shear junctions in the softer material multiplied by A_0 the indenting area perpendicular to the sliding direction. The adhesion component defined by s is the stress required to separate the mating surface junctions, and A the real contact area. This model demonstrates that friction is not dependent only on normal loads.

Under real-life conditions, asperities in surfaces are the real contact area, and they represent a very small fraction of the apparent surface contact area. This implies that stresses in asperities can easily be three orders of magnitude larger than the applied load (Yew et al., 2011).

Damage to polymer surfaces caused by the adhesive component of friction is sometimes higher than the ploughing component (Sinha and Briscoe, 2009b; Yew et al., 2011), due to intermolecular forces between mating surfaces being similar in strength to intermolecular forces in the bulk, resulting in commonly observed transfer films. Temperature and time are additional factors influencing the adhesive component response of polymer surfaces (Maeda et al., 2002; Lafaye et al., 2006).

The stick-slip phenomenon, measured at low speeds, in which friction alternates between a high-value F_s static and F_d dynamic friction, is attributed using molecular dynamics (MD) simulation models, to the detachment, slide, and reattachment cycles of adhesion between mating surfaces molecules (Yew et al., 2011). Other models based on atomic force microscopy (AFM) experiments attribute the static friction to the normal displacement required to reach the top of asperities, and dynamic friction to the sliding down from the top of the asperities part of the cycle (Israelachvili et al., 2005).

Wear is defined as the loss of material due to friction dissipation mechanisms. It is classified into three different categories: abrasive, adhesive, and fatigue wear (Sinha and Briscoe, 2009b). Abrasive mechanisms involve the interaction with hard surfaces, in which the hard surface asperities cause plastic deformation to the soft surface without (ploughing) or with (cutting) the presence of material removal. Adhesive wear involves the removal of material from the bulk, after intermolecular bonding has occurred between the mating surfaces, and when these forces are larger than cohesive forces at or near the interface. Wear by surface fatigue is caused by the growth of cracks underneath and at the surface, resulting from the cyclic stresses experienced near the surface.

Three stages are identified during wear: interaction of surfaces, changes in surface layer, and damage to surfaces (Totten and Liang, 2004). Interaction of surfaces involves both the mechanical and molecular components of friction mentioned earlier. Physical and chemical changes to surface layers occur easily in polymers due to their sensitivity to temperature and mechanical strains caused by low intermolecular bonding strength. Damage to surfaces take place by the normal loads acting against the surface, in which harder materials generate plastic deformation with eventual localized failure, or failure by fatigue when load induced deformations are mainly elastic.

TECHNIQUES FOR FRICTION CHARACTERIZATION

In order to further characterize frictional processes in polymer surfaces, several modern strategies have been developed. The mechanical approach is based on interrelations of mechanical properties of the bulk material. The molecular approach studies the effects of polymer morphology at the surface, often together with MD simulations to increase comprehension. The tribophysical approach analyzes surfaces and debris after tribological processes have occurred to identify physical and chemical changes in the material and relates those changes to specific conditions present at the surface while interactions occur.

MECHANICAL APPROACH

A comprehensive introduction to polymer friction and wear from a mechanical point of view is presented in Friedrich (1986b). It is recognized that friction dissipation processes occur markedly different within two discrete depth zones in the mating surfaces.

The first is the interface zone, located in the outer part of the surface, with thickness in the nanometer scale, where the adhesive component of friction is most influential. Temperature, contact pressure, strain magnitude, and rates are very high in this zone. In this approach, adhesion is caused by the interfacial shearing characteristics and conditions at the real contact area, which depend on surface topography and subsurface mechanical properties.

The cohesive zone, located just below the interface zone, has a thickness similar to the contact length, and is strongly influenced by the ploughing or deformation friction. Temperature, contact pressure, strain magnitude, and rates are low or moderate in this zone. The ploughing component creates stresses beneath the interface zone during every cycle of compression, shearing, and tension, at every real contact point while surfaces are sliding. This accumulates conformational changes while the surface viscoelastic history is generated, leading to the formation of defects underneath the surface.

Molecular Approach

With the introduction of the AFM and surface force apparatus, several experiments attempting to understand friction at the molecular level in the interface of two polymeric surfaces in contact became feasible. Equation 35.1 is derived from observations using this approach.

It is shown that the ploughing component of friction depends on the degree of flatness of the surfaces in contact. The concept of mathematically flat surfaces is replaced by atomically flat surfaces. Vertical displacements of only 25 pm required to move the surfaces sideways are already enough to increase the coefficient of friction of atomically flat surfaces to very high levels versus mathematical flat surfaces (Israelachvili et al., 2005).

It is recognized that the ploughing component of friction depends on the surface topography while the adhesive component is more dependent on molecular forces between the surfaces. For low-roughness polymer surfaces (Chen et al., 2005), or at very low loads (Yew et al., 2011), the adhesive component will dominate.

At the molecular scale, the adhesive frictional force resulting from van der Waals forces between molecules represents only a fraction of the force. Its main component is the adhesion hysteresis, which is the additional energy required to break the interfacial bonds formed during each adhesion cycle, in comparison to the energy used to create them. This extra energy to break the bonds is consumed by molecular rearrangements occurring at or near the polymer surface (Chen et al., 2005).

Experiments (Maeda et al., 2002; Chen et al., 2005) show that the main factor affecting adhesive friction is the presence of free chain ends at polymer surfaces as displayed in Figure 35.1. Below is a summary of their findings:

- Initially uncross-linked high-molecular-weight polymers present a low degree of interpenetration at the surface. Few chain coils and ends are free to penetrate into the mating surface. After several sliding cycles, the surface suffers severe changes in its conformational state, that is, adhesion hysteresis, and breakage of some polymer chains. These processes increase the number of free chains and chain ends at the interface, now available for surface interdiffusion.
- If polymer is cross-linked, the restricted movement of chains at the surface and the suppression of chain ends decreases friction forces a few orders of magnitude when compared to uncross-linked polymers (Maeda et al., 2002). However, cross-linked polymers display a very low toughness and as a result are not appropriate for tribological applications by themselves (Friedrich and Schlarb, 2013).
- While uncross-linked polymers required high speeds to stop surface interpenetration (smooth sliding regime), cross-linked polymer surfaces display the same smooth sliding behavior even at very low speeds because no chains are free to penetrate the mating surface.
- When surfaces are held in contact for long periods of time, molecular interdiffusion occurs between surfaces, increasing the energy required to separate the surfaces. Stiction spikes appear during the start of relative displacement between surfaces.

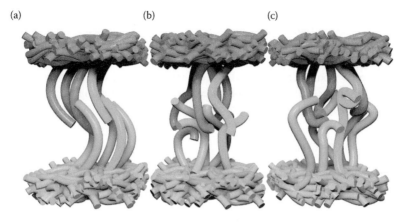

FIGURE 35.1 Interpenetration cases at the interface of polymer surfaces: (a) horizontal displacement, (b) vertical displacement, and (c) entanglement. (Adapted from Zeng, H. 2013. *Polymer Adhesion Friction and Lubrication*, pp. 699.)

- Uncross-linked polymers with different degrees of chain polarity displayed similar adhesion hysteresis at 25°C, suggesting minimal or no effect.

MD models can be constructed by defining two types of bonds: chemical or primary bonds and intermolecular or secondary bonds. It uses some degree of simplification by grouping several atoms into one single component to reduce computational complexity. Primary bonds can be modeled using a single harmonic potential, while intermolecular bonds can be modeled using the Lennard–Jonnes potential model (Yew et al., 2011; Kutvonen et al., 2012).

In a polyethylene (PE) thin-film MD simulation against the same material (Yew et al., 2011), a severe plastic deformation layer of depth 2.5 nm was identified at the interface. Underneath this layer, a 31-nm-thick layer displayed considerable shearing strain before surfaces actually displayed relative motion. The effects of molecular weight and polymer density were studied to understand their influence on frictional forces.

- At low molecular weights, the ploughing component was higher because the surface asperities increased and the adhesion component was lower because the contact area was smaller. At high molecular weights, the surface was flatter and the number of asperities decreased, reducing the ploughing component but increasing the adhesion component because of the resulting increased contact area.
- The low-density film allowed for a larger degree of molecular interdiffusion with the resulting increase in adhesive friction, while the lower number of asperities and relative compliance of material helped reduce the ploughing friction component. At high densities, the degree of molecular interdiffusion is reduced and therefore less adhesive friction results. The ploughing component increases from the increased number of surface asperities and their higher stiffness.

The study concludes that low friction between PE surfaces could be obtained using low-density and high-molecular-weight materials.

Tribophysical Approach

This relatively new approach focuses on physical changes taking place at the interface between sliding surfaces, influenced by heating produced from deformations and cyclic intermolecular bond shearing, that is, frictional forces. Several processes like polymerization, conformational changes, crystallization, thermal degradation, chemical reactions, and melting arise and should be considered when designing polymer surfaces (Sinha and Briscoe, 2009b). The former processes exclude the resulting abrupt changes in both static and dynamic mechanical properties of polymers when temperature increases from increased load, speed, stress history, or environmental temperature. Contact surfaces, subsurface, and wear debris are characterized using mainly surface spectroscopy and thermal analysis methods to identify processes present in the tribosystem.

EXPERIMENTAL METHODS

Three different types of semicrystalline polyamide (PA) materials were used for testing as reported in Table 35.1:

- Commercially available PA-6 extrusion grade natural material (PA6)
- Commercially available 30% by weight short glass fibers PA-66 extrusion grade black pigmented material (PA66GF30)
- Laboratory prepared 50% by weight short E-glass fibers coated with ZnO nanowires PA6 extrusion grade natural material (PA6GF50NW)

The PA6 and PA66GF30 materials were cut with a manual saw from an extruded rod. Parts from the PA6 material were cryogenically ground, and then mixed with the nanowire-coated glass fibers while fed into the mold. All materials were melted and then compression molded into 2.7-mm flat cylinders with length 10 mm.

Often, since material wear rate under real-life conditions is very slow, an increase in load or speed is introduced to the tribosystem to accelerate the test. These unrealistic conditions often completely change the phenomenon present affecting its response (Sinha and Briscoe, 2009a). Increased wear rate of polymeric surfaces under fatigue wear without introducing abrupt changes was achieved by increasing the frequency of mechanical deformations in the surface by using fine metal wire meshes. It has been concluded that wear under no transfer film formation in the mating surface can also be achieved with the same method. Wear debris do not form stable films on the surface of the small diameter wires of the mesh, and any loose particle is able to go into the holes, achieving considerable travel distance before the holes fill up with debris (Ratner, 1967; Friedrich, 1986b).

Following the above technique, a typical low stress fatigue wear process as those experienced by mining conveying surfaces was achieved using a pin-on-disk configuration, in which the polymeric flat cylinders were held fixed against a rotating 250-μm opening size, 0.19-mm rounded wire diameter, 30% opening area stainless-steel wire mesh. A 10-N load, linear speed of 0.4 m/s, and a center-line radius of 12 mm were used. Tests were run for 10,000 cycles or 752 m. Five tests were repeated for each material. After each test, the mesh was replaced, and specimens were ultrasonicated in

TABLE 35.1

Mechanical and Tribological Properties of Materials

Material	Elongation at Break (%)	Yield Stress (MPa)	Elastic Modulus (MPa)	Coefficient of Friction (μ)	Specific Volume Loss Rate (10^{-5} mm^3/Nm)
PA6	>100	45	1425	0.30	1.17
PA66GF30	10[a]	57[a]	2700	0.30	5.14
PA66GF50NW	3[a]	65[a]	4628[a]	0.35	3.73

[a] Extrapolated values from similar commercial materials using same test methods.

acetone for 2 min and weighted using a high-precision balance. The specific material volume loss rate was calculated as the volume loss divided into the product of applied force and traveled distance. The pin-on-disk tribometer (Brand: CSM Instruments SA, Model: Micro) reported the coefficient of friction.

IMPROVEMENT OF WEAR ENDURANCE: RESULTS AND DISCUSSION

The impossibility of mining and other transport systems to rely on protective transfer films impose severe requirements to these types of surfaces, because dissipation of friction energy occurs directly at the surface of the conveying material. Fatigue wear is introduced to the system during the loading, carrying, and unloading material cycles, adhesive wear occurs during material unloading and transfers, and small amounts of abrasive wear originate due to trapped material and contaminants present during operation.

The increase of life cycle by several orders of magnitude, without the need to alter the design of current conveyors, is feasible. Several methods to design improved surfaces, bulk morphologies, composites interfaces of conveying polymeric materials at the surface, together with samples from nature, will be described below.

INCREASING FRICTION DISSIPATION ABILITY AT THE SURFACE

The Ratner and Lancaster relation (Lancaster, 1968) for mild abrasive, single-pass, or nontransfer film formation conditions demonstrates the wear rate of a polymer W_{rate} is equal to the reciprocal of the product of ultimate tensile strength σ_{uts} and elongation at break ϵ_b. It is confirmed by several state-of-the-art publications.

$$W_{rate} = 1/\sigma_{uts}\epsilon_b \tag{35.2}$$

In general, large elongations will help concentrate the damage caused by friction to the interface zone, while the bulk is protected (Friedrich and Schlarb, 2013). A high-strength material withstands larger loads, and a considerable elastic modulus keeps the real contact area to acceptable values. Good balance of rigidity, high elongation at break, and high strength are normally found in semicrystalline thermoplastics, which explains its generally good performance in tribological applications. Amorphous thermoplastics are generally brittle and therefore do not display appropriate performance. Cross-linked polymers should perform very well because of their very low friction forces, but are extremely brittle and therefore not appropriated for tribological performance as bulk materials (Friedrich and Schlarb, 2013).

For prevalent fatigue wear conditions, considerable relation between the fatigue tension endurance of polymers and their wear rate has been demonstrated by measuring crack propagation rates for amorphous and semicrystalline polymers. Increased crack propagation and wear rates for amorphous PA versus semicrystalline PA, and for lower-molecular-weight versus high-molecular-weight polymers also displayed direct relations (Omar et al., 1986).

Based on the above correlation, a similar but upgraded Ratner–Lancaster relation was defined as the brittleness index B (Brostow et al., 2006), which is inversely proportional to a static property: elongation at break ϵ_b, and a dynamic response property: storage modulus E'. The storage modulus in this context is related to the material ability to display elastic responses during frictional processes.

$$B = 1/\epsilon_b E' \tag{35.3}$$

Sliding wear tests were performed by repetitive microscratch tests on the same groove at low speed and room temperature to prevent heat-induced changes at the surface. Strain hardening occurred beneath the groove surfaces, increasing the scratch resistance of the groove with increasing scratch number.

The viscoelastic recovery (f), defined as how much of the penetration depth was recovered or healed after a given scratch number, displayed an inverse correlation with the brittleness index (Brostow and Kovačevic, 2010). Amorphous, semicrystalline, and elastomeric thermoplastics were shown to follow this correlation (Brostow and Hagg Lobland, 2008).

A correlation between free volume and viscoelastic recovery has also been found (Brostow et al., 2006). As mentioned before, it was concluded that subsurface crack propagation is the main mechanism present in fatigue wear.

Using the mechanical properties from Table 35.1, the Ratner and Lancaster relation holds valid for the PA6 and PA66GF30 material. For the PA66GF50NW, an additional factor influenced the response, and will be covered later. It is important to highlight that the PA6 material was worn five times less than the PA66GF30 material, even when friction was same for all systems, implying that similar amounts of energy were introduced into each tribosystem. The worn surface of the PA6 material as shown in Figure 35.2 displays no signs of brittle material removal processes, but instead high amounts of shear yielding activity with some short cracks on the surface. Microyielding, shear band formation, and crystal slipping demonstrate their efficiency as energy dissipation processes, while the resulting conformational changes may include strain hardening of the surface layers, making it more wear resistant. This behavior was also observed in previous studies in heat-treated aluminum composites from one of the authors, in which wear rate was reduced by increased dispersion of smaller reinforcements, leading to more effective crack-arresting ability (Song et al., 1995).

Increasing the number of active yielding sites at the polymer surface will expand its energy dissipation ability. One route to achieve this is to manipulate the morphology of the polymer by dispersing particles of a different phase, displaying a strong interface with the matrix, in order to obtain a radially oriented crystalline structure of the matrix around the particles. If particles are separated by a distance less than the critical ligament thickness, the yield stress of the matrix will be reduced at those locations, since crystal slipping and deformations will be favored. Localization of defects is avoided by intense shear bands activity bridging the particles throughout all of the matrix as seen in Figure 35.3. Stiffness and strength remain constant, which favors the relations introduced earlier (Michler and Baltá-Calleja, 2005).

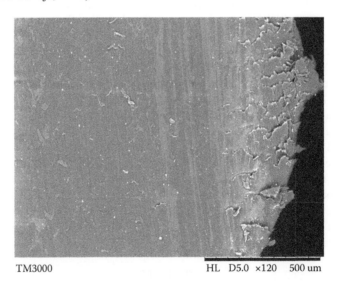

TM3000   HL   D5.0   ×120   500 um

FIGURE 35.2 Surface of PA6 material after fatigue wear testing. (Adapted from Gimenez, J. 2014. Fatigue wear of polymer composites having nano-engineered interfaces, https://www.researchgate.net/publication/271703673_Fatigue_Wear_of_Polymer_Composites_having_Nano-engineered_Interfaces?ev=prf_pub)

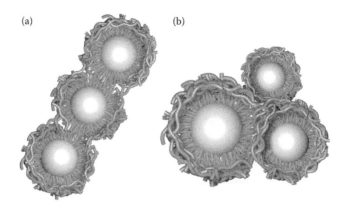

(a)   (b)

FIGURE 35.3 (a) A polymeric structure displaying continuous shear bands enabled by the critical ligament thickness (Ac). (b) Zones of amorphous arrangements interrupt continuous shear bands, leaving the structure prone to defect incubation. (Adapted from Michler, G. H. and F. J. Baltá-Calleja. 2005. In *Mechanical Properties of Polymers Based on Nanostructure and Morphology.* edited by G. H. Michler and F. J. Baltá-Calleja. CRC Press, Taylor & Francis Group.)

Matching the temperature of the tribosystem surface to the glass transition temperature (T_g) of the polymer is also expected to increase the energy dissipation capability of the surface. During nanoindentation creep tests at room temperature, creep rates showed a correlation when plotted against the difference between test temperature and material T_g. Surfaces near T_g hinder both of the creep causing mechanisms: excessive, and highly restricted chain mobility, which are the same mechanisms responsible for fatigue wear (Beake and Bell, 2007).

Ultra-high-molecular-weight PE (UHMWPE) displays very low wear rates due to its unique structure of smooth, very long chains, adopting semicrystalline structures. While shear yielding of lengthy macromolecules consumes large amounts of energy, their nonpolar nature permits chain slipping under low intermolecular forces. High-energy electron beam radiation on UHMWPE can produce cross-linked structures with reduced plastic deformation capacity, increasing wear resistance linearly at higher cross-link densities. While strength remains constant with increasing cross-link density, elongation at break and work to failure is highly reduced (Fu et al., 2011), resulting in lower fatigue resistance (Ho et al., 2007). In order to enhance the fatigue strength of the component while preserving its wear resistance, a gradient cross-linked surface is created by compression molding two layers of UHMWPE material having different concentrations of cross-linking radical scavengers, namely, antioxidants as shown in Figure 35.4. Under the same radiation dose, the outer layer with reduced antioxidant becomes highly cross-linked which provides high wear resistance up

FIGURE 35.4 Gradient from cross-linked surface (top) to uncross-linked core (bottom). (Adapted from Oral, E., A. Neils, and O. K. Muratoglu. 2014. *Journal of Biomedical Materials Research Part B: Applied Biomaterials*, doi:10.1002/jbm.b.33256.)

to the desired depth, transitioning to an increased antioxidant concentration zone which results in a low cross-link density zone displaying high fatigue strength (Oral et al., 2010). This type of surface structure mimics the skin of snakes, where a hard outer layer made of scales provides structural integrity, high strength, and stiffness, while an increasingly flexible material beneath the surface absorbs stresses and provides protection for defect initiation and propagation (Klein et al., 2010).

Reducing Hysteresis Loss in Thermoplastic Elastomers

Hysteresis loss or loss factor in viscoelastic materials is the ratio of energy dissipated to energy stored per cycle. It was shown earlier that the ability of these types of materials to dissipate energy via elastic responses instead of storing that energy in the form of defects accumulation is necessary for adequate wear resistance. Below, erosive wear is used to extrapolate fatigue wear in conveyor belts and other conveying surfaces made of thermoplastic elastomers (TPE), based on the presence of similar cyclic stresses and deformations, with, however, markedly different strain rates.

Erosive wear is caused by surface damage due to repetitive impacts or severe scratches. During a typical erosive wear process in TPE and highly ductile polymers, there is an incubation period in which minimal material loss is produced. At this stage, the material accumulates a high degree of plastic deformations (incubating defects, crazes, and shear bands in the surface and beneath it) which eventually exceed maximum strains. In a later stage, material removal starts increasing linearly with time, a result of the neighboring cracks intersection. Elastomeric and highly ductile polymers display high resistance to erosion when compared to other types of materials and brittle polymers due to their ability to dissipate large amounts of energy by elastic deformation, that is, large storage modulus, reducing the amount of energy available for plastic deformation, crack initiation, and propagation (Friedrich, 1986a). Specifically, thermoplastic polyurethanes (TPUs) are currently used as sacrificial surfaces in many mining applications to protect underlying structural and often less ductile materials, including linings in chutes and belt conveyors.

In order to identify the microstructure and viscoelastic factors that affect their response, erosive wear experiments were made over polyester- and polyether-based TPU materials. The polyether-based material displayed a brittle response in comparison to the polyester highly ductile response. Crystallization by fixation of chain positions due to strong intermolecular hydrogen bonding occurs at strains three times lower in polyether flexible chain segments than in polyester (Sigamani, 2010), which implies higher propensity to permanent deformation, and resulting lower wear resistance. Crystallization can be delayed by irregular chain configurations. Narrow molecular weight of rigid segments increases mechanical properties by creating finer domains.

Characterization was performed on TPU worn surfaces using Fourier transform infrared spectroscopy. For polyester-based TPU, it is suggested that flexible segments absorb large amounts of energy due to their ductility, and after large degrees of plastic deformation, they are removed from the surface, while hard segments do not display such mass reduction. In polyether-based TPU, a similar reduction of both hard and soft segments is observed, suggesting the weak bond between the hard and soft phase to be the prevalent failure mechanism, leading to equal segment mass loss (Sigamani, 2010).

One type of hybrid material used for high demanding applications utilizes TPU films to cover and protect underlying structural components like carbon fiber-reinforced polyetheretherketone (PEEK) composites, because of their brittle response to erosion and resulting high wear rates (Friedrich et al., 2013). In order to understand the effect of TPU film thickness on erosive wear rates at 30–40 impinging angles, experiments and simulations were performed with films attached to steel substrates (Zhang et al., 2013). Experimental results showed an initial decrease in wear rate up to some film thickness where mass loss started increasing. Computer simulations not including the effect of temperature predicted a decrease of wear rate with increasing film thickness due to larger flexibility of the film, with resulting decrease of stress concentration. In order to include the temperature effect, heat production at the surface due to each erosive particle friction with the surface, and from the resulting plastic deformation of the film, were used to estimate a total heat flux to the film, and confirmed experimentally at two different depths. Temperatures of 50°C were measured 1 mm below the surface. Since TPE are poor heat conductors, the surface temperature increases linearly with the film thickness, resulting in increased wear rates. It was concluded that the lowest wear rate thickness marks the point at which the weakening of mechanical properties due to the raise in surface temperature becomes more influential than the concentration of mechanical stresses at the surface.

Erosive wear experiments with several types of TPU, with hardness ranging from 83 to 88 shore A, but with different microstructures, displayed no correlation between wear rates and microstructural factors, including highly similar morphologies or interdomain spacing. Master curves for dynamic mechanical responses of all different grades were obtained by using the time–temperature superposition principle, from which the viscoelastic modulus was obtained at the same timescales that the erosive process occurred. The loss modulus displayed strong correlation to the mass loss for all types of TPU tested, independent of impingement angles (Arena et al., 2015). It was concluded that plastic deformations are directly related to the wear rate of TPU under erosive wear. Heat dissipation mechanisms, ductile hard phases, and very high elastic responses increase the material ability to dissipate frictional energy.

Reduction of the hysteresis loss in elastomers can also be achieved by incorporation of covalently bonded nanosprings into the polymer chain, or as a filler in the material as shown in Figure 35.5 (Liu et al., 2013). Using this approach, the viscoelastic response of TPE can be fine-tuned, since springs are able to store elastic energy during stress cycles, springs constants can be tailored, and polymer chains attached to the spring can follow the tension and compression displacements. Mechanical properties like Young's modulus and tensile strength can also be improved due to nanoreinforcement. Additional carbon-based nanostructures with high flexibility and reversible mechanical response include nanocoils, nanorings, and thin graphene sheets.

Carbon-based nanosprings have the ability to withstand tensile and compressive deformations and then return to their original shape, similar to macroscopic springs (Nakamatsu et al., 2005). MD simulations were performed to evaluate two different methods to integrate nanosprings into TPE. In the first method, nanosprings are mixed with the polymer matrix, the spring ends being uncross-linked or with different cross-link densities. An increase in cross-link density with surrounding polymer chains, from 0 links to 800, increases threefold the stress-to-strain ratio, and also reduces the permanent set in the same proportion due to increased volume interaction in the material. Changing the nanospring constant also displays improvements in the permanent set. In the second method, a self-assembly triblock copolymer composed of polymer chain, nanospring, and polymer chain, is studied, which achieves equilibrium in the simulation by adopting a lamellar morphology. The response in the spring extension axis displays an improved response in comparison to

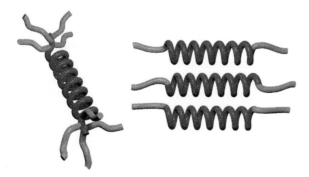

FIGURE 35.5 Polymers viscoelastic modification by inclusion of nanosprings. (a) Randomly mixed nanosprings cross-linked to nearby polymer chains. (b) Triblock composed of chain-nanospring-chain. (Adapted from Liu, J. et al. 2013. *Advanced Functional Materials*, 23: 1156–1163, doi:10.1002/adfm.201201438.)

the randomly mixed nanosprings, possibly due to enlarged volume interaction in the stress direction. Off axis responses, however, display a decrease in permanent set performance versus the randomly mixed method (Liu et al., 2013).

Role of Composite Interfaces

Micro- and nanocomposites based on polymeric matrices usually display lower levels of wear resistance when compared to unfilled polymers. For microcomposites, while the strength and modulus can be highly increased with many structural applications being opened for these type of materials, there is always a marked reduction in elongation at break, because fibers and other types of fillers act as crack nucleation points, interrupting energy dissipation mechanisms like plastic deformation, and ultimately leading to lower friction dissipation capabilities. In the case of nanocomposites, the effects are different. Usually properties like strength, stiffness, fatigue, and creep are increased, while elongation at break remains constant at low filling levels, starting to decrease at higher loadings. Using the essential work of fracture approach to evaluate the effect of toughness (Bárány et al., 2003), the amount of energy required for crack initiation is higher, due to restrictions of intermolecular displacement, and the properties achieved by the critical ligament thickness effect already discussed. Crack propagation energy, however, is lowered due to the stress concentration effects caused by nanoparticle aggregation after considerable plastic deformation has occurred.

When tribological conditions only include adhesive or fatigue mechanisms without the presence of sharp asperities, however, fiber-filled microcomposites can provide higher load-bearing capabilities, better stress distribution through the composite, lubrication in some cases, and heat dissipation in the case of carbonaceous fillers. Inclusion of nanoparticles in this type of composites helps distribute the loads to a further degree, reducing stress concentration, and creating anchoring points for macromolecules.

The main damaging mechanism of microcomposites under the adhesive or fatigue mechanism is failure of the interface between microfiller and the matrix, resulting in premature removal of the filler. Experiments on fatigue wear of fiber-reinforced composites show that the process starts with matrix and fiber wear, with resulting loss of mechanical properties in the fibers. Then, fiber cracking commences while the fiber–polymer interface is increasingly debilitated. Lastly, fiber–matrix separation occurs, increasing wear rate, due to large amounts of material being removed instantaneously (Friedrich and Schlarb, 2013). In this respect, nanocomposites have a clear advantage over microcomposites, since their surface-to-volume ratio is several orders of magnitude higher, and therefore, the interface performance is less critical for filler to matrix attachment. The interface, however, becomes increasingly influential on the matrix properties.

Both high strength and toughness are desirable in composites interfaces to keep microfillers attached to the matrix during most of the wear process, in order to exploit their benefits. Carbon fiber PEEK laminated composites obtained a 26% increase in oscillatory sliding wear resistance when treated with cold remote nitrogen oxygen plasma, because the oxygen and nitrogen functional groups attached to the fiber surface increased the strength interface with the matrix (Sharma et al., 2013).

In a different experiment, montmorillonite nanoparticles were compounded into a high-density PE (HDPE) matrix using either maleic anhydride (MA) or olefin elastomer (thermoplastic polyolefin [TPO]) as grafting molecules. MA is a high stiffness molecule while TPO is highly elastic. At 5% loading, the MA grafted composite displayed a slightly increased wear rate in comparison to HDPE, while the TPO grafted composite displayed half the wear rate of HDPE. Scanning electron microscope (SEM) images of TPO grafted composite worn surfaces displayed considerable plastic deformation when compared to the MA grafted composite, confirming the benefits of tough interfaces (Liu and Xu, 2011).

In the case of nanocomposites, grafting of polymer compatible molecules around nanoparticles ensures adequate anchoring to the matrix as displayed in Figure 35.6, while dispersion of the mix is also improved. Benefits include increased shear strength and resistance to irreversible

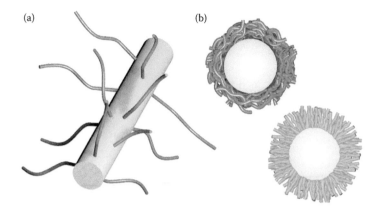

FIGURE 35.6 (a) CNTs attached to the surface of carbon fibers as a general reinforcing mechanism of carbon fiber polymer composites interfaces. (Adapted from Li, M., Y. Gu, Y. Liu, Y. Li, and Z. Zhang, *Carbon*, 52, 109–21, 2013.) (b) Arrangement of weak particle–polymer interfaces (top) in contrast to strong interfaces (bottom). (Reprinted from *Tribology of Polymeric Nanocomposites: Friction and Wear of Bulk Materials and Coatings*, Chapter 17, 2nd edn., Burris, D. L. et al., Polytetrafluoroethylene matrix nanocomposites for tribological applications, pp. 571–617, doi:10.1016/B978-0-444-59455-6.00017-9, Copyright 2013, with permission from Elsevier.)

morphological changes, reduced nanoparticle removal rates, and increased thermal conductivity when particles have this ability. Plasma-treated carbon nanotubes (CNTs) displayed considerable increase of oxygen groups in the surface with stronger adhesion to the matrix. When dispersed on UHMWPE, the coating displayed an increase in wear resistance of at least two orders of magnitude in comparison to pristine UHMWPE films (Samad and Sinha, 2011).

ZINC OXIDE NANOWIRES ON GLASS FIBER COMPOSITES

Owing to the limitations of particulate nanocomposites in comparison to long fiber microcomposites in terms of stiffness, strength, and impact resistance, methods to increase the wear resistance of fiber-filled microcomposites by increasing the interface strength are sought. Silanes are currently the most common type of coupling agents in glass fiber-reinforced polymers, and experiments in PA6 doubled the interlaminar shear strength when compared to untreated fibers (Mäder et al., 1994).

Since fibers are much harder and stiffer than polymeric matrices, fiber edges protruding from the surface and their interfaces are continually prone to impacts and high stresses during frictional processes, which rapidly damage the interface between matrix and polymer, resulting in the wear cycle mentioned above. Classical methods to enhance the interface strength include monolayers deposition, surface oxidation, and fiber whiskerization. Modern methods as shown schematically in Figure 35.6a include growth of CNTs (Lv et al., 2011) and zinc oxide (ZnO) nanowires (Ehlert et al., 2013) on carbon fibers, direct reaction of grafting groups on fiber surface (Laachachi et al., 2008), attachment of nanotubes to fiber surface grafted dendrimers (Peng et al., 2012), and cross-linkable polymeric sizing containing CNTs (Godara et al., 2010). All of the modern approaches demonstrate increase in interfacial shear strength (IFSS), some of them achieving a twofold increase. ZnO nanowires growth on carbon fibers doubled the IFSS of epoxy composites when measured using single fiber fragmentation testing, and a correlation was found between interface strength and ketone groups density in the fiber surface (Ehlert et al., 2013). However, none of them used thermoplastic polymeric matrices.

In our experiments, E-glass fibers were initially treated to remove the commercial sizing. Next, based on experiments over carbon fibers (Ehlert et al., 2013), etching of fibers was performed in aqueous solutions with high concentration of sodium hydroxide (NaOH). Subsequently, ZnO

(a)   (b)

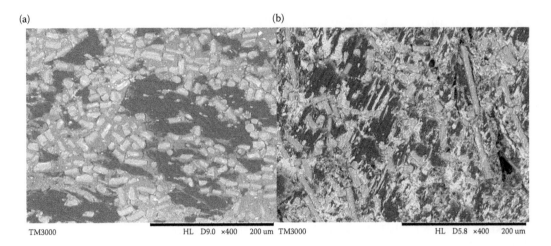

FIGURE 35.7 Surface of PA6GF50NW (a) PA66GF30 (b) materials after equivalent fatigue wear testing. (Adapted from Gimenez, J. 2014. Fatigue wear of polymer composites having nano-engineered interfaces, https://www.researchgate.net/publication/271703673_Fatigue_Wear_of_Polymer_Composites_having_ Nano-engineered_Interfaces?ev=prf_pub)

nanowires were grown on the fiber surface using a hydrothermal process. Details on preparation can be found in Gimenez (2014). ZnO nanowire-coated short glass fibers were mixed with PA6 and compression molded as described earlier in this document. Nanowires were 200 nm in diameter and 50 μm in length. A uniform coating was achieved throughout the fibers' surface, able to withstand ultrasonication for at least 10 min without noticeable damage, with no signs of delamination after fiber breakage. During molding, detachment of the coating occurred by the high shear forces of the viscous PA, observed in fibers on the outer molded surfaces, which hindered the possibility of characterizing their influence on fatigue wear.

During tribological experiments, however, the synergistic performance improvements of multiscale composites were confirmed. For the PA6GF50NW shown in Figure 35.7a, nanowires accumulated on the borders of glass fibers due to the rolling effect of some round-shaped nanoparticles, acting as rollers between the mating surfaces, and accumulating at microfibers interfaces, protecting them from shear stresses (Jiang and Zhang, 2013). When the surface of PA66GF30 shown in Figure 35.7b is compared to the PA6GF50NW multiscale composite, a much smoother surface, with reduced presence of severe scratches and ploughed material, and decreased premature fiber removal is observed in the multiscale composite. The specific wear rate of the multiscale composite was 38% lower than the microcomposite as described earlier.

High-Temperature and Corrosion-Resistant Systems

High temperatures and strong chemicals are present during mineral processing. Materials capable of resisting sodium hydroxide (NaOH) at high concentrations and temperature, in processes like bauxite ore processing, include semicrystalline polymers like polytetrafluoroethylene (PTFE) sold by Harrington Industrial Plastics, USA, tested to withstand NaOH (concentration 70%) at 170°C without accelerated degradation, and PEEK sold by Victrex PLC, UK, which remains unaffected at NaOH concentrations of 50% at 200°C, according to their product datasheets. In addition, the use of conventional lubricants in many extreme environments is not feasible due to their functional limited temperature range, risk of contamination, and periodic maintenance requirements among others. Solid lubricants in contrast display a larger functional temperature range, structural stability, low costs, and reduced environmental impact, making them attractive for those highly demanding tribosystems.

PTFE is a well-known solid lubricant, displaying chemical stability and low sliding friction conferred by its macromolecule configuration; however, a brittle and easily sheared internal structure results in very high wear rates when used in bulk for mechanical and tribological applications, ultimately restricting its range of applications. Methods to increase its wear resistance include inclusion of microparticles and nanoparticles.

In sliding against a 304 steel disk (Ra = 0.05 μm) at room temperature, using speed of 10 mm/s and contact pressure of 3.1 MPa, unfilled PTFE, micro- and nanocomposites were all filled at 5% with alumina particles. Microcomposites increased their wear resistance two orders of magnitude, and nanocomposites four orders of magnitude in comparison to the unfilled material. The inclusion of particles did not modify the coefficient of friction of the unfilled material. The morphology of transfer films on the counterface was coarse and uneven for the unfilled material, decreasing to a very fine and uniform film for the nanocomposites, suggesting different failure mechanisms. Energy dispersive x-ray spectroscopy (EDS) confirmed the accumulation of three times more microfillers on the worn surfaces in comparison to nanofillers, even when nanocomposites were tested over extended distances, and the rate of material removal was markedly slower. Mixing microparticles with nanoparticles in the same composite, using 5% as total content of alumina, and tested under the same conditions, led to the same response of microparticle-only composites (Burris et al., 2013). Fluorine surface-modified nanoparticles increased the wear resistance and displayed reproducible performance, in contrast to nonmodified nanoparticles, by creating homogeneous zones with crack-arresting and polymer chains anchoring capacities. The presence of nanoparticles during processing preserved the high elongation and tough aligned polymer chain morphologies achieved during jet milling processing deformation, and stabilized tough crystal structures instead of brittle phases (Burris et al., 2013).

While PTFE nanocomposites can increase the mechanical properties versus unreinforced material, the resulting properties are not enough for many demanding applications subjected to large stresses under high temperatures. Many blends of PEEK/PTFE are available commercially, which are able to withstand higher temperatures under load while maintaining self-lubrication capabilities; however, their tribological performance is only increased reasonably. An ultra-low-wear composite was prepared by grinding both materials, jet milling the PTFE, followed by compression molding. PEEK weight contents from 0% to 70% were used to evaluate the best formulation, and a 32 wt.% PEEK formula displayed the best performance by reducing the wear rate of unfilled PEEK by three orders of magnitude, and more than five orders of magnitude that of unfilled PTFE. Uniaxial compression tests of the composite found a continuous PEEK phase displaying an elastic modulus only 5% less than unfilled PEEK, and single edge notched tensile tests displayed strains at break in the order of 300%, much larger than the strains achieved by any of the components by itself. Characterization of the composite structure using SEM and EDS revealed a continuous PEEK phase, in which PTFE-oriented molecules, that is, nanofibrils, were anchored to the PEEK phase, creating a very strong interface between these dissimilar materials, conferring high strength and very large elongations before failure to the composite, as displayed in Figure 35.8 (Burris and Sawyer, 2006).

Natural (Bio) Systems Methods

The field of biotribology is faced with an overwhelmingly complex and extensive range of tribological systems, which include multiscale structures with interrelated mechanisms working in parallel, and tailored along thousands of years according to the wide range of needs of every living being, with dissimilar operating requirements, including life-threatening high-speed responses, and durability requirements of surfaces according to the life expectations of those organisms; all these without mentioning the complexity of characterization and experimental procedures.

One of the most studied systems is the articular joint of mammalians, which includes synergistic lubrication mechanisms with redundant systems to support the failure of primary systems,

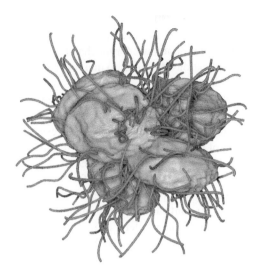

FIGURE 35.8   Structure of the PTFE (nanofibrils)/PEEK (bulk phase) composite. (Adapted from Burris, D. L. and W. G. Sawyer. 2006. *Wear*, 261(3–4): 410–418, doi:10.1016/j.wear.2005.12.016.)

producing time-dependent complex responses within different length scale structures, depending on stresses and rates (Zeng, 2013), as shown in Figure 35.9.

The articular cartilage is a porous network filled with interstitial fluid. The fluid inside the network is pressurized upon loading of the surface, creating a transitory load-carrying structure which highly reduces the stresses transfer to the collagen fibril network or solid phase, resulting in reduced friction forces and higher wear resistance (Greene et al., 2014). In addition, a considerable number of highly charged proteins entangled inside the porous network known as proteoglycans, produce two different effects that augment the length of time the fluid pressurization effect is active. By increasing the number of ions inside the cartilage, the pressure of the interstitial fluid is augmented in comparison to the surrounding, and the compressive stiffness of the network is enlarged by the action of the large repulsive forces opposing the collapse of the cartilage porous network.

While the latter mechanisms increase the creep resistance and reduce friction and wear rates, a redundant system is in place to provide boundary lubrication for situations of very large or prolonged loading: lubricin and hyaluronic acid, two types of macromolecules physically cross-linked. Lubricin molecules prevent solid-to-solid contact by providing the cartilage surface with a boundary lubrication

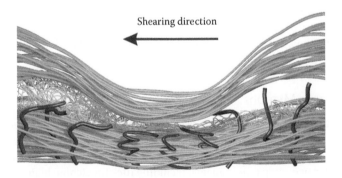

FIGURE 35.9   Multimodal lubrication mechanism in articular cartilage. Lubricin shown as the boundary layer concentrating at the front of the shearing direction. Hyaluronic acid shown as molecules entangled into the cartilage porous network. (Adapted from Greene, G. W. et al. 2011. *Proceedings of the National Academy of Sciences of the United States of America*, 108(13): 5255–5259, doi:10.1073/pnas.1101002108.)

layer that is dragged at large loading conditions and therefore concentrated at the front of the shearing direction protecting the surface. Hyaluronic acid molecules in contrast attach mechanically to the cartilage network when pores collapse, after the fluid pressurization effect has been lost, functioning as emergency boundary lubrication, avoiding premature wear of the cartilage (Zeng, 2013).

A synthetic material able to mimic the tribological performance of articular cartilage, displaying complex responses similar to the biosystem, confirmed the synergistic interaction of the three mechanisms just mentioned (Greene et al., 2014).

A different but equally interesting approach mimics the just mentioned boundary lubrication layer of articular cartilage by using polymer brushes grafted from cross-linked PE (CLPE), to reduce friction forces and wear rates of sliding surfaces in total hip arthroplasty bearings. Hydrated lubrication layers are achieved by using hydrophilic polymers with different charges: nonionic, anionic, cationic, and zwitterionic. Using a ball-on-plate tribometer and a hip simulator, frictional forces and wear rates were measured for CLPE liners sliding against a Co–Cr–Mo alloy ball, under lubricants simulating living beings fluids.

The anionic layer displayed mineralization when lubricated with a solution containing inorganic cations follow by high wear rates, while the cationic layer displayed large frictional forces caused by the attachment of proteins to the alloy ball and the polymeric layer, but showed low wear rates possibly by removal of the protein layer under high loads. The nonionic and zwitterionic layers displayed 50% and 75% reduction in frictional forces, respectively, versus the untreated surface. The nonionic, zwitterionic, and cationic layers presented reduced wear rates of at least 97% after three million cycles, versus the untreated CLPE surface.

While reduction in friction and wear rates are mainly attributed to the highly effective water hydration layer lubrication effect, a secondary mechanism of improvement is the repulsion of protein molecules from both sliding surfaces, reducing the adhesive component of friction at all times (Kyomoto et al., 2012).

CONCLUSIONS

Several methods to increase the wear endurance of materials applied to tribosystems similar to mining conveyors were reviewed. Their preliminary performance predicts increased life cycles of several orders of magnitude, which can compensate their higher costs, consequently creating realistic opportunities for footprint reduction in mineral processing plants. Most of materials reviewed are already mature, practical to manufacture, and demand short research cycles to realize their benefits.

Increasing the wear resistance of thermoplastic semicrystalline polymers under nontransfer film formation conditions requires the contemplation of a totally different range of material design parameters versus those studied for closed tribosystems relying on transfer film formation.

The internal configuration and conformation of macromolecules is the main mechanism responsible for improved wear resistance performance in this type of polymers. Innovative material morphologies enable materials with outstanding performance.

Further research is needed to generate strong and ductile interphases in fiber-filled thermoplastic semicrystalline composites, in order to achieve reduced wear rates under nonabrasive conditions.

Biotribological systems can provide all the insight, ideas, and examples required to develop the most advanced wear-resistant materials and systems.

REFERENCES

Arena, G., K. Friedrich, D. Acierno, E. Padenko, P. Russo, G. Filippone, and J. Wagner. 2015. Solid particle erosion and viscoelastic properties of thermoplastic polyurethanes. *Express Polymer Letters*, 9(3): 166–176, doi:10.3144/expresspolymlett.2015.18.

Bárány, T., T. Czigány, and J. Karger-Kocsis. 2003. Essential work of fracture concept in polymers. *Periodica Polytechnica, Mechanical Engineering*, 47(2): 91–102.

Beake, B. D. and G. A Bell. 2007. Nanoindentation creep and glass transition temperatures in polymers. *Polymer International*, 56(2): 773–778, May 2006, doi:10.1002/pi.

Brostow, W. and H. E. Hagg Lobland. 2008. Predicting wear from mechanical properties of thermoplastic polymers. *Polymer Engineering & Science*, 48(10): 1982–1985, doi:10.1002/pen.21045.

Brostow, W., H. E. Hagg Lobland, and M. Narkis. 2006. Sliding wear, viscoelasticity, and brittleness of polymers. *Journal of Materials Research*, 21(9): 2422–2428, doi:10.1557/jmr.2006.0300.

Brostow, W. and V. Kovačevic. 2010. Tribology of polymers and polymer-based composites. *Journal of Materials Education*, 32: 273–290, http://www.researchgate.net/publication/234840630_Tribology_of_polymers_and_polymer-based_composites/file/72e7e52b8fcc2e027c.pdf

Burris, D. L., K. Santos, S. L. Lewis, X. Liu, S. S. Perry, T. A. Blanchet, L. S. Schadler, and W. G. Sawyer. 2013. Polytetrafluoroethylene matrix nanocomposites for tribological applications. In *Tribology of Polymeric Nanocomposites: Friction and Wear of Bulk Materials and Coatings*, Chapter 17, 2nd edn. edited by K. Friedrich and A. K. Schlarb. Elsevier, UK, pp. 571–617, doi:10.1016/B978-0-444-59455-6.00017-9.

Burris, D. L. and W. G. Sawyer. 2006. A low friction and ultra low wear rate PEEK/PTFE composite. *Wear*, 261(3–4): 410–418, doi:10.1016/j.wear.2005.12.016.

Chen, N., N. Maeda, M. Tirrell, and J. N. Israelachvili. 2005. Adhesion and friction of polymer surfaces: The effect of chain ends. *Macromolecules*, 38(8): 3491–3503, http://pubs.acs.org/doi/abs/10.1021/ma047733e

Ehlert, G. J., U. Galan, and H. A. Sodano. 2013. Role of surface chemistry in adhesion between ZnO nanowires and carbon fibers in hybrid composites. *ACS Applied Materials & Interfaces*, 5(3): 635–645, doi:10.1021/am302060v.

Friedrich, K. 1986a. Erosive wear of polymer surfaces by steel ball blasting. *Journal of Materials Science*, 21: 3317–3332, doi:10.1007/BF00553375.

Friedrich, K. 1986b. *Friction and Wear of Polymer Composites*. Elsevier, Amsterdam.

Friedrich, K., X.-Q. Pei, and A. A. Almajid. 2013. Specific erosive wear rate of neat polymer films and various polymer composites, *Journal of Reinforced Plastics and Composites*, 32(9): 631–643, doi:10.1177/0731684413478478.

Friedrich, K. and P. Reinicke. 1998. Friction and wear of polymer-based composites. *Mechanics of Composite Materials*, 34(6): 503–514, http://link.springer.com/article/10.1007/BF02254659

Friedrich, K. and A. K Schlarb. 2013. *Tribology of Polymeric Nanocomposites: Friction and Wear of Bulk Materials and Coatings*, 2nd edn. Butterworth-Heinemann, Oxford, UK.

Fu, J., B. W. Ghali, A. J. Lozynsky, E. Oral, and O. K. Muratoglu. 2011. Wear resistant UHMWPE with high toughness by high temperature melting and subsequent radiation cross-linking. *Polymer*, 52(4): 1155–1162, doi:10.1016/j.polymer.2011.01.017.

Gimenez, J. 2014. Fatigue wear of polymer composites having nano-engineered interfaces, https://www.researchgate.net/publication/271703673_Fatigue_Wear_of_Polymer_Composites_having_Nano-engineered_Interfaces?ev=prf_pub.

Godara, A., L. Gorbatikh, G. Kalinka, A. Warrier, O. Rochez, L. Mezzo, F. Luizi, A. W. van Vuure, S. V. Lomov, and I. Verpoest. 2010. Interfacial shear strength of a glass fiber/epoxy bonding in composites modified with carbon nanotubes. *Composites Science and Technology*, 70(9): 1346–1352, doi:10.1016/j.compscitech.2010.04.010.

Greene, G. W., X. Banquy, D. W. Lee, D. D. Lowrey, J. Yu, and J. N. Israelachvili. 2011. Adaptive mechanically controlled lubrication mechanism found in articular joints. *Proceedings of the National Academy of Sciences of the United States of America*, 108(13): 5255–5259, doi:10.1073/pnas.1101002108.

Greene, G. W., A. Olszewska, M. Osterberg, H. Zhu, and R. Horn. 2014. A cartilage-inspired lubrication system. *Soft Matter*, 10: 374–3782, doi:10.1039/c3sm52106k.

Ho, Y.-C., F.-M. Huang, and Y.-C. Chang. 2007. Comparative fatigue behavior and toughness of remelted and annealed highly crosslinked polyethylenes. *Journal of Biomedical Materials Research Part B: Applied Biomaterials*, 83: 340–344, doi:10.1002/jbmb.

Israelachvili, J. N., N. Maeda, K. J. Rosenberg, and M. Akbulut. 2005. Effects of sub-ångstrom (pico-sale) structure of surfaces on adhesion, friction, and bulk mechanical properties. *Journal of Materials Research*, 20(08): 1952–1972, doi:10.1557/JMR.2005.0255.

Jiang, Z. and Z. Zhang. 2013. Wear of multi-scale phase reinforced composites. In *Tribology of Nanocomposites*, edited by D. J. Paulo, Chapter 5, Vol. 11. pp. 79–100, doi:10.1007/978-3-642-33882-3.

Klein, M. C. G., J. K. Deuschle, and S. N. Gorb. 2010. Material properties of the skin of the Kenyan sand boa *Gongylophis Colubrinus* (Squamata, Boidae). *Journal of Comparative Physiology A: Neuroethology, Sensory, Neural, and Behavioral Physiology*, 196: 659–668, doi:10.1007/s00359-010-0556-y.

Kutvonen, G. R. and T. Ala-Nissila. 2012. Correlations between mechanical, structural, and dynamical properties of polymer nanocomposites. *Physical Review E*, 85(4): 041803, doi:10.1103/PhysRevE.85.041803.

Kyomoto, M., T. Moro, K. Saiga, M. Hashimoto, H. Ito, H. Kawaguchi, Y. Takatori, and K. Ishihara. 2012. Biomimetic hydration lubrication with various polyelectrolyte layers on cross-linked polyethylene orthopedic bearing materials. *Biomaterials*, 33(18): 4451–4459, doi:10.1016/j.biomaterials.2012.03.028.

Laachachi, A., A. Vivet, G. Nouet, B. B. Doudou, C. Poilâne, J. Chen, J. B. Bai, and M. H. Ayachi. 2008. A chemical method to graft carbon nanotubes onto a carbon fiber. *Materials Letters*, 62(3): 394–397, doi:10.1016/j.matlet.2007.05.044.

Lafaye, S., C. Gauthier, and R. Schirrer. 2006. Analysis of the apparent friction of polymeric surfaces. *Journal of Materials Science*, 41(19): 6441–6452, doi:10.1007/s10853-006-0710-7.

Lancaster, J. K. 1968. Relationships between the wear of polymers and their mechanical properties. *Proceedings of the Institution of Mechanical Engineers, Conference Proceedings September 1968*, 183(16): 98–106. doi:10.1243/PIME_CONF_1968_183_283_02.

Li, M., Y. Gu, Y. Liu, Y. Li, and Z. Zhang. 2013. Interfacial improvement of carbon fiber/epoxy composites using a simple process for depositing commercially functionalized carbon nanotubes on the fibers. *Carbon*, 52: 109–121, February, doi:10.1016/j.carbon.2012.09.011.

Liu, J., Y. L. Lu, M. Tian, F. Li, J. Shen, Y. Gao, and L. Zhang. 2013. The interesting influence of nanosprings on the viscoelasticity of elastomeric polymer materials: Simulation and experiment. *Advanced Functional Materials*, 23: 1156–1163, doi:10.1002/adfm.201201438.

Liu, S.-P. and J.-F. Xu. 2011. Characterization and mechanical properties of high density polyethylene/silane montmorillonite nanocomposites. *International Communications in Heat and Mass Transfer*, 38(6), 734–741, doi:10.1016/j.icheatmasstransfer.2011.03.012.

Lv, P., Y.-Y. Feng, P. Zhang, H.-M. Chen, N. Zhao, and W. Feng. 2011. Increasing the interfacial strength in carbon fiber/epoxy composites by controlling the orientation and length of carbon nanotubes grown on the fibers. *Carbon*, 49(14): 4665–4673, doi:10.1016/j.carbon.2011.06.064.

Mäder, E., K. Grundke, H. J. Jacobasch, and G. Wachinger. 1994. Surface, interphase and composite property relations in fibre-reinforced polymers. *Composites*, 25(7): 739–744, http://www.sciencedirect.com/science/article/pii/0010436194902097

Maeda, N., N. Chen, M. Tirrell, and J. N. Israelachvili. 2002. Adhesion and friction mechanisms of polymer-on-polymer surfaces. *Science*, 297(5580): 379–382, doi:10.1126/science.1072378.

Michler, G. H. and F. J. Baltá-Calleja. 2005. Micromechanics of particle-modified semicrystalline polymers: Influence of anisotropy due to transcrystallinity and/or flow. In *Mechanical Properties of Polymers Based on Nanostructure and Morphology*. edited by G. H. Michler and F. J. Baltá-Calleja. CRC Press, Taylor & Francis Group, Boca Raton, FL, pp. 317–378.

Nakamatsu, K.-I., M. Nagase, H. Namatsu, and S. Matsui. 2005. Mechanical characteristics of diamond-like-carbon nanosprings fabricated by focused-ion-beam chemical vapor deposition. *Japanese Journal of Applied Physics, Part 2: Letters*, 44: 1–4, doi:10.1143/JJAP.44.L1228.

Omar, M. K., A. G. Atkins, and J. K. Lancaster. 1986. The role of crack resistance parameters in polymer wear. *Journal of Physics D: Applied Physics*, 19(2): 177, http://iopscience.iop.org/0022-3727/19/2/007

Oral, E., B. W. Ghali, S. L. Rowell, B. R. Micheli, A. J. Lozynsky, and O. K. Muratoglu. 2010. A surface crosslinked UHMWPE stabilized by vitamin E with low wear and high fatigue strength. *Biomaterials*, 31(27): 7051–7060, doi:10.1016/j.biomaterials.2010.05.041.

Oral, E., A. Neils, and O. K. Muratoglu. 2014. High vitamin E content, impact resistant UHMWPE blend without loss of wear resistance. *Journal of Biomedical Materials Research Part B: Applied Biomaterials*, 103(4): 790–797, doi:10.1002/jbm.b.33256.

Peng, Q., X. He, Y. Li, C. Wang, R. Wang, P. Hu, Y. Yan, and T. Sritharan. 2012. Chemically and uniformly grafting carbon nanotubes onto carbon fibers by poly(amidoamine) for enhancing interfacial strength in carbon fiber composites. *Journal of Materials Chemistry*, 22(13): 5928, doi:10.1039/c2jm16723a.

Ratner, S. B. 1967. In *Abrasion of Rubber*, edited by D. I. James. MacLaren. ISBN: 978-0853340027.

Samad, M. A. and S. K. Sinha. 2011. Mechanical, thermal and tribological characterization of a UHMWPE film reinforced with carbon nanotubes coated on steel. *Tribology International*, 44(12): 1932–1941, doi:10.1016/j.triboint.2011.08.001.

Sharma, M., J. Bijwe, E. Mäder, and K. Kunze. 2013. Strengthening of CF/PEEK interface to improve the tribological performance in low amplitude oscillating wear mode. *Wear*, 301(1–2): 735–739, doi:10.1016/j.wear.2012.12.006.

Sigamani, N. S. 2010. *Characterization of Polyurethane at Multiple Scales for Erosion Mechanisms Under Sand Particle Impact*. Texas A&M University, USA.

Sinha, S. K. and B. J. Briscoe, (eds.). 2009a. Part I bulk polymers. In *Polymer Tribology*, 724 p., Imperial College Press, London, UK

Sinha, S. K. and B. J. Briscoe. 2009b. *Polymer Tribology*. Imperial College Press, London, UK.

Song, W. Q., P. Krauklis, A. P. Mouritz, and S. Bandyopadhyay. 1995. The effect of thermal ageing on the abrasive wear behaviour of age-hardening 2014 Al/SiC and 6061 Al/SiC Composites. *Wear*, 185: 125–130.

Totten, G. E. and H. Liang. 2004. *Mechanical Tribology: Materials, Characterization, and Applications*. Marcel Dekker Inc., New York, NY.

Yew, Y. K., M. Minn, S. K. Sinha, and V. B. C. Tan. 2011. Molecular simulation of the frictional behavior of polymer-on-polymer sliding. *Langmuir: The ACS Journal of Surfaces and Colloids*, 27(10): 5891–5898, doi:10.1021/la201167r.

Zeng, H. 2013. *Polymer Adhesion Friction and Lubrication*, pp. 699. John Wiley & Sons, Inc., Hoboken, New Jersey.

Zhang, N., F. Yang, L. Li, C. Shen, J. Castro, and L. J. Lee. 2013. Thickness effect on particle erosion resistance of thermoplastic polyurethane coating on steel substrate. *Wear*, 303(1–2): 49–55, doi:10.1016/j.wear.2013.02.022.

36 Superhydrophobicity
From the Origin of the Concept to the Fabrication Methodologies for Applications in Pipeline Coatings

Shihao Huang, Sagar Cholake, Wen Lee,
Kenaope Maithakimako, Divya Eratte,
Sheila Devasahayam, and Sri Bandyopadhyay

CONTENTS

ABSTRACT

In industrial applications, the benefits of superhydrophobic surfaces are precisely obvious, including the improvement of antiwetting, anticorrosion, anticontamination features which improve wear-resistant and provide self-cleaning properties. The difference between theoretical models and case studies shows some gap that needs to be overcome to achieve cost and energy efficiency in the development of superhydrophobic materials. Application of hydrophobicity not only prevents blockage of gas and oil walls in a long pipeline but it also reduces 25% of the adhesion between fluid and pipeline walls. A range of preparation technologies for delivering superhydrophobic properties is discussed in this chapter with their advantages, problems, and contact angle results. Coal log is a typical area in the mining industry where the future research on superhydrophobicity needs to be focused. This chapter delivers a systematic and critical review from the origin of the concept to the most recent developments in the fabrication of superhydrophobic surfaces.

INTRODUCTION

(Super) hydrophobic surfaces have drawn immeasurable attention in both industry and the academic field, all contributed by the fact that superhydrophobic materials have self-cleaning, antiwetting, antisticking, anticorrosion, and anticontamination features.

In the mining industry, there are several applications used:

1. Coal log: Applications of superhydrophobic surfaces are widely used these days, mainly contributed to their amazing self-cleaning and antisticking features. Not only that, based on Wilson's research (Wilson et al., 1994), hydrophobicity can also improve the robustness and water resistance of coal logs. In his research, Wilson described the coal log pipeline transportation system that transported coal logs hundreds of miles before reaching coal-burning utility plants; owing to these long transportation distances, the robustness, water resistance, and abrasion resistance are important concerns of coal logs during their long journey. In order to meet these coal log performance requirements, it is important to select good hydrophobic characteristics to facilitate a coal log fabrication process, and to improve the robustness and water resistance of the coal logs. Future research needs to manufacture strong robust hydrophobic coal logs that are durable enough to withstand the erosion effects of long-distance transportation in a water-filled pipe.
2. Pipeline: Another example is given by cement-lined pipes for water-lubricated transport of heavy oil (Arney et al., 1996). Arney discovered different strategies for preventing oil from fouling the walls of core annular flow pipelines and also for restarting from an unexpected pipeline shutdown. Sulfonated plastics have remarkable oleophobic and hydrophilic properties (the plastic becomes durably wet after immersion in water and oil will not stick), but to use such plastics in a lubricated line will be problematic because they are too expensive, and may be too fragile for pipe applications.

As a wide range of superhydrophobic applications is used in the mining industry, the importance of understanding the superhydrophobic surface and the manufacturing process is even more mandatory. In this chapter, a concise and systematic overview beginning from the fundamental concept and mechanism of the "lotus effect" through some key theories of superhydrophobicity to the rationalization of relevant methodologies for the preparation of superhydrophobic surfaces are discussed in detail.

Natural Superhydrophobic Materials and an Overview of the Mechanism behind the Lotus Effect

There are various superhydrophobic materials (Gao and Jiang, 2004; Sun et al., 2005b; Neinhuis and Barthlott, 1997) in nature, including bird feathers, insect wings, animal furs, and plant leaves. The lotus leaf is the most well-known superhydrophobic natural material, resulting in its high water repellence, antisticking, and self-cleaning property. The contact angle of lotus leaf can be 160.7° with a contact angle hysteresis of 2°. Figure 36.1 is the illustration of a lotus leaf under scanning electron microscopy (SEM).

FIGURE 36.1 (a) Lotus leaves exhibit shows extraordinary water repellence on their upper surface. (b) Scanning electron microscopy (SEM) image of the leaf's upper surface showing hierarchical structure consisting of papillae, wax clusters, and wax tubules. (c) Wax tubules on the upper surface. (d) Leaf upper surface after critical-point drying with 15° tilt angle. (e) Leaf downside surface shows convex cells without stomata. (Adapted from Ensikat, H. J. et al. 2011. *Beilstein J. Nanotechnol.*, 2:152–161.)

The lotus leaf always stays clean from dirt mainly because of its highly water repellent surface. When water falls on the lotus leaf, it rolls out as beads carrying away dirt along with it. In earlier stages, it was believed to be due to the smoothness of the leaf surface. Then, the relationship between roughness and water repellency were studied by a number of researchers (Cassie and Baxter, 1944). Later, the lipid/wax layer with its crystalloid nature was considered as the major reason for the superhydrophobicity of lotus leaves (Jeffree, 1986). There is an interfacial layer between any plant surface and the surrounding atmosphere, which can be called the cuticle, which mainly consists of soluble lipids embedded in a polyester matrix (Holloway, 1994). Owing to its chemical lipid composition, they exhibit hydrophobic nature. Further investigation showed that two scales of waxy surface texture gave rise to the water repelling property of the lotus leaf—one at a micro-scale level and another at submicron or nanoscales (Bhushan and Jung, 2006; Bhushan et al.,

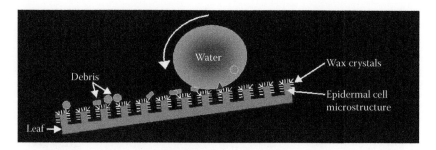

FIGURE 36.2 Schematic representation of "lotus effect". (http://www.thenakedscientists.com/HTML/articles/article/biomimeticsborrowingfrombiology. Reproduced by permission of The Royal Society of Chemistry.)

2009). The microscale roughness on the lotus leaf surface is formed by convex cell papillae, and the nanoscale roughness is formed by hydrophobic wax tubules (Figure 36.2). The extremely high contact angle, probably, in the range of 160°C and low contact angle hysteresis results in a tendency for water droplets to roll off the surface as opposed to slide, removing surface contaminants and dirt with them. Several researchers have observed the morphology and surface characteristics of the lotus leaf using advanced microscopical technologies such as SEM and have observed microscopic-level bump-like structures as well as nanolevel hair-like structures covering the leaf surface (Figure 36.3). Therefore, these two levels of roughness (micro and nano) allow the air to get entrapped in the space between and will not allow space for water to get in. The micro–nano–air composite surface is responsible for the high contact angle and thereby the superhydrophobicity of the lotus leaf. Recently, several attempts have been made to synthesize surface structures that mimic the superhydrophobic behavior of the lotus leaf (Shirtcliffe et al., 2010; Celia et al., 2013). Superhydrophobic, self-cleaning surfaces exhibiting the so-called "lotus effect" are of interest in many commercial and industrial applications (Wang et al., 2006, 2007, 2012; Zhu et al., 2006).

In summary, the morphology of the lotus leaf helps in understanding nature's superhydrophobicity which helps us in developing artificial superhydrophobic materials with similar advantages, which are suitable to many industrial applications such as antisticking, stain-resistant texture, as well as antisoiling architectural/mining coatings (Zielecka and Bujnowska, 2006).

FUNDAMENTALS OF SUPERHYDROPHOBICITY AND SOME RELATED THEORIES

Wettability is an important property for a solid surface (Shibuichi et al., 1996), and controlling the wettability may improve the performance of applications (De Coninck et al., 2001). Wettability can be expressed by the contact angle on the surface, and it can be divided into three categories, namely, hydrophilic, hydrophobic, and superhydrophobic.

When a liquid drop falls on a solid surface, it can take different shapes such as spherical, hemispherical, or flat. This is because different materials have different surface properties and surface

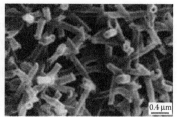

FIGURE 36.3 Morphological details of lotus leaf. (Adapted from Koch, K. and H. Ensikat. 2008. *Micron*, 39(7): 759–772.)

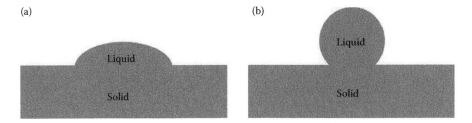

FIGURE 36.4 Illustration of the contact angle of a liquid on an ideal flat surface (Young's modulus). For water, if $\theta_{flat} < 90°$, it is an intrinsically hydrophilic surface (a); if $\theta_{flat} > 90°$, it is an intrinsically hydrophobic surface (b).

properties of any material actually depend on their chemical compositions. One important surface property is wetting behavior, which means how water interacts with surfaces. There are surfaces that spread water, which are mainly termed hydrophilic (water-loving) and those that do not spread water, but instead roll water beads, which are termed as hydrophobic (water-hating) in a most general way. The most important characterization parameter for determining this kind of surface wetting behavior is through the measurement of the contact angle of the liquid on the solid surface. The contact angle is defined as the angle between the tangent to the solid–liquid interface and the tangent to the liquid–fluid interface (Marmur, 2006). Contact angle analysis is an indirect technique to determine surface energy by placing a drop on the surface. It is generally accepted that if the contact angle is acute (<90°), the surface is hydrophilic, and if the contact angle >90°, the surface is considered hydrophobic (Figure 36.4). Superhydrophobic materials are those having contact angle >150°.

SURFACE ENERGY

Surface energy is an important physical parameter which controls a wide range of phenomena, including stress for brittle fracture as well as wettability (Owens and Wendt, 1969; Tyson and Miller, 1977). Many applications of materials require adherence to other substances while some do not. Adhesion is a manifestation of the attractive forces between all atoms.

Lower surface energy will enhance hydrophobicity and eventually achieve superhydrophobicity. Young's equation (36.1) is a good way to evaluate the wettability of a solid surface.

YOUNG'S EQUATION

The contact angle of a surface can be expressed by Young's equation as given below (Wolansky and Marmur, 1998; Coninck et al., 2002; Wang et al., 2014):

$$\cos\theta = \frac{\gamma_{SG} - \gamma_{SL}}{\gamma_{LG}} \tag{36.1}$$

where θ is the contact angle, γ_{SG} is the interfacial tension between solid and gas, γ_{SL} is the interfacial tension between solid and liquid, and γ_{LG} is the interfacial tension between liquid and gas.

Hydrophilic surfaces have a contact angle <90°, while hydrophobic surfaces have a contact angle >90°.

There are two types of contact angle values: (a) static contact angles and (b) dynamic contact angles. To determine the flat surface, the static contact angle is usually used as it is close to Young's angle. The method to measure this static contact angle is using the sessile drop technique, where a drop of water is vertically deposited onto the surface of the material and the value is determined by a goniometer. Dynamic contact angles are nonequilibrium contact angles measured by contact angle hysteresis ($\Delta\theta$), which is the difference between the growth and shrinkage of the droplet (Brzoska et al., 1994).

Young's equation can be applied for a perfectly flat and homogeneous smooth surface. Unfortunately, all real surfaces are chemically heterogeneous and rough. Therefore, the applicability of Young's equation in real time is more complex.

Contact Angle on Rough Surface

There are two models to calculate contact angles on rough surfaces: Wenzel's model (Wenzel, 1936) and Cassie–Baxter model (Cassie and Baxter, 1944).

Wenzel's Model

In Wenzel's model (Wenzel, 1936), the droplet penetrates into the asperities, and it is given by the equation

$$\cos\theta^w = \frac{r(\gamma_{SG} - \gamma_{SL})}{\gamma_{LG}} = r\cos\theta \tag{36.2}$$

It can be inferred from Wenzel's equation that if the contact angle on a smooth surface is acute ($\theta < 90°$), then the contact angle on the rough surface (θ_w) is lesser than the contact angle on the smooth surface (θ). On the other hand, if the contact angle on the smoother surface is obtuse ($\theta > 90°$), then the contact angle on the rough surface (θ_w) is greater than the contact angle on the smooth surface (θ)

Wenzel's theory states that when the droplet comes in contact with rough surface, θ will immediately change into θ^w. θ is the Young's contact angle on the smooth surface while r stands for the surface roughness factor, which is defined as the ratio between real surface area and apparent surface area as in the following equation:

$$r_s = \frac{A_{real}}{A_{apparent}} \times 100\% \tag{36.3}$$

where A_{real} stands for true value on surface and $A_{apparent}$ for apparent value on surface. Based on Equation 36.3, when $A_{apparent} = A_{real}$, $r_s = 1$, which means the surface is perfectly smooth; when $r_s > 1$, it should appear with a rough surface.

Wenzel's equation predicts that the wetting is enhanced with the degree of roughness when the surface is hydrophilic (contact angle <90°). On the other hand, the wetting is lessened by roughness when the surface is hydrophobic (contact angle <90°).

Cassie–Baxter Model

The Cassie–Baxter model assumes the presence of air bubbles trapped in the rough grooves. These air bubbles induce friction from droplets and air, causing the suspension of the droplet on top of the asperities.

The Cassie–Baxter model describes the contact as the sum of all surfaces, including the rough surface on top of the asperities and the air friction from air bubbles trapped in the grooves. The equation is given as

$$\cos\theta^c = f_1\cos\theta_1 + f_2\cos\theta_1 \tag{36.4}$$

where θ^c is the apparent contact angle, f_1 and f_2 are the surface fraction for air phase and solid phase, respectively, and θ_1 and θ_2 are the contact angle of air phase and solid phase, respectively

If the rough surface has only one asperity existing, then the fraction of solid will be f and the fraction of air will be $(1 - f)$. Because the contact angle of water in air is 180°, the equation will transform into

$$\cos\theta^c = f(1+\cos\theta)-1 \qquad (36.5)$$

Thus, the Cassie–Baxter model can be described by a sole function of solid fraction with its contact angle of solid phase.

In a cosine function, the value of cos θ will increase while θ decreases. So, to result in the highest contact angle, the solid phase needs to be extremely small and the contact angle on solid phase needs to be as large as possible which in this case will depend on the material properties (Elif Cansoy et al., 2011). Normally, the Cassie–Baxter model cannot determine the practical situation, but it can always be used to compare the results with the practice.

Figure 36.5 is an illustration of two behaviors for a liquid drop on a rough surface. On the left is Wenzel's theory, liquid penetrating into the spikes; on the right is Cassie–Baxter's theory, liquid suspending on the spikes (Li et al., 2007).

Both the Wenzel model and Cassie–Baxter models can determine the contact angle on rough surfaces only qualitatively and there is no clear evidence that either of them have an advantage over the other. Over the years, researchers in this field continued to improve these equations and even tried to invent new model to fit more practical situations for the contact angle test.

CONTACT ANGLE HYSTERESIS

The contact line will begin to advance and increase steadily after the volume of a droplet is placed on the surface. The advancing contact angle (θ_a) is the contact angle observed when it begins to move (Johnson and Dettre, 1964). When the droplets become stable and retract back until the contact line recedes, the contact line begins to move again. As a result, receding contact angle (θ_r) is the contact angle line set in motion by this process (Zisman, 1964; Wang et al., 2004). Contact angle hysteresis is also known as the sliding angle (α), and

$$\alpha = \theta_a - \theta_r \qquad (36.6)$$

The sliding angle is critical since it defines how easily a droplet will slide off a surface. A small sliding angle promotes the self-cleaning property of a superhydrophobic surface. It is an important parameter in defining superhydrophobicity (Figure 36.6).

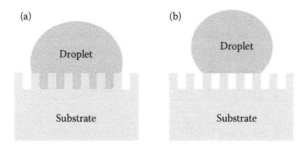

FIGURE 36.5 (a) Wenzel's theory, (b) Cassie–Baxter's theory.

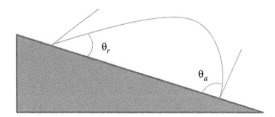

FIGURE 36.6   Illustration of contact angle hysteresis, showing advancing contact angle and receding contact angle when droplet slides on a surface.

SIGNIFICANCE OF CONTACT ANGLE THEORIES IN THE DEVELOPMENT OF SUPERHYDROPHOBIC SURFACES

Even though Young's theory is the basic theory, the Wenzel and Cassie–Baxter theory are found to be more realistic on solid rough surfaces in practical situations. Wenzel's model has actually explained homogeneous wetting stating that a liquid can completely penetrate into the grooves. However, homogeneous wetting may not happen in real time as vapor pockets may become entrapped under the liquid layer forming a composite regime. Therefore, the Cassie–Baxter model can be applied to heterogeneous wetting cases. Both the models have highlighted the geometrical configuration of solid surfaces as a significant feature in determining wettability. In most practical situations, Wenzel's concept and Cassie–Baxter's concept are simultaneously applied, as shown in Figure 36.5.

PREPARATION OF SUPERHYDROPHOBIC SURFACE

Both the Wenzel and Cassie–Baxter theories suggest that a rough surface would help improve hydrophobicity and hydrophilicity. Two approaches are given in the literature: (1) top-down approach and (2) bottom-up approach. The top-down approach involves lithographic and template-based techniques, whereas the bottom-up approach involves mostly self-assembly and self-organization, which includes chemical deposition, layer-by-layer (LBL) deposition, hydrogen bonding, and colloidal assemblies. At the same time, other methods have been discovered to combine both approaches; such as casting of polymer solution, phase separation, and electrospinning. These approaches are discussed in the following section in detail.

TOP-DOWN APPROACH

The top-down approach is based on microelectronic materials and devices manufactured in several ways, including carving, casting, machining, or by use of bulk materials tools and lasers. Template, micromachining, and plasma treatments have also been used over the past few years. Normally, a temptation process involves molding and replication steps, and then the template can be lifted off or dissolved, or removed by sublimation. Lithographic approaches involve a light through a mask having desired radiation characteristics to the substrate (silicon) with a photoresist. In the subsequent etching step, designed pattern with hydrophobicity will be transferred by silanization. Micromachining simply dices the surfaces into the desired texture. In the plasma treatment, the surface is anisotropically etched to produce a rough surface. In the following, each individual method is discussed in detail.

Template Technique

Temptation

The template technique is one of the soft lithography techniques (Qu et al., 2011). To replicate the same feature from the template, it can be done by molding and the subsequent lifting off the replica and dissolution of the templation. The template technique is very useful for preparing polymeric superhydrophobic surfaces. Many materials can be used as a template, ranging from natural lotus leaves to fish skin (Li et al., 2007; Qu et al., 2011).

In the Sun et al. (2005a) experiment, a lotus leaf was used as a template to reproduce a replicate using nanocasting. After the casting process, lifting off polydimethylsiloxane (PDMS), the negative replication of the lotus leaf was available.

This negative template is further used as the main source to replica the lotus leaf. The replica of the lotus leaf almost has the same nanostructure as a natural lotus leaf. Compared with a natural lotus leaf and positive PDMS replica, the average distance of the small papillae hills are the same (about 6 µm). The replica not only presents the same size of papillae hills, but also it brings perfect nano-textures between the hills and valleys with a closely identical shape. After the contact angle test, Sun's team found the positive replica had a contact angle of 160° which is the same value of a natural lotus leaf, while the negative replica has a contact angle value of 110°.

Peng et al. (2013) fabricated superhydrophobic surface with microcavities through the template method. These superhydrophobic surfaces with microcavities were produced by using taro leaves as a master template and dip coating was followed to modify the superhydrophobic surface. The authors claimed that this process is an improved template method (ITM) and the results obtained were significantly better than those achieved using traditional template methods. Additionally, the water repellence of the microcavities surface was significantly enhanced with a layer of polymerized *n*-octadecylsiloxane nanosheets (Figure 36.7).

Yuan et al. (2013) fabricated a superhydrophobic surface on a copper sheet by the combination of a PDMS template and chemical etching method and achieved the water contact angle and sliding angle of 153° and 7°, respectively. These surfaces can be used for the large-scale commercial production of superhydrophobic surfaces on metals. Wang et al. (2012) fabricated a novel flower-like superhydrophobic epoxy resin surface (Figure 36.8) with water contact angle (158 ± 1.7°) and sliding angle (3°) using ZnO powder agglomerations as template. These microflower-like superhydrophobic epoxy resin surface can be used for various applications such as anticorrosion, anti-icing, self-cleaning, etc.

Nanoimprint Lithography

One of the other template techniques is called nanoimprint lithography which is also another way to replicate patterns. This pattern replication process is completed by both heat and pressure with a

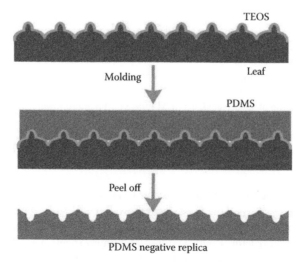

FIGURE 36.7 ITM on taro leaf. (Reprinted from *Journal of Colloid and Interface Science*, 395, Peng, P., Q. Ke, G. Zhou, and T. Tang, Fabrication of microcavity-array superhydrophobic surfaces using an improved template method, 326–328, Copyright 2013, with permission from Elsevier.)

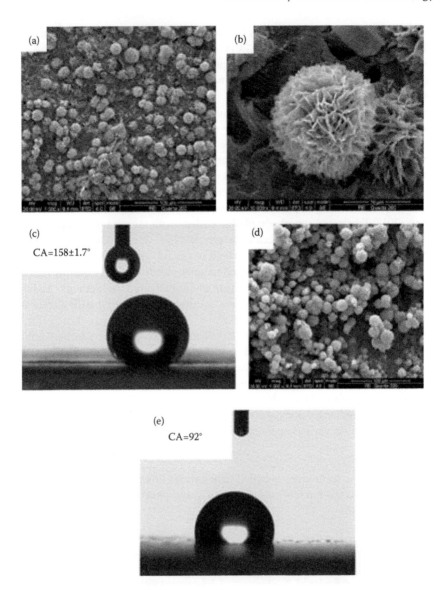

FIGURE 36.8 (a) SEM image of microflower-like epoxy resin surface before modification with stearic acid. (b) Higher magnification of (a). (c) Shape of a water droplet on the as-prepared microflower-like epoxy resin surface after the modification of stearic acid. (d) SEM image of the microflower-like epoxy resin surface after the modification of stearic acid. (e) Shape of a water droplet on the smooth epoxy resin surface modified by stearic acid. (Reprinted from *Materials Chemistry and Physics*, 135, Wang, C. et al., A novel method to prepare a microflower-like superhydrophobic epoxy resin surface, 10–15, Copyright 2012, with permission from Elsevier.)

hard master pressed onto the thermoplastic polymer layer above glass transition temperature of the polymer. After cooling down when the master is removed, the negative replica of the template will be available. An example of nanoimprint is deduced by Lee et al. (2004). His team used polystyrene to replicate different nanostructures with textured aluminum sheets and anodic aluminum oxide (AAO) membranes. After the heat and pressure process, nanostructure patterns are able to be transferred onto the polystyrene substrates. Then cooling to room temperature and releasing the pressure, the replication template can be removed from the polystyrene substrates by dissolution of AAO with saturated $HgCl_2$ solution. The diameters of polystyrene nanofibers can be controlled using AAO

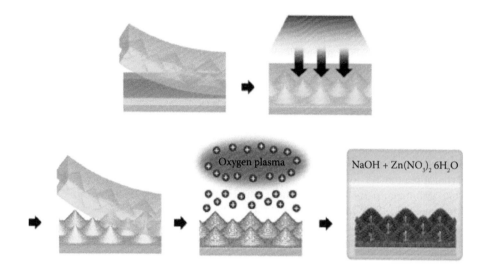

FIGURE 36.9 Fabrication of superhydrophobic and superoleophobic surfaces based on ZnO. (Reprinted from *Microelectronic Engineering*, 116, Jo, H. et al., Superhydrophobic and superoleophobic Surfaces Using ZnO nano-in-micro hierarchical structures, 51–57, Copyright 2014, with permission from Elsevier.)

replication temptation with different pore diameters while the diameters can be controlled by wet chemical etching using 5 wt.% H_3PO_4. Under this wet chemical etching condition, the pore diameter of the template increases at the rate about 1 nm/min. Then these replica surfaces have a water contact angle between 155.8° and 147.6° which reflects the high degree of water repellent property as successfully replicated by nanoimprint lithography.

Jo et al. (2014) fabricated superhydrophobic and superoleophobic surfaces based on ZnO nano-in-micro hierarchical structures on various large-area substrates through ultraviolet nanoimprint lithography combined with hydrothermal synthesis (Figure 36.9). These surfaces showed static water contact angle >160°, contact angle hysteresis <2°, and sliding angle of <1° and can be extended on different substrates such as Si wafer, glass, and flexible PET (polyethylene terephthalate) film.

Capillary Force Lithography

Capillary force lithography is a different imprint nanolithography. It needs a patterned elastomeric mold instead of a hard master. This patterned elastomeric mold is directly laid onto a spin-coated polymer film on a substrate. By increasing the temperature to about the glass transition temperature of the polymer, with solvent evaporation (temperature-induced capillarity), or by direct molding prior to solvent evaporation (solvent-induced capillarity), the negative replica of the mold is produced.

Suh and Jon (2005) used capillary lithography to control wettability of polyethylene glycol surfaces (PEG) depending on the geometry of the nanostructures. This method was used to fabricate PEG nanostructures by using functionalized polyurethane with acrylate group into an ultraviolet curable mold. In their report, there were two wetting states depending on the height of nanostructures. One was relatively low, which is <300 nm for 150 nm pillars with 500 nm spacing. The initial contact angle result was <80° and water content easily penetrated the surface into the grooves leading in a reduced contact angle at equilibrium (Wenzel state). By comparison, the other higher one showed a hydrophobic nature and did not have much change in contact angle result over time (Cassie–Baxter state). Both the wetting states were confirmed by dynamic wetting properties as contact angle hysteresis. However, due to PEG having a hydrophilic nature and not a hydrophobic nature, the contact angle observed was around 95°. The contact angle might be improved by swapping PEG to another polymer material with higher hydrophobicity itself or by changing the geometry of the mold.

All in all, templation is the method used very often for superhydrophobic surfaces. Templation will have a dominating effect on hydrophobicity. Using different sorts of templation, they can be fabricated with tunable surface morphology (Jin et al., 2005). The limitations for the templation approach include the following: (a) the material may not be suitable to prepare the superhydrophobic surface due to the temperature, (b) the attainable geometry is very limited in this approach, as well as all sorts of other issues.

Photolithography

Photolithography is a microfabrication method in which a light is irradiated through a mask with a designed pattern to substrates with a photoresist property. It uses the light to transfer a geometric pattern from a photomask to the photoresist chemical on the substrate. Photolithography can be divided into two major parts: (a) x-ray lithography (Heuberger, 1986, 1988) and (b) e-beam lithography (Marrian et al., 1994; Mendes et al., 2004).

Normally the preparation of superhydrophobic surfaces needs extra steps, but in photolithography, there are no such extra steps involved, and this easy operation is the reason why it is often used for surface molding.

X-Ray Lithography

Silicon wafers are normally manufactured by x-ray lithography; they have regular patterns of spikes.

The distance between the spikes and the distance between their width and height are varied, as shown by Fürstner et al. (2005). Table 36.1 shows the designed dimensions and actual dimensions in 18 different types of structures, measured by SEM, where the spikes' widths are measured at the top. With a large range of the pillars, the contact angle ranges from 113° to 161° depending on the width, clearance, and height of the spikes. Furstner concluded that the best cleaning surface would have width/clearance = 2 and height/width = 4 (Sun et al., 2005).

E-Beam Lithography

Martines et al. (2005) fabricated nanopatterns with increasing solid fraction (two samples with nanopillars and nanopits) on silicon wafers. The samples were modified by oxygen plasma for 30 min and the contact angle results on the samples were tested within 24 h. Experimental results in Table 36.2 show all dimensions of the nanopatterns on wafer with asperities (h), diameter (d), and length between pillars (l). The hydrophobic patterns were obtained by silanization of the hydrophilic surfaces with octadecyltrichlorosilane (OTS).

The Cassie–Baxter and Wenzel models give an accurate estimate of the advancing angle of the hydrophobic surface. The two models are very sensitive when dealing with the asperity profile.

TABLE 36.1
Hydrophobic Surface Dimensions

Sample Name	Dimensions		$h = 1\ \mu m$		$h = 2\ \mu m$		$h = 4\ \mu m$	
	d (μm)	a (μm)	d (μm)	a (μm)	d (μm)	a (μm)	d (μm)	a (μm)
d1a1	1	1	0.85	1.05	0.89	1.01	0.9	1.01
d1a1.5	1	1.5	0.77	1.59	0.85	1.55	0.84	1.53
d1a2	1	2	0.9	1.97	0.92	1.93	0.94	1.95
d1a3	1	3	0.88	2.89	0.92	2.89	0.96	2.9
d1a5	1	5	0.75	4.91	0.88	4.86	0.92	4.82
d2a2	2	2	1.94	1.88	1.94	1.89	2.02	1.91

Source: Adapted from Sun, T. L. et al. 2005. Bioinspired Surfaces with Special Wettability. *Accounts of Chemical Research*, 38: 644.

TABLE 36.2
Dimension of Nanopatterns

	H90	H83	P22	P21	P13	P12
d (nm)	105	138	157	156	124	117
h (nm)	116	114	239	286	268	792
l (nm)	300	300	300	300	300	300

Source: Adapted from Lee, W. et al. 2004. Nanostructuring of a polymeric substrate with well-defined nanometer-scale topography and tailored surface wettability. *Langmuir*, 20(18): 7665–7669.

Martinez (Kim et al., 2012) also confirmed that the forest is the most stable pillar of slender waterproof texture.

Nanosphere Lithography

Nanosphere lithography (Suh and Jon, 2005) is a comprehensive way for patterning nanosphere arrays in a large area. Spin coating onto the surface using a monodispersed polystyrene beads solution can easily self-organize a closed pack nanostructure which has been achieved by Hulteen and Van Duyne (1995). Haynes and Van Duyne (2001) suggested that for both double layer and single layer, closed-packed polystyrene arrays of more than few square centimeters can adjust the spinner's speed and concentrate the surfactant in the polystyrene solution. After the close-packed nanostructures are formed, oxygen plasma etching will tell the difference between liquid state and solid state for nanostructure surface fraction.

Therefore, the process should reduce the polystyrene beads' diameter while maintaining the separation distance as a constant. Octadecanethiol (ODT) is used to modify a 20-nm-thick gold film on a size-reduced polystyrene array, and then the process of water repellent surface has been made.

Plasma Treatment of Surfaces

Plasma treatment normally uses the plasma etching process with a high-speed stream of glow discharge and an appropriate mixed gas shot to the sample. The plasma source can be either a charged source (ions) or a neutral source (atoms and radicals). Plasma treatment of surfaces can cause a considerable change in the surface structure because of the anisotropic etching of the surface layers (Wu et al., 1997).

Fresnais et al. (2006a) used plasma treatment on low-density polyethylene (LDPE). The main characteristics of these superhydrophobic plasma-treated surfaces result in a high contact angle (170°) and low contact hysteresis (5°). Fresnais also tested a range of roughness from 20 to 400 nm. As a result of these samples he concluded that the range of roughness is very wide when using plasma treatment.

Balamurali et al. (Balu et al., 2008) demonstrated that superhydrophobic properties can also be applied on paper as well by a combination of deposition and plasma etching. He uses chemical vapor deposition (CVD) of pentafluoroethane (PFE) onto a fluorocarbon film which is around 100 nm on Si wafer; the gas he chose for the etching process was oxygen. As both CVD and plasma etching process only work on the surface, it is considered the bulk material remains the same. Furthermore, since this method disclosed the inherent roughness of the cellulose fibers has higher roughness on imparting superhydrophobic paper grafted polymer.

The material Balamurali (Heuberger, 1986) used is very different from Fresnais (Fresnais et al., 2006b). While Fresnais used a common inorganic polymer such as cellulose, which has the advantage of being renewable, flexible, biodegradable, a cheap biopolymer, and available in large quantity. After plasma-enhanced CVD, the superhydrophobicity for the material largely increased with the contact angle as high as 166.7°. When compare to the conventional polymers, grafted nanoparticle are being deposited to produce the roughness which divides all fibers and microscale roughness. The interior surface of the paper obtained the inherent roughness of Nanoscale cellulose are robust.

Bottom-Up Approach

The bottom-up approach is more suitable for large and complex objects by integration of smaller components. In general, the bottom-up approach includes involves self-organization and self-assembly. Self-organization (Lehn, 2002) is the driving force leading to the evolution of the biological from inanimate matter. Self-assembly (Whitesides and Grzybowski, 2002) is an integration method for components spontaneously assembled in solution or the gas phase reaching its stable structure minimum energy. There are many methods for the bottom-up approach to prepare a superhydrophobic surface, including chemical deposition, colloidal assembly, sol–gel method, and LBL deposition. In the following, these methods are discussed.

Chemical Deposition

Chemical deposition can be divided into three categories, (a) chemical bath deposition (CBD), (b) CVD, and (c) electrochemical deposition. These depositions involve chemical reactions producing a suitable substrate. The chemical deposition method is normally used for generating thin films mainly in crystalline inorganic materials such as InS, CdS, CuSe, etc.

Wu et al. (2005) used CBD to fabricate the superhydrophobic surface of ZnO, due to the suitable growth of ZnO crystalline forming an engineered surface, and a range of alkanoic acids being tested to tune the surface wettability. Through his experiment results, all the results on the contact angle are >150°, which really make a difference in the wetting mechanisms. The engineered surface showing a stable superhydrophobicity is mainly dominated by the Cassie–Baxter state when the chain length of alkanoic acid is more than 16. From the length of alkanoic acids between C8 and C14, it tends to have a Wenzel state instead and displays a great contact angle hysteresis.

Feng et al. (Zhang et al., 2005) also used zinc oxide (ZnO) to prepare the superhydrophobic surface. In Figure 36.10a and b field emission scanning electron micrographs (FE-SEM) for the surface in low and high magnification, respectively, show the nanorods are very densely packed with uniform arrays. Most of the nanorods have flat hexagonal crystallographic planes, while projecting out of the crystal seeds layer with range of the diameters from 50 to 150 nm. Figure 36.10c is an SEM image on a cross-sectional view of aligned ZnO nanorods which are all aligned vertically onto the surface with the height of 1.2 μm. Figure 36.10d shows an x-ray diffraction large peak in (002) plane while the surface nanorods are (001).

Huang et al. (2005) prepared aligned carbon nanotube-coated zinc oxide thin film. Contact angle measurements show that the coating of ZnO on CNT is superhydrophobic and the contact angle is 159°. Unlike the uncoated CNT surface, the ZnO-coated surface of carbon nanotubes showed signs of water leakage, even over an extended period of time. In this case, surface wettability is able to change from superhydrophobic to hydrophilic using dark storage or ultraviolet irradiation.

Other chemical deposition using gold clusters (Zhang et al., 2004) and silver aggregates (Zhao et al., 2005) have also been used to successfully produce superhydrophobic surfaces.

Colloidal Assemblies

Because of the interaction from van der Waals, monodispersed particles can form close-packed assemblies on the surface. The particle assemblies render roughness to the underlying substrates. Plasma etching is always introduced to colloidal assemblies due to its ability to improve roughness for substrates. The colloidal particles can vary from polymer beads to inorganic spheres. The main advantage and wide use of this method contribute to its cost efficiency due to the fact that it has no expense on lithographic techniques. It is also very easy to operate under laboratory conditions.

In the Zhang et al. (2005) research, superhydrophobic surfaces were derived from binary colloidal assemblies. $CaCO_3$ loaded hydrogel spheres and silica gel beads or polystyrene beats are

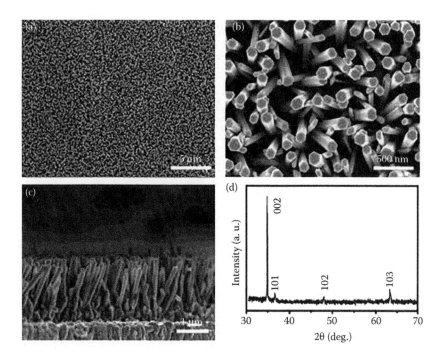

FIGURE 36.10 (a, b) FE-SEM top images of the as-prepared ZnO nanorod films at low and high magnifications, respectively. (c) Cross-sectional view of the aligned ZnO nanorods. (d) XRD pattern of the as-synthesized nanorod films. (Reprinted with permission from Feng, X. et al., 62–63. Copyright 2003 American Chemical Society.)

continuously dip-coated on the silicon wafer, the former a combination of recruitment silicon wafers and $CaCO_3$ loaded hydrogel spheres. Because silicon wafers and $CaCO_3$ loaded hydrogel spheres have different hydrophilicity, the selective region will localize the silica and polystyrene beats to an irregular dual structure with high roughness. Use of low surface energy molecules of subsequent modification produces a superhydrophobic surface. The heat treatment may increase the mechanical stability of the resulting dual structure to a large extent, which allows the reproduction superhydrophobic surface, in practice to provide good durability. Contact angle results are divided into two groups, as shown in Table 36.3.

TABLE 36.3

Water Contact Angle of the Surfaces of Silicon Wafers and Those Coated with Colloidal Self-Assemblies

Substrate Surfaces	Unmodified	Modified
Silicon wafer	<5°	114°
Silica colloidal self-assemblies	<5°	129°
$CaCO_3$-PNIPAM colloidal self-assemblies	44°	99°
Silica/$CaCO_3$-PNIPAM binary assemblies	45°	160°
PS colloidal self-assemblies	24°	123°
PS/$CaCO_3$-PNIPAM binary assemblies	<5°	156°

Source: Adapted from Jin, M. H. et al. 2005. *Advanced Materials*, 17(16): 1977–1981.

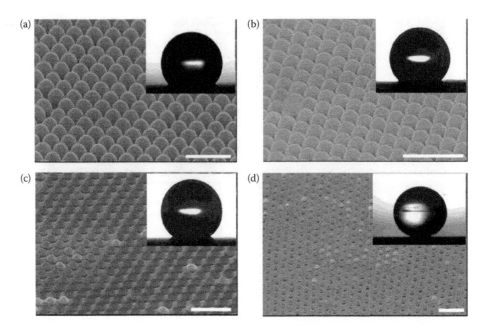

FIGURE 36.11   SEM images (60°) of the size-reduced polystyrene beads and the water contact angle measurement on the corresponding modified surfaces (insets). The diameters of polystyrene beads and water contact angles on these surfaces were measured to be (a) 400 nm, 135°, (b) 360 nm, 144°, (c) 330 nm, 152°, and (d) 190 nm, 168°. Bar: 1 μm. (Reprinted with permission from Heuberger, A., 107–121. Copyright 1988 American Chemical Society.)

The heating treatment may largely enhance the mechanical stability of the resulting binary structures, which allows regeneration of the surface superhydrophobicity, providing good durability in practice. Water contact angles are being examined when divided into two groups as in Table 36.3.

In Jau et al.'s (Shiu, 2004) research, spin coating is used in monodispersed polystyrene beads to form dense-packed superhydrophobic surfaces. In this case, it is also necessary to control the solid air fraction of the nanobeads, as is done by oxygen plasma etching. After surface treatment and coating with gold, their team observed contact angle with SEM images. Figure 36.11 shows SEM images of these size-reduced polystyrene arrays. During the experiment, 440 nm diameter polystyrene beads were used to form single-layer arrays. Compared between 36.11a and d with four images, decrease in the diameters of polystyrene beads due to oxygen plasma treatment results in increasing the contact angles. The apparent water contact angle of modified polystyrene surfaces changed from 132° with 440 nm diameter polystyrene arrays to 168° with 190 nm diameter polystyrene arrays. The contact angle result on the rough surface (ODT-modified gold surface) is much greater than on the flat surface. Jau et al. (cited from Zhang et al., 2005) concluded that it is the surface nanostructure and not the low surface energy molecules that governs the superhydrophobic behavior of surfaces.

Layer-by-Layer Deposition

LBL deposition normally takes advantage of the electrostatic charge interacting with different layers like polycation and polyanion (Li et al., 2007). The LBL method is easy to operate and easy to control on the layer thickness in nanoscale. Hydrophobization is always needed due to the fact that polyelectrolytes behave as hydrophilic.

Multilayer systems always need to have nanoparticles filled in order to increase the roughness effect. There are also some additional treatments that can increase the surface roughness.

Zhai et al. (2004) demonstrated superhydrophobic surfaces by creating a honeycomb-like poly-electrolyte multilayer surface coated with silica nanoparticles. Polyallylamine hydrochloride (PAH)/poly (acrylic acid) (PAA) multilayers is used via LBL deposition.

Zhai's superhydrophobic coatings achieved a very high texture of the multilayer surface using fluorinated silanes. Even if the surface is soaked in water for an extended time its superhydrophobic properties still remains. It was found that a combination of acid treatment and induced deposition of silica nano-microstructure nanostructured particles are both necessary to create a stable super-hydrophobic surface.

Zhao et al. (2005) describe the method using both electrochemical deposition and LBL assembly. His team using the matrix for electrodeposition was 6.5 bilayers of poly diallyldimethylammonium chloride (PDDA) and polystyrene sulfonate (PSS) multilayers fabricated onto the indium tin oxide (ITO) electrode by the LBL method. The repeated process included dipping ITO in PDDA aqueous solutions and in PSS solution for 5 min each as well as water rinsing and nitrogen drying. This process is used to ensure the deposition is LBL. They used the method prepared for potentiostatic electrodepositing in mixture solution of $NaNO_3$ and $AgNO_3$ on the ITO electrode adding with poly-electrolyte multilayer. Afterwards they used Ag aggregates immersed in an ethanol solution of n-dodecanethiol for 12 h and rinsed with pure ethanol the next morning. After this surface modification, the contact angle result showed a superhydrophobicity of 154° and with a 3° tilt angle.

Sol-Gel Methods

The basic principle behind the sol–gel method is to convert the precursor into a glassy material through a series of hydrolysis and polycondensation reactions. This technique can be utilized to produce a superhydrophobic surface with good thermal resistance (Celia et al., 2013). This process is mainly contributed by the conversion of the monomers into the colloidal solution sol as the precursor for an integrated network and gel formation. Silica is usually prepared by condensation ortho-silicate and hydrolysis. The sol may be administered directly or in combination with a filler such as silica nanoparticles. The surface property depends on how sol is modified and how functional group result on the gel as explained by Daoud et al., 2006.

In Hikita et al.'s (2005) experiment, sol–gel films were prepared using condensation of alkoxysilane compounds and hydrolysis. Fluoroalkylsilane and colloidal silica particles are the two keys in controlling surface energy and the roughness. The surface will start to become hydrophobic when the weight percentage for colloidal silica and fluoroalkylsilane were optimized. These films demonstrated a 150° water contact angle. These samples' other important features besides superhydrophobicity include transparency and durability.

Nakagawa and Soga (1999) successfully produced water repellent films which have high heat resistance of about 300°C. Yet, these films are not being used in the conventional sol–gel method. The sol–gel method is used with silicon as substrate and coated with ethyltrimethoxysilane and tetraethoxysilane. Then it uses a treatment of ammonia annealed to 300°C. However, the contact angle is only 110°.

Unlike Nakagawa, Doshi et al. (2005) produced superhydrophobic surface with a high contact angle of 160° using the sol–gel method. In this experiment, neutron reflectivity is used to probe the solid–liquid vapor interface of a porous superhydrophobic surface submerged in water. UV/ozone treatment is used to control the surface coverage in the silica skeleton of a hydrophobic organic ligand; thereby allowing the contact angle with water to be continuously changed when the range exceeds 160°.

Ma et al. (2012) fabricated a superhydrophobic surface on alumina film by using the sol–gel method. A "non-sticky" superhydrophobic surface was obtained upon self-assembly of fluoroalkyl phosphonic acid. Lin et al. (2013) fabricated a highly stable thermal (up to 400°C) and transparent superhydrophobic surface (transmittance close to 90%) by the sol–gel method using poly(methylhydrosiloxane) (PMHS) and tetraethoxysilane (TEOS) as precursors and the optimum coating achieved a maximum water contact angle of 164.7° and sliding angle of 2.7°; but with poor moisture resistance, which was further improved by cetyltrimethoxylsilane (CTMS). Lin et al. (2013) has fabricated superhydrophobic nanocoatings by a simple sol–gel dip-coating method and reported a good moisture resistance. They have used 3-aminopropyltriethoxylsilane (APTEOS) as the aggregated agent to give the nanocoatings transparency, and CTMS was used for improving the nanocoatings' superhydrophobicity

All in all, bottom-up approaches can control the chemical composition and thickness of the samples very precisely. But it is very hard to predict their contact angle and whether they are superhydrophobic or not.

CONCLUSION

There are several patents on using superhydrophobic materials/coatings in the mining industry over the years. Though these are not being largely produced due to cost efficiency, scientific research in this field have never been stopped. To maintain clean surfaces is quite important in the mining industry; this will not only reduce maintenance cost, but also keep the machines away from corrosion. One example is as follows: when machines are working in extreme cold conditions, ice will frost from water on the surface of a machine, but using a superhydrophobic surface will keep the water away. In other words, ice or water will not stick to the surface of the machine. It is urged that new-generation pipelines and coal logs be developed in the future to achieve higher energy efficiency for the mining industry.

REFERENCES

Arney, M. S. 1996. Cement-lined pipes for water lubricated transport of heavy oil. *International Journal of Multiphase Flow*, 22(2): 207–221.

Balu, B., V. Breedveld, and D. W. Hess. 2008. Fabrication of "roll-off" and "sticky" superhydrophobic cellulose surfaces via plasma processing. *Langmuir*, 24(9): 4785–4790.

Bhushan, B. and Y. C. Jung. 2006. Micro-and nanoscale characterization of hydrophobic and hydrophilic leaf surfaces. *Nanotechnology*, 17(11): 2758.

Bhushan, B., Y. C. Jung, and K. Koch. 2009. Micro-, nano- and hierarchical structures for superhydrophobicity, self-cleaning and low adhesion. *Philosophical Transactions of the Royal Society A: Mathematical, Physical and Engineering Sciences*, 367(1894): 1631–1672.

Brzoska, J. B., I. B. Azouz, and F. Rondelez. 1994. Silanization of solid substrates: A step toward reproducibility. *Langmuir*, 10(11): 4367–4373.

Cassie, A. B. D. and Baxter, S., 1944. Wettability of porous surfaces. *Transactions of the Faraday Society*, 40: 546–551.

Celia, E., T. Darmanin, E. T. D. Givenchy, S. Amigoni, and F. Guittard. 2013. Recent advances in designing superhydrophobic surfaces. *Journal of Colloid and Interface Science*, 402: 1–18.

Coninck, J. D., J. Ruiz, and S. Miracle-Sole. 2002. Generalized Young's equation for rough and heterogeneous substrates: A microscopic proof. *Physical Review E*, 65: 036139.

Daoud, W. A., J. H. Xin, and X. Tao. 2006. Synthesis and characterization of hydrophobic silica nanocomposites. *Applied Surface Science*, 252(15): 5368–5371.

De Coninck, J., M. J. de Ruijter, and M. Voue. 2001. Dynamics of wetting. *Current Opinion in Colloid and Interface Science*, 6: 49.

Doshi, D. A. et al. 2005. Investigating the interface of superhydrophobic surfaces in contact with water. *Langmuir*, 21(17): 7805–7811.

Elif Cansoy, C., H. Y. Erbil, O. Akar, and T. Akin. 2011. Effect of pattern size. *Journal of Colloids and Surfaces A*, 386: 116–124.

Ensikat, H. J. et al. 2011. Superhydrophobicity in perfection: The outstanding properties of the lotus leaf. *Beilstein Journal of Nanotechnology*, 2: 152–161.

Fresnais, J., L. Benyahia, and F. Poncin-Epaillard. 2006a. Dynamic (de)wetting properties of superhydrophobic plasma-treated polyethylene surfaces. *Surface and Interface Analysis*, 38(3): 144–149.

Fresnais, J., J. Chapel, and F. Poncin-Epaillard. 2006b. Synthesis of transparent superhydrophobic polyethylene surfaces. *Surface and Coatings Technology*, 200(18): 5296–5305.

Feng, X. et al. 2003. Reversible super-hydrophobicity to super-hydrophilicity transition of aligned ZnO nanorod films. *Journal of the American Chemical Society*, 126(1): 62–63.

Fürstner, R. et al. 2005. Wetting and self-cleaning properties of artificial superhydrophobic surfaces. *Langmuir*, 21(3): 956–961.

Gao, X. F. and L. Jiang. 2004. Biophysics: Water-repellent legs of water striders. *Nature*, 432: 36.

Haynes, C. L. and R. P. Van Duyne, 2001. Nanosphere lithography: A versatile nanofabrication tool for studies of size-dependent nanoparticle optics. *The Journal of Physical Chemistry* B, 105(24): 5599–5611.

Heuberger, A. 1986. X-ray lithography. *Microelectronic Engineering*, 5(1): 3–38.

Heuberger, A. 1988. X-ray lithography. *Journal of Vacuum Science & Technology B*, 6(1): 107–121.

Hikita, M. et al. 2005. Super-liquid-repellent surfaces prepared by colloidal silica nanoparticles covered with fluoroalkyl groups. *Langmuir*, 21(16): 7299–7302.

Holloway P. J. 1994. Plant cuticles: Physicochemical characteristics and biosynthesis. In *Air Pollutants and the Leaf Cuticle*, edited by K. E. Percy. Berlin, Heidelberg: Springer, pp. 1–13.

Huang, L. et al. 2005. Stable superhydrophobic surface via carbon nanotubes coated with a ZnO thin film. *Journal of Physical Chemistry B*, 109(16): 7746–7748.

Hulteen, J. C. and R. P. J. Van Duyne. 1995. Nanosphere lithography: A materials general fabrication process for periodic particle array surfaces. *Vacuum Science and Technology, A* 13: 1553.

http://www.thenakedscientists.com/HTML/articles/article/biomimeticsborrowingfrombiology

Jeffree, C. 1986. The cuticle, epicuticular waxes and trichomes of plants, with reference to their structure, functions and evolution. In *Insects and the Plant Surface*, edited by Juniper B. E. and T. R. E. Southwood. London: Edward Arnold, pp. 23–64.

Jin, M., X. Feng, L. Feng, T. Sun, J. Zhai, T. Li, and L. Jiang. 2005. Superhydrophobic aligned polystyrene nanotube films with high adhesive force. *Advanced Materials*, 17(16): 1977–1981.

Jo, H., J. Choi, K. Byeon, H. Choi, and H. Lee. 2014. Superhydrophobic and superoleophobic Surfaces Using ZnO nano-in-micro hierarchical structures. *Microelectronic Engineering*, 116(0): 51–57.

Johnson, R. E. and R. H. Dettre. 1964. Contact angle hysteresis. III. Study of an idealized heterogeneous surface. *Journal of Physical Chemistry*, 68(7): 1744–1750.

Kim, P. et al. 2012. Liquid-infused nanostructured surfaces with extreme anti-ice and anti-frost performance. *ACS Nano*, 6(8): 6569–6577.

Koch, K. and H. Ensikat. 2008. The hydrophobic coatings of plant surfaces: Epicuticular wax crystals and their morphologies, crystallinity and molecular self-assembly. *Micron*, 39(7): 759–772.

Lee, W. et al. 2004. Nanostructuring of a polymeric substrate with well-defined nanometer-scale topography and tailored surface wettability. *Langmuir*, 20(18): 7665–7669.

Lehn, J.-M. 2002. Toward self-organization and complex matter. *Science*, 295(5564): 2400–2403.

Li, X.-M., D. Reinhoudt, and M. Crego-Calama. 2007. What do we need for a superhydrophobic surface? A review on the recent progress in the preparation of superhydrophobic surfaces. *Chemical Society Reviews*, 36(8): 1350–1368.

Lin, J., H. Chen, T. Fei, C. Liu, and J. Zhang. 2013. Highly transparent and thermally stable superhydrophobic coatings from the deposition of silica aerogels. *Applied Surface Science* 273(0): 776–786.

Ma, W., H. Wu, Y. Higaki, H. Otsuka, and A. Takahara. 2012. A "non-sticky" superhydrophobic surface prepared by self-assembly of fluoroalkyl phosphonic acid on a hierarchically micro/nanostructured alumina gel film. *Chemical Communications*, 48(54): 6824–6826.

Marmur, A. 2006. Soft contact: Measurement and interpretation of contact angles. *Soft Matter*, 2(1): 12–17.

Marrian, C. et al. 1994. Low voltage electron beam lithography in self-assembled ultrathin films with the scanning tunneling microscope. *Applied Physics Letters*, 64(3): 390–392.

Martines, E. et al. 2005. Superhydrophobicity and superhydrophilicity of regular nanopatterns. *Nano Letters*, 5(10): 2097–2103.

Mendes, P. M. et al. 2004. Gold nanoparticle patterning of silicon wafers using chemical e-beam lithography. *Langmuir*, 20(9): 3766–3768.

Nakagawa, T. and M. Soga. 1999. A new method for fabricating water repellent silica films having high heat-resistance using the sol–gel method. *Journal of Non-Crystalline Solids*, 260(3): 167–174.

Neinhuis, C. and W. Barthlott. 1997. Characterization and distribution of water-repellent, self-cleaning plant surfaces. *Annals of Botany*, 79(6): 667–677.

Owens, D. K. and R. C. Wendt. 1969. Estimation of the surface free energy of polymers. *Journal of Applied Polymer Science*, 13(8): 1741–1747.

Peng, P., Q. Ke, G. Zhou, and T. Tang. 2013. Fabrication of microcavity-array superhydrophobic surfaces using an improved template method. *Journal of Colloid and Interface Science*. 395(0): 326–328.

Qu, B. et al. 2011. Bionic duplication of fresh Navodon septentrionalis fish surface structures. *Journal of Nanomaterials*, 2011: 4.

Shibuichi, S., T. Onda, N. Satoh, and K. Tsujii. 1996. Super water-repellent surfaces resulting from fractal structure. *The Journal of Physical Chemistry*, 100(50): 19512–19517.

Shirtcliffe, N. J., G. McHale, S. Atherton, and M. I. Newton. 2010. An introduction to superhydrophobicity. *Advances in Colloid and Interface Science*, 161(1–2): 124–138.

Shiu, J.-Y. et al. 2004. Fabrication of tunable superhydrophobic surfaces by nanosphere lithography. *Chemistry of Materials*, 16(4): 561–564.

Suh, K. Y. and S. Jon. 2005. Control over wettability of polyethylene glycol surfaces using capillary lithography. *Langmuir*, 21(15): 6836–6841.

Sun, M. H., C. X. Luo, L. P. Xu, H. Ji, O. Y. Qi, D. P. Yu, and Y. Chen. 2005a. Artificial lotus leaf by nanocasting. *Langmuir*, 21(19): 8978–8981.

Sun, T. L., L. Feng, X. F. Gao, and L. Jiang. 2005b. Bioinspired Surfaces with Special Wettability. *Accounts of Chemical Research*, 38: 644.

Tyson, W. R. and W. A. Miller. 1977. Surface free energies of solid metals: Estimation from liquid surface tension measurements. *Surface Science*, 62(1): 267–276.

Wang, C., Y. Song, J. Zhao, and X. Xia. 2006. Semiconductor supported biomimetic superhydrophobic gold surfaces by the galvanic exchange reaction. *Surf Science* 600(4): 38–42.

Wang, C., J. Xiao, J. Zeng, D. Jiang, and Z. Yuan. 2012. A novel method to prepare a microflower-like superhydrophobic epoxy resin surface. *Materials Chemistry and Physics* 135(1): 10–15.

Wang, M., N. Raghunathan, and B. Ziaie. 2007. A nonlithographic top-down electrochemical approach for creating hierarchical (micro-nano) superhydrophobic silicon surfaces. *Langmuir*, 23(5): 2300–2303.

Wang, X. S., S. W. Cui, L. Zhou, S. H. Xu, Z. W. Sun, and R. Z. Zhu. 2014. A generalized Young's equation for contact angles of droplets on homogeneous and rough substrates. *Journal of Adhesion Science and Technology*, 28(2): 161–170.

Wang, X.D. et al. 2004. Contact Angle hysteresis on rough solid surfaces. *Heat Transfer—Asian Research*, 33(4): 201–210.

Wenzel, R. N. 1936. Resistance of solid surfaces to wetting by water. *Industrial & Engineering Chemistry* 28: 988–994.

Wilson, J. W., Y. Ding, and B. Zhao. 1994. The effect of binders on the performance of coal logs in a pipeline transportation system. *Mech 94: International Mechanical Engineering Congress and Exhibition; Preprints of Papers; Resource Engineering; Volume 2 Conference 2; Bulk Solids Storage, Handling and Transportation in Mining and Mineral Processing.* Institution of Engineers, Australia, 1994.

Whitesides, G. M. and B. Grzybowski. 2002. Self-assembly at all scales. *Science*, 295(5564): 2418–2421.

Wolansky, G. and A. Marmur. 1998. The actual contact angle on a heterogeneous rough surface in three dimensions. *Langmuir*, 14: 5292–5297.

Wu, C. et al. 1997. Surface modification of indium tin oxide by plasma treatment: An effective method to improve the efficiency, brightness, and reliability of organic light emitting devices. *Applied Physics Letters*, 70(11): 1348–1350.

Wu, X., L. Zheng, and D. Wu. 2005. Fabrication of superhydrophobic surfaces from microstructured ZnO-based surfaces via a wet-chemical route. *Langmuir*, 21(7): 2665–2667.

Yuan, Z. et al. 2013. A novel fabrication of a superhydrophobic surface with highly similar hierarchical structure of the lotus leaf on a copper sheet. *Applied Surface Science* 285(Part B): 205–210.

Zielecka, M. and E. Bujnowska. 2006. Silicone-containing polymer matrices as protective coatings: Properties and applications. *Progress in Organic Coatings*, 55(2): 160–167.

Zisman, W. A. 1964. Relation of the equilibrium contact angle to liquid and solid constitution. In: *Advances in Chemistry*, Vol. 43, pp. 1–51, American Chemical Society: Washington, DC. Chapter DOI: 10.1021/ba-1964-0043.ch001.

Zhang, G. et al. 2005. Fabrication of superhydrophobic surfaces from binary colloidal assembly. *Langmuir*, 21(20): 9143–9148.

Zhang, X. et al. 2004. Polyelectrolyte multilayer as matrix for electrochemical deposition of gold clusters: Toward super-hydrophobic surface. *Journal of the American Chemical Society*, 126(10): 3064–3065.

Zhai, L. et al. 2004. Stable superhydrophobic coatings from polyelectrolyte multilayers. *Nano Letters*, 4(7): 1349–1353.

Zhao, N. et al. 2005. Combining layer-by-layer assembly with electrodeposition of silver aggregates for fabricating superhydrophobic surfaces. *Langmuir*, 21(10): 4713–4716.

Zhu, M., W. Zuo, H. Yu, W. Yang, and Y. Chen. 2006. Superhydrophobic surface directly created by electrospinning based on hydrophilic material. *Journal of Materials Science* 41(12): 3793–3797.

Index

Printed and bound by CPI Group (UK) Ltd, Croydon, CR0 4YY

01/11/2024

01782601-0017